# PHYSIOLOGIA

## Epicuro-Gaſſendo-Charltoniana:

### OR

## A FABRICK

### OF

## SCIENCE NATURAL,

Upon the Hypotheſis of

# ATOMS,

Founded ⎫ ⎧*EPICURUS,*
Repaired ⎬by⎨*PETRUS GASSENDUS,*
Augmented ⎭ ⎩*WALTER CHARLETON,*

Dr. in Medicine, and Phyſician to the late
*CHARLES*, Monarch of
Great-*Britain.*

---

## The FIRST PART.

---

Fernelius, in præfat. ad lib. 2. de Abditis rerum Cauſſis.
*Atomos veteres jam ridemus, miramurq; ut ſibi quiſquam perſuaſerit, Corpora quæ-*
*dam ſolida, atque individua, fortuita illa concurſione, res magnitudine immenſas,*
*varietate multitudineq; infinitas, omnemq; abſolutiſſimum hunc Mundi ornatum*
*effeciſſe. At certè, ſi Democritus mortem cum vita commutare poſſet, multo acriùs*
*hæc, quæ putamus Elementa, ſuo more rideret.*

---

## LONDON,

Printed by *Tho: Newcomb*, for *Thomas Heath*, and
are to be ſold at his ſhop in *Ruſſel*-ſtreet,
neer the *Piazza* of *Covent-Garden.*
1 6 5 4.

## TO THE
## *HONOURABLE*
## M<sup>rs</sup> ELIZABETH VILLIERS,
### WIFE
### TO THE HONORABLE
## ROBERT VILLIERS
### ESQUIRE.

## MADAM,

*He excellent Monſieur Des Cartes, I remember, in his Dedicatory Epiſtle of his Principles of Philoſophy, to that illuſtrious Lady, the Princeſs Elizabeth; ſhewed Himſelf ſo much a Courtier, as to profeſs unto Her Highneſs, that of all Perſons living, who had peruſed his former Writings, He knew none, that perfectly underſtood them, except Herſelf only. This, Madam, is ſomewhat more than*

A 2　　　　　*what*

*what I shall adventure to say to you, in this my humble
Addreſs. Not that I might not, with the Authority
of Truth, and the willing Teſtimonies of all judicious
Perſons, whom you have at any time dignified with
your incomparable Converſation, affirme; That A-
cuteneſs of Wit, and Soundneſs of Judgement are as
Eminent in you, as in any that I know, of either Sex.
But, that I conceive it to be more conſiſtent with my
Duty of Conformity to the ſtrict Laws of your Humility (which is ſupreme among your many Virtues, if
there can be Supremacy where All are Superlative)
only to ask you leave, ſo far to juſtifie My ſelf, in this
way of Devotion, as publikely to own my Aſſurance;
that of all my Readers, none will meet with fewer
Difficulties, or diſcover more Lapſes and Errors, than
your ſelf: nor could that Book be clearly underſtood by
the Author, when He wrote it, which you cannot eaſily underſtand, when you are pleaſed to read it; be the
Argument thereof of what kind ſoever, and the Language either* Italian, French, *or* English, *which are all
equally your own.*

*But, I have little reaſon to ſpeak of juſtifying this
my Devotion, to the World; when that, by the General Tribute of Admiration and Reverence, which
your Excellencies duely receive from it, is fully convinced, that I am not capable of declaring a greater
Prudence, in any action of my whole life, than in this
of laying down both my ſelf and this mean Oblation of my
Obſervance and Gratitude, at the feet of a Perſonage,
whoſe ſingle Name is acknowledged to define All the poſſible Perfections of Humanity: and, upon conſequence,
cannot fail to give to both Me and my Writings not on-*
*ly*

ly an *Estimation among Good Men*; but also a full *Protection from the Malevolence of Evil*. And, I have been very lately told by some (and Those such Eminent Witts too; as that very Noble Persons, to whom they have Dedicated their Labours, have thereby received no small Additions of Honour) that they seriously Envied the good fortune of my resolution of invocating your Patronage of this *Epicurean* Philosophy; forasmuch as they were confirmed, that I had taken the most certain course, to procure *Immortality* thereunto, by offering it up to the Favour of so great an Example of all Heroick Accomplishments, as that Her Memory must ever continue verdant and sacred to all Posterity since it could not be, while *Generous Minds* should conserve the Memorials of Her as the Mirrour in which Vertue used to dresse Herself, when she would appear Amiable and Graceful; but that they must often cast some glances of valew upon the Remains of Him, who had so deep a sentiment of Her goodness, as to have known no other Ambition, but that commendable one of making Himself eternally known for Her most humble and obsequious Votary.

That, which would more become me, were to make my Excuses for the exceeding Boldness of this my Application; and to prevent such Objections as may lye against the Rashness of my Zeal: in selecting such a way to express my Reverence, as cannot secure me from a suspect of Prophanation; and presenting to you such a Sacrifice of my Thankfulness, as, if estimated according to its own Unworthiness, must make it a question, whether I had any designe of being Thankful at all. And here, to the First, I might justly plead; that a great Part of
*this*

this Volume was compoſed in your Houſe (the chief
Manſion of well-order'd Hoſpitality) and All of it in
the ſtrength of your Inſpiration. That the Book comes
not into your hands, to Informe, but only Remember you
of many of thoſe Diſcourſes of Nature, which your Noble Husband and your ſelf have often ſuffered me to entertain (would to God, I might have ſaid, ſatisfy) your
eager Curioſity withal, at thoſe hours your induſtrious
Minds required Relaxation from the bent of more
grave and advantageous Thoughts. That, having
the Honour of ſo great a Truſt, as that of your moſt præcious Lives committed unto me; it highly concern'd me,
to ſtudy and purſue all ways of Demonſtrating my ſelf not
altogether uncapable thereof, and more eſpecially this of
Natural Philoſophy, which being the Grounds, is alſo the
Meaſure of a Good Phyſician. And, that when your
Husband being acquainted with my Purpoſe of Enquiring into the Nature of Souls, both Brutal and Human,
in a diſtinct Work, though but the Remaining Moity of
this Phyſiologie; had injoyned me to deliver the ſame into his hands, as ſoon as I ſhould have finiſhed it: I inſtantly apprehended, I had an opportunity of a Double Happineſs, the one of being equally Grateful to Two ſingular
Friends; the other, of Allying thoſe Two Treatiſes
by Conſecration, which would be of ſo neer Affinity in
their Subjects.

As for the Other; I might eaſily alleage, that Great
ſpirits uſe not to eſtimate Præſents that are brought them,
by the value they carry in themſelves: but the Affections of thoſe who offer them. That Thankfulneſs is the
Poor mans wealth, and makes him, in the eyes of Gene-
roſity, ſtand in competition, for reſpect, with the Rich.
That though this my Oblation hold no proportion to the

im-

*immense height of your Merit, yet it is equal to that of my Power, and, indeed, the best that my Gratitude was able to advance upon the slender stock of my Capacity. And, that I never intended it as a Retribution for your incompensable Favours; but only as an Homage, to testifie that I confess my self infinitely your Debtor.*

*But, Madam, for me to attempt to Excuse, unto your self, the Unfitnesse of this Act of my devotion; is no lesse unnecessary, than for me to justifie to the World, that I have placed it upon a most worthy Object: forasmuch as I have no more reason to doubt, that so transcendent a Charity, as is diffused through and surrounds your perfect Soul, can be large enough to dispense with the Rudeness of the Ceremonies, and Poverty of the Offering, where you are satisfied of the sincere Respects, and unalterable Fidelity of his Heart, who tenders it; than I have to fear, that the World should not most readily confirm my judgement, that your Deserts have rightfully entitled you to all the Demonstrations of Honour and Reverence, that can possibly be given to you.*

*The Chief part, therefore, yea the whole of my present Duty, is only humbly to Beg your benigne Acceptance of this Dedication, as the Best Expression I was able to make of those profound sentiments which as well your Goodness in General to others, as your Particular Favours to my self, have impressed upon my Soul. And this I now do, upon the Knees of my Heart; and solemnly vow, that as I esteem a perfect Friend, the greatest Treasure of my life, so I do and ever shall account you the most perfect of Friends: That I shall confess my self to have lost not only all Piety, but all*

*Hu-*

*Humanity also, when ever I shall willingly lose any the least opportunity of serving you: and that your own Good Angell (I speak familiarly, but at the same time believe you to be under the Tuition of a Legion of Good ones) cannot more fervently desire your complete Happiness, than, Incomparable Madam,*

Your Eternal Servant,

*London the 20 of July,*
*An. Dom. 1654.*

W. CHARLETON.

# THE
# CONTENTS, SERIES,
## AND ORDER
## OF THE WHOLE BOOK.

## BOOK THE FIRST.

### CHAP. I.

*All Modern Philosophers reduced to four general Orders; and the principal causes of their Dissention.* pag. 1.

#### SECT. I.

ARTIC.

1 THe principal Sects of the ancient Grecian Philosophers, only enumerated. pag. 1
2 The same revived among the Moderns, with encrease. 2
3 Who are reduced either to the Pedantique or Female Sect. 2
4 Or, to the Assertors of Philosophical Liberty. 3
5 Or, to the Renovators. 3
6 Or to the Electors. 4

#### SECT. II.

ARTIC.

1 THe principal causes of the Diversity of Philosophical Sects; and the chiefest among them, the Obscurity of Nature. 5
2 The Imperfection of our Understanding. 5
3 The Irregularity of our Curiosity. *A paradox.* 6

### CHAP. II.

*That this World is the Universe.* pag. 9.

#### SECT. I.

ARTIC.

1 THe Ambition of Alexander in affecting the Conquest, less vain then that of many ancient Philosophers in affecting the Knowledge of a Multitude of Worlds. 9
2 A reduction of those Philosophers to four distinct Sects; respective to their distinct perswasions: and the Heads of each Sect nominated. 9
3 The two main pillars on which the opinion of a Plurality of Worlds was anciently erected. 10

#### SECT. II.
### The Redargution.

ARTIC.

1 THe Question stated to be concerning the real Existence, not the possibility of an Infinity of Worlds. 11
2 Because the supposed Infinity of the Extramundan Spaces, is no impossibility. ibid.
3 Because an Infinity of Bodies is also possible as to the Omnipotence of God. ibid.
4 The Error of concluding the Esse, from the Posse of an Infinity of Worlds. 12
5 The first main Pillar of a Plurality of worlds subverted ibid.
6 The second Pillar found sophisticate, and demolished. 13
7 A Plurality of Worlds manifestly repugnant to Authority Divine 14
8 And Human. ibid.
9 The result of all; the Demonstration of the Authors Thesis, That this World is the Universe. ibid.
10 Extramundane Curiosity, a high degree of Madness. 15

### CHAP. III.
*Corporiety and Inanity,* p. 16
#### SECT. I.

ARTIC.

1 BOdy and Inanity, the two general Parts of the Universe.

(a) 2 These

# THE CONTENTS, SERIES,

2 *Three the most memorable Definitions of Corporiety extant among Physiologists, recounted and examined.* ibid.

3 *Four Descriptions of the nature of* Inanity, *by* Epicurus, Cleomedes, Empericus, Aristotle. 17

4 *Their importance extracted: and what is the formal or proper notion of a* Vacuum. 18

5 *The Existence of Bodies in the World, manifest by* Sense : *whose* Evidence *is perfect* Demonstration. ibid.

## Chap. IV.

### A Vacuum in Nature. p 21.

#### Sect. I.

Art.

1 *The Distinction of a* Vacuum *into* (1) *Natural and* (2) *Præternatural: and the one called* Disseminate, *the other* Coacervate. 21

2 *The nature of a* Disseminate *Vacuity, explained by the Analogy of a heap of Corn.* ibid.

3 *The first Argument of a Disseminate Vacuity, desumed from the evidence of Motion, in General: and* Aristotles *error concerning the Essence or Place, concisely detected, and corrected.* 22

4 *Motion demonstrated by* Sense : *and* Zeno's *enigmatical Argument, for an Universal Quiet, dissolved.* 23

5 *The Consequution of the Argument (if no Vacuum, no Motion) illustrated.* 24

6 *An Objection, that the Lococession of some Bodies, depends on their Rarity or Porosity; not on a Disseminate Vacuity: prevented.* ibid.

7 *No beginning of Motion, without Inanity interspersed.* 25

#### Sect. II.

Artic.

1 A *Second Argument of a Vacuity Disseminate, collected from the reason of Rarefaction and Condensation.* ibid.

2 *The eminent Phænomenon of an Aeroscloppet, or Wind Gun, solved by a Vacuity Disseminate among the incontiguous* (quoad totas superficies) *parts of air.* 26

3 *Experiment of an Æolipile, or Hermetical Bellows, attesting a Vacuity Disseminate.* ibid.

4 *Experiment of a Sulphurate Vapor, included in a Glass Vial, partly filled with Water: of the same importance.* 27

5 *No Combustible in Aer: and so the opinion of the Aristoteleans, that the Extinction of Flame imprisoned, is to be charged on the Defect of Aer for its sustentation; grosly erroneous.* 28

6 *A fourth singular and memorable* Experiment *of the Authors, of Ice at the nose of a large Reverberatory Furnace, charged with Ignis rotæ; evidencing a Vacuity interspersed in the Aer.* 29

7 *An inference from that Experiment; that Aer, as to its General Destination, is the Common Receptary of Exhalations.* ibid.

8 *A second Illation, that the Aer doth receive Exhalations at a certain rate, or definite proportion; which cannot be transcended without prodigious violence.* ibid.

9 *The Existence of Inane Incontiguities in the Aer, confirmed by two considerable Arguments.* ibid

#### Sect. III.

Artic.

1 THat Water *also contains Vacuola empty Spaces; demonstrated.* 31

2 *From the Experiment of the Dissolution of* Alum, Halinitre, Sal Ammoniac, *and* Sugar, *in Water formerly sated with the Tincture of Common Salt.* ibid.

3 *The verity of the Lord* Bacons *Assertion, that a repeated infusion of Rhubarb acquires as strong a virtue Cathartical, as a simple infusion of Scamony, in equal quantity: and why.* 32

4 *Why two Drachms of Antimony impregnate a pint of Wine, with so strong a vomitory Faculty at two ounces.* ibid.

5 *Why one and the same Menstruum may be enriched with various Tinctures.* ibid.

#### Sect. IV.

Artic.

1 TWo other Arguments of a Vacuity Disseminate inferrible from (1) the difference of Bodies in the degrees of Gravity: (2) the Calefaction of Bodies by the penetration of igneous Atoms into them. 33

2 *The Experiment vulgarly adduced to prove no vacuity in nature, so far from denying, that they confess a Disseminate one.* ibid.

3 *The grand Difficulty of the Cause of the Aers restitution of it self to its natural contexture, after rarefaction and condensation, satisfyed in brief.* ibid.

## Chap. V.

### A Vacuum præternatural. p.35.

#### Sect. I.

Artic.

1 WHat is conceived by a Coacervate Vacuity: and who was the Inventer of the famous Experiment of Quick-silver in a Glass Tube, upon which many modern Physiologists

ologists have erected their perswasion of the possibility of introducing it. 35
2 A faithful description of the Experiment, and all its rare Phænomena. 36
3 The Authors reason, for his selection of onely six of the most considerable Phænomena to explore the Causes of them. 37

## Sect. II.

ARTIC.
1 THe First Cardinal Difficulty. 37
2 The Desert space in the Tube argued to be an absolute Vacuum coacervate, from the impossibility of its repletion with Aer. ibid.
3 The Experiment presented in Iconism 38
4 The Vacuity in the Desert Space, not prevented by the insinuation of Æther. 40
5 A Paradox, that Nature doth not abhor all vacuity, per se; but onely ex Accidenti, or in respect to Fluxility. ibid.
6 A second Argument against the repletion of the Desert space by Æther. 41
7 The Vacuity of the Desert space, not prevented by an Halitus, or Spiritual Efflux from the Mercury: for three convincing reasons. 42
8 The Authors Apostacy from the opinion of an absolute Coacervate Vacuity in the desert space: in regard of ibid.
9 The possibility of the subingression of light. ibid.
10 Of the Atoms or insensible bodies of Heat and Cold: which are much more exile and penetrative then common Aer. 43
11 Of the Magnetical Efflux of the Earth: to which opinion the Author resigns his Assent. 44
12 No absolute plenitude, nor absolute Vacuity, in the Desert Space: but onely a Disseminate Vacuity. ibid.

## Sect. III.

ARTIC.
1 THe second Difficulty stated. 45
2 Two things necessary to the creation of an excessive, or præternatural Vacuity. ibid.
3 The occasion of Galilæos invention of a Brass Cylindre charged with a wooden Embol, or Sucker: and of Torricellius invention of the present Experiment. ibid.
4 The marrow of the Difficulty, viz. How the Aer can be impelled upward, by the Restagnant Quick silver, when there externally wants a fit space for it to circulate into. 46
5 The solution of the same, by the Laxity of the Contexture of the Aer. ibid.
6 The same illustrated, by the adequate simile of Corn infused into a Bushel. ibid.
7 A subordinate scruple, why most bodies are moved through the Aer, with so little resistence, as is imperceptible by sense? 47
8 The same Expeded. ibid.

9 A second dependent scruple concerning the Cause of the sensible resistence of the Aer, in this case of the Experiment: together with the satisfaction thereof, by the Gravity of Aer. ibid.

## Sect. IV.

ARTIC.
1 THe State of the Third Difficulty. 48
2 The Solution thereof in a Word. ibid.
3 Three precedent positions briefly recognized, in order to the worthy profounding of the mystery, of the Aers resisting Compression beyond a certain rate, or determinate proportion ibid.
4 The Æquiponderancy of the External Aer, pendent upon the surface of the Restagnant Mercury, in the vessel to the Cylindre of Mercury residuous in the Tube, at the altitude of 27 digits: the cause of the Mercuries constant subsistence at that point. 49
5 A convenient simile, illustrating and enforcing the same. 50
6 The Remainder of the Difficulty; viz. Why the Æquilibrium of these two opposite weights, the Mercury and the Aer, is constant to the præcise altitude of 27 digits: removed. ibid.
7 Humane Perspicacity terminated in the exterior parts of Nature, or simple Apparitions: which eluding our Cognition, frequently fall under no other comprehension, but that of rational Conjecture. ibid.
8 The constant subsistence of the Mercury at 27 digits, adscriptive rather to the Resistence of the Aer, then to any occult Quality in the Mercury. 51
9 The Analogy betwixt the Absolute and Respective Æquality of weights, of Quick-silver and Water, in the different altitudes of 27 digits and 32 feet. 52
10 The definite weights of the Mercury at 27 digits, and Water at 32 feet, in a Tube of the third part of a digit in diametre; found to be neer upon two pound, Paris weight. ibid.
11 Quære, Why the Æquilibrium is constant to the same point of altitude in a Tube of a large concave, as well as in one of a small; when the force of the Depriment must be greater in the one, then the other. 53
12 The solution thereof by the appropriation of the same Cause, which makes the descent of two bodies, of different weights, æquivelox. ibid.

## Sect. V.

ARTIC.
1 THe Fourth Capital Difficulty proposed. 54
2 The full solution thereof, by demonstration. ibid.
3 The same confirmed by the theory of the Cause of

of the Mercuries frequent Reciprocations, before it acquiesce at the point of Æquipondium. ibid.

### Sect. VI.

Artic.
1 THe Fifth Principal Difficulty. 55
2 Solved, by the Action of Restauration natural to each insensible particle of Aer. ibid.
3 The incumbent Aer, in this case, equally distressed, by two contrary Forces. 56
4 The motion of Restauration in the Aer extended to the satisfaction of another consimilar Doubt, concerning the subintrusion of Water into the Tube; if superaffused upon the restagnant Mercury. ibid.
5 A Third most important Doubt, concerning the nonapparence of any Tensity, or Rigidity in the region of Aer incumbent upon the Restagnant Liquors. ibid.
6 The solution thereof, by the necessary relation of a space in the vicine region of Lax aer, equal to that, which the Hand commoved possesseth in the region of the Compress. 57
7 A confirmation of the same Reason, by the adequate Example of the Flame of a Tapour. ibid.
8 By the Experiment of Urination. ibid.
9 By the Beams of the Sun, entring a room, through some slender crany, in the appearance of a White shining Wand, and constantly maintaining that Figure, notwithstanding the agitation of the aer by wind, &c. 58
10 By the constancy of the Rainbow, to its Figure, notwithstanding the change of position and place of the cloud and contiguous aer. ibid.
11 Helmonts Delirium, that the Rainbow is a supernatural Meteor: observed. ibid.

### Sect. VII.

Artic.
1 THe sixth and last considerable Difficulty. ibid.
2 The clear solution thereof, by the great disproportion of weight betwixt Quick-silver and Water. 59
3 A Corollary; the Altitude of the Atmosphere conjectured. ibid.
4 A second Corollary: the desperate Difficulty of conciliating Physiology to the Mathematicks: instanced in the much discrepant opinions of Galileo and Mersennus, concerning the proportion of Gravity that Aer and Water hold each to other. ibid.
5 The Conclusion of this Digression: and the reasons, why the Author adscribes a Cylindrical Figure to the portion of Aer impendent on the Restagnant Liquors, in the Experiment. 60

## Chap. VI.
## Of PLACE. p. 62.

### Sect. I.

Artic.
1 THe Identity Essential of a Vacuum and Place, the cause of the present Enquiry into the Nature of Place. ibid.
2 Among all the Quæries about the Hoti of Place; the most important is, Whether Epicurus or Aristotles Definition of it, be most adequate. ibid.
3 The Hypothesis of Aristotles Definition 63
4 A convenient supposition inferring the necessity of Dimensions Incorporeal. ibid.
5 The Legality of that supposition. ibid.
6 The Dimensions of Longitude, Latitude, and Profundity, imaginable in a Vacuum. 64
7 The Grand Peripatetick objection, that Nothing is in a Vacuum; ergo no Dimensions. ibid.
8 Des Chartes, and Mr. White seduced by the plausibility of the same. 65
9 The Peripateticks reduction of Time and Place to the General Categories of Substances and Accidents, the cause of this Epidemick mistake. ibid.
10 Place neither Accident nor Substance. 66
11 The precedent Giant-Objection, that Nothing is in a Vacuum; stab'd, ab ablow. ibid
12 Dimensions Corporeal and Incorporeal, or Spatial. 67
13 The former supposition reassumed and enlarged. ibid.
14 The scope and advantage thereof; viz. the comprehension of three eminent Abstrusities concerning the Nature of Place. ibid.
15 The Incorporiety of Dimensions Spatial, Discriminated from that of the Divine Essence, and other Substances Incorporeal. 68
16 This persuasion, of the Improduction and Independency of Place; preserved from the suspicion of Impiety. ibid.

### Sect. II.

Artic.
1 PLace, not the immediate superfice of the Body invironing the Locatum; contrary to Aristotle. 69
2 Salvo's for all the Difficult Scruples, touching the nature of Place; genuinely extracted from Epicurus his διάστημα. ibid
3 Aristotles ultimate Refuge. 70
4 The Invalidity thereof: and the Coexistibility, or Compatibility of Dimensions Corporeal and Spatial. 71

CHAP.

## CHAP. VII.
### *Of Time and Eterntiy.* p. 72.

### SECT. I.

ARTIC.
1 *The* Hoti *of Time more easily conceivable by the* Simple Notion *of the* Vulgar, *then by the* complex Definitions *of Philosophers.* ibid.
2 *The* General presumption *that Time is* Corporeal *, or an* Accident *dependent on* Corporeal Subjects; *the chief Cause of that Difficulty.* 73
3 *The* variety of opinions *, concerning it ; another Cause of the Difficulty : and* Epicurus *Description of its Essence, recited and explained.* ibid.
4 *Time defined to be* Cœlestial Motion, *by* Zeno, Chrysippus, &c. *and thereupon affirmed , by* Philo, *to be onely* Coævous to the World. 74
5 *Aristotles so much magnifyed Definition of Time to be the* Measure of Motion Cœlestial, *&c. perpended and found too light.* ibid.

### SECT. II.

ARTIC.
1 *Time, nor* substance*, nor* Accident *: but an* Ens more General, *and the* Twin-brother *of* Space. ibid.
2 *A* Paralelism *betwixt* Space *and* Time. ibid.
3 *Time,* Senior *unto, and* independent *upon* Motion *: and onely* accidentally *indicated by* Motion, *as the* Mensuratum *by the* Mensura. 76
4 *A* demonstration *of the* independence *of* Time

*upon* Motion *, from the miraculous* Detention *of the* Sun, *above the* Horison *, in the days of* Joshua. 77
5 *An* Objection *, that, during the arrest of the Sun, there was no Time , because no Hours ; satisfyed.* ibid.
6 *The* Immutability *of Time also asserted against* Aristotle. ibid.

### SECT. III.

ARTIC.
1 *The Grand Question, concerning the* Disparity *of Time and* Æternity *: stated.* 78
2 *Two* præparatory Considerations, *touchant the* æquivocal use *of the word* Æternity *: requisite to the cleer solution thereof.* ibid.
3 *Two* decisive Positions *thereupon inferred and established* 79
4 *The* Platonicks Definition *of* Eternity *, to be one* Everlasting Now ; *not intelligible , and therefore collusive.* 80
5 *Their* Assertors subterfuge, *that* Eternity *is* Coexistent *to* Time; *also unintelligible.* ibid.
6 *Our* Ecclesiastick Doctors, *taking* Sanctuary *in the* 3. Exod. *for the authorizing of their Doctrine, that the* Present Tense *is onely competent to* God, *and so that* Eternity *is one* permanent Instant, *without* Fusion *or* Succession *: not secure from the rigour of our* Demonstration. 81
7 *The* Objective Præsence *of all things at once, to the* Divine Intellect *; no ways impugned by our* contradiction *of the* Doctors theory. ibid.
8 *Nor the* Immutability *of the* Divine Nature, *against* Aristotle. 82
9 Coronis. 83

# The Second Book.

## CHAP. I.
### *The* Existence *of* Atoms*,* Evicted.
### p. 84.

### SECT. I.

ARTIC.
1 *The* right *of the* Authors Transition *from the* Incorporeal *to the* Corporeal *part of* Nature *: and a series of his subsequent speculations :* ibid.
2 Bodies *generally distinguished into* Principles *and* Productions, *with their* Scholastick Denominations *and* proprieties. 85

3 *The* right *of* Atoms *to the* Attributes *of the* First Matter. ibid.
4 *Their* sundry Appellations allusive *to their* three eminent proprieties. ibid.
5 *Two* vulgarly passant Derivations *of the word,* Atom, *exploded.* 86
6 *Who their* Inventor *: and who their* Nomenclator. 87
7 *Their* Existence demonstrated. 87
8 *That* Nature, *in her* dissolution *of* Concretions, *doth descend to the* insensible particles. 88
9 *That she can* run on *to* Infinity. ibid.
10 *But must consist in* Atoms, *the* Term *of* Exsolubility

b

*solubility.* ibid.

11 *A second* Argument *of their Existence, drawn from that of their Antitheton* Inanity. 89

12 *A third, hinted from the impossibility of the Production of* Hard Bodies *, from any other Principle.* ibid.

13 *A Fourth, from the* Constancy *of Nature in the specification and Determinate Periods of her Generations.* ibid.

### CHAP. II.

## No Physical Continuum, infinitely Divisible. p. 90.

### Sect. I.

ARTIC.

1 THe Cognation of this Theorem, to the Argument of the immediately precedent Chapter. ibid.
2 *Magnitude divisible by a continued progress through parts either* Proportional, *or* Aliquotal. ibid.
3 *The use of that Distinction in the present.* 91
4 *The verity of the* Thesis, *demonstrated.* ibid.
5 *Two detestable* Absurdities, *inseparable from the position of* Infinite parts *in a* Continuum. ibid.
6 Aristotles *subterfuge of* Infinitude Potential; 92
7 *Found openly* Collusive. 93
8 *A second subterfuge of the* Stoick; ibid.
9 *Manifestly disentaneous to* Reason. ibid.

### Sect. II.

ARTIC.

1 THe Absurdities, *by* Empericus, *charged upon the supposition of only* Finite parts *in a* Continuum. 94
2 *The sundry* Incongruities *and* Inconsistences, *by the Modern Anti-Democritans, imputed to the supposition of* Insecility. ibid.
3 *The full Derogation of them all together, by one single Response; that the minimum of* Atomists *is not Mathematical, but Physical, contrary to their presumption.* 95
4 *A seeming* Dilemma *of the Adversary, expeditely evaded.* 96
5 *A* Digression, *stating and determining that notable Question, Whether* Geometrical Demonstrations *may be conveniently transferred to the* Physical *or* sensible Quantity? ibid.

### CHAP. III.

## Atoms, the First and Universal Matter. p. 99.

### Sect. I.

ART.

1 THe introduction, *hinting the two general assumptions of the Chapter.* ibid.
2 Dmocritus *and* Epicurus *vindicated from the absurd admission of* Inanity *to be one Principle of Generables.* ibid.
3 *Atoms not inconsistent with, because the Principles of the four vulgar* Elements. 100
4 *The dissent of the Ancients, about the number of* Elements. 101
5 *No one of the four* Elements *sufficient to the production of either any of the other three, or of any Compound nature.* ibid.
6 *The four Elements, not the Protoprinciple of Concretions.* 102
7 *Atoms discriminated from the Homoiomerical Principles of* Anaxagoras. ibid.
8 *The principal Difficulties urged against the Hypothesis of* Atoms, *singularly solved.* 103
9 *A recapitulation of the premisses, introductory to the verification of the present thesis.* 106

### Sect. II.

ARTIC.

1 THe 4 notable opinions, *concerning the Composition of a* Continuum. 107
2 *A Physical Continuum cannot consist of* Points Mathematical. ibid.
3 *Nor of Parts and* Points Mathematical, *united.* 108
4 *Nor of a simple Entity, before division indistinct: but of Indivisibles.* ibid.
5 *A second Apodictical reason, desumed from the nature of Union, evincing that* Atoms *are the* First *and* Catholick Principle *of Concretions.* 109
6 *An objection prevented.* ibid.
7 *The reason of the Authors supercession of all other Arguments of the like importance.* ibid.

### CHAP. IV.

## The Essential Proprieties of Atoms: p. 111.

### Sect. I.

ARTIC.

1 THe two links connecting this to the precedent Chapter. ibid.
2 *The General Proprieties of* Atoms: *and the* Inseparabity *of each, demonstrated.* ibid.
3 *The*

3 *The* Resistence *of Atoms, no distinct proprie-*
*ty ; but pertinent to their Solidity or Gravity*
112

4 *The* specifical *Proprieties of Atoms.* ibid.

## Sect. II.

### Concerning the Magnitude of A-
### toms. p. 113.

Artic.

1 BY *the* Magnitude, *is meant the* Parvity *of*
Atoms. ibid.

2 *A consideration of the Grossness of our senses,*
*and the extreme subtilty of Nature in her o-*
*perations ; præparatory to our Conjectural ap-*
*prehension of the Exiguity of Her Materials,*
Atoms. ibid.

3 *The incomprehensible subtility of Nature, ar-*
*gued from the Artifice of an exquisite Watch,*
*contrived in a very narrow room.* 114

4 *The vast multitude of sensible particles, and*
*the vaster of Elemental Atoms, contained in*
*one grain of Frankinsense ; exactly calculated.*
ibid.

5 *The Dioptrical speculation of a Handworm,*
*discovering the great variety of Organical*
*Parts therein, and the innumerability of their*
Component Particles. 115

6 *A short Digressive Descant upon the Text of*
Pliny, *touching the multiplicity of parts in a*
Flea ; *hinting the possible perspicacity of Rea-*
son. ibid.

7 *The Exility of Atoms, conjectural from the*
*great diffusion of one Grain of Vermilion dis-*
*solved in Water.* 116

8 *The same, inferrible from the small quantity of*
*oil depredated by the Flame of a Lamp, in a*
*quarter of an hour.* ibid.

9 *The Microscope of great use, in the discern-*
*ment of the minute particles of Bodies : and*
*so advantageous to our Conjecture, of the exi-*
*lity of Atoms.* ibid

## Sect. III.

### Concerning the Figures of Atoms.
### p. 117.

Artic.

1 AN *Epitome of all that directly concerns*
*the Figures of Atoms in three General*

Canons. ibid.

2 *The* First *Canon* *explained and certified.* ibid.

3 *The Exility of Atoms, doth not necessitate their*
General Roundness ; *contrary to the common*
*conceit.* ibid.

4 *The Diversity of Figures in Atoms, evicted*
*from the sensible Dissimilitude of* Individuals,
*as well Animate as Inanimate.* 118

5 *A singular* Experiment, *autoptically demon-*
*strating the various Configurations of the mi-*
*nute Particles of Concretions.* 119

6 *A variety of Figures in Atoms, necessary to*
*the variety of all Sensibles.* ibid.

7 *The second Canon, explained and Certified.*
120

8 *The* Third *Canon, explained, and refuted.* 121

## Sect. IV.

### Concerning the Motions of Atoms.
### p. 121.

Artic.

1 TWo *introductory* Observables. ibid.

2 *The Motion of Atoms, according to the*
*General Distinction of the Ancients, Two-fold;*
viz. Natural, *and* Accidental : *and each of*
*these redivided into two different Species.* ibid.

3 *The summary of* Epicurus Figment, *of the*
*Perpendicular Motion of Atoms, without a*
*common* Centre. 122

4 *His* Declinatory *natural Motion of Atoms,*
*excused ; not justified.* ibid.

5 *The genuine sense of* Epicurus, *in his distincti-*
*on of the* Reflex *Motion of Atoms into* ex
Plaga, *and* ex Concussione. 123

6 *The several* Conceptions *of* Epicurus, *about*
*the perpetual Motions of Atoms.* 124

7 *The perpetual* Inquietude *of Atoms*, *even*
*in compact Concretions, adumbrated in*
melted Lead. ibid.

8 *The same more sensibly exemplified, in the spi-*
*rit extracted from* Mercury, Tin, *and* Sublimate. 125

9 *The Mutability of all Concretions, a good Ar-*
*gument of the perpetual intestine Commotion*
*of Atoms, in the most adamantine Compositi-*
ons. ibid.

10 *What we are to* explode, *and what* retain, *in*
*the opinion of* Epicurus, *touching the Motion*
*of Atoms.* ibid.

THE

# The Third Book.

### CHAP. I.

#### The Origine of Qualities. p. 127.

##### SECT. I.

ARTIC.

1 AN introductory Advertisement; of the obscurity of many thing to Reason, which are manifest to sense: and of the Possibility, not necessity of the Elementation of Concretions, and their sensible Qualities, from the Principles presumed.  1.7

2 The Authors Definition of a Quality, in general: and genuine exposition of Democritus mysterious Text, concerning the Creation of Qualities.  128

3 The necessary deduction of Qualities from Naked or Unqualified Principles.  130

4 The two primary Events of Atoms, viz. Order and Position, associated to their three essential Proprieties, viz. Magnitude, Figure, and Motion; sufficient to the Origination of all Qualities.  ibid.

5 The necessity of assuming the Magnitude and Motion of Atoms, together with their Order and Situation, as to their production of Qualities, evicted by a double instance.  131

6 The Figure, Order and Position of Parts in Concretions, alone sufficient to the Causation of an indefinite variety of Qualities, from the analogy of Letters.  ibid.

7 The same Exemplified in the wise of White Froth, on the Waves of the Sea.  132

8 The Nativity of Colours in General, explained by several obvious Examples.  ibid.

9 The Accension of Heat, from Concretions actually Cold, upon a meer transposition of their Component Particles; exemplified in sundry Chymical Experiments.  133

10 The Generation of all kinds of sensible qualities in one and the same Concretion, from the variegated positions of its particles: evidenced in the Example of a putrid Apple.  134

11 The assenting suffrage of Epicurus.  ibid.

### CHAP. II.

#### That Species Visible are Substantial Emanations. p. 136.

##### SECT. I.

ARTIC.

1 THe Visible Images of objects, substantial: and either corporeal Emanations from the superficial parts of Concretions; or Light it self, disposed into contextures, consimilar to the figure of the object.  ibid.

2 The position of their being Effluviaes, derived from Epicurus; and preferred to the common doctrine of the Schools of the Immateriality of Species Visible.  ibid.

3 Epicurus Text concerning the same.  137

4 The faithful Exposition thereof.  ibid.

5 The contents thereof reduced to four heads.  134

6 The Existence of Images visible, certified by autoptical Demonstration.  ibid.

7 Epicurus opinion, of the substantiality of Images Visible, consonant to the judgement of Plato and Empedocles.  139

8 The Aristoteleans Thesis, that Images optical are meer Accidents, recited: and  ibid:

9 Convicted of sundry Impossibilities, Inconsistences, and Absurdities.  ibid.

10 The grand Objection of Alexander, that a continual Efflux of substance must minorate the Quantity of the most solid Visible.  140

11 Solved by two Reasons; the possible Accretion of other particles; and the extreme Tenuity of the Emanent.  141

12 The Tenuity of Images visible, reduced to some degree of Comprehensibility, by conceiving them to be most thin Decortications.  ibid.

13 By Instance, in the Visible species of the Foot of a Handworm.  ibid.

14 By exemplifying in the numerous round Films of Wax, successively derepted from a Wax tapor by the flame thereof, in the space of an hour: and  142

15 In the innumerable Films of Oyl, likewise successively delibrated, by the flame of an Ellychnium, or Match, perpendicularly floating in a vessel of equal capacity with Solomons Brazen Sea, in the space of 48 hours.  ibid.

16 By the Analogy betwixt an Odorable and Visible Species.  ibid.

7 The

17 *The Manner and Reason of the Production of visible Images ; according to the hypothesis of* Epicurus.    143
18 *The Celerity of the Motion of visible Images, reasoned ; and compared to that of the* Light *of the Sun.*    144
19 *The Translation of a moveable from place to place, in an indivisible point of time, impossible : and why ?*
20 *The Facility of the Abduction, or Avolation of Images Visible, from solid Concretions ; solved by the Spontaneous Exsilition of their superficial Atoms : and the Sollicitation of Light incident upon them.*    ibid.
21 *That Objects do not emit their Visible Images, but when Illustrated :* a Conceit though paradoxical, yet not improbable.    145

### Sect. II.

Artic.
1 Visible Images Syßatical, *described ; and distinguißt from* Apostatical ones.    146
2 *Their* Existence *assured, by the testimony of* Diodorus Siculus : and    ibid.
3 Damascius, *together with the Autopsy of* Kircher.    ibid.
4 Kirchers *Description of that famous Apparition at* Rhegium, *called* Morgana Rheginorum: and    147
5 *Most ingenious* Investigation *of the Causes thereof.*    ibid.
6 *His admirable* Artifice *, for the exhibition of the like aereal Representation, in Imitation of* Nature.    148

### CHAP. III.

*Concerning the Manner and Reason of* VISION, p. 149.

### Sect. I.

Artic.
1 The Reason of Vision, *according to the opinion of the* Stoicks.    149
2 *Of* Aristotle.    150
3 *Of the* Pythagoreans.    ibid.
4 *Of* Empedocles.    ibid.
5 *Of* Plato.    ibid.
6 *Of* Epicurus.    ibid.
7 *Of* Monf. Des Chartes.    151
8 *The ingenuity of* Des Chartes *Conceit, acknowledged : but the solidity indubitated.*    152
9 *The Opinion of* Epicurus more satisfactory, *then any other : because more Rational, and less obnoxious to inexplicable Difficulties.*    ibid.
10 *The Two most considerable* Difficulties *opposed to* Epicurus position, *of the Incursion of* Sabßantial *Images into the Eye.*    153

### Sect. II.

Artic.
1 That the superfice of no body is perfectly *smooth : evicted by solid* Reason, *and* Autopsie.    ibid.
2 *That the visible Image doth consist of so many Rays as there are points designable in the whole superfice of the object : and that each Ray hath its line of Tendency direct , respective to the face of that particle in the superfice , from which it is emitted.*    154
3 *That the* Density *and* Union *of the Rays, composing the visible Image, is greater or less ; according to their less , or greater* Elongation *from the Object.*    ibid.
4 *That the* Visible Image *is neither total in the total medium ; nor total in every part thereof : but so manifold as are the parts of the medium from which the object is discernable.* Contrary to the Aristoteleans.    155
5 PARADOX, *That no man can see the same particle of an object , with both Eys at once ; nay, not with the same Eye, if the level of its* Visive Axe *be changed.*    ibid.
6 CONSECTARY. *That the Medium is not possessed with one simple Image ; but by an Aggregate of innumerable Images, deradiate from the same object : all which notwithstanding constitute but one entire Image.*    156
7 CONSECTARY. 2. *That Myriads of different Immages , emanant from different objects , may be Coexistent in the Aer ; without reciprocal penetration of Dimensions, or* Confusion *of particles : contrary to the* peripaticks.    ibid
8 *That the place of the visible Images ultimate* Reception, *and complete* Perception , *is the* Concave *of the* Retina Tunica.    157
9 *That the* Faculty *forms a judgement of the Conditions of the Object , according to the representation thereof by the Image, at its impression on the principal part of Vision, the* Amphibleßroides.    ibid.
10 CONSECTARY. *That the Image is the Cause of Objects apparence of this or that determinate* Magnitude.    158
11 CONSECTARY. 2. *That no Image can replenish the Concave of the Retina Tunica, unless it be deradiated from an object of an almost* Hemisphaerical *ambite.*    159
12 Why, *when the Eye is open there is always pourtrayed in the bottom thereof , some one* Total Image ; *whose various Parts, are the* Special Images *of the several things included in the visual* Hemisphere.    ibid.
13 PARADOX, *That the prospect of a shilling or object of a small diametre is as great, as the Prospect of the Firmament.*    160

(c)                        Why

14 *Why an object appears both greater in Dimensions and more Distinct in parts, neer at hand, than far off.* ibid.

15 *Why an object, speculated through a Convex Lens, appears both greater and more distinct; but through a Concave, less and more Confused: than when speculated only with the Eye.* 161

16 *DIGRESSION. What Figur'd Perspicils are convenient for Old: and what for Purblind persons.* 162

17 *That to the Disjudication of one of two objects, apparently Equal, to be really the Greater; is not required a greater Image: but only an Opinion of its greater Distance.* 163

18 *Des Cartes Opinion concerning the Reason of the Sights apprehending the Distance of an object:* 164

19 *Unsatisfactory; and that for two considerations.* ibid.

20 *And that more solid one of Gassendus (viz. that the Cause of our apprehending the Distance of an object, consisteth in the Comparation of the several things interjacent betwixt the object and the Eye, by the Rational Faculty) embraced and corroborated.* ibid.

21 *PARADOX. That the same Object, speculated by the same man, at the same distance, and in the same degree of light; doth alwayes appear greater to one Eye, than the other.* 165

22 *A second PARADOX. That all men see (distinctly) but with one Eye at once: contrary to that eminent Optical Axiom, that the Visive Axes of both eys concur, and unite in the object.* 166

23 *The three degrees of Vision, viz. most perfect, perfect, and imperfect: and the verity of the Paradox restrained onely to the two former Degrees.* 167

## Sect. III.

ARTIC.

1 *A Research into the Reason of the different Effects of Convex and Concave Glasses; as well Dioptrical, as Catoptrical.* ibid.

2 *A COROLLARIE. Hinting the Causes, why an Elliptical Concave reflects the incident rays, in a more Acute angle, than a Parabolical: and a Parabolical than a Spherical.* 170

3 *A CONSECTARY. Why a Plane Perspicil exhibits an object in genuine Dimensions; but a Convex, in Amplified, and a Concave in minorated.* 171

## Sect. IV.

ARTIC.

1 *A Recapitulation of the principal Arguments precedent: and summary of the subsequent.* 173

2 *The Eye Anatomized: and the proper use of each Part thereof, either absolutely Necessary, or onely Advantagious to Vision concisely demonstrated. viz. 1 The Diaphanity of the Horny Membrane, and the three Humors, Aqueous, Chrystalline, and Vitreous. 2 The Convexity of all its parts except the Amphiblestroides. 3 The Uvea Tunica, and Iris. 4 The Pupilla. 5 The Blackness of the inside of the Uvea Tunica. 6 The Tunica Arachnoides. 7 The Ciliary Filaments thereof. 8 The Chrystalline. 9 The Retina Tunica. 10 The six Muscles, viz. 1 The Direct, as the Attollent, Depriment, Adducent, Abducent. 2 And Oblique, as the 2 Circumactors, or Lovers Muscles. 173, to 177*

3 *Why the Situation of an object is perceived by the sight.* 177

4 *The Reason of the eversion of the Image, in the Amphiblestroides.* 178

5 *The same illustrate by an Experiment.* ibid.

6 *Why the Motion and Quiet of objects are discerned by the sight.* ibid.

7 *Why Catoptrical Images imitate the motions of their Antitypes or Originals.* ibid.

8 *Why the right side of a Catoptrical Image respects the Left of its Exemplar. And why two Catoptrick Glasses, confrontingly posited, cause a Restitution of the parts of the Image to the natural Form.* 180

## Chap. IV.
## The Nature of Colours. p. 182.

### Sect. I.

ARTIC.

1 *The Argument duly acknowledged to be superlatively Difficult, if not absolutely Acataleptical.* ibid.

2 *The sentence of Aristotle concerning the Nature of Colours: and the Commentary of Scaliger thereupon.* 183

3 *The opinion of Plato.* ibid.

4 *Of the Pythagorean and Stoick.* 184

5 *Of the Spagyrical Philosophers.* ibid.

6 *The reason of the Authors desertion of all these; and election of Democritus and Epicurus judgement, touching the Generation of Colours.* ibid.

7 *The Text of Epicurus, fully and faithfully expounded.* 185

### Sect. II.

ARTIC.

1 *A PARADOX. That there are no Colours in the Dark.* 186

2 *A familiar Experiment, attesting the Verity thereof.* ibid.

3 *The Constancy of all Artificial Tinctures, dependent*

pendent on the constancy of Disposition in the superficial Particles of the Bodies that wear them. 187

4 That so generally magnified Distinction of Colours into Inhærent, and meerly Apparent; redargued of manifest Contradiction. ibid.

5 The Emphatical, or Evanid Colours, created by Prisms; no less Real and Inhærent, than the most Durable Tinctures. 188

6 COROLLARY. The Reasons of Emphatical Colours, appinged on Bodies objected, by a Prism. 189

7 The true Difference of Emphatical and Durable Colours, briefly stated. ibid.

8 No Colour Formally inhærent in objects; but only Materially, or Effectively: contrary to the constant Tenent of the Schools. ibid

9 The same farther vindicated from Difficulty, by the tempestive Recognition of some præcedent Assumptions of the Atomists. 190

### SECT. III.

ARTIC.

1 The Nativity of White; or the reason of its perception by the sight. 191

2 Black, a meer Privation of Light. ibid.

3 The Genealogy of all Intermediate Colors. ibid.

4 The Causes of the Sympathy and Antipathy of some Colours 192

5 The intermistion of small shadows, among the lines of Light; absolutely necessary to the Generation of any Intermediate Colour. ibid.

6 Two eminent PROBLEMS concerning the Generation and Transposition of the Vermillion and Cærule, appinged on Bodies by Prismes. 193

7 The Solution of the Former: with a rational Conjecture of the Cause of the Blew, apparent in the Concave of the Heavens. 194

8 The Solution of the Latter. 195

9 The Reasons, why the Author proceeds not to investigate the Causes of Compound Colours in Particular. 196

10 He confesseth the Erection of this whole Discourse, on simple Conjecture: and enumerates the Difficulties to be subdued by him, who hopes to attain an Apodictical Knowledge of the Essence and Causes of Colours. ibid.

11 Des Cartes attempt to dissolve the chief of those Difficulties; unsuccessfull because grounded on an unstable Hypothesis. 197

## CHAP. V.

### The Nature of Light. p.198.

### SECT. I.

ARTIC.

1 The Clasp, or Ligament of this, to the præcedent Chapter. ibid.

2 The Authors Notion of the Rays of Light. ibid.

3 A Parallelism betwixt a stream of Water exsilient from the Cock of a Cistern, and a Ray of Light emanent from its Lucid Fountain. ibid.

PRÆCONSIDERABLES. 199

4 Light distinguisht into Primary, Secondary, &c. 199

5 All Light Debilitated by Reflection: and why. ibid.

6 An Example, sensibly demonstrating the same. 200

7 That light is in perpetual Motion; according to Aristotle. ibid.

8 Light, why Corroborated, in some cases, and Debilitated in others, by Refraction. 201

COROLLARY. Why the Figure of the Sun, both rising and setting, appears rather Elliptical, than Spherical. ibid.

9 PARADOX. That the proportion of Solary Rays reflected by the superior Aer, or Æther, toward the Earth, is so small, as not to be sensible. 202

10 That every Lucid Body, as Lucid, doth emit its Rays Sphærically: but, as Visible; Pyramidally. ibid.

11 That Light is invisible in the pure medium. 203

### SECT. II.

ARTIC.

1 The necessity of the Authors confirmation of the First Præconsiderable. 204

2 The Corporiety of Light, demonstrated by its just Attributes: viz. 1 Locomotion. 2 Resilition. 3 Refraction. 4 Coition. 5 Disgregation. 6 Ignity. 224, 225

3 Aristotles Definition of Light, a meer Ambage, and incomprehensible. 205

4 The Corporiety of Light imports not the Coexistence of two Bodies in one Place; contrary to the Peripatetick. 206

5 Nor the motion of a Body to be Instantaneous. ibid.

6 The Invisibility of Light in the limpid medium, no Argument of its Immateriality: as the Peripatetick presumes ibid.

7 The Corporiety of Light fully consistent with the Duration of the Sun: contrary to the Peripatetick. 207

8 The insensibility of Heat in many Lucent Bodies, no valid Argument against the present Thesis, that Light is Flame Attenuated. ibid.

CHAP.

## Chap. VI.

### The Nature of a Sound. p 208.

#### Sect. I.

ARTIC.

1 AN *Elogy of the sense of* Hearing : *and the* Relation *of this and the precedent* Chapter. ibid.

2 *The great Affinity betwixt* Visible *and* Audible *species ; in their representation of the superficial Conditions of Objects.* 209

3 *In the Causes and manner of their* Destruction. ibid.

4 *In their* Actinobolism, *or* Diffusion, *both* Sphærical *and* Pyramidal. 210

5 *In their certifying the sense of the* Magnitude, Figure, *and other Qualities of their Originals.* ibid.

6 *In the obscuration of* Less *by* Greater. 211

7 *In their* offence *of the organs, when* excessive. ibid.

8 *In their production of* Heat *by* Multiplication. ibid.

9 *In their* Variability, *according to the various disposition of the Medium.* ibid.

10 *In their chief Attributes, of* Locomotion, Easilition, Impaction, Resilition, Disgregation, Congregation. ibid.

#### Sect. II.

ARTIC.

1 THE *Product of the Premises, concerning the points of Consent, and Dissent of* Audible *and* Visible Species : viz. That *Sounds are* Corporeal. 213

2 *An obstruction of prejudice, from the generally supposed repugnant* Authorities *of some of the* Ancients ; *expeded.* ibid.

3 *An Argument of the Corporiety of* Sounds. 214

4 *A Second* Argument. ibid.

*COROLLARY.* ibid.

5 *The Causes of* Concurrent Echoes, *where the* Audient *is equally (almost) distant from the* Sonant *and* Repercutient. ibid.

*COROLLARY 2.* 215

6 *Why* Concaves *yield the strongest and longest* Sounds. ibid.

*COROLLARY. 3.* ibid.

7 *The reason of* Concurrent Echoes, *where the* Audient *is near the* Reflectent, *and remote from the* Sonant. ibid.

*COROLLARY. 4.* ibid.

8 *Why* Echoes Monophon *rehearse so much the fewer syllables, by how much nearer the audient is to the* Reflectent. ibid.

*COROLLARY. 5.* ibid.

9 *The reason of* Polyphon *Echoes.* ibid.

10 *A Third* Argument *of the* Materiality *of* Sounds. 216

11 *The necessity of a certain* Configuration *in a* Sound ; *inferred from the* Distinction *of one sound from another, by the* Sense. ibid.

12 *The same confirmed by the* Authority *of* Pythagoras, Plato, *and* Aristotle. ibid.

13 *And by the* Capacity *of the most subtle parts of the* Aer 217

14 *The Reason and manner of the* Diffusion *of* Sounds, *explicated by a congruous* Simile. ibid.

15 *The most subtle Particles of the* Aer *only, the* material *of* Sounds. 218

PARADOX. ibid.

16 *One and the same numerical voice, not heard by two men, nor both ears of one man.* ibid.

17 *A* PROBLEM *not yet solved by any Philosopher : viz. How such infinite Variety of Words is formed only by the various motions of the* Tongue *and* Lips. 219

18 *A Second (also yet unconquered)* Difficulty, viz. *the determinate* Pernicity *of the* Aers *motion, when exploded from the* Lungs, *in* Speech. ibid.

19 *All* Sounds *Created by* Motion, *and that either when that intermediate* Aer *is controlled by two solids mutually resistent ; or when the* aer *is percust by one* Solid ; *or when a* solid *is percust by the* Aer. ibid.

20 Rapidity *of motion necessary to the* Creation *of a* Sound, *not in the* First Case. 220

21 *But, in the* Second *and* Last. ibid.

22 *That all* Sounds *are of* equal Velocity *in the* Delation. ibid.

23 *The* Reason *thereof.* ibid.

24 *To measure the* Velocity *of great* Sounds. 221

25 Sounds, *not subject to* Retardation *from* adverse ; *nor* Acceleration, *from* Secund *Winds.* ibid.

#### Sect. III.

ARTIC.

1 THat *all* Sounds, *where the* Aer *is percussed by one* solid, *are created immediately by the* Frequency, *not the* Velocity *of* motion ; *demonstrated.* 222

2 *And likewise, where the* Aer *is the* Percutient. ibid.

3 *That all* Acute *sounds arise from the more, and* Grave *from the less* Frequent *percussions of the* aer, *demonstrated.* 223

4 *The* suavity *of* musical Consonances, *deduced from the more* frequent ; *and* Insuavity *of* Dissonances *from the less* frequent Union *of the* vibrations *of strings, in their* Terms. 224

5 *The same* Analytically *presented in* Scheme. 226

6 *A just and unanswerable* Exception *against the former* Harmonical Hypothesis. ibid.

7 PROBLEM 1. *In what instant, an* Harmonical

*monical Sound, resulting from a Chord percussed, is begun.* 227
8 *That a Sound may be created in a Vacuum; contrary to* Athanas. Kircher *in* Art.Magn. Conson & Dissoni lib.1.cap.6,Digres. 229
9 *Why all Sounds appear more Acute, at large, than at small distance.* 231
10 *Why Cold water falling, makes a fuller noise, than warm.* ibid.
11 *Why the voice of a Calf is more Base than than that of an Ox, &c.* 232
12 *Why a Dissonance in a Base is more deprehensible by the ear, than in a Treble voice.* ibid.

## CHAP. VII.
## Of Odours. p 233.

### SECT. I.

ARTIC.
1 *THat the Cognition of the Nature of Odours is very difficult; in respect of the Im-*perfection *of the sense of* Smelling, *in man: and* ibid.
2 *The contrary opinions of* Philosophers, *concerning it.* 234
3 *Some determining an Odour to be a* substance. ibid.
4 *Others a meer* Accident *or* Quality. 235
5 *The Basis of the Latter opinion, infirm and ruinous.* 235
6 *That all odorous Bodies emit* corporeal *Exhalations.* ibid.
7 *That Odours cause sundry* Affections *in our Bodies, such as are consignable onely to* substances. ibid.
8 *That the Reason of an Odour's affecting the sensory, consists only in a certain* Symbolism *betthe Figures and Contexture of its Particles, and the Figures and Contexture of the Particles of the Odoratory Nerves.* 236
9 *That the Diversity of* Odours *depends on the Diversity of Impressions made on the sensory, respondent to the various Figures and Contexture of their Particles.* 237
10 *Why some persons abhor those smells, which are grateful to most others.* ibid.
11 *Why, among* Beasts, *same species are offended at those scents in which others highly delight.* 238
12 *The Generation and Diffusion of Odours, due onely to Heat.* ibid.
13 *The Differences of Odours.* 239
14 *The Medium of Odours.* 240

## CHAP. VIII.
## Of Sapours. p. 241.

### SECT. I.

ARTIC.
1 *FRom the superlative Acuteness of the sense of Tasting,* Aristotle *concludes the cognition of the Nature of Sapours to be more easily acquirable, than the nature of any other sensible object: but refutes himself by the many Errors of his own Theory, concerning the same.* ibid.
2 *An Abridgment of his doctrine, concerning the Essence and Causes of a Sapour, in General.* 242
3 *And the Differences of Sapour, with the particular Causes of each.* ibid.
4 *An Examination and brief redargution of the same Doctrine.* 244
5 *The postposition thereof to the more verisimilous Determination of the sons of* Hermes, *who adscribe all Sapours to* Salt. ibid.
6 *But far more to that most profound and satisfactory Tenent of* Democritus *and* Plato; *which deduceth the Nativity of Sapours from the various Figures and contextures of the minute particles of Concretions.* ibid.
7 *The advantages of this sentence, above all others touching the same subject.* 245
8 *The Objections of* Aristotle *concisely, though solidly solved.* 246
9 *That the salivous Humidity of the Tongue serveth to the Dissolution and Imbibition of Salt in all Gustables.* 247

## CHAP. IX.
## Of Rarity, Density, Perspicuity, Opacity. p. 248.

### SECT. I.

ART.
1 *THis Chapters right of succession to the former.* ibid.
2 *The Divers acceptation of the term,* Touching. ibid.
3 *A pertinent (though short) Panegyrick on the sense of* Touching. 249
4 *Some Tactile Qualities, in common to the perception of other senses also.* ibid.
5 *A Scheme of all Qualities, or Commonly, or Properly appertaining to the Sense of Touching; as they stand in their several Relation to, or Dependencies on, the Universal Matter,* Atoms: *and so, of all the subsequent Capital Arguments to be treated of, in this Book.* 250
6 *The right of Rarity and Density to the Priority of consideration.* ibid.

d                                    SECT,

# THE CONTENTS, SERIES,

## Sect. II.

Artic.

1 THe Opinion of those Philosophers, who place the Reason of Rarity, in the actual Division of a Body into small parts; and the brief Refutation thereof. 251

2 A second Opinion, deriving Rarity and Density from the several proportions, which Quantity hath to its substance: convicted of incomprehensibility, and so of insatisfaction. ibid.

3 A third, deriving the more and less of Rarity in Bodies, from the more and less of VACUITY interceded among their particles: and the advantages thereof above all others, concerning the same. ibid.

4 The Definitions of a Rare, and of a Dense Body; according to the assumption of a Vacuity Disseminate. 252

5 The Congruity of those Definitions, demonstrated. ibid.

6 That Labyrinth of Difficulties, wherein the thoughts of Physiologists have so long wandered; reduced to a point, the genuine state of the Question. ibid.

7 That Rarity and Density can have no other Causes immediate, but the more and less of Inanity interspersed among the particles of Concretions; DEMONSTRATED. 253

8 Aristotles Exceptions against Disseminate Inanity; neither important nor competent. ibid.

9 The Hypothesis of a certain Æthereal substance to replenish the pores of Bodies, in Rarifaction; demonstrated insufficient, to solve the Difficulty, or demolish the Epicurean Thesis of small Vacuities. 254

10 The Facility of understanding the Reasons and Manner of Rarifaction and Condensation, from the Concession of small Vacuities; illustrated by a congruent Similitude. 255

11 PARADOX, That the Matter of a Body, when Rarified, doth possess no more of true Place, than when Condensed, and the Conciliation thereof to the proposed Definitions of a Rare and of a Dense Body. 256

12 PROBLEM, Whether Air be capable of Condensation to so high a rate as it is of Rarifaction: and the Apodictical solution thereof. ibid.

## Sect. III.

Artic.

1 THe opportunity of the present speculation, concerning the Causes of Perspicuity and Opacity. 258

2 The true Notions of a Perspicuum and Opacum. ibid.

3 That every Concretion is so much the more Diaphanous by how much the more, and more ample Inane Spaces are intercepted among its particles, cæteris paribus. ibid.

4 Why Glass though much more Dense, is yet much more Diaphanous, than Paper. 259

5 Why the Diaphanity of Glass is gradually diminished, according to the various degrees of its Crassitude. ibid.

6 An Apodictical Confutation of that popular Error, that Glass is totally, or in every particle, Diaphanous. 260

## CHAP. X.

### Of Magnitude, Figure; And their Consequents, Subtility, Hebetude, Smoothness, Asperity. 261

## Sect. I.

Artic.

1 THe Contexture of this Chapter, with the precedent. ibid.

2 That the Magnitude of Concretions, ariseth from the Magnitude of their Material Principles. ibid.

3 The present intention of the term, Magnitude. ibid.

4 That the Quantity of a thing, is meerly the Matter of it. 22

5 The Quantity of a thing, neither Augmented by its Rarefaction; nor diminished by its Condensation: contrary to the Aristotelians, who distinguish the Quantity of a Body from its Substance. ibid.

6 The reason of Quantity, explicable also meerly from the notion of Place. 263

7 The Existence of a Body, without real Extension; and of Extension without a Body: though impossible to Nature yet easie to God. ibid.

8 COROLLARY, That the primary Cause, why Nature admits no Penetration of Dimensions, is rather the Solidity, than the Extension of a Body. 264

9 The reasons of Quantity Continued and Discrete, or Magnitude and Multitude. ibid.

10 That no Body is perfectly Continued, beside an Atom. ibid.

11 Aristotles Definition of a Continuum, in what respect true, and what false. 265

12 Figure (Physically considered) nothing but the superficies, or terminant Extremes of a Body. ibid.

## Sect. II.

Artic.

1 THe Continuity of this, to the first Section. 266

2 Subtility and Hebetude, how the Consequents of Magnitude. ibid.

3 A considerable Exception of the Chymists (viz. that

that some Bodies are dissolved in liquors of grosser particles, which yet conserve their Continuity in liquors of most subtile and corrosive particles) prevented. ibid.

4 Why Oyle dissociates the parts of some Bodies, which remain inviolate in Spirit of Wine: and why Lightning is more penetrative, than Fire. 267

5 Smoothness and Asperity in Concretions, the Consequents of Figure in their Material Principles. ibid.

## CHAP. XI.

## Of the Motive Virtue, Habit, Gravity, and Levity of Concretions.
### 269

### SECT. I.

ARTIC.

1 THe Motive Virtue of all Concretions, derived from the essential Mobility of Atoms ibid.

2 Why the Motive Virtue of Concretions doth reside principally in their spiritual Parts. 270

3 That the Deviation of Concretions from motion Direct; and their Tardity in motion: arise from the Deflections and Repercussions of Atoms composing them. ibid.

4 Why the motion of all Concretions necessarily presupposeth something, that remains unmoved; or that, in respect of its slower motion, is equivalent to a thing Unmoved. ibid.

5 What the Active Faculty of a thing, is. 271

6 That in Nature every Faculty is Active: none Passive. ibid.

7 A Peripatetick Contradiction, assuming the Matter of all Bodies to be devoid of all Activity; and yet desuming some Faculties à tota substantia. 272

8 That the Faculties of Animals (the Ratiocination of man onely excepted) are Identical with their spirits. ibid.

9 The Reason of the Coexistence of Various Faculties in one and the same Concretion. ibid.

10 Habit defined. 273

11 That the Reason of all Habits in Animals, consisteth principally in the conformity and flexibility of the Organs, which the respective Faculty makes use of, for the performance of its proper Actions. ibid.

12 Habits, acquirable by Bruits: and common not onely to Vegetables, but also to some Minerals 274

### SECT. II.

ARTIC.

1 GRavity, as to its Essence or Formal Reason, very obscure. 275

2 The opinion of Epicurus good as to the Cause of Comparative: insufficient as to the Cause of Absolute Gravity. ibid.

3 Aristotles opinion of Gravity, recited. ibid.

4 Copernicus theory of Gravity, insatisfactory; and wherein. 276

5 The Determination of Kepler, Gassendus, &c. that Gravity is Caused meerly by the Attraction of the Earth: espoused by the Author. 277

6 The External Principle of the perpendicular Descent of a stone, projected up in the Aer; must be either Depellent, or Attrahent. ibid.

7 That the Resistence of the Superior Aer is the onely Cause which gradually refracteth, and in fine wholly oversometh the Imprest Force, whereby a stone projected, is elevated upward. ibid.

8 That the Aer, distracted by a stone violently ascending, hath as well a Depulsive, as a Resistent Faculty; arising immediately from its Elatericai, or Restorative motion. 279

9 That nevertheless, when a stone, projected on high in the Aer, is at the highest point of its mountee; no Cause can Began its Downward Motion, but the Attractive Virtue of the Earth. 280

10 Argument, that the Terraqueous Globe is endowed with a certain Attractive Faculty, in order to the Detention and Retraction of all its Parts. 281

11 What are the Parts of the Terrestrial Globe 282

12 A Second Argument that the Earth is Magnetical ibid.

13 A Parallelism betwixt the Attraction of Iron by a Loadstone, and the Attraction of Terrene bodies by the Earth. 283

14 That as the sphere of the Loadstones Allective Virtue is limited: so is that of the Earths magnetism. ibid.

15 An Objection of the Disproportion between the great Bulk of a large stone and the Exility of the supposed magnetique Rays of the Earth: Solved by three weighty Reasons. 284

16 The Reason of the Æquivelocity of Bodies, of different weights, in their perpendicular Descent: with sundry unquestionable Authorities to confirm the Hoti thereof. 285

17 That the whole Terrestrial Globe is devoid of Gravity: and that in the universe is no Highest, nor Lowest place. 286

18 That the Centre of the Universe is not the Lowest part thereof: nor the Centre of the Earth, the Centre of the World. 287

19 A Fourth Argument, that Gravity is onely Attraction. 289

20 Why a greater Gravity, or stronger Attractive force is imprest upon a piece of Iron by a Loadstone, than by the Earth. ibid.

21 A Fifth Argument, almost Apodictical; that

*that Gravity is the Effect of the Earths Attraction.* ibid.

### SECT. III.

ARTIC.

1 *L*Evity *nothing but less* Gravity. 290
2 *Aristotles Sphere of Fire, extinguisht.* 291
3 *That Fire doth not Ascend spontaneously ; but* Violently ; *i.e. is impell'd upward by the Aer.* ibid.

### CHAP. XII.

*Of Heat and Cold.* p. 293.

### SECT. I.

ARTIC.

1 *T*He *Connection of this to the immediately precedent Chapter.* ibid.
2 *Why the Author deduceth the 4 First Qualities, not from the 4 vulgar Elements ; but from the 3 Proprieties of Atoms.* ibid.
3 *The Nature of Heat is to be conceived from its* General Effect *; viz. the* Penetration, Discussion, *and* Dissolution *of the Bodies concrete.* ibid.
4 *Heat defined as no Immaterial, but a* Substantial *Quality.* 294
5 *Why such Atoms, as are comparated to produce* Heat *, are to be Named the* Atoms of Heat *; and such Concretions, as harbor them, are to be called* Hot, *either* Actually, *or* Potentially. ibid.
6 *The 3 necessary Proprieties of the Atoms of* Heat. ibid.
7 *That the Atoms of Heat are capable of* Expedition *or deliverance from Concretions, Two wayes; viz. by* Evocation *and* Motion. 296
8 *An* Unctuous *matter, the chief Seminary of the Atoms of Heat : and why.* 297
9 *Among Unctuous Concretions, Why some are more easily inflammable than others.* 298
10 *A CONSECTARY. That Rarefaction is the proper Effect of Heat.* ibid.
11 *PROBLEM 1. Why the bottom of a Caldron, wherein Water is boyling, may be touched by the hand of a man, without burning it: Sol.* 299
12 *PROBLEM 2. Why Lime becomes ardent upon the effusion of Water. Sol.* 300
13 *PROBLEM 3. Why the Heat of Lime burning is more vehement, than the Heat of any Flame whatever. Sol.* ibid.
14 *PROBLEM 4. Why boyling Oyl scalds more vehemently, then boyling Water. Sol.* 301
15 *PROBLEM 5. Why* Metals, *melted or made red hot, burn more violent than the Fire, that melteth or heateth them. Sol.* ibid.
16 *CONSECTARY 1. That, as the degrees of Heat, so those of fire are innumerably various.* ibid.

17 *That to the Calefaction, Combustion, or Inflammation of a body by fire, is required a certain space of time ; and that the space is greater or less, according to the paucity ; or abundance of the igneous Atoms invading the body objected ; and more or less of aptitude in the contexture thereof to admit them.* 301
18 *Flame more or less Durable, for various respects.* 303
19 *CONSECTARY 3. That the immediate and genuine Effect of Heat, is the* Disgregation *of all bodies, as well Homogeneous, as Heterogeneous : and that the* Congregation *of Homogeneous Natures, is only an* Accidental *Effect of Heat ; contrary to* Aristotle. 305

### SECT. II.

ARTIC.

1 *T*He *Link connecting this Section to the former.* 306
2 *That* Cold *is no Privation of* Heat *; but a* Real *and* Positive Quality *: demonstrated.* ibid.
3 *That the adequate* Notion *of* Cold, *ought to be desumed from its* General Effect, *viz. the* Congregation *and* Compaction *of bodies.* 307
4 *Cold, no Immaterial ; but a* Substantial *Quality.* ibid.
5 *Gassendus conjectural Assignation, of a* Tetrahedical *Figure to the Atoms of cold ; asserted by sundry weighty considerations.* ibid.
6 *Cold, not Essential to* Earth, Water, *nor* Aer. 309
7 *But to some* Special Concretions *, for the most part, consisting of* Frigorifick Atoms. 312
8 *Water, the chief* Antagonist *to* Fire *; not in respect of its* Accidental Frigidity, *but* Essential Humidity *: and that the* Aer *hath a juster title to the Principality of* Cold *; than either* Water, *or* Earth. 313
9 *PROBLEM : Why the breath of a man doth* Warm, *when expired with the mouth wide open ; and* Cool, *when efflated with the mouth contracted.* ibid.
10 *Three CONSECTARIES from the premisses.* 314

### CHAP. XIII.

*Fluidity, Stability, Humidity, Siccity.* p. 316.

### SECT. I.

ARTIC.

1 *W*Hy Fluidity *and* Firmness *are here considered before* Humidity *and* Siccity. ibid.
2 *The Latin Terms,* Humidum *and* Siccum, *too narrow to comprehend the full sense of* Aristotle. ibid.
3 *Aristotles*

2 *The Latin Terms*, Humidum *and* Siccum, *too narrow to comprehend the full sense of Aristotle.* ὑγρὸν καὶ ξηρὸν.                    ibid.
3 Aristotles *Definition of a Humid substance, not practise enough; but, in common also to a Fluid; and his Definition of a Dry, accommodable to a Firme.*                    317
4 Fluidity *defined*                    318
5 *Wherein the Formal Reason thereof doth consist.*                    ibid.
6 *The same farther illustrated, by the twofold Fluidity of Metals; and the peculiar reason of each.*                    319
7 *Firmness defined:*                    320
8 *And derived from either of 3 Causes.*    ibid.

## SECT. II.

ARTIC.
1 Humidity *defined.*                    321
2 Siccity *defined.*                    322
3 *Siccity rather* Comparative *than Absolute.* ibid.
4 *All moisture either* Aqueous *or* Oleaginous.                    ibid.
5 PROBLEM. *Why pure water cannot wash out oyl from a Cloth; which yet water, wherein Ashes have been decocted, or soap dissolved, easily doth? Solut.*                    323
6 PROBLEM 2. *Why stains of Ink are not to be taken out of cloaths, but with some Acid Liquor? Solut.*                    ibid.

## CHAP. XIV.

### Softness, Hardness, Flexility Tractility, Ductility, &c. p. 325.

## SECT. I.

ARTIC.
1 The Illation *of the Chapter.*                    ibid
2 Hard *and* Soft, *defined.*                    ibid.
3 *The Difference betwixt a Soft and Fluid.* 326
4 Solidity *of Atoms; the Fundament of Hardness and* Inanity *intercepted among them, the fundament of Softness, in all Concretions.* ibid.
5 *Hardness and Softness, no Absolute, but meerly* Comparative *Qualities; as ascriptive to* Concretions, *contrary to Aristotle.*                    327
6 Softness *in* Firme *things, deduced from the same cause, as Fluidity in Fluid ones.* ibid.
7 *The General Reason of the Mollification of Hard, and Induration of Soft bodies.*    ibid.
8 *The special manners of the Mollification of* Hard: *and Induration of Soft bodies.*    328
9 PROBLEM. *Why Iron is Hardned, by being immersed red-hot into Cold Water; and its* SOLUTION.                    ibid.
10 *The Formal Reasons of Softness and Hardness.*                    329
11 *The ground of Aristotles Distinction betwixt* Formabilia *and* Pressilia.                    ibid.

12 *Two* Axioms, *concerning, and illustrating the nature of Softness.*                    330

## SECT. II.

ARTIC.
1 Flexility, Tractility, Ductility, &c. *derived from* Softness, *and* Rigidity *from* Hardness                    331
2 PROBLEM. *What is the Cause of the motion of* Restoration *in* Flexiles? *and the* Solut.                    ibid.
3 *Two* Obstructions *expeded.*                    332
4 *Why* Flexile *bodies grow weak, by overmuch, and over frequent Bending.*                    333
5 *The Reason of the frequent Vibrations, or Diadroms of* Lutestrings, *and other* Tractile *Bodies; declared to be the same with that of the* Restorative *Motion of* Flexiles: *and demonstrated.*                    ibid.
6 PROBLEM. *Why the Vibrations, or Diadroms of a Chord distended and percussed, are* Æquitemperaneous, *though not* Æquispatial: *and the* SOLUT.                    335
7 PROBLEM. *Why doth a Chord of a duple length, perform its diadroms in a proportion of time duple, to a Chord of a single length; both being distended by equal force; and yet if the Chord of the duple length be distended by a duple force or weight, it doth not perform its Diadroms, in a proportion of time duple to that of the other; but onely if the Force or weight distending it, be quadruple to the First supposed: and its* SOLUT.                    336
8 *The Reasons of the vast* Ductility, *or Extensibility of* Gold.                    337
9 Sectility *and* Fissility, *the Consequents of* Softness.                    ibid.
10 Tractility *and* Friability, *the Consequents of* Hardness.                    338
11 Ruptility *the Consequent partly of* Softness, *partly of* Hardness.                    339
12 PROBLEM. *Why Chords distended, are more apt to break, neer the Ends, than in the middle? and its* SOLUT.                    ibid.

## CHAP. XV.

### Occult Qualities made Manifest.
### P. 341

## SECT. I.

ARTIC.
1 *That the Insensibility of Qualities doth not import their Unintelligibility; contrary to the presumption of the Aristoteleans.* ibid.
2 *Upon what grounds; and by whom, the Sanctuary of Occult Qualities was erected.*    342
3 Occult Qualities *and profest ignorance, all one.*                    ibid.
4 *The Refuge of* Sympathies *and* Antipathies, *equally*

(c)

equally obstructive to the advance of Natural Science, with that of Ignote Proprieties. 343

5 That all Attraction, referred to Secret Sympathy; and all Repulsion, ascribed to secret Antipathy, betwixt the Agent and Patient, is effected by Corporeal Instruments, and such as resemble those whereby one body Attracteth, or repelleth another, in sensible and mechanique operations. ibid.

6 The Means of Attractions sympathetical, explicated by a convenient Simile. 345

7 The Means of Abaction and Repulsions Antipathetical, explicated likewise by sundry similitudes. 346

8 The First and General Causes of all Love and Hatred betwixt Animals. 347

9 Why things Alike in their natures, love and delight in the Society each of other: and why Unlike natures abhor and avoid each other ibid

SECT. II.

ARTIC.

1 THe Scheme of Qualities (reputed) occult. 348

2 Natures Avoidance of Vacuity, imputed to the τυχυγια or Conspiration of all parts of the Universe; no Occult Quality. ibid.

3 The power and influence of Cælestial Bodies, upon men, supposed by Judicial Astrologers, inconsistent with Providence Divine, and the Liberty of mans will. 349

4 The Afflux and Reflux of the Sea, inderivative from any immaterial Influx of the Moon. ibid.

5 The Causes of the diurnal Expansion & conversion of the Heliotrope and other Flowers. ibid.

6 Why Garden Claver hideth its stalk, in the heat of the day. 350

7 Why the House Cock usually Crows soon after midnight; and at break of day. ibid.

8 Why Shell-fish grow fat in the Full of the moon, and lean again at the New. 352

9 Why the Selenites resembles the Moon in all her several Adspects. ibid.

10 Why the Consideration of the Attraction of Iron by a Loadstone, is here omitted. 353

11 The secret Amities of Gold and Quicksilver of Brass and Silver, unriddled. ibid.

12 A COROLLARY. Why the Granules of Gold and Silver, though much more ponderous then those of the Aqua Regis and Aqua Fortis, wherein they are dissolved, are yet held up, and kept floating by them. 354

13 The Cause of the Attraction of a Less Flame by a Greater. ibid.

14 The Cause of the Involation of flame to Naphtha at distance. ibid.

15 Of the Ascension of Water into the pores of a Spunge. 355

16 The same illustrated by the example of a Syphon. ibid.

17 The reason of the Percolation of Liquors, by a cloth whose one end lieth in the liquor, and other hangs over the brim of the vessel, that contains it. 356

18 The reason of the Consent of two Lutestrings, that are Æquison. ibid.

19 The reason of the Dissent betwixt Lutestrings of sheeps Guts, and those of Wools. 357

20 The tradition of the Consuming of all Feathers of Foul, by those of the Eagle; exploded. 358

21 Why some certain Plants befriend, and advance the growth and fruitfulness of others, that are their neighbours. ibid.

22 Why some Plants thrive not in the society of some others. 359

23 The Reason of the great friendship betwixt the Male and Female Palm-tree. 360

24 Why all wines grow sick and turbid, during the season wherein the Vines Flower and Bud. 361

25 That the distilled waters of Orange flowers, and Roses, do not take any thing of their fragrancy, during the season of the Blooming and pride of those Flowers; as is vulgarly believed. ibid.

SECT. III.

ARTIC.

1 WHy this Section considers onely some few select Occult Proprieties, among those many imputed to Animals. 362

2 The supposed Antipathy of a Sheep to a Woolf solved. ibid.

3 Why Bees usually invade Froward and Cholerick Persons: and why bold and confident men have sometimes daunted and put to flight, Lyons and other ravenous Wild-Beasts. 363

4 Why divers Animals Hate such men, as are used to destroy those of their own Species: and why Vermin avoid such Gins and Traps, wherein others of their kinde have been caught and destroyed. ibid.

5 The Cause of the fresh Cruentation of the Carcass of a murthered man, at the presence and touch of the Homicide. 364

6 How the Basilisk doth empoyson and destroy, at distance. 365

7 That the sight of a Woolf doth not cause Hoarsness and obmutescence in the spectator; as is vulgarly reported and believed. 366

8 The Antipathies of a Lyon and Cock: of an Elephant and Swine meerly Fabulous. 367

9 Why a man intoxicated by the venome of a Tarantula, falleth into violent fits of Dancing: and cannot be cured by any other means, but Musick. ibid.

10 Why Divers Tarantiacal Persons are affected and cured with Divers Tunes, and the musick of divers Instruments. 369

11 That

11 *That the venome of the Tarantula doth produce the same effect in the body of a man; as it
doth in that of the Tarantula it self: and
why.* ibid

12 *That the Venom of the Tarantula is lodged
in a viscous Humor, and such as is capable of
Sounds.* 371

13 *That it causeth an uncessant* Itching *and* Titillation *in the Nervous and Musculous parts
of mans body, when infused into it, and fermenting in it.* ibid.

14 *The cause of the* Annual Recidivation *of the
Tarantism, till it be perfectly cured.* 372

15 *A Conjecture, what kind of* Tunes, Strains,
*and* Notes *seem most accommodate to the cure
of* Tarantiacal Persons *in the* General. ibid

16 *The Reason of the* Incantation *of* Serpents,
*by a rod of the* Cornus. 373

17 *DIGRESSION. That the* Words, Spells,
Characters, &c. *used by Magicians, are of no
vertue or Efficacy at all, as to the Effect intended; unless in a remote interest, or as they
exalt the* Imagination *of* Him, *upon whom
they pretend to work the miracle.* ibid.

18 *The Reason of the* Fascination *of* Infants, *by
old women.* 374

19 *The Reason of the* stupefaction *of a mans
hand by a* Torpedo. 375

20 *That ships are not* Arrested *in their course, by
the Fish called a* Remora: *but by the Contrary impulse of some* Special Current *in the Sea.* ibid.

21 *That the Echineis, or Remora is not* Ominous. 377

22 *Why this place admits not of more than a* General *Inquest into the* Faculties *of* Poysons *and
Counterpoisons.* ibid.

23 Poysons *defined.* ibid.

24 *Wherein the* Deleterious *Faculty of poyson
doth consist.* ibid.

25 Counterpoisons *defined.* 378

26 *Wherein their* Salutiferous *Virtue doth consist.* ibid.

27 *How* Triacle *cureth the venome of* Vipers ibid

28 *How the body of a* Scorpion, *bruised and laid
warm upon the part, which it hath lately wounded and envenomed; doth cure the same.* 379

29 *That some* Poisons *are* Antidotes *against others
by way of direct* Contrariety. ibid

30 *Why sundry particular men, and some whole
Nations have fed upon* Poisonous *Animals and
Plants, without harm* 380

31 *The* Armary Unguent, *and* Sympathetick
Powder, *impugned.* ibid.

32 *The Authors* Retractation *of his quondam Defence of the* Magnetick Cure *of* Wounds,
*made in his* Prolegomena *to* Helmonts *Book
of that subject and title.* 381

## CHAP. XVI.

## The Phænomena of the Loadstone Explicated. p.383.

### SECT. I.

ARTIC.

1 THe Nature *and* Obscurity *of the Subject,
hinted by certain* Metaphorical *Cognomina, agreeable thereunto, though in divers relations.* ibid.

2 *Why the Author insisteth not upon the* (1) *several* Appellations, (2) Inventor *of the
Loadstone,* (3) *invention of the* Pixis Nautica. 384

3 *The Virtues of the Loadstone, in General, Two,
the* Attractive, *and* Directive. ibid.

4 Epicurus *his first Theory of the Cause and
Manner of the Attraction of Iron by a Loadstone; according to the Exposition of* Lucretius. ibid.

5 *His other solution of the same, according to the
Commentary of* Galen. 386

6 Galens *three Grand Objections against the
same, briefly Answered.* 387

7 *The insatisfaction of the* Ancients *Theory necessitates the Author to recur to the Speculations and Observations of the* Moderns, *concerning the Attraction of Iron by a* Magnes;
*and the Reduction of them all to a few* Capital observables. viz. 388

8 *A Parallelism betwixt the* Magnetique *Faculty of the Loadstone and Iron; and that of
Sense in Animals.* 389

9 *That the Loadstone and Iron interchangeably
operate each upon other, by the mediation of certain Corporeal Species, transmitted in* Rays:
*and the Analogy of the* Magnetick, *and* Luminous Rayes. 390

10 *That every Loadstone, in respect of the Circumradiation of its Magnetical Aporrha's
ought to be allowed the supposition of a Centre
Axis, and* Diametre *of an Æquator: and the
Advantages thence accrewing.* 391

11 *The Reason of that admirable* Bi-form, *or
Janus-like Faculty of Magneticks: and why
the Poles of a Loadstone are incapable, but those
of a Needle easily capable of transplantation
from one Extreme to the contrary.* 392

12 *An Objection, of the Aversion or Repulsion of the North Pole of one Loadstone, or
Needle, by the North Pole of Another: prævented.* 393

13 *Three principal Magnetick Axioms, deduced
from the same Fountain.* ibid.

14 *A DIGRESSION to the Iron Tomb of
Mahomet.* 394

15 *That the Magnetique Vigour, or Perfection
both of Loadstones and Iron, doth consist in either*

c

*their their* Native Purity *and Uniformity of Substance, or their Artificial Politeness.* 396

16 *That the Arming of a Magnet with polished Steel, doth highly Corroborate; but as much diminish the sphere of its* Attractive Virtue. ibid.

17 *Why a smaller or weaker Loadstone, doth snatch away a Needle from a* Greater, *or more* Potent *one; while the small or weak one is held within the sphere of the great or stronger ones Activity: and not otherwise.* 397

18 COROLLARY. *Of the Abduction of Iron from the Earth by a Loadstone.* 398

### Sect. II.

ARTIC.

1 THe Method, *and* Contents *of the Sect.* ibid.

2 *Affinity of the Loadstone and Iron.* ibid.

3 *The Loadstone conforms it self, in all respects, to the Terrestrial Globe; as a Needle conforms it self to the Loadstone.* 399

4 *Iron obtains a* Verticity, *not onely from the Loadstone, by affriction, or Aspiration; but also from the Earth it self: and that according to the laws of Position.* 400

5 *One and the same Nature, in common to the Earth, Loadstone and Iron.* 401

6 *The Earth, impraegnating Iron with a Polary Affection, doth cause therein a Local Immutation of its insensible particles.* 402

7 *The Loadstone doth the same.* 403

8 *The Magnetique Virtue, a Corporeal Efflux.* ib.

9 *Contrary* Objections, *and their* Solutions. 404

10 *A Parallelism of the* Magnetique Virtue, *and the* Vegetative Faculty *of Plants.* 405

11 *Why Poles of the same respect and name, are Enemies: and those of a Contrary respect and name, Friends.* 406

12 *When a Magnet is dissected into two pieces, why the Boreal part of the one half, declines Conjunction with the Boreal part of the other; and the Austral of one with the Austral of the other.* ibid.

13 *The Fibres of the Earth extend from Pole to Pole; and that may be the Cause of the firm Cohaesion of all its Parts, conspiring to conserve its Spherical Figure.* 407

14 *Reason of Magnetical* Variation, *in divers climates and places.* ibid.

15 *The Decrement of Magnetical Variation, in one and the same place, in divers years.* 410

16 *The Cause thereof not yet known.* ibid.

17 *No Magnet hath more than* Two Legitimate *Poles: and the reason of* Illegitimate *ones* 411

18 *The Conclusion, Apologetical; and an Advertisement, that the Attractive and Directive Actions of Magnetiques, arise from one and the same Faculty; and that they were distinguished onely ἡδυπαθλος χάριν, for convenience of Doctrine.* 412

# The Fourth Book.

## CHAP. I.

### Of Generation and Corruption.

### p. 415

### Sect. I.

ARTIC.

1 THe Introduction. ibid

2 *The proper Natives of* Generation *and* Corruption. 416

3 *Various opinions of the Antient Philosophers, touching the reason of Generation: and the principal Authors of each.* 417

4 *The two great opinions of the same Philosophers concerning the manner of the Commission of the Common Principles in Generation; faithfully and briefly stated.* 418

5 *That of* Aristotle *and the* Stoicks, *refuted: and* Chrysippus *subterfuge, convicted of 3 Absurdities.* 419

6 *Aristotles twofold Evasion of the Incongruities attending the position of the* Remanence *of things commixed, notwithstanding their supposed reciprocal Transubstantiation: found likewise meerly* Sophistical. 420

7 *That the* Forms *of things, arising in Generation, are no* New *substances, nor distinct from their matter: contrary to the* Aristoteleans. 422

8 *That the* Form *of a thing, is onely a certain* Quality, *or determinate Modification of its* Matter. 424

9 *An abstract of the theory of the* Atomists, *touching the same.* 425

10 *An illustration thereof, by a pregnant and opportune* Instance. viz. *the Generation of* Fire, Flame, Fume, Soot, Ashes, *and* Salt, *from* Wood *dissolved by Fire.* 426

SECT.

### Sect. II.

Artic.

1 *That in Corruption, no substance perisheth; but only that determinate Modification of substance, or Matter, which specified the thing.* 428

2 *Enforcement of the same Thesis by an illustrious Example.* 429

3 *An Experiment demonstrating that the Salt of Ashes was praexistent in Wood; and not produced, but onely educed by Fire.* ibid.

4 *The true sense of three General Axioms, deduced from the precedent doctrine of the Atomists.* 420

5 *The General Intestine Causes of Corruption, chiefly Two: (1) the interception of Inanimity among the solid particles of Bodies. (2) The essential Gravity and inseperable Mobility of Atoms.* 431

6 *The General Manners, or ways of Generation and Corruption.* 432

7 *Inadvertency of Aristotle in making Five General Modes of Generation.* 433

8 *The special Manners of Generation, innumerable; and why.* ibid.

9 *All sorts of Atoms, not indifferently competent to the Constitution of all sorts of things.* 434

### CHAP. II.
### Of Motion. P. 435.

### Sect. I.

Artic.

1 *Why the Nature of Motion, which deserved to have been the subject of the first speculation, was reserved to be the Argument of the Last, in this Physiology.* ibid.

2 *An Epicurean Principle, of fundamental concern to motion.* 436

3 *Aristotles Position, that the first Principle of motion, is the very Forme of the thing moved; absolutely incomprehensible: unless the Form of a thing be conceived to be a certain tenuious Contexture of most subtile and most active Atoms.* ibid.

4 *A second Epicurean Fundamental, concerning motion: and the state of the Difference betwixt Epicurus, Aristotle, and Plato, touching the same.* 437

5 *Epicurus's Definition of motion, to be the Remove of a body from place to place; much more intelligible and proper, than Aristotles, that it is the Act of an Entity in power, as it is such.* 438

6 *Empericus his Objections against that Definition of Epicurus: and the full Solution of each.* 439

7 *That there is motion; contrary to the Sophisms of Parmenides, Melissus, Zeno, Diodorus, and the Scepticks.* 441

### Sect. II.

Artic.

1 *Aristotles Definitions of Natural and Violent motion; incompetent: and more adequate ones substituted in the room of them.* 444

2 *The same deduced from the First Epicurean Principle of motion, premised: and three considerable Conclusions extracted from thence.* 445

3 *A short survey of Aristotles whole theory concerning the Natural motion of Inanimates: and the Errors thereof.* 446

4 *Uniformity, or Aequability, the proper Character of a Natural motion: and the want of uniformity, of a Violent.* 447

5 *The Downward motion of Inanimates, derived from an External Principle; contrary to Aristotle.* 449

6 *That that External Principle, is the Magnetique Attraction of the Earth* 450

7 *That the Upward motion of Light things, is not Accelerated in every degree of their Ascent as Aristotle praeariously affirmed: but, the Downward motion of Heavy things is Accelerated in every degree of their Descent.* ibid.

8 *The Cause of that Encrease of Velocity in Bodies descending; not the Augmentation of their Specifical Perfection as they approach neerer and neerer to their proper place: as Simplicius makes Aristotle to have thought.* 452

9 *Nor the Diminution of the quantity of Aer underneath them: as some Others conjectured.* ibid.

10 *Nor, the Gradual Diminution of the Force imprest upon them, in their projection upward: as Hipparchus alleadged.* 453

11 *But, the Magnetique Attraction of the Earth.* ibid.

12 *That the Proportion, or Ration of Celerity to Celerity, encreasing in the descent of Heavy things; is not the same as the Proportion, or Ration of Space to Space, which they pervade: contrary to Michael Varro the Mathematician.* 455

13 *But, that the moments or Equal degrees of Celerity, carry the same proportion, as the moments or equal degrees of Time, during the motion: according to the Illustrious Galilæo.* 456

14 *Galilæo's Grounds, Experience, and Reason.* 457

15 *The same Demonstrated.* 458

16 *The Physical Reason of that Proportion.* 460

17 *The Reason of the Equal Velocity of Bodies of very different weights, falling from the same*

*same altitude; inferred from the same Theory.*
ibid.

18 *Gravity Distinguish't into* Simple, *and Adjectitious.* 461

19 *The Rate of that superlative velocity with
which a Bullet would be carried, in case it
should fall from the Moon, Sun or region of the
Fixed stars, to the Earth : and from each of
those vast heights, to the* Centre *of the Earth.*
462

## Sect. III.

Artic.

1 W*Hat, and whence is that Force, or Virtue
Motive, whereby Bodies Projected are
carried on after their Dismission from the Projicient.* 463

2 *The Manner of the Impression of that Force.* 463

3 *That all Motion, in a free or Empty space,
must be Uniform, and Perpetual: and that
the chief Cause of the Inequality and Brevity
of the motion of things projected through the
Atmosphere, is the magnetique Attraction of
the Earth,* 466

4 *That, in the Atmosphere, no body can be projected in a Direct line; unless perpendicularly Upward, or Downward : And why.* 468

5 *That the Motion of a stone projected upwards
obliquely, is Composed of one Horizontal and
Perpendicular together.* ibid.

6 *Demonstration of that Composition.* 469

7 *That of the two different Forces, impressed up-*

*on a ball, thrown upward from the hand of a
man standing in a ship, that is under sayl; the
one doth not destroy the other, but each attains
its proper scope.* ibid.

8 *That the space of time, in which the Ball is
Ascending from the Foot to the Top of the
Mast: is equal to that, in which it is again
Descending from the top to the foot.* 470

9 *That, though the Perpendicular motion of a
stone thrown obliquely upward, be unequal,
both in its ascent and descent : yet is the Horizontal of Equal Velocity in all parts of space.*
ibid.

10 *The Reason and Manner of the Reflexion or
Rebounding motion of Bodies, diverted from
the line of their direction by others encountring
them.* 471

11 *That the Emersion of a weight appended to a
string, from the perpendicular, to which it had
reduced it self, in Vibration; is a Reflexion
Median betwixt No Reflexion at all, and the
Least Reflexion assignable; and the Rule of
all other Reflexion whatever.* 472

12 *The Reason of the Equality of the Angles of
Incidence and Reflexion.* ibid.

13 *Two Inferences from the premises: viz. (1)
That the oblique Projection of a Globe against
a plane, is composed of a double Parallel : and
(2) That Nature suffers no diminution of her
right to the shortest way, by Reflexion.* 474

14 *Wherein the Aptitude or Ineptitude of bodies
to Reflexion doth consist.* ibid.

BOOK

# READER.

THe Authors frequent Absence from the Towne, during this Impression; and the Division of the Copy among several Compositers, who could not all be Equally acquainted with His Hand; together with the multiplicity of Affaires, that diverted the Master Printer from the full discharge of his undertaking in the Correction of each sheet, before it was wrought off: have unhappily occasioned many Errata's in this Book. Of which such as consist only in the Misplacing, Duplication, Inversion, Omission, of Letters; or in the wrong position or Omission of Points, and other Pauses; these may be more easily Excused, than collected into a Catalogue. But, as for those less Venial ones, that seem either to trouble or invert the sense; or render the Authors care in Orthography suspected: you may please to Correct them ( so many of them, at least, as the Author observed in once reading over the Book ) thus.

## In the EPISTLE.

Page, 3. line 18. read value, for valew.

## In the TABLE.

PAg. 2. col. 1. line 18. read Essence of Place. col. 2. l. 14. r. Vacuola, or Empty spaces. & l. 23. r. in equal quantity, suspected. & l. 42. r. Inventor. p. 4. col. 2. l. 5. r. whether. & l. 10. r. Dimensions, & l. 16. r. Des Cartes. & l. 31. r. Dimensions. p. 5. col. 1. l. 5. r. generally distinguished: & col. 2. l. 44. omit the: & l. 45. r. cannot run on &c. Page. 6. col. 1. l. 23. r. dissentaneous to Reason, & l. ult. omit the. col. 2. l. 3. r. Democritus & l. 12. r. compound Nature; &. l. 25. r. mathematicall, Page 8. col. 1. l. 2. r. things: & l. 30. r. Accension. col. 2. l. 24. quantity of &c. P. 9. col. 1. l. 38. & 39. r. Des Cartes, p. 10. col. 1. l. 10. r. poreblind; p. 13. col. 1. l. 30. r. a symbolisme betwixt the &c. & col. 2. l. 38. in their several relations. p. 18. col. 1. l. 22. r. syzygia or &c. p. 21. col. 1. l. 32. r. the principal Authors of each.

# BOOK the FIRST.

## CHAP. I.

*All Modern Philosophers reduced to four general Orders; and the principal causes of their Dissention.*

### SECT. I.

IF we look back into the Monuments or Remains of *Antiquitie*, we shall observe as many several SECTS of *Philosophers*, as were the Olympiads in which *Greece* wore the Imperial Diadem of *Letters*; nay, perhaps, as many as she contained *Academies*, and publike *Professors* of Arts and Sciences: Each Master affecting to be reputed the principal Secretary of Nature; and his Disciples (their minds being deeply imbued with his principles) admiring him as the Grand Oracle of Divinitie, and the infallible Dictator of Scientifical Maxims. The chiefest, most diffused, and most memorable of these Sects, were the *Pythagorean*, the *Stoick*, the *Platonist*, the *Academick*, the *Peripatetick*, the *Epicurean*, and what, derided all the rest, the *Pyrrhonian*, or *Sceptick*; which feircely contended for the Laurel, by subtle disputations on the side of absolute Ignorance, and aspired to the Monarchy of Wisdom, by detecting the vanitie and incertitude of all Natural Science. As for the *Megarick*, *Eretrick*, *Cyreniack*, *Annicerian*, *Theodorian*, *Cynick*, *Eliack*, *Dialectick*, and others less famous; *Diogenes Laertius*, (*de vita Philosophor.*) hath preserved not only a faithful Catalogue of them, but hath also recorded their originals, declinations, periods, opinions.

*Art.* 1.
The principal Sects of the ancient Grecian Philosophers, only enumerated.

B      If

*Art. 2.*
*The same re-*
*vived among*
*the Moderns*
*with encrease.*

If we enquire into the *Modern* state of Learning, down even to our present age, we cannot but find not only the same Sects revived, but al. so many more New ones sprung up : as if Opinion were what mysterious Poets intended by their imaginary *Hydra* ; no sooner hath the sword of Time cut off one head, but there grows up two in the place of it ; or, as if the vicissitudes of Corruption and Generation were in common as well to Philosophy, as the subject of it ; Nature. Insomuch as that Adage, which was principally accommodated and restrained to express the infinite dissention of Vulgar and Unexamining Heads, *Tot sententiæ quot homines*, may now justly be extended also to the *Scholiarchs* and professed enquirers into the Unitie of Truth. To enumerate all these Modern dissenting Doctors (the most modest of all which hath not blushed to hear his pedantique Disciples salute him with the magnificent Attributes of a Despot in Physiologie, and the only Cynosure by which the benighted reason of man may hope to be conducted over the vertiginous Ocean of Error, to the Cape of Veritie) is neither useful to our Reader, nor advantageous or pertinent to our present Design. But, to reduce them to *four General Orders*, or range them into four principal Classes ; as it may in some latitude of interest, concern the satisfaction of those who are less conversant among Books : so can it in no wise affront the patience of those, whose studies have already acquainted them with the several kinds of Philosophy now in esteem.

*Art. 3.*
*Who are re-*
*duced either*
*to the Pedan-*
*tique or Fe-*
*male Sect.*

1 Some there are (and those not a few ) who in the minority of their Understandings, and while their judgments are yet flexible by the weak fingers of meer Plausibilitie, and their memories like Virgin wax, apt to retain the impression of any opinion that is presented under the specious disguise of Verisimilitie only, become constant admirers of the first Author, that pleaseth them, and will never after suffer themselves to be divorced from his principles, or to be made Proselytes to Truth ; but make it the most serious business of their lives to propugne their Tutors authoritie, defend even his very errors, and excogitate specious subterfuges against those, who have with solid Arguments and Apodictical reasons, clearly refuted him. These stifle their own native habilities for disquisition, believe all, examine nothing ; and, as if the Lamp of their own Reason were lent them by their Creator for no use at all, resign up their judgments to the implicite manuduction of some other ; and all the perfection they aim at, is to be able to compose unnecessary, and perhaps erroneous Commentaries upon their Masters text. This easie Sect may, without much either of incongruitie or scandal, be named *Secta Φιλαυθρωπίνης*, the FEMAL Sect ; because as women constantly retain their best affections for those who untied their Virgin Zone ; so these will never be alienated from immoderately affecting those Authors who had the Maiden-head of their minds. The chiefest Chair in this Classis ought to be consigned to our *Junior Aristoteleans*, who villifie and despise all doctrine, but that of the *Stagirite*, and confidently measure all mens deviations from truth, by their recessions from his dictates. This we say not to derogate from the honour due to so great a Clerk ; for we hold it our duty to pay him as large a tribute of Veneration, as any man that ever read his excellent Writings, without prejudice, and esteem him as one of the greatest and brightest stars in the sphere of Learning ; nay we dare assert, that He was the Centre in which all the choicest speculations

culations

culations and observations of his Prædecessors were united, to make up as complete a body of Natural Science, as the brain of any one single person, wanting the illumination of *Sacred Writ*, seems capable of, in this life of obscuritie: and that He hath won the Garland from all, who have laboured to invent and præscribe a general Method for the regulation and conduct of mens Cogitations and Conceptions. But, that I am not yet convicted, that his judgment was superior to mistake; that his Writings, in many places more then obscure, can well be interpreted by those who have never perused the Moniments of other Ancients; nor, that it can consist with Ingenuity to institute a Sacrament in Philosophy, (*i.e.*) to vow implicite vassalage to the Authoritie of any man, whose maxims were desumed from no other Oracle, but that of Natural Reason only; and to arrest all Curiositie, Disquisition, or Dubitation, with a meer ᾠντὸς ἔφα.

Hither may we refer also the patient Interpreters of *Scotus*; the vain Idolaters of *Raimund Lully*; but, above all, the stupid admirers of that Fanatick Drunkard, *Paracelsus*. In whose whole life, the only Rarities any sober man can discover. were his Fortune, and his Impudence. His *Fortune*, in that he being an absolute bankrupt in merit, could be trusted with so large a stock of Fame : his *Impudence*, in that, being wholly illiterate (for in stead of refining, He much corrupted his mother-tongue) He should prætend to subvert the Fundamentals of *Aristotle* and *Galen*, to reform the Common-weal of Learning, consummate the Arts and Sciences, write Commentaries on the *Evangelists*, and enrich the world with *Pansophy* in Aphorisms.

(2) Others there are (and those too few) whose brests being filled with true Promethean fire, and their minds of a more generous temper, scorn to submit to the dishonourable tyranny of that Usurper, Autority, and will admit of no Monarchy in Philosophy, besides that of Truth. These ponder the Reasons of all, but the Reputation of none; and then conform their assent, when the Arguments are nervous and convincing; not when they are urged by one, whose Name is inscribed in Golden Characters in the Legend of Fame. This Order well deserves the Epithite, Βεβαιωτικὸς, and therefore we shall Christen it, The Order of the ASSERTORS OF PHILOSOPHICAL LIBERTY; in regard, they vindicate the native privilege of our Intellectuals, from the base villenage of Præscription. Of this Order, Gratitude it self doth oblige us to account the Heroical *Tycho Brahe*, the subtle *Kepler*, the most acute *Galilæus*, the profound *Scheinerus*, the miraculous because universally learned *Kircherus*, the most perspicacious *Harvey*, and the Epitome of all, *Des Cartes*. In honour of each of these Hero's, we could wish (if the constitution of our Times would bear it) a Colossus of Gold were erected at the publick charge of Students; and under each this inscription :

*Amicus Plato, amicus Aristoteles, magis amica veritas.*

(3) The third Classis is possessed by such, who, without either totally neglecting or undervaluing the Inventions and Augmentations of the Modern; addict themselves principally to research the Moniments of the Ancients, and dig for truth in the rubbish of the *Grecian Patriarchs*. These are the noblest sort of *Chymists*, who labour to reform those once-excellent Flowers out of their Ashes : worthy *Geometricians*, that give us the true

*Art.4.*
Or, to the *Assertors of Philosophical Liberty.*

*Art.5.*
Or, to the *Renovators.*

B 2
dimensions

dimenfions of thofe Giant Wits, by the meafure of their Feet : and genuine fons of *Æfculapius*, who can revive thofe, whom the fleet chariot of
Time hath dragg'd to pieces, and recompofe their fcattered fragments into
large and complete bodies of Phyfiologie. The Courfe of thefe Worthies
in their ftudies doth denominate them Ἀνακαμψῶναι, RENOVATORS.

For, being of opinion, that Philofophy as well as Nature doth continually decline, that this is the Dotage of the World, and that the minds
of men do fuffer a tenfible decay of clarity and fimplicity; they reflect their
thoughts upon the πικὴν, or *Epoche* of *Phyfical Writings*, ranfack the urns
of *Athens* to find out the medal of fome grave Philofopher, and then with
invincible induftry polifh off the ruft, which the vitriolate dampnefs of Time
had fuperinduced; that fo they may render him to the greedy eyes of Po-
fterity in his primitive fplendor and integrity. The uppermoft feats in this
infinitely-deferving Claffis juftly belong to *Marcilius Ficinus*, who from
many mouldy and worm-eaten Tranfcripts hath collected, and interpreted
the femidivine Labors of *Plato* : to *Copernicus*, who hath refcued from the
jawes of oblivion, the almoft extinct Aftrology of *Samius Ariftarchus* : to
*Lucretius*, who hath retrived the loft Phyfiologie of *Empedocles* : to *Magnenus*, who hath lately raifed up the reverend Ghoft of *Democritus* : to *Merfennus*, who hath not only explained many Problems of *Archimed*; but reno-
vated the obfolete Magick of Numbers, and charmed the moft judicious
ears of Mufitians, with chiming *Pythagoras* Hammers, in an Arithmetick
Harmony: and to the greateft Antiquary among them, the immortal *Gaffendus*; who, out of a few obfcure and immethodical pieces of him, fcattered
upon the rhapfodies of *Plutarch* and *Diogenes Laertius*, hath built up the
defpifed *Epicurus* again, into one of the moft profound, temperate, and
voluminous among Philofophers.

*Art. 6.*
Or to the
*Electors.*
Our Fourth Claffis is to be made up of thofe, who indeed adore no Authority, pay a reverend efteem, but no implicite Adherence to Antiquity,
nor erect any Fabrick of Natural Science upon Foundations of their own
laying : but, reading all with the fame conftant Indifference, and æquanimity, felect out of each of the other Sects, whatever of Method, Principles, Pofitions, Maxims, Examples, &c. feems in their impartial judgments, moft confentaneous to *Verity*; and on the contrary, refufe, and, as
occafion requires, elenchically refute what will not endure the Teft of
either right *Reafon*, or faithful *Experiment*. This Sect we may call (as
*Potamon Alexandrinus*, quoted by *Diogenes Laertius*, long before us)
Ἐκλεκτικοὶ the ELECTING, becaufe they cull and felect out of all others,
what they moft approve.

Herein are Chairs provided for thofe Worthies, *Fernelius*, *Sennertus*,
and moft of the junior Patriots and Advancers of our Art. And the low-
eft room, we ask leave to referve for our felves. For. we profefs our felves
to be of his perfwafion, who faith ; *Ego quidem arbitror, re diu perpensâ,
nullius unquam fcientiam fore abfolutam, quin Empedoclem, Platonem, Ari-
ftotelem, Anaxagoram, Democritum adjungat Recentioribus, & ab unoquoque
quod verum eft, rejectis falfis, eligat. His enim Principibus peculiari ratione
Cælefte Lumen affulfit : & quamvis Corporis imbecillitate multa corruperint;
plurima tamen, quæ Fidei lumine difcernimus, fcripfére verifsima* He can
never make a good *Chymift*, who is not already an excellent *Galenift*, is
proverbial among us Phyficians : and as worthy the reputation of a Proverb is it among Profeffors in Univerfities: He can never clearly underftand
                              the

the *Moderns*, who remains ignorant of the doctrines of the *Antients*. Here to declare our selves of this Order, though it be no dishonour, may yet be censured as superfluous: since not only those Exercises of our Pen, which have formerly dispersed themselves into the hands of the Learned, have already proclaimed as much; but even this præsent Tractate must soon discover it.

---

## Sect. II.

TO explore the Cheif Grounds, or Reasons of this great Varietie of Sects in Philosophy; we need search no further, then the exceeding *Obscurity of Nature*, the *Dimness and imperfection of our Understanding*, the *Irregularity of our Curiosity*.

Of the *First*, they only can doubt, who are too stupid to enquire. For, Nature is an immense Ocean, wherein are no Shallows, but all Depths: and those ingenious Persons, who have but once attempted her with the founding line of Reason, will soon confess their despair of profounding her, and with the judicious *Sanchez* sadly exclaim; *Una Scientia sufficit toti orbi: nec tamen totus hic ei sufficit. Mihi vel minima mundi res totius vitæ contemplationi sat est superque: nec tamen tandem eam spero me nosse posse:* nor can they dislike the opinion of the *Academicks* and *Pyrrhonicks*, that all things are Incomprehensible. *Art.* 1. The principal causes of the Diversity of Philosophical Sects; and the chiefest among them, the Obscurity of Nature

And (as for the *second*) if Nature were not invelloped in so dense a Cloud of Abstrusity, but should unveil her self, and expose all her beuteous parts naked to our speculation: yet are not the Opticks of our Mind either clear or strong enough to discern them.    Men indeed fancy themselves to be Eagles; but really are grovelling Moles, unceffantly labouring for light: which at first glimpse perstringeth their eyes, and all they difcover thereby, is their own native Blindness. *Naturæ mysteria etiamsi mille facibus revelentur, arbitrantium oculis numquam tota excipientur: restabit semper quod quæras; & quo plus scies, eo plura à te ignorari miraberis.* This meditation, we confess, hath frequently stooped our ambitious thoughts, dejected us even to a contempt of our own nature, and put us to a stand in the midst of our most eager pursuit of Science: insomuch that had not the inhærent Curiosity of our Genius sharply spurred us on again, we had totally desisted, and sate down in this refolution; for the future to admire, and perhaps envy the happy serenity of their Condition, who never disquiet and perplex their minds with fruitless scrutiny, but think themselves wise enough, while they acquiesce in the single satisfaction of their Senses.    Nor do we look ever to have our Studies wholly free from this Damp: but expect to be surprifed with many a cold fit, even then when our Cogitations shall be most ardent and pleasing.    And to acknowledge our penfive fense of this Discouragement, is it that we have chosen this for our Motto:        *Quo magis quærimus, magis dubitamus.* *Art.* 2. The Imperfection of our Understanding.

But left this our despair prove contagious, and infect our Reader, and He either shut up our Book, or smilingly demand of us, to what purpose

purpose we wrote it; if (as we confess) Insatisfaction be the End of study, and (as we intimate) our Phisiology at most but ingenious Conjecture: we must divert him with the novelty of a Paradox, *viz.* that the *Irregularity of our Curiosity is one Cause of the Dissent of Philosophers.*

*Art. 3.*
*The Irregularity of our Curiosity. A paradox.*

That our desire of Truth should be a grand Occasion of our Error; and that our *First Parents* were deluded more by the instigation of their own essential C U R I O S I T Y, than by either the allurement of their Sensual Appetite, or the subtle Fallacies of the Serpent: is a conceit not altogether destitute of the support and warranty of Reason. For, the Human Soul (the only Creature, that understands the ἐξοχὴν, or transcendent Dignity of its Original, by reflecting upon the superlative *Idea*, which it holds of its Creator) from the moment of its immersion into the cloud or opacity, of flesh labours with an insatiable Appetence of Knowledge; as the only means, that seems to conduce to the satisfaction of its congenial Ambition of still aspiring to Greater and Better things: and therefore hath no Affection either so Essential, or Violent, as the Desire of Science; and consequently, lyeth not so open to the deception of any Objects, as of those which seem to promise a satisfaction to that desire. And obvious it is from the words of the Text; that the Argument which turned the scales, *i.e.* determined the Intellect, and successively the Will of our Grandmother *Eve*, from its indifferencie, or æquilibration, to an Appetition, and so to the actual Degustation of the Forbidden Fruit, was this: *Desiderabilis est arboris fructus ad habendam scientiam.* Besides, though we shall not exclude the Beauty of the fruit, transmitted by the sight to the judicatory Faculty, and so allecting the Sensual Appetite, from having a finger in the Delusion; yet can we allow it to have had no more then a finger; and are perswaded, that in the syndrome or conspiracy of Causes, the most ponderous and prævalent was the Hope of an accession or augmentation of *Knowledge.* Since it cannot but highly disparage the primitive or innocent state of man, to admit, that his Intellect was so imperfect, as not to discern a very great Evil, through the thin Apparence of Good, when the utmost that Apparence could promise, was *no more*, than the momentany pleasure of his Palate or Gust: Or, that the express and pœnal Interdiction of God, yet sounding in his ears, could be over-balanced by the light species of an object, which must be lost in the Fruition.

Nor is this *Curiositie* to be accused only of the First Defection from Truth, but being an inseparable Annex to our Nature, and so derived by traduction to all *Adams* posteritie, hath proved the procatarctick Cause of many (some contemplative Clerks would have adventured to say of All) the *Errors* of our judgments. And, though we have long cast abour, yet can we not particular any one Vicious inclination, or action, whose Scope or End may not, either directly or obliquely, proximly or remotely, seem to promise an encrease of *Knowledge* in some kind or other. To instance in one, which appears to be determined in the Body, to have no interest beyond the Sense, and so to exclude all probabilitie of extending to the Mind, as to the augmentation of its Science. Whoever loves a beutiful Woman, whom the right of Marriage hath appropriated to another, ardently desires to enjoy her bed; why, not only for the satisfaction of his sensual Appetite, because that might be acquired by the act of carnality

with

with some other less beutiful, and Beuty is properly the object of the Mind: but because that Image of Beuty, which his eye hath transmitted to his mind, being præsented in the species or apparition of Good and Amiable, seems to contain some Excellence, or comparitively more Good, then what He hath, formerly understood. If it be *objected*, that if so, one enjoyment must satisfie that Desire; and consequently, no man could love what He hath once enjoyed, since Fruition determineth Desire: We *Answer*, that there is no such necessitie justly inferrible, when Experience assures, that many times Love is so far from languishing, that it grows more strong and violent by the possession of its Object. The Reason is, because the passionate Lover, apprehending no fruition total, or possession entire, supposeth some more Good still in the object, then what his former enjoyment made him acquainted withall. And if it be *replyed*, that the Lover doth, in the perseverance of his Affection, propose to himself meerly the *Continuation* of that Good, which He hath formerly enjoyed: we are provided of a sufficient *Rejoynder*, *viz.* that whoso wisheth the Continuation of a Good, considers it not as a thing præsent, but to come; and consequently as a thing which yet He doth not know: for, no man can know what is not.

Other Instances the Reader may be pleased to select from among the *Passions*; tracing them up to their first Exciting Cause, in order to his more ample satisfaction: it being digressive and only collateral to our Scope. Good thus being the only proper Object of our Affections (for Evil exhibited naked, *i. e.* as Evil, never Attracts, but ever Averts our Will, or Rational Appetite: as we have clearly proved in our Discourse of the *Liberty Elective of mans Will.*) if we mistake a real evil præsented under the disguise of a Good: this mistake is to be charged upon the account of our Rational or Judicatory Faculty, which not sufficiently examining the Reality of the species, judgeth it to be good, according to the external Apparence only; and so misguideth the Will in its Election. Now, among the Causes of the Intellects erroneous judicature (we have formerly touched upon its own *Native Imperfection*, or *Cæcity*, and *Præjudice*,) the chiefest and most general is the *Impatience*, *Præcipitancy*, or *Inconsiderateness* of the Mind; when, not enduring the serious, profound, and strict examen of the species, nor pondering all the moments of Reason, whi h are on the Averting part of the Object, with that impartialivy requisite to a right judgment; but suffering it self, at the first occursion or præsentation thereof, to be determined, by the moments of Reason apparent on the Attracting part, to an Approbation thereof: it misinformeth the Will, and ingageth it in an Election and prosecution of a Falsity, or Evil, couched under the specious semblance of a positive Truth, or Good.

Now, to accommodate all this to the interest of our *Paradox*; if Good, real or apparent, be the proper and adæquate object of the Intellect; and the chief reason of Good doth consist in that of Science, as the principal end of all our Affections: then, most certainly, must our præcedent assertion stand firm, *viz. that our understanding lyeth most open to the delusion of such objects, which by their Apparence promise the most of satisfaction to our Desire of Science*; and, upon consequence, by how much the more we are spurred on by our *Curiosity*, or Appetence of Knowledge, by so much the more is our mind impatient of their strict examen, and æquitable perpension. All which we dayly observe experimented in our
selves.

selves. For, when our thoughts are violent and eager in the pursuit of some reason for such or such an operation in Nature; if either the discourse, or writings of some Person, in great esteem for Learning or Sagacity, or our own meditations furnish us with one, plausible and verisimilous, such as seems to solve our Doubt: how greedily do we embrace it, and without further perpension of its solidity and verity, immediately judge it to be true, and so set up our rest therein?    Now, it being incontrovertible, that Truth consists in a Point, or Unity; it remains as incontrovertible, that all those judgements, which concur not in that Point, must be erroneous: and consequently that we ought ever to suspect a multiplicity of dissenting judgments, and to suppose that Phænomenon in Nature to be yet in the dark, *i. e.* uncomprehended, or not understood, concerning whose solution the most various opinions have been erected.

And thus have we made it out; that our Curiosity is the most frequent Cause of our Minds Impatience or Præcipitancy: that Præcipitancy the most frequent Cause of our Erroneous judgments, concerning the Verity or Falsity of Objects: those Erroneous judgments alwayes the Cause of the Diversity of Opinions: and the Diversity of Opinions alwayes the Cause of the Variety of Sects among Philosophers.

CHAP.

# CHAP. II.

## That this World is the Universe.

### Sect. I.

Mong those Fragments of Antiquity, which History hath gathered up from the table of sated Oblivion, we find two worthy the entertainment of our Readers memory, though, perhaps, not easie to be digested by his Belief. The *one* that *Alexander* the Great grew melancholy at the lecture of *Anaxarchus* his discourse of an *Infinity of Worlds*, and with tears lamented the confinement of his Ambition to the Conquest of One: when yet, in truth, the wings of his Victory had not flown over so much as a third part of the Terrestrial Globe; and there remained Nations more then enough to have devoured his numerous Armies at a breakfast, to have learned him the unconstancy of Fortune, the instability of Empire, and the vanitie of Pride, by the experiment of his own overthrow, and captivity in a narrow prison. The *Other*, that there were whole Schools of *Philosophers*, who fiercely contended for a *Plurality* of *Worlds*, and affected the honour of invincible Wits, by extending their disquisitions beyond the Extrems or Confines of this adspectable World to a multitude of others without it, as vast, as glorious, as rich in variety of Forms: when, indeed, their Understandings came so much short of conquering all the obvious Difficulties of this one, that even the grass they trod on, and the smallest of Insects, a Handworm, must put their Curiosity to a stand, reduce them to an humble acknowledgment of their Ignorance, and make them sigh out the Scepticks Motto, *Nihil Scitur*, for a Palinodia. Whether His or Their Ambition were the greater, is not easie to determine; nor can we find more wildness of Phansy, or more insolent Rhodamontadoes in Camps, than Academies, nay if we go to Absurdities, *Cedunt Arma Togæ*, the Sword must give place to the Gown. But, that his Error was more venial then theirs, is manifest from hence; that He had conquered all of the World that he knew: but they could not but find themselves foiled and conquered by every the most minute and sensible part of the world, which they had attempted to know.

This *Genus* of Philosophers doth naturally divide it self into two distinct *species*. The *First* of which doth consist of those, who assert only a *Plurality* of Worlds: the *Second* of those, who have been so bold as to ascend even to an *Infinity*. Those who assert only a Plurality may be again

C                                           sub-

subdistinguished into two subordinate divisions : (1) Such as held a Plurality of Worlds *Coexistent* ; among whom the most eminent was *Plutarch*, who *(in lib. de Oracul. defect.)* affirms, that to have many Worlds at once, was consistent with the majesty of the Divine Nature, and consonant to Human Reason ; and (in 1. *placit.* 5.) earnestly labours to dissolve the contrary Arguments of *Plato* and *Aristotle* for the Unity of the World. Nor were these all of one Sect ; for *some* opinioned that there were many other Worlds synchronical in the Imaginary space, or on the outside of this : and *others* would admit of nothing, beyond *Trismegistus* Circle, or without the convex part of the *Empyræum* ; but conceived that every Planet, nay, every Star, contained in this, was an intire and distinct World. Among these the Principal were *Heraclides*, the *Pythagoreans*, and all the Sectators of *Orpheus* : as they are enumerated by *Plutarch* (2. *Placit.* 13.)

(2) Such as held a Plurality of worlds, not coexistent or synchronical, but *successive* ; i.e. that this præsent world, Phœnix-like, sprung up from the ruines of another præcedent ; and that the Ashes of this shall produce a Third, the Cinders of that a Fourth, &c. of this perswasion were *Plato*, *Heraclitus*, and all the *Stoicks*.

The *Second* species is made up of those, who dreamt of an *Infinity* of Worlds coexistent in an infinite space : and the chief seats in this *Classis* belong to *Epicurus* and *Metrodorus*, upon the last of which this peremptory saying is commonly fathered ; Ἄτοπον τὸ ἐν μεγάλῳ πεδίῳ ἕνα φύεσθαι στάχυν καὶ ἕνα κόσμον ἐν τῷ ἀπείρῳ. *Tam absurdum esse in Universo infinito unum fieri mundum, quàm in magno agro unam nasci spicam.* And below them shall sit *Anaximander*, *Anaximenes*, *Archelaus*, *Xenophon*, *Diogenes*, *Leucippus*, *Democritus*, and *Zeno Eleates*, as may be collected from the records of *Stobæus (Ecl. Physic. l. 9.)* That *Epicurus* was a grand Patron of this Error, is confest by himself (*in Epist. ad Herodotum, apud Laertium*) in these words : Ἀλλὰ μὴν καὶ κόσμοι ἄπειροί εἰσιν οἵθ᾽ ὅμοιοι τούτῳ ὄντ᾽ ἀνόμοιοι, *Cæterum in universitate, seu natura rerum, infiniti sunt mundi, alij quidem similes isti quem nos incolimus, alij verò dissimiles.*

*Art. 3.*
The two main pillars on which the opinion of a Plurality of Worlds was anciently erected.

The Reasons, or rather the Apparences of Reason, which seduced the Understandings of so many and great Philosophers into a judgment, that there was an Infinity of Worlds ; are comprehended under these Two.

(1) *Quòd Causæ sunt infinitæ. Nam si hic quidem mundus sit, finitus Causæ verò, ex quibus est, fuère omninò infinitæ : necesse est mundi etiam sint infiniti. Prorsus enim, ubi sunt Causæ, Effectus quoque ibi sunt.* That Worlds there are infinite in multitude, is manifest from hence, that there are infinite Causes for Worlds : for, since this World is finite, and the Causes of which it was made, were infinite ; necessary it is that there be infinite Worlds. Insomuch as where are Causes, there also must be Effects. This *Epicurus* more then intimated, when He argued thus : *Quippe Atomi, cum sint infinitæ, per infinitatem spatiorum feruntur, & alibi alia, ac procul ab hoc ad fabricationem mundorum infinitorum variè concurrunt.* Consule *Plutarchum*, (1. *Placit.* 5.) & *Lucretium.* (*lib.* 2.)

(2) *Quòd nulla sit specialis res, cui non suo sub genere sint singularia multa similia :* That there is no one thing special, to which under that kind, many singulars are not alike. Upon this sand was it that *Plutarch* erected his feeble structure of a Plurality of Worlds ; for (*in defect. Oracul.*) he expresseth

presseth it at large, in these words, *Videmus naturam ipsis generibus, specie-busque, quasi quibusdam vasculis aut involucris seminum, res singulares con-tinere. Neque enim res ulla est numero una, cujus non sit communis ratio, neque ulla certam denominationem nunciscitur, quæ singularis cum sit, non eti-am communem qualitatem habeat. Quare & hic mundus, ita singulariter di-citur, ut communem tamen rationem, qualis atemque mundi obtineat : singula-ris autem conditionis sit, ex differentia ab alijs quæ ejusdem Generis sunt ; Et certe non unicus Homo, non unicus Equus, non unicum Astrum, non unicus Deus, non unicus Dæmon in rerum natura est : quid prohibit, quo minus plures, non unicum mundum Natura contineat, &c.*

## Sect. II.

## *The Redargution.*

THat our *Redargution* of this vain *Error* may obtain the more both of *Perspicuity* and *Credit*, we are to advertise that the *Question* is not concerning the *Possibility*, but the real or actual *Existence* of an *Infinity* of *Worlds*. For, of the *Possibility*, no man, imbued with the principles of Phy-siology, or Theology, can doubt.

*Art.* 1.
The Question stated to be concerning the real *Existence*, not the *possibi-lity* of an Infi-nity of Worlds

(1) Because, to the most profound and nice Enquirers into that ab-struse point, no Argument, whether simple or complex, hath appeared weighty enough to disswade them from admitting an *immense Tohu*, or *in-finite Vacuum*, without the extremities of this World. For, not a few, nor the least judicious part of even our Christian Doctors have asserted those *Extramundane spaces* calling them IMAGINARY, because we can ima-gine the same Dimensions of Longitude, Latitude, and Profundity, to be in them, as are in those real Spaces, wherein Bodies are included in this world ; and since all men, acknowledging the Omnipotence of God, con-clude, that He might, had He so pleased, have created this World larger and larger even to infinity ; necessary it is, that they also admit a larger and larger *space* or Continent, for the Reception of that enlarged World. Which may with equal Truth be accommodated also to an *Infinity* of Worlds, insomuch as all, who acknowledge Gods *Omnipotence*, readily condescend, that He could, had it seemed good in the eye of his Wisdom, have created more and more Worlds even to Infinity: necessary it is, that they understand those Worlds must be received in proportionate spaces, which ought to be over and above that space, which this World possesseth. For, whereas some have conceived, that if God would create more Worlds besides this, He must also create more spaces to contain them; undoubtedly they entangle themselves in that inextricable Difficulty which is objected upon them, concerning the space interjected between any two Worlds; since that space may be brought under the laws of Mathematical Com-mensuration, and clearly explained by a greater or less Distance.

*Art.* 2.
Because the supposed *Infini-ty* of the *Extra-mundan Spaces* is no impossi-bility.

(2) Because, it is found no *Unjust*, or desperate Difficulty to defend a *Possible Infinity of Bodies*. For the Fathers of our Church have delivered it as Canonical, that God might have created any thing Actually Infi-

*Art.* 3.
Because an *I-finity* of *Bodies* is also possible, as to the Omni-potence of God.

nite

nite not only in Magnitude, but also in Multitude. Only they reserve the infinity of *Essence*; which since it can be competent to none but the *Divine Essence*, and comprehends all perfections whatever in a most transcendent or Eminent manner: it is as absolutely impossible that any thing should be Created *Actually Infinite in Essence*, as that God should be created. Which we conceive to be the ground of that Truth; that *to imagine God to be able to create any thing equal to Himself: is to suppose an Imperfection in his Nature.* Nor have They, without good Cause, deserted the conduct of *Plato* and *Aristotle*, when they would seduce them into an opinion, that Infinity is only *Potential*, not Actual, *i. e.* that nothing *in Rerum Natura* can be infinite *in Actu*, but only *in Potentia*; insomuch as though a *Continuum* may be either divided, or Augmented even to Infinity: yet cannot that *Continuum* either by Division, or Augmentation, ever become Actually infinite. For, since even *Aristotle* himself describes an *Infinite* to be, *non cujus extra nihil est, sed ex quo accipientibus semper aliquid accipiendum restat,* that from which though nere so much be abstracted, yet still there shall more remain undeducted; which is, in the sum or importance, to say that the Essence of Infinity is *Inexhauribility*: it seems possible to admit not only many, but even infinite infinities in an Infinite. Thus we say, and truly, that in an infinite *Number* are comprehended not only infinite *Unities*, but also infinite *Binaries*, infinite *Ternaries*, infinite *Denaries, Centenaries, &c.* which is the reason of that Axiom, *That all the parts of an Infinite are infinite.*

**Art. 4.**
The Error of concluding the *Esse*, from the *Posse* of an Infinity of Worlds.

Now though to be able, by perfect *Demonstration*, to evince that there are no more Worlds but this one, which we inhabit, is that of which to despair can be no dishonour to the most acute and Mathematical Wit in the world; since none ought to doubt, but God might have created, and may yet at his pleasure create others innumerable, because neither can His *Infinite Power* ever be exhausted, nor that *Abyss* of *Nothing*, out of which the Energie of his Word instantly educed this World, not afford or space or matter for them : yet notwithstanding to affirm, that because 'tis *possible* therefore there *are* many other Worlds actually coexistent; is a manifest *inartificial Argument*, and a Conclusion repugnant to all the inducements of Persuasion.

For, albeit we readily concede, that there is an *Infinite Inanity* or Ultramundan Space, yet can it not follow of necessity, that there are *Infinite Atoms* contained in that Ultramundane Space; as *Democritus* and *Epicurus* præposterously infer : insomuch as it sounds much more concordant to reason, that there are no more Atoms, then those of which this single World was compacted.

**Art. 5.**
The *first* main Pillar of a Plurality of Worlds subverted.

And when they Argue thus; *Since the vacuity or ultramundane Space is infinite in Magnitude or Capacity, necessary is is that the Abyss of Atoms included therein be also Infinite in Extent; because otherwise they could never have convened, and coalesced in that Form, which the World now holds :* we admit their *Induction* for natural and legitimate, but detest their *supposition* as absurd and impossible. For, They take it for granted, that the Chaos of Atoms was not only eternal and *Increate*, but also that it disposed, and compacted it self into that Form, which constitutes the World, by the spontaneous motion inhærent in Atoms, and their fortuitous coalescence in such and such respective Figures : when to a sober

judgment

judgment it appears the highest *Impossibility* imaginable, that either the
Chaos of Atoms could be eternal, self-principiate, or increate, or dispose and fix it self into so vast, so splendid, so symmetrical, so universally
harmonical, or Analogical a structure, as this of the World. For, as the
*Disposition* or Dispensation of the Chaos of Atoms into so excellent
a form, can be ascribed to no other Cause, but an *Infinite Wisdom*: so
neither can the *Production* or Creation of the same Chaos be ascribed to
any other Cause, but an *Infinite Power*, as we have formerly demonstrated
in our *Darkness of Atheism, cap. 2.*

And therefore, since it is most probable that Atoms were the *Materia Prima*, or material Principle of the World; as we shall clearly enunciate in a singular Chapter subsequent: we may adventure to affirm,
that God created exactly such a proportion of Atoms, as might be
sufficient to the making up of so vast a Bulk, as this of the World,
and that there remained no one superfluous. 'Tis unworthy a Philosopher to acknowledge any superfluity in Nature: and consequently a dangerous solœcism to say the *God* of Nature knowing not how to proportion the quantity of his materials to the model or platform of his structure, created more Atoms, then were necessary, and left an infinite
Residue to be perpetually hurried too and fro in the ultramundane space.
If they shall *urge* upon us, that no man was privy to the Councel of
God at the Creation, and consequently no can know, whether He created
either more Atoms then were requisite to the amassment of this World,
or more Worlds then this one: we may justly *retort* the Argument upon
them, and conclude, that since no man was privy to the Councel of
God, they have no reason to pretend to know, that God created either
more matter, or more Worlds; and so the whole substance of the Dispute must be reduced only to this: That they have no more Reason
for the support of their opinion of a Plurality of Worlds then we have for
ours of the Unity of the World. Nay the greatest weight of Reason hangs
on our end of the scale; for, we ground our Opinion upon that stable
Criterion, our *sense*, and asserting the singularity of the world, discourse of
what our sight apprehends: but They found theirs upon the fragil reed of
wild Imagination, and affirming a Plurality discourse of what neither the
information of their sense, nor solid reason, nor judicious Authority, hath
learned them enough to warrant even Conjecture.

And, as to their second *Argument*, viz. *That there is in Nature no one
Thing special, to which under the same kind, there are not many singulars alike*:
we *Answer*, that All those *singulars*, which we observe to be multiplied under one and the same kind, are such which perish in the *Individual*, and
therefore cannot but be lost, if not conserved by the multitude of Succes-
sors; and not such as are not obnoxious to destruction by Corruptibility,
for they, constantly existing in the Individual, need not Multiplicity to their
conservation. For which cause, one Sun, and one Moon are sufficient, and in al
probability of this sort is the World; for though it be conceived obnoxious
to corruption, and shall once confess a *Period*: yet is this no valid reason to
justifie the necessity of a multitude of worlds, since the Dissolution of the
World shal be synchronical to the Dissolution of Nature, when Sun, Moon,
and all other kinds of Creatures, as well single as numerous shall be blended
together in one common ruine; and then the same *Infinite Cause* which hath
destroyed them, can, with as much facility as he first Created them, repair
                                                                their

*Art. 6.*
The second Pil-
lar found so-
phisticate, and
demolished.

their ruines, educe them out of their second Chaos, and redintegrate them into what Form His Wisdom shall design.

*Art. 7.*
*A Plurality of Worlds manifestly repugnant to Authority Divine,*

Nor is this opinion of a Plurality of Worlds only destitute of, but even *è diametro* repugnant to the principal *Inducements* of Belief. For, if we consider *Authority Divine*; in *Moses* inæstimable Diary or Narrative of the Creation can be found no mention at all of a Multitude of Worlds, but on the contrary a positive assertion of *one* world; and the express declarement of the manner how the *Fiat* of Omnipotence educed the several Parts thereof successively out of the Chaos, disposed them into subordinate Piles, and endowed them with exquisite configurations respective to their distinct destinations, motions and uses: and in all the other Books of Sacred Writ, whatever concerns the Providence of God, the Condition of man, the mysteries of his Redemption, means of salvation, &c. doth more then intimate the singularity of the World; nor is there any one word, if rightly interpreted, which can be produced as an excuse for the opposite Error.

*Art. 8.*
*And Human.*

If *Humane Authority*; we may soon perceive, that those Ancient Philosophers, who have declared on our side, for the Unity of the World, do very much exceed those *Pluralists* nominated in our præcedent Catalogue, both in *Number* and *Dignity*. For, *Thales, Milesius, Pythagoras, Empedocles, Ecphantus, Parmenides, Melissus, Heraclitus, Anaxagoras, Plato, Aristotle, Zeno* the Stoick, attended on by all their sober Disciples, have unanimously rejected and derided the Conceit of many Worlds, not only as vain and weak, but as extremly Hypochondriack, and worthy a whole acre of Hellebor. Nor, indeed, are we persuaded, that so great Wits as those of *Democritus* and *Epicurus*, did apprehend it as real; but only Imaginary, proposing it as a necessary Hypothesis, whereon to erect their main Physical Pillar, τὸ ὅλον ἀ'ξμιλον ἔναι καὶ ἀφξαφον, *Vniversum esse ortus interitusque expers*, That the Universe is nonprincipiate and indissoluble. For, having mediated thus; Whatever is Finite, is circumscribed by an External Space, from which a cause may come and invading destroy it, and into which the matter thereof, after the dissolution of its Form, may be received: now this World, being Finite, must be environed by a circumambient space, from which a Cause may invade and destroy it; and into which the matter thereof, after the dissolution of its Form, may be received; must of necessity therefore be dissoluble: They inferred, that, unless they would concede the Universe to be dissoluble; which could never consist with their Principles, they must affirm it to be Infinite, *i. e.* without which no space can be, from whence any Cause might invade it, and into which the matter thereof after the destruction of its Form, might be received: and thereupon concluded to suppose an Infinity of Worlds Coexistent.

Which seems to be the Reason also that induced *Epicurus* and *Metrodorus* to opinion, that the *Vniverse* was not only Ἀμετάβλητον *Immutable*; but also ἀκίνητον *Immoveable*; as may be collected from these words of *Plutarch* quoted by *Eusebius* (1.præpa. Evang.5.) concerning *Metrodorus, Is inter cætera non moveri universum dixit quoniam non est quo migrare possit, nam si*

*Art. 9.*
*The result of all the Demonstration of the Authors Thesis, That this World is the Universe.*

*posset quidem, vel in plenum, vel in vacuum; atqui universum continet quicquid hujusmodi est, quia si non contineret, minimè foret Vniversum.*

Having thus amply refuted the Dream of a Plurality of Worlds, both by detecting the exceeding invalidity of those two Cardinal Reasons, on which

which the Authors and Abettors of it had raſhly fixed their Aſſent ; and by convicting it of manifeſt Repugnancy to Authority Divine and Human: we may ſafely præſume , the underſtanding of our Reader is ſufficiently præpared to determine his judgment to an Approbation of our Theſis, the Argument and Title of this Chapter, *viż.* That this Adſpectable world is the τὸ πᾶν *Omne,* τὸ ὅλον *Univerſum,* the *All in Rerum Natura,* the large Magazine wherein all the wealth and treaſure of Nature is included , and that there is Nothing Quantitative, but meerly Local, beyond the Convex extremity, or (as *Ariſt.*) τὴν ὑσίαν τ̄ ἐϛάτη; ὃ πᾶσαϛ ϖεριοϛᾶϛ, *ſubſtantiam quæ eſt in ultima Cæli converſione* ; the outſide of the *Empyræum.* Thus much *Ariſtotle,* though upon the conviction of other Arguments , ſeems fully to have both underſtood and embraced, when in poſitive terms He affirmed, μήτε εἶναι μεδὲν ἔξω Σῶμα τ̄ ἀϛανῦ, μήτε ἐσδὲ ενεϛ̄ται γίνεϛται *Extra cælum neque eſt quicqnam Corpus, neque eſſe omninò poteſt (de cælo l.i. c.9.)* As alſo whenſoever He uſed thoſe two words, τὸ ὅλον & ὁ κόσμοϛ, *Univerſum* & *Mundus,* as perfect ſynonymæes, indifferently ſignifying one and the ſame thing: which was moſt frequent not only to him, but to P*lato* alſo , and moſt of the moſt judicious ſort of Philoſophers.

*Art.*18.
*Extramundane
Curioſity,* a
high degree of
*Madneſs.*

If any Curioſity be ſo immoderate, as to tranſgreſs the Limits of this All, break out of *Triſmegiſtus* Circle, and adventure into the Imaginary Abyſs of Nothing, vulgarly called the *Extramundan Inanity* ; in the Infinity (or, rather, *Indefinity* ) of which many long-winged VVits have, like ſeel'd Doves, flown to an abſolute and total loſs : the moſt promiſing Remedy we can præſcribe for the reclaiming of ſuch wildneſs; is to advertiſe , that a ſerious Diverſion of thought to the ſpeculation of any the moſt obvious and ſenſible of ſublunary Natures, will prove more advantagious to the acquiſition of Science, then the moſt acute metaphyſical Diſcourſe, that can be hoped from the groveling and limited Reaſon of man, concerning that imperveſtigable Abſtruſity ; of which the more is ſaid, the leſs is underſtood ; and that the moſt inquiſitive may find Difficulties more then enough within the Little VVorld of their own Nature, not only to exerciſe, but empuzle them. To which may be annexed that judicious Corrective of *Pliny,* (*l.*2. *Nat.Hiſt.c.*1.) *Furor eſt, profecto furor eſt egredi ex hoc mundo,& tanquam interna ejus cuncta planè jam ſint nota, ita ſcrutari Extera. Quaſi verò menſuram ullius poſsit agere, qui ſui neſciat : aut mens Hominis videre, quæ mundus ipſe non capiat.* And that facete ſcoff of the moſt ingenious Mr. *White* (*in Dialog.*1. *de mundo.*) That the Extramundan Space is inhabited by *Chymæra's* which there feed, and thrive to Giants upon the dew of *Second Intentions.*

CHAP:

## CHAP. III.

*Corporiety and Inanity.*

### SECT. I.

*Art. 1.*
*Body and Inani-*
*ty, the two ge-*
*neral Parts of*
*the Universe.*

THE Universe, or this adfpectable World (henceforth Synonymaes) doth, in the general, confift of
only two Parts, *viz.* Something and Nothing, or
*Body* and *Inanity.* Τὴν τῶν ὄντων φύσιν σωμα[q. εἶναι,
κỳ κενον, *Naturam rerum effe Corpora & Inane,* was the
Fundamental pofition of *Epicurus (apud Plutarch.
adverf. Colot.)* which his faithful Difciple *Lucretius*
hath ingenuofly rendred in this Diftich:

*Omnis, ut eft igitur per fe, Natura duabus*
*Confiftit rebus; quæ Corpora funt, & Inane.*

The All of Nature in two Parts doth lye,
That is, in Bodies and Inanity.

*Art. 2.*
*Three the moft*
*memorable De-*
*finitions of*
*Corporiety ex-*
*cant among*
*Phyfiologifts,*
*recounted and*
*examined.*

Concerning the nature or effence of a BODIE, we find more then
one Notion among Philofophers.

(1) Some underftanding the root of *Corporiety* to be fixt in *Tangibility:*
as *Epicurus (apud Empericum adverf. Phyfic.)* faith, κỳ ἄ χρονισμόν Σχήμα[ος
τέ κỳ μηγέθ[ος, κỳ ἀντιτυπίας, κỳ βάρυς, ᾧ Σῶμα νενοείδαι: *intelligi Corpus ex*
*congerie figuræ magnitudinis, refiftentiæ (feu foliditatis ac impenetrabilitatis*
*mutuæ & gravitatis*; that by Bodie is to be underftood a congeries of figure, magnitude, refiftence (or folidity and impenetrability mutual) and
gravity.

To which *Ariftotle* feems to allude *(in 4. Phyfic. 7.)* where He faith of
thofe who affert a Vacuum, Σῶμα ᵹ πᾶν ἅπαν οἱονìαι εἶναι ἁπλόν: they
conceive all Bodies to be Tangible: and *Lucretius, Tangere enim & tangi*
*fine Corpore nulla poteft res.* Here we are, *per tranfennam,* to hint; that
the Authors of this Notion, do not reftrain the *Tangibility* of Bodies only
to the Senfe of *Touching* proper to Animals; but extend it to a more general importance, *viz.* the *Contact of two Bodies reciprocally occurring*
each to other *fecundum fuperficies*; or what *Epicurus* blended under
the word, Ἀντιτυπίας, Refiftence mutual arifing from Impenetrability.

(2) Others

(2) Others placing the Essential Propriety of a Body in its *Extension into the three Dimensions of Longitude, Latitude, and Profundity.* Thus *Aristotle* ( *Nat. Auscult.* 4. *cap.* 3. ) strictly enquiring into the Quiddity of Place, saith most profoundly; διαστημᾶτ(. μὲν ἂν ἐχεῖ τρία, μῆκος, ἢ πλατῦῷ, ἢ βάθος, οἶς ὁρίζελαι Σῶμα πᾶν: *Sane Dimensiones tres habet, longitudinem, latitudinem, & altitudinem, quibus omne Corpus definitur.* And thus *Des Cartes* ( *princip. Philos. Part.* 2. *Sect.* 4. ) *Naturam materiæ, sive Corporis in universum spectati, non consistere in eo quod sit res dura, vel ponderosa, vel colorata, vel aliquo alio modo sensus afficiens, sed tantum in eo, quòd sit res extensa in longum latum & profundum:* that the Essence of matter, or a Body considered in the General doth not consist in its hardness, weight, colour, or any other relation to the senses; but only in its Extension into the three Dimensions.

And (3) Others, by an excessive acuteness of Wit, dividing the *Substance* of a Body from the *Quantity* thereof, and distinguishing *Quantity* from *Extension.* Of this *immoderately subtle* Sect are all those, who conceived that most Bodies might be so rarified and condensed, as that by *Rarefaction* they may acquire *more*, and by *Condensation less* of Extension, then what they have before in their native dimensions. We say *immoderately subtle*, because whoever shall with due attention of mind profound the nature of Rarefaction and Condensation, must soon perceive; that by those motions a Body doth suffer no more then a meer *Mutation of Figure*, but its Quantity admits of neither Augmentation, nor Diminution. So as those Bodies may be said to be *Rare*, betwixt whose parts many Intervals or Interstices, repleted with no Bodies, are interspersed; and those Bodies affirmed to be *Dense*, whose parts mutually approaching each to other, either diminish, or totally exclude all the Intervals or intercedent Distances. And when it eveneth, that the Intervals betwixt the distant parts of a Body, are totally excluded by the mutual access, convention and contact of its parts: that Body must become so absolutely, or ( rather ) superlatively Dense, as to imagine a possibility of greater Density, is manifestly absurd. But yet notwithstanding, is not that Body thus extremly Dense, of less *Extension*, then when having its parts more remote each from other, it possessed a larger space: in respect, that whatever of Extension is found in the Pores, or Intervals made by the mutually receiving parts, ought not to be ascribed to the Body rarified, but to those small Inanities that are intercepted among the dissociated particles. For instance; when we observe a Sponge dipt in Liquor to become turgent and swell into a greater bulke; we cannot justly conceive, that the Sponge is made more Extense in all its parts, then when it was dry or compressed: but only, that it hath its pores more dilated or open, and is therefore diffused through a greater space. But we may not digress into a full examen of the nature of Rarefaction and Condensation; especially since the Syntax of our Physical Speculations will lead us hereafter into a full and proper consideration thereof.

Of the nature of the other ingredient of the Universe, INANITY, there are several Descriptions:

D　　　　　　　　　　　　　(1) *Epi-*

*Art.* 3.
Four Descriptions of the nature of Inanity, by *Epicurus, Cleomedes, Empiricus, Aristotle.*

(1) *Epicurus* names it ỳ κυψϱ, ỳ ἀϕϕι ϕύϵιν, a *Region*, or *Space*, and *a Nature that cannot be touched* : thereby intimating the direct Contrariety betwixt the essential notion of Corporiety and Inanity ; which Antithesis *Lucretius* plainly expresseth in that Verse, *Tactus coporibus cunctis intactus Inani.*

(2) *Cleomedes* describes a Vacuum to be, ỳ ἐϰϕῖϵν ἀϕώμϕαϑν, *ex sua natura incorporeum* : adding for further explanation, *siquidem est incorporeum, tactionque fugit, & neque figuram habet ullam, neque recipit, & nequepatitor quicquam, neque agit, sed probet subnimodo liberum per seipsum corporibus motum* ; it is incorporeal, because it cannot be touched, hath no figure of its own, nor is capable of any from others, neither suffers nor acts any thing, but only affords free space for the motion of other bodies through it.

(3) *Empiricus* (2. *adverf Phyfic.*) descanting upon *Epicurus* description of Inanity, faith ; *Natura eadem corpore destituta, appellatur Inane, occupata vero a corpore, Locus dicitur, pervadentibus ipsam corporibus evadit Regio* ; the same Nature devoid of all body, is called a *Vacuum*, if possessed by a body, 'tis called a *Place*, and when bodies pervade it, it becomes a *Region*.

And (4) *Aristotle* (3.*Phyfic*.7.) defines a *Vacuum* to be *Locus in quo nihil est*, a Place wherein no body is contained.

**Art. 4.**
Their importance extracted ; and what is the formal or proper notion of a *Vacuum*.

　　　　Now if we faithfully extract the importance of all these several Descriptions of Inanity, we shall find them to concurr in this common Notion. As according to vulgar sense, a Vessel is said to be empty, when it being capable of any, doth yet actually contain no body ; so, according to the sense of Physiology, that Place, that Region, or that Space, which being capable of bodies, doth yet actually receive or contain none, is said to be a *Vacuum* or Emptiness. Such would any Vessel be if upon remove of that body, whereby its capacity was filled, no other body, the Aer, nor ought else, should succeed to possess it : or such would that Space be, which this Book, that Man, or any other Body whatever doth now actually replenish, if after the remove of that Tenent, neither the circumstant Aer, nor ought else should succeed in possession, but it should be left on every side as it were limited by the same concave superficies of the circumambient, wherein the body, while a Tenent, was circumscribed and included.

**Art. 5.**
The Existence of Bodies in the world, manifest by Sense ; whose Evidence is perfect Demonstration.

　　　　Of the Existence of *Bodies* in the World, no man can doubt, but He who dares indubitate the testimony of that first and grand Criterion, *SENSE*, in regard that all *Natural Concretions* fall under the perception of some one of the Senses : and to stagger the Certitude of Sense, is to cause an Earthquake in the Mind, and upon consequence to subvert the Fundamentals of all Physical Science. Nor is Physiology, indeed, more then the larger Descant of Reason upon the short Text of Sense : or all our *Metaphysical* speculations (those only excluded, which concern the Existence and Attributes of the Supreme Being, the *Rational Soul* of man, and *Spirits* : the Cognition of the two former being desumed from proleptical or congenial impressi-

ons

ons implantate in, or coeffential to our mind, and the belief of the last being founded upon Revelation supernatural ) other then Commentaries upon the Hints given by some one of our External senses. Which Confideration caused *Epicurus* to erect these two Canons, as the Base of Logical Judicature.

( 1 )

*Opinio illa vera est, cui vel suffragatur, vel non refragatur sensus evidentia.*

( 2 )

*Opinio illa falsa est, cui vel refragatur vel non suffragatur sensus evidentia.*

That Opinion is true, to which the Evidence of Sense doth either assent, or not dissent: and that false, to which the evidence of Sense doth either not assent, or dissent.

By the *suffragation* or Assent of the Evidence of Sense, is meant an Assurance that our Apprehension or Judgment of any Object occurring to our sense, is exactly concordant to the reality thereof; or, that the Object is truly such, as we, upon the perception of it by our sense, did judge or opinion it to be. Thus *Plato* walking far off towards us, and we seeing him conjecture or opinion, as confidently as the great distance will admit, that it is *Plato*, whom we see coming toward us : but when, by his nearer approach, the great impediment of Certitude, Distance is removed; then doth the evidence of sense make an Attestation or suffragation of the verity of our opinion, and confirm it to be *Plato*, whom we saw.

The *Non-refragation* of Sense, intends the Consequution of some Inevident thing, which we suppose or præsume to be, with reflection upon something sensibly evident, or apparent. As when we affirm that there is a *Vacuum*; which taken singly, or speculated. in its own obscure nature, is wholly inevident, but may be demonstrated by another thing sufficiently evident, *viz. Motion*: for if no Vacuum, no Motion; since the Body to be moved must want a Place, wherein to be received, if all Places be already full and crouded. Hence comes it that the thing Evident doth not *Refragari* to the Inevident. And thus the Suffragation and Nonrefragation of the Evidence of sense, ought to be understood as one *Criterion*, whereby any Position may be evicted to be true.

D 2                    Hither

Hither also may be referred that *Tetrastick* of *Lucretius*, (*lib.*1.)

*Corpus enim per se communis deliquat esse*
*Sensus: quo nisi prima fides fundata valebit,*
*Haud erit, occultis de rebus, quò referentes*
*Confirmare Animi quicquam ratione queamus.*

That *Bodies* in the World existent are,
Our *Senses* undeniably declare :
Whose Certitude once quæstion'd ; we can find
No judge to solve nice scruples of the *Mind.*

It remains therefore only that we prove (1) *That there is a Vacuum in Nature.* (2) *That there is in the Universe no Third Nature besides that of Body and Inanity.*

CHAP.

# CHAP. IV.

## A Vacuum in Nature.

### SECT. I.

IN order to our more prosperous Evacuation of that Epidemick Opinion, *Vacuum non dari in rerum natura*, that there is no Vacuity or Emptiness in the World; it is very requisite, that we præmise, as a convenient Præparative, this short advertisement.

Among the speculations of many Ancient Phyfiologists, and especially of *Aristotle* (4. *Physic. 6*) we find a *Vacuum* distinguished into κατ φύσιν, & παρὰ φύσιν, *Secundum naturam*, & *Eternaturam*, a Vacuum consistent with, and a Vacuum totally repugnant to the fundamental constitutions of Nature. According to which proper distinction, we may consider a *Vacuum* (1) as παρεσπαρμένον, *Disseminatum, Intersperfed*, or of so large diffusion as variously to interrupt the Continuity of the parts of the World. 2 As Ἀθρόον, *Coacervatum, Coacervate* or separate from all parts of the World, such as the Ultramundan Space is conceived to be. Now, if we respect the *First* consideration or acception of a Vacuum, the Quæstion must be, *An detur vacuum Disseminatum*? Whether there be any small Vacuity in nature, or more plainly, Whether among the incontinued particles of Bodies there be any minute insensible Spaces intermixed, which are absolutely empty, or unpossessed by any thing whatever? If the *second*; then the doubt is to be stated thus: *An detur vacuum intra mundanum Coacervatum*? Whether within the World (for of the extramundane Inanity, the difficulty is not great, as may be collected from the contents of our *Second Chapter* præcedent) there can be any great or sensible Vacuity, such as we may imagine possible, if many of the small or interspersed Vacuities should convene and remain in one entire coacervate Inanity.

*Art. 1.*
The Distinction of a Vacuum into (1) Natural, and (2) Præternatural: and the one called Disseminate, the other Coacervate.

Concerning the *First* Problem, we cannot state the Doubt more intelligibly, then by proposing it under the analogy of this *Example*. Let a man intrude his hand into a heap of Corn, and his hand shall possess a certain sensible space among the separated Grains: his hand again withdrawn, that space doth not remain empty, but is immediately repossessed by the mutuall confluent grains, whose Confluxibility, not impeded, causeth

*Art. 2.*
The nature of a Disseminate Vacuity, explained by the Analogy of a heap of Corn.

feth their inftant convention. And yet betwixt the Grains mutually convened there remaine many intercepted or interpofed Spaces or Intervalls,
unpoffeffed by them; becaufe the Grains cannot touch each other fo *fe-*
*cundum totas fuperficies,* according to all parts of their fuperficies, as to be
contiguous in all points.    Exactly thus, when any Body is intruded into
Aer, Water, or any fuch rare and porous nature, betwixt whofe incontinued parts there are many Interftices varioufly diffeminated, it doth poffefs
a certain fenfible fpace proportionate to its dimenfions: and when that
Body is withdrawne, the fpace cannot remain empty, becaufe the infenfible
or atomical particles of the Aer, Water, &c. agitated by their own native
Confluxibility, inftantly convene and repoffefs it.    And yet, betwixt the
convened particles, of which the Aer, and Water, and alfo all porous
Bodies are compofed, there remain many empty fpaces (analogous to
thofe Intervalls betwixt the incontingent Grains of Corn) fo minute or
exiguous, as to be below the perception and commenfuration of fenfe.
Which is the very Difficulty, concerning which there are fo many Controverfies extant, as their very Lecture would be a Curfe to the greateft
Patience.    However, we conceive our felves fufficiently armed with Arguments to become the Affertors of a *Vacuum Diffeminatum,* or empty
Intervalls betwixt the particles of Rare, Porous, or Incontinued Bodies.

*Art 3.*
*The firft Argu-*
*ment of a Diffe-*
*minate Vacuity,*
*defumed from*
*the evidence*
*of Motion, in*
*General; and*
*Ariftotle's error*
*concerning the*
*Effence or*
*Place, concisely*
*detected, and*
*corrected.*

Our Firft Argument is that Reafon given for a Vacuum by *Epicurus* :
Ɛἰ ᾖ μὴ ἦν ὁ κενὸν, ὐκ ἂν ἔιχε τὰ σώματα ὅπε ἔιι, ἔδὲ δὶ ὁ ἐκινῶτο, καὶ
βάπερ φαίνεζαι κινοῦδμα, *Nifi effet Inane, non haberent Corpora neque ubi*
*effent, neque qua motus fuos obirent, cùm moveri ea quidem manifeftum fit :*
Unlefs there were a Vacuum, Bodies could have neither where to confift,
nor whither to be moved; and manifeft it is, that they are moved.    Which
folid Reafon, though feemingly perfpicuous, hath in it fo many receffes of
obfcurity, as may not only excufe, but efflagitate a curfory paraphrafe.
Firft, we are to obferve that, in the theory of *Epicurus,* the Notions of
*Inanity* and *Locality* are one and the fame *effentially,* but not *refpectively* :
*i e.* that the fame fpace when replenifhed with a Body, is a *Place,* but when
devoid or deftitute of any Tenent whatever, then it is a *Vacuum.*    Secondly, that *Ariftotle* did not fufficiently profound the Quiddity of *Place,* when
He taught, *that the Concave fuperficies of the Circumambient did conftitute*
*the Effence thereof.*    For, when it is generally conceded that the *Locus*
muft be adæquate to the *Locatum ;* it is truly præfumed, that the internal
fuperficie of the Circumambient or Place, ought to be adæquate to the external fuperficies of the Locatum or Placed ; but not to its *Profundity,* or
Internal Dimenfions.    And, fince it is of the formal reafon of Place, that
it be *Immoveable,* or uncapable of Tranflation ; for, otherwife any thing
might, at one and the fame time, be immote and yet change place : it is
evident, that the fuperficies of the Circumambient is not Immoveable, fince
it may both be moved, the Locatum remaining unmoved, and *è contrà,*
perfift unmoved, when the Locatum is removed.    And, therefore, the
Concave fuperficies of the Circumambient may, indeed, obtain the reafon
of a Veffel, but not of a Place.    And, upon confequence, we conclude,
that the *Space comprehended within the fuperficies of the Circumambient,* is
really and effentially what is to be underftood by *Place*    Since that Space
is adæquated perfectly to its Locatum in all its internal Dimenfions, and is
alfo truly Immoveable ; in regard that upon the remove of the Locatum,

it remains fixt, unchanged, unmoved ; in the fame ftate as before its occupation , it perfevers after its defertion.   And when the Body removed
poffeffeth a new Space: the old Space is inftantly poffeffed by a new Body.  Thirdly, that this argument defumed from the Evidence of Motion,
was propofed by *Empiricus, (adverf. Geometr.)* more Syllogiftically, thus,
Ἐι ἐϛι κίνησις, ἔϛι κενὸν, ἀλλα μὴν ἔϛι κίνησις, ἔϛι ἄρα κενόν. *Si Motus
eft , Inane eft ; atqui Motus eft , eft ergo Inane.*  If there be Motion ,
there muft be Inanity ; but *Motion there is , therefore there is a
Vacuum.*

*Art. 4.*
Motion de-
monftrated by
*Senfe*: and *Ze-
no's* ænigmati-
cil Argument ,
for an Univer-
fal *Quiet*, diffol-
ved.

That there is Motion, is manifeft from fenfe.  And as for that memorable Argument of *Zeno* againft Motion , though we judge that he
affected it more for the fingularity, then folidity thereof , and only propofed it as a new Paradox to gain fome credit to Scepticifm, of which he
was a fierce Affertor ; and that no man did ever admit it to a competition with the Authority of his Senfe : yet , fince many have reputed it
indiffoluble , we conceive the folution thereof muft become this
place.

> *Motus non poteft fieri per fpatium quodvis . uifi priis mobile pertranfeat
> minus, quam majus ; fed quamcunque affignes partem, alia eft minor,
> & alia minor in infinitum : Ergo non poteft fieri motus, nunquam enim
> incipiet.*   No Motion can be made through any fpace whatever,
> unlefs the Moveable firft pafs through a lefs, before a greater fpace ;
> but, what part of fpace foever you fhall pleafe to affign, ftill there
> will be another lefs part, and another lefs then that, and fo up to in-
> finity : therefore can there be no motion at all, fince it can never
> begin at a fpace fo little as that no lefs can remain.

### Solution.

The Fallacie lyeth in the *Minor,* which we concede to be true *ratione
Mathematica,* in the Mathematical acceptation thereof ; and fo no folution
can be fatisfactory to the Argument, unlefs we admit an infinite Divifibility in the parts of a Continuum : But deny it *ratione Phyfica,* in the proper
Phyfical acceptation, and fo we may folve the riddle by proving the parts
of a Continuum not to be divifible *ad infinitum,* and Motion is to be confidered *penes realem rerum exiftentiam.*  Now, that Space is divifible *ad
infinitum* only *Extrinfece* and *Mathematice,* not *Phyfice,* may be thus evinced.   If Motion be divifible *in infinitum,* the parts of a flow Motion
will be as many as the parts of a fwift Motion : but 'tis indubitate, that
two parts of a fwift motion are coexiftent to one of a flow : therefore
either that one part muft be permanent, fince it exifteth in two times, or
all Motions are equall in velocity and tardity, which is repugnant to experience.  And *Motion, Space,* and *Time,* are perfectly Analogous, *i.e. Proportional :* for there is no part of Motion , to which there may not be
affigned a Part of Space and Time fully refpondent.  Befides, fhould we
allow the Argument to be too clofe for the teeth of *Reafon* ; yet no man
can affirm it to be too hard for the fword of *Senfe* , and therefore it ought
not to be reputed inextricable : fince thofe objects which fall under the fincere judicature of the fenfe , need no other Criterion to teftifie their Verity.  Upon which the judicious *Magnenus* happily reflected *( p.* 162.
*Democritis*

*Democriti revivifcent.)* when He layed down this for a firm Principle: *Sensibilia per sensus sunt judicanda, nam illius potentia est judicare de re, per quam res cognoscitur; neque fides omnis sensibus deneganda.*

**Art. 5.**
The *Consequution* of the Argument (*If no Vacuum, no Motion*) illustrated.

This short Excursion ended, we revert to our *Fourth* observable, *viz.* the *Consequution* or Inference of *Epicurus*, in his argument for a Vacuum: *If no Vacuum, no Motion.* Which seems both natural and evident; for what is full, cannot admit a second tenent: otherwise nothing could prohibit the synthesis or Coexistence of many Bodies in one and the same place; which to imagine, is the extremest Absurdity imaginable.

For Illustration, let us Imagine, that the Uuniverse (having nothing of Inanity interspersed among its parts) is one Continued Mass of Bodies so closely crouded, ramm'd, and wedged together, that it cannot receive any the least thing imaginable more: and keeping to this Hypothesis, we shall soon deprehend, whether any one Body among those many disposed within this compact or closely crouded Mass may be removed out of its own to invade the place of another. Certainly, if all places be full, it must extrude another body out of its place, or become joint-tenant with it and possess one and the same place. Extrude a body out of its possession it cannot, because the Extruded must want a room to be received into; nor can the Extruded dispossess a third, that third expel a fourth, that fourth eject a fifth, &c. Since the difficulty sits equally heavy on all: and therefore, if the invaded doth not resign to the invading, there can be no beginning of Motion, and consequently no one Atome *in the* Universe can be moved. And, as for its becoming synthetical or joint-tenant, that is manifestly impossible: because a Collocation of two Bodies in one and the same place, imports a reciprocal Penetration of Dimensions, then which nothing can be more repugnant to the tenor of Nature: and therefore it remains, that every part of the Universe would be so firmly bound up and compacted by other parts, that to move those Cochles, Snails, or Insects, which are found in the ferruminated womb of Rocks, and incorporated to the heart of Flints, would be a far more modest attempt, then to move the least atome therein.

**Art. 6.**
An *Objection,* that the *Loco-cession* of some Bodies, dependes on their *Rarity* or *Porosity,* not on a Disseminate Vacuity, prevented.

Nor can the *Dissenting* evade the compulsion of this Dilemma, by prætending, that in the Universe are Bodies of *rare, porous,* and *fluxible* Constitutions, such as are more adapted to Lococession, or giving place upon their invasion by other Bodies, then are Rocks or Flints. Because, unless their *Rarity, Porosity, Fluxibility,* or yeeldingness be supposed to proceed from *Inanity disseminate;* or, that all the particles of those Bodies are contiguous, or mutually contingent *secundum totas superficies:* doubtless, they must be so Continued, as that it can make no difference, whether you call them Bodies of Flint or Aer. For, neither shall the Aer possess a place less absolutely then a Flint: because how many particles soever of place you shall suppose, no one of them can remain unpossessed; it being of the Essence of Place, that it be adæquate to *its* Tenent in all its internal Dimensions, *i.e.* in the number and proportion of Particles: nor a Flint more perfectly then *t*Aer, whose insensible Particles are præsumed to be reciprocally contingent in all points, and so to exclude all Interspersed Inanity.

We

We said, *without Inanity interfperfed, there can be no Beginning of Motion.* Which to explain, let us fuppofe that a Body, being to be moved through the Aer, doth in the firft degree of motion propel the contiguous aer, the fpace of a hairs bredth,    Now, the Univerfe being abfolutely full, that fmall fpace of a hairs bredth muft be præpoffeffed, and fo the Body cannot be placed therein, untill it hath thence depelled the incumbent Aer. Nor can the contiguous Aer poffeffing that fpace of a hairs bredth be depelled *per latera* to a place behind: becaufe that place alfo is replete with Aer. Infomuch, therefore, as the body to be moved, cannot progrefs through fo fmall a fpace, as that of an hairs bredth, becaufe of the defect of place for the reception of the Aer replenifhing that fpace: it muft of neceffity remain bound up immoveably in that place, wherein it was firft fituate.    But if we conceive the Aer to have fmall Inane *Vacuolas*, or Spaces (holding an analogy to thofe fpaces interceding betwixt the Grains of a Heap of Corn or Sand) varioufly interpofed among its minute infenfible particles: then may we alfo conceive, how the Motion of a Body through the Aer is both begun and continued: *viz.* that the Body moved, doth by its fuperfice protrude the particles of the contiguous Aer, thofe protruded particles being received into the adjacent empty interftices, prefs upon the next vicine particles of aer, and likewife protrude them, which received alfo into other adjacent empty fpaces become contiguous to, and urgent upon other next particles of Aer, and fo forward untill, upon the fucceffive continuation of the Compreffion by protrufion, and the confequent dereliction of a place behind, the lateral particles of the Aer, compreffed by the anterior parts diffilient, are effufed into it: and fo, how much of Aer is compreffed and impelled forward, fo much recurrs backward *per latera*, and is dilated.    The fame alfo may be accommodated to the Lococeffion of the Parts of *Water*, allowing it this prærogative, that being propelled by a Body movent, it doth by its particles more eafily propel the contiguous particles of the Aer, then its own; becaufe the empty minute fpaces of the aer incumbent upon the Water, are larger, which may be the reafon, why water propelled forwards, becomes tumid and fwelleth fomewhat upwards in its fuperfice, and is depreffed proportionately backward.    Now according to this theory, ought we to underftand the Reafon of *Epicurus* for a Vacuum, defumed from the neceffity of motion.

*Art.* 7.
No beginning of Motion, without Inanity interfperfed.

---

## Sect. II.

AS the nature of Motion confidered in the General, hath afforded us our Firft Argument, for the comprobation of a Vacuity Diffeminate: fo likewife doth the nature of *Rarefaction* and *Condenfation*, which is a fpecies of Local Motion, fpeculated in particular, readily furnifh us with a *Second*.    Examine we therefore, with requifite fcrutiny, fome of the moft eminent *Apparences* belonging to the *Expanfion* and *Compreffion* of *Aer* and *Water*: that fo we may explore, whether they can be falved more fully by our hypothefis of a Diffeminate Vacuity, then by any other, relating to an Univerfal *Plenitude.*

*Art.* 1.
A fecond Argument of a Vacuity Diffeminate, collected from the reafon of *Rarefaction* and *Condenjation.*

E                            Take

*Art.* 2.
The eminent
Phænomenon
of an *Æolipile,*
*pet, or Wind*
*Gun,* solved by
a Vacuity Dif-
feminate a-
mong the in-
contiguous
*(quoad totas*
*superficies )*
parts of aer.

Take we a *Pneumatique* or *Wind-Gun*, and let that part of the Tube, wherein the Aer to be compressed is included, be four inches long (the diameter of the bore or Cavity being supposed proportionate: ) now if among the particles of that aer contained in the four inched space of the Tube, there be no empty Intervals, or minute Inanities ; then of necessity must the mass of Aer included be exactly adæquate to the capacity or space of four inches, so as there cannot be the least particle of place, wherein is not a particle of aer æqual in dimensions to it, *i. e.* the number of the particles of aer is equal to the number of the particles of the Cavity. Suppose we then the number of particles common to both, to be 10000. This done, let the aer, by the Rammer artificially intruded, be compressed to the half of the space ( not that the compression may not exceed that rate, for *Mersennus* ( *in præf. ad Hydraulicam Pneumaticam Artem.* ) hath by a most ingenious demonstration taught, that Aer is capable of Compression even to the tenth part of that space, which it possessed in the natural disposition, or open order of its insensible particles: ) and then we demand, how that half space, *viz.* two inches, can receive the double proportion of Aer, since the particles of that half space are but 5000. Either we must grant that, before compression, each single particle of Aer possessed two particles of space, which is manifestly absurd : or, that after Compression, each single particle of space doth contain two of aer, which is also absurd, since two bodies cannot at once possess the same place : or else, that there were various Intervals Inane disseminate among the particles of Aer, and then solve the Phænomenon thus. As the Grains of Corn, or Granules of Sand, being powred into a vessel up to the brim, seem wholly to fill it, and yet by succussion of the vessel, or depression of the grains upon the imposition of a great weight, may be reduced into a far less space ; because from a more lax and rare, they are brought to a more close and constipate congeries, or because they are reduced from an open, to a close order, their points and sides being more adapted for reciprocal contact *quoad totas superficies*, nor leaving such large Intervals betwixt them as before succussion or depression. So likewise are the particles of aer included in the four-inched space of the Tube, by Compression or Coangustation reduced downe to the impletion of onely the half of that space ; because from a more lax or rare Contexture they are contracted into a more dense or close, their angles and sides being by that force more disposed for reciprocal Contingence, and leaving less Intervals, or empty spaces betwixt them then before.

*Art.* 3.
Experiment
of an *Æolipile,*
or *Hermetical*
Bellows, arre-
sting a Vacuity
Disseminate.

Our *Second* Experiment is that familiar one of an *Æolipile* which having one half of its Concavity replete with Water, and the other with Aer, and placed in a right position near the fire : if you will not allow any of the spaces within it to be empty, pray, when the Water by incalescence rarefied into vapours, issues out with thundering impetuosity through the slender perforation or exile outlet of its rostrum, successively for many hours together, how can the same Capacity still remain full ? For, if before incalefaction the particles of Water and Aer were equal to the number of the particles of space contained therein ; Pray, when so many parts both of Water and Aer,

consociated

confociated in the form of a vapour, are evacuated through the
Orifice, muſt not each of their remaining parts poſſeſs more parts of
the capacity, and ſo be in many places at once? If not ſo, were there
not, before the incaleſcence, many parts of Water and Aer crouded
into one and the ſame part of ſpace, and ſo a manifeſt penetration of real
dimenſions? Remains it not therefore more veriſimilous, that, as an
heap of duſt diſperſed by the Wind, is rarefied into a kind of cloud
and poſſeſſeth a far larger ſpace then before its diſperſion; becauſe the
diſgregated Granules of Duſt intercept wider ſpaces of the ambient
aer: ſo the remaining parts of Water and Aer in the cavity of the
Æolipile poſſeſs all thoſe Spaces left by the exhaled parts, becauſe they
intercept more ample empty Spaces, being diſpoſed into a more lax
and open contexture. And that this is cauſed by the particles
of Fire, which intruding into, and with rapid impetuoſity agitated
every way betwixt the ſides of the Æolipile, ſuffer not the parts
of Aer and Water to quieſce, but diſperſe and impel them variouſly:
ſo that the whole ſpace ſeems conſtantly full by reaſon of the rapidity of
the Motion.

*Art.* 4.
Experiment of
a *Sulphurate
Tapor*, included
in a Glaſs Vi-
al, partly filled
with Water
of the ſame
importance.

The *Third* Mechanick Experiment, which may juſtifie the ſub-
miſſion of our aſſent to this Paradox, is this. Having præpared a ſhort
Tapor of Wax and Sulphur groſly powdered, light and ſuſpend
it by a ſmall Wier in a Glaſs Vial of proportionate reception,
wherein is clean Fountain Water ſufficient to poſſeſs a fifth part, or
thereabout, of its capacity: and then with a Cork fitted exactly to
the Orifice, ſtop the mouth of the Vial ſo cloſely, that the eruption of the moſt ſubtle Atom may be prevented.    On this you
ſhall perceive the flame and fume of the Sulphur and Wax inſtantly to diffuſe and in a manner totally poſſeſs the room of the Aer,
and ſo the fire to be extinguiſhed: yet not that there doth ſucceed either any diminution of the Aer, ſince that is impriſoned, and all poſſibility of evaſion præcluded; or any aſcent of the
Water, by an obſcure motion in vulgar Phyſiology called *Suction*,
ſince here is required no ſuction to ſupply a vacuity upon the deſtitution of aer. But if you open the orifice, and enlarge the impriſoned
Aer, you ſhall then indeed manifeſtly obſerve a kind of obſcure ſuction,
and thereupon a gradual aſcention of the Water: not that the flame
doth immediately elevate the water, as well becauſe it is extinct, and
the water doth continue elevated for many hours after its extinction, as
that, if the flame were continued, can it be imagined that it would with
ſo much tenacity adhære to the tapor, as is requiſite to the elevation of
ſo great a weight of water; but rather, that upon the Coaguſtation
or compreſſion of the aer reduced to a very cloſe order in the mutual contact of its inſenſible particles, the empty ſpaces formerly intercepted betwixt them being repleniſhed with the exhalations of the tapor; when the
orifice is deobturated, there ſenſibly ſucceeds a gradual expiration of the
atoms of Fire, as the moſt agile, volatile and prepared for motion, and
then the aer, impelled by its own native Fluxibility, re-expands or
dilates it ſelf by degrees. But ſince the narrowneſs of the Evaporato-
ry, or orifice prohibits the ſo ſpeedy reflexion or return of the compreſſed particles of the aer to their naturall contexture or open order, as the renitency of their fluxibility requireth, ſo long as there

E 2                                        remain

remain any of the atoms of Fire in poſſeſſion of their Vacuities, as long continues the reexpanſion of the Aer; and that reexpanſion preſſing upon the ſides of the water, cauſeth it to aſcend, and continue elevated. And no longer, for ſo ſoon as the aer is returned to its native contexture, the water by degrees ſubſideth to the bottom, as before the accenſion of the Tapor : and ſo that motion commonly called a *Suction in avoidance of Vacuity,* is more properly a *Protruſion,* cauſed by the expanding particles of aer compreſſed.

*Art 5.*
*No Combuſtible in Aer : and ſo the opinion of the Ariſtotelians, that the Extinction of Flame impriſoned, it to be charged on the Defect of Aer, for its ſuſtentation; groſly erroneous.*

If any præcipitous Curioſity ſhall recur to this Sanctuary, that in the Subſtance of the Aer is contained *Aliquid Combuſtibile,* ſome combuſtible matter, which the hungry activity of the flame of the Tapor doth prey upon, conſume and adnihilate : He runs upon a double abſurdity; (1) That in Nature is a ſubſtance, which upon the accidental admotion of Fire, is ſubject to abſolute *Adnihilation,* which to ſuppoſe, ſmels of ſo great a wildneſs of Imagination as muſt juſtifie their ſentence, who ſhall conſign the Author of it to ſeven years diet on the roots of White Hellebor, nor durſt any man but that *Elias Artium Helmont,* adventure on the publique Patronage of it. (2) That the Aer is the *Pabulum,* or Fewel of Fire : which though no private opinion, but paſſant even among the otherwiſe venerable Sectators of *Ariſtotle* (who unjuſtly refer the Extinction of flame impriſoned, to the *Defection of Aer* : as intimating that the deſtruction of Fire, like that of Animals, doth proceed from the deſtitution of Aliment) is yet openly inconſiſtent to Reaſon and Experiment. To *Reaſon,* becauſe the Aer, conſidered ſincerely as Aer, without the admixture of vapours and exhalations, is a pure, ſimple and Homogeneous ſubſtance; whoſe parts are conſimilar : not a compoſition of heterogeneous and diſſimilar, whereof ſome ſhould ſubmit to the conſumptive energie of Fire, and other ſome (of the invincible temper of Salamandes Wool, or Muſcovy Glaſs,) conſerve their originary integrity inviolable in the higheſt fury of the flames. Again, Themſelves unanimouſly approve that Definition of *Galen (lib. 1. de Element. cap. 1.) Elementa ſunt natura prima & ſimpliciſſima corpora, quæque in alia non amplius diſſolvi queant :* that it is one of the eſſential Proprieties of an *Element* as to be ingenerable, ſo alſo *Indiſſoluble* : and as unanimouſly conſtitute the Aer to be an Element. To *Experiment,* becauſe had the Fire found (and yet it is exceedingly inquiſitive, eſpecially when directed by Appetite, according to their ſuppoſition) any part of the Aer inflamable ; the whole Element of aer had been long ſince kindled into an univerſal and inextinguable conflagration, upon the accenſion of the firſt focal fire : nor could a flaſh of Lightning or Gunpowder, be ſo ſoon extinct if the flame found any maintenance or ſuſtentaculum in the Aer, but would enlarge it ſelf into a Combuſtion more prodigious and deſtructive then that cauſed by the wild ambition of Phaeton. Moſt true it is, that Fire deprived of aer, doth ſuffer immediate extinction : yet not in reſpect of Aliment denyed (for *Nutrition* and *Vitality* are ever convertible ) but of the *want of room* ſufficient to contain its igneous and fuliginous Exhalations, which therefore recoiling back upon the flame, coarctate, ſuffocate, and ſo extinguiſh it. For upon the exceſſive and impetuous ſuddain afflation of aer, Flame doth inſtantly periſh, though not impriſoned in a glaſs : the cauſe is, that the flame, not with tenacity ſufficient adhæring to the body of the tapor, or lamp, is eaſily blown off, and being thus diſlodged hath no longer ſubſiſtence in the aer. And Heat, beating upon
the

the outside or convex part of a Glass, seems sensibly to dilate the Aer imprisoned within; as is manifest upon the testimonie of all Thermometres, or Weather-Glasses, those only which contain Chrysulca, or Aqua Fortis in stead of Water, at least if the experiment be true, excepted : but Fire in the Concave or inside of the Glass violently compresseth the aer, by reason of its fuliginous Emissions, which wanting vacuities enough in the aer for their reception, recoil and suffocate the fire.

The *Fourth*, this. Being in an intense frost at *Droitwich* in *Worcestershire*, and feeding my Curiosity with enquiring into the Mechaniek operations of the *Wallers* (so the *Salt-boylers* are there called ) I occasionally took notice of *Yce*, of considerable thickness, in a hole of the earth, at the mouth of a Furnace very great and charged with a Reverberatory fire, or *Ignis rota.* Consulting with my Phylosophy, how so firm a congelation of Water could be made by Cold at the very nose of so great a fire ; I could light on no determination, wherein my reason thought it safe to acquiesce, but this. That the ambient Aer, surcharged with too great a cloud of exhalations from the fire, was forced to a violent recession or retreat, and a fresh supply of aer as violently came on to give place to the receding, and maintain the reception of fresh exhalations ; and so a third, fourth and continued relief succeeded : and that by this continued and impetuous afflux, or stream of new aer, loaden with cold Atoms, the activity of the cold could not but be by so much the more intense at the mouth of the furnace, then abroad in the open aer, by how much the more violent the stream of cold aer was there then elsewhere.    To complete and assure the Experiment, I caused two dishes, of equal capacity, to be filled with river Water ; placed one at the mouth of the furnace, the other *sub Dio :* and found that near the furnace so nimbly creamed over with Yce, as if that visibly-freezing Tramontane Wind, which the Italian calls *Chirocco*, had blown there, and much sooner perfectly frozen then the other.    And this I conceive to be also the reason of that impetuous suction of a stream of aer, and with it other light and spongy bodies, through the holes or pipes made in many Chimneys, to prævent the repercussion of smoke.

*Art. 6.*
A fourth singular and memorable Experiment of the Authors, of Yce at the nose of a large Reverberatory Furnace, charged with *Ignis rotæ* evidencing a Vacuity interspersed in the Aer.

From these observations equitably perpended and collated, our meditations adventured to infer

( 1 ) That the Aer ; as to its principal and most universal Destination was created to be the Aτμοδοχεῖον, or common RECEPTARY of Exhalations : and that for the satisfaction of this End, it doth of necessity contain a *Vacuum Desseminatum* in those minute and insensible *Incontiguities* or Intervals betwixt its atomical Particles ; since Nature never knew such gross improvidence, as to ordain an *End*, without the codestination of the *Means* requisite to that End.    To prævent the danger of misconstruction in this particular, we find our selves obliged to intimate ; that in our assignation of this Function or Action to the Aer, we do not restrain the aer to this use alone : since Ignorance it self cannot but observe it necessarily inservient to the Conservation of Animals endowed with the organs of Respiration, to the transvection of Light, the convoy of odours, founds, and all Species and Aporrhæas, &c. but that, in allusion to that Distinction of Anatomists betwixt the *Action* and *Use* of a Part, we intend ; that the grand and most General *Action* of the Aer, is the Reception or entertainment

*Art. 7.*
An Inference from that Experiment ; that Aer, as to its General Destination, is the Common Receptary of Exhalations.

entertainment of Vapours and Exhalations emitted from bodies situate in or near the Terraqueous Globe. And in this acception, allowing the Aer to be constituted the General *Host* to admit; we insinuate that it hath rooms wherein to lodge the arriving Exhalations : insomuch as the necessity of the one, doth import as absolute a necessity of the other ; the existence of the *Final* ever attesting the existence of the Conductive, or *Mediatory* Cause.

*Art. 8.*
A second illation, that the Aer doth receive Exhalations at a certain rate, or definite proportion ; which cannot be transcended without prodigious violence.

( 2 ) That , though the Aer be variously interspersed with empty Interstices , or minute Incontiguities, for the reception of Exhalations : yet doth it receive them at a just *Rate, Tax,* or *determinate Proportion,* conform to its own Capacity, or *Extensibility* ; which cannot without Reluctancy and Violence be exceeded   For when the Vacuities, or Holds have taken in their just portage, and equal fraught, the compressed aer hoyseth sail, bears off, and surrenders the Scene to the next adventent or vicine aer ,which acteth the like part successively to the continuation of the motion. This may be exemplified in the experiment of the Furnace and Chimneys newly mentioned, but more manifestly in that of the *Sulphurate Taper* in the Vial: where the Aer, being overburthened with too great a conflux of fuliginous Exhalations , and its recession impeded by the stopping of the Vial, it immediately recontracteth it self, and in that renitency extinguisheth by suffocation the rude Flame, which oppressed it with too copious an afflux.   As also in those of *Canons* and *Mines* ; which could not produce such portentous effects , as are dayly observed in Wars , if it were not in this respect, that the Receptaries in the Aer suffer a rack or extension beyond their due Capacities   For, when the Powder fired in them is, in the smallest subdivision of time, so much subtiliated, as to yeeld a Flame (according to the compute of *Mersennus*) of 10000 parts larger in extension, then it self, while its Atoms remained in the close order and compact form of Powder ; and the Aer, by reason of its imprisonment, is not able to recede, and bear off so speedily, as the velocity of the motion requires : for avoidance of a mutual Penetration of Dimensions among the minute particles of the Fire, smoke, and its own, it makes an eruption with so prodigious an impetuosity, as to shatter and evert all solid bodies situate within the orb of impediment.

*Art. 9.*
The Existence of Inane Incontiguities in the Aer, confirmed by two considerable Arguments.

For the further Confirmation of our *First Thesis, viz.* That the Aer is interspersed with various Porosities, or Vacuities, by reason of the Incontiguity of its insensible Particles ; and that these serve to the reception of all Exhalations : we shall superadd these two considerable Arguments. ( 1 ) If this *Vacuum Disseminatum* of the Aer be submoved, and an absolute Plenitude in the Universe from a Continuity of all its parts supposed ; then must every the smallest motion, with dangerous violence run through the whole Engine of the World , by reason of that Continuity. ( 2 ) If the Aer were not endowed with such Porosities, other Bodies could never suffer the dilatation or rarefaction of themselves; since,upon the subtiliation or dilatation of their minute particles, *i e.* the remove of their Atoms from a close to an open contexture, they possess 1000 times larger Capacities : and so there would be no room to entertain the continual *Effluviums,* expiring from all bodies passing their natural vicissitudes and degenerations.

SECT.

### Sect. III.

TO these *Four* eminent *Experiments*, we might have annexed others numerous enough to have swelled this Chapter into a Volume; but conceiving them satisfactory to any moderate Curiosity, and that it can be no difficulty to a Phyſiological Meditation, to ſalve any *Apparence* of the ſame nature, by this *Hypotheſis* of a *Vacuum Diſſeminatum* in the Aer, as the *Cauſſa ſine qua non* of its *Rarefaction* and *Condenſation* : we judged it more neceſſary to addreſs to the diſcharge of the reſidue of our duty, *viz.* to præſent it as veriſimilous; that in the *Water* alſo are variouſly diſperſed the like *Vacuola*, or empty ſpaces, ſuch as we have not unfitly compared to thoſe διας-ηματα, or *Intervals* betwixt the Granules of Sand in a heap, in thoſe parts where their ſuperficies are not contiguous, in reſpect of the ineptitude of their Figures for mutual contact in all points. And this ſeems to us ſo illuſtrious a Verity, as to require neither more atteſtation, nor explanation, then what this one ſingular Experiment imports.

*Art. 1.*
That *Water* alſo contains *Vacuola*, empty Space; demonſtrated.

'Tis generally known, that *Water* doth not diſſolve *Salt* in an indefinite quantity, but *ad certam taxam*, to a certain determinate proportion; ſo as being once ſated with the Tincture thereof, it leaves the overplus entire and undiſſolved. After a long and anxious ſcrutiny for a full ſolution of this *Phenomenon*, our thoughts happily fixed upon this; That, the Salt being in diſſolution reduced (*Analyſi retrograda*) into its moſt minute or Atomical Particles, there ought to be in the Water Conſimilar or adæquate Spaces for their Reception; and that thoſe Spaces being once repleniſhed, the Diſſolution (becauſe the Reception) ceaſeth. Not unlike to a full ſtomach, which eructates and diſgorges all meats and drink; ſuperingeſted : or full veſſels, which admit no liquor affuſed above their brim.    Hereupon, having firſt reflected upon this, that the Atomical Particles of common Salt are *Cubical*; and thereupon inferred, that, ſince the *Locus* muſt be perfectly adæquate to the *Locatum*, they could only fill thoſe empty ſpaces in the water, which were alſo *Cubical*: we concluded it probable, that in the water there ought to be other empty ſpaces *Octohedrical, Sexangular, Spherical,* and of other Figures, which might receive the minute particles of other Salts, ſuch as *Alum, Sal Ammoniac, Halinitre, Sugar, &c.* after their diſſolution in the ſame Water. Nor did Experiment falſifie our Conjecture. For, injecting *Alum* parcel after parcel, for many dayes together, into a veſſel of Water formerly ſated with the tincture of common Salt; we then, not without a pleaſant admiration, obſerved that the Water diſſolved the Alum as ſpeedily, and in as great quantity, as if it altogether wanted the tincture of Salt; nor that alone, for it likewiſe diſſolved no ſmall quantities of other Salts alſo. Which is no obſcure nor contemptible Evidence, that water doth contain various inſenſible *Loculaments, Chambers,* or *Receptaries* of different *Figures*: and that this variety of thoſe Figures doth accommodate it to extract the *Tinctures* of ſeveral Bodies injected and infuſed therein. So as it is exceedingly difficult, to evince by Experiment that any Liquor is ſo ſated with precedent Tinctures, as not

to

*Art. 2.*
From the Experiment of the Diſſolution of *Alum, Halinitre, Sal Ammoniac,* and *Sugar,* in Water formerly ſated with the Tincture of *Common Salt*

to be capable of others also : especially while we cannot arrive at the exact knowledge of the Figure of the Atomical Particles of the body to be infused, nor of the Figures of those minute spaces in the liquor, which remain unpossessed by the former dissolutions.

*Art. 3.*
*The verity of the Lord Verulam's Assertion, that a repeated infusion of Rhubarb acquires as strong a virtue Cathartical, as a simple infusion of Scamony, in equal quantity; and why.*

Upon which reason, we are bold to suspect the truth of the *Lord S. Albans* assertion ; (*Centur. 1. Nat. Hist.*) *that by repeating the infusion of* Rhubarb *several times, letting each dose thereof remain in maceration but a small time* (*in regard to the Fineness and volatility of its Spirits, or Emanations*) *a medicament may be made as strongly Cathartical or Purgative, as a simple infusion of* Scamony *in the like weight.* For (1) when the empty spaces in the Menstruum, or Liquor, which respond in Figure to the Figure of the Atomical particles of the Rhubarb, are replenished with its Tincture ; they can admit no greater fraught, but the Imbibition of Virtue ceaseth : and that two or three infusions , at most, suffice to the repletion of those respective spaces , may be collected from hence , that the Rhubarb of the fourth infusion loseth nothing of its Purgative Faculty thereby, but being taken out and singly infused in a proportionate quantity of the like liquor, it worketh as effectually as if it had never been infused before. (2) Experience testifieth the Contrary, *viz.* that a Drachm of Scamony singly infused in an ounce and half of White wine, doth operate (*cæteris paribus*) by 15 parts of 20, more smartly then 5 drachms of Rhubarb successively infused in the like quantity of the same or any other convenient Liquor.

*Art. 4.*
*Why two Drachms of Antimony impregnate a pint of Wine, with so strong a vomitory Faculty, as two ounces.*

Here also is the most probable Cause, why two Drachms of *Antimony* crude, or *Crocus Metallorum*, give as powerful a Vomitory impregnation to a Pint of Sack, or White wine, as two ounces : *viz.* because the *menstruum* hath no more Vacuities of the same Figure with the Atomical *Effluviums* of the Antimony, then what suffice to the imbibition or admission of the two Drachms. For the Certitude of this, we appeal to the experience of a *Lady* in *Cheshire*, who seduced by an irregular Charity , and an opinion of her own skill, doth prætend to the cure of the sick, and to that purpose præpares her Catholique Vomitory, consisting of four Drachms and an half of crude *stibium* infused all night in 3 or 4 ounces of White wine, and usually gives it ( without respect to the individual temperament of the *Assument* for one dose to the sick : and yet, as our selves have more then once observed, the infusion doth work with no greater violence , in some persons, then as much of our common *Emetique Infusion* præscribed in the reformed Dispensatory of our Venerable *College.* Nay more then this, our selves have often reduced the Dose of the same Emetique Infusion down only to 4 *Scruples*, and yet found its operation come not much short of the usual Dose of an *ounce.*

*Art. 5.*
*Why one and the same Menstruum may be enriched with various Tinctures.*

Hence also may be desumed a satisfactory reason for the imprægnation of one and the same *Menstruum* with various Tinctures : for Example, Why an Infusion of *Rhubarb*,sated with its tincture,doth afterward extract the tinctures of *Agarick*, *Senna*, the *Cordial Flowers*, *Cremor Tartari*, &c. injected according to the præscript of the judicious Physician, in order to his confection of a Compound Medicament requisite to the satisfaction of a *Complex Scope* or Intention.

SECT.

### Sect. IV.

*A* Third Argument, for the comprobation of a *Vacuum Disseminatum*, may be adferred from the Cause of the Difference of Bodies in the degrees of *Gravity*, respective to their *Density* or *Rarity*, (*i. e.*) according to the greater or less Inane Spaces interspersed among their insensible Particles. And a *Fourth* likewise from the reason of the *Calefaction* of Bodies by the subingress or penetration of the Atoms of Fire into the empty Intervals variously disseminate among their minute particles. But, in respect that we conceive our *Thesis* sufficiently evinced by the *Præcedent Reasons*; and that the consideration of the *Causes* of *Gravity* and *Calefaction*, doth, according to the propriety of Method, belong to our succeeding Theory of *Qualities*: we may not in this place insist upon them.

*Art. 1.*
Two other Arguments of a Vacuity Disseminate inferrible from (1.) the difference of Bodies in the degrees of Gravity : (2) the Calefaction of Bodies by the penetration of igneous Atoms into them.

And as for those many *Experiments* of *Water-hour-glasses, Syringes, Glass Fountains, Cuppinglasses*, &c. by the inconvincible Assertors of the *Peripatetick Physiology* commonly objected to a *Vacuity*: we may expede them altogether in a word. We confess, those experiments do, indeed, demonstrate that Nature doth abhorr a *Vacuum Coacervatum*; as an heap of Sand abhors to admit an Empty Cavity great as a mans hand extracted from it: but not that it doth abhor that *Vacuum Disseminatum*, of which we have discoursed; nay, they rather demonstrate that Nature cannot well consist without these small empty Spaces interspersed among the insensible Particles of Bodies, as an heap of Sand cannot consist without those small Interstices betwixt its Granules, whose Figures prohibit their mutual contact in all points. So that our Assertion ought not to be condemned as a Kænodox inconsistent to the laws of Nature, while it imports no more then this; that, as the Granules of a heap of Sand mutually flow together to replenish that great Cavity, which the hand of a man by intrusion had made, and by extraction left, by reason of the *Confluxibility* of their Nature: so also do the Granules, or Atomical Particles of Aer, Water, and other Bodies of that *Rare* condition, flow together, by reason of the *Fluidity* or *Confluxibility* of their Nature, to prævent the creation and remanence of any considerable, or *Coacervate Vacuum* betwixt them. To instance in one of the *Experiments* objected. Water doth not distil from the upper into the lower part of a *Clepsydra*, or *Water-hour-glass*, so long as the Orifice above remains stopped; because all places both above and below are ful,nor can it descend until,upon unstopping the hole,the aer below can give place, as being then admitted to succeed into the room of the lateral aer, which also succeeds into the room of that which entered above at the orifice, as that succeeds into the room of the Water descending by drops, and so the motion is made by succession, and continued by a kind of Circulation. The same also may be accommodated to those Vessels, which *Gardners* use for the irrigation of their Plants, by opening the hole in the upper part thereof, making the water issue forth below in artificial rain.

*Art. 2.*
The Experiments vulgarly adduced to prove no vacuity in nature, so far from denying, that they confess a Disseminate one.

*Art. 3.*
The grand Difficulty of the Cause of the Aers restitution of it self to its natural contexture, after rarefaction and condensation; satisfied in brief.

It only remains, therefore, that we endeavour to solve that Giant *Difficulty*, proposed in defiance of our *Vacuum Disseminatum*, by the mighty *Mersennus*

F

*Mersennus (in Phænomen. Pneumatic. propos. 31.)* thus. *Quomodo Vacuola, solitò majora in rarefactione, desinant, aut minora facta in condensatione crescant iterum : quænam enim Elateria cogunt aerem ad sui restitutionem?* How do those Vacuities minute in the aer, when enlarged by rarefaction, recover their primitive exility ; and when diminished by condensation, re-expand themselves to their former dimensions : What *Elaters* or *Springs* are in the aer, which may cause its suddain restitution to its natural constitution of insensible particles ?

We *Answer*; that, as it is the most catholique Law of Nature, for every thing, so much as in it lies, to endeavour the conservation of its originary state ; so, in particular, it is the essential quality of the Aer, that its minute particles conserve their natural Contexture, and when forced in Rarefaction to a more open order, or in Condensation to a more close order, immediately upon the cessation of that expanding, or contracting violence, to reflect or restore themselves to their due and natural contexture. Nor need the Aer have any Principle or Efficient of this Reflection, other then the *Fluidity* or *Confluxibility* of its Atomical Parts: the essence or *Quiddity* of which Quality, we must reserve for its proper place, in our ensuing theory of Qualities.

CHAP.

# CHAP. V.

## *A Vacuum Præternatural.*

### Sect. I.

 E fides a Natural, or Diſſeminate Vacuity frequently intercepted betwixt the incontiguous Particles of Bodies
(the Argument of our immediately precedent Chapter)
not a few of the higheſt form in the ſchool of *Democritus* have adventured to affirm not only the poſſibility,
but frequent introduction of a *Præternatural* or *Coacervate Inanity:* ſuch as may familiarly be conceived, if we
imagine many of thoſe minute inane ſpaces congrega-
ted into one ſenſible void ſpace.   To aſſiſt this Paradox, the autoptical
teſtimony of many Experiments hath been pleaded ; eſpecially of that
*Glaſs Fountain* invented by *Hero* (*præf.in Spirit.*) and fully deſcribed by
the learned and induſtrious *Turnebus* (*in lib. de calore*) and of that Braſs
Cylindre, whoſe concave carries an *Embolus,* or ſucket of wood, concerning which the ſubtle *Galilæo* hath no ſparing diſcourſe in the firſt of his
*Dialogues :* but, above all, of that moſt eminent and generally ventilated
one of a *Glaſs Cylindre,* or Tube filled with Quickſilver, and inverted ;
concerning which not long after the invention thereof by that worthy
Geometrician, *Torricellius,* at Florence, have many excellent Phyſicomathematical Diſcourſes been written by Monſieur *Petit,* Dr. *Paſchal Merſennus , Gaſſendus, Stephanus Natalis.* Who, being all French, ſeemed
unanimouſly to catch at the experiment, as a welcom opportunity to challenge all the Wits of Europe to an æmulous combat for the honour of perſpicacity.    Now albeit we are not yet fully convinced, that the chief
Phænomenon in this illuſtrious Experiment doth clearly demonſtrate the
exiſtence of a *Coacervate* Vacuity, ſuch as is thereupon by many conce.led,
and with all poſſible ſubtlety defended by that miracle of natural Science,
the incomparable *Merſennus* (*in reflexionib. Phyſicomathemat.*) yet, inſomuch as it affords occaſion of many rare and ſublime ſpeculations, whereof ſome cannot be ſolved either ſo fully, or perſpicuouſly by any Hypotheſis, as that of a *Vacuum Diſſeminatum* among the inſenſible particles of
Aer and Water ; and moſt promiſe the pleaſure of Novelty, if not the
profit of ſatisfaction to the worthy conſiderer ; we judge it no unpardonable Digreſſion, here to preſent to our judicious Reader, a faithful Tran-
ſcript of the Experiment, together with the moſt rational ſolutions of all

*Art. 1.*
What is con-
ceived by a *Co-
acervate Vacuity:*
and who was
the *Inventor* of
the famous Ex-
periment of
*Quickſilver* in a
*Glaſs Tube,* up-
on which ma-
ny modern
Phyſiologiſts
have erected
their perſwaſi-
on of the poſſi-
bility of in-
troducing it.

*Experientiam
appenam , cujus
inventionem etſi
neſcio qui alii
ambitioſius ſibi
arrogent , certo
tamen mihi con-
ſtat, primam à
Torricellio ,
nobili magni
Ducis Æturiæ
Mathematico
detectam, &c.
Athanaſ Kir-
cherus, in Artis
magn. Conſoni
& Diſſoni l. 1.
p. 11. in ſingu-
lari Digreſſione*

the admirable Apparences obſerved therein, firſt by *Torricellius* and the
reſt beyond Sea, and ſince more then once by our ſelves.

## The Experiment.

*Art.* 2.
A faithful de-
ſcription of
the *Experiment*,
and all its rare
*Phænomena.*

  *Having prepared a Glaſs Tube (whoſe longitude is 4 feet, and the diameter
of its concavity equal to that of a mans middle finger ) and ſtopped up one of
its extremities, or ends, with a ſeal Hermetical : fill it with Quickſilver, and
ſtop the other extreme with your middle finger. Then, having with a moſt
ſlow and gentle motion (leſt otherwiſe the great weight of the Quickſilver break
it ) inverted the Tube, immerge the extreme ſtopt by your finger into a Veſſel
filled with equal parts of Quickſilver and Water, not withdrawing your finger
untill the end of the Tube be at leaſt 3 or 4 inches deep in the ſubjacent Quickſilver : for, ſo you prævent all inſinuation or intruſion of Aer. This done, and
the Tube fixed in an erect or perpendicular poſition ; upon the ſubduction of
your finger from the lower orifice , you may obſerve part of the Quickſilver
contained in the Tube to deſcend ſpeedily into the reſtagnant or ſubjacent
Quickſilver, leaving a certain ſpace in the ſuperior part of the Tube, according to apparence at leaſt, abſolutely Void or Empty : and part thereof (after
ſome Reciprocations or Vibrations ) to remain ſtill in the Tube, and poſſeſs its
cavity to a certain proportion, or altitude of 27 digits, or 2 feet, 3 digits and
an half (proxime) conſtantly. Further, if you recline, with a gentle motion
alſo, the upper extreme of the Tube, untill the lower, formerly immerſed in the
Quickſilver, ariſe up into the region of the Water incumbent on the ſurface of
the Quickſilver : you may perceive the Quickſilver remaining in the Tube to
aſcend by ſenſible degrees up to the ſuperior extreme thereof, together with part
of the Water ; both thoſe liquors to be confounded together ; and, at length,
the Quickſilver wholly to diſtill down in parcels, ſurrendring the cavity of the
Tube to the poſſeſſion of the Water. Likewiſe, if you recline the ſuperior extreme of the Tube, untill its altitude reſpond to that of 27 digits, ſtill retaining the oppoſite extreme in the region of the ſubjacent Quickſilver in the veſſel : then will the Quickſilver be ſenſibly impelled up again into the Tube, untill that ſpace formerly vacated be repleniſhed. Finally, if, when the Quickſilver hath fallen down to the altitude of 27 digits, the Tube be ſuddainly
educed out of the ſubjacent Quickſilver and Water, ſo as to arrive at the confines of the Aer; then doth the Aer ruſh into the Tube below, with ſuch impetuoſity , as to elevate the Quickſilver and Water contained in the Tube, to the
top ; nay, to blow up the ſealed end thereof, and drive out the liquors
4 or 5 feet perpendicular up in the aer ; not without ſome terror, though not
much danger to the Experimentator, eſpecially if he do not expect it.*

  Now though it be here præſcribed , that the Tube ought to be 4 feet
in length, and the amplitude of its Cavity equal to that of an ordinary
mans finger: yet is neither of theſe neceſſary : For, whatever be the
longitude, and whatever the amplitude of the Tube, ſtill doth the Quickſilver, after various reciprocations, acquieſce and ſubſiſt at the ſame ſtan-
dard of 27 digits ; As Dr. *Paſchal* junior found by experience in his Tube
15 feet long, which he bound to a ſpear of the ſame length, ſo to prevent
the fraction thereof , when it was erected perpendicularly , replete with
Quickſilver, (*in libro cui titulus , Experiences Novelles touchant le Vuide.*)

Among

Among those many (*Natalis* reckons up no less then 20) stupendious Magnalities, or rare *Effects*, which this eminent Experiment exhibits to observation; the least whereof seems to require a second *Oedipus* more perspicacious then the first, for the accommodation thereof though but to plausible and verisimilous Causes, and might had *Aristotle* known it, have been reputed the ground of his despair, with more credit then that petty Problem of the frequent and irregular Reciprocation of *Euripus*: we have selected only *six*, as the most considerable, and such whose solution may serve as a bright tapor to illuminate the reason of the Curious, who desire to look into the dark and abstruce *Dihoties* of the rest.

*Art.3.* The Authors. reason, for his selection of only six of the most considerable *Phænomena* to explore the Causes of them.

<br/>

## S E C T. II.

## *The First Capital Difficulty.*

**VV**Hether that Space in the Tube, betwixt the upper extreme thereof and the Quicksilver delapsed to the altitude only of *27 digits*, be really an entire and absolute Vacuity?

*Art. 1.* The First Cardinal Difficulty.

Concerning this, some there are who confidently affirm the space between the superfice of the Quicksilver defluxed and the superior extreme of the Tube, to be an absolute COACERVATE VACUITIE: such as may be conceived, if we imagine some certain space in the world to be, by Divine or miraculous means, so exhausted of all matter or body, as to prohibit any corporeal transflux through the same. And the *Reasons*, upon which they erect their opinion, are these subsequent.

This space, if possessed by any Tenent, must be replenished either with common *Aer*, or with a more pure and subtle substance called *Æther*, which some have imagined to be the Universal *Cament* or common *Elater*, by which a general Continuity is maintained through all parts of the Universe, and by which any Vacuity is prævented: or by some *exhalation* from the mass of Quicksilver included in the Tube.

*Art. 3.* The Defect space in the Tube argued to be an absolute Vacuum coacervate, from the impossibility of its repletion with Aer.

First, that it is not possessed by *Aer*, is manifest from several strong and convincing reasons.

(1) Because the inferior end of the Tube, *D*, is so immersed into the subjacent mass of Quicksilver below the line *E F*, that no particle of aer can enter thereat.

(2) Because, if there were aer in the Tube filling the deserted space *C K*, then would not the circumambient or extrinsecal aer, when the Tube is educed out of the restagnant Quicksilver, and Water, rush in with that violence, as to elevate the remainder of the Quicksilver in the Tube, from *K* to *D*, up to the top *C*, and break it open; as is observed: in regard, that could not happen without a penetration of bodies. So that, if we suppose any portion of aer to have slipped into the Tube below, at the subduction of the finger that closed the orifice: then would not the Mercury reascending (upon the

the inclination of the Tube down to the horizontal line *K M*, ) rise up quite to the top *C*, but subsist at *O P*. But the contrary is found upon the experiment.

(3) When the Tube, after the deflux of the Mercury to *K*, is reclined so as the extreme *G*, be of the same horizontal altitude with the point *K*, as is visible in the Tube *L M*: then doth the Mercury in the subject vessel reascend into the same, and again possess the desert Space *K C*, or *N M*. This being so, Whither can the aer, if any the least portion of it were resident, in the space *N M*, retreat, since the extreme *M*, is hermetically closed, and so no way for its egression can be prætended.

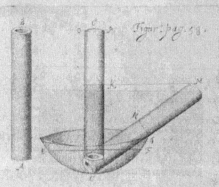

AB, *A Tube of Glass, replete with Quicksilver.*

A, *The lower extreme thereof, hermetically sealed.*

B, *The upper extreme thereof; open.*

DC, *The same Tube inverted, and perpendicularly erected in a vessel full of Quicksilver: so as the orifice D, be not unstopped, untill it be immersed in the subjacent Quicksilver.*

HG I, *A vessel filled up to the line* EF, *with Quicksilver: and thence up to the brim* H I, *with Water.*

CK, *The Vacuum, or Space deserted*

by the Quicksilver *descended*

OCP, *The quantity of Aer supposed to have insinuated it self at the subduction of the finger from the inferior orifice* D.

KM, *A Line parallel to the Horizon.*

LM, *The same Tube again filled with Quicksilver, and reclined untill the upper extreme thereof become parallel to the same horizontal altitude with* K.

N, *The distance of 27 inches from* L, *as K from* D.

(4) If any portion of Aer chance to intrude into the cavity of the Tube; which may come to pass either if, when the superior orifice of the Tube is inverted, it be not exactly obturated by the finger of the Experimentator; or, if at the extraction of his finger the lower extreme be not immersed deep enough in the subjacent Mercury, to prevent the subingress of some aer; or, if the orifice of the

Tube educed out of the region of the subjacent Mercury and Water be not wholly deobturated at once, but so as there is only some slender inlet of Aer: We say, if in any of these Cases it happen, that some small portion of aer be admitted into the cavity of the Tube; we have the evidence of our sense, and the most infallible one too, that the aer so admitted doth not ascend to the top *C*, but remaine visible in certain small Bubbles (such as usually mount up to the surface of seething water) immediately upon the superfice of the Mercury at the altitude of 27 digits *K*. As if, indeed, the aer were attracted, and in a manner chained down by the *Magnetical Effluviums* of the earth, together with the pendent Quicksilver: which having more *Ansulæ* or *Fastnings*, whereon the small Hooks of the Magnetical Chains exhaling from the Globe of the Earth, may be accommodately fixed, is therefore attracted downward more forcibly, and, in that respect, is reputed to have the greater proportion of *Gravity*. Again, If upon the inclination of the Tube, and the succeeding repletion of the same by the regurgitating Mercury, that portion of aer formerly entered be propelled up to the top of the Tube, *C*, and then the Tube again reduced to its perpendicular, so as the Quicksilver again deflux to *K*: in this case the aer doth not remain at *C*, but sinks down as formerly to *K* also, and there remains incumbent upon the face of the Quicksilver. Which Descent of the aer cannot be more probably referred to any Cause, then the Attraction of the Magnetick streams of the Earth.

(5) Having admitted some few Bubbles of aer to slide up by the margine of the Mercury into the desert Space *K C*; and then reclined the Tube to the altitude of the horizontal line *K M*: you may perceive the delapsed Quicksilver not to be repelled up again quite to the top, as before the irreption of aer, but to make a stand when it arrives at the confines of the included aer at *O P*, leaving so much space, as is requisite for the reception of it. Nor can it do otherwise, without a penetration of Dimensions, by the location of two Bodies in one and the same place.

(6) Moreover, after the acquiescence of the Quicksilver at *K*, if you stop the inferior extreme *D*, with your finger, while it remains immersed in the restagnant Quicksilver *E F*, so as to præclude the irreption of any more aer; and then invert the Tube again: the Scene of the Desert Capacity *C K*, will be changed to the contrary extreme stopt by your finger, and yet without the least sign of aer pervading the mass of Quicksilver in a kind of small stream of Bubbles, contrary to what evene's, when aer is admitted into the Tube in a small quantity, for in that case, upon the inversion of the Tube, you may plainly behold an intersection between the descending Quicksilver and the ascending aer, which mounts up through it in a small stream or thread of Bubbles.

(7) To those, who conceive that a certain portion of the *Circumstant* Aer, being forced by the compression of the restagnant Mercury in the Vessel, rising higher, upon the deflux of the Mercury contained in the Tube, doth penetrate the sides of the Tube, and so replenish the desert Capacity therein: we answer; that though we deny not but aer may penetrate the pores or Incontiguities of Glass, since that is demonstrable in Weather Glasses, and in the experiment of

Sir *Kenelm*

Sr. *Kenelm Digby*, of making a sensible transudation of Mercury mixt with Aqua Fortis in a Bolt-head, through the sides thereof, if gently confricated with a Hares-foot on the outside; yet cannot it be made out, that therefore the Desert Capacity in the Tube is possessed with Aer, for two inoppugnable reasons. (1) Because though the Tube be made of Brass, Steel, or any other Metal, whose contexture is so close, as to exclude the subtlest aer, yet shall the Experiment hold the same in all Apparences, and particularly in this of the deflux of the Quicksilver to the altitude of 27 digits. (2) Because, if the desert Cavity were replete with aer; the incumbent aer could not rush in to the Tube, at the eduction of its lower end *D*, out of the restagnant Mercury and Water, with such violence; since no other cause can be assigned for its impetuous rushing into the Tube, but the regression of the compressed parts of the ambient aer to their natural laxity, and to the repletion of the violent or forced Vacuity. Since, if the whole Space in the Tube were possessed, *i.e.* if there were as many particles of Body, as Space therein: doubtless, no part of place could remain for the reception of the irruent aer.

*Art. 5.*
The Vacuity in the Desert Space, not prevented by the insinuation of *Æther.*

*Secondly,* As for that most subtile and generally penetratrive substance, ÆTHER, or pure Elementary Fire, which some have imagined universally diffused through the vast Body of Nature principally for the maintenance of a Continuity betwixt the parts thereof, and so the avoidance of any Vacuity, though ne're so exile and minute; we do not find our selves any way obliged to admit, that the *Desert Space* in the Tube is repleted with the same, untill the *Propugnators* of that opinion shall have abandoned their Fallacy, *Petitio principii,* a præcarious assumption of what remains dubious and worthy a serious dispute, *viz. That Nature doth irreconcileably abhor all vacuity, per se.* For, until they have evinced beyond controversie, that Nature doth not endure any Emptiness or solution of Continuity, *quatenus* an *Emptiness,* and not meerly *ex Accidenti,* upon some other sinister and remote respect: their Position, that she provided that subtile substance, *Æther,* chiefly to prevent any Emptiness, is rashly and boldly anticipated, and depends on the favour of Credulity for a toleration. Nor is it so soon demonstrated, as affirmed, that all Vacuity is repugnant to the fundamental constitution of Nature.

*Art. 6.*
A Paradox, that Nature doth not abhor all vacuity, *per se,* but only *ex Accidenti,* or in respect to Fluidity.

*Naturam abhorrere Vacuum,* is indeed, a maxim, and a true one: but not to be understood in any other then a *metaphorical* sense. For, as every Animal, by the instinct of self-conservation, abhors the solution of Continuity in his skin, caused by any puncture, wound, or laceration; though it be no offence to him to have his skin pinkt or perforated all over with insensible pores: so also by the indulgence of a Metaphor, may Nature be said to abhor any great or sensible vacuity, or solution of Continuity, such as is imagined in the Desert Space of the Tube; though it be familiar, nay useful and grateful to her, to admit those insensible inanities, or minute porosities, which constitute a *Vacuum Disseminatum.* We say, by the *indulgence of a Metaphor*; because we import a kind of *sense* in Nature, analogous to that of Animals. And, tollerating this Metaphorical Speech, that Nature hath a kind of sense like that of Animals; yet, if we allow for the vastity of her Body can it be conceived no greater

greater trouble or offence to her, to admit such a solution of Continuity, or Emptiness, as this supposed in the Desert space of the Tube, then to an Animal, to have any one pore in his skin more then ordinarily relaxed and expanded for the transudation of a drop of sweat. This perpended, it can seem no *Antiaxiomatisme*, to affirm, that nature doth not abhor Vacuity, *per se*, but onely *ex Accidenti*: i. e. upon this respect, that in Nature is somewhat, for whose sake she doth not, without some reluctany, admit a Coacervate or sensible Vacuity. Now that somewhat existent in Nature *per se*, in relation to which, she seems to oppose and decline any sensible Vacuity, can be no other then the *Fluxility* of her Atomical Particles, especially those of *Fire*, *Air*, and *Water*. And, for ought we poor Haggard Mortals do, or can, by the Light of Nature, know to the contrary, all those vast spaces from the margent of the Atmosphere, whose altitude exceeds not 40 miles (according to *Mersennus* and *Gassendus*) perpendicular, up to the Region of the fixed Stars ; are not only Fluid, but *Inane* ; abating only those points, which are pervaded by the rayes of the Sun and other Celestial Bodies. But, why should we lead the thoughts of our Reader up to remote objects, whose sublimity proclaims their incertitude ; when from hence only, that the Aer is a *Fluid* substance : it is a manifest, direct and unstrained consequence, that the immediate cause of its avoidance of any sensible or coacervate Vacuity, is the Confluxibility of its Atomical particles ; which being in their natural contexture contiguous in some, though not all points of their superficies, must of necessity press or bear each upon other, and so mutually compel each other, that no one particle can be removed out of its place, but instantly another succeeds and possesses it ; and so there can be no place left empty, as hath been frequently explained by the simile of a heap of Sand ? Now, if the Confluxibility of the insensible particles of the aer, be the immediate and *per se* Cause of its avoidance of any aggregate sensible solution of Continuity : we need no farther justification of our position, that Nature doth oppose vacuity sensible not *per se*, but only in order to the affection of Confluxibility, *i. e. ex Accidenti*

Again, should we swallow this præcarious supposition of the *Æther*, with no less pertinacity, then ingenuity asserted by many Moderns, but professedly by *Natalis*, in both his Treatises ( *Physica Vetus & Nova, & Plenum experimentis novis confirmatum* ) and admit, that Nature provided that most tenuous and fluid substance chiefly to prævent Vacuity : yet cannot the Appetite of our Curiosity be satisfied, that the Desert space in the tube is replenished with the same, penetrating through the glass ; untill they have solved that Apparence of the violent irruption of the ambient Aer into the orifice of the tube, so soon as it is educed out of the subjacent liquors, the Quicksilver and Water, by the same Hypothesis. Which whether they have done, so as to demonstrate, that the sole cause of the Aers impetuous rushing into the canale of the Tube, and prodigiously elevating the ponderous bodies of Quicksilver and Water residuous therein, is not the Reflux of the incumbent aer, by the ascention of the restagnant Quicksilver in the vessel, compressed to too deep and diffused a subingression of its insensible Particles, to recover its natural laxity, by regaining those spaces, from which it was expelled and secluded ; and to supply the defect of this reason, by substituting some other syntaxical to their hypothesis of the Æther, which shall

*Art. 7.*
A second Argument against the replenion of the Desert space by *Æther*

G                                                                              be be more verisimilous and plausible: this we ought to refer to the judgment of those, who have attentively and æquitably perused their Writings.

*Art. 8.*
*The Vacuity*
*of the Desert*
*space, not pre-*
*vented by an*
*Halitus, or Spi-*
*ritual Efflux*
*from the Mer-*
*cury: for three*
*convincing*
*reasons.*

Lastly, as for the third thing supposed to replenish the Desert space in the Tube, *viz.* A certain spiritual *Efflux*, or *Halitus*, in this exigent, educed out of the Mass of Quicksilver, by a secret force of Nature, which makes any shift to avoyd that horrid enemy of hers, Inanity; we deny not the possibility of extracting or exhaling a spiritual substance from Quicksilver, fine enough to possess such a space, without obnubilating it: but cannot conceive in this case, what should be the efficient of that Extraction; for who can acquiess in that General, *a secret Force of Nature?* (2) What becomes of that Exhalation, when the Tube, meerly upon its reclination to the altitude of the Horizontal line, *K. M.* is repossessed with Mercury; for, to admit its reduction to what it was before separation, is to suppose a second secret force in Nature syncritical, or Conjunctive, Antagonist to the former Diacritical or Separative, which operateth without Heat, as the other without Cold: and to admit, its expiration through the pores or incontiguities of the Glass, is either to suppose the same portion of Quicksilver rich enough in spirit to replenish that Desert space a thousand times successively, in case the Tube be so often elevated and reclined; for if all the spiritual substance be once exhausted, then must that Fox, Nature, recur to another expedient, or else tollerate a vacuity Coacervate; or to suppose that the same exhalation doth again return into the Glass, by the same slender ways it expired, which is a Fancy worthy the smile of *Heraclitus.* (3) How this *Halitus*, in respect it is præsumed more rare and subtile, then the aer admittible by the orifice of the Tube, upon its reseration, can consist without *Inanity Disseminate*: which implicateth an Universal Plenitude.

And these are the *Reasons*, which at first inclined our judgement to determine on their part, who opinion the *Desert space* in the Tube to be an absolute *Coacervate Vacuity.*

*Art. 9.*
*The Authors*
*Apostacy from*
*the opinion of*
*an absolute*
*Coacervate Va-*
*cuity in the*
*desert space;*
*in regard of*

But, it was not long, before our second and more circumspect cogitations, assisted by time, which insensibly delivered our mind from that pleasant enchantment of novel conceptions, and reduced it to that just temper of indifferency, requisite to sincere discernment and æquitable arbitration; perpending also the Arguments impugning the former perswasion of a Coacervate Vacuity, and diminishing it down onely to a *Disseminate* one in the Desert space of the Tube: found them, by incomparable excesses, to preponderate the former, and with many more grains or moments of Verisimilty to counterpoyse our judgement to their end of the balance. And the Arguments Negative, are these.

*Art. 10.*
*The possibili-*
*ty of the subin-*
*gression of*
*light.*

(1) Manifest it is even to the most critical of our senses, that LIGHt penetrating the sides of the Glass Tube, doth totally pervade the Desert Space: therefore it cannot be an absolute sensible Vacuum. Now, that Light is a Body, or that the rayes of Light are certain i Corporeal, though most minute Effluviums transmitted from the luminous Body, or Focus; is a Truth so universally embraced by all Knowing men, and upon such apodictical commendations, that here to demonstrate it, would not only be an unseasonable Digression, but a criminal Parergy.

(2) Though

*Art.* 2
Of the Atoms
or insensible
bodies of *Heat*
and *Cold*,which
are much more
exile and pe-
netrative then
common Aer.

(2) Though the Tube might be made of some metal, or other material, whose contexture of Atomical Particles is so dense and compact, as not to permit the trajection of the beams of Light; and though the Experiment would be the same, in all Apparences, if made in the dark: yet may the Desert Space be possessed by the subtle *Atoms* of *Heat*, or *Cold*, proceeding from the ambient aer, and insinuating themselves through the incontiguities of the Tube. That the Atoms of Heat and Cold ordinarily transfix Glass, is evident from the Experience of Weather-glasses: in which the cause of the descent of the Water included, is the Rarefaction of the aer therein by the Heat, and the cause of the ascent of the water in cold Weather, is the Condensation of the same aer by Cold; neither of which were possible, if the subingression of Cold and Hot Atoms through the Glass were excluded. And, that the aer incarcerated in a Thermometre, or Temperamental organ of Silver, Coper, or Brass, is subject to the same mutations of qualities, upon the same vicissitude of Causes: hath been so frequently experimented, as to cut off all prætext of diffidence. Which is also a sufficient manifest, that the Atoms of Heat and Cold are more exile and penetrative, then those of the common Aer of use to Animals in Respiration: insomuch as they insinuate themselves through such bodies, whose almost continued parts interdict the intrusion of the grosser particles of Aer, which cannot permeate through ordinary Glass. (1) Because, if you shut your self in a closet, or chamber, that hath but one small window consisting of one entire pane of Glass, and that so cæmented into Lead, as that no chinke is left between; and whose cranies as well in the door, as elsewhere are all damm'd up: you cannot hear the voice of another person, though speaking very loud and near the Glass on the outside, notwithstanding you lay your ear close thereunto. Now, since a *Sound* (at least the *Vehicle* of a sound ) can be nought else, but a subtle portion of the aer modified; as shall be professedly commonstrated, when time hath brought us so far on our præsent journey, as the proper place for our Enquiry into the Nature of Sounds: and yet this so subtle and fine a portion of the aer cannot penetrate Glass of an ordinary thickness: we have the auctority of no weak nor obscure Reason, to countenance this our Conjecture, that the Atoms of Cold and Heat, are more exile and penetrative, then the common Aer. (2) If you include small Fishes in a large vial of the thinnest Glass, filled with River water; they may live therein for many months, provided the orifice of the Glass remain open and free to the aer: but, if you once stop it, so as to exclude the aer, they shall expire in few moments. Whence we may conclude, that however Fishes seem to have an obscure kind of Respiration, such as may be satisfied with that small portion of Aer, which is commixt with Water: yet is not that thin and subtile aer, supposed to penetrate Glass, the same they (or any other Animal) use in Respiration. Which had those grand Masters of mysterious Disquisitions, *Mersenus* and *Robervallius* animadverted; they might have soon divined, what would be the event of their intended Experiment, of including some small Animal, as a Mouse or Grashopper, in a Glass of sufficient capacity, and luting on the same on the top of the Tube, where the Desert Space useth to be, in the Experiment of Mercury, so to try whether the vital organs thereof could keep on their motions in a place devoid of aer: insomuch as that purer substance diminant from the region of the circumjacent Aer, is not corporeal enough to serve the necessity of Respiration in any Animal, though ne're so minute. The manner of

G 2                                    making

making this Experiment, is, by *Mersennus* (p.50. *reflect. physicomathemat.*) præscript; thus : *Porro, operæ pretium foret aliquam muscam admodum vegetam & robustam, v.c. Crabronem, aut Vespam, in tubo includere, priusquam Mercurio impleretur, ut post depletionem ad altitudinem 27 digit. proximè, videretur num in eo Vacuo, aut, si mavis, æthere viveret, ambularet, volaret, & num Bombus à volante produceretur.*

*Art.* 12.
Of the *Magnetical Efflux* of the *Earth:* to which opinion the Author resigns his Assent.

( 3 ) Deducting the possibility of both these, there yet remains a *Third* substance, which may well be conceived to prævent a Coacervate Vacuity in the forsaken space of the Tube: and that's the MAGNETICAL EFFLUX of the Earth. For (1) that the Terraqueous Globe is one great *Magnet*, from all points of whose superfice are uncessantly deradiated continued Threads or beams of subtle insensible *Aporrhæa's,* by the intercession whereof all Bodies, whose Descent is commonly adscribed to Gravity, are attracted towards its Centre; in like manner as there are continually expired from the body of the Loadstone invisible Chains, by the intercession whereof Iron is nimbly allected unto it : is so generally conceded a position among the Moderns, and with so solid reasons evicted by *Gilbert, Kircher, Cartesius, Gassendus* and others, who have professedly made disquisitions and discourses on that subject; that we need not here retard our course, by insisting on the probation thereof.

( 2 ) That, as the Magnetical expirations of the Loadstone, are so subtle and penetrative, as in an instant to transfix and shoot through the most solid and compact bodies, as Marble, Iron, &c. without inpediment; as is demonstrable to sense, the interposition of what solid body soever, situate within the orb of energy, in no wise impeding the vertical or polory impregnation of a steel Needle by a Magnet loricated, or armed : so also the Magnetical Effluvias of the Globe of Earth do pervade and pass through the mass of Quicksilver contained both in the Tube, and the Vessel beneath it, and fixing their Uncinulæ or hamous points, on the Ansulæ, or Fastnings of the Quicksilver therein, attract it downward perpendicularly toward the Centre : is deduceable from hence, that if any Bubbles of aer chance to be admitted into the Tube together with the Quicksilver, that aer doth not ascend to the top of the Tube, but remains incumbent immediately upon the summity of the Quicksilver, as being, in respect of its cognation to the Earth, attracted and as it were chained down by the Magnetical, Emanations of the Earth transmitted through al interjacent bodies, and hooked upon it. For we shall not incur the attribute of arrogance, if we dare any man to assign the incumbence of the aer upon the Mercury, to any more probable Cause. It being, therefore most Verisimilous, that the Earth doth perpetually exhale insensible bodies from all points of its surface, which tending upward in direct lines, penetrate all bodies situate within the region of vapors, or Atmosphere without resistence; and particularly the masses of Quicksilver in the Tube and subjacent vessel : we can discover no shelf, that can disswade us from casting anchor in this serene Haven; *That the magnetical Exhalations of the Earth, do possess the Desert space in the Tube, so as to exclude a sensible Vacuity.*

*Art.* 13.
No absolute plenitude, nor absolute Vacuity, in the Desert Space: but only a Disseminate Vacuity.

We said, so as to exclude a *sensible Vacuity*, thereby intimating that it is no part of our conception, that either the Rayes of *Light*, or the Atoms of *Heat* and *Cold*, or the *Magnetical Effluvia's* of the Earth, or all combined together, do so enter and possess the Desert space, as to cause an absolute folute Plenitude therein. For, doubtlefs, were all thofe fubtle Effluxions coadunated into one denfe and folid mafs; it would not arife to a magnitude equal fo much as to the 10ᵗʰ, nay the 40ᵗʰ part of the capacity abandoned by the delapfed Mercury. But fill it to that proportion, as to leave only a *Vacuity Diſſeminate :* fuch as is introduced into an Æolipile, when by the Atoms of fire entered into, and varioufly difcurrent through its Concavity, the infenfible Particles of Aer and Water therein contained, are reduced to a more lax and open order, and fo the inane Incontiguities betwixt them ampliated. And this we judge fufficient concerning the folution of the *Firſt Difficulty.*

---

## Sect. III.

### *The Second Capital Difficulty.*

WHat is the immediate *Remora*, or Impediment, whereby the *Aer*, which in refpect of the natural Confluxibility of its infenfible particles, fo ftrongly and expeditely preventeth any exceſſive vacuity, in all other cafes, is forced to fuffer it in this of the Experiment ?

*Art.* I.
The ſecond Difficulty ſtated.

### The Solution.

Infomuch as the *Fluidity*, or *Confluxibility* of the Atomical or infenfible particles of the Aer, is the proxime and fole Caufe of Natures abhorrence of all fenfible Vacuity; as hath been proved in the præcedent Section: Manifeft it is, that whofoever will admit a Vacuity exceſſive, or againft the rite of Nature, muft, in order to the introduction or Creation thereof, admit alfo two diftinct Bodies; (1) One, which being moved out of its place, muft propel the contiguous aer forward. (2) Another, which interpofed, muft hinder the parts of the circumftant aer, propulfed by the parts of the aer impelled by the firft movent, from obeying the Confluxibility of their Figure, and fucceeding into the place deferted by the body firft moved.

*Art.* 2.
Two things neceſſary to the creation of an exceſſive, or præternatural Vacuity.

Which is the very fcope, that the profound *Galilæo* propofed to himfelf, when He invented a wooden Cylindre, as an Embolus or Sucker to be intruded into another concave Cylindre of Brafs, impervioufly ftopped below; that by the force of weights appended to the outward extreme, or handle thereof, the fucker might be gradually retracted from the bottom of the Concave, and fo leave all that fpace, which it forfaketh, an entire and coacervate Vacuum. Upon which defign *Torricellius* long after meditating, and cafting about for other means more conveniently fatisfactory to the fame intention; He moft happily lighted upon the præfent Experiment : wherein the Quickfilver became an accommodate fubftitute to *Galilæo*'s wooden fucker, and the Glafs Tube to the Brafs concave Cylindre.

*Art.* 3.
The occaſion of *Galilæus* invention of a *Braſs Cylindre* charged with a wooden *Embol,* or Sucker : and of *Torricellius* invention of the præſent Experiment.

The

*Art. 4.*
The narrow
of the Difficul
ty &c. How
the Aer can be
impelled up-
ward by the
Restagnant
Quickfilver,
when there
externally
wants a fit
space for it to
circulate into.

The remaining part of the Difficulty, therefore, is only this relative *Scru-*
*ple ; How the Aer can be propelled by the wooden fucker, downward, or by*
*the reftagnant Quickfilver in the Veffel, upward, when externally there is pro-*
*vided no void fpace for its reception.* For, indeed in the ordinary Tranflation
of bodies through the aer, it is no wonder that the adjacent aer is propelled
by them; fince they leave as much room behind them, as the aer propelled
before them formerly poffeffed, whereinto it may and doth recur: but in this
cafe of the Experiment, the condition is far otherwife, there being, we confefs, a place left behind, but fuch as the aer propelled before cannot retreat
into it, in regard of the interpofition of another denfe folid & impervious body. Upon which confideration, we formerly and pertinently reflected when
reciting fome of thofe Experiments vulgarly objected to a *Vacuum Diffemi-*
*natum*, we infifted particularly upon that of a *Garden Irrigatory:* fhewing, that
the Reafon of the Waters fubfiftence, or pendency therein, fo long as the
orifice in the Neb remains ftopped, is the defect of room for the aer preffed upon by the bafis of the Water to recur into upon its refignation of
place; becaufe all places being full, there can be none whereinto the inferior aer may recede, until upon deobftruction of the hole above, the circumjacent aer enters into the cavity of the Veffel, and refignes to the aer
preffed upon below, and fo the motion begins and continues by a fucceffive furrender of places. For, though the aer contiguous to the bottom
of the Irrigatory, be not fufficient to refift the compreffure of fo great a
weight of water, by the fingle renitency of the Confluxibility of its atomical particles; yet the next contiguous aer, poffeffing the vicine fpaces, and
likewife wanting room to recede into, when compelled by the firft aer, aggravates the refiftence : which becomes fo much the greater, by how much
the farther the preffure is extended among the parts of the circumjacent
aer; and by fo much the farther, is the preffure of the circumjacent aer
extended, by how much the greater is the preffure of the next contiguous
aer; and that preffure is proportionate to the degrees of Gravity and velocity in the body defcendent. Which is manifeftly the reafon, why the
water doth not defcend through the perforated bottom of the Veffel, *viz.*
becaufe the Gravity thereof is not fufficient to counterpoyfe fo diffufed,
prolix, and continued refiftence, as is made and maintained by the confluxibility of the parts of the circumambient aer fucceffively uniting their
forces.

*Art.* 5.
The folution of
the fame, by
the Laxity of
the Contex-
ture of the Aer

Notwithftanding this feeming plenitude, we may abfolve our reafon
from the intricacy of the *fcruple*, by returning : that, though all places
about the Tube are filled with aer, yet not without fome *Laxity*. So,
though there be, indeed, no fenfible or coacervate fpace, wherein there
are not fome parts of the aer : yet are there many *infenfible* or *diffeminate*
fpaces, or *Loculaments* varioufly interfperfed among the incontiguous
(in all points) particles of the aer, which are unpoffeffed by any Tenent
at all. For the familiarizing of this Nicety, let us have recourfe once a-
gain to our fo frequently mentioned example of a heap of Corne.

*Art.* 6.
The fame illu-
ftrated, by the
adæquate fimi-
le of Corne in-
fufed into a
Bufhel.

When we have poured Corne into a Bufhel up to the brim thereof; the
capacity feems wholly poffeffed by the Graines of Corne, nor is there
therein any fpace, which fenfibly contains not fome Graines : yet if we
fhake the bufhel, or deprefs the Corne, the Graines fink down in a clofer
pofture, and leave a fenfible fpace in the upper part of the bufhel, capable
of

of a confiderable acceſs or addition. The reaſon is, that the Grains, at their firſt infuſion, in reſpect of the ineptitude of their Figures for mutual contact, in all points of their ſuperficies, intercept many empty ſpaces betwixt them; which diſperſed minute inane ſpaces are reduced to one great and coacervate or ſenſible ſpace, in the ſuperior part of the Continent, when, by the ſuccuſſion of the veſſel, the Grains are diſpoſed into a cloſer poſture, *i. e.* are more accommodated for mutual contingency in their ends and ſides. Thus alſo may aer be ſo compreſſed, as the Granules, or inſenſible particles of it, being reduced to a more cloſe or denſe order, by the ſubingreſſion of ſome particles of the aer neareſt to the body Compreſſing, into the incontiguities of the next neighbouring aer; may poſſeſs much leſs of ſpace, then before compreſſion; and conſequently ſurrender to the body propelling or compreſſing, leaving behind a certain ſpace abſolutely devoid of aer, at leaſt, ſuch as doth appear to contain no aer.

But this Difficulty, *Hydra*-like, ſends out two new Heads in the room of one cut off. For, Curioſity may juſtly thus expoſtulate.

(1) Have you not formerly affirmed, that no body can be moved, but it muſt compel the aer forward, to ſuffer a certain *ſubingreſſion* of its inſenſible particles into the pores, or Loculaments of the next contiguous aer, ſuch as is requiſite to the leaving of a ſpace behind it for the admiſſion of the body moved? And, if ſo; how comes it, that when moſt bodies are moved through the aer, with ſo much facility, and therefore cauſe the parts thereof before them to intrude themſelves into the incontiguities of the next vicine aer, with a force ſo ſmall, as that it is altogether inſenſible: yet in this caſe of the *Experiment*, is required ſo great a force to effect the ſubingreſſion and mutual Coaptation of the parts of the aer?

*Art. 7.*
A ſubordinate ſcruple, why moſt bodies are moved through the Aer, with ſo little reſiſtence, as is imperceptible by ſenſe?

The *Cauſe* ſeems to be this. In all common motions of bodies through the liberal aer, there is left a Space behind, into which the parts of the aer may inſtantly circulate, and deliver themſelves from compreſſion; and ſo there is a ſubingreſſion and Coaptation of only a few parts neceſſary, and conſequently the motion is tolerated without any ſenſible Reſiſtence: but in this Caſe of the Experiment, in regard there is no place left behind by the Propellent, into which the compreſſed parts of the aer may be effuſed; neceſſary it is that the parts of aer immediately contiguous to the body Propellent, in their retroceſſion and ſubingreſſion compreſs the parts of the next contiguous aer; which though they make ſome reſiſtence (proportionate to their meaſure of Confluxibility) do yet yeild, retrocede, and intrude themſelves into the incontiguities of the next contiguous aer; and thoſe making alſo ſome reſiſtence, likewiſe yeild, retrocede, and inſinuate themſelves into the Loculaments of the next, which acts the like part upon the next, and ſo ſucceſſively. So that a greater force then ordinary is required to ſubdue this gradually multiplied reſiſtence ſucceſſively made and maintained by the many circumfuſed parts of the aer; and to effect, that the retroceſſion, ſubingreſſion and coaptation of the parts of the aer be propagated farther and farther, untill convenient room be made, for the reception of the body Propellent.

*Art. 8.*
The ſame Expeded.

(2) Whence do you derive this *Reſiſtence* of the Aer?

From its *Gravity.* For, the Aer of its own nature is Heavy, and can be ſaid

*Art. 9.*
A ſecond dependent ſcruple concerning the Cauſe of the *ſenſible* reſiſtence of the Aer, in this caſe of the Experiment; together with the ſatisfaction thereof, by the *Gravity* of Aer

said to be *Light* only *comparatively*, or as it is lefs ponderous then Water and Earth: nor can there be given any more creditable reafon of the Aers tendency upward here below near the convexity of the Earth, then this; that being in fome degree ponderous in all its particles, they defcend downwards from the upper region of the Atmofphere, and in their defcent bear upon and mutually compel each other, untill they touch upon the furface of the Earth, and are by reafon of the folidity and hardnefs thereof repercuffed or rebounded up again to fome diftance: fo that the motion of the Aer upwards near the face of the Earth, is properly *Refilition*, and no natural, but a *violent* one. Now, infomuch as the Aer feems to be no other, but a common Mifcelany of minute bodies, exhaled from Earth and Water and other concretious fublunary, and proportionately to their Craffitude or Exility, emergent to a greater or lefs altitude: it can be no illegal procefs for us to infer, that all parts thereof are naturally endowed with more or lefs Gravity proportionate to their particular bulk; whether that Gravity be underftood to be (as common Phyfiology will have it) a Quality congenial and inhærent, or (as Verifimility) their conformity to the magnetick Attraction of the Earth. And, infomuch as this *Gravity* is the caufe of the mutual *Depreffion* among the particles of aer in their tendency from the upper region of the Atmofphere down to the furface of the Earth: we may well conceive, that the *Depreffion* of the inferior parts of the aer by the fuperior incumbent upon them, is the origine immediate from whence that *Reluctancy* or *Refiftence*, obferved in the Experiment, upon the induction of a præternatural Inanity between the Parts thereof. But a farther profecution and illuftration of this particular, depends on the folution of the next Problem.

## Sect. IV.

### *The Third Capital Difficulty.*

Art. 1.
The State of the Third Difficulty.

**W**Hat is the Caufe of the Quickfilvers *not defcending below that determinate Altitude, or Standard of 27 digits?*

#### Solution.

Art. 2.
The Solution thereof in a Word.
Art. 3.
Three præcedent pofitions briefly recognifed, in order to the worthy profounding of the myftery, of the Aers refifting Compreffion beyond a certain rate, or determinate proportion.

The *Refiftence* of the parts of the aer, which endures no compreffion, or *fubingrefs* of its infenfible particles, beyond that *certain proportion*, or determinate rate.

To profound this myftery of Nature to the bottom, we are to requeft our Reader to endure the fhort recognition of fome paffages in our præcedent difcourfes. (1) That upon the ordinary tranflation of bodies through the Aer, the refiftence of its infenfible parts is fo fmall, as not to be difcoverable by the fenfe; becaufe the fubingreffion of its contiguous parts into the loculaments of the next vicine aer, is only perexile, or fuperficial: and that we may fafely imagine this fuperficial fubingreffion not to be extended beyond the thicknefs of a fingle hair; nay, in fome
cafes

cafes, perhaps, not to the hundreth part thereof.   *So ftupendiaufly fubtle are the fingers of Nature in many of her operations.*   But, that the refiftence obferved in the prefent Experiment, for the enforcing of a præternatural Vacuum, is therefore deprehenfible by the fenfe, becaufe in refpect of a defect of place behind the body propellent, into which the parts of the aer compelled forward may circulate, the fubingreffion muft be more profound; and fo the refiftence being propagated farther and farther by degrees, muft grow multiplied, and confequently fenfible.  (2) That the Force of the body propellent is greater, then the force of the next contiguous aer protruding the next, and the force of the third protruded wave of the aer (for a kind of Undulation may be afcribed to aer) greater on the Fourth, then that of the Fourth upon the Fifth, and fo progreffionally to the extrem of its diffufion or extenfion: fo that the Force becomes fo much the weaker and more oppugnable, by how much the farther it is ex_tended; and dwindles or languifhes by degrees into a total ceffation. (3) That, as upon the fuccuffion, or fhock of a Bufhel apparently full of Corn, is left a certain fenfible fpace above, unpoffeffed by any part or Grain thereof; which coacervate empty fpace refponds in proportion to thofe many Diffeminate *Vacuola*, or Loculaments intercepted among the incontingent fides of the Grains, before their reduction to a more clofe order by the fuccuffion of the Bufhel: fo likewife, upon the impulfe of the aer by a convenient body, is left behind a fenfible fpace abfolutely empty, as to any part of aer; which Coacervate empty fpace muft refpond in proportion to thofe many Diffeminate fpaces intercepted among the incontiguous parts, or Granules of the aer, before their reduction to a more clofe order, or mutual fubingreffion and coaptation of fides and points, by the body comprefing.

Thefe Notions recogitated, our fpeculations may progrefs with more advantage to explore the proxime and proper Caufe of the Mercuries conftant fubfiftence at the altitude of 27 digits, in the Tube perpendicularly erected.  For, upon the credit of their importance, we may juftly affume, that upon the compreffion of the circumambient Aer by a fmall quantity of Quickfilver (fuppofe only of two or three inches) impendent in the concave of the tube, can be caufed, indeed, fome fmall fubingreffion of the particles thereof; but fuch, as is only *fuperficial* and *infenfible*: in refpect the weight of fo fmall a proportion of Quickfilver is not of force fufficient to propel the parts of the aer to fo great a craffitude that the fpace detracted from the Aggregate of Diffeminate Vacuities fhould amount to that largnefs, as to become vifible above the Quickfilver in the Tube, fince the quantity of the Quickfilver being fuppofed little, the force of Reluctancy, or Refiftence in the parts of the aer, arifing from their inhærent Fluidity, muft be greater then the force of compreffion arifing from Gravity; and therefore there fucceeds no fenfible Deflux of the Quickfilver.  But, being that a greater and greater mafs of Quickfilver may be fucceffively infufed into the Tube, and fo the compreffive force of its Gravity be refpectively augmented; and thereupon the aer become lefs and lefs able fucceffively to make refiftence: 'tis difficult not to obferve, that the proportion of Compreffion from Gravity in the Quickfilver, may be fo equalized to the Refiftence from Gravity in the Aer, as that both may remain *in ftatu quo*, without any fenfible yeilding on either fide.  Hence comes it, that at the æquipondium of thefe two Antagonifts, the fpace in the

*Art. 4.* The *Æquiponderancy* of the External Aer, pendent upon the furface of the Reftagnant Mercury, in the veffel, to the Cylindre of Mercury refiduous in the Tube, at the altitude of 27 digits: the caufe of the Mercuries conftant fubfiftence at that point.

H                                    Tube

Tube detracted from the Aggregate of minute Inanities disseminate in the aer, is so small as not to be commensurated by sense: and at the cessation of the Æquilibrium, or succeding superiority of the encreased weight of the Quickfilver, the parts of the Aer being compelled thereby to a farther retrocession and subingression; the space detracted from the Aggregate of disseminate Vacuities in the aer, becomes larger, and consequently sensible, above the Quickfilver in the upper region of the Tube.

This may be most adæquately illustrated, by the *simile* of a strong man, standing on a plane pedestal, in a very high wind. For, as He by a small afflation or gust of wind, is in some degree urged or prest upon, though not so much as to cause him to give back; because the force of his resistence is yet superior to that of the Wind assaulting and impelling him; nor, when the force of the Wind grows upon him even to an Æquilibrium, is He driven from his station, because his resistence is yet equal to the impulse of the wind; but when the force of the Wind advances to that height, as to transcend the Æquilibrium, then must the man be compelled above the rate of his resistence, and so be abduced from the place of his station: so likewise, while there is only a small quantity of Quickfilver contained in the Tube, though, by the intervention or mediation of the Quickfilver restagnant in the subjacent vessel, it press upon the parts of the incumbent aer, in some degree; yet is not the aer thereby urged so, as to be compelled to retrocede, and permit the restagnant Quickfilver to ascend higher in the vessel; and therefore the Quickfilver impendent in the Tube cannot descend, because the restagnant wants room to ascend. But, when the quantity, and so the Gravity of the Quickfilver contained in the Tube is so augmented, as to exceed the Resistence of the aer; then is the aer compelled or driven back, by the restagnant Quickfilver rising upwards, to a sensible subingression of its atomical particles, and the Quickfilver in the Tube instantly defluxeth into the place resigned by the restagnant, until it arriveth at that point of altitude, or standard, where the resistence of the aer becomes again equal to the force compressing it, and there subsisteth, after various reciprocations up and down in the Tube.

Now concerning the remaining, and, indeed, the most knotty part of the Difficulty, *viz. Why the Æquilibrium of these two opposite Forces, is constant to the certain præcise altitude of 27 digits?* of this admirable Magnale no other cause seems worthily assignable, but this; *that such is the nature of aer, in respect both of the atomical particles of which it is composed, and of the disseminate vacuities variously interspersed among them, as that it doth resist compression at such a determinate rate, or definite proportion, as exactly responds to the altitude of 27 digits.* Should it be demanded of us, Why He, who stands on a plane, doth resist the impulse of a mighty wind to such a determinate rate or height, but not farther: we conceive our Answer would be satisfactory to the ingenious, if we returned only, that such is the exact proportion of his strength, resulting from the individual temperament of his body.

We are *Men, i.e. Moles*; whose weak and narrow Opticks are accommodated only to the inspection of the *exterior* and *low* parts of *Nature*, not perspicacious enough to penetrate and transfix her *interior* and *abstruse* Excellencies: nor can we speculate her glorious beauties in the direct and
incident

incident line of *Essences* and *Formal Causes*, but in the refracted and reflected one of *Effects*; nor that, without so much of obscurity, as leaves a manifest incertitude in our Apprehensions, and restrains our ambition of intimate and *apodictical Science*, to the humble and darksome region of mere superficial *Conjecture*. Such being the condition of our imperfect Intellectuals; when we cannot explore the profound recesses, and call forth the *Formal Proprieties* of some Natures, but find our disquisitive Faculties terminated in the some *Apparences*, or *Effects* of them : it can be no derogation to the dignity of *Humanity*, for us to rest contented, nay thankful to the *Bounty* of our *Creator*, that we are able to erect verisimilous Conjectures concerning their causation, and to establish such rational Apprehensions or Notions thereupon, as may, without any incongruity, be laudably accommodated to the probable solution of other consimilar Effects, when we are required to yeild an account of the manner of their arise from their proper originals. Thus, from our observation of other things of the like condition, having extracted a rational Conjecture, that this so great *Gravity* of the *Quicksilver* doth depend upon the very Contexture of its insensible particles, or minute bodies, whereof it doth consist, by which they are so closely and contiguously accommodated each to other in the superficies of their points and sides, as no body whatever (*Gold* only excepted) doth contain more parts in so small a bulk, nor consequently more *Ansulæ*, or Fastnings, whereon the *Magnetique Hooks* of the Earth are fixable, in order to its attraction downward : and on the contrary, that the so little *Gravity* of the *Aer*, depends on a quite dissimilar Contexture of its insensible particles, of which it is composed, by which they are far less closely and contiguously adapted each to other, and so incomparably fewer of them are contained in the like space, and consequently have incomparably fewer *Ansulæ* or Fastnings, whereon the Hooks of the Magnetick Chains of the Earth may be fixed : having, we said, made this probable conjecture, what can be required more at our hands, then to arrest Curiosity with this solution; that the Aer is of such a Nature, *i.e.* consisteth of such insensible particles, and such Inane Spaces interspersed among them, as that it is an essential propriety of it, to resist compression, to such a determinate rate, and not beyond? Had we bin born such *Lyncei*, as to have had a clear and perspect Knowledge of the Atoms of Aer, of their Figure, magnitude, the dimensions of the Inane spaces intercepted among them, of the facility or difficulty of their reciprocal adaptation, of the measure of their Attraction, the manner and velocity of their Tendency, &c. then, indeed, might we, without any complex circumambage of Discourse, have rendered the express and proper Reason, why the Aer doth yeild præcisely so much, and no more to the Gravity of the Quicksilver compressing it. Since we were not, it may be reputed both honour and satisfaction, to say; that it is essential to the Natures of Mercury and Aer, thus and thus opposed, to produce such and only such an Effect.

*Art. 8.*
The constant subsistence of the Mercury at 27 digits, ascriptive rather to the *Resistence* of the Aer, then to any occult Quality in the Mercury.

However, that we may not dismiss our Reader absolutely jejune, who came hither with so great an Appetite; we observe to him, that the constant subsistence of the Mercury at the altitude of 27 digits, doth seem rather to proceed from the manifest Resistence of the Aer, then from any secret Quality in the Mercury, unless its proportion of Gravity be so conceived. This may be collected from hence; that *Water* infused into the

H 2                                                     Tube

Tube doth also descend to the point of *Æquipondium*, and stops at the altitude of 32 Feet, nor more, nor less; and in that altitude becomes æquiponderant to the Mercury of 27 digits. So that it is manifest, that with what Liquor soever the Tube be filled, still will the Aer resist its deflux at a certain measure: provided only, that the Tube be long enough to receive so much of it, as the weight thereof may equal that of the Mercury at 27 digits, or the Water at 32 feet.

*Art 9.*
The Analogy betwixt the Absolute and Respective Æquality of weights, of Quickfilver and Water, in the different altitudes of 27 digits and 32 feet.

Here we meet an opportunity also of observing to Him, by how admirable an Analogy this respective Æquality of the weights of Quickfilver and Water, in these so different altitudes, doth consent with the absolute weight of each. When, as the weight of Quickfilver carries the same proportion to the weight of Water, of the same measure or quantity, as 14 to 1 : so reciprocally doth the Altitude of 32 feet, carry the same proportion to 27 digits, as 14 to 1. And hence comes it, that, if *Water* be ⅟ peraffused upon the restagnant Quickfilver in the vessel under the Tube; the Quickfilver doth instantly ascend above the standard of 27 digits, higher by a 14th. part of the water superaffused. Which truly, is no immanifest argument, that the Aer, according to the measure of its weight, or the præcise rate of its resistence, becomes æquilibrated to the Mercury at the altitude of 27 dig. since the superaffused Water doth no more then advance the Æquilibrium according to the rate of its weight, or proportion of resistence. Besides, it is farther observable, that because the Tube is replenished by a 14th part in 27 dig. of the altitude, above the first Æquilibrium (a proportionate access to the Mercury in the Tube, being made by a like part of that in the subject vessel, impelled into it) therefore is the *Vacuum* above the Mercury in the Tube, diminished also by one 14th. part; and the compression of the Aer, impendent on the surface of the restagnant Mercury, relaxed and diminished also by a 14th part. So that if the vessel underneath the Tube be large enough to admit an addition of Water successively affused, until so much of the restagnant Mercury, as formerly descended, shall be again propelled up into the Tube : then must the whole Tube be replenished, and so the whole Vacuity disappear, for then all Compression of the incumbent aer ceaseth, and so much space as was possessed before the Experiment, both without and within the Tube, by the Mercury, Water, Aer, is again repleted.

*Art.* 10.
The definite weight of the Mercury at 27 digits, and of Water at 32 feet, in a Tube of the third part of a digit in diametre; found to be near upon two pound, *Paris* weight.

* *Confulendus Mersennus, in tract. de Mensuris & ponderibus, cap 1. & reflexion. physicomathemat. p.* 229.

If you shall still insist, and urge us to a præcise and definite account of the weight of the Quickfilver contained in the Tube to the altitude of 27 digits, and of the Water of 32 feet; which makes the Æquilibrium with the opposite weight of the circumstant Aer : our *Answer* is, that the exact weight of neither can be determined, unless the just Diameter or Amplitude of the Tube be first agreed upon. For albeit neither the Longitude nor the Amplitude of the Tube makes any sensible difference in this Phænomenon of the Experiment, the Æquilibrium being still constant to the same altitude of 27 digits, for the Mercury, and 32 feet for Water : yet, according as the Cavity of the Tube is either smaller, or greater, must the weight of the Liquors contained therein be either less, or more. Since therefore, we are to explore the definite weight of the Liquor contained, by the determinate Amplitude of the Tube containing; suppose we the Diametre of the cavity of the Tube to be one third part of a * Digit, and we shall find the weight of the Quickfilver, from the base to the altitude

of

of 27 digits, to be near upon two pound, *Paris* weight : and upon confequence the weight of Water in the fame Tube, of 32 feet in altitude, to be the fame ; and the weight of the Cylindre of Aer, from its bafe incumbent on the furface of the reftagnant Quickfilver, up to its top at the fummity of the Atmofphere, to be alfo the fame ; otherwife there could be no Æquilibrium. Here, as a *Corollary*, we may add, that infomuch as the force of a body Attrahent may be æquiparated to the weight of another body fpontaneoufly defcending or attracted magnetically by the Earth : thereupon we may conclude, that the like proportion of weight appended to the handle of the wooden Sucker, may fuffice to the introduction of an equal vacuum, in *Galileo's* Brafs Cylindre.

But, perhaps, you'l *object*, that this feems rather to entangle then diffolve the Riddle. Since by how much the larger the cavity of the Tube, by fo much the greater the quantity, and fo the weight of the Quickfilver contained : and by how much the greater the weight, or force of the Depriment, by fo much the more muft the Depreffed yeild, and confequently, fo much the lower muft the Æquilibrium be ftated.

To extricate you from this Labyrinth, we retort ; that the caufe of the Æquilibriums conftancy to the point of 27 digits, whatever be the quantity of the Mercury contained in the Tube, is the fame with that, which makes the defcent of two bodies of the fame matter, but different weights, to be *Æqually Swift*: for a bullet of Lead of an ounce, falls down as fwiftly as one of 100 pound. For, in refpect, that a Cylindre of Quickfilver contained in a Tube of a large diametre, doth not defcend more fwiftly, then a Cylindre of Quickfilver contained in a Tube of a narrow diametre : therefore is it, that the one doth not prefs the bottom, upon which as its Bafe, it doth impend, more violently then the other doth prefs upon its Bafe ; and confequently, the reftagnant Quickfilver about the larger Cylindre doth not, in its elevation or rifing upward, more comprefs the Bafis of the impendent Cylindre of Aer, then what is reftagnant about the leffer Cylindre. Whereupon we may conclude, that a great Cylindre of Aer refifting a great Cylindre of Quickfilver, no lefs then a fmall doth refift a fmall : therefore ought the Æquilibrium betwixt the depreffure of the Quickfilver, and the refiftence of the circumftant Aer, to be conftant to the altitude of 27 digits, afwell in a large, as a narrow Tube. Which reafon may alfo be accommodated to *Water* and all other *Liquors*.

*Art.* 11.
*Quære*,
Why the Æquilibrium is conftant to the fame point of altitude in a Tube of a large concave, as well as in one of a fmall, when therefore of the Depriment muft be greater in the one, then the other.
*Art.* 12.
The folution thereof by the appropriation of the fame Caufe, which makes the defcent of two bodies, of different weights, æquivelox.

S E C T.

## SECT. V.

### *The Fourth Capital Difficulty.*

VVHy is the deflux of the Quickfilver alwayes ftinted at the altitude of 27 digits, though in Tubes of different longitudes ? when it feems more reafonable, that according to the encreafe or enlargement of the Inanity in the upper part of the Tube, which holds proportion to the Longitude thereof; the Compreffion, and fo the Refiftence of the Aer circumpendens, ought alfo to be encreafed proportionately: and confequently, that the Æquilibrium ought to be fo much the higher in the Tube, by how much the greater Refiftence the Aer makes without; becaufe, by how much a larger Space is detracted from the Aer, by fo much more diffufed and profound muft the fubingreffion of its Atomical Particles be, and fo the greater its refiftence.

### Solution.

Certain it is, afwel upon the evidence of fenfe, as the conviction of feveral demonftrations excogitated chiefly by *Merfennus* (in *Phænom. Hydraulic.*) that a Cylindre of any Liquor doth with fo much the more force or Gravity impend upon its Bafe, or bottom, by how much the higher its perpendicular reacheth, or, by how much the longer it is : and confequently, having obtained a vent, or liberty of Exfilition below at its Bafe, iffues forth with fo much the more rapidity of motion. And this fecret reveals what we explore. For, according to the fame fcale of Proportions, we may warrantably conceive; that, by how much the higher the Cylindre of Quickfilver is in the Tube, by fo much the more forcibly it impendeth upon its Bafe, in the Reftagnant Quickfilver; and fo having obtained a vent below, falleth with fo much the more rapidity of motion or exfilition thereupon : and upon confequence, by fo much the more violently is the incumbent Aer compreffed by the reftagnant Quickfilver afcending, its refiftence overcome, and the fubingreffion of its infenfible particles into the inane Loculaments of the vicine aer, propagated or extended the farther; and the fpace detracted from the Aggregate of Diffeminate Inanities, fo much the larger, and confequently the *Coacervate Vacuum* apparent in the fuperior region of the Tube, becomes fo much the greater. And, becaufe the Refiftence made againft the fubingreffion, dilating or diftending it felf, is in the inftant overcome, by reafon of a greater impulfe caufed by the Cylindre of Mercury defcending from a greater altitude; and that refiftence remains, which could not be overcome, by the remnant of the Mercury in the Tube, at the height of 27 digits : therefore, is this Remaining Degree of refiftence, the manifeft Caufe, why the Mercury is Æquilibrated here at the point of 27 digits, afwell when it falls from a high as a low perpendicular.

*Art.* 3.
The fame con-
firmed, by the
theory of the
Caufe of the
Mercuries fre-
quent *Recipro-
cations*, before
it acquiefce at
the point of
Æquipondi-
um.

This may receive a degree of perfpicuity more, from the tranfitory obfervation of thofe frequent *Reciprocations* of the Quickfilver, at the firft

deflux

deflux of it into the reftagnant, before it acquiesce and fix at the point of *Æquiponderancy*: no otherwise then a Ball bounds and rebounds many times upon a pavement, and is by fucceffive fubfultations unceffantly agitated up and down, untill they gradually diminifh and determine in a ceffation or quiet. The *Caufe* of which can be no other then this; that the extreme or remoteft fubingreffion of the infenfible particles of the Aer, is (we confefs) propagated fomewhat farther, then the neceffity of the *Æquipondium* requireth, by reafon of a new accefs of Gravity in the Quickfilver; but, inftantly the infenfible particles of the Aer, being fo violently and beyond the rate of fubingreffibility preft upon, and made as it were more powerful by their neceffary Reflexion, then the refidue of Quickfilver remaining in the Tube; refult back to their former ftation of liberty, with that vehemency, as they not only prævent any further fubingreffion, and reduce the even-now-fuperior and conquering force of the Quickfilver to an equality; but alfo repell the Quickfilver delapfed up again into the Tube above the point of the Æquipondium: and again, when the Quickfilver defluxeth, but not from fo great an altitude, as at firft; then is the Aer again compelled to double her files in a countermarch, and recede from the reftagnant Quickfilver, though not fo far, as at firft charge. And thus the force of each being by reciprocal conquefts gradually decreafed, they come to that Equality, as that the Quickfilver fubfifts in that point of altitude, wherein the Æquilibrium is.

---

## Sect. VI.

## *The Fifth Capital Difficulty.*

WHat Force that is, whereby the Aer, admitted into the lower orifice of the Tube, at the total eduction thereof out of the reftagnant *Quickfilver and Water; is impelled fo violently, as fufficeth not only to the elevation of the remaining Liquors in the Tube, but even to the difcharge of them through the fealed extreme, to a confiderable height in the Aer?*

Art. 1.<br>The Fifth<br>Principal Dif-<br>ficulty.

### Solution.

The immediate *Caufe* of this impetuous motion, appears to be only the *Reflux*, or *Refilition* of the fo much compreffed Bafis of the Cylindre of Aer, impendent on the furface of the Reftagnant Liquors, Quickfilver and Water, to the natural Lavity of its infenfible particles upon the ceffation of the force Compreffive: the Principle, and manner of which *Reftorative* or *Reflexive* Motion, may be perfpicuoufly deprehended, upon a ferious recognition of the Contents of the laft Article in the præcedent Chapter of a *Diffeminate Vacuum*; and moft accommodately Exemplified in the difcharge or explofion of a bullet from a *Wind-Gun*. For, as the infenfible particles of the Aer included in the Tube of a Wind-Gun, being, by the Embolus or Rammer, from a more lax and rare contexture, or order, reduced to a more denfe and clofe (which is effected, when they are made more contiguous in the points of their fuperfice, and fo compelled to diminifh

Art. 2.<br>Solved, by the<br>Motion of Re-<br>ftauration natu-<br>ral to each in-<br>fenfible parti-<br>cle of Aer.

minifh the inane fpaces interjacent betwixt them, by fubingreffion ) are, in a manner fo many Springs or Elaters, each whereof, fo foon as the external Force, that compreffed them, ceafeth (which is at the remove of the Diaphragme or Partition plate in the chamber of the Tube ) reflecteth, or is at leaft reflected by the impulfe of another contiguous particle: therefore is it, that while they are all at one and the fame inftant executing that Reftorative Motion, they impel the Bullet, gaged in the canale of the Tube, before them with fo much violence, as enables it to transfix a plank of two or three digits thickneſs. So alfo do the infenfible Particles of the Bafe of the Cylindre of Aer incumbent on the furface of the Reftagnant Liquors, remain exceedingly compreffed by them, as fo many Springs bent by external Force: and fo foon as that Force ceafeth (the Quickfilver in the Tube, after its eduction, no longer preffing the Reftagnant Maſs of Quickfilver underneath, and fo that by his tumefaction no longer preffing the impendent Aer ) they with united forces reflect themfelves into their natural rare and liberal contexture, and in that Reftorative motion drive up the remainder of Quickfilver in the canale of the Tube to the upper extreme thereof with fuch violence, as fufficeth to explode all impediments, and fhiver the glaſs.

*Art. 3.*
*The incumbent Aer, in this cafe, equally diffuſed, by two contrary Forces.*

For, in this cafe, we are to conceive the Aer to be æqually diftreffed betwixt two oppofite Forces; on *one* fide by the Gravity of the long Cylindre of Aer from the fummity of the Atmofphere down to the Bafe impendent on the fuperfice of the Reftagnant Liquors; on the *other*, by the afcendent Liquors in the fubjacent veffel, which are impelled by the Cylindre of Quickfilver in the tube, defcending by reafon of its Gravity: and confequently, that fo foon as the *obex, Barricade,* or impediment of the Reftagnant Quickfilver, is removed, the diftreffed Aer inftantly converteth that refiftent force, which is inferior to the Gravity of the incumbent aereal Cylindre, upon the remainder of the Quickfilver in the Tube, as the now more fuperable Opponent of the two; and fo countervailing its Gravity by the motion of Reflexion or Reftoration, hoyfeth it up with fo rapid a violence, as the eafily frangible body of the Glaſs cannot fuftain.

*Art. 4.*
*The motion of Reftauration in the Aer, extended to the fatisfaction of another confimilar Doubt, concerning the fubintrufion of Water fuperaffuſed into the Tube, if fuperaffuſed upon the reftagnaut Mercury.*

Which Reafon doth alfo fatisfie another *Collateral Scruple,* viz. *Why Water, fuperaffuſed upon the Reftagnant Quickfilver, doth intrude it felf as it were creeping up the fide of the Tube, and replenifh the Defert Space therein;* fo foon as the inferior orifice of the Tube is educed out of the Reftagnant Quickfilver, into the region of Water. For, it is impelled by the Bafe of the Aereal Cylindre exceedingly compreffed, and relaxing it felf: the refiftence of it, which was not potent enough to prævail upon the greater Gravity of the Quickfilver in the Tube, fo as to impel it above the point of Æquiponderancy; being yet potent enough to elevate the Water, as that whofe Gravity is by 13 parts of 14 leſs then that of the Quickfilver.

*Art. 5.*
*A Third moft important Doubt, concerning the non-apparence of any Tenfity, or Rigidity in the region of Aer incumbent upon the Reftagnant Liquors.*

Here the Inquifitive may bid us ftand, and obferve a fecond fubordinate *Doubt,* fo confiderable, as the omiffion of it together with a rational folution, muft have rendred this whole Difcourfe not only imperfect, but a more abfolute Vacuum, *i.e.* containing leſs of matter, then the Defert Space in the Tube; and that is: *How it comes, that during the Æqui-*

*librium*

*librium betwixt the weight of the Quicksilver in the Tube on the one hand, and the long Cylindre of Aer on the other; even then when the Base of the Cylindre of Aer is compressed to the term of subingression; we find the aer as Fluxile, soft, and yeilding, (for, if you move your hand transversly over the Restagnant Quicksilver, you can deprehend none the least Tensity, Rigidity, or Urgency thereabout) as any other part of the Region of Aer not altered from the Laxity of its natural contexture?*

We reply, that though nothing occurr in the whole Experiment more worthy our absolution; yet nothing occurrs less worthy our admiration then this. For, if my hand, when moved toward the region of the compressed Aer, did leave the space, which it possessed before motion, absolutely Empty, so as the aer impelled and dislodged by it could not circulate into the same; in that case, indeed, might I perceive, by a resistence obvening a manifest Tensity or Rigidity in the compressed aer: but, insomuch as when my hand leaves the region of the lax aer, and enters that of the compressed, there is as much of space left in the lax aer for the compressed to recurr into, as that which my hand possesseth in the region of the compressed; and when it hath passed through the region of the compress'd, and again enters the confines of the lax, there is just so much of the lax aer propelled into the space left in the compressed, as responds in proportion to the space possessed by it in the lax: therefore doth my hand deprehend no sensible difference of *Fluxility* in either, and yet is the Urgency or Contention of the Base of the Cylindre of aer impendent upon the restagnant Quicksilver, constantly equal, though it may be conceived to suffer an *Undulation* or Wavering motion by the traversing of my hand to and again, by reason of the propulse and repulse.

*Art.* 6.
The solution thereof by the necessary reliction of a space in the vicine region of Lax aer, equal to that, which the Hand commoved possesseth in the region of the Compress.

This may be enforced by the *Example* of the *Flame* of a Candle, which though ascending constantly with extreme pernicity, or rapidity of motion, and made more crass and tense by the admixture of its own fuliginous Exhalations: doth yet admit the traversing of your finger to and fro through it so easily, as you can deprehend no difference of Fluxility between the parts of the Flame and those of the circumvironing Aer; the cause whereof must be identical with the former.

*Art.* 7.
A confirmation of the same Reason, by the adæquate Example of the Flame of a Taper.

*Secondly*, by the Experience of *Urinators* or Divers; who find the Extension and contraction of their arms and legs as free and easie at the depth of 20 fathoms, as within a foot of the surface of the Water; notwithstanding that water comes many degrees short of Aer, in the point of * Fluidity.

*Art.* 8.
2 By the Experiment of Urination.

* Quam ob causam, corpus hominis ad

*quantitatis immersum aquæ profunditatem nullum incumbentis aquæ pondus sentiat, legat Petar ex Mersenno plenidum Hydraulis. propol. 49 p.203.*

I

*Thirdly*

*Thirdly,* by the *Beams of the Sun*; For, when these insinuate themselves through some slender hole or crany into a chamber, their stream or Thread of Solary Atoms appears like a *white shining wand* (by reason of those small Dusty bodies, whose many faces, or superficies making innumerable refractions and reflections of the rayes of Light towards the Eye) and constantly maintains that figure, though the wind blow strongly transverse, and carry off those small dusty bodies, or though with a fan you totally dispel them: why? Because fresh Particles of Dust succeeding into the rooms of those dispelled, and æqually refracting and reflecting the incident radii of light toward the Eye, conserve the Apparence still the same. So though the wind blow off the first Cylindre of comprest aer, yet doth a second, a third, &c. instantly succeed into the same Space, so as that region, wherein the Base thereof is situated, doth constantly remain comprest: because the compression of the insensible Particles of the Aer and Wind, during their Continuation in that region, continues as great as was that of the particles formerly propulsed and abduced.

And *Fourthly,* by the *Rainbow*; which persisteth the same both in the extent of its Arch, and the orderly-confused variety of Colours: though the Sun, rapt on in his diurnal tract, shifts the angle of incidence from one part of the confronting Cloud to another, every moment; and the Wind change the Scene of the Aer, and adduce consimilar small bodies, whose various superficies making the like manifold Refractions and Reflexions of the incident lines of Light, dispose them into the same colours, and præsent the eye with the same delightful Apparition.

Which had the Hairbrain'd and Contentious *Helmont* in the least measure understood; he must have blush't at his own most ridiculous whimsy, that the Rainbow, is *a supernatural Meteor,* or *Ens extempore created* by Divinity, as a sensible symbol of his Promise no more to destroy the inhabitants of the Earth by Water, having no dependence at all on Natural Causes: especially when the strongest Argument He could excogitate, whereby to impugn the common Theory of the Schools, concerning the production thereof, by the refraction and reflection of the rayes of the Sun incident upon the variously figured parts of a thin and rorid Cloud in opposition diametrical; was only this. *Oculis, manibus, & pedibus cognovi istius figmenti falsitatem. Cùm ne quidem simplex Nubes esset in loco Iridis. Neque enim, etsi manu Iridem finderem, eamque per colores Iridis ducerem, sensi quidpiam, quod non ubique circumquaque in aere vicino: imo non proin Colores Iridis turbabantur, aut confusionem tollerabant. (in Meteoron Anomalon.)*

---

### Sect. VII.

### *The Sixth and last Capital Difficulty.*

Upon the *eduction of the lower extreme of the Tube out of the region of the Restagnant Quicksilver, into that of Water superaffused; wherefore doth the Water instantly intrude into the Tube, and the Quicksilver residuous therein by sensible degrees deflux, until it hath totally surrendred unto it?*

Solution.

## Solution.

This *Phænomenon* can have for its *Cause* no other but the great *Disparity of weight betwixt those two Liquors*.    For, insomuch as the subsistence of the Quicksilver in the erected Tube, at the altitude of 27 digits, justly belongs to the Æquipondium betwixt it and the circumpendent Cylindre of Aer; and the proportion of Weight which Quicksilver holds to Water, is the same that 14 holds to 1: it must as manifestly, as inevitably follow, that the Water, being by so much less able, in regard of its so much minority of Weight, to sustain the impulse of the Aer unceffantly contending to deliver it self from that immoderate Compreffion, muft yeild to the descending Base of the aereal Cylindre, and so ascend by degrees, and poffefs the whole Space; every part of Quicksilver that delapfeth, admitting 13 parts of Water into the Tube.

*Art. 2.*
The clear *folution* thereof, by the great difproportion of weight betwixt *Quickfilver* and *VVater*.

Here occurs to us a fair opportunity of erecting, upon the præmifed foundation, a rational *Conjecture* concerning the *perpendicular Extent of the Region of Aer from the face of the Terraqueous Globe*.    For, if Aer be 1000 times (according to the compute of the great *Merfennus* (reflect. *phyficomath. pag.* 104.) who exceedingly differs from the opinion of *Galilæo* (*Dialog. al. moviment. pag.* 81.) and *Marinus Ghetaldus* (*in Archimed. promot.*) both which demonftrate Aer to be only 400 times) lighter then Water, and Water 14 times lighter then Quickfilver: hence we may conclude (1) That Aer is 14000 times lighter then Quickfilver; (2) That the Cylindre of Aer æquiponderant to the Cylindre of Quickfilver of the altitude of 27 digits, is 14000 times higher; and (3) That the altitude of the Cylindre of Aer amounts to 21 Leucæ, or Leagues. Since 14000 times 27 digits (*i. e.* 378000 digits) divided by 18000 digits (so many amounting to a French League, that confifteth of 15000 feet) the Quotient will be 21.

*Art. 3.*
A *Corollary*, the *Altitude* of the *Atmofphere* conjectured.

From the fo much difcrepant opinions of these fo excellent Mathematitians, and moft ftrict Votaries of Truth, *Galilæo* and *Merfennus*; each of which conceived his way for the exploration of the exact proportions of *Gravity* betwixt *Aer* and *Water*, absolutely *Apodictical*: we cannot omit the opportunity of obferving; how insuperable a difficulty it is, to conciliate *Ariftotle* to *Euclid*, to accommodate those Axioms, w<sup>ch</sup> concern Quantity *abftract* from Matter, to Matter united in one notion to Quantity, to erect a folid fabrick of *Phyfiology* on Foundations *Mathematical*. Which Difficulty the ingenious *Magnenus* well refenting, made this a chief præparatory Axiom to his fecond Difputation concerning the Verifimility of *Democritus* Hypothefis of Atoms: *Non funt expendendæ Actiones Phyficæ regulis Geometricis*; fubnecting this ponderous Reafon, *Cum Demonftrationes Geometricæ procedant ab Hypothefi, quam probare non eft Mathematici, fed alterius Facultatis, quæ eam refellit; ideo lineis Mathematicis, regulifque ftricte Geometricis, Actiones Phyficæ non funt expendendæ.* ( *Democrit. Reviviscent. p.* 318.)

*Art. 4.*
A fecond *Corollary*; the defperate} Difficulty of conciliating *Phyfiology* to the *Mathematicks*: inftanced in the much difcrepant opinions of *Galilæo* and *Merfennus*, concerning the proportion of Gravity that Aer and *VVater* hold each to other.

And now at length having run over theſe ſix ſtages, in as direct a courſe, and with as much celerity, as the intricacy and roughneſs of the way would tolerate; hath our Pen attained to the end of our *Digreſſion*: wherein, whether we have gratified our Reader with ſo much either of ſatisfaction, or Delight, as may compenſate his time and patience; we may not præſume to determine. However, this præſumption we dare be guilty of, and own; that no *Hypotheſis* hitherto communicated, can be a better *Clue* to extricate our reaſon from the myſterious Labyrinth of this Experiment, by ſolving all its ſtupendious *Apparences*, with more veriſimility, then this of a *Diſſeminate Vacuity*, to which we have adhæred. But, before we revert into the ſtraight tract of our Phyſiological journey, the præcaution of a ſmall *ſcruple* deduceable from that we have conſigned a *Cylindrical Figure* to the portion of Aer impendent on the ſurface of the Reſtagnant Liquors; adviſeth us to make a ſhort ſtand, while we advertiſe; That though we confeſs the Diametre of the Sphere of Aer to be very much larger then that of the Terraqueous Globe, and ſo, that the Aer, from the Convex to the Concave thereof incumbent on the ſurface of the Reſtagnant Liquors in the veſſel placed on the Convex of the Earth, doth make out the *Section* or *Fruſtum* of a *Cone*, whoſe Baſis is in the ſummity of the Atmoſphere; and point at the Centre of the Earth (as this *Diagram* exhibiteth.)

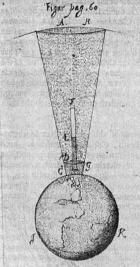

Figur pag. 60

C I K, *The Terraqueous Globe.*
B, *The Centre thereof.*
C D G, *A veſſel ſituate on the ſuperficie thereof.*
C, *The lower region of the veſſel, filled with Quickſilver.*
G, *The upper region poſſeſſed by Water.*
F E D, *The Tube perpendicularly erected in the Veſſel.*
E, *The point of Æquilibrium, at 27 dig. to which the Cylindre of Quickſilver hath deſcended.*
A B H, *A Cone extending from the Centre of the Earth to the convex ſuperficie of the region of Aer.*
A D G H, *A Fruſtum, or part of that Cone extending from the Convex to the Concave of the Aer, impendent on the ſurface of the Reſtagnant Liquors in the veſſel D C G.*

Note that neither Earth, Aer, Veſſel, nor Tube, are delineated according to their due proportions: ſince ſo, the Earth would have appeared too great, and the reſt too ſmall, for requiſite inſpection.

Yet

Yet, infomuch as the Aer is Æquipønderant to the Cylindre of Quick-filver contained in the Tube (the only requifite to our præfent purpofe) no lefs in the Figure of a Cone, then in that of a Cylindre; and fince both *Merfennus* and *Gaffendus* (to either of which we are not worthy to have been a meer *Amanuenfis*) have waved that nicety, and declared themfelves our Præcedents, in this particular: we have thought our felves excufable for being conftant to the moft ufual Apprehenfion, when the main intereft of Truth was therein unconcerned.

OF

# PLACE

# CHAP. VI.

## OF

# PLACE.

### SECT. I.

The Identy
*Essential* of a
*Vacuum* and
*Place*, the
cause of the
præsent En-
quiry into the
Nature of
*Place*,

Hat *Inanity* and *Locality* bear one and the same Notion, *Essentially*, and cannot be rightly apprehended under different conceptions, but *Respectively*; or, more expresly, that the same *Space*, when possessed by a Body, is a *Place*, but when left destitute of any corporeal Tenent whatever, then it is a *Vacuum*: we have formerly insinuated, in the *third Article*, *Sect. 1. of our Chap. concerning a Vacuum in Nature*. Which essential *Identy*, or only relative *Alterity* of a *Vacuum* and *Place*, is manifestly the Reason, why we thus subnect our præsent Enquiry into the Nature or Formality of Place, immediately to our præcedent Discourse of a Vacuum: we conceiving it the duty of a Physiologist, to derive his Method from *Nature*, and not to separate those Things in his Speculation, which she hath constituted of so near Affinity in Essence.

*Art. 2.*
Among all the
Queries, about
the *Hoti* of
Place; the most
important is,
Whether Epi-
curus, or Ari-
stotles Defini-
tion of it, be
most adæ-
quate.

Among those numerous and importune *Altercations*, concerning the *Quiddity* or formal reason of *Place*, in which the too contentious *Schools* usually lose their Time, their breath, their wits, and their Auditors attention; we shall select only one *Question*, of so much, and so general importance, that, if rightly stated, calmly and æquitably debated, and judiciously determined, it must singly suffice to imbue the mind of any the most Curious Explorator, with the perspicuous and adæquate Notion thereof.

*Epicurus (in Epist. ad Herodot.)* understands Place to be, τὸ διάςημα, *Intervallum illud, quod privatum Corpore, dicitur INANE, & oppletum corpore*

*corpore, LOCUS :* That Interval, or Space, which being destitute of any body, is called, a *Vacuum,* and possessed by a body, is called Place.

And *Aristotle* (in 3. *Auscult. Natur. cap. 6.*) thinks He hath hit the white, when He defines Place to be, τὸ τῶ περιέχουθω. πέρας ἀκίνητον πρῶτον, *Circumdantis Corporis extremum immobile primum ; Concava nempe, seu proxima immediataque, & ipsum locatum contingens corporis ambientis superficies :* the concave, proxime, immediate superfice of the body circumambient, touching the *Locatum.*

Now the *Difficulty* in Quæstion, is only this : Whether this Definition of *Aristotle,* or that modest Description of *Epicurus,* doth with the greater measure of verisimility and perspicuity respond to the nature of what we ought to understand, in propriety of conception, signified by the word, *Place.*

In order to our impartial perpension of the moments of reason on each side, requisite it is, that we first strictly ponder the *Hypothesis,* or Ground, on which *Aristotle* erected his assertion, which is this ; *Præter dimensiones Corporis locati, & ipsam ambientis superficiem, nullas alias dari (in 4. Physic. 1.)* that in nature are none but *Corporeal* Dimensions : for, if we can discover any other Dimensions, abstruct from Corporiety, such wherein the formal reason of Space may best and most intelligibly be radicated ; it can no longer remain in the suspence of controversie, how unsafe it is for the Schools to recurr to that superstructure, as a Sanctuary imprægnable, whose Foundation is only sand, and depends for support upon no other but a præcarious supposition.

*Art.* 3.
The Hypothesis of Aristotles Definition.

Imagine we, therefore, that God should please to adnihilate the whole stock or mass of Elements, and all Concretions resulting there-from, *i. e.* all Corporeal Substances now contained within the ambite, or concave of the lowest Heaven, or Lunar Sphere : and having thus imagined, can we conceive that all the vast *Space,* or *Region* circumscribed by the concave superfice of the Lunar Sphere, would not remain the same, in all its Dimensions, after as before the reduction of all bodies included therein to nothing ? Undoubtedly, that conceipt cannot endure the test of Reason, which admits, that this sublunary Space can suffer any other alteration, but only a privation of all Bodies that possessed it. Now, that it can be no Difficulty to God, at pleasure, to adnihilate all things comprehended within it ; and yet at the same time to conserve the Sphere of the Moon entire and unaltered : cannot be doubted by any, but those inhumane Ideots, who dare controvert his Omnipotence.

*Art.* 4.
A convenient supposition inferring the necessity of Dimensions Incorporeal.

Nor can it advantage our Dissenting Brother, the *Peripatetick* to plead ; that we suppose, what ought not to be supposed, an absolute *Impossibility,* as to the Firm and fundamental Constitutions of Nature, which knows no such thing, as *Adnihilation* of Elements : since, though we allow it impossible to Nature, yet can no man be so steeled with impudence, as to deny it facile to the *Author* and *Governour* of Nature ; and should we conced it impossible to *Him* also, yet doth not the impossibility of any Effect interdict the *supposition* thereof as possible, in order to the appropinquation of a remote, and explanation of an obscure verity, not invalidate that *Illation* or assumption, which by genuine cohærence depends thereupon.

*Art.* 5.
The *Legality* of that supposition.

Besides,

Besides, 'tis no Novelty, nor singularity in us, upon the same consideration, to suppose *Natural Impossibilities*: insomuch as nothing is more
usual, nor laudable amongst the noblest order of *Philosophers*, then to take
the like course, where the abstruse condition of the subject puts them upon
it; and even *Aristotle* Himself hath been more then once our Præcedent
and Exemplar therein. For, when He had demonstrated the Necessity of
the motion or circumgyration of the Cœlestial Orbs; He yet requires of
us, that we suppose them to *quiesce* constantly: that so we may the more
satisfactorily apprehend the truth of that position, at which his whole discourse was collineated; *viz. that the Cause of the Earths Quiet, is not*, as
some dreamed, *the rapid motion of the Heavens*; for, having cleared the
eye of his Readers mind from all the dust of præsumption, with this supposition, He then with advantage demands of him, *Ubinam terra moraretur?*
(*2 de Cælo.*) Nay, even concerning this our Argument, need we not want
the Authority of *Aristotle* to justifie the lawfulness of this our supposition;
for, attempting to enforce, that in a large imagined Vacuum, in part whereof a Cube of Wood is conceived to be situate, there can be no Dimensions but those of the Cube; He admits them conceiveable as clearly abstracted from the mass or bulk of wood, and devested of all corporeal Accidents; wherein (under favour) He more then seems to incur an open
Contradiction of his own dear *Tenet*, that it is absurd to imagine any *Dimensions Incorporeal*. Nor is the *Facility* of our supposition less manifest
then the *Lawfulness* thereof: since we dare our Opponents to produce any
contemplative Person, who shall conscientiously attest, that He could not,
when He fixed his thoughts thereupon, clearly and easily imagine the same;
What therefore can remain to impede our progress to the *Use*, or scope of
this our supposition?

*Art.6.*
The Dimensi-
ons of Longi-
tude, Lati-
tude, and Pro-
fundity, imagi-
nable in a Va-
cuum,
Having, therefore, imagined the whole sublunary Region to be one
continued and entire Vacuum: we cannot but also imagine, that from any
one point designed in the concave superfice of the Lunar Sphere, to another
point *è diametro* opposite in the same, there must be a certain *Distance*, or
*Intercedent Space*. If so; must not that Distance import a *Longitude*, or
more expresly an incorporeal and invisible Line? (2) If so; must not the
*medium* of that Line be the *Central point* of the empty Space, the same which
stood for Centre to the Terraqueous Globe, before its annihilation?
(3) If so; may we not conceive *How much* of that voyd Region was formerly possessed by the mass of Elements: and with mental Geometry
commensurate how much of that Space did once respond to the
superfice, how much to the profundity of each of those Bodies?
(4) If so; must we not allow the Dimensions of *Longitude, Latitude*, and
*Profundity* imaginable therein? undoubtedly, yea: since we can no
where conceive a *Distance*, or intercedent Space, but we must there also conceive a *Quantum*; and Quantity imports *Dimensions*, nor is
there any Distance, but of determinate extent, and so commensurable.

*Art.7.*
The Grand
*Peripatetick* ob-
jection, that
*Nothing* is in a
Vacuum; ergo
no Dimensions
From the pressure of this Socraticism, hath our *Peripatetick* retreated to
that ruinous sanctuary of the Term, *Nothing*: retarding our pursuit, with
this Sophism. When you suppose the sublunary Region to be an absolute Vacuum, you expresly concede, that *Nothing* is contained therein; and
upon consequence, that those Dimensions by you imagined therein, are
                                                                    Nothing,

Nothing, and so that therein are no Dimensions at all. Why; because Dimensions consist essentially and so inseparably in *Quantity*; and all Quantity is inseparable from *Corporiety.* Wherefore, supposing no Body existent in that Empty Space: you implicitely exclude all Quantity, and consequently all Dimensions from thence.

This Evasion, we confess, is plausible; nor hath it imposed only upon young and pædantique Pretenders to Science, such as having once read over some Epitome of the Commentaries upon *Aristotles Physicks,* and learned to cant in Scholastick Terms, though they understand nought of the Nature of the Things signified, believe themselves wise enough to rival *Solomon* : but even many grey and sage Enquirers, such who most sedulously digged for the jewel of Knowledge in the Mine of Nature, and emancipated their intellectuals betimes from the slavery of Books. For, among the most celebrated of our Modern Physiologists, we can hardly find two, who have judged it safer to abide the seeming rigour of this Difficulty, then to run upon the point of this Paradox; that, if all Bodies included in the ambite of the Lunar Heaven, were annihilated, then would there be no Distance at all betwixt the opposite sides of the same : and the Reason they depend upon, is this; Necessary it is that those points should not be distant each from other, but be contiguous, betwixt which Nothing doth intercede. Nay, even *Des Cartes* himself cannot be exempted : since, 'tis confest by him (*in Princip. Philosoph. articul.* 18.) that He subscribed the same common Mistake, in these Words : *si quæratur, quid fiet, si Deus auferat omne corpus, quod in aliquo vase continetur, & nullum aliud in ablati locum subire permittat? Respondendum est vasis latera hoc ipso fore contigua. Cum enim inter duo corpora nihil interjacet, necesse est, ut se mutuò tangant, ac manifestè repugnat, ut distent, & tamen ut distantia illa sit Nihil : quia omnis Distantia est modus Extensionis, & ideo sine substantia extensa esse non potest.* To him also may we associate Mr. *White* (*in Dialog.* 1. *de Mundo.*)

The most direct and shortest way to the *Redargution* of this Epidemick Errour, lyes in the detection of its grand and procatarctick *Cause*; which is the Præoccupation of most Scholers minds by the Peripatetick Institutions, that limit our Notions to their imperfect Categories, and explode those Conceptions as Poetical and extravagant, that transcend their classical Distinction of all *Entities* into *Substance* and *Accident.* For, first, insomuch as in the Dialect of the Schools, those three Capital Terms, *Ens, Res, Aliquid,* are mere Synonyma's, and so used indiscriminately; it is generally concluded, that whatever is comprehensible under their signification, must be referred either to the Classis of *Substances,* or that of *Accidents* : and upon illation, that what is neither Substance, nor Accident, can pretend to no *Reality,* but must be damned to the predicament of Chimæra's, or bee excluded from *Being.* Again, having constituted *one* Categorie of all *Substances,* they mince and cantle out poor thin *Accident* into *Nine,* accounting the first of them *Quantity,* and subdividing that also into (1) *Permanent,* i. e. the Dimensions of Longitude, Latitude, Profundity; and so make Place to consist if not in all three, yet at least in one of them, *viz.* Latitude

      or

*Art.* 8.
*Des Cartes,* and Mr. *White* seduced by the plausibility of the same.

*Art.* 9.
The *Peripateticks* reduction of *Time* and *Place* to the General Categories of *Substance* and *Accidents,* the Cause of this Epidemick mistake.

or the superficies of a Body: (2) *Succeſsive, i.e.* Time and Motion, but especially Time, which may be otherwise expreſſed by the Term, *Durati-on.* Hereupon, when they deliver it as oraculous, that Quantity is a *Corporeal Accident:* they confidently inferr, that if any Quantity, or Permanent, or Succeſsive, be objected, that is not or ſeparately, or conjunctly Corporeal, it ought to be exploded, as not *Real,* or an abſolute *Nothing.*

Now this their *Scheme* is defective. (1) Becauſe it fails in the General Diſtribution of *Ens,* or *Res,* into *Subſtance* and *Accident:* in regard, that to thoſe two Members of the Diviſion there ought to be ſuperadded other two, more general then thoſe; viz. *Place* and *Time,* Things moſt unreducible to the Categories of Subſtance and Accident. We ſay, *more General then thoſe Two;* becauſe as well all Subſtances as Accidents whatever, have both their Exiſtence in ſome Place, and their Duration in ſome Time; and both Place and Time are, even by thoſe who diſpute whether they are Accidents, or not, willingly granted to perſever conſtantly and invariately the ſame.

(2) Becauſe it offends Truth in the confinement of all *Quantity,* or Dimenſion, and ſo of that of Place and Time, to the Category of *Accidents,* nay even of *Corporeal* ones: when there wants not a ſpecies of Quantity, or Extenſion having Dimenſions, that is not *Corporeal;* for, nor Place, nor Time, are Corporeal. Entities, being no leſs congruous to Incorporeal, then Corporeal Beings. Upon which conſideration, 'tis a genuine and warrantable Inference; that albeit Place and Time are not pertinent to the Claſſis either of ſubſtances, or Accidents: yet are they notwithſtanding *Realities, Things,* or *not-Nothings;* inſomuch as no ſubſtance can be conceived exiſtent without Place and Time. Wherefore, when any Cholerick Bravo of the *Stagirites* Faction, ſhall draw upon us with this Argument; *Whatever is neither Subſtance, nor Accident,* is a downright Nothing, &c. we need no other buckler then to except Place and Time.

*Art.* 10.
Place, neither
*Accident* nor
*Subſtance.*

To authenticate this our Schiſm, and aſſert our Affirmation; we muſt now evince, that Place is neither Accident, nor ſubſtance: which to effect, we need not borrow many moments of its Twin-brother, Time, to hunt for Arguments in. For (1) though it be objected, that Place is capable of Acceſſion to, and ſejunction from the *Locatum,* without the impairment, or deſtruction thereof; and in that relation ſeems to be a mere Accident: yet cannot that juſtifie the conſignation of Place to the Category of Accidents; becauſe Place is uncapable of *Acceſs* and *Receſs,* and 'tis the *Locatum* to which in right we ought to aſcribe *Mobility.* So that when various Bodies may be ſucceſſively ſituate in one and the ſame Place, without cauſing any the leaſt mutation therein: we muſt allow the force of this Argument, to bring it neareſt to the propriety of a ſubſtance. (2) A *ſubſtance* it cannot be; becauſe the Term, *Subſtance* imports ſomething, that doth not only *exiſt per ſe,* but alſo, and principally, what is *Corporeal,* and either *Active* or *Paſſive:* and neither Corporiety, nor Activeneſs, nor Paſſiveneſs, are Attributes competent to Place: *Ergo.*

*Art.* 11.
The prace-
dent Giant-
*Objection,* that
Nothing is in
a Vacuum;
ſtabb'd, at a
blow.

Now, to leave our roving, and ſhoot level at the mark; the Extract

of

of thefe præmifed Confiderations, will eafily and totally cure the defpe-
rate *Difficulty* objeded. For, when it is urged, that betwixt the oppofite
fides of a veffel fuppofed to be abfolutely devoyd of any Body whatever,
nothing doth intercede, and confequently that they are Contiguous;
we need no other *folution* but this: that (indeed) nothing *Corporeal*
doth interced, betwixt the diametrally oppofite fides of a voyd concave,
that is either Subftance, or Accident; but yet there doth intercede
*fomething Incorporeal*, fuch as we underftand by *Spatium, Intercape-
do, Diftantia, Intervallum, Dimenfio*, which is neither Subftance
nor Accident.    But, alas! that Thing you call *Space* is, according to
your own fuppofition, an abfolute *Vacuum*: What though? it muft not
therefore be *Nothing*, unlefs in the fenfe of the *Peripatetick*: becaufe it hath
a *Being (fuo modo)* and fo is *fomething*.

The fame alfo concerns thofe *Dimenfions*, which we conceive, and
the Schools deny to be in our imaginary Vacuum: For of them
it may be likewife truly faid, that they are *Nihil Corporeum*, but
not that they are *Nihil Incorporeum*, or more emphatically, *Nihil
SPATIALE*, Nothing *Spatial*. Hence, according to the diftinction of Things into Corporeal, and Incorporeal; we may, on the defign of Perfpicuity, difcriminate Dimenfions alfo into (1) *Corporeal*, fuch as are competent to a *Body*, wherein we underftand Longitude, Latitude, Profundity: (2) *Spatial*, fuch as are congruous
to *Space*, wherein we may likewife conceive Longitude, Latitude, and
Profundity. And fo we may conclude, that thofe Dimenfions, which
muft remain in that fuppofed *Inane Region* circumfcribed by the concave
of the Lunar Orb, in cafe God fhould adnihilate the whole mafs of Elements, and all their off-fprings, included therein; are, in truth, not Corporeal, but *Spatial*.

*Art. 12.*
Dimenfions
*Corporeal* and
*Incorporeal*, or
*Spatial.*

Let us fkrew our *fuppofition* one pin higher, and farther imagine,
that God, after the Adnihilation of this vaft machine, the Univerfe, fhould
create another, in all refpeds confimilar to this, and in the fame part of
Space, wherein this now confifteth: and then fhall our thoughts be tuned
to a fit key for the fpeculation, nay the comprehenfion of *Three* notorious
*Abftrufities, viz;*

*Art. 13.*
The former
fuppofition re-
affumed and
enlarged.

(1) That as the Spaces were *Immenfe*, before God created the
World; fo alfo muft they eternally perfift of infinite Extent, if He
fhall pleafe at any time to deftroy it: that He, according to the
counfel of his own *Beneplacit*, eleded this determinate *Region* in the
infinite Spaces, wherein to ered or fufpend this huge Fabrick of the
World; leaving the refidue which we call *Extramundan* Spaces, abfolutely voyd: and that as the whole of this determinate Region of
Space is adæquately competent to the whole of the World; fo alfo is each part thereof adæquately competent to each part of the
World; *i. e.* there is no part of the World, Great or Small, to
which there is not a part of Space exadly refpondent in all dimenfions.

*Art. 14.*
The fcope and
advantage
thereof, viz.
the compre-
henfion of
three eminent
Abftrufities
concerning
the Nature of
Place.

(2) That thefe immenfe Spaces are abfolutely *Immoveable*. And
therefore fhould God remove the World into another determinate region

of them, yet would not this Space wherein it now persisteth ; accompany it, but remain immote, as now. In like manner, when any part of the World is translated from one place to another ; it leaves the part of Space, which it formerly possessed, constant and immote, and the Spaces through which it passeth, and wherein it acquiesceth, continue also immote.

(3) That, in respect the Dimensions of these Spaces are Immoveable, and Incorporeal : therefore are they every where *Coexistent*, and *Compatient* (we speak in the dialect of the Schools) with Corporeal Dimensions, without reciprocal repugnancy ; so as in what part soever of Space any Body is lodged, the Dimensions of that part of Space, are in all points respondent to the Corporeal Dimensions thereof. In this case, therefore, 'tis far from an Absurdity, to affirm, that *Nature doth not abhor a Penetration of Dimensions.* To bring up the rear of these advantages resulting from our supposition, we may from thence deprehend, Why *Aristotle* hath not cleft a hair in his position, that there is in the Universe no *Interval*, nor *Dimensions*, but what are *Corporeal.*

*Art.* 15.
The *Incorporiety* of Dimensions *Spatial*, Discriminated from that of the *Divine Essence,* and other *Substances Incorporeal.*

To discriminate the *Incorporiety* of these *Dimensions Spatial*, from that adscribed to the *Divine Nature, Intelligences Angelical,* the *Mind of Man*, and other (if there be any) Incorporeal substances ; we advertise, that the term *Incorporeal* bears a double importance. (1) It intends not only a simple *Negation* of Corporiety, and so of corporeal Dimensions ; but also a true and germane *substance,* to which certain *Faculties* and *Operations* essentially belong ; and in that sense it is adscriptive properly to God, Angels, the Souls of men, &c. spiritual Essences. (2) It signifies a *mere Negation of Corporiety,* and so of corporeal Dimensions, and not any positive Nature capable of Faculties and Operations ; and in this sense only is it congruous to the Dimensions of Space, which we have formerly intimated to be neither *Active,* nor *Passive,* but to have only a general *Non-repugnancy,* or *Admissive Capacity,* whereby it receives Bodies either *permanenter,* or *transeunter.*

*Art.* 16.
This persuasion, of the *Improduction* and *Independency* of *Places* præserved from the suspicion of *Impiety.*

Here we discover our selves in danger of a nice *scruple,* deductive from this our Description of *Space,* viz, that, according to the tenor of our Conceptions , *Space must be unproduced by, and independent upon the original of all Things,* God. Which to prævent, we observe, that from the very word *Spatial* Dimensions, it is sufficiently evident, that we understand no other Spaces in the World, then what most of our Ecclesiastical Doctors allow to be on the outside thereof, and denominate *Imaginary* : not that they are meerly *Phantastical,* as Chimæra's ; but that our Imagination can and doth apprehend them to have Dimensions, which hold an analogy to the Dimensions of Corporeal substances, that fall under the perception and commensuration of the sense. And, in that respect, though we concede them to be *improduct* by, and *independent* upon God ; yet cannot our Adversaries therefore impeach us of impiety, or distort it to the disparagement of our theory : since we consider these Spaces, and their Dimensions to be *Nihil Positivum, i.e.* nor Substance, nor Accident, under which two Categories all works of the Creation are comprehended. Besides, this sounds much less harsh in the ears of the Church, then that which not a few of her Chair-men have adventured to patronize ; viz. *that the Essences of Things are Non-principiate, Improduct , and Independent :*
insomuch

insomuch as the Essence being the noblest, constitutive, and denominative part of any Thing, Substance or Accident; to hold it uncreat and independent, is obliquely to infer God to be no more then an *Adopted Father* to Nature, a *Titular* Creator, and Author of only the material, grosser and unactive part of the World.

## S E C T.  II.

BY the discovery of Dimensions independent upon Corporiety, such wherein the Formal reason of Space appears most intelligibly to consist, have we fully detected the weakness of *Aristotles* Basis, *præter dimensiones Corporis locati, & ipsam ambientis superficiem, nullas alias dari*: it remains only, that we demolish his thereupon-erected *Definition* of *Place*, in which his legions of *Sectators* have ingarrisoned their judgments, as most impregnable.

That *Place* is not the immediate and contiguous *superfice* of the body invironing the *Locatum*, may by the single force of this Demonstration be fully evicted. *Immobility* is essential to Place, as *Aristotle* well acknowledged; for if Place were moveable, then would it follow of inevitable necessity, that a body might be translated without mutation of place, and *è converso*, the place of any thing might be changed, while the thing it self continues immote; both which are *Absurdities* so manifest, as no mist of Sophistry can conceal them even from the purblind multitude: Now the superfice of the Circumambient can in no wise prætend to this propriety of place, Immobility; as may be most conveniently argued from the example of a Tower; for that space, which a Tower possesseth, was there before the structure, and must remain there the same in all dimensions after the ruine thereof; but the superfice of the contiguous Aer, the immediate Circumambient, is removed, and changed every moment, the whole mass of Aer being uncessantly agitated more or less, by winds and other violences: *Ergo.*    So numerous are the shifts and subterfuges of the distressed Disciples of *Aristotle*, whereby they have endeavoured to *Fix* this *Volatile* superfice of the Circumambient: that should we insist upon only the commemoration of them all; we might justly despair of finding any Charity great enough, to pardon so criminal an abuse of leasure.

Besides, from *Epicurus διάσυσμα*, or *Space*, we may extract *Salvo's* for all those *Scruples*, which are commonly met with by all, who worthily enquire into the nature of Place. For, when it is questioned (1) How a body can persist invariately in the same place, though the circumambient be frequently, nay infinitely varied? (2) How a body can change place, though the Circumambient accompany it in its remove? (3) Why one body can be said to be thus or thus far, more or less distant from another? we may easily satisfie all with this one obvious Answer, that all mobility is on the part of the Locatum, all Space continuing constant and immote. Further, hence come we to understand, in what respect Place is commonly conceived to be exactly *adæquate* to the *Locatum*: for, the Dimensions of all Space possessed, are in all points respondent to those of the body possess-

sing

sing; there being no part of the body, profound or superficial, to which there is not a part of Space respondent in æqual extent; which can never be made out from the mere superfice of the Circumambient, in which no one of the Profound or Internal parts of the Locatum, but only the *superficial* are resident. Moreover, hence also may we understand, How *Incorporeal substances*, as God, Angels, and the Souls of men, may be affirmed to be *in loco*. For, when God, who is infinite, and therefore uncapable of Circumscription, is said to be in Place; we instantly cogitate an infinite Space: which is more then we can do of Place, if accepted in *Aristotles* Notion, which imports either that God cannot be in any place, or else He must be circumscribed by the contiguous superfice thereof: which how ridiculous, we need not observe. For *Angels* likewise, who dares affirm an Angel to be in a place, that considers his Incorporiety, and the necessity of his circumscription by the superfice of the Circumambient, if *Aristotles* Definition of Place be tolerable? To excuse it with a distinction, and say that an Angel may be conceived to be in a determinate place, not *Circumscriptivè*, but *definitivè*, i. e. So *Here*, as no *where* else: is implicitely and upon inference, to confess the truth of our assertion; Since that *Here*, designs a certain part of Space, not the superfice of any circumambient. For, though you reply, that an Angel, being an incorporeal substance, wants as well *internal* and *profound* Dimensions, by which his substance may respond to *Space*, as those *superficial* ones, that respond to *Place*: yet cannot that substance hath any evasion, since if his substance hath any *Diffusion* in place, as is generally allowed; and though it be constituted *in puncto*, as is also generally conceived: nevertheless, doth that Diffusion as necessarily respond to a certain æqual part of Space, as a point is a determinate part of space. This perhaps, is somewhat abstruse, and therefore let us conceive an Angel to be resident in some one point of that *Inane Region* circumscribed by the concave of the Lunar orb, formerly imagined: and then we may without any shadow of obscurity understand, How his substance may respond to a certain part, or point of the Inane Space, so as He may be said to be *Here*, not *There*, in this but no other place: but impossible it is, to make it out, How the substance of an Angel constituted *in puncto* of an empty space, can respond to the superfice of a Body Circumambient, because all Bodies formerly included in that sublunary Region are præsupposed to be adnihilated. Lastly, by the Incorporiety of Space we are præserved from that Contradiction, which *Aristotle* endeavouring to prævent, præcipitated himself upon no small Absurdity, *viz.* that the *supreme Heaven, or Primum mobile is in no Place*. For, if we adhere to his opinion, that place is the superfice of a body circumambient; the *Primum mobile* being the extreme or bounds of the World, we deny any thing of *Corporiety* beyond it, and so exempt it from *Locality*: but if we accept space to be the same without and within the world, we admit the *Primum mobile*, the noblest, largest, and most useful of all Bodies in the World, to enjoy a Place proportionate to its dimensions, and motion, as adæquately as any other. The necessity of which concession, *Thales Milesius* well intimated, when interrogated, What *Thing was greatest*? He answered, *Place*: because, as the World contains all other Bodies, so Place contains the World.

*Art. 3.*
*Aristotles ulti-*
*mate Refuge.*

Reduced to these straights, *Aristotle*, among sundry other Sophisms, entrusteth the last part of his Defence, to this slight *Objection*; If *Place*

*were*

*were a certain Space, constant in three dimensions ; then would it inevitably follow, that the* Locatum *and the* Locus *must reciprocally penetrate each others dimensions, and so the parts of each be infinitely divided : which is manifestly absurd, since Nature knows nor penetration of Dimensions, nor infinity of corporeal division.*

To this *Induction* we could not refuse the attribute of *Probability*, no more then we do now of *Plausibility*, had we not frequently prævented it, and openly by our Distinction of Dimensions into *Corporeal* and *Incorporeal*, and appropriating the *last* to Space. For, indeed, the Fundamental Constitutions of Nature most irrevocably prohibite the substance of one Body to penetrate the substance of another, through all its Dimensions : but, alas ! Place is (κυρίως ἀσώματος) properly and altogether *Incorporeal*; and therefore may its dimensions Incorporeal be *Coexistent*, or *Compatient* with the Corporeal Dimensions of any Body, without mutual repugnancy, the Spatial Dimensions not excluding the Corporeal, nor those extruding the spatial. This cannot be a *diaphanous*, or ænigmatical to those, who concede *Angels* to be *Incorporeal*, and therefore to penetrate the Dimensions of any the most solid Bodies, so that the whole substance of an Angel may be *simul & semel*, altogether and at once in the same place with that of a stone, a wall, the hand of a man, or any other body whatever, without any necessity of mutual *Repugnancy*. Nor to those, who observe the Synthesis, or Collocation of *Whiteness*, *Sweetness*, and *Qualities* in the substance of Milk : for as those are conceived to pervade the whole substance of Milk, without any reciprocal repugnancy of Dimensions, so are we to conceive that the Dimensions of Space are totally pervaded by the whole Body of the *Locatum*, without *Renitency*.

*Art. 4.*
The Invalidity thereof: and the *Coexistibility*, or *Compatibility* of Dimensions *Corporeal and Spatial.*

CHAP.

# CHAP. VII.

## OF

# TIME

## AND

# ETERNITY.

### SECT. I.

**Art. I.**
*The Noti of Time more easily conceivable by the Simple Notion of the Vulgar, then by the complex Definitions of Philosophers.*

**S**ome *Texts* there are in the *Book of Nature*, that are best interpreted by the sense of the *Vulgar*, and become so much the more ænigmatical, by how much the more they are commented upon by the subtile discourses of the *Schools*: their over-curious *Descants* frequently rendring that *Notion* ambiguous, comp'ex and difficult, which accepted in its own genuine *simplicity*, stands fair and open to the discernment of the unpræjudicate, at the first conversion of the acies of the Mind thereupon. Among these we have just cause to account *TIME*; since if we keep to the popular and familiar use of the word, nothing can be more easily understood: but if we range abroad to those vast Wildernesses, the Dialectical Paraphrases of Philosophers thereupon, and hunt after an adæquate *Definition*, bearing its peculiar *Genus*, and essential *Difference*; nothing can be more obscure and controversial. This the sacred *Doctor* (*August.* 11. *Confess.*14.) both ingenuously confessed, and most emphatically expressed, in his, *Si nemo ex me quærat, quid sit Tempus, scio; si quærenti explicare velim, nescio*: intimating that the *Mind* may, indeed, at first glance speculate the nature of Time by a proper Idea; but so pale and fine a one, as

not

nor *Tongue*, nor *Pen* can ever pourtray a lively representation thereof. And *Cicero* ( 1. *de invent.* ) is bold to list it among the most desperate Difficulties, *Tempus definire Generaliter.* To which we may annex that saying of one quoted by *Stobæus* ( *Eccl. Phys.* 11. ) *Tempus esse* Νόημα ἐφ' ὑποστασιν, *Quidpiam non re, sed cogitatione constans.* As also that of *Aristotle*, who not only injoyns, that we discourse of Time in a certain key of thought far different from that wherein we use to consider things, which have a real inhærence *in subjecto*; as if Time had no other subject of inhærence but the Mind, were only a mere *Ens Rationis*, extrinsecal Denomination, and could expect no exacter a description, then His *Numerus, qui absque ratione numerante est nullus*: but adviseth, if any shall demand, what Time is, to afford him no other but *Democritus* Answer; *Tempus esse* ἡμεριδὲς χ vαχθὲς φάντασμα, *quale spatium diei noctisque appuret.*

If we research profoundly into the *Original* of this Difficulty, of acquiting a clear and perspect theory of the *Quiddity* of Time, from the Lecture of those prolix Treatises, whose plausible Titles promise satisfaction concerning it: we shall soon find the chief *Cause* to be this; that most Philosophers have præsupposed Time to be some *Corporeal Ens*, or at least some certain *Accident* inexistent in and dependent on Corporeal Subjects; when (in verity) if it be any thing at all it seems to be the *Twin-brother* of *Space*, devoyd of all relation to Corporiety, and absolutely independent on the Existence of any Nature whatever. For, to Him, who shall, in abstract and attentive meditation, sequestre Time from all Bodies, from their motions, successive alternations, and contingent vicissitudes insequent upon those motions; *i. e.* all Years, Months, Weeks, Dayes, Hours, Minutes, Seconds, and all Accidents or Events contingent therein: it will soon appear most evident, that Time ( *in suo esse*) owes no respect at all to Motion, its constancy, variety, or measure; since the understanding must deprehend Time to continue to be what it ever was and is, whether there be any Motion or Mutation in the World, or not, nay, whether there be any World or not. For, examining what is meant by the term *Duration*, and what by the term *Motion*, in their single importances apart: we discover, that Motion holds no relation to Duration, nor *è converso*, Duration to Motion, but what is purely *Accidental*, and *Mental*, i. e. imagined by man, in order to his commensuration of the one by the other.

*Art.* 2.
The Generall præsumption that Time is *Corporeal*, or an *Accident* dependent on Corporeal Subjects; the chief Cause of that Difficulty

Another *Cause* of this Difficulty, may be the irreconcileable *Discrepancy* of judgments concerning it, even among the most Venerable of the Ancients. For (1) *Epicurus* hath a complex and periphrastical Description of the Essence of Time, when He concludes it to be, Σύμπτωμα Συμπτωμάτων, παρεπόμῃνον ἡμέραις τε, χ νυξί, χ ὥραις χ παθεσιν, χ ἀπαθείαις, χ κινήσεσι, χ μοναῖς, an *Accident of Accidents, or Event of Events*, consequent to dayes and nights, and hours to passions and indolency, motion and quiet. The reason of which *Empiricus* (2. *adverf. Physic.*) by way of explanation, thus renders: Days and Nights are Accidents supervenient upon the ambient Aer, the one being caused by the præsence, the other by the absence of the Sun; Hours are also accidents, as being parts of day or night; but Time is coextended to each day, night & hour, & therefore we say, that this day is long, this night short, while our thoughts are constantly pointing

*Art.* 3.
The variety of opinions, concerning it; another Cause of the Difficulty: and Epicurus Description of its Essence, recited and explained.

at

at Time in that respect supervenient; Passions likewise and Indolences or Dolours and Pleasures, are Accidents not without Time evenient; lastly, Motion and Quiet are Accidents contingent in Time, and therefore by it we commensurate the Celerity and Tardity of Motion, the long or short duration of Quiet: therefore is Time the Accident of Accidents. And *Lucretius* alluding to the same opinion of *Epicurus*, translates his Ἀσώματον ἢ χρόνον ὑπάρχειν, *Tempus esse incorporeum*, into *Tempus item per se non est, &c. lib. 1.*

*Art. 4.*
*Time defined to be Cælestial Motion, by Zeno, Chrysippus &c. and thereupon affirmed, by Philo, to be only Coævous to the World.*

(2) *Zeno, Chrysippus, Apollodorus, Posidonius*, and their Sectator *Philo*, define Time to be, *Motus cælestis, sive mundani intervallum*, understanding as well all particular Conversions, as the Generality of Motion from the beginning to the end of the World. Whereupon *Philo* would inferr, that Time was *coævous* to the World, *i. e.* before the World there was no Time, nor should be any after: though the *Stoicks* unanimously defend the *Infinity* of Time, in regard they affirmed an *infinitie of Worlds successive*, the second springing up, Phœnix-like, from the ashes of the first, the third from the second, &c. (3) *Pythagoras*, according to the Records of *Plutarch (in quæstion. Platonic.)* to one interrogating him concerning the Essence of Time, calls it *Animam Cœli*, the soul of Heaven. To which *Plotinus (En. 3. lib. 7. cap. 10.)* seems to have alluded, when interpreting *Plato's* saying, that Time was the *Image of Eternity (in Timæo)* He make *Æternity* to be the very soul of the World, as considered *in se*, in its own simple essence; and *Time* to be the same soul of the world, considered, *prout varias mutationes suscipit*, as it admits various mutations.

*Art. 5.*
*Aristotle's so much magnified Definition of Time, to be the Measure of Motion Cælestial, &c. perpended and found too light.*

(4) And *Aristotle*, as every Pædagogue hath heard, after a long and anxious scrutiny, positively and magisterially determines Time to be, *Numerum Motus (cælestis ac primi) secundum prius & posterius*, the Number of the first Cœlestial Motion, according to former and later, *i. e.* insomuch as in Motion we may observe parts *Antecedent* and *Consequent* by a perpetual succession. At the first word of this eminent Definition, some superficial *Criticks* have sawcely nibbled, urging (forsooth) that it sounds solœcistical, because *Number* is Quantity *Discrete*, but *Time Continued*; and therefore that the Word *Measure* ought to be its substitute: but alas! had they read His whole discourse of the nature of Time, they could not have been ignorant, that *Aristotle* intended nothing less, then that Time should be reputed a Quantity *Discrete*; when both in his præcedent and subsequent lines He expresly teacheth, that *Motion is continued, in respect of Magnitude, and Time in respect of Motion.* Had They Excepted against the whole, indeed, their Quarrel had bin justifiable, and our selves might safely have espoused it; because, if Time be the Measure of Cœlestial Motion, then must it follow, that if there were a Plurality of Worlds, or *Prima Mobilia's*, there would also be a Plurality of Times, because a Plurality of Motions. To those of His Disciples, who reply, that in case there were many *First Moveables*, and consequently many distinct Motions; yet would there be but *one* Measure of them all: we rejoyn, if it be supposed that some of the many Motions are swifter then others, then of necessity must they have many *Prior* and *Posterior* Parts; and if so,
how

how can all those, more or less difcrepant in velocity and tardity, fall under one and the fame meafure? or, what fober man can admit, that
there would be but one Time, where muft be many diftinct *fubjects* of Motion, and fo of Time? Nor can it more avail them to diftinguifh Time
INTERNAL from EXTERNAL, affigning to each particular *Primum
Mobile* a proper or *Internal* Time within its ambite, and one General or
*External* Time to them all in common: becaufe it is a manifeft *Adynaton,*
that there fhould be a General Time, without a General Motion, whofe
parts being prior and pofterior, in refpect of perpetual fucceffion, muft be
the common *Norma,* or Rule of obfervation to all the reft; nor, indeed, can
we admit, that a Flux of ten hours at once, or together, is poffible, where
ten Spheres are in one hour moved. And, therefore, though *Ariftotle* feems
to have had fome Hint of the true nature of Time, in his *Objection* againft
thofe, who opinioned it to be *Cœleftial Motion:* yet he loft it again, when
He defined it to be the *Meafure of Cœleftial Motion.* For, Reafon atte-
fteth the contrary, it being evident that the Cœleftial Motion is rather the
Meafure of Time: infomuch as the meafure ought to be more known then
the thing meafured; and Time is a certain Flux no lefs independent upon
Motion then Quiet. Which thofe Worthies well underftood, who confeft Time to be IMAGINARY, fuch as flowed infinitely in duration before the Creation, and fhall continue its flux infinitely after the Diffolution of the World.

## Sect. II.

FAiling of fatisfaction concerning the Nature of Time, from the *Defini-
tions* of others: it remains only, that we feduloufly imploy our own
Cogitations in queft of fome competent *Defcription* of it. *Seneca* (in
*Epift.* 58.) defcanting upon *Plato's* General Diftinction of all Entities into
fix *Claffes,* faith thus: *Sextum Genus eft eorum, quæ quafi funt, tanquam
Inane & Tempus,* the fixth Genus contains only thofe things, which have
*as it were* a being, as INANITY and TIME: which we thus expound,
*Space and Time are things more General then to be comprehended under the
Categories of Subftance and Accident.* With this Text we had not long
exercifed our thoughts, before we conceived, that the moft hopeful way
for exploring the myfterious *Quiddity* of Time, lay in the ftrict examen of
the *Affinity* or *Analogy* betwixt it and the fubject of our immediately præcedent Chapter, *Space.* Nor did our Conjecture prove abortive; for,
having confronted their proprieties in all points, we foon found their Natures fully correfpondent: fo as the Notion of one feems involved in that
of the other; as is manifeft in this *Paralellifm.*

*Art.* I.
Time, nor fub-
ftance, not Ac-
cident but an
*Ens* more Ge-
neral, and the
Twin-brother
of Space.

(1) As Place, or Space, in the total, is *illimitate* and *immenfe*: fo is
Time, in its totality, *non-principiate* and *interminable.* (2) As every *Moment* of Time is the *fame* in all places: fo is every *canton* or part of Place
the *fame* in all times. (3) As Place, whether any, or no Body be
*collocated* therein, doth ftill perfift the fame immoveable and invariately:
fo doth unconcerned Time flow on eternally in the fame calm and

*Art.* 2.
A *Paralelifm*
betwixt Spa-
and Time.

equal tenor, whether any or nothing hath *duration* therein, whether any thing be moved or remain quiet. (4) As Place is uncapable of *expansion, interruption* or *discontinuity,* by any Cause whatever: so is Time uncapable of *acceleration, retardation,* or *suspension*; it moving on no less, when the Sun was arrested in the midst of its race in the dayes of *Joshua,* when the Hebrews vanquished & pursued the Amorrhites, then at any time before, or since. (5) As God was pleased, out of the *Infinite Space* to elect a certain determinate *Region* for the *situation*: so hath He, out of *Infinite Time,* elected a determinate *part* for the *Duration* of the World. (6) And therefore, as every Body, or Thing, in respect to its HERE or THERE, enjoyes a proportionate part of the *Mundane Space*: So likewise doth it, according to its NOW, or THEN of Existence, enjoy a proportionate part of the *Mundane Duration.* (7) As, in relation to Place, we say, *Everywhere*, and *Somewhere*: so, in relation to Time, we say, *Alwayes*, and *Sometimes.* Hence, as it is competent to the *Creature* to be only somewhere, in respect of Place, and sometimes, in respect of Time: so is it the prærogative of the *Creator,* to be Everywhere as to place, and Forever, as to time. And therefore those two illustrious Attributes, *Immensity,* whereby He is præsent in all places, and *Æternity,* whereby He is existent at all Times, are proper only to God. (8) As Place hath Dimensions *Permanent,* whereby it responds to the Longitude, Latitude, and Profundity of *Bodies*: so hath Time Dimensions *successive,* to which the *Motions* of Bodies may be adæquated. Hence comes it, that as by the Longitude, of any standing measure (V.G.) of an Ell, we commensurate the longitude of Place: so by the flux of an Horologe do we commensurate the flux of Time. And, insomuch as no motion is more General, Constant and Observed, then that of the Sun: therefore do we assume its motion for a *General Horodix,* by it regulate all our computations, and confide in it as an universal Directory, in our Mensuration of the flux of Time. Not that the Feet of Time are chained to the Chariot of the Sun, so as the Acceleration or Retardation of the motion of that should cause an equal Velocity, or Tardity in the progress of this: but that Custom hath so prævailed, as we compute the flux of Time by the diurnal and annual revolution of the Sun. For, in case the motion of the Sun were doubly swifter, then now it is, that of Time would not therefore be doubly swifter also; but only the space of two dayes would then be equal to the space of one, as now during the præsence of the Sun to our Hemisphere: nor, on the contrary, if the motion of the Sun were doubly slower, would the pace of Time be likewise doubly slower; but only the Space of one day, would be equal to that of two. And, therefore, He that will defend *Empedocles* conceit, that in the beginning of the World, the length of the dayes did by six parts in seven exceed that of our dayes: must demonstrate that the urnal Arch of the Sun was then by six of seven larger then now, or its motion so much slower.

*Art. 3.*

Time, *Senior unto, and independent upon Motion: and only accidentally indicated by Motion, as the Mensuratum by the Mensura.*

From this *Paralellism* 'tis difficult not to conclude, *that Time is infinitely elder then Motion,* and consequently *independent* upon it: as also, that Time is only indicated by Motion, as the *Mensuratum* by the *Mensura.* For, insomuch as it had been otherwise impossible for Man to have known how much of Time He had spent either in action, or rest; therefore did He fix his observation upon the Cœlestial motion, and compute the quantity of Time præterlapsed by the Degrees of the Suns motion in the Heavens.

vens. And becaufe the obfervation of the Suns motion was eafie and familiar; therefore did the Ancients invent feveral inftruments, as *Water*
and *Sand Hour-glaffes*, and *Sun-dials*, and the Neotoricks *Trochiliack Ho-
rodixes*, circumgyrated by internal fprings, or external weights appenfed;
and fo artificially adæquated them to the motion of the Sun, that defines
the day by its præfence, and might by its abfence, as having fubdivided
their horary motions into equal fmaller parts, at laft they defcended to the
defignation of each ftep in the progrefs of Time, *i.e.* to the computation
even of Minutes and Seconds.

If any yet doubt (which we cannot fuppofe, without implicite fcandal)
of the *Independence* of *Time* on *Cæleftial Motion*; or, that old *Chronos*
muft ftand ftill, in cafe the *Orbs* fhould make a Halt: we advife him feri-
oufly to perpend that *fupernatural Detenfion* of the Sun in the day of battle
betwixt the Ifraelite and the Amorrhite; affuring our felves that his
thoughts will foon light upon this *Apodictical Argument*. Either there
was no Time during the *Ceffation* of the Suns motion on that day; or
elfe Time kept on its conftant flux: for one of thefe pofitions muft be true.
That the *Firft* is *falfe*, is manifeft from the extraordinary *Duration* of the
day, the *Text* pofitively expreffing, *that no day was ere*, *nor fhould be fo long
as that*; and the word *Long* undeniably importing a *Continued flux* of
time: *Ergo*, the *fecond* muft be moft *true*; and upon Confequence,
though the *Detention* of the Sun was *miraculous*, yet was the *Duration*
of the day *Natural*, becaufe Time hath no dependence on Cæleftial
Motion.

*Art.* 4.
A *demonftration*
of the inde-
pendence of
Time upon
Motion, from
the miraculous
*Detention of the
Sun*, above the
Horifon, in the
days of *Jofhua*.

Nor do they at all infirm the news of this *Dilemma*, who *object*; *that
there was then no Time*, *becaufe there were no Hours*: fince Hours are no
more Effential to Time then *Spring, Summer, Autumn*, and *Winter*, which
are only fucceffive mutations of the temperament of Aer, convenient to
the confervation and promotion of feminalities; and as for *Dayes*, they
likewife are abfolute Aliens to Time, fince while our Hemifphere enjoyes
the illumination of the Sun, the fubterraneous one wants it; and fo our
day is night to the Antipodes inhabiting the oppofite part of the Globe
Terreftrial; but Time is conftantly the fame through the Univerfe. Befides, there were Hours during the arreft of *Don Phœbus*; in this refpect,
that the fpace of Time, in which he ftood ftill, was defignable by the
flux of Hour-glaffes, or any other Temporary Machine: nor ought we to
fay, there are no hours but thofe which we commenfurate. And therefore, we incur no *Solœcifm* when we fay, that God, had it feemed good in
the eye of his Wifdom, might have created the World many thoufands of
millions of *years* fooner then He did: becaufe fuch was the præcedent Flux
of Time as might be computed by Spaces of Duration in longitude refpondent to that determinate fpace of Time, which the Sun in its progrefs through the Zodiack annually doth fulfill; not that before the Creation, there were real years, diftinct and defined by the repeated Converfi-
ons of the Sun.

*Art.* 5.
An Objection,
that, during
the arreft of
the Sun, there
was no *Time*,
becaufe no
Hours; fatisfi-
ed.

Further, As Time hath no Dependence on, fo can it receive no
*Mutation* from Motion. *Ariftotle*, indeed, accufeth it of *Mutability*,
merely becaufe we ufe to connect that Time in which we fall afleep,
to

*Art.* 6.
The *Immutabi-
lity* of Time
alfo afferted,
againft *Arifto-
tle*.

to that in which we awake, lofing that of which the ceffation of
our fenfes operation makes us infenfible : But alas! this looks like
too weak a conceit to be the mature iffue of fo ftrong a brain as His;
infomuch as albeit we concede fome Mutations to be neceffary, as
to our *perception* of the flux of Time, yet doth it not follow,
that therefore thofe Mutations are neceffary; as to the *Flux of Time
it felf.* True it is alfo, that we ufe to meafure various Mutations
by Time: but if we examine the matter profoundly, we fhall ani-
madvert, that the Time, during which thofe Mutations laft, is rather meafured by Motion then the contrary; for though that motion be not obferved in the Heavens, yet may it be æquivalent indicated by Hour-Glaffes, or any other *Chronodix.* Which *Ariftotle* himfelf feems to acknowledge ( in 1 2. *de Cælo* ) when He
affirms, *that as Motion may be meafured by Time, fo may time by
Motion.*

---

## SECT. III.

*Art.* 1.
The Grand
Quæftion, con-
cerning the
*Difparity* of
*Time* and *Eter-
nity;* ftate I.

IF Time be, as our Defcription imports, *Non-principiate* and *Infinite :* how can we *Difcriminate* it from *Æternity?* Should we
refolve, that *Æternity,* in the ears of an unpræjudicate underftanding, founds no more then PERPETUAL DURATION,
or Time that never knew beginning, nor can ever know an end :
we are inftantly affaulted with this Difficulty; that *Time* hath Dimenfions fucceffive, comprehends Priority and Pofteriority of parts,
and effentially confifteth in a certain perpetual Flux; but *Eternity*
is radicated in one permanent point, falls under none but the Præfent Tenfe, and is only a certain conftant *ὃ νῦ,* or intranfible NOW;
or, as *Boætius* defines it, *Interminabilis vitæ tota fimul & perfecta
poffefsio,* an interminable and perfect poffeffion of life altogether,
*i. e.* without præterite and future, or, *Forever at once.* To extricate
our felves from this feeming Confufion of two things, whofe Natures
appear fo irreconcileably difparate; we are to begin at two prævious Con-
fiderables.

*Art.* 2.
Two præpara-
tory Confide-
rations, touch-
ant the æquivo-
cal ufe of the
word *Æterni-
ty;* requifite to
the clear folu-
tion thereof.

(1) That *Plato* ( out of whofe *Timæus* that eminent Definition of
*Boætius* was extracted, which hath received the approbation and prai-
fes of moft of our Ecclefiaftick Patriarchs ) afferting his opinion,
that *Immutable and Eternal Natures are not fubject to Time*; to which A-
*riftotle* alfo affented; doth not intend the word, *Æternity,* abftractly
and præcifely, to fignifie a fpecies of *Duration*: but Concretely, for
*fomething whofe Duration is Eternal,* viz. the *Divine Subftance,* which He
otherwife calls, the *Soul of the World.* This may be, without violence or
finifter perverfion, collected from hence, that He diflikes the incongruous
conference of both and either of thofe Tenfes, *Fuit* and *Erit,* as well
upon *Eternity* or interminable Duration, abftractly confidered; as
ὅτι τὴν αἰδίον οὐσίαν, upon the *Eternal Subftance.* And *Plotinus*
( *En.* 3. *lib.* 7. *cap.* 1. ) more then once exprefly declares as much:
and moft ingenioufly infinuates the fame both when He derives the
word

word Æternity, ἡ αἰῶνα, ἀπὸ τὸ ἀεὶ ὀνΘ., *ab eo quod semper est*; and when he excludes all real Alterity, or difference from τὸ ὄν, *quod est*, and τὸ ἀεὶ ὄν, *quod semper est*, importing that *Is* and *Eternity* are *Identical*.

(2) That when *Plato* denieth the Congruity of *Preterite* and *Future*, but allowes that of the *Present* Tense, or *Est*, to the *Eternal Substance*; He only aims at this, that, saying of the Eternal Substance, *Fuit*, *it hath been*, we do not understand it the same with *Non amplius est*, *it is no more*; and also when we say of it, *Erit*, *it shall be*, we do not understand it as *Nondum est*, *it is not yet*: but not that *Fuit* is incompetent to the Eternal Substance, provided we intend that it doth now continue to be the same it ever hath been, nor *Erit*, while we conceive it shall be to all Eternity the same, that it ever hath been, and now is. It being manifest from the Syntax and purport of all his Dialogue, that his cardinal scope was only to prævent the dangerous adscription of those *temporary Mutations* to the *Eternal Being*, which are properly incident to *Generable* and *Corruptible* Natures: and to demonstrate, that we ought to conceive God, ἔτε πρεσβύτερον, ὅτε νεώτερον, *neque seniorem*, *neque juniorem*. In a word, *Plato* doth judge, that the Tense *Est* is proper only to the *Divine Nature*, because it is ever the same, or invariably possesseth the same perfections; nor is there any moment in the vast amplitude of Eternity, wherein it can be justly said, Now it hath some Attribute, which it had not formerly, or which it shall not have in the future: since the progress of Time can neither add any thing unto, nor detract any thing from it, as it doth to other Natures, that are obnoxious to mutation; so that God may well be called, in *Plato's* Phrase, ἔχων ἀκινήτος, *Habens se immobiliter*.

These remora's of ambiguity removed, we may uninterruptedly advance to inference, and without further hæsitancy determine, (1) That when Æternity is said to be, *Quidpiam totum simul*, something wanting succession or flux of parts, as in the memorable Definition of *Boetius*, then is it to be accepted, not *abstractly* for *Duration*, but *Concretely* for the *Divine Substance*, whose Duration is sempiternal. (2) That *Time* and *Eternity* differ each from other, in no other respect, then that Eternity is an infinite Duration, and Time (according to the Vulgar intent of the word) a certain *part* of that infinite Duration, commensing at the Creation, and determining at the Dissolution of the World.

This *Cicero* rightly apprehended, and emphatically expressed, in his sentence, *Tempus est pars quadam Æternitatis*, *cum alicujus anni*, *menstrui*, *diurni*, *nocturnive spatii certa significatione*. In this respect, Eternity is said to be Duration *Non-principiate* and *Interminable*; which is proper only to *God*: and *Time* is said to be Duration *Principiate* and *Terminable*; which is competent to all *Caduce*, *Mutable*, and *Corruptible* Natures: as also that part of Eternity, which the *Neotericks* by a special idiome name *Ævum*, is Duration *Principiate*, but *Interminable*, which is adscriptive to *Angeleal* or *Intellectual* Natures: and

to

to the *Rational Soul* of man; for thus we understand that frequent Bipar-
tition of *Eternity* into *à parte ante*, & *à parte post*, invented by the
Schoolmen.

*Art. 4.*
The *Platonicks*
Definition of
Eternity, to be
one Everlasting
*Now*; not in-
relligible, and
therefore col-
lusive.

These Positions being indisputable, the remaining subject of our præsent
Disquisition, is only *Whether the Platonicks spake rationally and intelligi-*
*bly, when they defined Eternity to be one everlasting* NOW, *or a Duration*
*void of succession, or flux of parts?*

Concerning this grand *Doubt*, we profess, would Truth have con-
nived, we could most willingly have past it by untoucht; because
most of our *Christian Doctors* have fully assented unto them in this
particular : but, since the convulsion of this their opinion doth stagger no Principle of Faith, or Canonical Document made sacred
and established by the Authority of the Church; we shall not deserve Excommunication, nor suffer the expurgatory Spunge of *Rome*,
if we quæstion the Congruity of that Definition, and affirm that
*No man can understand it.* For, what Wit is so acute and sublime, as to conceive, that a thing can have *Duration*, and that
Duration can be as a *point* without *Fusion* and *Continuation* from
one moment to another, by intervenient or mediate moments? Easie enough, we confess, it is to conceive, that the *Res durans* is altogether at once, or doth retain the sameness of its Nature, without
mutation, diminution, or amission of any Perfection : but that, in
this *Perseveration*, there is not many *Nows*, or many *Instants*, of
which, compared among themselves, some are *Antecedent*, and others
*Consequent*; is to us absolutely incomprehensible.

Nor can we understand, why it may not be good Christian Phrase,
to say; God *WAS* in the time of the First Man, and *SHALL*
be in the time of the Last : or why it is not more Grammatical and
proper for us to say, God Created the World HERETOFORE,
and will both destroy and renovate the World HEREAF-
TER; then, that God doth NOW Create, destroy and renovate.

*Art. 5.*
Their *Assertors*
subterfuge,
that Eternity is
*Coexistent* to
*Time*, also un-
intelligible.

To this the Common *Answer* is, that the Reason why these
Anthropopathical Phrases are tolerable, is because *Eternity is Coexi-*
*stent to our Time :* but this is *Ignotum explanare per ignotius;* for
the manner of that supposed *Coexistence* hath been never explained,
and seemeth laid by till the advent of Elias. That an *Instant*, i. e. what
wants succession, can be Coexistent to a *successive thing* ; is as manifest an impossibility, as that a *Point*, i. e. what wants Longitude,
can be Coexistent or Coextensive to a *Line.* Indeed, They have endeavoured to wave the Difficulty, by subnecting, that the Instant
of Eternity is of such peculiar *Eminency*, as that it is Æquivolent to
Time though Successive : But as to the Formal Reason, and manner
of this peculiar Eminency, they have left it wholly to our Enquiry also. Nor did they bestow one serious thought upon the consideration of it; for had they, doubtless they must have found their
Wit at a loss in the Labyrinth of Fancy, and perceived themselves reduced to this Exigent : either that they had fooled themselves in
                                        trifling

trifling with words not well underſtood; or that they had præcariouſly
uſurped the Queſtion; or that the ſame Inſtants are in Eternity, that are
in our Time, but with ſuch Eminency, that infinitely *more* are contained
in Eternity, then in our Time.    How much better were it ſaid, that
we are Coexiſtent with God; or, that we are exiſtent in a ſmall
part of that Duration, in which God infinitely exiſteth ?  For, while
we are, certainly, we cannot imagine *Two* diſtinct Durations; but
*one*, which in reſpect to our Nature, that is principiate, mutable, and
terminable, doth contain deſignable Terms; and in reſpect of the Divine Nature, which is nonprincipiate, immutable, interminable, hath
its Diffuſion or Extenſion infinitely long before, and as long after us.
This may receive ample juſtification from that ſpeech of the *Hebrew*
*Poet*, whoſe Inſpirer was the Holy Ghoſt, ( *Pſal.* 101. ) *Thou ſhalt*
*Change them and they ſhall be changed; but thou*, O *God* ! *art the ſame*
*forever, and thy years ſhall not fail.*   For here YEARS are attributed to God, but not any mutation of Subſtance : ſo that when our
years are exhauſted, in a ſhort, or ſpan-like flux of Time, the Glaſs of
His Duration is alwayes full.  And, therefore, the Expreſſion is only
Tropological, when it is ſaid, *that the years of our life make but*
*a Day in the Almanack of Divinity :* for the life of the *Hemerobii*
compared to ours of threeſcore years and ten, holds ſome proportion; but the life of *Methuſalem*, compared to the Duration of the
Life of our lives, the *Divine Eſſence*, holds none at all.    Upon
this conſideration, it was more then a Heathen obſervation of *Plu-*
*tarch ( in Conſolat. ad Apollon. )* that there is no difference betwixt a long and a brief time, in reſpect of *Eternity :* ſince, as
*Simonides*, a thouſand, nay a million of years make but a point,
nor ſo much as the leaſt part of a point in the line of infinite Duration.

Convicted thus by Reaſon, our *Doctors* convert to *Scripture*,
urging that God (*Exod.* 3.) indicates his Beeing only in the *Præ-*
*ſent* Tenſe, as peculiar to his Eternity, ſaying, *I am*, *that I*
*am*, *and I am hath ſent thee* to Moſes.    But this Objection admits of a threefold evaſion.   ( 1 ) The *Hebrew* Text doth not, in
that place, uſe the Præſent, but the *Future* Tenſe, *I ſhall be*, *what*
*I ſhall be, and I ſhall be hath ſent thee.*   ( 2 ) We can oppoſe
many other Texts, which adſcribe to God as well Præterite and
Future, as Præſent time; and moſt eminently in the *Revelation*, He
is deſcribed, ὁ ὢν, ἦ ὁ ἦν, ἦ ὁ ἐρχόμενος, *He that is*, *and was*,
*and is to come.*   ( 3 ) God Himſelf doth frequently enunciate many
actions, not that He now doth, but that He hath formerly done,
and will do in the future, in that moment of opportunity, which His
Wiſdom hath prædetermined.  Hence alſo expulſed, They fly to their
laſt fortreſs, *viz.*

*If Eternity be not one permanet Now, then cannot all things be*
*præſent to God, objectively.*   But vain is their hope of ſecurity in
this alſo.    For, many things, if we reſpect the *when* of their exiſtence, have already been, and as many are not yet; but, becauſe
the Omniſcience of God pervades as well the darkneſs of *paſt*, as of
*præſent*

                        M

*Art. 6.*
Our Eccleſia-
ſtick *Doctors*,
taking San-
ctuary in the
3 *Exod.* for the
authorizing of
their doctrine,
that the Præ-
ſent Tenſe is
only compe-
tent to God,
and ſo that E-
ternity is one
permanent In-
ſtant, without
Fuſion or Suc-
ceſſion : not
ſecure from
the rigour of
our Demon-
ſtration.

*Art. 7.*
The *Objective*
*Præſence of all*
*things at once,*
to the Divine
In ellect, no
wayes impug-
ned by our
contradiction
of the *Doctors*
theory.

*præsent* Time, and alwayes speculates all things most clearly and distinctly: therefore do we say, that all things are objects to His Opticks, or, that all things are præsent to His Cognition; not that He knows, all things to be præsent at once altogether, but that He hath before Him at once all the diversities of Times, and as perfectly contemplates them Future and Præterite, as Præsent. For, the *Divine* Intellect doth not apprehend Objects, as the *Humane*, one after another, or in a successive and syntactical series; but grasps all things *together* in one entire act of Cognition, and comprehends in one simple intuition whatever hath been, or may be known. And, therefore, our opinion is not at all impugned by that sacred sentence; *All things are open and naked to His eyes, and He calls upon those things, that are not, as if they were*. Hereupon some have, with unpardonable temerity and incogitancy, inferred; that ONCE there was no Time; for in this their very denial, they openly confess, that Time hath ever been: it being all one as if they had said, *There was a Time when there was no Time.*

Lastly, as the Omniscience of God cannot be indubitated by our persuasion of the *Identity of Eternity and Time*, so neither can His *Immutability*, as *Aristotle* would have it, only for this Reason (forsooth) that Time, or that Duration, which hath successive, and so prior and posterior parts, is the General *Cause* of *Corruption*. For, our præcedent Discourse hath left no room for the intrusion of that futile Objection; insomuch as it rather commonstrateth the *Divine* Nature to be so Constant and Perfect, that in the eternal flux of Time it can know nothing of Innovation or Corruption. Besides, Time, or the succession of Duration, is not the Cause, that induceth Corruption: but the *Native Imbecillity* of compound Natures, invaded and subdued by some *Contrary Agent*; and God is a *Pure, Simple, Homogeneous* substance, and so not subject to the invasion of any Contrary. Evident it is, therefore, that *Aristotle*, when He urged this Sophism, spoke more like a Poet, then a Philosopher; since Poets only use to give Time the Epithite of *Edax rerum:* nor could He be so absurd, as to dream, that Time was a vast Animal, with sharp teeth, an insatiate appetite, and a belly inexplebile, or an old man armed with a Sithe, as the Poets describe *Saturn,* making κρόνος and χρόνος, *Saturn* and *Time* one and the same thing. For, Time really doth neither Eat not Mow down any thing; and the *Dissolution* of all Create compound Natures can be imputed to no other Cause, but the Domestick Hostility of their Heterogenieties, or the uncessant intestine warr of their Elements, from whose commixture their Compositions, or Concretions did first result. With this qualification, therefore, we are not angry at that of *Periander*, in *Stobæus, Tempus est Causa omnium rerum:* because in the process of Time all things have their origin, state, and declination. In this restrained sense we also tolerate the saying of *Thales Milesius*, quoted by *Laertius, Tempus est sapientisimum:* since Time produceth Experience, and Experience Prudence. And that Antitheton of *Pharon* the *Pythagorean*, recited by *Aristotle, Tempus est Ineruditisimum:* because in process of Time the Memory of all things is obliterated, and so *oblivion* may well be called the *Hand-maid of Time*, that perpetually follows at the heels of her Mistriss.

Our

Our *Clue* of thoughts concerning *Time* is now wholly unravelled; *Art. 9.*
and though we may not præsume, that we have therewith led the mind *Coronis.*
of our Reader through all the mysteries of its Nature: yet may we hope,
that it may serve as a conduct to those, who have a more ample stock of
Learning and Perspicacity for the support and encouragement of their
Curiosity; at least that the Attentive and Judicious may easily collect
from thence, that we have, upon no Interest but that main one of Verity, withdrawn our assent from the common Doctrine of the Schools,
that Eternity is one permanent Now, without Succession, or Priority
and Posteriority of Moments.

# The Second Book.

## CHAP. I.

### *The Existence of Atoms, Evicted.*

#### Sect. I.

*Art.* I.
The right of
the Authors
Transition
from the *Incorporeal* to the
*Corporeal* part
of Nature: and
a series of his
subsequent
speculations:

A Mong infinite other hypochondriack Conceits of the *Teutonick* (rather, *Fanatique*) Philosophers, they frequently adscribe a *Dark*, and a *Light* side to God; determining the Essence of *Hell* in the one, and that of *Heaven* in the other. Whether the expression be proper and decent enough to be tolerated; requires the arbitration of only a mean and vulgar judgment. We shall only affirm, that had they accommodated the same to the shadow, or Vicegerent General of God, to *Nature*; their Dialect had been, as more familiar to our capacity, so more worthy our imitation. For, that the INCORPOREAL, and therefore Invisible part of the Universe, the *Inane Space*, may bear the name of the DARK; and the CORPOREAL and visible part of the LUMINOUS side of Nature: seems consentaneous to reason. On the *First*, hath the eye of our Mind been thus long levelled; taking in by collateral and digressive glances the Essential Proprieties of *Place* and *Time*; the one of which is absolutely *Identical*, the other perfectly *Analogous* to *Inanity*: on the *other* we are now to convert it, and with more then common attention, therein to speculate the Catholique *Principles, Motions* and *Mutations*, or *Generation* and *Corruption* of BODIES.

All

All Bodies, by an univerſal Diſtinction, are either (1) τὰ ἐξ ὧν αἱ
Συγκρίσεις, ſuch, from the convention and coalition of which all Concretions reſult; familiarly called by Phyſiologiſts, *Principia, Primordia,
Componentia,* but moſt commonly, *Elements,* and *Materia Prima.* Or
(2) τὰ Συγκρίματα, ſuch as conſiſt of the former coacervated, and coaleſced: or ſuch as are compoſed of many ſingle particles Component.
The *Former* were made by *Creation,* and are ſuperiour to Corruption:
the *Later* are produced by *Generation,* and reducible by Corruption. The
*Firſt* are *Simple* and *Originary*; ſuch as *Plato* intends (*in Phædro*) when he
ſaith, *Principii nullam eſſe originem, quoniam ex ipſo principio oriuntur omnia*: the *other, Compound* and *Secondary*; ſuch as *Lucretius* (*lib.* 1.) underſtands by his *Concilio qua conſtant principiorum.*

*Art.* 2.
*Bodies general-
ly diſtinguiſh-
ed into Prin-
ciples* and *Pra-
duſtions*, with
their Scholaſtick Denomi-
nations and
proprieties.

What theſe *Firſt, Simple, Ingenerable, Incorruptible, Univerſally Component Bodies* are, or to ſpeak in the Dialect of the Vulgar, What is the *General Matter* of all *Concretions* (it is no ſolœciſm in Phyſiology, to transfer
a word abſtractly importing a Natural Action upon the thing produced by that action) hath been by more Diſputed, then Determined, in all Academies. That there muſt be ſome one *Catholique
Material Principle*, of which all Concrete Subſtances are compoſed; and into which they are again, at length by Corruption reſolved:
is unanimouſly confeſſed by all. And, conſequently, that this Matter is
*Incorruptible*, or the Term wherein all Diſſolution ceaſeth; hath been indubitated by none, but thoſe, who, upon a confuſion of Geometrical with
Phyſical Maxims, run upon the point of that dangerous Abſurdity, *that
the infinite diviſion of a real Continuum is poſſible.* Inſomuch therefore, as
the Eſſential reaſon or Formality of *Corporiety* doth ſolely conſiſt in *Exten-
ſibility*, or the Dimenſions of Longitude, Latitude, and Profundity real;
as our *Third Chapter* præcedent hath demonſtrated, and as the Patriarch of
the Schools doth expreſly confeſs (*Natur. Auſcult.* 4. *cap.* 3.) and inſomuch as nothing can be the *Root* or beginning of Material or Phyſical
Extenſion, but τὶ ἀδιαίρετον, *Aliquid indiſſolubile*, ſomething ſo minute
and ſolid, that nothing can be conceived more exiguous and impatible in
Nature (for, as the *Radix* of *Mathematick*, or Imaginary Continuity, is
a *Point*: ſo muſt that of *Phyſical* or ſenſible Continuity be a *Body of the
ſmalleſt Quantity*) ſuch as are the ATOMS of *Democritus, Epicurus,*
and other their *Sectators*; and the *Inſenſible Particles* of *Carteſius*: therefore, from manifeſt neceſſity, may we determine, *that no Principle can juſtly challenge all the Proprieties, or Attributes of the Firſt Univerſal Matter,
but* Σώματα ἀδιαίρετα, *Indiviſible Bodies*, or *Atoms.* Which fundamental Poſition clearly to eſtabliſh by demonſtration; is a chief part of our
difficult Province: having, for method and prevention of obſcurity, firſt
briefly inſiſted upon their various *Appellations*, with the Etymological
relation of each, traced them up to their πηγὴ, or *Invention*, and evicted
their *Exiſtence.*

*Art.* 3.
The right of
*Atoms* to the
Attributes of
the *Firſt Mat-
ter.*

(1) As for their various DENOMINATIONS; they natu-
rally reduce themſelves to *three General Imports,* bearing a congruous and
emphatick reſpect to their three moſt eminent *Proprieties.* For,

*Art.* 4.
Their ſundry
*Appellations* al-
luſive to their
three eminent
proprieties.

(1) In relation to their *Corporiety,* they are called, τὰ Σώματα, *Bodies*, by way of tranſcendency: becauſe they are ἀμετοχὴ κενᾶ,
devoyd

devoyd of all *Incorporiety*, *i.e.* they contain nothing of *Inanity*, as do all Concretions emergent from them, there being in all Compound Bodies more or lefs of Inanity diffeminate among their particles. For which reafon, they are alfo named, πληρη, *Plena.*

(2) In regard of their affording *Matter* to all Concretions, they are denominated, Ἀρχαι, *Principles*, στοιχεῖα, *Elements*, Πρῶτα σωματα, *First Bodies*, Πρῶτα Μεγεθη, *First Magnitudes*, τῶν οντων ὑλη, the *Matter of all things*, and Πανσπερμια, *Genitalia femina rerum*, the feminaries of all productions: becaufe all material things are compofed of them. In which concern alfo, by a Pythagorical Epithite, they are ftiled, Μοναδες, *Unities*; becaufe, as all Numbers arife from Unities, fo all Compofitions from them.

(3) To denote their *Indiffolubility*, they are moft frequently known by the term, Ἀτομα, and Ἀτομοι, *Atoms*; either becaufe they are incapable of *Section*, as *Ifodor, Plutarch, Servius, Budæus, Scapula*, &c. or δια την ἀλλων σεμνοτητα, *ob indiffolubilem foliditatem*, for their indiffoluble folidity. For, all Concrete Bodies, infomuch as they came fhort of abfolute folidity, having fomewhat of Inanity intermixt, may be divided, and fubdivided until their ultimate refolution into thefe, their component parts: but Atoms admit of no divifion below themfelves. Wherefore they are ufually chriftned, ἀδιαιρετα, ἀμερη, *Individual, Insectile, Impartible*; as likewife, ἀοραζα, λογω θεωρητα, *Invisible*, and *by the mind only perceptible*, Bodies, *i. e.* fo exile as no man can conceive a real Exility beyond theirs.

*Art.5.*
Two vulgarly paffant Derivations of the word, *Atom*, exploded.

Hence are we affured, that *Two* vulgarly paffant *Derivations* of the word, *Atome*, are ingenuine and extorted. (1) That of *Hefychius*, with too much femblance of approbation mentioned by the Reviver of the great Democritus, *Magnenus*, (*de Atom. difput. 2. cap. 2.*) which would have it a fprigg of that root, Ἀτμος, *Fumus*; becaufe (forfooth) from all bodies, in their reverfion from mixtion to diffolution, their Elements difperfe by *Exhalation*: as if this Etymologie were fo adæquate and important, as to compenfate the defect of an *omicron*, in the fecond fyllable. (2) That embraced not only by many pædantique *Grammarians*, but even acute *Philologers*, who interpret the word *Atomus* to fignifie a *Defect of Parts*; as if an Atom were deftitute of all *Magnitude*, or no other then a mere Mathematical Point: when, indeed, the *Nomenclator* had his eye fixt only on their *Solidity*, Hardnefs, or Impatibility, which is fuch, as excludes all poffibility of Fraction, Section, Divifion. Thus much *Epicurus* himfelf expreffeth, in moft perfp cuous and unpervertible terms (*apud Plutarch.* 1. *placit.* 3. ) thus, *Dicitur Atomus, non quod minima fit, vel inftar puncti, fed quod non pofsit dividi; cùm fit patiendi incapax, & inanis expers.* And *Galen* (1 *de Elem.*) recounting their doctrine, who affirmed the Principles of all Bodies to be Atoms, faith of *Epicurus, Fecit Atomas,* ἀθραυστους ὑπο σκληροτητος, He made them Infrangible in refpect to their *folidity.*

(2) Concerning their INVENTION; if we reflect upon them as *in Re*, before their reception of any constant Denomination; we have the tradition not only of *Possidonius* the Stoick, related by *Empiricus* (*adverf. Physic. lib. ib.*) but also of *Strabo*, to assure the honour thereof upon one *Moschus*, a *Phaenician*, who flourished not long before the ruine of *Troy* by the Graecians. Allowing this for Authentique, we have some cause to judge *Magnenus* to have been too favourable to his Grand Master, *Democritus*, when (*in testimon. de Democrito. pag. 32.*) He enricheth his Panegyrick of him with, *Effluvia Corporum Atomosque comperit, & invexit omnium primus : ex Laertio quod unum tanti apud me est, ut congestas omnium Philosophorum laudes vel exaequet vel superet.* Besides, to do *Laertius* right, He finds *Leucippus*, not *Democritus*, to have been the Founder of this incomparable Hypothesis : as his records lye open to testifie (*in vita Leucippi.*) But, if we reflect upon them only as *in Nomine*, enquiring who was their Godfather, that imposed the most convenient name, *Atoms*, upon them; we need not any more ancient, or faithful monuments to silence all competition about that honour, then those of *Theodoret* : who rightly sets the Laurel on the deserving front of *Epicurus*, in this text ; Ἐπίκουρος ὁ Νεοκλέους, 'ῷ ὑπ' ἐκείνων νεφη, κ̀ ἀδιαίρετα δὴ κληθεῖς, Ἄτομα προσηγόρευσεν; *Epicurus, Neoclis filius*, dicta illis (meaning *Democritus* and *Metrodorus Chius*) *Nasta & Adiareta, appellavit Atomos.* We are not ignorant, that *Sidonius Apollinaris* (*carmin. 15.*) adscribes the imposition of this name, to *Archelaus* in these Verses :

> *Post hos, Archelaus divina, mente paratam*
> *Concipit hanc molem, confectam partibus illis,*
> *Quos Atomos vocat, ipse leveis, &c.*

But how unjustly, even S. *Augustine* (8. *de Civit. Dei, cap* 3.) sufficiently declares; saying, that *Archelaus* deduced all things, *non ex Atomis, sed ex Particulis dissimilibus.* And therefore, though we may not file up the first Discovery of this noble Principle, Atoms (of all others, hitherto excogitated, the most verisimilous, because most sufficient to the solution of all Natures Phaenomena) among those many benefits, which the Commonweal of Philosophy owes to the bounteous Wit of *Epicurus* : yet hath his sagacity in accommodating them with so perfectly congruous an Appellation, and successful industry in advancing and refining their Theory, in the General, worthily entituled him to the homage of a grateful Estimation equal to that, which the merit of their *Inventor* claims.

(3) Concerning their EXISTENCE; that there are such Things, as Atoms, or Insectile Bodies, *in Rerum Natura*; cannot be long doubted by any judicious man, who shall thus reason with himself.

( 1 ) *Nature can produce Nothing out of Nothing ; nor reduce any thing to Nothing :* is an Axiome, whose tranquility was never yet disturbed ; no not by those who have invaded the Certitude of even First Notions, and accused Geometry of delusion. If so ; there must be some *Common Stock*, or an Universal Something, *Ingenerable*, and *Incorruptible* , of which being praeexistent, all things are Generated, and into which being

being indiſſoluble, all things are, at the period of their duration, again reſolved.

That Nature doth diſſolve Bodies into exceeding minute, or inſenſible particles ; Her ſelf doth undeniably manifeſt, as well in the *Nutrition* of *Animate* (their Aliment being volatilized into ſo many inſenſible particles, as thoſe whereof the Body nouriſhed doth conſiſt ; otherwiſe there could be no General Appoſition, Accretion, Aſſimilation ) as the *Incineration* of *Dead* Bodies. Which ground *Des Cartes* rightly apprehended to be ſo firm and evident, that he thought the exiſtence of his Inſenſible Particles ſufficiently demonſtrable from thence. *Quis dubitare poteſt* (ſaith He) *quin multa Corpora ſint tam minuta, ut ea nullo ſenſu deprehendamus, ſi tantum conſideret, quidnam ſingulis horis adjiciatur iis quæ lentè augentur, vel quid detrahatur ex iis quæ ſenſim minuuntur ? Creſcit enim arbor quotidiè, nec poteſt intelligi majorem illam reddi quam prius ſuit, niſi ſimul intelligatur aliquod corpus eidem adjungi. Quis autèm unquam ſenſu deprehenderit, quanam ſint illa corpuſcula, quæ in una die arbori creſcenti acceſſerunt, &c.* (*princip. Philoſ. part.4. articul.* 201.)

That ſhe cannot in her Diſſolution of Bodies, proceed to *Infinity,* but muſt conſiſt in ſome *definite Term,* or *extreme,* the loweſt of *Phyſical Quantity* ; is demonſtrable from hence, that *every real Magnitude is uncapable of interminable Diviſion.* For, ſince to an infinite proceſs is required an infinite Time ; ſhe could never Generate any thing New, becauſe the old would require an infinite time and proceſs to their Diſſolution. Convicted by this apodictical Argument, *Ariſtotle* (1 *Phyſ.* 9. ) deteſting the odious Abſurdity of (εἰς ἄπειρον εἶναι) running on to Infinity ; ſolemnly concludes (ἀνάγκη δὴ ϛῆναι) that there muſt be an Extreme Matter, wherein all Exolution is terminated : only herein He recedes from the ſuppoſition of *Democritus, Epicurus,* and other Patrons of the ſame Doctrine, that they terminated all Exolution in the *Inſectility* of *Atoms* ; but He deſcribes no ſuch Extreme, or point of Conſiſtence, his *Materia Prima* being ſtated rather *Potential,* then *Actual,* and abſolutely devoid of all *Quantity* ; then which we know no more open and inexcuſable a *Contradiction.* Again, if the Exolution of Bodies were not Definite, and that Nature knowing no *ne ultra,* did progreſs to *Adnihilation* : then muſt it inevitably follow, that the Matter of all things, that have been formerly, is totally Adnihilated ; and the matter of all things now Exiſtent, was educed out of Nothing. Two moſt intolerable Abſurdities ; ſince *Adnihilation* and *Creation* are terms not to be found in the Dictionary of *Nature,* but proper only to *Omnipotence :* nor is there any ſober man, who doth not underſtand the Common Material of Things to be conſtantly the *ſame,* through the whole flux of Time, or the duration of the World ; ſo as that from the Creation thereof by the *Fiat* of God, no one particle of it can periſh, or vaniſh into Nothing, until the total Diſſolution of Nature, by the ſame Metaphyſical power ; nor any one particle of new matter be ſuperadded thereto, without miracle. The Energy of Nature is definite and præſcribed : nor is ſhe Commiſſioned with any other Efficacy, then what extends to the moulding of *Old Matter* into *New Figures* ; and ſo, the nobleſt Attribute we can allow her, is that of a *Tranſlator.*

Now, to extract the ſpirit of all this, ſince there muſt be an Extreme, or ultimate

ultimate Term of Exolubility, beyond which can be progress; since this Term can be conceived no other but the lowest degree of Physical Quantity; and since, beyond the Insectility of Atoms, no Quantity Physical can be granted: what can the genuine Consequent be, but that *in Nature there are extremly minute Bodies*, Ἄτομα ᾗ ἀμετάβλητα, *Indivisible* and *Immutable*?

(2) For Confirmation; as in the Universe there is, *Aliquid Inane*, something so purely *Inane*, as that it is absolutely devoyd of all Corporiety: so also must there be *Aliquid Corporeum*, somewhat so purely *Corporeal*, or *solid*, as to be perfectly devoyd of all Inanity; to which peculiar solidity nothing but *Atoms*, in regard of their *Indivisibility*, can pretend: therefore is their *Existence* to be confessed. This Reason *Lucretius* most elegantly thus urgeth;

Art. 11. A second Argument of their Existence, drawn from that of their Antitheton, Inanity.

> *Tum porrò, si nil esset, quod* INANE *vacaret,*
> *Omne foret solidum; nisi contrà* CORPORA *cæca*
> *Essent, quæ loca complerent quæcunque tenerent,*
> *Omne quod est spatium, Vacuum constaret Inane, &c.*    Lib. 1.

(3) Evident it is to sense, that in the World are two sorts of Bodies, *Soft* and *Hard*; now, if we assume the Principles of all things to be exquisitely Hard, or Solid; then do we admit the production of not only Hard, but also of soft Bodies to be possible, because softness may arise to a Concretion of Hard Principles, from the Intermistion of Inanity: but, if we assume soft Principles, then do we exclude all possibility of the production of Hard Bodies, that Solidity, which is the Fundament of Hardness, being substracted: Therefore is the Concession of *Atoms* necessary.

Art. 12. A third, hinted from the impossibility of the Production of Hard Bodies, from any other Principle.

(4) Nature is perpetually *Constant* in all her specifical Operations, as in her Production and Promotion of Animals to the determinate periods of their Increment, Stature, Vigour, and Duration; and, more evidently, in the impression of those marks, whereby each species is discriminated from other. Now, to what Cause can this Her Constancy be, with greater probability, referred, then to this, that her *Materials* are *Certain, Constant*, and inobnoxious to Dissolution, and consequently to mutation: and such are *Atoms* præsumed to be? *Ergo*, they are *Existent*.

Art. 13. A Fourth, from the Constancy of Nature in the specification and Determinate Periods of her Generations.

N                    CHAP.

# CHAP. II.

## *No Physical Continuum, infinitely Divisible.*

### SECT. I.

*Art.* 1.
The Cognation of this Theorem, to the Argument of the immediately præcedent Chapter.

He Grand Base on which the whole Fabrick of the Atomists, *i. e.* our Physiology is supported, confesseth it self to be this; that Nature cannot extend her Dissolution of Bodies beyond τι ἀτομον καὶ ἀδιαίρετον, somewhat that is *Firm* and *Inexsoluble.* And the rock on which that adamantine Base is fixt, is soon understood to be this; that *the Parts of no Physical Continuum, or Magnitude, are subdivisible to Infinity.* The *Former,* we conceive so clearly comprobated by Reasons of evidence and certitude equal to that of the most perfect Demonstration in Geometry, that to suspect its admission for an impregnable Verity, by all, who have not, by a sacramental subscription of *Aristotles* Infallibility, abjured the ingenious Liberty of estimating Philosophical Fundaments more by the moments of Verisimility, then the specious Commendums of Authority, were no less then implicitely to disparage the Capacity of our Reader, by supposing Him an incompetent judge of their importance and validity. And that the *Other* is equally noble in its alliance to Truth, and so secure from subversion by the minds of the acutest Sophistry, that may oppose it; is the necessary Theorem of this præsent *Exercitation.*

*Art.* 2.
Magnitude divisible by a continued progress through parts either *Proportional,* or *Aliquotal.*

To usher in this Verity with the greater splendor, we are required to advertise

( 1 ) That Philosophers have instituted two distinct Methods, for the regular Division of Magnitude. For, their Divisions are continued by a progression through Parts either ( 1 ) PROPORTIONAL; which is when a Physical Continuum is divided into two parts, and each of those parts is subdivided again into two more, and each of those into two more; or when the whole of any magnitude is divided into 10 equal parts

parts, and each of thofe into 10 more, and each of thofe into 10 more, and fo forward, obferving the fame decimal proportions through the whole divifion: or (2) ALIQUOTAL; *i.e.* when a Continuum is divided into fuch parts, as being divers times repeated, are æquated to the whole, or into fo many parts as feem convenient to the Divifor, provided they hold equal proportions among themfelves, whether they be Miles, Furlongs, Fathoms, Feet, Digits, &c. Which Diftinction *Ariftotle* feems to allude unto, when he declares (*3.phyfic 7.*) that the Difference betwixt Magnitude and Number doth confift in this, that by the Divifion of *Numbers* we arrive at laft, ἐπὶ τὸ ἐλάχιςον, *ad Minimum*, at the *Leaft*; but of *Magnitude*, ἐπὶ τὸ ἐλάττον, *ad Minus*, only to a *Lefs*.

(2) That when *Democritus*, *Epicurus*, and other Ancients of the fame *Antiftoical* Faction, treating of the Divifion of Magnitude, determine it ἐπὶ τ ἐλάττον; they did chiefly intend that Methodical Divifion, which is made *in partes Proportionales*; infomuch as every part made by a fecond divifion muft be lefs then that made by a firft.

## The Demonftration.

*If in a Finite Body, the number of Parts, into which it may be divided, be not Finite alfo; then muft the Parts comprehended therein be really Infinite: and, upon Confequence, the whole Compofition refulting from their Commixture, be really Infinite; which is repugnant to the fuppofition.*

So perfectly Apodictical, and fo inoppugnably victorious, is this fingle Argument, that there needs no other to the juftification of our inftant Caufe: nor can the moft obftinate and refractory Champion of the Peripateticks, refufe to furrender his affent thereto, without being reduced to a moft difhonourable exigent. For, He muft allow either that the whole of any Body is fomething befides, or diftinct from the Aggeries, or Mafs of Parts, of which it is compofed: or, that all the Parts, together taken, are fomewhat greater then the whole amaffed by their convention and coalefcence. If fo, there muft be as many parts in a grain of Muftard feed, as in the whole Terreftrial Globe: fince in either is fuppofed an equal Inexhauribility; which is contrary to the Firft Notion of *Euclid*, *Totum eft majus fua parte*. And if any mans skull be fo foft, as to admit a durable impreffion of an opinion fo openly felf-contradictory, as this, that *the Whole is lefs then its Parts*; we judge him a fit Scholer for *Chryfippus*, who blufht not publiquely to affirm, that one drop of Wine was capable of commiftion with every particle of the Ocean, nay, diffufive enough to extend to an union with every particle of the Univerfe, were it 100000 times greater, then now it is. Nor, need we defpair to make him fwear, that *Arcefilas* did not jeer the Difciples of *Zeno*, when he exemplified the inexhauible divifion of Magnitude, in a mans Thigh, amputated, putrified, and caft into the Sea; ironically affirming the parts thereof fo infinitely fubdivifible, that it might be incorporated *per minimus*, to every particle of Water therein; and confequently, that not only *Antigonus* Navy might fail at large through the thigh, but also that *Xerxes* thousand two hundred ships might freely maintain a Naval fight with 300 Gallies of the Greeks, in the compass of its dispersed parts. We deny not, but *Zeno's* Argument against Motion, grounded on the supposition of interminable Partibility in Magnitude, is too hard and full of Knots, to be undone by the teeth of common reason: yet who hath been so superlatively stupid, as to prefer the mere plausibility thereof to the contrary Demonstration of his sense, and thereupon infer a belief, that there is no Motion in the World? What Credulity is there so easie, as to entertain a conceit, that one granule of sand (a thing of very small circumscription) doth contain so great a number of parts, as that it may be divided into a thousand millions of Myriads; and each of those parts be subdivided into a thousand millions of Myriads; and each of those be redivided into as many; and each of those into as many: so as that it is impossible, by multiplications of Divisions, ever to arrive at parts so extremely small, as that none can be smaller; though the subdivisions be repeated every moment, not only in an hour, a day, a month, or a year, but a thousand millions of Myriads of years? Or, What Hypochondriack hath been so wild in Phansie, as to conceive that the vast mass of the World may not be divided into more parts then the Foot of a Handworm, a thing so minute as if made only to experiment the perfection of an Engyscope? And yet this must not be granted, if we hearken to the spels of *Zeno* and the Stoicks; who contend for the Divisibility of every the smallest quantity into infinite parts: since, into how many parts soever the World be divided, as many are assumable in the Foot of a Handworm, the parts of this being no less inexhaustible, nor more terminable by any continued division, then the parts of that, according to the supposition of Infinitude. And, hereon may we safely conclude, that albeit the Arguments alledged in defence of Infinite Divisibility of every Physical Continuum, were (as not a few, nor obscure Clerks have reputed them) absolutely indissoluble: yet notwithstanding, since we have the plain Certificate of not only our Reason, but undeluded sense also to evidence the Contrary, ought we to more then suspect them of secret Fallacy and Collusion; it being a rule, worthy the reputation of a First Notion, that in the examination of those Physical Theorems, whose Verity, or Falsity is determinable by the sincere judicature of the sense, we ought to appeal to no other Criterion, but to acquiesce in the Certification thereof; especially where is no Refragation, or Dissent of Reason.

Notwithstanding the manifest necessity of this apodictical Truth, yet have there been many *Sophisms* framed, upon design to evade it : among which we find only *Two*, whose plausibility and popular approbation seem to præscribe them to our præsent notice.

*Art. 6.*
*Aristotles (subterfuge of Infinitude Potential.*

The *First* is that famous one of *Aristotle* (*de insecabil. lineis*) *Non creari propterea infinitum actu ex hujusmodi partibus infinitis, quoniam tales partes non actu, sed potestate duntaxat infinita sunt; adeo proinde ut creent solùm infinitum potestate, quod idem sit actu finitum:* that the division of a finite body into infinite parts doth not make it actually infinite, because the parts are not actually, but only potentially infinite; so as they render it infinitely divisible only potentially, while it still remains actually Finite.

The

The Collusion of this Distinction is not deeply concealed. For, every   *Art. 7.*
Continuum hath either *no* parts *in actu*, or *infinite* parts *in actu*. Since, if Found openly
by parts *in actu*, we understand those that are *actually divided*: then hath Collusive.
not any Continuum so much as two or three parts; the supposed *Conti-*
*nuity* excluding all *Division*. And if we intend, that a Continuum hath
therefore two parts actually, because it is *capable* of division into two parts
actually: then is it necessary, that we allow a Continuum to have parts
actually infinite, because we presume it capable of division into infinite
parts actually; which is contradictory to *Aristotle*. Nor can any of his
*Defendants* excuse the consequence by saying; that the Division is never
finishable, or terminable, and that his sense is only this, that no Continuum can ever be divided into so many parts, as that it may not be again
divided into more, and those by redivision into more, and so forward
without end. Since, as in a Continuum two parts are not denyed to exist,
though it be never divided into those two parts: so likewise are not infinite parts denied to exist therein, though it be never really divisible into
infinite parts. Otherwise, we demand, since by those requisite divisions
and subdivisions *usque ad infinitum*, still more and more actuall parts are
discovered; can you conceive those parts, which may be discovered to be
of any *Determinate Number*, or not? If you take the *Affirm.* then will
not there be parts enough to maintain the division to infinity: if the *Ne-*
*gat.* then must the parts be actually infinite. For, how can a Continuum
be superior to final exhaustion, unless in this respect, that it contains infinite
parts, *i.e.* such whose Infinity makes it Inexhaustible. Because, as those
parts, which are deduced from a Continuum, must be præexistent therein
before deduction (else whence are they deduceable?) so also must those,
which yet remain deduceable, be actually existent therein, otherwise they
are not deducible from it. For, Parts are then Infinite, when more and
more inexhaustibly, or without end, are conceded Deducible.

The other, with unpardonable confidence insisted on by the *Stoicks*, is   *Art. 8.*
this; *Continuum non evadere infinitum; quoniam illud proprie resultat non* A second sub-
*ex Proportionalibus, sed ex Aliquotis partibus, quas constat esse Definitas, cùm* terfuge of the
*inter extrema Corporis versentur:* that [by admitting an infinity of parts Stoick;
in a Finite Continuum] a Continuum doth not become infinite; because
that results properly not from *Proportional*, but *Aliquotal* parts, which are
therefore confess'd to be *Definite*, because they relate only to the *Extremes*
of a Body.

First, this subterfuge is a mere *Lusus Verborum*, sounding nought at all   *Art. 9.*
in the ears of Reason. For since every thing doth consist of those parts, Manifestly dis-
into which it may be at last resolved; because every Continuum is at last sentaneous to
resolved into, therefore must it consist of Proportional Parts. Again, since Reason.
every one of Aliquotal parts is Continuate, each of them may be divided
into as many Aliquotal parts, as the whole Continuum was first divided
into, and so upwards infinitely: so as at length the Division must revert into Proportional Parts, and the Difficulty remain the same.

SECT.

## Sect. II.

THe impoſſibility of Dividing a Phyſical Continuum into parts inter-
minably ſubdiviſible, being thus amply Demonſtrated; and the So-
phiſtry of the moſt ſpecious Receſſes, invented to aſſiſt the Contrary opinion, clearly detected: the reſidue of this Chapter belongs to our Vindi-
cation of the ſame Theſis from the guilt of thoſe *Abſurdities* and *Incongruities*, which the Diſſenting Faction hath charged upon it.

*Art. I.*
*The Abſurdi-*
*ties, by Empi-*
*ricus, charged*
*upon the ſup-*
*poſition of ſon-*
*ly Finite parts*
*in a Continu-*
*um.*

     *Empiricus,* with great Virulency of language inveighing againſt the Pa-
trons of Atoms, accuſeth them of ſubverting all Local Motion, by ſuppoſing that not only Place and Time, but alſo Natural Quantity indiviſible beyond Inſectile Parts. To make this the more credible, He Objects (1) That if we aſſume a Line, conſiſting of nine Inſectils, and imagine two inſectile Bodies to be moved, with equal velocity, from the oppoſite extremes thereof toward the middle; it muſt be, to their mutual
occurſe, and convention in the middle, neceſſary that both poſſeſs the median part of the median, or Fifth Inſectile place (there being no cauſe, why
one ſhould poſſeſs it more then the other) when yet both the Places and
Bodies therein moved, are præſumed Inſectile, *i. e.* without parts.
( 2 ) That all Bodies muſt be moved with equal celerity; for, the pace of
the Sun and that of a Snail muſt be æquivelox, if both move through an
inſectile ſpace, in an inſectile Time. (3) That, if many Concentrical Circles be deſcribed by the circumduction of one Rule, defixed upon one of
its extremes, as upon a Centre; ſince they are all delineated at one and
the ſame time, and ſome are greater then others: it muſt follow, that unequal portions of Circles are deſcribed in the ſame individual point of
Time, and conſequently that an Inſectile of an Interior Circle muſt be
æquated to a ſectile of an Exterior.

*Art. 2.*
*The ſundry*
*Incongruities &*
*Inconſiſtences,*
*by the Modern*
*Anti-Democri-*
*tans, imputed*
*to the ſuppoſi-*
*tion of Inſectili-*
*lity.*

     To theſe our Modern *Anti-Epicureans* have ſuperadded many other
Ἀσύμμετρα, or *Inconciſtencies*, as dependent on the poſition of *Inſectility.*
*viz.* (1) That a Line of unæqual Inſectiles, ſuppoſe of 3. 5. 9. or 11.
cannot be divided into two equal halfs: when yet, that any Line whatever
may be exactly bipartited, is demonſtrable to ſenſe. (2) That a leſs line
cannot be divided into ſo many parts, as a Greater: though the Contrary be concordant to the maximes of Geometry. (3) That though lines
drawn betwixt all the points of the Leggs of an Iſoſcelis Triangle, parallel to its Baſe, are leſs then its Baſe; yet will they be found greater: becauſe, ſuppoſing the Baſe to be of five points, and the Leggs of 10; it muſt
follow, that the leaſt Line, or the neareſt to the Vertex, doth conſiſt of
only two points, the ſecond of 3, the third of 4, the fourth of 5, the fifth
of 6, the ſixth of 7, the ſeventh of 8, and the greateſt, or neareſt to the
Baſe, of 9; then which nothing can be more abſurd. (4) That the Dia-
gone of a Quadrate would be commenſurable in longitude with the ſide
thereof: one and the ſame point being the meaſure common to both;
though the Contrary is demonſtrated by *Euclid.* (5) That the ſame Di-
agone of a Quadrate could not be greater then, but exactly adæquate to
the

the fide thereof : becaufe each of all its points muft be poffeffed by juft fo many, nor more nor fewer lines, then may be drawn betwixt the points of the oppofite fides; which is highly abfurd. (6) That, with the danger of no lefs abfurdity, would not a femicircle be greater then its Diametre; fince to every point in the femicircle there would refpond another in the Diametre, and there would be in both as many points, on which as many perpendicular Lines, deduced from them, might be incident. (7) That, according to the fuppofition of Infectility, of many Concentrick Circles the Exterior would not be greater then the Interior; infomuch as all the Lines drawn from all the points of it toward the Centre, muft pafs through as many points of the other. Many other Exceptions lye againft our Infectility; but being they are of the fame Nature with thefe, rather Mathematical, then Phyfical, and that one common folution will ferve them all: we may not abufe our leafure in their recitation.

That there have been hot and fcarce ingenious Altercations among the graveft and leading Philofophers, in all ages; and even about thofe Arguments, which wear the proper Characters of Truth fairly engraven on their Fronts; can be efteemed no wonder; becaufe the general cuftom of men to fpeculate the Fabrick of Nature, through the deceivable Glafs of Authority, doth amply folve it. But, that fo many Examples of Sagacity and Difquifition, as have condemned the Hypothefis of Atoms, fhould think their Choler againft the Patrons of it excufable only by the allegation of thefe light and impertinent Exceptions: cannot be denyed the reputation of a Wonder, and fuch a one as no plea, but an ambitious Affectation of extraordinary fubtilty in the invention of Sophifms (wherein Fallacy is fo neatly difguifed in the amiable habit of right Reafon, as to be charming enough to impofe upon the incircumfpection of common Credulity, and caft difparagement upon the moft noble and evident Fundamentals.) can palliate. For, certainly, They could not be ignorant, that they corrupted the ftate of the Quæftion; the *Minimum*, or *Infectile* of *Atomifts*, being not *Mathematicum*, but *Phyficum*, and of a far different nature from that Leaft of Quantity, which Geometricians imagining only, denominate a *Point*. And therefore, what *Cicero* (1. *de finib.*) faid againft *Epicurus*; *Non effe ne illud quidem Phyfici, credere aliquid effe minimum*: may be juftly converted into, *Effe præfertim Phyfici, naturale quoddam minimum afferere*; fince Nature in her Exolutions cannot progrefs to infinity. We fay, *Phyfici*; becaufe it is the *Naturalift*, whofe enquiries are confined to fenfible objects, and fuch as are really Exiftent in Nature: nor is He at all concerned, to ufe thofe *Abstractions* (as they are termed) from *Matter*; the Mathematician being the only He, who cannot, with fafety to his Principles, admit the Tenet of Infectility, or Term of Divifibility. For to Him only is it requifite, to fuppofe and fpeculate Quantity abftract from Corporiety; it being evident, that if He did allow any Magnitude divifible only into Individuals, or that the number of poffible parts, or points in a Continuum, were definite: then could he not erect Geometrical, or exquifite Demonftrations. And hence only is it, that He fuppofeth an Infinitude of points in every the leaft Continuum, or (in his own phrafe) that every Continuum is divifible into parts infinitely fubdivifible: not that He doth, or can really underftand it fo; but that many Convenient Conclufions, and no confiderable Incongruities, follow upon the Conceffion thereof. This confidered, we need no other evidence,

that

*Art. 3.*
The full Derogation of them all together, by one fingle Refponfe; that the *minimum* of Atomifts is not Mathematical, but Phyfical, contrary to their præfumption.

that all the former Objections, accumulated upon *Epicurus* by the malicious Sophistry of *Empiricus* and others, concern only the *Mathematicians*, not the *Physiologist*, who is a stranger to their supposition of interminable Divisibility.

*Art. 4.*
A seeming Dilemma of the Adversary, expeditely evaded.

If this *Response* prævail not, and that we must 'yet sustain this seeming Dilemma; Either the suppositions of the Mathematicians are *True* or *False* : if *true*, then doth their verity hold, when accommodated to Physical Theorems, by the assumption of any sensible Continuum, or real Magnitude; if *false*, then are not the Conclusions Necessary, that are deduced from them, but the contrary is apparent in their demonstrations; Therefore, &c. Our *Expedient* is, that, though we should concede those suppositions to be False, yet may they afford true and necessary Conclusions: every Novice in Logick well knowing how to extract undeniable Conclusions out of most false propositions, only supposed true, as may be Instanced in this Syllogism. *Omnes arbores sunt in cœlo* (that's false) *sed omnia sydera sunt Arbores* (that's false) *Ergo, omnia sydera sunt in cœlo* (that's indisputable). Besides, 'tis evident, that of those many *Hypotheses* celebrated by *Astronomers*, either no one is absolutely true, or all except one, are false: yet Experience assures, that from all, at least from most of them the Motions of Cœlestial Bodies may be described, and respective Calculations instituted with equal *Certude.*

## Digression.

*Art. 5.*
A Digression, stating and determining that notable Quæstion, Whether Geometrical Demonstrations may be conveniently transferred to Physical or sensible Quantity?

Here, because our Reader cannot but perceive us occasionally fallen into the mouth of that eminent Quæstion; *An liceat in materiam physicam, sive sensibilem, transferre Geometricas Demonstrationes?* Whether it be convenient to transfer Geometrical Demonstrations to Physical or sensible Quantity? Since they, who accept the *Negative*, seem to admit the *Negative*, seem to annihilate the use of Geometry: we need not deprecate his impatience, though we digress so long, as to præsent him the summary of our thoughts concerning it.

First, we conceive it not justifiable, alwayes to expect the eviction of Physical Theorems, by Geometrical Demonstrations. This may be authorized from hence, that Geometricians themselves, when they fall upon the theory of those parts of the Mathematicks, which are *Physicomathematical*, or of a mixt and complex Consideration, are frequently necessitated to convert to suppositions, not only different from, but directly and openly repugnant to their own proper and establisht maxims. Thus, in Opticks, *Euclid* concedes a *Least Angle*; and *Vitellio* admits a *Least Light*, such as being once understood to be divided, hath no longer the act of Light, *i. e.* wholly disappears: which is no less, then in Opticks to allow a Term, or point of Consistence to the Division of Quantity, which yet in Geometry they hold capable of an infinite process. We are provided of a most pertinent Example, for the illustration of the whole matter. The *Geometrician* Demonstrateth the Division of a Line into two equal segments, to be a thing not only possible, but most easie : and yet cannot the *Physiologist* be induced to swallow it, as really performable.
For

For He considers (1) That the superfice of no body can be so exactly smooth and polite, as to be devoyd of all uneveness or asperity, every common Microscope discovering numerous inæqualities in the surface of even the best cut Diamonds, and the finest Chrystal, Bodies, whose Tralucency sufficiently confesseth them to be exceeding polite: and consequently, that there is assumable thereon no Line so perfectly uniform, as not to be made unequal by many *Valleculæ* and *Monticulæ*, small pits and protuberances frequently interjacent. (2) That the Edge of no Dissecting Instrument can be so acute, as not to draw a line of some Latitude. (3) That should the edge of the acutest Rasor be laid on the foot of a Handworm, which may be effected by the advantage of a good Magnifying Glass, and a steady hand: yet is that composed of many Myriads of Atoms, or insensible particles of the First universal Matter. And thence Concludes that no real Line drawn upon the superfice of any the smoothest Body, can be practically divided into two Halfs, so exactly, as that the section shall be in that part, which is truly the median to both extremes. Since, that part, which appears, to the sense, to be the median, and is most exiguous; doth yet consist of so many Myriads of particles, as that though the edge of the Rasor be imposed by many Myriads of particles aside of that, which is truly in the middle, yet will it seem to the eye still to be one and the same. This duely perpended, we have no cause to fear the *section of an Atome*, though the edge of a knife were imposed directly upon it: Since the edge must be gross and blunt, if compared to the exility of an Atome: so that we may allow it to divide an Assembly, or Heap of Atoms, but never to cut a single one.

Secondly, We judge it expedient in some cases to accommodate suppositions Geometrical to Subjects merely Physical; but to this end only, that we may thereby acquire *majorem ἀκρίβειαν*, a greater degree of *Acuteness*, or advance our speculations to more Exactness. Thus the soul of the Mathematicks, *Archimed,(de Arenarum num.)* supposed the Diametre of a grain of Poppy seed to consist of 10000 particles; not that He conceived that any Art could really discern so vast a multitude of parts in a body of so minute circumscription: but that, by transferring the same reason to another body of larger dimensions, He might attain the certitude of his Proposition by so much the nearer, by how much the less he might have erred by neglecting one of those many particles. Thus also is it the custom of Geometricians, in order to their exactness in Calculations, to imagine the Semidiametre, or Radius of any Circle, divided into many Myriads of Parts; not that so many parts can be really distinguished in any Radius, but that, when comparation is made betwixt the Radius, and other right lines, which in parts Aliquotal, or such as are expressed by whole numbers, do not exactly respond thereunto, particles may be found out so exile, as though one, or the fraction of one of them be neglected yet can no sensible Error ensue thereupon. And this (in a word) seems to be the true and only Cause, why Mathematicians constantly suppose every Continuum to consist of Infinite parts: not that they can, or ought to understand it to be Really so; but that they may conserve to themselves a liberty of insensible Latitude, by subdividing each division of Parts into so many as they please; For, they well know, that the Physiologist is in the right, when He admits no Infinity, but only an Innumerability of parts in natural Continuum. Lastly, if these Reasons appear not weighty enough to counterpoise the Contrary Persuasion; we can aggravate them with a Grain of noble Authority.    For, no meaner a man then *Plato*, who seems to have understood Geometry as well as the Ægyptian *Theuth*, the suppo-sed Inventor thereof ( *vide Platon. in Phædro* ) and to have honoured it much more in a solemn Panegyrick (*9. dialog. de Rep.*) sharply reprehends *Eudoxus*, *Archytas*, *Menæchonus*, &c. for their errour in endeavouring to adjust Geometrical speculations to sensible objects: subnecting in positive termes, that (διαφθείρεϑαι τὸ γεωμετρίας ἀγαθὸν) thereby the good of Geometry was corrupted. (*Lege Marsil. Ficin. in Compend. Timæi. cap. 19.*)

CHAP.

# CHAP. III.

*Atoms, the First and Universal Matter.*

### SECT. I.

NO man so fit to receive and retain the impressions of *Truth*, as He, who hath his Virgin mind totally dispossessed of *Prejudice* : and no *Thesis* hath ever, since the Envy of *Aristotle* was so hot, as to burn the Volumes of *Democritus* and most of the Elder Philosophers, which might have conserved its lustre, been more Eclipsed with a præsumption of sundry *Incongruities*, then this noble one, that *Atoms are the First and Catholique Principle of Bodies*. Requisite it is therefore that this Chapter have, *Janus* like, two faces : *one* to look backward on those *Impediments* to its general admission, the *Inconsistences* charged upon, and sundry *Difficulties* supposed inseparable from it ; the *other* to look forward at the plenary *Remonstrance* of its *Verity*.

*Art. 1.*
The introduction, hinting the two general assumptions of the Chapter.

*Superbissimo furore ambitiosus nominis Aristoteles, in Philosophorum Principes est debacchatus, unoque incendio congestas triginta sex seculis tot sapientiæ divitias absumpsit, & si quæ voluit superesse suncris, ca omnium ludibris, dicteriisque lacessenda tradidit posteris, dum in optimorum bona invehitur, abscissis perditisque sapientiæ statuarum capitibus, suum imposuit singulis : ut Magnenas, in Democrit. Script. Elench. ex Plinio in præfat. ad D. Vespasianum Imp.*

*Art. 2.*
*Democritus* & *Epicurus* vindicated from the absurd admission of Inanity to be one Principle of Generables.

In obedience to this necessity, therefore, we advertise, *first* ; that it hath proved of no small disadvantage to the promotion of the Doctrine of Atoms, that the Founders thereof have been accused of laying it down for a main Fundamental, *that there are two Principles of all things in the Universe*, BODIE and INANITY; importing the necessary Concurrence of the *Inane Space* to the constitution of Bodies complex, as well as of *Atoms*. This Absurdity hath been unworthily charged upon *Epicurus* by *Plutarch*, in these words ; *Principia esse Epicuro Insuitatem & Inane :* and upon *Leucippus* and *Democritus* by *Aristotle* (1. *Metaphys.* 4.) in these ; *Plenum & Inane Elementa dicunt.*

To vindicate these *Mirrors* of Science from so dishonourable an Imputation, we plead ; that though they held the Universe to consist of two *neral Parts, Atoms* and *Vacuity :* yet did not they, therefore, affirm, that

all things were composed of those two, as *Elementary* Principles. That
which imposed upon their *Accusers* judgment, was this, that supposing
*Atoms* and the *Inane Space* to be *Ingenite* and *Incorruptible*, they conceived
the whole of Nature to arise from them, as from its two universal Parts;
but never dreamt so wild an Alogy, as that all *Concretions*, that are produced by Generation, and subject to destruction by Corruption, must derive their Consistence from those two, in the capacity of Elements, or *Componentia*. For, albeit in some latitude and liberty of sense, they may be
conceded Elements, or Principles of the Universe : yet doth it not naturally follow, that therefore they must be equal Principles, or Elements of
*Generables*; since Atoms only fulfill that title, the Inane Space affording
only *Place* and *Discrimination*. Nor is it probable, that those, who had
defined Vacuity by *Incorporiety*, should lapse into so manifest a Contradiction, as to allow it to be any Cause of Corporiety, or to constitute one
moiety of Bodies. Besides, neither can *Epicurus* in any of those Fragments
of his, redeemed from the jaws of oblivion by *Laertius*, *Cicero*, *Empiricus*, *Plutarch*, *&c*. nor his faithful Disciple and Paraphrast, *Lucretius*, in all
his Physiology, be found, to have affirmed the Contexture of any Concretion from Inanity, but of all things simply and solely from Atoms. And for
*Democritus*, him doth even *Aristotle* himself wholly acquit of this Error;
for (*in 1.Phys*.) enumerating the several opinions of the Ancients concerning the Principles, or Elements of all things, He saith of him; *Fecit principiorum Genus unicum, Figuras verò differentes*. All therefore that lyeth
against them in this case, is only that they asserted the interspersion or dis-
semination of Inanity among the incontingent particles of Bodies concrete,
as of absolute necessity to their peculiar Contemperation : which we conceive our selves obliged to embrace and defend, untill it shall be proved unto us, by more then paralogistical arguments, that there is any one Concretion in the world so perfectly solid, as to contain nothing of the Inane
Space intermixt; which till it can be demonstrated that a Concretion may
be so solid, as to be Indissoluble, we have no cause to expect.

*Art. 3.*
Atoms not in-
consistent
with, because
the Principles
of the four
vulgar Ele-
ments.

*Secondly*, That the Patrons of Atoms do not (as the malice of some, and
incogitancy of others hath prætended, to cast disparagement upon their
Theory) deny the Existence of those four Elements admitted by most
Philosophers : but allow them to be *Elementa Secundaria*, Elements
Elementated, *i. e*. consisting of Atoms, as their First and Highest Principles. Thus much we may certifie from that of *Lucretius* (*2. lib*.) treating
of Atoms;

* *Accipitur pro
Igne seu Æthe-
re, quem dictum
Anaxagoras
censuit, ἀπὸ ṫ
αἴθειν, ab u-
rendo.*

> *Unde mare, & Terra possent augescere, & unde
> Adpareret spatium Cæli* * *domus, altaq, tecta,
> Tolleret à terris procul, & consurgeret Aer, &c.*

Nor can the most subtle of their Adversaries make this their Tenet bear an
action of trespass against right Reason; especially when their Advocate shall
urge, the great Dissent of the Ancients concerning both the Number and
Original of Elements, the insufficiency of any one Element to the Production of Compound Natures, and that the four vulgar Elements cannot
justly be honoured with the Attributes of the First Matter.

(1) *The*

(1) *The Dissent of the Ancients about the number of Elements* cannot be unknown to any, who hath revolved their monuments and taken a list of their several opinions; their own, or their Scholiasts volumes lying open to record, that of those who fixt upon the four Vulgar Elements, *Fire, Aer, Earth, Water,* for the universal Principles, some constituted only *one* single first Principle, from which by Consideration and Rarefaction, the other three did proceed, and from them all Elementated Concretions: among which are *Heraclitus,* who selected Fire; *Anaximenes,* who pitched upon Aer; *Thales Milesius,* who præferred Water; and *Pherecydes,* who was for Earth. Others supposed only *Two* primary, from which likewise, by Condensation and Rarefaction the other two secondary were produced: as *Xenophanes* would have Earth and Water; *Parmenides* contended for Fire and Earth; *Oenopides Chius* for Fire and Aer; and *Hippo Rheginus* for Fire and Water. Others advanced one step higher, and there acquiesced in *Three;* as *Onomacritus* and his Proselytes affirmed Fire, Water, and Earth. And some made out the *Quaternian,* and superadded also Aer; the Principal of which was *Empedocles.* Now, to him who remembers, that there can be but one Truth; and thereupon justly inferrs, that of many disagreeing opinions concerning one and the same subject, either all, or all except one must be false; and that it is not easie which to prefer, when they are all made equally plausible by a parity of specious Arguments: it cannot appear either a defect of judgment, or an affectation of singularity in *Democritus* and *Epicurus* to have suspected them all of incertitude, and founded their Physiology on an Hypothesis of one single Principle, Atoms, from the various transposition, configuration, motion, and quiescence of whose insensible Particles, all the four generally admitted Elements may be derived, and into which they may, at the term of Exsolubility, revert without the least hazard of Absurdity or Impossibility; as will fall to our ample enunciation in our subsequent Enquiries into the Originals of Qualities, and the Causes of Generation and Corruption.

Art. 4.
The dissent of the Ancients, about the number of Elements.

(2) *That one of the four Elements cannot singly suffice to the production of any Compound Nature;* needs no other eviction but that Argument of *Hippocrates* (*de Natur. Hominis*) *Quo pacto, cùm unum existat, generabit aliquid, nisi cùm aliquo misceatur?* Instance we in *Heraclitus* Proto-Element, Fire; from which nothing but Fire can be educed: though it run through all the degrees of those fertile Modifications of Densefcence and Rarescence. (2) To suppose Rarefaction and Condensation, without the more or less of Inanity intercepted; as they do: is to usurp the concession of an Impossibility. (3) 'Tis absurd, to conceive Fire transformable, by Extinction, into any other Element: because a simple substance cannot be subject to essential transmutation. So that, if after its extinction any thing of Fire remain, as must till Adnihilation be admitted; its surviving part must be the Common Matter, such as Atoms, which according to the various and respective addition, detraction, transposition, agitation, or quiet of them, now put on the form of Fire, then of Aer, anon of Water, and lastly of Earth; since, in their original simplicity, they have no actual, but a potential Determination to the forms of all, indiscriminately. And, what is here urged, to evince the impossibility of Fires being the sole Catholique Element, carrieth the same proportion of reason and evidence, (the two pathognomick characters of Verity) to subvert the supposition of any of the other three for the substantial Principle of the rest.

Art. 5.
No one of the four Elements sufficient to the producti- on of either any of the o- ther three, or of any Com- pound nature.

*Art. 6.*
The four Elements, not the Protoprinciple of Concretions.

(3) *That though the four vulgar Elements may be the Father, yet can they not be the Grandfather Principle to all Concretions*, is evidencible from hence. (1) They are Contrary each to other, and so not only Asymbolical or Disharmonious, but perfectly Destructive among themselves, at least uncapable of that mutual correspondence requisite to peaceful and durable Coalescence. (2) They are præsumed to coalesce, and their Concretions to consist without Inanity interspersed among their incontiguous particles: which is impossible. (3) Their Defendants themselves concede a degree of Dissolution beyond them: and consequently that they know a Principle Senior. (4) Their Patrons must grant either that they, by a prævious deperdition of their own nature, are changed into Concretions, which by mutation of Forms escheat again into Elements; in which case Elements can be no more the Principle of Concretions, then Concretions the Principle of Elements, since their Generations must be vicissitudinary and Circular, as that of Water and Ice: or, that, conserving their own natures immutable, they make only confused Heaps, and confer only their visible Bulks to all productions; in which case, nothing can *revera* be said to be generated, since all Generations owe their proprieties and peculiar denominations to their Forms. (5) Whoso admits a reciprocal or symbolical Transmutation of Elements: must also admit one Common, and so a Former Matter, which may successively invest it self in their several Forms; For Contraries, while Contraries, cannot unite in the assumption of the same nature. (6) That *Achilles,* or Champian Objection, that Vegetables and Animals owe their Nutrition and Increment to the four Elements, is soon conquered by replying; that Elements are not therefore the First Principles, but rather those from whose respective Contexture they borrowed the nature of Elements, and so derived an aptitude, or qualification requisite to the condition of Aliment.

*Art. 7.*
Atoms discriminated from the *Homoiomerical* Principles of *Anaxagoras.*

*Thirdly,* that the Principles of *Democritus, Epicurus,* &c. are *toto cælo,* by irreconcileable disparities, different from those of *Anaxagoras,* called ὁμοιομερῆ, CONSIMILAR Parts, or abstractly, ὁμοιομέρεια, SIMILARITY (ἀπὸ τῷ ὁμοια ᾗ μέρη ἔιναι τοῖς γρωύδοσι) because they are supposed to be parts in all points consimilar to the Things generated of them, according to the paraphrase of *Plutarch* ( 1.*placit.*3. ) who there explains it by the Example of Aliment. Wherein, whether it be Wine, Water, Bread, Flesh, Fruits, &c. notwithstanding the seeming difference in the outward form, there are actually contained some Sanguineous, some Carnous, other Osseous, other Spermatick Parts, which, upon their sequestration, and selection by the Nutritive Faculty are discretely apposed to the sanguineous, carnous, osseous, and spermatick parts præexistent in the body nourisht. And the *Disparity* doth chiefly consist herein; that *They* endow their Atoms with only three congenial Qualities, *viz. Magnitude, Figure,* and *Gravity*: but *He* investeth his *Similarities* with as great variety of essential Proprieties, as there is of Qualities, nay Idiosyncrasies in Bodies.

Which to suppose, is to dote: (1) Because if the nature of the whole be one and the same with that of its Parts: then must the Principles, no less then the Concretions consisting of them, be obnoxious to Corruption. (2) Because, if it be assumed, that Like are made of Like, or that Concretions are absolutely Identical to their Elements; it cannot be denyed; that there are Laughing and Weeping Principles concurrent to the generations

of

of Laughing and Weeping Compofitions. (3) Becaufe from hence, that (concordant to *Anaxagoras*) all things are actually exiftent in all things, and that the difference refteth only in the external Apparence, arifing from the prædominion of fuch or fuch over fuch or fuch parts of the Confimilar Principles: it neceffarily enfues (as *Ariftotle* argueth againft Him, 1 *Phyfic.* 4.) that in the contufion, fection, or éctrition of Fruits, Herbs, &c. there muft frequently appear Blood, Milk, Sperm, &c. as being thereby enfranchifed from the tyranny of thofe parts, which ruled the coit in the induction of the outward apparence, and emergent out of thofe Clouds, which concealed and difguifed them. All which are Abfurdities fo palpable that a blind man may thereby Diftinguifh the rough and fpurious Hypothefis of *Anaxago-ras*, from the fmooth and genuine Principle of *Democritus* and his *Sectators.*

*Fourthly* and laftly, that the *Difficulties*, which many Diffenters, and more eminently their moft potent and declared Opponent, *Lactantius* (*in lib. de Ira Dei, cap.* 10.) have pofted up againft the fuppofition of Atoms for the Catholick Principle of Bodies Concrete, thereby to prævent their further approbation, and admiffion into the Schools; carry not moments enough of reafon to inflect and determine the judgment of an æquitable Arbiter to a fufpition, much lefs a pofitive negation of its verifimility. Of this we defire our Reader to be judge, when he hath made himfelf competent, by a patient hearing, and upright perpenfion of the pleas of both parties, here præfented.

(1) *Anti-Atomift*; Whence had thefe minute and indivifible Bodies, called Atoms, their original? or, out of what were they educed?

*Atomift*; This inappofite Demand lyeth open to a double refponfe. As a mere *Philofopher* I return; that the affumption of Atoms for the *Firft Matter* doth exprefly prævent the pertinency of this Quære. Nor would *Ariftotle*, *Plato*, or any other of the *Ethnick* Philofophers, who would not hear of a *Creation*, or production of the Firft Matter out of Nothing, but contumacioufly maintained its *Ingeneration* and *Eternity*, have had Gravity enough to fupprefs the infurrection of their fpleen againft the abfurdity thereof: fince to enquire the Matter of the *Firft* Matter, is a *Contradiction in terminis*. As a proficient in the facred School of *Mofes*, I may anfwer; that the fruitful *Fiat* of God, out of the *Tohu*, or infinite fpace of Nothing, called up a fufficient ftock of the Firft Matter, for the fabrication of the World in that moft excellent Form, which He had Idea'd in his own omnifcient intellect from Eternity.

(2) *Anti-Atomift*; If Atoms be fmooth and fphærical, as their Inventors fuppofe; it is impoffible they fhould take mutual hold each of other, fo as by reciprocal adhæfion and coalition to conftitute any Concretion. For, what power can mould an heap of Millet-feed into a durable figure, when the Lævitude or politenefs, and roundnefs of the Grains inexcufably interdict their Coition into a Mafs?

*Atomift*; This Objection difcovers the rancour, no lefs then the præcedent Interrogation did the weaknefs of the propofers. For, they could not be ignorant, that the Defendants of Atoms do not fuppofe them to be all
fmooth

smooth and globular, but of *all sorts of figures* requisite to mutual *Applicati-on, Coalition, Cohærence.* And therefore they could not but expect this solution. That, though polite and orbicular Atoms, cannot by mutual apprehension and revinction each of other, compact themselves into a Mass; yet may they be apprehended and retained by the Hooks, and accommodated to the Creeks and Angles of other Atoms, of Hamous and Angular figures, and so conspire to the Coagmentation of a Mass, that needs no other Cæment besides the mutual dependence of its component particles, to maintain its Tenacity and Compingence. This may receive light, from observation of the successive separation of the dissimilar Parts of Bodies, by Evaporation. For, first those Atoms, which are more smooth, or less angular and hamous, easily extricate themselves, and disperse from the Concreted Mass; and then, after many and various Evolutions, circumgyrations, and change of positions, the more rough, hamous, and angular, they expede themselves from reciprocal concatenation, and at last, being wholly disbanded, pursue the inclination of their inhærent Motive Faculty, and disappear. Experience demonstrating, that by how much more Unctuous and Tenacious any Consistence is, by so much a longer time do the particles thereof require to their Exhalation. Thus is Water much sooner evaporated, then Oyl: and Lead then Silver.

(3) *Anti-Atomist*; If Atoms be unequal in their superfice, and have angular and hamous processes; then are they capable of having their rugosities planed by detrition, and their hooks and points taken off by amputation: contrary to their principle propriety, Indivisibility.

*Atomist*; the hooks, angles, asperities, and processes of Atoms are as insecable and infrangible as the residue of their bodies, in respect an equal solidity belongs to them, by reason of their defect of Inanity interspersed, the intermixture of Inanity being the Cause of all Divisibility.

*Hæc, quæ sunt rerum primordia, nulla potest vis*
*Stringere, nam solido vincunt ea corpore demum.*

(4) *Anti-Atomist*; That Bodies of small circumscription, such as grains of sand, may be amassed from a syndrome, and coagmentation of Atoms; seems, indeed, to stand in some proportion to probability: but to conceive a possibility, that so vast a Bulk, as the adspectable World bears may arise out of things but one degree above nothing, such insensible materials convened and conglobated; is a symptome of such madness, as Melancholy adust cannot excuse, and for which Physitians are yet to study a cure.

*Atomist*; To doubt the possibility, nay dispute the probability of it: is certainly the greater madness. For, since a small stone may be made up of a Coagmentation of grains of Sand; a multitude of small stones, by coacervation, make up a Rock; many Rocks by aggregation, make a Mountain; many Mountains, by coaptation, make up the Globe of Earth; since the Sun, the Heavens, nay the World may arise from the conjunction of parts of dimensions equal to the Terrestrial Globe: what impossibility doth he incurr, who conceives the Universe to be amassed out of Atoms? Doubtless, no Bulk can be imagined of such immense Dimensions, as that the

                                                                        greatest

greateſt parts thereof may not be divided into leſs, and thoſe again be ſubdivided into leſs; ſo that, by a ſucceſſive degradation down the ſcale of Magnitude, we may not at laſt arrive at the foot thereof, which cannot be conceived other then Atoms. Should it appear unconceivable to any that a Piſmire may perform a perambulation round the terreſtrial Globe; we adviſe him to inſtitute this Climax of Dimenſions, and conſider, firſt that the ambite of the Earth is defined by miles, that miles are commenſurated by paces, paces conſiſt of feet, feet of digits, digits of grains, &c. and then He may ſoon be convinced, that the ſtep of a Piſmire holds no great diſproportion to a grain, and that a grain holds a manifeſt proportion to a digit, a digit to a foot, a foot to a pace, a pace to a perch, a perch to a furlong, a furlong to a mile, and ſo to the circumference of the whole Earth, yea by multiplication to the convexity of the whole World. If any expect a further illuſtration of this point, it can coſt him no more but the pains of reading the 45. page of our Treatiſe againſt *Atheiſm*; and of *Archimeds* book *de Arenarum Numero.*

( 5 ) *Anti-Atomiſt*; If all peices of Nature derived their origine from Individual Particles; then would there be no need of *Seminalities* to ſpecifie each production, but every thing would ariſe indiſcriminately from Atoms, accidentally concurring and cohæring: ſo that Vegetables might ſpring up, without the præactivity of ſeeds, without the aſſiſtance of moyſture, without the fructifying influence of the Sun, without the nutrication of the Earth; and all Animals be generated ſpontaneouſly, or without the prolification of diſtinct ſexes.

*Atomiſt*; This inference is ingenuine, becauſe unneceſſary, ſince all Atoms are not Conſimilar, or of one ſort, nor have they an equal aptitude to the Conformation of all Bodies. Hence comes it, that of them are firſt compoſed certain Moleculæ, ſmall maſſes, of various figures, which are the ſeminaries of various productions; and then, from thoſe determinate ſeminaries do all ſpecifical Generations receive their contexture and Conſtitution, ſo præciſely, that they cannot owe their Configuration to any others. And, therefore, ſince the Earth, imprægnated with Fertility, by the ſacred Magick of the Creators Benediction, contains the ſeeds of all Vegetables; they cannot ariſe but from the Earth, nor ſubſiſt or augment without roots, by the mediation of which, other ſmall conſimilar Maſſes of Atoms are continually affected for their nutrition; nor without moyſture, by the benefit of which, thoſe minute maſſes are diluted, and ſo adapted for tranſportation and final aſſimilation; nor without the influence of the Sun, by vertue whereof their vegetative Faculty is conſerved, cheriſhed and promoted in its operations. Which Reaſon is æquivalent alſo to the Generation, Nutrition and Increment of *Animals.*

( 6 ) *Anti-Atomiſt*; If your Proto-Element, Atoms, be the Principle of our 4 common Elements, according to the various Configurations of it into Moleculæ, or ſmall maſſes; and that thoſe are the Semina-

P                                                                    ries ties of all things : then may it be thence inferred, that the *Seeds* of *Fire* are invisibly contained in Flints, nay more, in a Sphærical Glass of Water, exposed to the directly incident rayes of the Sun ; our sense convincing, that Fire is usually kindled either way.

*Atomist* ; Allowing the legality of your Illation, we affirm, that in a Flint are concealed not only the Atoms, but *Molecule*, or Seeds of Fire, which wanting only retection, or liberty of Exsilition, to their apparence in the forme of fire, acquire it by excussion, and pursuing their own rapid motion *undiquaque*, discover themselves both by affecting the sight and accension of any easily combustible matter, on which they shall pitch, and into whose pores they shall with exceeding Celerity penetrate. Nor can any man solve this eminent Phænomenon so well, as by conceiving, that the body of a Flint, being composed of many igneous (i. e. most *exile, spherical,* and *agile*) Atoms, wedged in among others of different dimensions and figures ; (which contexture is the Cause of its Hardness, Rigidity and Friability) upon percussion by some other body conveniently hard, the insensible Particles thereof suffering extraordinary stress and violence, in regard it hath but little and few Vacuola, or empty spaces intermixt, and so wanting room to recede and disperse, are conglomorated and agitated among themselves with such impetuosite, as determinately causeth the constitution of Fire. It being manifest, that violent motion generateth Heat: and confessed even by *Aristotle* ( 1. *Meteor.* 3.) that Fire is nothing but the *Hyperbole* or last degree of Heat. Secondly, That the seeds of Fire are not contained either in the sphærical Glass or the the Water included therein ; but in the Beams of the Sun (whose Composition is altogether of Igneous Atoms) which being deradiated in dispersed lines, want only Concurse and Coition to their investment in the visible form of Fire ; and that the Figure of the Glass naturally induceth, it being the nature of either a Convex, or Concave Glass to transmit many Beams variously incident towards one and the same point, which the virtue of Union advanceth to the force of Ignition.

Art. 9.
A recapitula-
tion of the
præmises, i-
troductory to
the verificati-
on of the præ-
sent thesis.

Having thus vindicated our Atoms from the supposed Competition of the Inane Space, in the dignity of being one Principle of Bodies ; reconciled them to the 4 Peripatetick Elements ; discriminated them from the Consimilar Particles of *Anaxagoras* ; solved the most considerable of the Difficulties charged upon them ; and thereby fully performed our assumption of removing the principal prætexts of *Præjudice* : we may now, with more both of perspicuity, and hopes of persuasion, advance to the Demonstration of our *Thesis*, the Title and Argument of this Chapter.

### Sect. II.

BEsides the manifest Allusion of Reason, we have the assent of all Philosophers, who have declared their opinions concerning the Composition of a Continuum, to assure a necessity, that it must consist either (1) *of Mathematical Points*; or (2) *of Parts and Mathematical points, united*; or (3) *of a simple Entity, before actual division, indistinct*; or (4) *of Individuals, i. e.* Atoms.

*Art.* 1.
The 4 notable opinions, concerning the Composition of a Continuum

(1) *Not of Mathematical Points*; because Σημεῖον, *Punctum*, in the sense of *Euclid*, is *Cujus nulla sit pars*, in respect it wants all Dimensions, and consequently all Figure: which is the ground of *Aristotles* Axiom, *Punctum puncto additum non potest facere majus*. To render the absurdity of this opinion yet more conspicuous, let us remember, that the Authors and Defendants of it have divided themselves into three distinct *Factions*. (1) Some have admitted in a Continuum, points Finite *simpliciter & determinatè*; (2) Others allow points also Finite, but not *simpliciter, sed & τι secundum quid*; (3) And others contend for points *Infinite, simpliciter, & absolute*. The *First and Second* endeavour to stagger the former Axiom of *Aristotle*, by an illegal transition from Quantity Continued, to *Discrete*, alledging this instance, that one Unity added to another makes a greater quantity. The *Last* recur to *Plato's* Authority, who concedeth two Infinites, a Greater and Less, commemorated by *Aristotle* (3. *phys.* 27.) Now, for a joint redargution of all, we demand, how they can divide a Line consisting of 5 insectiles into two equal segments? For, either they must cast off the intermediate insectile, or annex it to one division: if the first, they split themselves upon that rock, our *supposition*; if the last, they clash with the 9. *proposit.* 1. *lib. Euclid.* To evade the force of this Dilemma, they have invented many subterfuges: but how unsuccessfully, may be enquired of *Aristotle* (*in 6. physicor.*) who there convicts them all of either Falsity, or Impossibility; where, having præmised an excellent enunciation of the Analogy between Motion, Time, and Place, He apodictically concludes, that, if a Continuum did consist of points Mathematical, all Motions would be equally swift. Notwithstanding this, such was the contumacy of *Arriaga*, that in hopes to elude this insoluble Difficulty, He prætends to discover a new kind of Motion, distinguished by certain *Respites*, or *Pauses* intercedent; thereupon inferring that all things are moved, during their motion, with equal Celerity; but because the motion of one thing is intercepted with many pauses, and the motion of another with few, therefore doth the motion of this seem swift, and the motion of that slow; as if the degrees of *Celerity* and *Tardity* did respond to the *Frequency* and *Rarity* of *Respites* interceeding. If this be true, then must a Pismire move slower then an Eagle only because this distinguisheth its motion by shorter pauses, and that by longer: nor can a Faulcon overtake a Partridge, since our eyes assure, that a Partridge strikes six strooks at least with his wings, while its pursuer strikes one. *Macgravius* (*in histor. Animal.*

*Art.* 2.
A Physical Continuum cannot consist of Points Mathematical.

P 2    *Brasiliens*

*Brasiliens*) tells of an Animal, which from the wonderful tardigradous inceffion of it, is named by the Portugals *P R I G U I Z A*, or *Lubart*: becaufe though goaded on, it cannot fnail over a ftage of 10 paces in 48 hours. Had *Arriaga* beheld this *floth*, either He muft have difavowed his nicery, or held it an equal lay which fhould have fooner run over a four mile courfe, that, or the fleeteft Courfer in the Hippodrome at Alexandria: becaufe the Paufes, which intercept the conftant progreffion of the one, in the fpace of 10 paces, cannot be more then thofe that interrupt the continuity of the others motion, in the fpace of four miles. Thefe confiderations therefore enable us to conclude, that thofe who conftitute a Continuum of points Mathematical, abfurdly maintain, (1) That a point added to a point makes an augmentation of quantity; (2) That no Motion is fucceffive, but only Difcrete; (3) That all motions are of equal velocity, *funt enim puncta minimum quod pertranfiri pofsint* : and *Arriaga's* Quiet, imagined to be in motions, is no part of Motion. (4) That a Wheel is diffolved, when circumrotated upon its Axis; for, fince the Exterior Circle muft præcede the Interior, at leaft, by one point, it follows that the fame points do not correfpond to the fame points; which is abfurd and incredible. Therefore is not a Continuum compofed of Mathematical points.

**Art. 3.**
Nor of Parts and Points Mathematical, united.

(2) *Not of Parts and Mathematical points, united.* Becaufe (1) Parts cannot be conceived to be united or terminated, unlefs by an adæquation of Points to them; (2) Since thofe points, which are imagined to concur to the conjunction of parts, are even by *Suarez* the chief Patron of them, (*in Metaphyf. Difput. de quantitat.*) named *Entia Modalia*; it muft thence follow, that Parts, which are *Entia Abfoluta*, cannot confift without them; which is ridiculous. (3) Since they allow no Laft Part, how can there be a Laft, *i.e.* a Terminative Point? (4) Either fomething, or nothing is intermediate between one Indivifible and other Indivifibles: if fomething, then will there be a part without points; if nothing, then muft the whole confift of Indivifibles, which is the point at which we aim.

**Art. 4.**
Nor of a fimple Entity, before divifion indiftinct: but of Indivifibles

(3) *Not of a fimple Entity before Divifion, Indiftinct*; as not a few of our Modern Metaphyficians have dreamt, among whom *Albertinus* was a Grand Mafter. Who, that He might palliate the Difficulty of the Diftinction of Parts, that threatned an eafie fubverfion of his phantaftick pofition, would needs have that all Diftinction doth depend *ab Extrinfeco*, i.e. arifeth only from *mental Defignation*, or actual Divifion. But, O the Vanity of affected fubtilty! all that He, or his whole faction hath erected upon this foundation of Sand, may be blown down with one blaft of this fingle Argument. Thofe things which can exift being actually feparate, are really diftinct: but Parts can exift being actually feparate, therefore are they really diftinct, even before divifion. For Divifion doth not give them their peculiar Entity and Individuation, which is effential to them and the root of Diftinction. The *Major* is the general and only Rule of Diftinctions, which whofo denyes cannot diftinguifh *Plato* from *Ariftotle*, nor *Albertinus* from *Therfites*. The *Minor* holds its verity of fenfe, for the part A, is exiftent without the part B, though being before conjoyned, they both confpired to the conftitution of one Continuum. And that the Propriety of Entity, is the Bafe of Diftinguibility, is verified by that ferene Axiome, *Per idem res diftinguitur ab omni alia, per quod conftituitur in fuo effe.* Therefore cannot a Continuum confift of a fimple Entity before divifion indiftinct :

indistinct : but of *Individuals*, or *Atoms*, which is our scope and Conclusion.

Our *second* Argument flowes from the nature of *Union*. For the decent introduction of which, we are to recognize, that a *Modal Ens* cannot subsist without conjunction to an *Absolute* ; as, to exemplifie, *Intellection* cannot be without the *Intellect*, though on the reverse, the Intellect may be without the act of Intellection : so likewise cannot *Union* be conceived without *Parts*, though on the contrary, Parts may be without Union. And hence we thus argue :

*Art. 5.*
A second Apodictical reason, defumed from the nature of Union, evincing that Atoms are the First and Catholick Principle of Concretions.

That only which is made *independentèr à subjecto*, or holds its essence *ex proprio*, is the Term of Creation ; but Union is not independent *à subjecto*: therefore is not Union the Term of Creation. Since therefore the Term of Creation in the First Matter is devoid of Union ; it must consist of Individuals, for Division proceeds from the solution of Union. This derives Confirmation from hence ; that the subject from whence another is deduced, must be præcedent in nature to that which is derived : now the Parts of the First Matter are the Subject from whence Union is derived ; *Ergo*, are the Parts of the First Matter in nature præcedent to all Union ; and consequently they are Individuals, i.e. *Atoms*.

If it be objected, that the understanding cannot apprehend the First Matter to consist without some implicite Union we appeal to that Canon, *Quod non est de essentia rei, non ingreditur ejus conceptum :* For, Union not being of the essence of the parts of the First Matter, ought not to fall under the comprisal of that Idea, by which we speculate them. And, upon consequence, if they are conceived without implicite Union : certainly they are conceived as *Individuals*, or Atoms. The *Major* is justified by that common Principle ; *Ex eo quod res est, vel non est, dici potest vel esse, vel non esse ; conceptus enim mensura est rei Entitas, mensura autem vocis est conceptus.* And the Certitude of the *Minor* results from that Metaphysical Canon, *Nullus modus actualis est de Essentia rei.*

*Art. 6.*
An objection prævented.

Upon these *Two* Arguments might we have accumulated sundry others of the like importance, such as are chiefly insisted upon by the Modern Redeemers of *Democritus* and his noble *Principles* from that obscurity and contempt, which the Envy of Time and the Peripatetick had introduced, *Sennertus* (*in Hypomnemat. de Atomis.*) and *Magnenus* (*in cap. 2. disput. 2. de Atomis.*) and, in imitation of their ample model, have explicated the mystery of our Thesis, from the *Syncritical* and *Diacritical* Experiments of Chymistry, (whereby all Bodies are sensibly dissolved into those *Molecula*, or First Conventions of Atoms, which carry their specifical seminaries ; and the Heterogeneous parts of diverse Concretions, after dissolution, coagmentated into one mass, and united *per minimas*) but most eminently from that natural miracle, the *Tree of Hermes*, made by an artificial Resuscitation of an entire Herb from the Atoms of it in a Glass, honestly effected by a Polonian Physitian in the præsence of *Gaffarel*, as himself records (*in Curiositat. inaudit.*) asserted by *Quercetan* (*in defens. contra Anonym. cap. 23*) and to the life described by *Hierom. Cornarius*, famous for his long profession of Philosophy and Medicine at *Brandenburgh*, in an Epistle to the great *Libavius*, which he therefore made an Appendix to his acute dissertation

*Art. 7.*
The reason of the Authors supercession of all other Arguments of the like importance.

de

*de Refuscitatione Formarum ex cineribus plantarum (syntagm. Arcan.Chymic. lib.1. cap. 22.)* But having upon an upright and mature perpension of their weight, found it such, as warrants our adscription of them to the golden number of those Reasons, that are λόγοι ἀναγκαῖοι, ἢ ἐκ ἐυπορει διαλύειν (as *Aristotle* speaks of other Arguments concerning the same subject, *in de Generat. & Corrupt. cap. 2.*) such as urge and compel the mind to an assent, and bid defiance to all solution : we judged our præsent Fundamental sufficiently firm, though erected upon no other but those two pillars ; especially when we remembred that *supererogation* is a kind of *Deficiency.*

CHAP.

# CHAP. IV.

## *The Essential Proprieties of Atoms.*

### Sect. I.

THat our Theory of those *Qualities*,
which being congenial to, and inseparable from Atoms, fulfil the necessary *Attributes* of the First Universal Matter, may, according to
the Method requisite to perspicuity,
immediately succeed to our Demonstration of their Existence, and
the impossible Elementation of
Concrete substances from any other
general Principles ; and that the
expectance raised in our Reader by
our frequent transitory mention of
the *Proprieties* of Atoms, may be
opportunely sated by a profess Enumeration and Enunciation thereof : are
the two reasons that justifie our subnection of this to our præcedent Discourse.

*Art. 1.*
The two links
connecting
this to the
præcedent
Chapter.

The PROPRIETIES of our Atoms difference themselves into
*General* and *Specifical*. The General are (1) *Consimilarity of Substance*,
for all Atoms being equally Corporeal and solid, must be substantially identical, or of one and the same nature, knowing no disparity of Essence. Thus
much *Aristotle* intimates (1. *Physic*. 2.) when He infers *Democritus* holding, *esse principiorum τὸ ἓν, Genus unicum*, or τὴν φύσιν μίαν, *Naturam unam* , that the Principles of all things are of one Kind, or of one
Nature. In respect of this, there is no difference among Atoms. (2) *Magnitude*, or Quantity, which they cannot want, since they are not Mathematical Insectiles, but Material Realities, and Quantity or Extension is the
proper and inseparable affection of Matter ; and therefore every thing hath
so much of Extension, as it hath of Matter. (3) *Figure*, which is the essential Adjunct of their Quantity. For, insomuch as Atoms are most minute
Bodies, and stand diametrally opposed to Points Imaginary ; therefore must
they have dimensions real, and consequently a termination of those dimensions in their extreme or superfice, i. e. determinate Figure. Which is
the ground of *Magnenus* 3. *Postulate* ( de *Atoms* , disput. 2.) *Quicquid*
*magnitudinevi*

*Art. 2.*
The General
Proprieties of
Atoms ; and
the *Inseparabi-
lity* of each,
demonstrated.

*magnitudinem habet, finitamque extensionem, si pluribus dimensionibus subsistet, concedatur illi suam inesse Figuram* ; and perhaps also of *Euclids* definition of Figure, *Figura est, qua sub aliquo, vel sub aliquibus terminis comprehenditur.* Nor have they only a Plain figure, but a *solid* one, according to that of *Euclid* (*lib. 2. def. 1.*) *solidum est, quod longitudinem, latitudinem, & crassitudinem habet.* (4) *Gravity*, or Weight ; which is also coessential to them in respect to their solidity , and the principle of their Motion. And therefore *Epicurus* had very good cause to add his ʼἐ βάρος, to *Democritus* μεγεϑὸς τέ ἢ ϲχήμα : which *Plutarch* (*1.placit. 3.*) expresly renders thus, ἀναϗκὴ γαϸ κινεῖωϳ τὰ ϲώμαϗα τῆ τῶ βάϸυς πληϑῇ, *quia necesse est Corpora moveri ipso impetu Gravitatis.* For, having supposed that *Motion* was essentially competent to Atoms, it must have been no venial defect, not to have assigned them a certain special *Faculty* , or Virtue for a *Cause* to that motion præsumed ; and such must be their inhærent Gravity , or the tendency of weight. Now, in respect to either of these three last Proprieties, Atoms may be conceived to admit of difference among themselves ; for, in regard of Magnitude, some may be greater then others, of Figure, some may be sphærical, others cubical, some smooth, others rough, &c. and of Gravity, some may be more, and others less ponderous, though this can cause no degrees of Velocity or Tardity in their Motion, it being formerly demonstrated, that two bodies of different weights are æqually swift in their descent.

*Art. 3.*
The *Resistence* of Atoms, no distinct propriety ; but pertinent to their Solidity or Gravity.

To these 4 Essential Attributes of Atoms, *Empiricus* hath superadded a Fifth, *viz.* Ἀντιτυπία, *Renitency*, or Resistence. But, by his good leave, we cannot understand this to be any distinct Propriety ; but as τι ὑποκείμενον, something resilient from and dependent on their *solidity*, which is the formal reason of Resistence : besides , we may confound their Renitency with their Gravity , insomuch as we commonly measure the Gravity of any thing , by the renitency of it to our arms in the act of Elevation. Which may be the reason, why *Aphrodisæus* (*lib. 1. Quæst. cap. 2.*) enumerating the proprieties of Atoms, takes no notice at all of their Gravity ; but blends it under the most sensible effect thereof, *Resistence*.

*Art. 4.*
The *specifical* Proprieties of Atoms.

The *specifical* are such as belong to Atoms of particular sorts of Figure, as *Smoothness, Acuteness, Angularity*, and their Contraries, *Asperity, Obtuseness, Orbicularity*, &c. These, in the dialect of *Epicurus*, are συμφυῆ, *Cognata Proprietates.* Now all these Proprieties, both Generical, and Specifical, or Originary and Dependent, are truly αχώϸιϲα, as *Plutarch* (*1. adv. Colot.*) calls them, *Congenial*, and *inseparable.* Other Proprieties there are adscriptive to Atoms, such as their *Concurse, Connexion, Position, Order, Number*, &c. from which the Qualities of Compound Bodies do emerge ; but since they are only *Communia Accidentia*, Common Accidents , or (as *Lucretius*) *Atomorum Eventa*, the fortuitous Events of Atoms considered as complex and coadunated in the Generation of Concretions, and not in the intire simplicity of their Essence ; and consequently *seperable* from them : therefore may we hope, that our Reader will content himself with our bare mention of them in this place, which is designed for the more advantagious Consideration of only the *Essential* and *Inseparable.*

Sect.

## SECT. II.

### *Concerning the Magnitude of Atoms.*

*Art.* 1.

By the *Magnitude*, is meant the *Parvity* of *Atoms.*

**M**agnitude and Atoms, though two terms that make a graceful Consonance to ears acquainted with the moſt charming harmony of Reaſon, may yet ſound harſh and diſcordant in thoſe of the Vulgar, which is accuſtomed to accept Magnitude only *Comparatively*, or as it ſtands Antithetical to *Parvity*: and therefore it concerns us to provide againſt miſapprehenſion by an early advertiſement; that in our aſſumption of Magnitude as the firſt eſſential Propriety of Atoms, we intend not that they hold any *ſenſible* bulk, but that, contrary to Inſectiles, or Points Mathematical, they are *Entities Quantitive ſimply*, i.e. Realities endowed with certain corporeal Dimenſions, though moſt minute, and conſiſting in the loweſt degree of phyſical quantity; ſo that even thoſe of the largeſt ſize, or rate, are much below the perception and diſcernment of the acuteſt Opticks, and remain commenſurable only by the finer digits of rational Conjecture. And ſomewhat the more requiſite may this Præmonition ſeem, in reſpect that no meaner an Author then *Theodoret* hath, through groſs inadvertency, ſtumbled at the ſame block of ambiguity. For (*in Serm. 4. therapeut.*) He poſitively affirms, that *Democritus, Metrodorus,* and *Epicurus*, by their exile Principles, Atoms, meant no other but thoſe ſmall pulverized fragments of bodies, which the beams of the Sun, tranſmitted through lattice Windows, or chincks, make viſible in the aer: when according to their genuine ſenſe, one of thoſe duſty granules, nay, the ſmalleſt of all things diſcernable by the eyes of *Linceus*, though advantaged by the moſt exquiſite Engyſcope, doth conſiſt of Myriads of Myriads of thouſands of true Atoms, which are yet corporeal and poſſeſs a determinate extenſion.

*Art.* 2.

A conſideration of the *Groſſeneſs* of our *ſenſes*, and the extreme *ſubtility* of *Nature*, in her operations; præparatory to our Conjectural apprehenſion of the Exiguity of Her Materials, *Atoms.*

To avert the Wonder impendent on this nice aſſertion, and tune our thoughts to a key high enough to attain the Veriſimility thereof; We are firſt to let them down to a worthy acknowledgment of the exceeding *Groſſneſſe* and *Dulneſſe* of our *Senſes*, when compared to the ſuperlative *Subtility*, and *Acuteneſs* of *Nature* in moſt of her Operations: for that once done, we ſhall no longer boaſt the perſpicacity of our Opticks, nor circumſcribe our Intellectuals with the narrow line of our ſenſible diſcoveries, but learn there to ſet on our Reaſon to hunt, where our ſenſe is at a loſs. Doubtleſs, the ſlender Crany of a Piſmire contains more diſtinct Cellules, then that magnificent Fabrick, the Eſchurial, doth rooms; which though imperceptible to the eye of the body, are yet obvious to that of the mind: ſince no man can imagine how, otherwiſe, the Faculties of ſenſe and voluntary Motion can be maintained, a perpetuall ſupply of Animal (or, as Dr. *Harvey* will have them, Vital) ſpirits being indiſpenſably neceſſary to the continuation of thoſe actions; and therefore there muſt be Elaboratories for the præparation and confection, Treaſuries for the conſervation, and various Conduits for the emiſſion, and occaſional tranſvection of them

Q         into

into the Nerves and Muscles of that industrious and provident Animal. The due resentment of which prægnant Instance, is alone sufficient to demonstrate the incomputable degrees of distance betwixt the sensible Capacity of man, and the curious Mechanicks of Nature : and make the acutest of us all call for a Table-book to enroll this Aphorism ; *Ubi humana industria subtilitasque desinit, inde incipit industria subtilitasque Naturæ.* The wings of our Arrogance being thus clipt, let us display those of our Discoursive Faculty, and try how near we can come to deprehend the Magnitude, *i. e.* the *Parvity* of Atoms, by an ingenious Conjecture.

**Art. 3.**
The incomprehensible subtility of Nature, argued from the Artifice of an exquisite *Watch*, contrived in a very narrow room.

Consider we, first, that an exquisite Artist will make the movement of a Watch, indicating the minute of the hour, the hour of the day, the day of the week, moneth, year, together with the age of the Moon, and time of the Seas reciprocation ; and all this in so small a compass, as to be decently worn in the pall of a ring : while a bungling Smith can hardly bring down the model of his grosser wheels and balance so low, as freely to perform their motions in the hollow of a Tower. If so ; well may we allow the finer fingers of that grand Exemplar to all Artificers, Nature, to distinguish a greater multiplicity of parts in one Grain of *Millet seed*, then ruder man can in that great Mountain, *Caucasus* ; nay, in the whole *Terrestrial Globe.*

**Art. 4.**
The vast multitude of sensible particles, &c. the vaster of *Elemental Atoms*, contained in one grain of *Frankinsense*, exactly calculated.

Consider we, with *Magnenus*, that one grain of Frankinsense being fired, doth so largely diffuse it self in fume, as to fill a space in the aer, more then *seven hundred millions* of times greater then it possessed before combustion. For, to the utmost dispersion of its fume, the space might easily have received of grains of Frankinsense, equal in dimensions to the seed of a Lupine,

|  | | |
|---|---|---|
| according to its Altitude | | 720 |
| according to its Latitude | | 900 |
| in the | Longitude | 1200 |
| | Superfice of the whole figure | 5184000 |
| | Superfice of the end only | 648000 |
| | Area, or whole enclosure | 777600000 |

Since, therefore, our nostrils ascertation, that in all that space of Aer, there is no one particle which is not impregnated with the fragrant exhalations of that combust grain of Frankinsense, which, while it was entire might be by a steddy hand, a sharp incision knife, and a good magnifying Glass, or by that shorter way of trituration, divided at least into a thousand sensible particles : it must follow, in spite of Contradiction, that the sensible odorous particles of it do fulfil the number of 777600000000. And, insomuch as each of these sensible Particles, is mixt, it being lawful and commendable according to the subtile speculations of *Archimed (in Arenar.)* to assume that the smallest of them is composed of a Million of Elemental Atoms : therefore by the same rule, must there have been in the whole Grain of Elemental Atoms 777600000000000000, at least. If so ; we have but one step lower to Insectility, and so may guess at the Exiguity of a single Atome.

Consider

Confider we the delicate Contexture of Atoms, in the body of that smallest of Animals, a *Handworm*. First, if we speculate the *outside* of that organical tenement of life, a good Engyscope will præsent out eye with not only an oval-head, and therein a mouth, or prominent snout, armed with an appendent proboscis, or trunk consisting of many villous filaments contorted into a cone, wherewith it perforates the skin, and sucks up the bloud of our hands; but also many thighs, leggs, feet, toes, laterally ranged on each side; many hairy tufts on the tail, and many asperities, protuberances, and rugosities in the skin. Then our Reason if we contemplate the *inside* thereof, will discover a great variety of Organs necessary to the several functions of an Animal. For *Nutrition*, there must be Gullet, Stomach, Intestines, Liver, Heart, Veins; or at least parts in their offices and uses perfectly analogous thereto: For *Vitality*, there must be Lungs, and Heart for the præparation and confection, and Arteries for the general diffusion of Spirits; for *Locomotion* voluntary and *sensation*, there must be Brain, Spinal Marrow, Nerves, Tendons, Muscles, Ligaments, Articulations; and for the support and firmitude of all these, there must be some more solid *stamina*, or a kind of Bones and Cartilagineous contextures; in a word, there must be all members requisite to entitle it to Animation, with a double skin for the investiment of the whole Machine. Now, if we attentively compute, how many particles go to the composure of each of those organical parts, and how many Myriads of Atoms go to the contexture of each of those particles (for even the Spirits intervient to the motion of one of its toes, are compositions consisting of many thousands of Atoms), as we shall think it no wonder, that the exile and industrious fingers of Nature have distinguished, sequestred, selected, convened, accommodated, coadunated, and with as much aptitude as decorum disposed such an incomprehensible multitude of Parts, in the structure of so minute an Animal; so may we, in some latitude of analogy, conjecture the extreme Parvity of Her common Material, Atoms. On this ingenious pin hung the thoughts of *Pliny*, when (*in lib.* 11. *cap.* 1 & 2.) He exclaimed, *Nusquam alibi Natura artificium spectabilius est, quàm in Insectis: in magnis siquidem corporibus, aut certè majoribus, facilis officina sequaci materia fuit. In his verò tam parvis, atque tam nullis; quæ ratio, aut quanta vis, tanquam inextricabilis perfectio? ubi tot sensus collocavit in Culice? & sunt alia dictu minora. Sed ubi visum in ea pratendit? ubi Gustatum applicavit? ubi odoratum inseruit? ubi truculentam illam, & proportione maximam vocem ingeneravit? Qua subtilitate pennas adnexuit, prælongavit pedum crura, disposuit jejunam caveam, uti alvum, avidam sanguinis, & potissimum humani sitim accendit? Telum verò perfodiendo tergori, quo spiculavit ingenio? atque cùm præ exilitate pene non videatur, ita reciproca generavit arte, ut fodiendo acuminatum, pariter sorbendoque fistulosum esset, &c.*

Here had we haulted awhile, and wondered, how *Pliny* could, without the assistance of a *Magnifying Glass* (an Invention, whose Antiquity will hardly rise above the last revolution of Saturn) deprehend so vast a multiplicity of Parts in the machine of an Insect, of so small circumscription, that to commensurate the Base of the visive Cone, by which its slender image is transmitted to the pupil of the eye, would trouble a good Master

*Art. 5.*
The Dioptrical speculation of a *Handworm*, discovering the great variety of Organical Parts therein, and the innumerability of their Component Particles.

*Art. 6.*
A short Digressive Descant upon the Text of *Pliny*, touching the multiplicity of parts in a *Flea*; hinting the possible perspicacity of *Reason*.

Master in Opticks, and drive him to the *Minimus Angulus* of *Euclid* : but that it soon came into our thoughts, that He speculated the same by the subtiler Dioptrick of *Reason* ; which indeed is the best Engyscope of the Mind, and renders many things perspicuous to the *Understanding*, whose exceeding Exility is their sufficient Darkness.

**Art. 7.**
*The Exility of Atoms, conjectural from the great diffusion of one Grain of Vermillion dissolved in Water.*

To put more weights into the Scale of Conjecture, let us moreover observe ; how great a quantity of Water may be tinged with one grain of Vermillion ; how many sheets of Paper may be crimsoned with that tincture ; how innumerable are the points, by the apex of a needle, designable on each of those sheets : and when 'tis manifest that many particles of Vermillion are found in each of those points ; who can longer doubt, that the particles comprehended in the compass of that grain are indefinable by the exactest Arithmetique.

**Art. 8.**
*The same, infellible from the small quantity of oil depredated by the Flame of a Lamp, in a quarter of an hour.*

Again, (for we could be content, to let the Almund tree bud, before we take off our cogitations from this pleasant Argument) consider we, how small a portion of oyl is consumed by the flame of a Lamp, in a quarter of an hour ; and yet there is no moment passeth, wherein the stock of flame is not wasted and as fast repaired, which if it could be conserved alive all at once, would fill not only whole rooms, but even ample Cities : and if so, what need we any further eviction of the extreme *Exiguity* of those Parts, of which all Concretions are material'd ?

**Art. 9.**
*The Microscope of great use, in the discernment of the minute particles of Bodies: and so advantageous to our Conjecture, of the exility of Atoms.*

Had the Ancients, indeed, been scrupulous in this point ; their want of that useful Organ, the Engyscope, might in some part have excused their incredulity : but for us, who enjoy the advantages thereof, and may, as often as the Sun shines out, behold the most lævigated Granule of dissolved Pearl, therein præsented in the dimensions of a Cherry stone, together with its various faces, planes, asperities, and angles, (such as before inspection we did not imagine) most clear and distinct, longer to dispute the possible Parvity of Component Principles, is a gross disparagement to the Certitude of Sense, when it is exalted above deception, and all possible impediments to its sincere judicature are prævented.

Conclude we therefore, since the Diametre of a granule of Dust, when speculated through a good Engyscope, is almost Centuple to the diametre of the same, when lookt on meerly by the eye, on a sheet of Venice Paper : we may safely affirm, with *Archimed* (*in arenario.*) that it is conflated of ten hundred thousand millions of insensible Particles ; which is enough to verifie our præsent Assumption.

## *Concerning the Figures of Atoms.*

IN all the sufficiently prolix Discourses of the Ancient Assertors of A-toms, concerning their FIGURE, and the no sparing Commentaries of the Moderns thereupon; whatever seems either worthy our serious animadversions, or in anywise pertinent to our Designation: may be, without perversion, or amission of importance, well comprized under one of these 3 *Canons*. (1) *That Atoms are, in their simple essence, variously figurate;* (2) *That the distinct species of their Figures are Indefinite, or Incomprehensible, though not simply, or absolutely Infinite;* (3) *That the Number of Atoms retaining unto, or comprehended under each peculiar species of Figure, is not only indefinite, but simply Infinite.*

*Art.1.*
An Epitome of all that directly concerns the Figures of Atoms in 3 General Canons.

Concerning the FIRST; we advertise, that no man is to conceive them to have supposed the Figure of Atoms deprehensible by the Sight, or Touch, no more then their Magnitude, the termination whereof doth essence their figure, according to that definition of *Euclid*, lately alledged; but such, as being inferrible from manifold reasons, is obvious to the perception of the Mind. Which *Plutarch* (1. *placit.* 2.) personating *Epicurus*, expresly declares in his, ἰδία ἔχ*ειν* σχήμα*τ*α, λόγω *θ*εωρη*τ*ά, *Atomos proprias habere, sed ratione, seu mente contemplabiles Figuras.* To avouch the verity hereof, we need no other argument but this; insomuch as every Atome hath some determinate Quantity, or Extension, and that all Quantity must be terminated in some certain Figure: therefore is it necessary, that however exile the dimensions of an Atome are, yet must the superfice thereof be or plane, or sphærical, or angular, or Cubical, &c. i. e. of some figure either regular, or irregular.

*Art.2.*
The First Canon, explained and certified.

Doth any incline to believe, that the extreme Exility of Atoms may necessitate their general *Roundness*; and the rather because he perceives all those dusty fragments of bodies, visible in the act by Sunshine, (which are the Atoms of the Vulgar) to be clad in that figure: We advise him to collect a multitude of them, on a clean sheet of the finest white Paper, and then speculate any the smallest granules among them with a perfect Engyscope. For, in so doing He will acquire autoptical satisfaction, that none of them are exactly orbicular and perpolite, but all of various angular figures, pyramidal, pentahedrical, cubical, trapezian, heptahedrical, octahedrical, dodecahedrical, icosihedrical, &c. nay of so many irregular and dissimilar apparences, as must refute his error with a delightful Wonder. Though, in troth, it can be no wonder to him that considers the Defect of any Cause, that should break off the angles from those fragments volatile, after their detrition from hard bodies, and so tornate them into smooth sphærules: observation ascertaining, that when hard bodies are broken into large pieces, those pieces are always angular, and extremely discrepant in the parts of their superfice; and Reason
thence

*Art. 3.*
The *Exility* of Atoms, doth nor necessitate their *General Roundness;* contrary to the common conceit.

thence inferring, that lesser pieces must confess the like irregularity and disparity of figures among themselves. True it is, they enter the eye in a perfect sphear, because of the exiguity of their Angles; for every small, or remote Icosahedrical body, nay even Oblong and Cylindrical, posited at excessive distance, the extremities of their images being, in their long trajection through the aer, confracted, refused, and so entering the *Retina tunica* in a lesser angle; always appear orbicular. Thus, if we speculate any star, which is not of a spherical figure, as *Saturn*, which both *Kircher* and *Hevelius*, having beheld it with their excellent Telescopes, describe in this apparence

*(In Phositima Corporum cælestium, & See-nographia.)* ☊ it will deradiate its species in a pyramid, which hath so many distinct faces, as are comprehended in the Section, made from the position of the eye, in right lines drawn to the circumference thereof; and yet in the decurse of the angle, they all become so refused, as that the image of the Starr is received by the eye in a figure perfectly spherical. And, as the excessive Remotion, so likewise doth the immoderate Exiguity of objects cause our sense not to discern their genuine Figure and so to delude the common judicatory Faculty, by giving in dissimilar images: as is demonstrable from the reason, whereby *Magnifying Glasses* meliorate the sight, i. e. their enlarging the basis of the *Radius Visorius*, according to the theory of *Kircherus* (in *Magia Catoptrica.*) and *Scheinerus* (in *Fundam. Optic lib. 3. part. 2.*). Thus, if he credit the single information of his eye, who doth not judge a *Handworm* to be exactly round? and yet let him but behold it through an Engyscope, and he shall at first inspection discern the several divarications of its Members, Leggs, Feet, Tail, and other parts, no less diverse in proportion, then those observed in multipedous Insects, of farr greater bulk.

*Art. 4.*
*The Diversity of Figures in Atoms, evidenced from the sensible Dissimilitude of individuals, as well Animate, as Inanimate.*

To guard this Assertion of the variety of Figures in Atoms, with other Arguments of its Verisimility, let us Consider, that all Individuals, as well Animate, as Inanimate, are distinguishable each from other of the same species, by some peculiar signature of disparity visible in the superficial parts of their Bodies: and Reason will thereupon whisper us in the ear, that they are also different in their Configurations; and that the Cause of that sensible Dissimilitude, must be a peculiar, or idiosyncritical Contexture of their insensible Component particles. For *Animals*, we may instance in the noblest species. Among the Myriads of swarms of men, who can find any two Persons, so absolute Twinns in the aer of their faces, the lines of their hands, the stature of their bodies, proportion of their members, &c. as that Nature hath left no impression, whereby not only their familiar friends, but even strangers comparing them together, may distinguish one from the other? For *Inanimates*; doth it not deserve our admiration, that in a whole Bushel of Corn, no two Grains can be found so exquisitely respondent in similitude, as that a curious eye shall not discover some disparity betwixt them: and yet we appeal to strict observation, for the verity thereof. If our leasure and patience will bear it, let us conferr many Leaves, collected at one time from the same Tree; and try whether among them all we can meet with any two perfectly consimilar in magnitude, colour, superfice, divarications of filaments, equality of stemms, and other external proportions. If not, we must assent to a variety of

Con-

Configurations in their parts, and consequently admit no less, but indeed a farr greater variety of Figures in the particles of those parts, their Atoms.

To these it concerns us to annex one singular *Experiment*, easie, delightful, and satisfactory. Exposing a vessel of Salt water, to the Sun, or other convenient heat, so as the aqueous parts thereof may be gently evaporated, we may observe all the Salt therein contained, to reside in the bottome, conformed into Cubical Masses. And, if we do the like with Alum Water, the Alum will concrete in Octohedrical figures. Nay, the Cubes generated of Salt, will be so much the larger, by how much the more and deeper the Water, wherein it was dissolved; and *è contra*, so much the smaller, by how much shallower the Water: so that from a large vessel will arise saline Cubes in dimensions equal to those of a Gamesters Die, but from a small we shall receive Cubes, by five parts of six, lesser, and if we drop a small quantity of brine upon a plane piece of Glass, the Cubical Concretions thereon fixing, will be so minute, as to require the help of an Engyscope to their discernment. Now, as to that part of this Experiment, which more directly points at our præsent scope, we may perceive the greater Cubes to be a meer Congeries or assembly of small ones, and those small ones to be coagmentated of others yet smaller, or certainly composed of exiguous Masses bearing the figure of Isoscele Triangles, from four of which convened and mutually accommodated, every Cube doth result. Hence is it obvious to Conjecture, that those small Cubes, discernable only by an Engyscope, are contexed of other smaller, and those again of smaller, until by a successive degradation they arrive at the exility of Atoms, at least of those Moleculæ, which are the Seminaries of Salt, and, according to evident probability, of either exactly Quadrate, or Isoscele Triangular figures. Now, insomuch as the same, allowing the difference of Figure, is conjectural also concerning Alum, Sugar, Nitre, Vitriol, &c. Saline Concretions: why may we not extend it also to all other Compositions, especially such as have their Configurations certain and determinate, according to their specifical Nature.

*Art. 5.*
*A singular experiment, au-topically demonstrating the various Configurations of the minute particles of Concretions.*

Again, whoso substracts a diversity of Figures from Atoms: doth implicitely destroy the variety of sensibles. For, what doth cause the Odoratory Nerves of man to discriminate a Rose from Wormwood? but the different Configurations of those Moleculæ, *Flores Elementorum*, or Seminaries of Qualities, which being conflated of exceeding fine and small congregations of Atoms, do constitute the odorable species; and so make different impressions upon them. What makes a Dog, by the meer sagacity of his nose, find out his Master, in the dark, in a whole host of men? but this, that those subtle *Effluvia*, or Expirations, emitted insensibly from the body of his Master, are of a different Contexture from those of all others, and so make a different impression upon the mamillary processes, or smelling Nerves of the Dog. The like may also, with equal reason, be demanded concerning those wayes of Discrimination, whereby all Animals agnize their own from others young, and Beasts of prey, in their difficult venations, single out the embossed and chased; though blended together with numerous Herds of the same species.

Nor

*Art. 6.*
*A variety of Figures in Atoms, necessary to the variety of all Sensibles.*

Nor doth the Verifimility hereof hold only in objects of the fight and fmelling; but diffufeth to thofe of the Hearing, Tafting, and Touching: as may be foon inferred by him, who fhall do us the right, and himfelf the pleafure to defcend to particulars. Thefe things jointly confidered, we are yet to feek, what may interdict our Conception of great Diverfity of Figures in the Principles of Concretions, Atoms.

Art. 7.
The fecond Ca-
non, explained
and Certified.

Concerning the SECOND, *viz.* Ἐῖναι τὰ ϛ⟨…⟩ τῶ͂ Ἀτόμων ἀπειϡάντα, ἆ χ ἄπειϱα, *effe Figuræ Atomorum incomprehenfibiles, non infi_nitas,* that the figures of Atoms are fo various, as to be incomprehenfible, though not fimply infinite: this can be nor Problem, nor Paradox. For, though the fpecies of Regular Figures be many, of Irregular more, and of thofe that are producible from both regular and irregular, according to all the poffible wayes of their Commixture and Tranfpofition, fo amufingly various; as that the mind of man, though acquainted with all the my-fteries of Arithmetique and Algebra, cannot attain to a definite compute, nor præcife defcription of them all: yet do they not run up to abfolute Infinity, fo as that there can be no extreme and terminating fpecies. That the variety of Figures competent to Atoms, ought to be held *only Incom-prehenfible*; thefe Reafons evince (1) Since Atoms are circumfcribed and limitate in Magnitude, that Configurations in diverfity infinite fhould arife from that finite magnitude, is clearly impoffible. For, every diftinct figu-ration præfuppofeth a diftinct pofition of parts; and the parts of finite Mag-nitude may be tranfpofed fo many feveral wayes, as no further way of tranf-pofition can remain poffible: otherwife there would be new and new parts inexhauftibly, and fo magnitude would become infinite. (2) If the Diver-fity of figures were infinite, then could not the Qualities arifing to concre-tions from the various Contexture of their parts, be certain and determinate: fince, allowing an inexhauftible novelty of Configurations, their infenfible particles might be fo variegated, as that a better then the beft, and a worfe then the worft of Configurations might be produced; which is no obfcure abfurdity. (3) All things are determined by Contrary Qualities, which are fo extreme, that they admit many mediate or Inclufive degrees, but none Exclufive, or without their boundaries. (4) That only a Finite va-riety is fufficient to that incomprehenfible diverfity of figures, obferved in nature.

That the variety of Figures allowable to Atoms, is *Incomprehenfible*; may be thus familiarized. Thinke we, what great multiplicity of words may be compofed of only a few Letters varioufly tranfpofed. For, if we affume only Two Letters, of them we can create only two words; if three, 6; if four, 24; if five, 120; if fix, 720; if feven, 5040; if eight, 40320; if nine, 362880; if ten, 3628800: fo that before we fulfil the 24 Letters, the number of words componible of them, according to all the poffible wayes of pofitions, will fwell above our computation. This done, let us no more but exchange Letters for Figures, and affuming only Round, Oblong, Oval, Eliptick, Lenticular, Plane, Gibbous, Turbinate, Hamous, Polite, Hifpid, Conical, Obtufe, Tetrahedical, Pentahedrical, Hexahedrical, Heptahedri-cal, Dodecahedrical, Icofahedrical, Striate or fkrewed, Triangular, Cylin-drical Atoms: caft up to what an inaffignable number the Figures produci-ble from them, according to the feveral wayes of their Compofition and tranfpofition, may amount. Doubtlefs, we fhall difcover fo great variety, as

to

to elude our comprehension. If so, how much more incomprehensible must that Diversity be, which is possible from the assumption, and complication of all the Regular and Irregular figures, that a good Geometrician can conceive, and which it is justifiable for us to allow existent in Nature?

But as for the L A S T; viz. *that the number of Atoms, retaining to each distinct species of Figures, ariseth to Infinity*, i.e. that there are infinite Oval, infinite Pyramidal, infinite Sphærical, &c. Atoms: from this we must declare our Dissent. Because, how great a number soever be assigned to Atoms, yet must the same be Defined by the Capacity of the World, i.e. of the Universe, as hath been formerly intimated. And, therefore, the common Objection, that if so, the summe of things existent in the World, would be Finite; is what we most willingly admit, there being no necessity of their Infinity, and a copious syndrome of reasons, that press the Contrary. And as it is unnecessary to Nature: so likewise to her Commentator, the Physiologist; to whom it sufficeth, having exploded this delirium of Infinity, to suppose (1) that the material Principles of the Universe are essentially Figurate, (2) that the species of their figures are incomprehensible, as to their Variety, (3) that the Number of indivisible Particles comprehended under each difference of Figures, is also incomprehensible, but not inexhaustible, as *Epicurus* inconsiderately imagined.

*Art. 8.*
The Third Canon, explained, & refuted.

---

## Sect. IV.

## Concerning the Motions of Atoms.

T O give the more light to this dark Theorem, we are to præpossess our Reader with *Two* introductory *Observables*; (1) that our præsent insistence upon only the M O T I O N of Atoms, doth not suppose our omission of their G R A V I T Y; but duely include the full consideration thereof: since their Motion is the proper Effect of their Gravity, and that which doth chiefly bring it within the sphære of our Apprehension. (2) That the genuine Atomist doth worthily disavow all Motion, but what *Plutarch* in the name of *Epicurus*, hath defined to be, Μεταβασις απο τουτε εις τουτον, *Migratio de loco in locum*, the translation of a thing from one place to another. The suspicion of a *Chasme* in our Discourse, and the Ambiguity imminent from the Æquivocality of the term, Motion, thus maturely prævented: we may more smoothly progress to our short Animadversions on the Conceptions of the Ancients, touching the Last General Propriety of Atoms, their Congenial and intestine Motion.

*Art. 1.*
Two introductory Observables.

*Art. 2.*
The Motion of Atoms, according to the General Distinction of the Ancients, Two-fold; viz. *Natural*, and *Accidental*: & each of these redivided into two different *Species.*

Herein we are to recognize their opinions, that concern (1) the *Multiplicity*, (2) the *Perpetuity* of motions essentially competent to Atoms.

As to the F I R S T; they have, according to a General Distinction, assigned to Atoms a Two-fold Motion; (1) *Natural*, whereby an Atom, according

R

ing to the tendency of its essential weight, is carried directly downward:
(2) *Accidental*, whereby one Atom justling or arienating against another, is
diverted from its perpendicular descendence, and repercussed another way.
The Former, they called *Perpendicular*, the other, *Reflex*. The Natural or
Perpendicular *Epicurus* hath doubled again into κατὰ σιβυλω, *ad per-
pendiculum*, or as *Cicero* (*de fato*) interprets it, *ad Lineam*: and κατὰ
παμάτκλιον, *ad Declinationem*. The Accidental, or Reflex hath also,
according to the tradition of *Plutarch*, (1. *placit*.12.) been by him subdivided into κατὰ πληγω, *ex plaga, seu ictu*; and κατὰ πληγω, *ex concussione*, or rather, *ex Palpitatione*. So that, according to this special Distinction, there must be four different sorts of motions assignable to Atoms.

*Art. 3.*
The summary
of *Epicurus*
Figment, of
the *Perpendi-
cular* Motion
of Atoms,
without a
common *Cen-
tre.*

For the *Perpendicular* Motion, we advertise; that *Epicurus* therein
had no respect to any *Centre* either of the *World*, or the *Earth*; for
He conceded none such possible in the Universe, which He affirmed of
infinite extent: but to two contrary Regions allowable therein, the
one *Upward*, from whence, without any *terminus a quo*, Atoms flowed;
the other *Downward*, toward which, without any *terminus ad quem*, in
a direct line they tended. So that, according to this wild dream, any
coast from whence Atoms stream, may be called *Above*, and any to
which they direct their course, *Below*; insomuch as He conceited the
superfice of the Earth, on which our feet find the Centre of Gravity
in standing or progression, to be one continued plane, and the whole Horizon above it likewise a continued plane running on in extent not only to
the Firmament, but the intire immensity of the Infinite Space. According to which Delirament, if several weights should fall down from the firmament, one upon *Europe*, another upon *Asia*, a third upon *Africa*, a
fourth upon *America*; and their motion be supposed to continue beyond
the exteriors of the terrestrial Globe: they could not meet in the Centre
thereof, but would transfix the four quarters in lines exquisitely parallel,
and still descend at equal distance each from other, untill the determination
of their motion in the infinite Space, by the occurse and resistence of other
greater Weights.

*Art. 4.*
His *Declinatory*
natural Moti-
on of Atoms,
excused; not
justified.

For the *Declinatory* Motion, we observe, that *Epicurus* was by
a kind of seeming necessity constrained to the Fiction thereof, since
otherwise He had left his fundamental Hypothesis manifestly imperfect,
his Principles, destitute of a Cause for their Convention, Confliction,
Cohærence, and consequently no possibility of the emergency of Concretions from them. And, therefore, to what *Cicero* (*in* 1. *de fin.*) objects
against him, viz. that he acquiesced in a supposition meerly *præcarious*, since
he could assign no *Cause* for this motion of Declination, but usurped the
indecent liberty of endowing his Atoms with what Faculties he thought advantagious to the explanation of Natures Phænomena in Generation and
Corruption: we may modestly respond, by way of excuse not justification,
that such is the imbecillity of Human understanding, as that every Author of a
physiological Fabrick, or mundane Systeme, is no less obnoxious to the same
objection, of præsuming to consign Provinces (for the phrase of *Cicero*,
is *dare provincias principiis*) to his Principles, then *Epicurus*. For,
in Concretions or Complex Natures, to determine on a reason for
this or that sensible Affection, is no desperate difficulty, since the condition

of

of præassumed Principles may afford it: but, concerning the originary Causes of those Affections inhærent in and congenial to the Principles of those Concretions, all we can say, to decline a downright confession of our ignorance, is no more then this, that such is the necessity of their peculiar Nature; the proper and germane δὶ ὅτι remaining in the dark to us, and so our Curiosity put to the shift of simple Conjecture, unless we level our thoughts above Principles, and acknowledge no term of acquiescence. And even the acute and perspicacious *Cicero*, notwithstanding his reprehension of it in *Epicurus*, is forced to avow the inevitability of this Exigent, in express words, thus, *Ne omnes à Physicis irrideamur, si dicamus quicquam fieri sine Causa distinguendum est, & ita dicendum; ipsius Individui hanc esse naturam, ut pondere & gravitate moveatur, eamque ipsam esse Causam, cur ita feratur, &c.* Nor is this Crime of consigning provinces to his Principles, proper only to *Epicurus*, but common also to the *Stoick*, *Peripatetick*, &c. since none of them hath adventured upon a reason of the Heat of Fire, the Cold of Water, the Gravity of Earth, &c. Doubtless, had *Cicero* been interrogated, Why all the Starrs are not carried on in a motion parallel to the Æquator, but some steer their course obliquely; why all the Planets travel not through the Ecliptick, or at least in a motion parallel thereto, but some approach it obliquely: the best answer He could have thought upon, must have been only this, *ita Naturæ leges ferebant*; which how much beseeming the perspicacity of a Physiologist more then to have excogitated Fundamentals of his own, endowed with inhærent Faculties to cause those diverse tendencies, we referr to the easie arbitration of our Reader.

Concerning the *Accidental*, or *Reflex* Motion, all that is worthy our serious notice, is only this, that when *Epicurus* subdivideth this Genus into two species, namely κατὰ πληγὴ, *ex plaga*, and κατὰ παλμὸν, *ex concussione*, and affirmeth that all those Atoms which are (ἄνω κινουμένα) moved *upward*, pursue both sorts of this Reflex tendency; we are not to understand him in this sense, that both these kinds of Reflex motion are opposite to the Perpendicular, since it is obvious to every man, that Atoms respective to their Direct, or Oblique incidence in the different points of their superfice, may make, or rather suffer or direct, or oblique resilitions, and *Epicurus* expresly distinguisheth the Motion from Collision or Arietation into that which pointeth *upward*, and that which pointeth *sidewayes*; but in this, that he might constitute a certain Generical Difference, whereby both the species of Reflex motion might be known from both the species of the Perpendicular. For the further illustration of this obscure Distinction, and to prævent that considerable Demand, which is consequent thereto, viz. *whether all the possible sorts of Reflex Motion are only two, the one directly Upward, the other directly Lateral:* we advertise, that *Epicurus* seems to have alluded to the most sensible of simple Differences in the Pulse of Animals. For, as *Physitians*, when the Pulsifick Faculty distends the Artery so amply, and allows so great a space to the performance of both those successive contrary motions, the Diastole and Systole, as that the touch doth apprehend each stroke fully and distinctly, denominate that kind of Pulse, σφυγμὸς; and on the contrary, when the vibrations of the Artery are contracted into a very little space as well of the

R 2                    ambient

*Art. 5.*
The genuine
Rule of *Epicurus*, in his distinction of
the *Reflex* Motion of Atoms
into *ex Plaga*
& *ex Concussione.*

ambient, as of time, so as they are narrow and confusedly præsented to the
touch, they call it πλλμος: so likewise *Epicurus* terms that kind of Rebound,
or Resilition, which by a strong and direct incurse and arietation of one
Atom against another, is made to a considerable distance, or continued
through a notable interval of space, χξ' πλησjω; and, on the contrary, that
which is terminated in a short or narrow interval (which comes to pass,
when the resilient Atom soon falls foul upon a second, and is thereby revi-
berated upon a third, which repercusseth it upon a fourth, whereby it is again
bandied against a fifth, and so successively agitated, until it endure a perfect
*Palpitation*) he styles χξ' πιλμgσ. Upon this our Master *Galen* may be
thought to have cast an eye, when he said (*lib. de facult. nat.*) it was the opinion
of *Epicurus, Omnes attractiones per resilitiones atque implexiones Atomorum
fieri* that all Attractions were caused by the Resilitions and Implexions
of Atoms. Which eminent passage in *Galen*, not only assisted, but interpreted by another of *Plato* (*Magnetem non per Attractionem, sed Impulsio-
nem agere, in Timæo*) of the same import; hath given the hint to *Des Cartes, Regins*, Sir *K. Digby*, and some other of our late Enquirers, of supposing
the Attractive, rather *Impulsive* Virtue of the Loadstone, and all other bodies Electrical, to consist in the Recess, or *Return* of those continued Effluvia,
or invisible filaments of streated Atoms, which are uncessantly exhaled from
their pores. Nor doth He much strain these words of *Gilbert* [ *Effluvia
illa tenuiora concipiunt & amplectuntur corpora, quibus Uniuntur, & Ele-
ctris, tanquam extensis brachiis, & ad fontem prope invalescentibus effluviis,
deducuntur* ] who hath charged them with the like signification.

*Art. 6.*
The several
Conceptions
of *Epicurus*, a-
bout the per-
*petual* Motions
of Atoms.

As to the SECOND, *viz.* the *Perpetuity* of these Motions adscribed to Atoms; we think it not a little material to give you to understand,
at least to recognize that the conceptions of *Epicurus* concerning this particular, are cozen Germans to Chimæras, and but one degree removed
from the monstrous absurdities of Lunacy. For, He dreamt, and then believed, that all Atoms were from all Eternity endowed, by the charter, of
their uncreate and independent Essence, with that ingenite Vigour, or internal Energy, called Gravity, whereby they are variously agitated in the
infinite space, without respect to any Centre, or General term of Consistence: so as they could never discontinue that natural motion, unless they
met and encountred other Atoms, and were by their shock or impulse deflected into another course. That the Dissilient or deflected Atoms, whether rebounding upwards directly, or *ad latus* obliquely, or in any line intercedent betwixt those two different regions, would also indefinently pursue
that begun motion, unless they were impeded and diverted again by the occurse and arietation of some others floating in the same part of space. And,
that because the Revibrations, or Resilitions of Atoms regarding several
points of the immense space, like Bees variously interweaving in a swarm,
must be perpetual: therefore also must they never quiesce, but be as variously and constantly exagitated even in the most solid or adamantine of
Concretions, though the sense cannot deprehend the least inquietude or
intestine tumultuation therein; and the rather in respect of those Gro-
tesques or minute Inanities densely intermixed among their insensible particles.

*Art. 7.*
The perpetual
*inquietude* of
Atoms, even
in compact
Concretions,
adumbrated in
*melte d Lead.*

To explicate this Riddle, we must præsent some certain adumbration
of this intestine æstuation or commotion of Atoms in Concretions; and
this

this may moſt conveniently be done in melted Mettals, as particularly in Lead yet floating in the Fuſory veſſel. To apparence nothing more quiet and calm: yet really no quickſand more internally tumultuated. For, the inſenſible particles of Fire having penetrated the body of the crucible, or melting pan, and ſo permeating the pores of the Lead therein contained; becauſe they cannot return back upon the ſubjacent fire, in regard they are unceſſantly impelled by other ingeneous particles continually ſucceeding on their heels, therefore are they ſtill protruded on, untill they diſunite all the particles of the Lead, and by the pernicity and continuation of this their ebullition, hinder them from mutual revinction and coaleſcence: and thereby make the Lead a fluid, of a compact ſubſtance, and ſo keep it, as long as the ſuccuſſion of igneous particles is maintained from the fire underneath. During this act of Fuſion, think we, with what violence or pernicity the Atoms of Fire are agitated up and down, from one ſide to another, in the ſmall inanities interſperſed among the particles of the Lead; otherwiſe they could not diſſolve the compact tenour thereof, and change their poſitions ſo as to introduce manifeſt Fluidity: and, ſince every particle of the Lead, ſuffers as many various concuſſions, repercuſſions, and repeated vibrations, as every particle of Fire; how great muſt be the Commotion on both ſides, notwithſtanding the ſeeming quiet in the ſurface of the Lead?

But, becauſe our ſenſe, as well as our Reaſon, may have ſome ſatisfaction, touching the perpetual Commotion of Atoms, even in Compoſitions; we offer to Exemplifie the ſame either in the *Spirit* of *Halinitre*, or that which Chymiſts uſually extract from *Crude Mercury*, *Tin*, and *Sublimate* codiſſolved in a convenient menſtruum: For, either of theſe Liquors being cloſe kept in a luted glaſs, you may plainly perceive the minute moleculæ, or ſeminarie conventions of Atoms, of which it doth conſiſt, to be unceſſantly moved every way, upward, downward, tranſverſe, oblique, &c. in a kind of fierce æſtuation, as if goaded on by their inhærent Motor, or internal impulſive Faculty, they attempted ſpeedy emergency at all points, moſt like a multitude of Flyes impriſoned in a glaſs Vial.

*Art. 8.*
The ſame more ſenſibly exemplified, in the ſpirit extracted from *Mercury*, *Tin*, and *Sublimate*.

Now, the *Argument* that ſeems to have induced *Epicurus* to concede this perpetual Inquietude of Atoms, was the *inevitable mutation of all Concrete Subſtances*, cauſed by the continual Acceſs and Receſs of their inſenſible particles. For, indeed, no Concretion is ſo compact and ſolid, as not to contain within it ſelf the poſſible Cauſes of its utter Diſſolution; yea, though it were ſo immured in Adamant, as to be thought ſecure from the hoſtile invaſion of any Extrinſecal Agent whatever. And the ruine of ſolid bodies (i. e. ſuch whoſe parts are of the moſt compact Contexture allowable to Concretions,) cannot be ſo reaſonably adſcribed to any Cauſe, as this; that they are compacted of ſuch Principles, as are indeſinently motive, and in perpetual endeavour of Emergency or Exſilition: ſo that never deſiſting from internal evolutions, circumgyrations, and other changes of poſition; they at length infringe that manner of reciprocal Coaptation, Coheſion, and Revinction, which determined their ſolidity, and thereby diſſolving the Compoſitum, they wholly emancipate themſelves, obey their reſtleſs tendency at randome, and diſappear.

*Art. 9.*
The Mutability of all Concretions; a good Argument of the perpetual inteſtine Commotion of Atoms, in the moſt adamantine Compoſitions.

This fæculent Doctrine of *Epicurus*, we had occaſion to examine and refine all the droſs either of Abſurdity, or Atheiſm, in our Chapter concerning

*Art. 10.*
What we are to explode, and what retain, in the opinion of *Epicurus*, touching the Motion of Atoms.

cerning the *Creation of the World ex nihilo*, in our Book against *Atheism*. However, we may not dismiss our Reader without this short Animadversion. The Positions to be exploded are (1) *That Atoms were Eternally existent in the infinite space*, (2) *that their Motive Faculty was eternally inherent in them, and not derived by impression from any External Principle*, (3) *that their congenial Gravity affects no Centre*, (4) *that their Declinatory motion from a perpendicular, is connatural to them with that of perpendicular descent, from Gravity.* Those which we may with good advantage substitute in their stead, are (1) *That Atoms were produced* ex nihilo, *or created by God, as the sufficient Materials of the World, in that part of Eternity, which seemed opportune to his infinite Wisdom*; (2) *that, at their Creation, God invigorated or impregnated them with an Internal Energy, or Faculty Motive, which may be conceived the First Cause of all Natural Actions, or Motions,* (for they are indistinguishable) *performed in the World*; (3) *that their gravity cannot subsist without a Centre*; (4) *that their internal Motive Virtue necessitates their perpetual Commotion among themselves, from the moment of its infusion, to the expiration of Natures lease.* For, by virtue of these *Correctives*, the poisonous part of *Epicurus* opinion, may be converted into one of the most potent *Antidotes* against our Ignorance: the *Quantity* of Atoms sufficing to the Materiation of all Concretions; and their various *Figures* and *Motions* to the Origination of all their *Qualities* and *Affections*, as our immediately subsequent Discourse doth professedly assert.

The

# The Third Book.

## CHAP. I.

### *The Origine of Qualities.*

#### Sect. I.

That the sounding Line of Mans Reason is much too short to profound the *Depths*, or Channels of that immense Ocean, *Nature*, needs no other evictment but this, that it cannot attain to the bottom of Her *Shallows*. It being a discouraging truth, that even those things, which are familiar and within the sphere of our *Sense*, and such to the clear discernment whereof we are furnished with Organs most exquisitely accommodate, remain yet ignote and above the Moon to our *Understanding*. Thus, what can be more evident to sense, then the *Continuity* of a Body: yet what more abstruse to our reason, then the *Composition* of a *Continuum?* What more obviously sensible then *Qualities*: and yet what problem hath more distracted the brains of Philosophers, then that concerning their *Unde*, or *Original?* Who doth not know, that all Sensation is performed by the Mediation of certain Images, or Species: yet where is that He, who hath hit the white, in the undoubted determination of the Nature of a species, or apodictically declared the manner of its Emanation from the Object to the Sensorium, what kind of insensible-sensitive impression that is, which it maketh thereupon, and how being from thence, in the same

*Art.* 1.
An introductory Advertisement, of the *obscurity* of many things to *Reason*, which are *manifest to sense*: and of the *Possibility*, not necessity of the Elementation of *Concretions*, and their sensible *Qualities*, from the Principles praelibated.

same instant transmitted to that noble *something* within us, which we understand not, it proves a lively *Transumpt*, or type, and informs that ready judge of the Magnitude, Figure, Colour, Motion, and all other apparences of its *Antitype* or Original? or, what hath ever been more manifest or beyond dubitation, then the reality of *Motion?* and yet we dare demand of *Galilæo* himself, what doth yet remain more impervestigable, or beyond apodictical decision, then the *Nature* and *Conditions* thereof.

Concerning the *First* of these 4 ænigmatical Quæstions, we have formerly præsented you no sparing account of our Conjectural opinion: which we desire may be candidly accepted in the latitude of *Probability* only, or how it *may* be, rather then how *it is*, or *must* be; i.e. that it is, though most *possible* and *verisimilous* that every Physical Continuum should consist of Atoms; yet not absolutely *necessary*. For, insomuch as the true Idea of Nature is proper only to that *Eternal Intellect*, which first conceived it: it cannot but be one of the highest degrees of madness for dull and unequal man to prætend to an exact, or adæquate comprehension thereof. We need not advertise, that the Zenith to a sober Physiologists ambition, is only to take the copy of Nature from her shadow, and from the reflex of her sensible Operations to describe her in such a symmetrical Form, as may appear most plausibly satisfactory to the solution of all her Phænomena. Because 'tis well known, that the eye of our grand Master *Aristotles* Curiosity was levelled at no other point, as himself solemnly professeth (*in Meteorolog. lib.*1. *cap.* 7. *initio*) in these words: Ἐπεὶ δὲ περὶ τῆς ἀφωνίων τῇ αἰσθήσει νομίζομεν ἱκανῶς ἀποδεδεῖχθαι κατὰ τ λόγον, ἐὰν εἰς τὸ δυνατὸν ἀναγάγωμεν, ἐκ τε τῆς νῦν φαινομένων ὑπολάβοι τις ἂν, ὧδε περὶ τύτων μάλιςα συμβαίνειν: i.e. *Cum autem de hisce, quæ sensui pervia non sunt, satis esse juxta rationem demonstratum putemus, si ad id quod fieri possit ea reduxerimus, ex hisce quæ in præsentia dicuntur, existimaverit quispiam de hisce maxime ad hunc modum usu venire.* And evident it is that *Monf. Des Cartes* never was more himself, that is, profoundly ingenious, then when he crowned his excellent Principles of Philosophy with this advertisement: *at quamvis forte hoc pacto intelligatur, quomodo res omnes naturales fieri potuerint; non tamen ideo concludi debet, ipsas reverà sic factas esse: & satis à me præstitum esse putabo, si tantum ea quæ scripsi, talia sint, ut omnibus Naturæ Phænomenis accurate respondeant; hoc enim ad usum vitæ sufficiet.*

And, concerning the *other three*, which according to the natural order of their dependence, are successively the Arguments of our next ensuing Exercitations; we likewise deprecate the same favourable interpretation, in the General: that so, though our attempts perhaps afford not satisfaction to others, yet they may not occasion the scandal of *Arrogance* and *Obstinacy* in opinion to our selves.

*Art.* 2.

The Authors Definition of a Quality, in general: and genuine exposition of *Democritus* mysterious Text, concerning the *Creation* of *Qualities.*

By the *Quality* of any Concretion, we understand in the General, no more but that *kind of Apparence, or Representation, whereby the sense doth distinctly deprehend, or actually discern the same, in the capacity of its proper Object.* An *Apparence* we term it, because the *Quale* or *Suchness* of every sensible thing, receives its peculiar determination from the relation it holds to that sense, that peculiarly discerns it: at least from the judgment made in the mind according to the evidence of sensation. Which doubtless

was

was the genuine intent of *Democritus* in that remarkable and mysterious text, recorded by *Galen* (*in lib.* 1. *de Element. cap.* 2.) thus: Νόμῳ χροιή, νόμῳ πικρόν, νόμῳ γλυκύ; ἐτεῇ δ᾽ ἄτομον, ᾗ κενόν ὁ Δημόκριτος φησὶν ἐκ τ᾽ ζωίωδε τῶν ἀτόμων γίνεσθαι νομίζων ἀπάσας τας αἰσθητὰς ποιότητας, ὡς πρὸς ἡμᾶς τὰς αἰσθανομένας αὐτῶν, φύσει δ᾽ ἀδὲν ἔιναι λδκον, ἤ μέλαν, ἤ ξανθον, &c. *Lege enim Color, lege amaror, lege dulcor ; revera autem Atomus, & Inane, inquit Democritus, exiſtimans omneis Qualitates ſenſibileis ex Atomorum concurſu gigni, quatenus ſe habent ad nos; qui ipſarum ſenſum habemus : Natura autem nihil candidum eſſe, aut flavum, aut rubrum, &c.*
The importance of which may be fully and plainly rendred thus ; that ſince nothing in the Univerſe ſtands poſſeſſed of a Real or True Nature, i.e. doth conſtantly and invariately hold the præciſe Quale, or Suchneſs of their particular Entity, to Eternity ; Atoms (underſtand them together with their eſſential and inſeparable Proprieties, lately ſpecified. ) and the Inane Space only excepted : therefore ought all other things, and more eminently Qualities, in regard they ariſe not from, nor ſubſiſt upon any indeclinable neceſſity of their Principles, but depend upon various tranſient Accidents for their exiſtence, to be reputed not as abſolute and entire Realities, but ſimple and occaſional Apparences, whoſe ſpecification conſiſteth in a certain modification of the Firſt Matter, reſpective to that diſtinct Affection they introduce into this or that particular ſenſe, when thereby actually deprehended. Not that *Democritus* meant, in a litteral ſenſe, that their production was determinable *ex inſtituto hominum*, by the opinionative laws of mans Will ; as moſt of his Commentators have inconſiderately deſcanted : but in a *Metaphorical*, that as the juſtice, injuſtice, decency, turpitude, culpability, laudability of Human actions, are determined by the Conformity or Diſformity they bear to the Conſtitutions Civil, or Laws generally admitted, ſo likewiſe do the whiteneſs, blackneſs, ſweetneſs, bitterneſs, heat or cold, of all Natural Concretions receive their diſtinct eſſence, or determination from certain poſitions and regular ordinations of Atoms. And this eaſily hands us to the natural ſcope of that paſſage in *Laertius*, Ἀρχὰς ἔιναι τῶν ὅλων ἀτόμες, ᾗ κενόν, τὰ ᾗ ἄλλα πάντα νενομίσθαι, *Eſſe Atomos & Inane Univerſorum principia, cætera omnia Lege ſanciri :* as alſo of another in *Empiricus* (1. *hypot.* 30.) ἐτον Ἀτόμα ᾗ κενόν, *VERE eſſe Inſectilia ac Inane.* However, if any pleaſe to prefer the expoſition of *Magnenus*, that *Democritus* by that unfrequent and gentiliritious phraſe, *Nemo eſſe Qualitates*, would have the determinate nature of any Quality to conſiſt *in certa quadam lege, & proportione inter agens & patiens*, in a certain proportion betwixt the Agent and Patient, or object and ſenſorium ; we have no reaſon to proteſt againſt his election. For we ſhall not deny, but what is *Hony* to the palate of one man, is *Gall* to another ; that the moſt delicious and poynant diſhes of *Europe*, are not only inſipid but loathſome to the ſtomachs of the *Japones*, who in health eat their Fiſh boyled, and in ſickneſs raw, as *Maſſeus* ( *in libro de Japonum moribus* ) reports ; that ſome have feaſted upon Rhubarb, Scammony, and Eſula, which moſt others are ready to vomit and purge at the ſight of ; that Serpents are dainties to Deer, Hemlock a perfect Cordial to Goats, Hellebora choyce morſel to Quails, Spiders reſtorative to Monkeys, Toads an Antidote to Ducks, the Excrements of man pure Ambre Griſe to Swine, &c. All which moſt evidently declare the neceſſity of a certain proportion or Correſpondence betwixt the object and particular organ of ſenſe, that is to apprehend and judge it.

S                                          But,

But since the Notion of a *Quality* is no rarity to common apprehension, every Clown well understanding what is signified by *Colour, Odour, Sapour, Heat, Cold, &c.* so far as the concernment of his sense we are no longer to suspend our indagation of their possible ORIGINE, in the general.

*Art. 3.*
The necessary deduction of Qualities from Naked or Unqualified Principles.

Which, were our Atoms identical with the *Homoiomerical* Principles of *Anaxagoras* formerly described, and exploded, might be thought a task of no difficulty at all: in regard those Consimilarities are supposed actually to contain all Qualities, in the simplicity of their nature, or before their Convention and Disposition into any determinate Concretion; i. e. that Colour, Odour, Sapor, Heat, Cold, &c. arise from Colorate, Odorate, Sapid, Hot, Cold particles of the First Catholique Matter. But, insomuch, as *Atoms*, if we except their three congenial Proprieties, *viz.* Magnitude (which by a general interest, retains to the Category of Qualities) Figure, and Motion; are unanimously assumed to be *Exquales*, seu *Qualitatis Expertes*, absolutely devoid of all Quality: it may seem, at first encounter, to threaten our endeavors with infelicity, and damp Curiosity with despair of satisfaction. And yet this Giant at distance, proves a mere Pygmie at hand. For, the *Nakedness*, or *Unqualifiedness* of Atoms, the point wherein the whole Difficulty appears radicated; to a closer consideration must declare it self to be the basis of our exploration, and indispensably necessary to the Deduction of all sensible Qualities from them, when disposed into Concrete Natures. Because, were any Colour, Odour, &c. essentially inhærent in Atoms; that Colour, or Odour must be no less intransmutable then the subject of its inhæsion: and that Principles are Intransmutable, is implied in the notion of their being Principles; for it is of the formal reason of Principles, constantly to persever the same in all the transmutations of Concretions. Otherwise, all things would inevitably, by a long succession of Mutations, be reduced to clear Adnihilation. Besides, all things become so much the more Decoloured, by how much the smaller the parts are into which they are divided; as may be most promptly experimented in the pulverization of painted Glass, and pretious stones: which is demonstration enough, that their Component Particles, in their Elementary and discrete capacity, are perfectly destitute of Colour. Nor is the force of this Argument restrained only to Colour, as the most eminent of Qualities sensible: but extensible also to all others, if examined by an obvious insistence upon particulars.

*Art. 4.*
The two primary *Events* of Atoms, *viz. Order* and *Position*, associated to their three essential *Proprieties, viz. Magnitude, Figure, and Motion*; sufficient to the Origination of all Qualities.

Now, having taken footing on the necessary Incompetence of any sensible Quality to the Material Principles of Concretions: we may safely advance to our Investigation of the Reason, or Manner how Colour, and all other Qualities may be educed from such naked and unqualified Principles. And first we must have recourse to some few of the most considerable E V E N T S consignable to Atoms, as well as to their 3 inseparable Proprieties. The primary, and to this scope, most directly pertinent Events of Atoms, are only two, *viz.* τάξις καὶ χέσις, ORDER and SITUATION. That *Leucippus* and *Democritus*, besides those two eminent events, Cύνκρισις καὶ διάκρισις, *Concretion*, and *Secretion*, from which the *Generation* and *Corruption* of all things are derived; have also attributed unto Atoms, two other as requisite to all *Alteration*, i. e. the procreation of various Qualities, namely *Order* and *Position*: is justifiable upon the testimony

mony of *Aristotle* (*in lib. de ortu & interitu*) however He was pleased (*in 8. Metaphys. cap. 2.*) interpreting the Abderitane terms of *Democritus*, to adnumerate το ρυσμα, *Figure*, unto them, and thereupon inferr that Atoms are different, ή ρυσμω, ο ϛι σχημα, ή ξοπη, ο ϛι θεσις, ή διαθγη, ο ϛι ταξις, i. e. aut *Rhysmo*, quod est *Figura*; aut *Trope*, quod est *situs*; aut *Diathege*, quod est *ordo*: & (*in Metaphys.* 1. *cap.* 4.) to exemplifie this difference in Letters of the Alphabet; saying that A and N differ in *Figure*; A N, and N A, in order; and Z N, in situation. Which is the same with what *Empiricus* (2. *adverf.phys.*) reports to have been delivered by *Epicurus*. True it is, his Difciple *Lucretius*, exceeded him in the number of *Events* assignable to Atoms, in order to the emergency of all fenfible Qualities from them; for he composing this Diftich

> *Intervalls, Vie, Connexus, Pondera, Plaga,*
> *Concurfus, Motus, Ordo, Pofitura, Figura,*

confounds both *Events* and *Conjuncts* together: wherein He feems to have had more regard to the finoothnefs of his Verfes, then the Methodical traction of his Subject. For, *Motion, Concurfe,* and *Percuffion* are the natural Confequents of *Gravity*: and *Distance* and *Connexion* are included in *Position*; and *Wayes* or Regions belong to *Order*, as may be exemplified in the former Letters, which refpective to their remote or Vicine Pofition, and their Change from the right to the left hand, exhibite to the fenfe various faces or apparences.

That those two Conjuncts, *Magnitude* and *Motion*, are necefsarily to be affociated to Order and Pofition; is evident from hence, that if it be enquired, why there is in Light fo great a fubtility of parts, as that in an inftant it penetrates the thickeft Glafs; but fo little in Water, as that it is terminated in the fuperfice thereof: what more verifimilous reafon can be alledged to explain the Caufe of that difference in two fluid bodies, then this, that the Component Particles of Light are more minute, or have lefs of Magnitude, then those of Water? And if it be enquired, why the Aer, when agitated by the wind, or a fan, appears Colder, then when quiet; what folution can be more fatisfactory, then this, that by reafon of its motion it doth more deeply penetrate the pores of the skin, and fo more vigoroufly affect the fenfe? However, if we confine our affumption only to these three Heads, *Figure, Order,* and *Pofition*; we fhall yet be able, without much difficulty, to make it out, how from them, either fingle, or diverfly commixt, an infinite Multiplicity of Qualities may be created; as may be moft appofitely explained by the Analogy which Letters hold to Atoms. For as *Letters* are the *Elements* of *Writing*, and from them arife by gradation, Syllables, Words, Sentences, Orations, Books: fo proportionately are *Atoms* the *Elements* of *Things*, and from them arife by gradation, moft exile Moleculæ, or the Seminaries of Concretions, then greater and greater Maffes fucceffively, until we arrive at the higheft round in the fcale of Magnitude.

But we are reftrained to an infiftence only upon our 3 Heads affumed. As Letters of divers Figures, U, G, A, E, O, when præfented to the eye, carry 3 different fpecies, or afpects; and when pronounced, affect the Ear with as many diftinct founds: exactly fo do Atoms, refpectively to the variety

S 2                                                           riety

*Art. 5.*
The neceffity of affuming the *Magnitude* and *Motion* of Atoms, together with their *Order* and *Situation*, as to their production of Qualities, evicted by a double inftance.

*Art. 6.*
The *Figure, Order* and *Pofition* of Partes in Concretions, alone fufficient to the Caufsation of an indefinite variety of Qualities, from the analogy of Letters.

riety of their Figures, and determinate Contexture into this or that species, occurring to the Organs of Sight, Hearing, Smelling, Tasting, Touching, make divers impressions thereupon, or præsent themselves in divers
Apparences, or (what is tantamount) make divers Qualities. (2) As one
and the same Letter diversly posited, is divers to the Sight, and Hearing, as
may be instanced in Z, N, y, ⋏, b, d, p, q : so likewise doth one and the
same Atom, according to its various positions, or faces, produce various
affections in the Organs of Sense. For instance, if the Atome assumed be
Pyramidal : when the Cone is obverted to the sensory Organ, it must make
a different impression upon it, from that which the Base, when obverted
and applyed, will cause. (3) As the same two three or more Letters, according to their mutation of Site, or Antecession and Consequution, impart divers words to the eye, divers sounds to the ear, and divers things to
the mind; as ET, TE, IS, SI, SUM, MUS, ROMA, AMOR, MARO,
RAMO, ORAM, MORA, ARMO, &c. so also may two three, or
more Atoms, according to their various positions and transpositions, affect
the sense with various Apparences, or Qualities. (4) And as Letters,
whose variety of Figures exceeds not those of the Alphabet, are sufficient
only by the variety of order, to compose so great diversity of words, as
are contained in this, or all the Books in the World : so likewise, if there
were but 24 diverse Figures competent to Atoms, they alone by variety of
Order, or transposition, would suffice to the constitution of as incomprehensible a diversity of Qualities. But, when the diversity of their Figures is
incomparably greater: how infinitely more incomprehensible must that variety of Qualities be, which the possible changes of their Order may
produce ?

*Art. 7.*
The same Ex-
emplified in
the arise of
*White Froth,* on
the Waves of
the Sea.

Thus in the Water of the Sea, when agitated into a white froth, no other
mutation is made, save only the situation and differing contexture of the
parts thereof disposed by the included aer into many small bubbles; from
which the incident rayes of Light ( which otherwise would not have been
reflected in united ) and direct streams to the eye, and so creat a whiteness
continued, which is but paler, or weaker light, which must disappear immediately upon the dissolution of the bubbles, and return of the parts of the
water to their natural constitution of fluidity.

*Art. 8.*
The Nativity
of *Colours* in
General, ex-
plained by se-
veral obvious
Examples.

And since we are fallen upon that eminent Quality, *Colour*; we shall illustrate the obscure nativity thereof, in the general, by a most prægnant example. Immerge into a Glass Vial of clean fountain Water, set upon warm
embers, half an ounce (more or less, according the quantity of Water) of the
leaves of Senna; and after a small interval of time, instill into the infusion
a few drops of the oil of Tartar made *per Deliquium,* which done, you shall
perceive the whole mixture to become Red. Now, seeing that no one of
the three ingredients, in their simple and divided state, do retain to that species of Colour, in the remotest degree of affinity; from what original can
we derive this emergent Redness ? Doubtless, only from hence; that the
Water doth so penetrate, by a kind of Discussion separate, and educe the
smaller particles of that substance, whereof the leaves of Senna are composed, as that the particles of the oyl of Tartar subtily permeating the infusion,
totally alter the Contexture thereof, and so commove and convert its minute dissolved particles, as that the rayes of Light from without falling upon
them,

them, suffer various refractions and reflections from their several obverted faces, and præsent themselves to the eye in the apparence of that particular Colour.     And to confirm you herein, you need only instead of oyl of Tartar, infuse the like proportion of oyl of Vitriol into the same Tincture of Senna: for, thereupon no such redness at all will arise to the composition.     Which can be solved by no better a reason than this, that the oyl of Vitriol wants that virtue of commoving and converting the educed particles of the Senna into such positions and order, as are determinately requisite to the incidence, refraction, and reflection of the rayes of Light to the eye, necessary to the creation of that Colour.     On the Contrary, instead of Senna, infuse Rose leaves in the Water, and superaffuse thereto a few drops of the Spirit of Vitriol: and then the infusion shall instantly acquire a purple tincture, or deep scarlet; when from the like or greater quantity of oyl of Tartar instilled, no such event shall ensue.     Both which Experiments collated are Demonstration sufficient, that a Red may be produced from simples absolutely destitute of that gloss, only by a determinate Commixture, and position of their insensible particles: no otherwise then as the same Feathers in the neck of a Dove, or train of a Peacock, upon a various position of their parts both among themselves, and toward the incident Light, præsent various Colours to the eye; or as a peice of Changeable Taffaty, according as it is extended, or plicated, appears of two different dyes.     The same may also be conceived of the Cærule Tincture caused in White Wine by *Lignum Nephriticum* infused when the Decoction thereof shall remain turbid and subnigricant.

Moreover, left we leave you destitute of Examples in the other 4 orders of Qualities, respondent to the 4 remaining senses, to illustrate the sufficiency of Figure, Order and Situation, to their production; be pleased to observe.

First, that *Lead* calcined with the *spirit* of the most eager *Vinegre*, so soon as it hath imbibed the moysture of the ambient aer, or be irrigated with a few drops of Water, will instantly conceive so intense a *heat*, as to burn his finger that shall touch it.     Now, since both the Calcined Lead and Water are actually Cold, and no third Nature is admixt, and nothing more can be said to be in them when commixt, that was in them during their state of separation; whence can we deduce that intense Heat, that so powerfully affecteth, indeed, misaffecteth the sense of *Touching?*     Quæstionless, only from this our *triple fountain*, i. e. from hence, that upon the accession of humidity, the acute or pointed particle of the spirit of Vinegre, (whereby the fixed salt of the Lead was, by potential Calcination, dissolved; and the Sulphur liquated) change their order and situation, and after various convolutions, or the motions of Fermentations, obvert their points unto, and penetrate the skin, and so cause a dolorous Compunction, or discover themselves to the Organ of Touching in that species of Quality, which men call Heat.     The reason of this Phænomenon is clearly the same with that of a heap of *Needles*, which when confused in oblique, transverse, &c. irregular positions, on every side prick the hand that

*Art. 9.*
The Accensification *of Heat*, on *of Heat*, from Concre. tions actually Cold, upon a meer transposition of their Component Particles; exemplified in sundry Chymical Experiments.

that graspeth them : but if disposed into uniform order, like sticks in a Fagot, they may be laterally handled without any asperity or puncture : or that of the *Bristles* of an *Urchine*, which when depressed, or ported, may be stroked from head to tayle, without offence to the hand; but when erected or advanced, become intractable.

By the same reason also may we comprehend, why *Aqua Fortis*, whose Ingredients in their simple natures are all gentle and innoxious, is so fiery and almost invincible a poyson to all that take it : why the Spirit of *Vitriol*, freshly extracted, kindles into a fire, if confused with the *Salt* of *Tartar* : why the Filings of *Steel* when irrigated with Spirit of *Salt*, suffer an æstuation, ebullition, and dissolution into a kind of Gelly, or Paste : with all other mutations sensible, observed by Apothecaries and Chymists, in their Compositions of Dissimilar natures, from which some third or neutral Quality doth result.

*Art.* 10.
The Generation of all kinds of sensible Qualities in one and the same Concretion, from the variegated positions of its particles: evidenced in the Example of a *putrid Apple.*

Secondly, that in the parts of an Apple, whose one half is rotten, the other found, what strange disparity there is in the points of Colour, Odour, Sapour, Softness, &c. Qualities. The found half is sweet in taste, fresh and fragrant in smell, white in Colour, and hard to the touch : the Corrupt, bitter, earthy or cadaverous, duskish, or inclinining to black, and soft. Now to what Cause can we adscribe this manifest dissimilitude, but only this : that the Particles of the Putrid half, by occasion either of Contusion, or Corrosion, as the Procatarctick Cause, have suffered a change of position among themselves, and admitted almost a Contrary Contexture, so as to exhibite themselves to the several Organs of Sense in the species of Qualities almost contrary to those resulting from the found half; which upon a farther incroachment of putrefaction, must also be deturbed from their natural Order, and Situations in like manner, and consequently put on the same Apparences, or Qualities.　For, can it be admitted, that the found moity, when it shall have undergone Corruption, doth consist of other Particles then before? if it be answered, that some particles thereof are exhaled, and others of the aer succeeded into their rooms; our assertion will be rather ratified, then impugned : because it præsumes, that from the egression of some particles, the subingression of others of aer, and the total transposition of the remaining, Corruption is introduced thereupon; and thereby that general change of Qualities, mentioned.

*Art.* 11.
The assenting suffrage of *Epicurus.*

These *Instances*, and the insufficiency of any other *Dihoties*, to the rational explanation of them, with due attention and impartiality perpended, we cannot but highly applaud the perspicacity of *Epicurus*, who constantly held, τὴν μεταβλητικὴν κίνησιν, εἶδός ἐναι τῆς μεταβάσεως, that the Motion of *Mutation* was a species of *Local Transition* : and τὸ γὰρ μεταβάλλον κ᾿ ποιότητα σύγκριμα αὐτὸς; κατὰ τὴν τῶν συγκειμένων, αὐτῷ λόγῳ θεωρητῶν, σωμάτων, τὸ πλεῖον τε, κ᾿ μεταβατικὴν κίνησιν μεταβάλλ᾿ : *Concretum, quod secundum Qualitatem mutatur, omnino mutatur Locali & transitivo motu eorum corporum, ratione intelligibilium, quæ in ipsum concreverint.* Which *Empiricus* (2.

*advers.*

*adverf. Phyf.*) defcanting upon, faith thus; *Exempli cauffâ, ut ex dulci fiat aliquid amarum, aut ex albo nigrum; oportet moleculas, feu Corpufcula quæ ipfum conftituunt, tranfponi, & alium, vice alterius, ordinem fufcipere: Hoc autem non contigerit, nifi ipfa molecule, motione tranfitus, moveantur. Et rurfus, ut ex molli fiat quid Durum, & ex duro molle; oportet eas, quæ illud conftituunt, particulas fecundum locum moveri: quippe earum extenfione mollitur, coitione vero & condenfatione durefcit, &c.* All which is moft adæquately exemplified in a rotten Apple.

And this, we conceive, may fuffice in the General for our Enquiry into the poffible Origine of fenfible Qualities.

CHAP.

# CHAP. II.

## *That Species Visible*

## are

## SUBSTANTIAL EMANATIONS.

### SECT. I.

*Art. 1.*
The Visible
Images of objects, *substantial*: and either
*corporeal Ema-
nations* from
the superficial
parts of Con-
cretions; or
*Light* it self,
disposed into
contextures,
consimilar to
the figure of
the object.

*Ensus non suscipere SUBSTANTIAS,*
though the constant assertion of *A-
ristotle,* and admitted into his Definition of Sense, αἴσθησίς ἐςὶ τὸ
δεκτικὸν τῶς αἰσθητῶν εἰδῶν ἀνὰ τ̃
ὕλης, *Sensus est id, quod est capax
sensibilium specierum sine materia;*
(*lib.* 2. *de Anima, cap. ultim.*) and
swallowed as an Axiome by most
of his. *Commentators* : is yet so far
from being indisputable, that an intent examination of it by reason
may not only suspect, but convict
it of manifest *absurdity.* Witness
only one, and the noblest of Senses, the SIGHT: which discerns
the exterior Forms of Objects, by the reception either of certain *Substantial,* or *Corporeal Emanations,* by the sollicitation of *Light* incident
upon, and reflected from them, as it were Direpted from their superficial parts, and trajected through a diaphanous Medium, in a direct line
to the eye: or, of *Light it self,* proceeding in streight lines from Lucid
bodies, or in reflex from opace, in such contextures, as exactly respond in
order and position of parts, to the superficial Figure of the object, obverted to the eye.

*Art. 2.*
The position
of their being
*Effluvies,* de-
rived from *Epi-
curus*; and
p referred to
the common
doctrine of
the Schools of
the Immateria-
lity of Species
Visible.

For the FIRST of these Positions, *Epicurus* hath left us so rational a
Ground, that deserves, besides our admiration of His Perspicacity, if
not our plenary Adhærence, yet at least our calm Allowance of its *Verisimility,* and due prælation to that jejune and frothy Doctrine of the *Schools,*
that *Species Visible are Forms without Matter, and immaterial not only in their
admission into the Retina Tunica, or proper and immediate Organ of sight;
but even in their Trajection through the Medium interjacent betwixt the*
*object*

*object and the eye.* Which Argument, since too weighty, to be entrusted to the support of a *Gratis*, or simple *Affirmation*; we shall endeavour to prop up with more then one solid *Reason.*

And this that we may, with method requisite to perspicuity, effect: we are to begin at the faithful recital of *Epicurus* Text, and then proceed to the Explanation, and Examination of it.

*Reputandum est, esse in mundo quasdam Effigies, ad Visionem inservien-teis, quæ corporibus solidis delineatione consimiles, superant longè sua tenuita-te quicquid est rerum conspicabilium. Neq; enim formari repugnat etiam in medio aere circumfusove spatio, hujusmodi quasdam Contexturas : uti neque repugnat, esse quasdam in ipsis rebus, & maxime in Atomis, dispositiones, ad operandum ejusmodi spectra, quæ sunt quasi quædam mera inanesq; Cavitates, & superficiales, soliditatisve expertes tenuitates. Neq; præterea repugnat, fieri ex Corporibus extimis Effluxiones quasdam Atomorum continenter à volanti-um in quibus idem positus, idemq; ordo, qui fuerit in solidis, superficiebusve ipsorum, servetur : ut tales proinde Effluxiones sint quasi Formæ, sive Effi-gies, & Imagines Corporum, à quibus dimanant. Tales autem Formæ sive Ef-figies & Imagines sunt, quas moris est nobis, ut Idola, seu simulachra appellite-mus. Ex lib. 10. Diogen. Laertij. & versione Gassendi.*

The importance of which, and the remainder of his judgment, concerning the same theorem, may be thus concisely rendred. Without repugnan-cy to reason, it may be conceived (1) That in the Universʃity of Nature are certain most tenuious Concretions, or subtle Contextures, holding an exquisite analogy to solid bodies. (2) That by these, occurring to the sense, and thence to the Mind, all Vision, and Intellection is made : for they are the same that the Græcian Philosophers call Ειδωλα, ᷴ φαισμαᷥα, and the Latine *Imagines, Spectra, Simulachra, Effigies,* and most frequently *Species Intenti-onales.* (3) That among all the sundry possible wayes of the generation of these Species Visible, the two primary and most considerable are (1) by their *Direption* from the superficial parts of Compound bodies, (2) by their *Spontaneous Emanation,* and Concretion in the aer ; and therefore those of the First sort are to be named Απορροᷥ, and those of the second Συϛασᷥ. (4) That those Images, which are direpted from the extreams of solid bo-dies, do conserve in their separated state the same order and position of parts, that they had during their united. (5) That the ineffable or insuperable *Pernicity,* whereby these Images are transferred through a free space, de-pends upon both the *Pernicity* of the Motion of *Atoms,* and their *Tenuity* or *Exility.* For, the motion of Atoms, while continued through the Inane Space, and impeded by no retundent, is supposed to be inexcogitably swift : nor are we to admit, that when an Atom is repercussed by another directly arietating against it, and afterward variously bandied up and down by the re-tusion of others encountring it ; these partial or retuse motions are less swift, i.e. are performed in a space of time more assignable or distinguishable by thought, then if they were extended into one direct, simple, or uninterrupt-ed motion. And for the second Fundament, the extreme Tenuity of A-toms ; insomuch as these Images are præsumed to be no more but certain superficial Contextures of Atoms : it cannot seem inconsequent, that their Pernicity can know no remora. And thus much of *Epicurus Text,* and the competent *Exposition* thereof.

*Art.* 5. The Contents thereof reduced to 4 *Heads*

It succeeds that we examine the relation it bears to *Probability*; refering the consideration of his *spontaneous* and *systatical* Images, to the Last Section: and reducing our thoughts concerning the *Direpted* and *Apostatical* (which are, indeed, the proper subject of our præsent disquisition) to four capital points, viz. (1) their *An sint*, or Existence; (2) their *Quid sint*, or proper Nature; (3) their *Unde*, or Production; (4) their *Celerity* of Transmission.

*Art.* 6. The *Existence* of Images visible, certified by autoptical Demonstration.

Of the FIRST, namely the EXISTENCE of Species Visible; this is sufficiently certified by the obvious experience of Looking-glasses, Water, and all other Catoptrick or Speculary bodies: which autoptically demonstrate the Emission of Images from things objected. For, if the object be removed, or eclipsed by the interposition of any opace body, sufficiently dense and crass to terminate them, the Images thereof immediately disappear; if the object be moved, inverted, expansed, contracted, the Image likewise is instantly moved, inverted, expansed, contracted; in all postures conforming to, and so undeniably proclaiming its necessary dependence upon its Antitype. Thus also, when in Summer we shade our selves from the intense fervor of the Sun, in green Arbours, or under Trees; we cannot but observe all our cloaths tincted with a thin Verdure, or shady Green: and this from no other Cause, but that the Images or Species of the Leaves, being as it were stript off by the incident light, and diffused into the vicine Aer, are terminated upon us, and so discolour our vestiments. Not, as *Magirus* would solve it, *qualitate, i.e. immateriali forma, qua aer, corpus* διαφανὲς, *à folijs arborum viridibus imbuitur, tingitur, pingitur*; (*Comment. in Phylologiam Peripat. lib. 6. cap. 6. num.* 27.) And thus are the bodies of men sitting, or walking in a large room, infected with the Colours of the Curtains or Hangings, when the Sun strikes upon them: Of which *Lucretius* thus,

> *Nam jacier certè, atq; emergere multa videmus,*
> *Non solum ex alto, penitusque, ut diximus ante;*
> *Verum de summis ipsum quoq; sæpe Colorem.*
> *Et vulgo faciunt id lutea, russaq; vela,*
> *Et ferruginea, cum magnis intenta theatris*
> *Per malos volgata, trabeisq; frementia flutant.*
> *Namq; ibi concessum caveai subter, & omnem*
> *Scendi speciem patrum, matrumque, Deorumque,*
> *Inficiunt, coguntq; suo fluitare Colore.*
> *Ergo lintea de summo cum Corpore fucum*
> *Mittunt, Effigias quoq; debent mittere tenueis*
> *Res quæque, ex summo quoniam jaculantur utraq; &c.* Lib. 4.

Upon which Reason also the admirable *Kircher* hinted his parastatical Experiment, of Glossing the inside of a Chamber, and all things as well Furniture as Persons therein contained, with a pleasant disguise of grass Green, Azure, Crimson, or any other light Colour (for Black cannot consist in any Liquor, without so much density, as must terminate the Light:) only by disposing a capacious Vial of Glass, filled with the Tincture of Verdegrease, Lignum Nephriticum, or Vermilion, &c. in some aperture of the Window respecting the incident beams of the Sun. (*Art. Magn. Lucis, & Umbræ, lib.* 10. *part.* 2. *Magiæ, parastaticæ Experimento* 5.)

Concerning

Concerning the SECOND, *viz.* the NATURE of Images Visible; we obferve Firft, that *Epicurus* feems only to have revived and improved the notion of *Plato*, and *Empedocles*, who pofitively declared the fenfible Forms, or Vifible fpecies of things, to be Ἀτʋργόια, *Effluxiones quædam fubftantiales*: in that He denominates them *Aporrhea*, and defines them to be moft thin and only fuperficial Contextures, of Atoms effluxed from the fuperficial parts of Bodies, and *jugi fluore*, by a continued ftream emaning from them into all the circumfufed fpace.

*Art. 7.*
*Epicurus* opinion, of the *fubftantiality* of Images Vifible; confonant to the judgment of *Plato* and *Empedocles.*

Secondly, that the Common Opinion, moft pertinacioufly patronized by *Alexander* the Peripatetick, and *Scaliger*, with the numerous herd of *Ariftoteleans* (whom it is as eafie to convert, as nominate) is, that vifible fpecies are *mera Accidentia*, fimple pure Accidents, that neither poffefs, nor carry with them any thing of *Matter*, or Subftance; and yet being tranfmitted through a diaphanous Medium from folid objects, they affect the organ of Sight, are reflected from polite and fpeculary bodies, &c. Here we are arrefted with wonder, either how thefe great Mafters of Learning could derive this wild conceit from their Oracle, *Ariftotle*; when introth all they could ground upon his Authority of this kind, is defumable only from thefe words of his, *Colorem rei Vifibilis movere perfpicuum actu, quod deinceps oculum moveat*: or how they could judge it confentaneous to reafon, that thofe Affections fhould be attributed to meer *Accidents*, which are manifeftly Competent only to meer *Subftances*. For, to be moved or to be the fubject of Local Motion, to be impinged againft, and reflected from, or permeate a body; to be dilated, contracted, inverted, &c. cannot confift, nor indeed by a fober man be conceived, without abfolute *fubftantiality*. Some there are, we confefs, who tell us, that they kindled this Conceit from fundry fcattered fparks blended both in his general Difcourfes of Motion and Alteration, and particular Enquiries into the nature of Dreams, and Sounds, in his Problems: and thefe, thereupon, moft confidently ftate the whole matter, thus. That the Vifible Object doth firft Generate a Confimilar Species in the parts of the aer next adjacent; that this Embryon fpecies doth inftantly Generate a fecond in the parts of the aer next to it, that generates a third, that third a fourth, and fo they generate or fpawn each other fucceffively in all points of the Medium, untill the laft fpecies produced in the aer contiguous to the Horny membrane of the eye, doth therein produce another; which præfents to the Optick Nerve the exact delineations and pourtraiture of the Protoplaft, or Object. To Cure the Schools of this Delirium, our advice is, that they firft purge off that fæculent humor of Pædantifm, and implicite adhærence to Authority; and then with clean ftomachs take this effectual *Alterative.*

*Art. 8.*
The *Ariftotelean* Thefis, that Images optical are meer Accidents, recited: and

If the *Vifible Species* of Objects be, as they define; meer *Accidents*, i. e. *immaterial*: we Demand (1) What doth *Creat* them? Not the *Object*; fince that hath neither power, nor art, nor inftruments, to pourtray its own Counterfeit on the table of the contiguous aer. (2) What doth *Conferve* and *Support* them when pourtray'd? Not the *Aer*; fince that is varioufly agitated, and difpelled by the wind, and commoved every way by Light pervading it: and yet the Species of objects are alwayes tranfmitted in a direct line to the eye. (3) What can *Tranfport* them? Neither Aer, nor *Light*: fince it is of the formal reafon of an *Accident*, not to be removed or tranfmitted but in the arms of it Subject. Nor can the fame numerical

*Art. 9.*
Convicted of Sundry *Impoffibilities, Incoufiftences,* and *Abfurdities.*

species be extended through the whole space of the Medium; because it is repugnant to their supposition: and themselves affirm the transmigration of an Accident from one subject to another, impossible. (4) Is the species changed and multiplied by *Propagation*? That's if not an impossibility absolute, yet a Difficulty inexplicable; first because no man ever hath, nor can explain the *Modus Propagationis*, the manner of their Propagation: Secondly, since the parts of space intermediate betwixt the Object and the Eye, though but at a small distance removed, are innumerable; and a fresh propagation must be successively in each of those parts; and the space of Time required to each single propagation is a moment; certainly it must be long before the propagation could attain to so small a part of space, as is æqual to one Digit. If so, how many hours would run by, after the Suns Emergency out of an Eclipse, before the light of it would arrive at our eye? since, as the moments, or points of space betwixt it and us are more then innumerable; so likewise must the moments, or points of Time, while a fresh species is generated in each point of that vast space, be more then innumerable: and yet we have the Demonstration of the most Scientifick of our senses, that the light of the Sun is darted through that immense space, in one single moment. (5) What is the material of these species, or Whether is the Adam or First species educed out of *Nothing*? That's manifestly absurd, because above the power of Nature: and to recur to any other power superior to Hers, is downright madness. (6) Or, *ex Materia Potentia*, out of some secret Energie of the matter of the Medium? That's Unconceivable; for we dare the whole world to define, what kind of Power that is, supposed inhærent in the Medium (Aer, Water, Glass, or any other ω διαφανὲς) that can be actuated so expeditely into the production of infinite several species, in a moment. From one and the same part of Aer, in one and the same moment, how can be educed the different species not only of the Sun and a Stone, of a Man and a Stock, of a Head and a Foot; but even of two absolute Contraries, Snow and Pich? (7) If Visible Species contain nothing of Matter; how can they with such insuperable Velocity be projected on a speculary body, and recoyl back from it to so great a distance, as is commonly observed, even in the Repercussion, or rather Reflection of a Species from a Concave Glass: How consist of Various Parts, and conserve the order and position of them invariate, and the Colours of each clearly inconfused, through the interval of the Medium? How be really ampliated, contracted, deflected, inverted, &c. All which are properly and solely Congruent to Bodies or Entities consisting of Matter? (8) But all these and many more as manifest Incongruities and open Absurdities may be prævented by the assumption of the more durable and satisfactory Hypothesis of *Epicurus*: for conceding the Visible Species of Objects to be *Substantial Effluxes*, it can be no difficulty to solve their Trajection, Impaction, Refraction, Reflexion, Contraction, Diduction, Inversion, &c.

**Art. 10.**
*The grand Objection of Alexander, that a continual Efflux of substance must minorate the Quantity of the most solid Visible.*

Nor is it oppugnable by the objection of any *Difficulty* more considerable, then that so insultingly urged by *Alexander* the Peripatetick: *quanam ratione fieri possit, ut ex tot, tantisque effluentibus particulis, unumquodque adspectabilium non celeriter absumatur?* How can it consist with reason, since the Visible Species are præsumed to be substantial Effluviæs, that any the most solid and large adspectable body should not in a short time be minorated, nay wholly exhausted by the continual deperdition of so

many

many particles? (*in Comment. in lib. de Sensu & Sensili, cap. 3. & Epist. 56. ad Dioscor.*)

*Art. 11.*
Salved by two Reasons; the possible Accretion of other particles, and the extreme Tenuity of the Emanant.

Which yet is not so ponderous, as not to be counterpoysed by these two Reasons, (1) *ἐπιτρέχκρρῶδαι ὑπταις ἄλλα*, *Accrescere ipsis adspectabilibus advenientia ex opposito corpuscula alia*; that the decay is prævented by the apposition and accretion of other minute particles succeeding into the rooms of the effluxed; so that how much of substance decedes from the superficial parts of one body towards others, as much accedes to it by the advent of the like Emanations from others, and thereupon ensues a plenary Compensation. Nor can it diminish one grain of the weight of this solution, to rejoyn; that the Figures of adspectables must then be changed: because the substantial Effluxes which Accede, cannot be in point of Figure, Order, and Position of parts exactly consimilar to those which Recede. For, though there be a dissimilitude in Figure, betwixt the Deceding and Acceding particles; yet, in so great a tenuity of particles, as we suppose in our substantial species, that can produce no mutation of Figure in the object deprehensible by the sense: for many things remain invariate to the eye, which are yet very much changed as to Figure, in the judgment of the understanding; as may most eminently be exemplified in the Change that every Age insensibly stealeth upon the face of man. (2) *λεπτομερῶν ἱδέων ὄντων Ἀνυπερβλητον*, *Tenuitatem simulachrorum esse omnem modum excedentem*, the Tenuity of these Emanant Images is Extreme; and therefore the uninterrupted Emission of them, even for many hundreds of years, can introduce no sensible either mutation of Figure, or minoration of Quantity in the superficies of the Emittent. Which *Averrhoes* (at least the Author of that Book, *Destructionis Destructionum*, fathered upon him) had respect unto, when He said; *Neminem agniturum decrementum in Sole factum, tametsi ab eo circum deperierit quantitas pa'mi, aut etiam major.*

*Art. 12.*
The Tenuity of Images visible, reduced to some degree of Comprehensibility, by conceiving them to be most thin *Decortications.*

To approach some degrees nearer in our Comprehension to the almost Incomprehensible TENUITY of these substantial Emanations, that essence the Visible Images of Objects; Let us First, conceive them, with *Lucretius*, to be, *Quasi Membrana summo de Corpore rerum Derepta*, Certain *Extorticationes*, or a kind of most thin *Films*, by the subtle fingers of *Light*, stript off from the superficial Extremes of Bodies; for *Alexander* himself calls them *ὑπῤροῖ εἰς, & φλοιοῖ εἰς, Pellicula & Membranula*, & *Apuleius Exuviæ*, because as the *slough* or spoil of a Snake, is but a thin integument blancht off the new skin, and yet representing the various Spots, Scales, Magnitude, Figure &c. thereof: so likewise do the Visible Species, being meer *Decorticationes*, or *Sloughs* blancht off from Bodies, carry an exact resemblance of all Lineaments and Colours in the Exteriours thereof.

*Art. 13.*
By instance, in the Visible species of the Foot of a Handworm.

Secondly, assume the smallest of things Visible, the Foot of an *Handworm*, for the Object. For conceding the species Emanant from it, which is deprehensible by a Microscope, to consist only of those Atoms, which cohæring only *Secundum Latera*, and *non* ዪ *βάθος*, *Laterally* and not *Profoundly*, constitute the superficies: and then we cannot deny, that this species must be by many Myriads of Myriads of Atoms thinner then the Foot, or Object it self.

Thirdly,

*Art.* 14.
By *Exemplify-
ing in the nu-
merous round
Films of Wax,
successively
derepted from
a Wax vapour
by the flame
thereof, in the
space of an
hour: and*
Thirdly, Exemplifie the ineffable Tenuity of these Excortications, in those round Films of Wax that are successively lickt off by the Flame of a Tapour accended. For, having supposed, that one inch of a Wax Candle may suffice to maintain its flame, for the space of an hour: let us thus reason. Since the Diminution of that inch, perpendicularly erected, is unceffant, i. e. that there is no diftinguishable moment of time, wherein there is not a diftinct round of Wax taken off the upper part thereof, by the depredatory activity of the flame: how many muft the Round Films of Wax be, that are successively ditepted? Certainly, as many as there are diftinguishable points, or parts in the 24 part of the Æquator, or ambite of the Primum Mobile, successively interjacent toward the Meridian. And if, in ftead of that vaft Heaven, the Primum Mobile, you think it more convenient to affume the Terreftrial Globe (whose Magnitude, in comparison of the other, amounts not above a point) obferve what may be thence inferred. Since, according to the fupputation of *Snellius* and *Gaffendus*, the ambite of the Earth is commenfurable by 26255 Italian miles; and the 24 part thereof makes 1094 miles, and fo 1094000 paces, and fo 5470000 feet, each whereof is again fubdivifible into 1000 fenfible parts: it follows, that as the product, or whole number of thefe parts in the 24 part of the Circumference of the Globe Terreftrial arifeth to 5470000000; fo likewife muft the diftinct membranules of Wax successive derepted from the inch of Candle in the space of an hour fulfil the fame high number of 5470000000. And if fo, pray how incomprehenfible thin muft each of them be?

*Art.* 15.
In the innu-
merable Films
of Oyl, like-
wife succes-
fively delibra-
red, by the
flame of an El-
lychnium, or
March, per-
pendicularly
floating in a
veffel of equal
capacity with
*Solomons Bra-
zen Sea*, in the
space of 48
hours.
If this Example feem too grofs to adumbrate the extreme Tenuity of our species; be pleafed to exchange the Wax Tapour of an inch diameter, for *Solomons Brafen Sea*, filled with oyl, and an inch of Cotten Weeck perpendicularly immerfed, and at the upper extreme accenfed, in the middle thereof. For, infomuch as the Decrement of the oyl in altitude muft be unceffant, as is the exhaufting activity of the flame, there being no inftant of time, wherein its diminution is interrupted; and that, fhould the flame conftantly adhære to the Weeck for 48 hours, without extinction, the space of the oyls defcent from the margin of the veffel could not in craffitude equal that of a piece of Lawn, or a Spiders Web: certainly the number of Rounds of oyl fuccessively delibrated by the flame, in that conftitute time, muft require a far greater number of Cyphers to its Calculation. Which would you definitely know; 'tis but computing the diftinguifhable points of time in 48 hours, during which the flame is fuppofed to live, and you have your defire; and we ours, as to the conjectural apprehenfion of the Tenuity of each of them.

*Art.* 16.
By the Ana-
logy betwixt
an *Odorable* &
*Vifible* Species.
Laftly, let us argue *à fimili*, and guefs at the Tenuity of a Vifible, from that of an *Odorable* Species. How many *Aromaticks* are there, that for many years together, emit fragrant exhalations, that replenifh a confiderable space of the ambient aer; and gratefully affect the noftrils of all perfons, within the orb of projection: and yet cannot, upon the exacteft ftatick experiment, or trutination of the Scate, be found to have amitted one grain of Quantity? Now if we confider, how Crafs the Emanation of an Aromatick, or an odorous Anathymiafis, is comparatively to the fubftance of a Vifible Species (for no meaner a Philofopher then *Gaffendus*, whofe name founds all the Liberall Sciences, hath conceived; that the Vifible Images effluxing from an Apple in a whole year, if all caft into one bulk, would not
exceed

exceed that of the odorous vapour exhaled from it in one moment ) we shall not gainsay, but a solid Body may constantly maintain an Emanation of its Images Visible, for many hundreds of years, from its superficial parts, without any sensible abatement of Quantity, or variation of Figure. To which we shall superadd only this; that should we allow these substantial Effluxes, that are supposed to constitute the Visible Species, to amount in many hundred years, to a mass deprehensible by sense, in case the collection of them all into one were possible: yet would it be so small, as to elude the exactest observation of man; for, who that hath perchance weighed a piece of Marble, or Gold, and set down the præcise gravity thereof in his life time, can obtain a parrol from the grave and return to complete his experiment; after the deflux of so many Ages, as are required to fulfill the sensibility of its minoration?

Concerning the THIRD, *viz.* the PRODUCTION of Species Visible; *Epicurus* Text may be fully illustrated by this Exposition. That a solid Body, so long as environed with a rare or permeable space, may be conceived without Alogie, freely to emit its Images: because it hath Atoms ready in the superfice, that being actuated by their coessential motive Faculty, uncessantly attempt their Emancipation, or Abduction; and those so exile, that the Ambient cannot impede their Emanation. (2) That in regard they conserve the Delineations both of the Depressed and Eminent parts in the superfice of the Antitype, or Object, after their Efflux therefrom: therefore do the Images deceding from it become Configurate of Atoms coha͛rently exhaling in the same Order and Position that they held among themselves, during their Contiguity, or Adhæsion. Which also satisfies for the præsumed meer superficiality, i. e. *Improfundity* of the species: because it is derided only from the Extremities of the Object. (3) That, forasmuch as no Cause can be alledged, why the particles of the Image should, in their progress through a pervious medium to considerable distance, be deturbed or discomposed from that Contexture, or order and situation, which they obtained from the Cortex or outward Film of their solid original: therefore do they invariately hold the same Configuration, untill their arrival at the eye. Which to familiarize, we are to reflect upon a position or two formerly conceded, viz. that Atoms are, by the impulse of their ingenite Motion, variously agitated even in Concretions most compact; and yet cannot without difficulty expede themselves from the Interior or Central parts, because of their mutual Revinction, or Complication: but for those in the Exterior or superficial parts, they may, upon the least evolution disingage themselves, having no Atoms without to depress, but many within to express or impel them. (4) That, since the Motion of all Atoms, when at liberty to pursue the Tendency of their Motive Faculty, is *Æquivelox:* hence is it, that those Atoms which exhale from the Cavities or Deprest parts of the superficies of any Concretion, and those which exhale from the Prominencies, or Eminent Parts, are transferred together in that order, that they touch not, nor crowd each other, but observe the same distance and decorum, that they had in their Contiguity to, and immediate separation from the superficies. So that the Antecedent Atoms cannot be overtaken, or prævented by the Consequent: nor those farther outstrip these, then at the first start. (5) That the Emanation of Visible Images is *Continent,* i. e. that one succeeds on the heels of another, *jugi quodam Fluore,* in a continued stream more swiftly then that thought can distinguish any

interme-

*Art. 7.*
The Manner and Reason of the *Production* of visible Images; according to the hypothesis of *Efficaces.*

intermediate diſtance. So that, as in the Exſiltion of Water from the Cock of a Ciſtern perpetually ſupplied by a Fountain, the parts thereof ſo cloſely ſuccede each other, as to make one Continued ſtream, without any interruption obſervable: are we to conceive the Efflux of Images to be ſo Continent, that the Conſequent preſs upon the neck of the Antecedent ſo contiguouſly, as the Eye can deprehend no Diſcontinuity, nor the Mind diſcern any Interſtice in their Flux. And this uſhers us to the reaſon, why *Apuleius*, diſcourſing in the Dialect of *Epicurus*, ſaith, *Profectas à nobis Imagines, velut quaſdam exuvias jugi fluore manare.* (6) And laſtly, that a Viſible Image doth not ſo conſtantly retain its Figure, and Colours, as not to be ſubject to Mutilation and Confuſion, if the interval betwixt its original and the eye be immoderately large: as may be exemplified in the ſpecies of a ſquare Tower, which by a long trajection through the aer, hath its Angles retuſed, ſo that it enters the eye in a Cylindrical Figure. This *Epicurus* expreſly admitted in his ἐνίοτε συγγεομβϗθμου ὑπάρχης, *confuſam interdum evadere imaginem.* Which ought to be interpreted not only of the detriment ſuſtained in its long progreſs through the Medium, but alſo of that which may ariſe from ſome perturbation cauſed in the ſuperfice of the Exhalant.

*Art.* 18.
The Celerity of the Motion of viſible Images, reaſoned; and compared to that of the Light of the Sun.

Concerning the FOURTH, *viz.* the CELERITY of their Motion; this will *Epicurus* have to be Ἀνωπέρβλητον, *Inexſuperabilem*, ſwift in the higheſt degree: and his Reaſon is, becauſe ſuch is the Pernicity of Atoms, when enfranchiſed from Concretions, and upon the Wings of their Gravity. *Lucretius* moſt appoſitely compares the Celerity of Images in their Trajection, to that of the beams of the Sun, which from the body thereof are darted to the ſuperfice of the Earth in an inſtant, or ſo ſmall a part of time, as none can be ſuppoſed leſs. And this we may clearly comprehend, if we obſerve that moment when the Sun begins its Emergency from the Diſcus of the Moon, in an Eclipſe; for in the ſame moment, we may diſcern the Image of its cleared limbus, appearing in a veſſel of Water, reſpectively ſituate.

*Art.* 19.
The Tranſlation of a moveable from place to place, in an indiviſible point of time, impoſſible: and why?

And yet we ſay, the Celerity of their Trajection, not, with the Vulgar, the *Inſtantaneous Motion*: becauſe we conceive it impoſſible, that any *Moveable* ſhould be transferred to a diſtant place, in an indiviſible moment, but in ſome ſpace of time, though ſo ſhort as to be imperceptible; becauſe the Medium hath parts ſo ſucceſſively ranged, that the remote cannot be pervaded before the vicine.

*Art.* 20.
The Facility of the Abduction, or Avolation of Images Viſible, from ſolid Concretions; ſolved by the Spontaneous Exſilition of their ſuperficial Atoms: and the Sollicitation of Light, incident upon them.

And thus have we conciſely Commented upon the 4 *Conſiderables* comprehended in the Text of *Epicurus*, touching Apoſtatical Images Viſible, and thereupon accumulated thoſe Reaſons, which juſtifie our prælation of this His Opinion, to that not only leſs probable, but manifeſtly impoſſible one of the *Ariſtoteleans*: ſo that there ſeems to us only one Conſideration more requirable to complete its Veriſimility, and that is touching the FACILITY of the ABDUCTION of Viſible Images from ſolids.

We confeſs, that *Epicurus* ſuppoſition, of the ſpontaneous Evolution and conſequent Avolation of Atoms from the extremes of ſolid Concretions; is not alone extenſible to the ſolution of this Difficulty: and therefore

ʋe

we must lengthen it out with that confentaneous Pofition of *Gaffendus* (*de apparente magnitudine folis humilis & fublimis*, Epift. 2. pag. 24.) *Lucem follicitare fpecies*, that *Light* doth follicite and more then excite the Vifible fpecies of Objects, as well by agitating the fuperficial Atoms of Concretions, as by Carrying them off in the arms of its reflected rayes. For, that Light is intinged not only with Colours, which it pervades, but alfo with thofe, which it only fuperficially toucheth upon, provided the Colorate body be compact enough to repercufs it; all opace and fpeculary bodies, on which its beams are either trajectly, or reflextly impinged, fenfibly demonftrate. And though it may be objected, that the follicitation of *Light* is not neceffary to the Dereption, or Abduction of Images Vifible; becaufe it is generally præfumed, that they continually Emane from Objects, and fo as well in the thickeft Darknefs, as in the Meridian light: it muft notwithftanding be confeft, that they are unprofitable to Vifion, unlefs when they proceed from an object *Illuftrate*; and confequently that they flow hand in hand with the particles of Light reflected from it fuperfice. Which truly is the reafon why the Eye that is pofited in the dark doth well difcern Objects pofited in the Light; but that which is in the light hath no perception at all of objects in the dark.

And therefore whofo fhall affirme, that Vifible Species are not Emitted from bodies, unlefs Light ftrike upon them, and being repercuffed, carry their fuperficial Atoms, which conftitute the Vifible Species, off from them, in direct lines towards the eye: though He may perhaps want a Demonftration, yet not the evidence of Experience and probability, to credit his Paradox. Nor is there, why we fhould opinion, that only the Primary, or firft incident Light is reflected; becaufe Light emaneth from the Lucid, in a continued Fluor, fo that the præcedent particles are ftill contiguoufly purfued by the confequent: and hence is it that Light is capable of repercuffions even to infinity, if folid and impervious bodies could be fo difpofed, as that the firft oppofed might repercufs it on the fecond, the fecond reflect it to the third, the third to the fourth, &c. fucceffively, fo long as the Fluor fhould be continued, and no Eclipfe intervene. For, the reafon, why Light, formerly diffufed, doth immediately difappear, upon the intervention of any body, that interfects it ftream, is really the fame with that, wherefore Water exfilient from the Tube of a Ciftern, in an arched ftream, doth immediately droop and fall perpendicularly, upon the fhutting of the Cock: the fucceffive flux of thofe parts of Water, which, by a clofe and forceable preffure on the back of the præcedent, maintained the Arcuation of the ftream, being thereby prævented, and the effluxed committed to the tendency of their Gravity. And the reafon, why by the mediation of a fmall remainder of light, after the interfection of its fluor from the Lucid fountain, we have an imperfect and obfcure difcernment of objects; is no more then this: that only a few rayes, here and there one, are incident upon and fo reflected from the fuperfice thereof, having touched upon only a few fcattered particles, and left the greater number untouched; which therefore remain unperceived by the eye, becaufe there wanted Light fufficient to the illuftration of the whole, and fo to the Excitement and Emiffion of a perfect fpecies.

*Art.* 21.
That Objects do not emit their Vifible Images, but when *Illuftrated*: a Conceit though paradoxical, yet not improbable.

U                                        S e c t.

## SECT. II.

THere is yet a *ſecond* ſort of Images Viſible, which though conſiſtent of the ſame *Materials* with the Former; are yet different in the reaſon of their production, according to the theory of *Epicurus.* For, as the former are perfectly ſubſtantial, being Corporeal Effluviaes, by a kind of Dereption as it were blancht from the Extremes of Concretions: ſo likewiſe are theſe of the ſecond Genus, perfectly ſubſtantial, being certain Concrements or Coagmentations of Atoms in the aer, repreſenting the ſhapes of Men, Beaſts, Trees, Caſtles, Armies, &c. not cauſed by an immediate Dereption from ſuch ſolid Prototypes, but a SPONTANEOUS convention and coheſion of convenient particles. So that if we only call them, *Spontaneous Syſtatical Repreſentations*; we ſhall not only import the Diſparity of their Creation to that of the *Derepted Apoſtatical* ones, but alſo afford a glimpſe of their abſtruſe *Nature.* Of theſe, all that can be brought to lye in lines parallel to our præſent Theorem, doth concern only their *Exiſtence:* and that may be evicted by the conſpiring teſtimonies of many Authors, whoſe pens were not dipt in the fading ink of meer Tradition, nor their minds deluded with the affectation of Fabulous Wonders. Among which our leaſure will extend to the quotation of only *Two,* moſt pertinent and ſignificant.

*Diodorus Siculus (lib.3.)* ſpeaking of certain *Spectraes*; ſpontaneouſly conceited, and at ſet ſeaſons of the year exhibiting themſelves to Travellers in the regions of Africa, beyond the Quick-ſands and Cyrene; ſaith thus: περὶ γὰρ τίνας καιρὰς, ἡ μάλιςα κγ̃ὰ τὰς νlωφεμίας, συρἰοὺς ὁρᾶν]αι κγτὰ τ̀ ἄεερα παντξοίων ζ́ώων ἰδέας ἐμφαίνεσαι; *Quandoque, ac præſertim vigente tranquillitate aeris, conſpiciuntur per aerem Concrementa quædam, formas Animalium omnis generis referentia. Ipſorum nonnulla quietè ſe habent, nonnulla verò motionem ſubeunt. Quinetiam interdum inſequentes fugiant, interdum fugientes inſequuntur, &c.*

And *Damaſcius (in Vita Iſidori Philoſophi, apud Photium)* declaring the common report about that memorable τέρεᾶlog, or *Prodious* Aereal Repreſentation, annually beheld in the lower region of the aer, imminent upon that arm of the Adriatick Sea, that runs up betwixt Meſſana in Sicily, and Rhegium Julium in Calabria; delivers it thus: *Noſtra tempeſtate narrarunt homines bonæ fidei, juxta Siciliam in campo nominato Tetrapyrgio, & in aliis non paucis locis, videri Equitum pugnantium ſimulacra; idq, maximè æſtatis tempore, cum ardentiſſimus eſt meridies; &c.* Concerning the verity of this report, the moſt Curious *Athanaſius Kircherus* having ſome doubt; purpoſely takes a long journey from Rome to Meſſana and thence croſſeth over to Rhegium, at the opportune time for its obſervation. Where what He beheld, and by what Phyſical reaſons he ſolved the wonderment; we have thought worthy your patient notice, to extract from his excellent diſcourſe thereupon *(in cap.1. Magiæ Paraſtaticæ, paraſtaſi 1. Natura.)*

MORGANA

# MORGANA RHEGINORUM.

*Art. 4.*
*Kirchers De-*
*scription of*
*that famous*
*Apparition at*
*Rhegium, cal-*
*led Morgana*
*Rheginorum: &*

In the midst of Summer, when the Sun boyls the Tyrrhene Ocean with most fervent rayes, then is it, that wanton Nature entertains the wondring eyes of the inhabitants of Rhegium, a Town in Calabria most ancient and no less famous for having been the seat of many Philosophers, with a prodigious spectacle in the aer. There may you, whether with more delight, or wonder, is not soon determined, behold a spacious Theatre in the vaporous aer, adorned with great variety of Scenes, and Catoptrick representations; the Images of Castles, Palaces, and other Buildings of excellent architecture, with sundry ranges of Pillars, præsented according to the rules of Perspective. This Scene withdrawn, upon the sayling by of the Cloud, there succeeds another, wherein, by way of exquisite Landskip, were exhibited spacious Woods, Groves of Cypress, Orchards with variety of trees, but those artificially planted in Uniform rows like a perfect Phalanx, large Meadows, with companies of men, and herds of beasts walking, feeding, and couching upon them: and all these with so great variety of respondent Colours, so admirable a commixture of Light and Darkness, and all their motions and gestures counterfeited so to the life, that to draw a Landskip of equal perfection seems impossible to human industry.

It may well be conceived, though not easily exprest, how much this Parastatical Phantasm (which the Inhabitants of Rhegium call *Morgana*) hath excruciated the greatest Wits of *Italy*, while they laboured to explore a reason for the apparence of such things in the Cloud, as were not found either on the shore, or adjacent fields. This much encreased the ardor of Curiosity in me, so that crossing over from Messana to Rhegium, at the usual time of the Apparition, I examined all the Circumstances thereof, together with the situation of the place, the nature and propriety of the soyl, and the constitution of the vapours arising from the Sea: and examining my observations by Physical and Optical reasons, I soon detected the Causes of the whole Phænomenon. First I observed the Mountain called Tinna, on the Sicilian side, directly confronting Rhegium, to run along in a duskish obscure tract upon Pelorus; and the shores subjacent, as also the bottom of the Sea, to be covered with shining sand, being the fragments of *Selenites, Antimony*, and other *pellucid* Concretions, devolved from the eminent parts of the land, the contiguous Hills, that are richly fraught with veins of those Minerals. Then I observed that these translucid sands, being, together with vapors from the Sea and Shore, exhaled into the aer, by the intense fervor of the Sun; did coalesce into a Cloud, in all points respondent to a perfect Polyedrical, or Multangular Looking-glass: the various superficies of the resplendent Granules, making a multiplication of the species; and that these, being opacated behind by crass and impervious vapours, directly facing the Mountains, did make reflection of the various Images of objects respective to their various positions to the eye. The several Rows of Pillars in the aereal Scene are caused by one single Pillar, erected on the Shore; for being by a manifold reflection from the various superficies of the tralucent particles, opacated on the hinder part by dense Vapours, in the speculary Meteor, it is multiplyed even to infinity. No otherwise then as one single Image, posited betwixt two polyedrical Looking-glasses, confrontingly

*Art. 5.*
*Most ingeni-*
*ous Investiga.*
*tion of the*
*Causes thereof.*

ingly diſpoſed, is ſo often repercuſſed or reflected from ſuperſice to ſuperſice, that it exhibiteth to the eye almoſt an infinite multitude of Images exactly conſimilar.   Thus alſo doth one man ſtanding on the ſhore, become a whole Army in the Cloud, one Beaſt, a whole Herd, and one Tree a thick-ſet Grove.   As for the vaniſhing of this firſt Scene, and the ſucceſſion of a ſecond, adorned with the repreſentations of Caſtles, and other magnificent ſtructures; the Cauſe hereof is this: ſince the eye of the Spectator hath its ſight variouſly terminated in the ſeveral ſpeculary ſuperficies of the Cloud, that is in perpetual motion according to the impulſe of the Wind; it comes to paſs, that according to the rules of the Angles of Incidence and Reflection, divers Species are beheld under the ſame conſtitute Angle, and as the ſpeculary Vapour doth reflect them toward the eye, which divers ſpecies are projected from objects conveniently ſituate; and particularly from the Caſtle on the aſcent towards Rhegium from the place of our proſpect.

Some, perhaps, may judge our affirmation, of the Elevation of thoſe ſhining Grains of Vitreous Minerals into the aer, by the meet attraction of the Sun; and the Coalition of them there with the Cloud of Vapours: to be too large a morſel, to be ſwallowed by any throat, but that Cormorant one of Credulity.   If ſo, all we require of them, is only to conſider; that Hairs, Straws, grains of Sand, fragments of Wood, and ſuch like Feſtucous Bodies, are frequently found immured in Hailſtones: which doubtleſs, are ſufficient arguments, that thoſe things were firſt elevated by the beams of the Sun, recoyling from the earth, into the middle region of the aer, and there coagmentated with the vapours condenſed into a Cloud, and frozen in its deſcent.

*Art. 6.*
*His admirable Artifice, for the exhibition of the like aereal Repreſentation, in Imitation of Nature.*   Now this ſolution of the *Morgana,* acquires the more of Certitude and Auctority from hence; that in imitation of this Natural Prodigious Oſtent, or Aereal Repreſentation, *Kircher* invented a way of exhibiting an Artificial one, by the Fragments of Glaſs, Selenites, Antimony, &c. ſtewed in an iron trough, and vapours aſcending from Water ſuperaffuſed, and terminated by a black Curtain ſuperextended.   The full deſcription of which Artifice, He hath made the Subject of his 2. *paraſtaſis in Magia Paraſtat. cap.1.*

CHAP.

# CHAP. III.

### CONCERNING THE

## MANNER and REASON

### OF

# VISION.

---

### SECT. I.

Mong the many different Conceptions of Philofophers, both Ancient and Modern, touching the Manner and Reafon of the Difcernment of the Magnitude, Figure, &c. of Vifible Objects by the Vifive Faculty in the Eye, the moft Confiderable are thefe.

(1) The STOICKS affirmed, that certain Vifory Rayes deradiated from the brain, through the flender perforations of the Optick Nerves, into the eye, and from *Art. 1.* The Reafon of Vifion, according to the opinion of the *Stoicks.*
thence in a continued fluor to the object; do, by a kind of Procufion, and Compreffion, difpofe the whole Aer intermediate in a direct line, into a Cone, whofe Point confifteth in the fuperfice of the Eye, and Bafe in the fuperfice of the Object. And that, as the Hand by the mediation of a ftaff, impofed on a body, doth, according to the degrees of refiftence made thereby either directly, or laterally, deprehend the Tactile Qualities thereof, i.e. whether it be Hard, or Soft, Smooth or Rough, whether it be Clay, or Wood, Iron, or Stone, Cloth, or Leather, &c. So likewife doth the Eye, by the mediation of this Aereal ftaff, difcern whether the Adfpectable Object, on which the Bafis of it refteth, be White or Black, Green or Red, Symetrical or Afymetrical in the Figure of its parts, and confequently Beautiful or Deformed.

(2) ARISTO=

*Art. 2.*
Cf *Aristotle.*

(2) ARISTOTLE, though his judgment never acquiesced in any one point, as to this particular, doth yet seem to have most constantly inclined to this; that the Colour of the Visible doth move the *Perspicuum actu*, i. e. that *Illustrate Nature* in the Aer, Water, or any other *τὸ διαφανές*, *Transparent* body; and that, by reason of its *Continuity* from the extremes of the Object to the Eye, doth move the Eye, and by the mediation thereof the *Internal Sensorium* or Visive Faculty, and so inform it of the visible Qualities thereof. So that, according to the Descant of those, who pretend to be his most faithful Interpreters, we may understand Him, to have imagined the *Colour* of the object to be as it were the *Hand*; the diaphanous *Medium*, as it were the *Staff*; and the *Eye* as it were the *Body* on which it is imposed and imprest: *è diametro* opposite to the conceit of the *Stoicks*, who suppose the *Eye* to supply the place of the *Hand*; the *Aer* to analogize the *Staff*; and the *object* to respond to the *Body* on which it is imposed and imprest.

*Art. 3.*
O. the *Pytha-goreans.*

(3) The PYTHAGOREANS determined the reason of Vision on the Reflexion of the Visive Rayes, in a continued stream emitted from the internal Eye, to the visible, back again into the eye; or, more plainly, that the radious Emanations from the Eye, arriving at the superfice of the object, are thereby immediately Repercussed in an uninterrupted stream home again to the eye, in their return bringing along with them a perfect representation thereof, as to Colour, Figure and Magnitude.

*Art. 4.*
Of *Empedocles.*

(4) EMPEDOCLES, though admitting (as we hinted in the next præceding Chapter) substantial Effluxes, from the Visible to the Organ of Sight; doth also assume the Emission of certain Igneous or Lucid Spirits from the Organ to the Object: supposing the Eye to be a kind of Glass Lantern, illustrate, and illustrating the Visible, by its own Light.

*Art. 5.*
Of *Plato.*

(5) PLATO, though He likewise avouched the Emanation of Corporeal Effluviæs from the Object; doth not yet allow them to arrive quite home at the Eye: but will have them to be met half way by rayes of Light extramitted from the Eye: and that these two streams of External and Internal Light encountring with some Renitency reciprocal, do recoyl each from other, and the stream of Internal Light resilient back into the eye, doth communicate unto it that particular kind of Impression, which it received from the stream of Extradvenient Light, in the encounter; and so the Sentient Faculty comes to perceive the adspectable Form of the object, at which the Radius of Internal Light is levelled. This we judge to be sense of his words (*in Timæo, circa finem tertiæ partis*) *Simulachrorum, quæ vel in speculis oboriuntur, vel in perspicua, læviq; cernuntur superficie; facilis assecutio est. Nam ex utriusq; ignis, tam intimi, quam extra positi Communione, ejusq; rursus consensu, & congruentia, qui passim terso, læviq; corpori accommodatus est; necessario hæc omnia oriuntur, quam ignis oculorum cum eo igne, qui est è conspecto effusus, circa læve nitidumq; Corpus sese confundit.*

*Art. 6.*
O *Epicurus.*

(6) EPICURUS, tacitely subverting all these, foundeth the Reason of Vision, not in any Action of the intermediate Aer, as the *Stoicks* and *Aristotle*; nor in any Radious Emanation from the Eye to or toward the Object,

Object, as the *Pythagoreans*, *Empedocles*, and *Plato :* but, in the Derivation of a substantial Efflux from the Object to the Eye.

(7) And as for the opinion of the excellent *Monsieur Des Cartes*, which with a kind of pleasant violence, hath so ravisht the assent of most of the Students of Physiology, in the præsent Age, especially such as affect the accommodation of Mechanick Maxims to the sensible operations of Nature; that their minds abhor the embraces of any other : those, who have not heedfully perused his *Dioptricks*, may fully comprehend it in summary, thus. *Art.* 7. Of *Monf. Des Cartes.*

For *Sensation in Common*, He defines it to be a simple *Perception*, whereby a certain *Motion*, derived from a body conveniently objected, communicated, by Impression, to the small Fibres, or Capillary Filaments of a Nerve, and by those, in regard of their Continuity, transmitted to the Tribunal, or Judicatory Seat of the Soul, or Mind (which He supposeth to be the *Glandula Pinealis*, in the centre of the Brain) and there distinctly apprehended, or judged of. So that the Divers Motions imprest upon the slender threads of any Nerve, are sufficient to the Causation of divers perceptions ; or, that we may not eclipse his notion by the obscurity of our Expression, that the Impulse, or stroke given to the Nerve, doth, by reason of the Continuity of its parts, cause another Motion, in all points answerable to the first received by the External Organ, to be carried quite home to the Throne of the Mind, which instantly makes a respective judgment concerning the Nature of the Object, from whence that particular Motion was derived. In a word, that only by the Variety of Strokes given to the External Organ, thence to the filaments of the Nerve annexed thereto, thence to the Præsence Chamber of the Soul : we are informed of the particular Qualities, and Conditions of every Sensible ; the variety of these sensory Motions being dependent on the variety of Qualities in the Object, and the variety of judgments dependent on the variety of Motions communicate.

And for the sense of *Seeing*, in *special* ; He conceives it to be made, not by the mediation of Images, but of certain *Motions* (whereof the Images are composed) transmitted through the Eye and Optick Nerve to the Centrals of the Brain : præsuming the Visible Image of an Object to be only an exact representation of the motions thereby impressed upon the External Sensorium ; and accordingly determining the Reason of the Minds actual Discernment of the Colour, Situation, Distance, Magnitude, and Figure of a Visible, by the Instruments of Sight, to be this. (1) The Light desilient from the adspectable Body, in a direct line, called by the Masters of the Opticks, the Axe of Vision, percusseth the diaphanous fluid Medium, the Æther, or most subtile substance (by Him assumed to extend in a Continuate Fluor through the Universe, and so to maintain an absolute Plenitude, and Continuity of Parts therein.) (2) The Æther thus percussed by the Illuminant, serving as a Medium betwixt the Object and the Eye ; conveyeth the impression through the outward Membranes and Humors, destined to Refraction, to the Optick Nerve most delicately expansed into the *Retina Tunica*, beyond the Chrystalline. (3) The Motion thus imprest on the outward Extreme of the Optick Nerve, runs along the body of it to the inward Extreme, determined in the substance of the Brain. (4) The Brain receiving the impression, immediately gives notice thereof to its Noble Tenent;

nent, the Soul; which by the Quality of the ſtroke judgeth of the Quality of the Striker, or Object. In ſome proportion like an Exquiſite *Muſitian*, who by the tone of the ſound thereby created, doth judge what Cord in a Virginal was ſtrook, what jack ſtrook that ſtring, and what force the jack was moved withall, whether great, mean, or ſmall, ſlow or quick, equal or unequal, tenſe or lax, &c.

*Art. 8.*
*The ingenuity of Des Cartes Conceit, acknowledged: but the ſolidity indubitated*

This you'l ſay, is a Conceit of ſingular Plauſibility, invented by a Wit tranſcendently acute, adorned with the elegant dreſs of moſt proper and ſignificant Termes, illuſtrate with appoſite ſimiles and prægnant Examples, and diſpoſed into a Method moſt advantageous for perſuaſion; and we ſhould betray our ſelves into the Cenſure of being exceedingly either ſtupid, or malicious, ſhould we not ſay ſo too: but yet we dare not (ſo ſacred is the intereſt of Truth) allow it to be more then ſingularly *Plauſible*; ſince thoſe Arguments, wherewith the ſage *Digby* (*in the 32. chap. of His Treatiſe of Bodies*) hath long ſince impugned it, are ſo exceedingly præponderant, as to over-ballance it by more then many moments of Reaſon; nor could *Des Cartes* himſelf, were He now Unglorified, ſatisfie for his Non-Retraction of this Error; after his examination of their Validity, by any more hopeful Excuſe, then this, that no other opinion could have been conſiſtent to His Cardinal Scope of *Solving all the Operations of Senſe by Mechanick Principles.*

*Art. 9.*
*The Opinion of Epicurus, more ſatisfactory, then any other: becauſe moreRational, and leſs obnoxious to inexplicable Difficulties.*

Now, of all theſe Opinions recited, we can find, after mature and æquitable examination, none that ſeems, either grounded on ſo much Reaſon, or attended with ſo few Difficulties, or ſo ſufficient to the veriſimilous Explanation of all the Problems, concerning the Manner of Viſion, as that of *Epicurus*; which ſtateth the Reaſon of Viſion in the INCURSION of ſubſtantial Images into the Eye. We ſay

FIRST, *Grounded on ſo much Reaſon.* For, inſomuch as it is indiſputable, that in the act of Viſion there is a certain *Sigillation* of the figure and colour of the object, made upon that part of the Eye, wherein the Perception is; and this ſigillation cannot be conceived to be effected otherwiſe then by an *Impreſſion*; nor that Impreſſion be conceived to be made, but by way of *Incurſion* of the Image, or Type: it is a clear Conſequence, that to admit a Sigillation without Impreſſion, and an Impreſſion without Incurſion of the Image, is a manifeſt *Alogy*, an open Inconſiſtence. And upon this conſideration is it, that we have judged *Epicurus* to have ſhot neareſt the White, in his Poſition that Viſion is performed, διὰ εἰδώλων ἐμπτώσεως, *per ſimulachrorum Incurſionem, ſive Incidentiam*: which *Agellius* (*lib.5.cap.16.*) deſcanting upon, ſaith expreſly, *Epicurus aſſuere ſemper ex omnibus corporibus ſimulachra quaedam ipſorum, eaque ſeſe in oculos inferre, atque ita fieri ſenſum videndi putat.*

SECONDLY, *Encumbred with ſo few Difficulties.* For, of all that have been hitherto, either by *Alexander* (*2. de Anima 34.*) *Macrobius* (*7.Saturnal.14.*) *Galen* (*lib.7. de Conſenſu in Platonicis, Hippocratici ſque Decretis*) or any other Author, whoſe leaves we have revolved, objected againſt it; we find only *Two*, that require a profound exerciſe of the Intellect to their Solution: and they are theſe.

(1) *Obvious*

(1) *Obvious it is even to sense, that every Species Visible is wholly in the whole space of the Medium, and wholly in every part thereof ; since in what part soever of the Medium, the Eye shall be admoved, in a position convenient, it shall behold the whole object, represented by the species : and manifest it is, that to be total in the total Space, and total in every part thereof, is an Affection proper only to Incorporeals ; therefore cannot Vision be made by Corporeal Images incurrent into the Eye.*

*Art.* 10.
The Two most considerable Difficulties opposed to *Epicurus* position, of the Incursion of *Substantial* Images into the Eye.

(2) *In the intermediate Aer are coexistent the Images of many, nay innumerable Objects ; which seems impossible, unless those Images are presumed to be Incorporeal : because many Bodies cannot coexist in one and the same place, without reciprocal penetration of Dimensions, Ergo, &c.*

## SECT. II.

TO dispel these Clouds, that have so long eclipsed the splendor of *Epicurus* Assertion, of the *Incidence* of Images Visible into the Eye (for we shall not here dispute, whether he intended the sigillation to be made in that *Convex Speculum*, the *Chrystalline Humour* ; or that *Concave* one, the *Retina Tunica*) and explicate the abstruse nature of Vision : we ask leave to possess you with certain necessary *Propositions* : We assume therefore,

## Assumption the First.

*That the superfice of no Visible is so exquisitely smooth, polite, or equal, as not to contain various Inequalities, i. e. Protuberant and Depress parts, or certain (Monticuli and Vallecule) small Risings and Fallings :* which in some bodies being either larger, or more, are discoverable by the naked intuition of the Eye ; and in others, either smaller, or fewer, require the detection of the Microscope.

*Art.* 1.
That the superfice of no body is perfectly smooth : evicted by solid Reason, and Autopsie.

This is neither Præcarious, nor Conjectural : but warranted by Reason, and autoptical Demonstration. For, if the object assumed be polisht *Marble* ; since that apparent Tersness in the surface thereof is introduced by the detrition of its grosser inæqualities by Sand, and that Sand is nothing but a multitude of Polyedrical solid Grains, by the acuteness and hardness of their Angles cutting and derasing the more friable particles of the Marble : it must follow, that each of the grains of Sand must leave an impression of its edge, and so that the whole superfice must become scarified by innumerable small incisions , variously decussating and intersecting each other. If *Steel* of a speculary smoothness, such as our common Chalybeat Mirrours ; since the Tersness thereof is artificial, caused by the affriction of Files , which cut only by the acuteness of their teeth , or lineal inæqualities : it is not easie to admit, that they leave no scratches, or exarations on the surface thereof ; and where are many Incisions , each whereof must in Latitude respond to the thickness of the Tooth in the File, that made it, there also must

X                                        be

be as many Eminences or small Ridges intercepted among them. And if *Glass*, whose smoothness seems superlative; since it is composed of Sand and Salts, not so perfectly dissolved by liquation, as not to retain various Angles: it cannot be unreasonable to inferr, that those remaining points or angular parts must render the Composition in its exteriors full of Asperities. And, as for Autoptical Evidence; that Marble, Steel, and Glass are unequal in their superfice, is undeniable not only from hence, that a good Engyscope, in a convenient light, doth discover innumerable rugosities and Cavities in the most polisht superfice of either: but also from hence, that Spiders and Flyes do ordinarily run up and down perpendicularly on Venice Glass, which they could not do, if there were not in the surface thereof many small Cavities, or Fastnings for the reception of the Uncinulæ, or Hooks of their Feet. To which may also be added, the Humectation of Glass by any Liquor affused; for, if there were no Fosses and Prominences in the superfice thereof, whereon the Hamous particles of the Liquid might be fastned, it would instantly run off without leaving the least of moisture behind. And hence

## Assumption the Second.

*Art.* 2.
That the visible Image doth consist of so many Rayes as there are Points designable in the whole superfice of the object: and that each Ray hath its line of Tendency direct, respective to the face of that particle in the superfice, from which it is emitted.

　　*That as the whole Visible Image doth emane from the whole superfice of the Object; so do all the parts thereof emane from all the parts of the Object: i.e.* that look how many Atoms are designable in the superfice, from so many points thereof do Atoms exhale, which being contiguously pursued by others and others successively deceding, make continued Rayes, in direct lines tending thitherward, whither the faces of the particles point, from which they are deradiated.

　　For, insomuch as in the superfice no particle can be so minute to the sense, as, in respect to the Asperity, or Inæquality of its surface, not to have various Faces, by which to respect various parts of the Medium: it must inevitably follow, that all the rayes effluxed from an object, do not tend one and the same way, but are variously trajected through the Medium, some upward, others downward, some to the right, others to the left, some obversly or toward, others aversly or fromward, &c. So that there is no region or point of the compass designable, to which some rayes are not direct. And from this branch shoots forth our

## Third Assumption.

*Art.* 3.
That the Density and Union of the Rayes, composing the visible Image, is greater or less; according to their less, or greater Elongation from the Object.

　　*That every visible Image is then most Dense and United, when it is first abduced from the Object: or, that by how much the neerer the visible Species is to the Body, from which it is delibrated, by so much the more Dense and United are the rayes of which it doth consist; and so much the more Rare or Disgregate, by how much the farther it is removed from it.* This may be exemplified in lines drawn from the Centre of a Circle to the Circumference; for by how much the farther they run from the Centre, by so much the greater space is intercepted betwixt them: and by how much the larger space is intercepted betwixt them, by so much the greater must their Rarity be, the degrees of Rarity being determinable by the degrees of intercepted space.
Thus

Thus also muſt the rayes of the Viſible Image, in their progreſs mutually recede each from other, and according to the more or leſs of their Elongation from the point of abduction, become more or leſs Rare and ſcattered, into the amplitude of the Medium. However, we deny not the neceſſity of their innumerable *Decuſſations*, and *Interſections*; in reſpect to the various Faces, and Confrontings of the parts of the ſuperfice, from which they are emitted. And hence we extracted our

## Fourth Aſſumption.

*That the Viſible Image, though really diffuſed through the ſpace of the medium within the ſphear of Projection; is notwithſtanding neither total in the total ſpace, nor total in every part thereof, as is ſuppoſed in the Firſt Objection: but ſo Manifold, as there are parts of the Medium, from which the Object is adſpectable.*

*Art.* 4.
That the Viſible Image is neither total in the total medium; nor total in every part thereof but ſo manifold as are the parts of the medium from which the object is diſcernable. Contrary to the Ariſtoteleans.

Here may we introduce a *Paradox*, which yet doth not want a conſiderable proportion of Veriſimilitude to juſtifie the ſobriety and acuteneſs of his Wit, that firſt ſtarted it; which is, *That of divers men, at the ſame time, ſpeculating the ſame object, no one doth behold the ſame parts thereof, that are beheld by another;* nay more, *that no man can ſee the ſame parts of an Object, with both eyes at once;* nay more, *not the ſame parts with the ſame eye,* if he remove it never ſo little, becauſe the level of the Viſive Axe is varied. This may be verified by a ſingle reflection on the Cauſe hereof, which is the Inequality, or Aſperity of the ſuperfice of Bodies, ſeemingly moſt polite: for, in reſpect of that, it is of neceſſity, that various Rayes, proceeding from the various parts thereof, variouſly convene in the parts of the Medium, and inſomuch as each of thoſe rayes doth repreſent that particle only, from which it was effuſed, and no other, in their concurſe they cannot but repreſent other and other parts, according to the reſpective places or regions of the Medium, in which the Eye is poſited, that receives them. However, we ſhall familiarize it by *Example*. Let two men at once behold a Third, one before, the other behind: and both may be ſaid to behold the ſame man, but, truly, not the ſame parts of him; becauſe the eyes of one are obverted to his Anterior, and thoſe of the other to his Poſterior parts. Take it yet one note higher. Let the Face of a man be the Object, on which though divers perſons gaze at the ſame time, one on the right a ſecond on the left ſide, a third confrontingly, a fourth and a fifth obliquely betwixt the other three; and all may be ſaid to have an equal proſpect of the face: yet can it not be aſſerted, that they do all ſee the ſame parts thereof, but each a particular part. Whence it may be inferred, that albeit we may allow them all to behold his Fore-head, Eyes, Noſe, Cheeks, Mouth, &c. yet can we not allow them all to ſee the ſame parts of Forehead, Eyes, Noſe, Cheeks, &c. becauſe of their unequal ſituation, which Cauſeth that the whole ſpecies pro ſient from the face, doth not tend into the whole medium, but into various parts of it, reſpective to the various faces of the deradiant parts. Moreover, becauſe this præſumed Inæqualitv is not competent only to the greater parts of the face, ſuch as the Eyes, Noſe, Mouth, Chin, &c. but as juſtly conſiderable in the very *Skin*, which hath no deſignable place; wherein are not many ſmaller and ſmaller Eminencies and Depreſſions, deprehenſible (if not by the Opticks of the body, yet) by the acies of the

*Art.* 5.
PARADOX.
That no man can ſee the ſame particle of an object, with both Eyes at once; nay, not with the ſame Eye, if the level of its Viſive Axe be changed.

X 2    Mind:

Mind: hence is, that having imagined the Eyes of the Five Spectators
to move their visive Axes from part to part succesively, and as slowly as
the shadow of the Gnomon steals over the parts of a Dial, untill they have
ranged over the whole face; we may comprehend the necessity, of the
discovery of a fresh part by every new aime or levell of each eye, and the
baulking of others; as if in Particles of devex Figure, no Particels can be
detected a new, but as many of those formerly discerned must be lost, and
as many, nay more remain concealed.

And this Consideration smoothly ushers in two *Consectaries*

(1) That to say, *one simple species doth replenish the whole Medium*, is not,
in the strict Dialect of Reason, so proper, as to say, *the Medium is possessed by
an Aggeries, or Convention of innumerable species:* which being divers in
respect to the divers parts of the Object, from which they were deradiated,
must also be divers in their Existence, and Diffusion through the several
parts of the Perspicuum.    And yet must they be allowed to constitute but
*one entire* species; and this in respect to their Emanation from one Object:
because as the single parts of the species represent the single parts of
the object, so doth the whole of the species represent the whole of the
Object.

**Art. 6.**
CONSECTA-
RY.
That the Me-
dium is not
possessed with
one simple I-
mage; but by
an *aggregate*
of innumera-
ble Images, de-
radiate from
the same ob-
ject : all which
notwithstand-
ing constitute
but one entire
Image.

**Art. 7.**
CONSECTA-
RY 2.
That Myriads
of different
Images, ema-
nant from dif-
ferent objects,
may be COEX-
istent in the
Aer; without
reciprocal pe-
netration of Di-
mensions, or
*Confusion* of
particles, con-
trary to the
*Perspicuicks*.

(2) *That many, nay Myriads of different Species may be Coexistent in the
Common Medium, the Aer;* and yet *no necessity of the Coexistence of many
Bodies in one and the same place;* it being as justifiable to affirm, that they
reciprocally penetrate each others dimensions, as that the Warp and Woof,
or intersecting threads in a Cloth, do mutually penetrate each other: because the Aer is variously intersperfed with Inanities, or small empty *Roads,*
convenient to the inconfused transmission of all those swarms of Rayes, of
which the species consist. Have you not frequently observed, when many
Candles were burning together in the same room, how, according to the
various interposition of opace bodies, various degrees of Shadows and Light
have been diffused into the several quarters of the same? and can you give
any better reason of those various Intersections and Decussations of the several Lights, then this; that the rayes of Light streaming from the diverse
Flames, are directly and inconfusedly trajected through the several inane
Receptaries of the Aer, respective to the position of each Candle, without
reciprocal impediment; the rayes of one, that are projected to the right
hand, in no wise impeding the passage of those of another, that are projected
to the left, in the same sensible part of the Aer. Exactly so do the rayes
of divers Species Visible, in their progress through the aer, pass on in direct
and uninterrupted lines, without Confusion: and though they may seem
to possess the same sensible part of the medium, yet will not reason allow
them to possess the same Insensible particles thereof; in regard the
distinct transmission of each clearly demonstrateth, that each possesseth a
distinct place. Nor doth this their *Juxta-position,* or extreme Nearness
necessitate their *Confusion;* since we daily observe that Water and Wine
may be so Commixt in a Vial, as therein can be assigned no sensible part,
wherein are not some parts of both Liquors: and yet most certain it is, that
the particles of Wine possess not the same *Invisible Loculaments,* or Receptaries, that are replete with the particles of Water, but others absolutely distinct; because otherwise there would be as much of Water, or Wine
alone, in the Vial, as there is of both Water and Wine; which in that Continent

tinent is impossible. And hereupon we Conclude, that to admit every diſtinct ſpecies to repleniſh the whole medium, is no leſs dangerous, then to admit, that each of two Liquors confuſed doth ſingly repleniſh the whole Capacity or the Continent: the parity of reaſons juſtifying the *Paralleliſm.*

### Aſſumption the Fifth.

*That the viſible Image, being trajected through the Pupil, and having ſuffered its ultimate refraction in that Convex Mirror, the Chryſtalline Humor, is received and determined in that principal ſeat of Viſion,* (which holds no remote analogy to a Concave Mirror) *the Retina Tunica, or Expanſion of the Optick Nerve in the bottom of the eye; and therein repreſents the Object from whence it was deradiated, in all particulars to the life, i.e. with the ſame Colour, Figure, and Situation of parts, which it really beareth;* provided the Diſtance be not exceſſive.

*Art. 8.* That the viſible Images ultimate Reception, and complete Perception, is the Concave of the *Retina Tunica.*

The *Firſt part* of this eminent Propoſition, that excellent Mathematician, *Chriſtopher Scheinerus,* hath ſo evicted by Phyſical Reaſons, Optical Demonſtrations, and ſingular Experiments; as no truth can ſeem capable of greater illuſtration, and leſs oppoſition: and therefore the greateſt right we can do our ſelves, or you, in this point, is to remit you to the obſervant lecture of his whole Third Book, *de Fundament. Opticis;* which we dare commend with this juſt Elogie, that it is the moſt Elaborate and Satisfying inveſtigation of the Principal Seat of Viſion, that ever the World was enriched with, and He who ſhall deſire a more accompliſht Diſcourſe on that (formerly) abſtruſe Theorem, muſt encounter the cenſure of being either ſcarce Ingenious enough to comprehend, or ſcarce Ingenuous enough to acknowledge the convincing Energy of the Arguments and Demonſtrations therein alledged, for the confirmation of his Theſis, *Radij formaliter viſorij nativam ſedem eſſe tunicam retinam.*

And the other is ſufficiently evincible even from hence; That *the Sight,* or (if you pleaſe) *the Interior Faculty doth alwayes judge of the aſpectable form of an Object, according to the Condition of the Image emanant from it,* at leaſt, *according as it is repreſented by the Image, at the impreſſion thereof on the principal viſory part.* Which is a poſition of Eminent Certitude. For, no other Cauſe can be aſſigned, why the Viſive Faculty doth deprehend and pronounce an object to be of this, or that particular Colour: but only this, that the Image impreſt on the Net-work Coat doth repreſent it in that particular Colour, and no other. Why, when half of the Object is ecliſed, by ſome opake body interpoſed, the eye can ſpeculate, nor the faculty judge of no more then the unobſcured half: but only this, that the Image is mutilated, and ſo conſiſteth of onely thoſe radii, that are emitted from the unobſcured half, and conſequently can inferr the ſimilitude of no more.

*Art. 9.* That the Faculty forms a judgment of the Conditions of the Object, according to the repreſentation thereof by the Image, at its impreſſion on the principal part of Viſion, the *Amphibleſtroides.*

Why

Why an Object, of whatever Colour, appeareth Red, when speculated through Glass of that Tincture : but only because the Image, in its trajection through that Medium, being infected with rednefs, retains the fame even to its figillation on the Expanfion of the Optick Nerve. Why the fight, in fome cafes, efpecially in that of immoderate diftance, and when the object is beheld through a Reverfing Glafs, deprehends the object under a falfe figure : but becaufe the Image reprefents it under that diffimilar figure, having either its angles retufed, by reafon of a too long trajection through the Medium, or the fituation of its parts inverted, by decuffation of its rayes in the Glafs.

## Consectary the First.

*Art.* 10.
CONSECTA-
RY.
That the I-
mage is the
Caufe of Ob-
jects apparence
of this or that
determinate
*Magnitude*.

Now, it being no lefs Evident, then Certain, that the Image is the fole caufe of the Objects apparence under fuch or fuch a determinate Colour, and of this or that determinate Figure : it is of pure Confequence, that the Image muft alfo be the Caufe of the Objects appearance in this or that determinate *Magnitude*; efpecially fince Figure is effenced in the Termination of Magnitude, according to *Euclid.*(*lib.*1.*def.*14.)*Figura eft,qua fub aliquo, vel aliquibus terminis comprehenditur.* For, why doth the object appear to be of great, fmall, or mean dimenfions; if not becaufe the Image arriving at the fentient, is great, fmall, or mean? Why doth the whole object appear greater then a part of it felf; unlefs becaufe the whole Image is greater then a part of it felf? To fpeak more profoundly, and as men not altogether ignorant of the Myfteries in Opticks; demonftrable it is, that the Magnitude of a thing fpeculated may be commenfurated by the proportion of the Image deradiated from it, to the diftance of the Common Interfection. For as the Diametre of the Image, projected through a perfpective, or Aftronomical Tube, on a fheet of white paper, is in proportion to the Axis of the Pyramid Everfed; fo is the diameter of the bafis of the Object to the Axis of the Pyramid Direct. And hereby alfo come we to apprehend the *Diftance* of the Object from the Eye, for having obtained the Latitude of the object, we cannot want the knowledge of its Diftance : and by converfion, the knowledge of its diftance both affifts and facilitates the comprehenfion of its Magnitude. Which comes not much fhort of abfolute neceffity; fince as *Des Cartes* (*Dioptrices cap.*6.) hath excellently obferved, in thefe words : *Quoniam autem longitudo longius decurrentiam radiorum non exquifite falis ex modo impulfus cognofci poteft, præcedens Diftantia fcientia hic in auxilium eft vocanda. Sic, ex Gr. fi diftantia cognofcatur effe magna, & Angulus vifionis fit parvus; res objecta longius diftans judicatur magna : fin vero diftantia fciatur effe parva, & angulus Vifionis fit magnus; objectum judicatur effe parvum, fi vero diftantia objecti longius difiiti fit in cognita, nihil certi de ejut magnitudine decerni poteft :* if the Diftance of an object far removed be unknown, the judgment concerning the magnitude thereof muft be uncertain.

## CONSECTARY the Second.

Again, insomuch as the Receptary of the Visible Image, is that Concave Mirrour, the Retina tunica (we call it a *Concave Mirrour*, not only in respect of its Figure and Use, but also in imitation of that grand Master of the Opticks, *Alhazen*, who *(in lib.*1. *cap.* 2.) saith thus; *Et sequitur ex hoc, at corpus sentiens, quod est in Concavo Nervi (retina nimirum) sit aliquantulùm Diaphanum, ut appareant in eo formæ lucis & coloris, &c.*) Hence is it, that no Image can totally fill that Receptary, unless it be derived from an object of an almost *Hemispherical* ambite, or Compass; so that the rayes, tending from it to the eye, may bear the form of a Cone, whose Base is the Hemisphere, and point (somewhat retused) the superfice of the Pupil. This perfectly accords to *Keplers* Canon; *Visionem fieri, cum totius Hemisphæriy mundani, quod est ante oculum, & amplius paulò, idolum statuitur ad album subrufum Retinæ cavæ superficiei parietem. \in Paralipomen. ad Vitellion. cap.*5. *de modo Vision. num.*1.) Not that either He, or we, by the Optical Hemisphere, intend only the Arch of the Firmament; but any Ambite whatever, including a variety of things obverted to the open eye, partly directly, partly obliquely, or laterally, and Circumquaq; in all points about.

And this being conceded, we need not long hunt for a reason, why, when the eye is open, there alwayes is pourtraied in the bottom of the eye some one *Total Image*; whose various parts may be called the *Special Images* of the diverse things at once objected. For, as the whole Hemisphere Visive includes the reason of the whole Visible: so do the parts thereof include the reason of the special Visibles, though situate at unequal distance. And, since, the Hemisphere may be, in respect either of its whole, or parts, more Remote, and more Vicine; hence comes it, that no more Rayes arrive at the Eye from the Remote, than the Vicine: because in the Vicine, indeed, are less or fewer bodies, than in the Remote, but yet the Particles, or Faces of the particles of bodies, that are directly obverted to the Pupil, are more. Which certainly is the Cause, why of two bodies, the one Great, the other Small, the Dimensions seem equal; provided the Great be so remote, as to take up no greater a part of the Visive Hemisphere, than the small: because, in that case, the rayes emanant from it, and in direct lines incident into the pupill of the Eye, are no more then those deradiate from the small, and consequently cannot represent more parts thereof, or exhibit it in larger Dimensions. Whereupon we may conclude that the Visive Faculty doth judge of the Magnitude of Objects, by the proportion that the Image of each holds to the amplitude of the Concave of the Retina Tunica : or, that by how much every special Image shall make a greater part of the General Image, that fills the whole Hemisphere Visive, and so possess a greater part of the Concave of the Retina Tunica; by so much the greater doth the Faculty judge the quantity thereof to be: and *è Contra*. And, because a thing, when near, doth possess a greater part of the Visive Hemisphere, than when remote : therefore doth the special Image thereof also possess a greater part of the Concave in the Retina Tunica, and so exhibit in greater Dimensions; and it decreaseth, or becometh so much the less, by how much the farther it is abduced from the eye, For it then makes room
for

*Art.* 11.
CONSECTA-
RY 2.
That no I-
mage can re-
plenish the
Concave of the
Retina Tuni-
ca, unless it be
deradiated
from an object
of an almost
*Hemispherical*
ambite.

*Art.* 12.
Why, when
the Eye is o-
pen there is al-
wayes pour-
trayed in the
bottom there-
of, some one
*Total Image*;
whose various
Parts, are the
*Special Images*
of the several
things inclu-
ded in the vi-
sual Hemi-
sphere.

for another Image of another thing, that is detected by the abduction of the former, and enters the space of the Hemisphere obverted. And hereupon may we ground a

## PARADOX.

*Art.* 13.
PARADOX.
That the prospect of a shilling or object of a small diametre is as great, as the Prospect of the *Firmament.*

*That the Eye sees no more at one prospect then at another: or, that the Eye beholds as much when it looks on a shilling, or any other object of as small diameter, as when it speculates a Mountain, nay the whole Heaven.*

Which though obscure and despicable at first planting, will yet require no more time to grow up to a firm and spreading truth, than while we investigate the Reasons of Two Cozen-German optical *Phænomena's.*

(1) Why an Object appears not only greater in dimensions, but more distinct in parts, when lookt upon near at hand; than afarr off?

(2) Why an Object, speculated through a *Convex* Glass, appears both *larger* and more *distinct*; than when beheld only with eye: but through a *Concave,* both *Smaller,* and more *confused?*

*Art.* 14.
Why an object appears both greater in Dimensions and more Distinct in parts, near at hand, than far off.

To the solution of the *First*, we are to reflect on some of the præcedent Assumptions. For, since every Visible diffuseth rayes from all points of it superfice, into all regions of the medium, according to the second *Assumption;* and since the superfice of the most seemingly smooth and polite body, is variously interspersed with Asperities, from the various faces whereof, innumerable rayes are emitted, tending according their lines of Direction, into all points of medium circularly; according to the first *Assumption;* and since those swarms of Emanations must be so much the more Dense and Congregate, by how much the less they are elongated from their fountain, or body exhalant; and *è Contra,* so much the more Rare and Disgregate, by how much farther they are deduced, according to the third *Assumption:* Therefore, by how much nearer the eye shall be to the object, by so much a greater number of Rayes shall it receive from the various parts thereof, and the particles of those parts; and *è Contra:* and Consequently by how much a greater number of rayes are received into the pupill of the eye, by so much greater do the dimensions of the object, and so much the more distinct do the parts of it superfice appear. For it is axiomatical among the Masters of the Opticks, and most perfectly demonstrated by *Scheinerus* (*in lib.* 2. *Fundament. Optic. part.*1. *cap.*13.) that the Visive Axe consisteth not of one single raye, but of many concurring in the point of the pyramid, terminated in the concave of the Retina Tunica: and as demonstrable, that those rayes only concur in that conglomerated stream, which enters the Pupil, that are emitted from the parts of the object directly obverted unto it; all others tending into other quarters of the medium. And hence is it, that the image of a remote object, consisting of rayes (which though streaming from distant parts of the superfice thereof, do yet, by reason of their concurse in the retused point of the visive Pyramid, represent those parts as Conjoyned) thin and less united, comparatively; those parts must appear as Contiguous in the visifical Representation, or Image, which are really Incontiguous or seperate in the object: and upon consequence, the object must

muſt be apprehended as Contracted, or Leſs, as conſiſting of fewer parts, an d alſo Confuſed , as conſiſting of parts not well diſtinguiſht. This may be truly, though ſomewhat groſly, *Exemplified* in our proſpect of two or three Hills ſituate at large diſtance from our eye, and all included in the ſame Viſive Hemiſphere ; for, their Elongation from the Eye makes them appear Contiguous, nay one and the ſame Hill, though perhaps they are, by more then ſingle miles, diſtant each from other : or, when from a place of eminence we behold a ſpacious Campania beneath, and apprehend it to be an intire Plane ; the Non-apparence of thoſe innumerable interjacent Foſſes, Pits, Rivers, &c. depreſt places, impoſing upon the ſenſe, and exhibiting it in a ſmooth continued plane.

And to the ſolution of the *ſecond Problem,* a conciſe enquiry into the Cauſes of the different Effects of *Concave* and *Convex* Perſpicils, in the repreſentation of Images Viſible, is only neceſſary. A Concave Lens, whether Plano-concave, or Concave on both ſides, whether it be the ſegment of a great, or ſmall Circle, projects the Image of an Object, on a paper ſet at convenient diſtance from the tube that holds it, Confuſed and inſincere ; becauſe it refracts the rayes thereof even to Diſgregation, ſo that never uniting again, they are tranſmitted in divided ſtreams and cauſe a chaos, or perpetual confuſion. On the Contrary, a Convex Lens refracts the rayes before divided, even to a Concurſe and Union, and ſo makes that Image Diſtinct and Ordinate, which at its incidence thereon was confuſed and inordinate. And ſo much the more perfect muſt every Convex Lens be, by how much greater the Sphere is, of which it is a Section. For, as *Kircher* well obſerves ( *in Magia paraſtatica.* ) if the Lens be not only a portion of a great ſphere, V. Gr. ſuch a one , whoſe diametre contains twenty or thirty Roman Palms ; but hath its own diametre conſiſting of one, or two palmes: it will repreſent objects of very large dimenſions, with ſo admirable ſimilitude, as to inform the Viſive Faculty of all its Colours, Parts, and other diſcoverables in it ſuperfice. Of which ſort are thoſe excellent Glaſſes, made by that famous Artiſt, *Euſtachio Divini,* at Rome ; by the help whereof the Painters of Italy uſe to draw the moſt exquiſite Chorographical, Topographical, and Proſopographical Tables, in the World. This Difference betwixt Concave and Convex Perſpicils is thus ſtated by *Kircher* ( *Art. Magnæ Lucis & Umbræ lib.* 10. *Magiæ part.* 2. *Sect.* 5. ) *Hinc patet differentia lentis Convexæ & Concavæ ; quod illa confuſam ſpeciem acceptam tranſmiſſamque ſemper diſtinguit, & optimè ordinat: illa verò eandem perpetuo confundit ; unde officium lentis Convexæ eſt, eaſdem confuſè acceptas, in debita diſtantia, ſecundum ſuam potentiam, diſtinguere & ordinare.* And by *Scheinerus* ( *in Fundam. Optic. lib.* 3. *part.* 1. *cap.* 11. ) thus ; *Licet in vitro quocunque refractio ad perpendicularem ſemper accidat, quia tamen ipſum ſuperficie cava terminatur, radij in aerem egreſſi potius diſperguntur, quàm colliguntur : cujus contrarium evenit vitro Convexo, ob contrariam extremitatem. Rationes ſumuntur à Refractionibus in diverſa tendentibus, vitri Convexi & Concavi, ob contrarias Extremitatum configurationes. Concavitas enim radios ſemper magis divergit : ſicut Convexitas amplius colligit, &c.*

*Art.* 15.
Why an object, ſpeculated through a *Convex Lens,* appears both greater and more *diſtinct* ; but through a *Concave, leſs* and more *Confuſed* : than when ſpeculated only with the Eye.

Y                                    Now

Now, to draw thefe lines home to the Centre of our problem ; fince the Rayes of a Vifible Image trajected through a *Convex* Perfpicil, are fo refracted, as to concurr in the Vifive Axe : it is a clear confequence, that therefore an object appears both larger in dimenfions, and more diftinct in parts, when fpeculated through a Convex Glafs, than when lookt upon only with the Eye ; becaufe more of the rayes are, by reafon of the Convexity of its extreme obverted to the object, conducted into the Pupil of the Eye, than otherwife would have been. For, whereas fome rayes proceeding from thofe points of the object, which make the Centre of the Bafe of the Vifive Pyramid, according to the line of Direction, incurr into the Pupil ; others emanant from other parts circumvicine to thofe central ones, fall into the Iris ; others from other parts circumvicine fall upon the eyelids ; and others from others more remote, or nearer to the circumference of the Bafe of the Pyramid, ftrike upon the Eyebrows, Nofe, Forehead, and other parts of the face : the Convexity of the Glafs caufeth, that all thofe rayes, which otherwife would have been terminated on the Iris, eyelids, brows, nofe, forehead, &c. are Refracted, and by refraction deflected from the lines of Direction, fo that concurring in the Vifive Axe, they enter the Pupil of the Eye in one united ftream, and fo render the Image impreft on the Retina Tunica, more lively and diftinct, and encreafed by fo many parts, as are the rayes fuperadded to thofe, which proceed from the parts directly confronting the Pupil. On the Contrary ; becaufe an Image trajected through a *Concave* Perfpicill, hath its rayes fo refracted, that they become more rare and Difgregate : the object muft therefore feem lefs in dimenfions, and more confufed in parts ; becaufe many of thofe rayes, which according to direct tendency would have infinuated into the Pupill, are diverted upon the Iris, Eyelids, and other circumvicine parts of the face.

*Art.* 16.
DIGRESSI-
ON.
What Figur'd
*Perfpicils* are
convenient for
Old : and what
for *Purblind*
perfons.

Here opportunity enjoyns us to remember the duty of our Proffeffion, nor would Charity difpenfe, fhould we, in this place, omit to prefcribe fome General Directions for the Melioration of fight, or natively, or accidentally imperfect. The moft common Diminutions of Sight, and thofe that may beft expect relief from Dioptrical Aphorifms, and the ufe of Glaffes ; are only Two : *Presbytia*, and *Myopia*. The *Firft*, as the word imports, being moft familiar to old men, is (*Vifus in perfpiciendis objectis propinquis obfcuritas ; in remotis verò integrum acumen* ) an imperfection of the fight, by reafon whereof objects near hand appear obfcure and confufed, but at more diftance, fufficiently clear and diftinct. The Caufe hereof generally, is the defect of due Convexity on the outfide of the Chryftalline Humor ; arifing either from an Error of the Conformative Faculty in the Contexture of the parts of the Eye, or (and that moftly) from a Confumption of part of the Chryftalline Humour by that Marafmus, Old Age : which makes the common Bafe of the Image Vifible to be trajected fo far inwards, as not to be determined precifely in the Centre of the concave of the *Retina Tunica.* And therefore, according to the law of Contrariety, the Cure of this frequent fymptome is chiefly, if not only to be hoped from the ufe of *Convex* Spectacles, which determine the point of Concurfe exactly in the Centre of the Retina Tunica ; the rayes, by reafon of the double Convexity, viz. of the Lens and Chryftalline Humor, being fooner and more vigoroufly united, in the due place.

The *Other*, being Contrary to the firft, and alwayes *Native*, commonly
named

named *Purblindnefs*, Phyfitians define to be *Obfcuritus vifus in cernendis rebus diftantibus ; in propinquis vero integrum acumen :* a Dimnefs of the fight in the difcernment of Objeſts, unlefs they be appropinquate to the Eye. The *Caufes* hereof generally are either the too fpherical Figure of the Chryftalline Humor ; or, in the Duſtus Ciliares, or fmall Filaments of the Aranea Tunica (the proper inveftment of the Chryftalline) a certain ineptitude to that contraſtion, requifite to the adduſtion of the Chryftalline inwards towards the retina tunica, which is neceffary to the difcernment of objeſts at diftance : either of thefe Caufes making the common Bafe of the Image to be determined in the Vitrious Humor, and confequently the Image to arrive at the retina tunica, perturbed and confufed. And therefore our advice is to all Purblind Perfons, that they ufe *Concave* Speſtacles: for, fuch prolong the point of concurfe, untill it be convenient, i. e. to the concave of the retina tunica.

## Affumption the Sixth and laft.

Since all objeſts fpeculated under the fame Angle, feem of equal Magnitude (according to that of *Scheinerus, ficut oculus rem per fe parvam, magnam arbitratur, quia fub magno angulo, refraſtionis beneficio, illam apprehendit : & magnam contrario parvam ; fundament. Optic. lib.* 2. *part.* 2. *cap.* 5.) and are accordingly judged, unlefs there intervene an Opinion of their unequal Diftance, which makes the Speſtator præfume, that that Objeſt is in it felf the Greater, which is the more Remote, and that the Lefs, which is the lefs Remote : therefore, to the appehenfion and Dijudication of one of two objeſts, apparently equal, to be really the greater, is not required a greater Image, than to the apprehenfion and dijudication of an objeſt to be really the lefs ; but only an opinion of its greater Diftance.

*Art.* 17. That to the Dijudication of one of two objeſts, apparently Equal, to be really the Greater ; is not required a greater Image : but only an *Opinion* of its greater *Diftance.*

This may receive both Illuftration and Confirmation from this eafie *Experiment.* Having placed horizontally, in a valley, a plane Looking Glafs, of no more then one foot diametre ; you may behold therein, at one intuition the Images of the firmament, of the invironing Hills, and all other things circumfituate, and thofe holding the fame magnitude, as when fpeculated direſtly, and with the naked eye : and this only becaufe, though the Image in Dimenfions exceed not the Area of the Glafs, yet is it fuch, as that together with the things feen, it doth alfo exhibit the Diftance of each from other. Exaſtly like a good Landfkip, wherein the ingenious Painter doth artificially delude the eye by a proportionate diminution and decurtation of the things præfented, infinuating an opinion of their Diftance. And therefore, the Reafon, why the Images of many things, as of fpacious Fields, embroydered with rowes of Trees, numerous Herds of Cattle, Flocks of Sheep, &c. may at once be received into that narrow window, the Pupill of the eye, of a man ftanding on an Hill, Tower, or other eminent place, advantageous for profpeſt: is only this, that to the Speculation of the Hemifphere comprehending all thofe things, in that determinate magnitude, is required no greater an Image, than to the Speculation of an Hemifphere, whofe diametre is commenfurable

only

only by an inch. Since neither more rayes are derived from the one to the Pupil of the Eye, than from the other: nor to the judication of the one to be ſo much Greater than the other, is ought required, beſide an Opinion that one is ſo much more Diſtant than the other. And this we conceive a ſufficient Demonſtration of the Verity of our laſt *Paradox,* viz. that the Eye ſees as much, when it looks on a ſhilling, or other object of as ſmall diametre; as when it looks on the greateſt Ocean.

Here moſt opportunely occurs to our Conſideration that notorious PROBLEM, *Quomodo objecti diſtantia deprehendatur ab oculo?* How the Diſtance of the Object from the eye is perceived in the act of Viſion?

Art. 18.
Des Cartes Opinion, concerning the Reaſon of the Sights apprehending the Diſtance of an object:

This would *Des Cartes* have ſolved (1) By the *various Figuration of the Eye.* Becauſe in the Conſpection of Objects remote, the Pupil of the Eye is expanded circularly, for the admiſſion of more Rayes; and the Chryſtalline Humor ſomewhat retracted toward the Retina Tunica, for the Determination of the point of Concurſe in the ſame, which otherwiſe would be ſomewhat too remote: and on the contrary, in the conſpection of objects vicine, the Pupil is contracted circularly, and the Chryſtalline Lens protruded ſomewhat outwardly, for the contrary reſpects. (2) By the *Diſtinct,* or *Confuſed* repreſentation of the object; as alſo the *Fortitude,* or *Imbecillity* of Light illuſtrating the ſame. Becauſe things repreſented confuſedly, or illuſtrated with a weak light alwayes appear Remote: and on the contrary, things præſented diſtinctly or illuſtrate with a ſtrong light, ſeem vicine.

Art. 19.
Unſatisfactory; and that for two Conſiderations;

But all this we conceive unſatisfactory. (1) Becauſe, unleſs the variation of the Figure of the Eye were *Gradual,* reſpective to the ſeveral degrees of diſtance intercedent betwixt it and the object; it is impoſſible the light ſhould judge an object to be at this or that Determinate remotion: and that the variation of the Figure of the Eye is not Gradual reſpective to the degree of diſtance, is evident even from hence; that the Pupil of the Eye is as much Expanded, and the Lens of the Chryſtalline Humor as much Retracted toward the Retina Tunica, in the conſpection of an object ſituate at one miles diſtance, as of one at 2, 3, 4, or more miles; there being a certain Term of the Expanſion of the one part, and Retraction of the other. (2) Becauſe though Viſion be Diſtinct, or Confuſed, both according to the more or leſs illuſtration of the object by light, and to the greater or leſs Diſtance thereof from the Eye; yet doth this Reaſon hold only in mean, not large diſtance: ſince the orbs of the Sun and Moon appear greater at their riſing immediately above the Horizon, that is, when they are more Remote from the Eye, than when they are in the Zenith of their gyre, that is, when they are more Vicine to the Eye; and ſince all objects illuſtrate with a weak light, do not appear Remote, nor *è contra,* as common obſervation demonſtrateth.

Art. 20.
And that more ſolid one of *Gaſſendus* (viz. that the Cauſe of our apprehending the Diſtance of an object, conſiſteth in the Comparation of the ſeveral things interjacent betwixt the object and the Eye, by the Rational Faculty) embraced and corroborated.

And therefore allowing the Acuteneſs of *Des Cartes* Conceit, we think it more ſafe, becauſe more reaſonable to acquieſce in the judgment of the grave *Gaſſendus;* who (*in Epiſt.* 2. *de Apparente Magnitud. ſolis humilis & ſublimis*) moſt profoundly ſolves the Problem, by deſuming the Cauſe of our apprehending the diſtance of an Object, in the act of Viſion,

from

from a *Comparison of the thing interjacent between the object seen, and the Eye.* For, though that *Comparation* be an act of the *Superior Faculty*; yet is the connexion thereof to the sense, necessary to the making a right judgment, concerning the Distance of the Visible. And, most certainly, therefore do two things at distance seem to be Continued, because they strike the Eye with cohærent, or contiguous Rayes. Thus doth the top of a Tower, though situate some miles beyond a Hill, yet seem Contiguous to the same, nay to the visible Horizon ; and this only because it is speculated by the Mediation of Contiguous Rayes : and the Sun and Moon, both orient and occident, seem to cohære to the Horizon because though the spaces are immense, that intercede betwixt their Orbs and the Horizon, yet from those spaces doth not so much as one single Raye arrive at the Eye, and those which come to it from the Sun and Moon are contiguous to those which come from the Horizon. And hence is it, that the Tower, Hill, and Horizon seem to the sight to be equidistant from the Eye, because no other things are interposed, at least, seen interposed, by the comparison of which, the one may be deprehended more than the other. Besides, the distance of the Horizon it selfis not apprehended by any other reason, but the diversity of things interjacent betwixt it and the Eye : for, look how much of Space is possessed valleys and lower grounds interjacent, so much of Space is defalcated from the distance ; the sight apprehending all those things to be Contiguous, or Continued, whose Rayes are received into the Eye, as Contiguous, or Continued, none of the spaces interjacent affording one raye. Of which truth *Des Cartes* seems to have had a glimpse, when (*in Dioptrices cap. 6. Sect. 15.*) he conceds ; *objectorum, quæ intuemur, præcedaneam cognitionem, ipsorum distantiæ melius dignoscendæ inservire :* that a certain præcognition of the object doth much conduce to the more certain dignotion of its Distance.

Art. 21.

And on this branch may we ingraft a PARADOX ; *that one and the same object, speculated by the same man, in the same degree of light, doth alwayes appear greater to one Eye, than to the other.* The truth of this is evincible by the joint testimony of those incorruptible Witnesses of Certitude, Experience and Reason. (1) Of *Experience,* because no man can make the vision of both his eyes equally perfect ; but beholding a thing first with one eye, the other being closed, or eclipsed, and then with the other, the former being closed or eclipsed ; shall constantly discover it to be greater in dimensions in the apprehension of one Eye, than of the other : and *Gassendus,* making a perfect and strict Experiment hereof, testifies of himself, (*in Epist. 2. de Apparent. Magnitud. Solis, &c. Sect. 17.*) that the Characters of his Book appeared to his right Eye, by a fifth part, greater in dimensions, though somewhat more obscure, than to his left. (2) Of *Reason ;* because of all *Twin Parts* in the body, as Ears, Hands, Leggs, Testicles, &c. one is alwayes more vigorous and perfect, in the performance of its action, than the other. Which Inæquality of Vigour, if it be not the Bastard of Custom, may rightfully be Fathered upon either this ; that one part is invigorated with a more liberal *afflux* of *Spirits,* than the other : or this, that the *Organical Constitution* of one Part is more perfect and firm, than that of the other. And, therefore, one Eye having its Pupill wider ; or the figure of the Chrystalline more Convex, or the Retina Tunica more concave, than the other ; must apprehend an object to be either larger in Dimensions, or more Distinct in Parts, than the other, whose parts are of a different configuration

guration

*Art.* 21.
PARADOX.
That the same
Object, specu-
lated by the
same man, at
the same di-
stance, and in
the same de-
gree of light ;
doth alwayes
appear greater
to one Eye ,
than the other

guration: either of these Causes necessitating a respective Disparity in the Action.

**Art. 22.**
A second PARADOX. That all men see (distinctly) but with one Eye at once: contrary to that eminent Optical Axiom, that the *Visive Axes* of both eyes concurr and unite in the object.

If this sound strange in the ears of any man, how will he startle at the mention of that much more Paradoxical Thesis of *Joh. Baptista Porta* (lib. 6. de *Refraction*. cap. 1.) That *no man can see* (distinctly) *but with one eye at once?* Which though seemingly repugnant not only to common perswasion, but also to that high and mighty Axiom of *Alhazen*, *Vitellio*, *Franc. Bacon. Niceron*, and other the most eminent Professors of the Optiques, *That the Visive Axes of both eyes concurr and unite in the object speculated :* is yet a verity, well worthy our admission, and assertion. For, the Axes of the Eyes are so ordained by Nature, that when one is intended, the other is relaxed, when one is imployed, the other is idle and unconcerned; nor can they be both intended at once, or imployed, though both may be at once relaxed, or unimployed: as is Experimented, when with both eyes open we look on the leaf of a Book; for we then perceive the lines and print thereof, but do not *distinctly* discern the Characters, so as to read one word, till we fix the Axe of one eye thereon; and at that instant we feel a certain suddain subsultation, or gentle impulse in the Centre of that eye, arising doubtless from the rushing in of more spirits through the Optick Nerve, for the more efficacious performance of its action. The *Cause* of the impossibility of the intention of both Visive Axes at one object, may be desumed from the *Parallelism* of the *Motion* of the Eyes; which being most evident to sense, gives us just ground to admire, how so many subtle Mathematicians, and exquisite Oculists have not discovered the Coition and Union of the Visive Axes in the object speculated, which they so confidently build upon, to be an absolute Impossibility. For, though man hath two Eyes; yet doth he use but one at once, in the case of *Distinct* inspection, the right eye to discern objects on the right side, and the left to view objects on the left: nor is there more necessity, why he should use both Eyes at once, than both Arms, or Leggs, or Testicles, at once. And for an *Experiment* to assist this Reason, we shall desire you only to look at the top of your own Nose, and you shall soon be convicted, that you cannot discern it with both eyes at once; but the right side with the right eye, and afterward the left side with the left eye: and at the instant of changing the Axe of the first eye, you shall be sensible of that impulse of Spirits, newly mentioned. Nor, indeed, is it possible, that while your right eye is levelled at the right side of your nose, your left should be levelled at the left side, but on the contrary averted quite from it: because, the motion of the eyes being *Conjugate*, or *Parallel*, when the Axe of the right eye is converted to the right side of the nose, the Axe of the left must be converted toward the left Ear. And, therefore, since the Visive Axes of both Eyes cannot Concurr and Unite in the Tipp of the Nose; what can remain to perswade, that they must Concurr and unite in the same Letter, or Word in a book, which is not many inches more remote than the Nose? And, that you may satisfie your self, that the Visive Axes doe never meet, but run on in a perpetual Parallelism, i. e. in direct lines, as far distant each from other, as are the Eyes themselves; having fixed a staff or launce upright in the ground, and retreated from it to the distance of 10 or 20 paces, more or less: look as earnestly as you can, on it, with your right eye, closing your left, and you shall perceive it to eclipse a certain part of the wall, tree, or other body situate beyond it. Then look on it again with your left eye, closing your right; and you shall

ſhall obſerve it to eclipſe another part of the wall: that ſpace being intercepted, which is called the Parallaxe. This done, look on it with both eyes
open; and if the Axes of both did meet and unite in the ſtaff, as is generally ſuppoſed, then of neceſſity would you obſerve the ſtaff to eclipſe either both parts of the Wall together, or the middle of the Parallaxe: but
you ſhall obſerve it to do neither, for the middle ſhall never be eclipſed;
but only one of the parts, and that on which you ſhall fix one of your eyes
more intently than the other. This conſidered, we dare ſecond *Gaſſendus*
in his promiſe to Gunners, that they ſhall ſhoot as right with both eyes
open, as only with one: for levelling the mouth of the Peece directly at
the mark, with one eye, their other muſt be wholly unconcerned therein, nor is it ought but the tyrannie of Cuſtome, that can make it
difficult.

Here, to prevent the moſt formidable *Exception*, that lyes againſt this
Paradox, we are to advertiſe you of two Conſiderables. *Firſt*, that as
well Philoſophers, as Oculiſts unanimouſly admit three *Degrees*, or gradual
Differences of ſight. (1) *Viſus Perfectiſſimus*, when we ſee the ſmalleſt
(viſible) particles of an object, moſt diſtinctly: (2) *Perfectus*, when we ſee
an object diſtinctly enough, in the whole or parts, but apprehend not the
particles, or minima viſibilia thereof: (3) *Imperfectus*, when beſides the
object directly obverted to the Pupil of the eye, we alſo have a glimmering
and imperfect perception of other things placed *ad latera*, on the right and
left ſide of it. *Secondly*, that the verity of this Paradox, that we ſee but
with one eye at once, is reſtrained only to the *Firſt* and *Second* degrees of
Sight, and extends not to the *Laſt*. For, Experience aſſures, that, as many
things circumvicine to the principal object, on which we look only with
one eye open, præſent themſelves together with it, in a confuſed and obſcure
manner: ſo likewiſe, when both eyes are open, many things, obliquely incident into each eye, are confuſedly, and indiſtinctly apprehended. So that
in confuſed and Imperfect Viſion, it may be truly ſaid, that a man doth ſee
with both eyes at once: but not in Diſtinct and Perfect.

*Art. 23.*
The three De-
grees of Viſi-
on, *viz. moſt
perfect, perfect,*
and *imperfect:*
and the verity
of the Para-
dox reſtrained
only to the
*two former* De-
grees.

## SECT. III.

TO entertain Curioſity with a ſecond Courſe, we ſhall here attempt the
Conjectural Solution of thoſe ſo much admired Effects of *Convex*
and *Concave* Glaſſes; that is, Why the Rayes of Light, and together with
them thoſe ſubſtantial Effluxes, that eſſence the Viſible Images of Objects, being trajected through a Convex Glaſs, or reflected from a Concave,
are *Congregated* into a perpendicular ſtream: and likewiſe, why the Rayes
of Light, being trajected through a Concave, or reflexed from a Convex,
are *Diſgregated* from a perpendicular radius.

*Art. 1.*
A reſearch in-
to the Reaſon
of the different
Effects of Con-
vex and Con-
cave Glaſſes, as
well Dioptrical,
as Catoptrical.

Firſt, inſomuch as Glaſs, of the moſt polite and equal ſuperfice is full of
inſenſible *Pores*, or Perforations, and ſolid impervious *Granules*, alternately
interſperſed; we may upon conſequence conceive, that each of thoſe ſolid
Granules is as it were a certain *Monticle*, or ſmall Hillock, having a ſmall top,
and ſmall ſides circularly declining toward thoſe little Valleys, the Pores.
                                                                        This

This conceded; if a Glaſs, whoſe ſuperfice is *Plane*, be obverted to the Sun, ſince the ſmall Pores thereof tend from one ſuperfice to the other in direct and parallel lines, for the moſt part; it muſt be, that all the Rayes incident into the Pores, paſs through in direct and parallel lines, into the Aer beyond it: and ſo can be neither Congregated, nor Diſgregated, but muſt conſtantly purſue the ſame direct courſe, which they continued from the body of the Sun, to their incidence on the ſurface of the Glaſs. But if the Extream of the Glaſs, reſpecting the Sun, be of a *Convex* figure; then, becauſe one Pore (conceive it to be the Central one) is directly obverted to the Sun, and all the others have their apertures more oblique and, pointing another way; therefore it comes to paſs, that one ray, falling into the directly obverted pore, is directly trajected through the ſame, and paſſeth on into the aer beyond it in a direct line; but another ray, falling on the ſide of the Hillock next adjacent to the right pore, is thereby Refracted and Deflected, ſo that it progreſſeth not forward in a line parallel to the directly trajected ray, but being conjoined to it, paſſeth on in an united ſtream with it. And neceſſary it is, that the Angle of its Refraction be by ſo much the more *obtuſe*, by how much nearer the point of the Hillock, from which it was refracted, is to the direct or perpendicularly tranſmitted ray; and, on the contrary, by ſo much the more *Acute*, by how much the more remote: becauſe *There* the ray falls more deeply into the obvious pore, and ſtrikes lower on the adjacent Hillock, whoſe Protuberancy therefore doth leſs Deflect it; but *Here* the ray falls higher on the ſide of the Hillock, and ſo by the Protuberancie, or Devexity thereof is more deflected. But if the Extreme of the Glaſs confronting the Sun, be of a *Concave* figure; in that caſe, becauſe one pore being directly open, others have their apertures more obliquely reſpecting the Sun, it comes to paſs, that the ray falling into the direct pore, is directly trajected, and paſſeth through the aer in a perpendicular; but another ray falling on the ſide of the next adjacent Hillock, is thereby refracted and deflected, ſo that it doth not continue its progreſs in a line parallel to the directly-tranſient ray, but is abduced from it, and that ſo much the more, by how much the farther it paſſeth beyond the Glaſs. And neceſſary it is, that the Angle of its Refraction be alſo ſo much the more obtuſe, by how much nearer the point of its incidence on the ſide of the Hillock, is to the Aperture of the Direct pore; becauſe it falls deeper into it, and ſtrikes lower on the devex ſide of the Hillock: and on the contrary, ſo much the more Acute, by how much more remote its point of incidence is to the Aperture of the Direct pore; for the contrary reſpect. And this is the ſumm of our Conjecture, touching the reaſons of the different Trajection of Rayes through Convex and Concave Glaſſes.

As for the other part of our Conception, concerning *Reflexed* Rayes; if the Glaſs obverted to the Sun be *Plane* in it ſuperfice, then, becauſe all the Topps of the ſolid and impervious *Hillocks*, are directly obverted to the Sun, therefore muſt it be, that all the rayes incident upon them become *Reflected* back again toward the Sun, if not in the ſame, yet at leaſt in Contiguous lines. But if the face of the Glaſs obverted to the Sun, be *Convex*; then, becauſe the topp of one Hillock is directly obverted, and thoſe of others obliquely reſpecting the Sun; it comes to paſs, that one ray being directly Reflected, the others are reflected obliquely in lines quite different: and this in an Angle by ſo much more Acute, by how much nearer the Topps of the obliquely refpecting Hillocks are to that of the directly refpecting one ; and by fo much the more obtufe, by how much the more Remote.   And, if the fide of the Glafs turned toward the Sun, be *Concave* ; becaufe the Top, of one Hillock is directly, and thofe of others obliquely obverted to the Sun ; hence comes it, that the Ray incident on the directly-obverted one, is directly Reflected, and thofe that fall on the topps of the obliquely-obverted ones, are accordingly reflected obliquely, toward the Directly reflected ; fo that at a certain diftance they all Concurr and Unite with it in that point, called the *Term of Concurfe*: and this in an Angle fo much more Acute, by how much nearer the Topps of the obliquely-reflecting Hillocks are to that of the Directly-reflecting one ; and *è contra*.

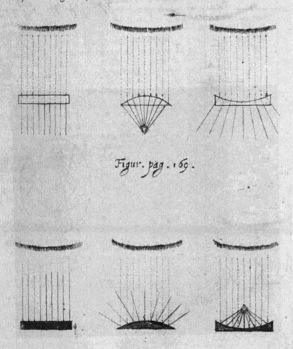

*Figur. pag. 169.*

*Art.* 2.
A COROL-
LARIE.
Hinting the
Causes, why
an Elliptical
Concave re-
flects the inci-
dent rayes, in
a more Acute
angle, than a
*Parabolical* :
and a Parabo-
lical than a
*Spherical.*

These things clearly understood, we need not want a perfect Demonstration of the Causes, why a Concave Glass, whose Concavity consisteth of the segment of an *Ellipsis*, reflecteth the rayes of the Sun in a more *Acute* Angle, and consequently burneth both more vigorously, and at greater Distance, then one whose Concavity is the segment of a *Parabola*: and why a Parabolical Section reflecteth them in an Angle more *Acute*, and so burneth both at greater distance, and more vigorously, than the Section of *Circle*. Especially if we familiarize this theory by the accommodation of these *Figures.*

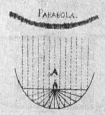

PARABOLA.

*Figur. pag.* 190.

ELLIPSIS.

B

CIRCVLVS.

Thus

Thus have we, in a short Discourse, not exceeding the narrow limits of a single Article, intelligibly explicated the Cause of that so much admired Disparity in the Effects of *Plane*, *Convex*, and *Concave*, Glasses; as well *Dioptrical*, or Trajecting the rayes of Light into the Aer beyond them, as *Catoptrical*, or Reflecting them back again from their obverted superfice. And we ask leave to encrease our Digression only with this CONSECTARY.    Because the Rayes of Light, and the rayes of visible Images are Analogical in their nature, and flow hand in hand together into the Eye, in the act of Vision; therefore is it, that to a man using a *Plane Perspicil*, an object always appears the *same*, i. e. equal in dimensions, and distinction of parts, as it doth to his naked Eye: by reason the Angle of its Extreams is the same in the Plane Glass, as in the Eye.    But, to a man using a *Convex Perspicil*, an object appears *Greater*; because the Angle of its Extreams is amplified: and through a *Concave*, *Less*; because the Angle is diminished.    In like manner, the Image of an object *reflected* from a *Plane Mirrour*, appears the *same* to the Spectator, as if Deradiated immediately, or without reflexion, from the object it self; because the Reflex Angle is equal to the Direct: but the Image of an Object Reflected from a *Convex Mirrour* appears *Less*; because the Angle of its Reflection is less than that of its Direction: and from a *Concave*, *Greater*; because the Reflex Angle is greater than the Direct:    This may be autoptically Demonstrated thus.    If you admit the Image of a man, or any thing else, through a small perforation of the wall, into an obscure chamber, and fix a Convex Lens in the perforation, with the Convex side toward the Light; you shall, admoving your eye thereto, at Convenient distance, observe the transmitted Image to be *Amplified*: but, receiving the Image on a sheet of white Paper, posited where your Eye was, you shall perceive it to be *Minorated*: the Contrary Effect arising from a *Concave Lens*, posited in the hole, with its Concave side toward the Light.    And this, because the Convex Congregating the rayes into the Pupill of the Eye (and so making the φασις, or Apparition Greater, for the cause formerly exposited) doth also Congregate them on the Paper; and therefore the Image cannot but appear Contracted, or Minorated: but on the contrary, the Concave Disgregating the rayes from the Pupil (and so making the φασις, or Apparition less in the Retina of the Optick Nerve) doth also Disgregate, or diffuse them largely on all parts of the Paper, and so the Image thereon received cannot but appear much Amplified.

*Art. 3.*
A CONSEC-
TARY.
Why a *Plane
Perspicil* exhibits an object in genuine Dimensions; but a *Convex*, in *Amplified*, and a *Concave*, in *minorated*.

Z 2                    FIGURE

## SECT. IV.

Hitherto we have in some degree of satisfaction, explicated the Manner, how, by the Incursion of substantial Images, deradiated from the object to the Eye, the Visive Faculty comes to apprehend the *Colour, Figure, Magnitude, Number,* and *Distance* of objects: and therefore it remains only, to the Complement of our present Designation, that we explore the Reasons of the Perception of the *Situation, Quiet,* and *Motion* of objects, by the sight. To our more perspicuous solution of which notable Difficulty, and to the illustration of many passages precedent in the two last Sections: it must be confest not only ornamental, or advantageous, but simply necessary, that we here Anatomize the whole Eye, and consider the proper Uses of the several parts thereof; those especially, that are either immediately and primarily instrumental, or only secundarily inservient to Vision.

*Art. 1.*
A Recapitulation of the principal Arguments precedent: and summary of the subsequent.

*Figur. pag.* 173

In the Conformation of the Eye, or minor Microcosm, as *Casserius Placentinus* calls it, in respect to the admirable Constructure thereof; the *First* observable is, that it is composed of many *Diaphanous,* or Transparent Parts, as the *Horny Membrane* (B C B) the *Aqueous Humour* (E F K F E) the *Chrystalline* (L) and the *Vitreous* (M G H M I N): and the intention of that Unimitable Mistress of the Optiques, Nature, herein was, that the Visive Rayes might not be Reflected from, but easily *Trajected* through them, into the Amphiblestroides, or

*Art. 2.*
The Eye Anatomized: and the proper use of each Part thereof, either absolutely Necessary, or only Advantagious to Vision, concisely demonstrated. viz. of

1.
The *Diaphanity* of the Horny Membrane, and the three Humors, *Aqueous, Chrystalline,* and *Vitreous.*

Net-work Coat, The *Second* is it *Convex Figure*; wherein the Providence of Nature had respect to the necessary Congregation and Unition of most of the rayes incident on the Area of the Eye, so that the Visive Axe might consist of many more rayes, than otherwise, i.e. had the figure of the Eye been Plane, or Concave, it would have done: for, being by this Convexity refracted, they convene in a Cone determined in the Centre of the *Amphiblestroides.* For the Convexity is so exactly proportionate to the Distance of the Retina Tunica from the Chrystalline, that most of the Visive Rayes, emanant from the several points of the object, and incident upon the several points of the Horny Membrane, may, after various Refractions, have their Rendezvouz, or point of Concurse exactly in the middle of the Retina Tunica: because, should their point of Concurse

2.
The *Convexity* of all its parts, except the *Amphiblestroides.*

curfe be either fhort of, or beyond the Retina Tunica; of neceffity the Image could not be at all, or, at moft, but very obfcurely prefented therein, as confifting of Difperfed, and mutually Interfecting rayes. The *Third*

3.
The *Uvea Tunica*, and *Iris*.

is the *Uvea Tunica*, or anterior part of the *Choroides*, whofe exterior fuperfice ( E F, F E ) being Diverficolor , or of various Colours, is called the *Iris* or Rainbow : which *Galen, Cafferius Placentinus* , and *Riolanus* will have to confift of a fix-fold Circle, but *Plempius* only of a Three-fold, the Two outmoft at the white of the Eye being more narrow in latitude, and the Third refpecting the Pupil of the Eye more ample, and illuftrate with the conftant colour on the Limbus of the *Uvea Tunica*, which in fome bears Sables , in others Azure, in others Sables and Argent confufed : whence the Difference of *Black, Blewifh,* and *Grey Eyes.* In the middle of

4.
The *Pupilla.*

this Coat is a Perforation, called the Pupil, ( F K F ) and by the vulgar, the *Apple* of the eye ; of fuch a Conftitution, that by Dilatation and Conftriction, as if it were a Sphincter Mufcle, it might be made wider, or narrower : and this for the Moderation of the incurrent rayes, which being fometimes more, fometimes fewer , and fometimes ftrong , fometimes weak , require a certain Moderation proportionate to the Faculty of the recipient and terminating fenfory. For, infomuch as an excefs of Light is deftructive, and the Defect of it infufficient to diftinct Vifion ; therefore did the Eternal Wifdome in the Entrance into the Chryftalline, contrive this Window capable of Dilatation and Conftriction : in Dilatation to admit fo much of the weaker Light as is required to perfect and diftinct Vifion ; in Contraction to exclude fo much of the Exceffive, as would offend, if not perifh the Organ. Yet in many the Amplitude of the Pupil varies, and thofe who have it very narrow, are ftrong and acute fighted ; but thofe, who have it more dilated conftantly, fee but weakly and obtufely.

5.
The *blacknefs* of the infide of the *Uvea Tunica.*

The interior fuperfice of this Membrane is obduced, or lined with a certain *Fuligineus* fubftance that gives it the Colour of a blackifh Grape, fully ripe : but to what end Nature provided this opacating Tincture, hath been a quæftion, that, even from *Galens* dayes to ours , hath made the Schools both of Anatomifts and Profeffors of the Optiques, ring again with Controverfies. Some affirming the defign of Nature therein to be, that the Chryftalline being veyled over with this obfcure parget, might have its own fplendor more intenfe by Congregation : becaufe, according to the pofition of *Alhazen* (*lib.*1. *propof.* 33.) as a fmall light in a dark obfcure place is better perceptible, as diffufing a brighter luftre, than in a wide, luminous place ; and confequently makes the circumjacent parts more vifible : fo doth the internal fplendour of the Chryftalline become more illuftrious, becaufe the inner circumference of the whole Uvea Tunica is lined with this footy matter, the rayes deradiating from it by reflection from the oppofite opacity of the Membrane, becoming reaffembled and united in a more vigorous luftre. *Others* conceiving the intention of it to be, the Recreation or Refection of the Vifive Spirits ; becaufe when ever the Chryftalline is offended, or rather the Amphiblestroides, with too vehement a Light, we ufe, for prefent remedy, to clofe our eyes, and the fpirits recoyling upon the Chryftalline from the natural darknefs of this Coat, are reaffembled, and fo refrefhed. And *others* contefting that the only ufe of it is , the Interception of Light ; for, fince the Pupil , or anterior perforation of the Uvea Tunica, is the only Aperture, or portal framed for the intromiffion of the Vifible Images, and there ought to be no other paffage, whereat Light might intrude it felf into the concave of the Eye : what could

wife

wise Nature have thought on more convenient to the Exclusion of unnecessary light, than the interjection of this sable Curtain? Experience evincing that nothing intercepts and shuts out Light, than opace Bodies interposed. These, indeed, are ingenious and plausible Conceits, but if truth be to be preferred to Acuteness; we may determine, that the only and proper use of this Atramentous or footy Blackness is, that the Rayes of Light, incident on the Concave of the Amphibleftroides, (G H I) and thence resilient back to the Concave of the Uvea Tunica, might by the Blackness of its lining be extinguisht, i. e. absolutely terminated: left thence again Reflected to the Amphibleftroides, they might perturb the Visible Image, and consequently the fight. The *Fourth* observable, is the *Tunica Arachnoides*, in its middle containing the moft pretious of Gemms, the *Chryftalline* Humor, whofe Figure alfo is Convex (but whether of a Parabolical, Elliptical, or Sphærical Section, is a noble problem, becaufe not yet determined.) on both fides, though fomewhat more on that fide refpecting the Retina Tunica, and manifeftly oblong, or inclining to an oval. This Coat, by the Mediation of the *Ciliary Proceffes*, or flender Filaments (B N, N B) difperfed from the Tunica Arachnoides, doth move the Chryftalline either nearer to, or farther from the Retina Tunica, as the greater or lefs Diftance of the object requires. For, in the Chryftalline, by reason of its greater both Denfity and Convexity, the rays of the fpecies are more ftrongly Refracted and more clofely United, than in any other part of the Eye: which juftifieth their opinion who make it the Primary Medium of Vifion. Becaufe, as a Convex Lens pofited in a hole of the wall, admits the fpecies into an obfcure room and alfo collect the rayes of it: fo doth the Chryftalline both admit and congregate them. And becaufe it is Diaphanous, therefore are not the fpecies terminated therein, as *Galen*, and after him moft *Anatomifts* have dreamt: fince otherwife no reason can be alledged, why the fpecies fhould not be as well terminated in the Horny Membrane, the Vitreous, or Aqueous Humour. Wherefore, Vifion is not made in the Chryftalline but the Retina Tunica: becaufe the fpecies are therein Terminated, as in an opace body. *Scheinerus* opinioned, that the fpecies, which otherwife, by reason of feveral refractions before their arrival at the Chryftalline, would have been exhibited in *Reverfe* pofitions, are therein refracted, and Rectified. But, from the Obfervation of *Franc. Sylvius*, *Franc. Vander Schagen*, *Joh. Wallæus*, and *Athanas Kircherus*, the *tunica Choroides* being fublated from the hinder part of the Eye, and then the *Sclirotica*, and laftly the *Amphibleftroides*; all objects appear inverfed in the Chryftalline: and in a fmaller form by much in the Eye of an Oxe, than in the Eye of a Man. The fame hath *Plempius* demonftrated by the Experiment of an Artificial Glafs Eye, placed in the fmall Aperture of a Window: all things externally objected appearing therein Inverft, as alfo on a fheet of paper pofited before the decuffation of the rayes. And, doubtlefs, it is neceffary, that the fpecies be inverted, at their termination on the Retina Tunica; fince otherwife we fhould have apprehended the object as inverft: which *Kepler* demonftrates from hence, that (*in paffione Patientia Agentibus è regione effe oppofita debere*) in Paffion the Patients muft be on the contrary region to the Agents. Some, we confefs would have it, that the *judicatory Faculty* doth correct the depraved Figure of the fpecies: becaufe (forfooth) it difcerns the juft magnitude of objects and their fituation, by moft fmall Images; as a good Geometrician doth judge of the dimenfions of *Hercules* whole body, by commenfurating thofe of his Heel. And others confign

that

*6.*
The Tunica
*Arachnoides.*

*7.*
The Ciliary
Filaments
thereof

*8.*
The Chryftalline.

*9.*
The Retina
Tunica.

that office to the *Common Sense*, which looking (*retro & desuper*) on the inverted species, apprehends them in a right position. And lastly, others desume the right judgment, from the *rectitude* of the *line*, by which the species are imprest. And thus poor man aggravates the Difficulties in Nature, though to his own greater disquiet and perplexity.       The *Last* of Parts in the Eye, immediately necessary to Vision, is the *Retina Tunica*, or Net-work Coat (G H I) in the bottom of the Eye; contexed of an innumerable multitude of Filaments, or thread-like Expansions of the Optick Nerve : and this is that noble sensory, formed for the Last Reception and Sigillation of the Image, which from hence by the Continuity of the Optick Nerve, is communicated ἐν Ἡγεμονικῷ, to the *Principal Faculty*, residing in the Brain.

10.<br>The six Muscles, viz.

But, because the Axe of the Visive Pyramid is a perpendicular line, beginning in the Extrems of the object, and ending in the Amphiblestroides; had the Eye been nailed or fixt in its orbita, we must have been necessitated to traverse the whole Machine of the body, for a position thereof convenient to Vision, since it can distinctly apprehend no object, but what lyes *è directo* opposite ; or have had this semi-rational sense, whose glory builds on Variety, restrained to the speculation of so few things, that we should have received more Discomfort from their Paucity, than either Information, or Delight from their Discernment : therefore, that we might enjoy a more enlarged Prospect, and read the whole Hemisphere over in one momentany act of Vision, Nature hath furnished the Eyes with *Muscles*, or Organs of agility; that so they may accommodate themselves to every visible, and hold a voluntary versility to the intended object ;

Parvula sic magnum pervisit Pupula Cœlum.

1.<br>The Direct, as the

And of these *Ocular Muscles* there are in Man, just so many, as there are kinds of Motion, 4 *Direct*, and 2 *Oblique* or Circular ; all situate within the Orbita, and associated to the Optick Nerve, and conjoining their Tendons, at the Horny Membrane, they constitute the *Tunica Innomitata*, so named by *Columbus*, who arrogates the invention thereof to himself, though *Galen* (*lib. 10. de usu part. cap. 2.*) makes express mention of it.

Atollens,

The *First* of the four *Direct* Muscles, implanted in the superiour part of the Eye, draweth it *Upward* ; whence it is denominated *Atollens*, the Lifter up, and *Superbus*, the Proud : because this is that we use in Haughty and sublime looks.

Deprimens,

The *Second*, situate in the inferiour part of the Eye, and Antagonist to the former, stoops the Eye Downward ; and thence is called *Deprimens*, the Depressor, and *Humilis*, the Humble : for this position of the eye speaks the Dejection, and Humility of the Mind.

Adducent,

The *Third*, fastned in the Major Canthus, or great angle of the Eye, and converting it toward the Nose; is therefore named *Adducens*, the Adducent, and *Bibitorius*, for in large draughts we frequently contract it.

The

The *Fourth*, oppofite both in fituation and office to the former, abduceth the Eye laterally toward the Ear; and is therefore named *Abducens*, and *Indignatorius*, the fcorning mufcle: for, when we would caft a glance of fcorn, contempt, or indignation, we contract the Eye towards the outward angle, by the help of this mufcle.

*Abducent.*

If all thefe Four work together, the Eye is retracted inward, fixt, and immote: which kind of Motion Phyfitians call *Motus Tonicus*, and in our language, the *Sett*, or *Wift* Look.

Of the *Oblique* Mufcles, the *Firft*, running betwixt the Eye, and the tendons of the Second and Third Mufcles, by the outward angle afcends to the fuperior part of the Eye, and inferted near to the Rainbow, circumgyrates the Eye downward.

*2. And Oblique, as the 2 Circumactors, or Lovers Mufcles.*

The *Second*, and fmalleft, twifted into a long tendon, circumrotates the Eye toward the interior angle, and is called the *Trochlea*, or Pully. Thefe two Circumactors are firnamed *Amatorij*, the Lovers Mufcles; for thefe are they that roul about the eye in wanton or amorous Glances.

And thus much of the Conformation of the Eye.

Now, as to the Solution of our Problem, *viz.* How the S I T U A-T I O N of an object is perceived by the fight? Since it is an indifputable Canon, *Omnem fenfum deprehendere rem ad eam regionem, è qua ultimo directa motione feritur*, that every fenfe doth apprehend its proper object to be fituate in that part of Space, from whence, by direct motion, it was thereby affected: we may fafely inferr, that the Vifible Object alwayes appears fituate in that part of fpace, from whence the Image thereof in a direct line invadeth the Eye, and enters the Pupil thereof. Which is true and manifeft not only in the intuition of an object by immediate or *Direct* rayes; but alfo in the infpection of Looking-Glaffes, that reprefent the object by *Reflex*: and a pure Confequence, that a Vifible Object, by impreffion of its rayes proceeding from a certain place, or region, muft of neceffity be perceived by the fight, in its genuine pofition, or *Erect* Form; though we have the teftimony both of Reafon and Autopfie, that the Image of every Vifible is pourtraid in the Amphibleftroides, in an unnatural pofition, or *Everfe* Form.

*Art. 3. Why the Situation of an object is perceived by the fight,*

*Art.* 4.
The Reason of the Evasion of the Image, in the *Amphible-stroides.*

As for that of *Reason*, it is thus Demonstrated. Suppose the Eye to be C D; the bottome thereof to be E; the object illustrate by the Sun A B. *viz.* a Cross painted on a Wall; the Pupil of the Eye, G H; and the Centre of the Pupill, I; Now, the Image of this Cross emanant therefrom, and entering the Pupill, in the lines A T, B S, must arrive at the bottome of the Eye, S T, in an Everse, or præposterous Form: because the narrowness of the Pupil, together with the prævious Refraction, makes the rayes concurrent at the point I, to Decussate, or mutually intersect each other; so that the raye proceeding from A, falls upon the part of the Retina Tunica, T, and the ray B falls upon S. Which makes it of absolute

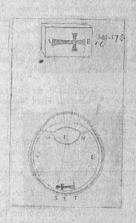

necessity, that the upper part or Head of the Cross, A, be depicted in the lower part of the Concave of the Retina Tunica, T: and the lower part, or Foot, B, in the upper part of the Concave of the Retina Tunica, S.

*Art.* 5.
The same illustrate by an *Experiment.*

And, as for that of *Autopsie,* or Ocular Experiment; Take the Eye of an Oxe, or (if the Anatomick Theatre be open) of a man, for in that the species are represented more to the life, than in the Eye of any other Animal, as *Des Cartes* (*in dioptrices cap.* 5. *Sect.* 11.) and having gently stript off the three Coats in the bottome, in that part directly behind the Chrystalline, so that the Pellucidity thereof become visible, place it in a hole of proportionate magnitude, in the wall of your Closet, made obscure by excluding all other light, so that the Anterior part theaeof may respect the light. This done, admoving your Eye towards the denudated part of the Chrystalline; you may behold the Species of any thing obverted to the outside of the Eye, to enter through the Chrystalline to the bottom thereof, and there represented in a most lively figure, as if pourtrayed by the exquisite Pencil of *Apelles*; but wholly Eversed: as in this following *Iconisme.*

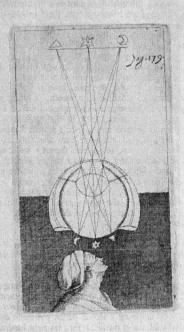

Finally, an object appears either in *Motion*, or *Quiet*, according as the Image thereof, represented on the Retina Tunica, is moved: or Quiet: only because, according to the Canon, in the præcedent Article, touching the reason of the perception of the situation of an object, the Visible is alwayes judged to be in that part of Space, from which, in a direct line, the last impression is made upon the *Sensorium*.

*Art. 6.*
Why the *Motion* and *Quiet* of objects are discerned by the fight.

And this Reason is of extent sufficient to include the full Solution also of that PROBLEM, by *Alexander* (2. *de Anima* 34.) so insulting proposed to the Defendants of *Epicurus* Material Actinobolisme Visive, or the Emanation of substantial Images from the Object to the Eye: *viz. Why doth the Image of a man move, when reflected from a Mirrour, according as the man moves?* For, this Phænomenon we are to referr to the Variation of the parts of the Mirrour, from each of which it is necessary that a fresh Reflexion of the Species be made into the Eye: and consequently, that the Image appear moved,

*Art. 7.*
Why *Catoptrical* Images imitate the motions of their Archetypes or Originals.

A a 2                                    ved,

ved, according to the various motions of the object. The necessity of this is evident from hence, if you stand beholding your face in a Glass, and there be divers others standing by, one at your right hand, another at your left, a third looking over your head in the same Glass, they shall all behold your image, but each in a distinct part of the Glass. Whence you may also understand, that in the Looking-glass is not only that Image, which you behold, but also innumerable others; and those so mutually communicant, that in the same place, where you behold your nose, another shall see your chin, a third your forehead, a fourth your mouth, a fifth your Eyes, &c. and yet doth no one see other then a simple and distinct Image. Moreover you may hence inferr, that in the medium is no point of Space, in which there is not formed a perfect Image of the rayes concurring therein, and advenient from the same object; though not from the same parts, or particles thereof: and consequently that in the whole Medium there are no two Images perfectly alike; as also, that what the Vulgar Philosophers teach, that the whole Image is in the whole Space or Medium, and whole in every part thereof, is a manifest Falsity. For, though it may be said justly enough, that the whole Image, i. e. the *Aggregate* of all the Images, is in the whole Space: yet is there no part of that Space, in which the whole Image can be.

*Art. 8.*
Why the right side of a Catoptrical Image respects the *Left* of its Exemplar. And why two Catoptrick Glasses, confrontingly posited, cause a *Restitution* of the parts of the Image to the natural Form.

To this place belongs also that PROBLEM; *Why doth not the right hand of the Image respond to the right of the object: but contrariwise, the left to the right, and right to the left?*

The *Cause* whereof consisteth onely in the Images *Confronting* the Object: or, as *Plato* (*in Timæo*) most perspicuously expresseth it, *quia contrarijs visus partibus ad contrarias partes sit contactus.* Understand it by supposing a second person posited in the place of the Mirrour, and confronting the first: for, his right hand must be opposed to the others left.

Nor is the reason of the Inversion of the parts of the Image other than this, that the rayes emitted from the right side of the object, are reflected on the left, and *è Contra.* Just as in all Impressions, or Sigillations, the right side of the Antitype responds to the left of the type. *Consule Aquilonium, lib. 1. opt. proposit. 46.* And, as for the reason of the *Restitution* of the parts of the Image to the right position of the parts of the object, by two Mirrours confrontingly posited: it may most easily and satisfactorily be explained by the *Decussation* of the reflected rayes.

To Conclude. We need not advertise, that the *Optical Problems* referrible to this place, are, (if not infinite) so numerous, as to require a larger Volume to their orderly Proposition and Solution, than what we have designed to the whole of this our *Physiology.* Nor remember you, that our principal Scope in this Chapter, was only to evince the Præeminence of *Epicurus* Hypothesis above all others, concerning the Reason and Manner of Vision; and this by accommodating it to the Verisimilous *Explanation* of the most Capital *Difficulties,* occurring to a profound inquest into that abstruse subject.

All

All therefore that remains unpaid of our præsent Debt, is modestly to refer it to your equitable Arbitration; Whether we have deserted the Doctrine of the *Aristoteleans*, touching this theorem, and addicted ourselves to the Sect of the *Epicureans*, on any other Interest, but that sacred one of *Verity*: which once to decline, or neglect, upon the sinister prætext of vindicating any Human Auctority; is an unpardonable Profanation of Reason, and high treason against the state of Learning.

CHAP.

# CHAP. IV.

## THE

# NATURE

## OF

# COLOVRS.

### SECT. I.

*Art.* 1.
The *Argument*
duely ac-
knowledged
to be superla-
tively *Difficult*,
if not abso-
lutely *Acata-
leptical*.

He *Rabbins*, whenever they encounter any Problem; that seems too strong for their Reason; to excuse their despair of conquering it, they instantly recurr to that proverbial Sanctuary, *Reservatur in adventum Eliæ*, it belongs to the Catalogue of secrets, that are reserved for the revealment of *Elias*. And, ingenously, if any *Abstrusity* in Nature be so imperveſtigable, as to juſtifie our open profeſſion of Incapacity, and neceſſitate our oppreſt Underſtanding to retreat to the ſame common Refuge; it muſt be this of the NATURE OF COLOURS, to the conſideration whereof the Clue of our Method hath now brought us. For, though all Philoſophers unanimouſly embrace, as an indubitable verity, that the object of Sight in *General*, is τὸ ὁρατὸν, *Viſible*, whatever is deprehenſible by that Senſe ; and that, in P*articular*, the Proper and Adequate object thereof, is τὸ χρῶμα, *Colour*, becauſe nothing is viſible but under the gloſs or verniſh of Colour, nor doth Light it ſelf ſubmit to the diſcernment of the eye, *quatenus Lux*, in the capacity of its Form, or meerly as Light, but *inſtar Albedinis*, as it retains to Whiteneſs; all which *Merſeanus* (*optica part.* 2. *theorem.* 1.) hath judiciouſly contracted into this one Theorem, *objectum viſus præcipuum eſt Lux & Color, vel Lux colorata, aut Color lucidus :* we ſay, notwithſtanding this their Ground-work be laid in the rock of manifeſt Certitude , yet when they attempt to erect thereon an eſtabliſht and permanent Theory of the Eſſence of Colours, either in their ſimple and firſt

Natures,

Natures, or complex and secondary Removes; they find the eye of their Curiosity so obnubilated with dense and impervious Difficulties, that all of certainty they can discover, is only this; that their most subtle indagations were no more but anxious Gropings in the dark, after that, whose Existence is evidenced only by, and Essence consisteth chiefly in Light. But, this Infelicity of our Intellectuals will be more fully commonstrated by our abridged rehearsal of the most memorable Opinions of others, and the declarement of our own, concerning this Magnale.

The *Despot* of the Schools (*in lib. de sensu & sensili, cap.3.*) defines Colour to be, ẟ ἐν τοῖς σώμασι διαφάνες τὸ ἔσχατον, *the Extremity of a Diaphanum, or transparent body terminated:* subjoining that Colour appertains to all things, *ratione Perspicuitatis,* and consequently, that the extremity of a perspicuous body terminated is the *Subject* of Colour. Which that we may clearly understand, let us consult the great *Scaliger*, who (*in Exercit.* 325.) thus concisely Comments thereupon. If the Perspicuum (saith He) suffer condensation so far as to the amission of its Transparency, and so prohibit the trajection of the Visible Species; it instantly becomes Colorate, and ought to be accounted Terminate, because it bounds or limits the Visive rayes. Wherefore, the law of Consequence injoineth, that we explore the Essence of Colours, in the Gradual Termination of the Diaphanum; and derive that Termination (1) from meer *Condensation,* without the admixture of any other thing to the Diaphanum; as may be instanced in the *Starrs,* for they become visible, though of a Lucid nature, only because they are of a Compact or Dense contexture. (2) From the *Admission of an Opace with a Translucid body;* as is exemplified in our *Culinary Fire,* which though in the simplicity of its most perspicuous, doth yet appear Red, because commixt and in some degree obnubilated with fuliginous Exhalations, from the pabulum or Fewel thereof, or compound body in combustion. The same likewise is to be understood of Aer and Water; for, those three Elements are all perspicuous, though in divers degrees: Fire being most perspicuous, Aer possessing the next degree, and Water coming behind them both, as seeming to be a Medium betwixt Perspicuity and Opacity. And, therefore, from the admission of the parts of that Opace Element, Earth, to any other of the three Diaphanous; one or other Colour among the many must arise. But, the Perspicuum passeth first into Whiteness, and therefore is it that Perspicuity, Light and Whiteness, are of the same nature, cozen Germans once removed, and discriminate only by Degrees: as, on the contrary, an Opacum, Darkness, and Blackness are also cognate. This being the original of the Two Father, or Ground Colours: it can be no Difficulty to attain the specifical Causes of all others, since they are only Intermediate, i. e. they arise from the various Complexion or Contemperation of the two Extrems. And this is the sense of *Aristotles* Text, if we admit the interpretation of *Scaliger.*

*Plato,* being either unable, or unwilling to erase out of the table of his mind some of the ingravements of *Democritus*; understands Colour to be *Flammula quædam, sive Fulgor, è singulis corporibus emicans, partes habens visui accommodatas* (*in Timæo*). For, having held, as *Diogenes Laertius* (*lib. 3.*) hath well observed, and we may easily collect from that discourse of his, in the name of *Timæus Locrus*; that the world consisteth of the four Elements, of Fire, as it is Visible, of Earth, as Tangible, of Aer and

Water

Water, *ut proportione non vacet* : left he fhould apoftate from his Fundamentals, He affirmed, *Corpora videri propter Ignem, & propter Terram
tangi,* that the Vifibility of all things was radicated in their participation
of Fire, and their Tangibility in their fhare of Earth ; and confequently
that the Colour of bodies was nothing but an 'Εκλαμ-ψις, or *Emicancy of
their internal Fulgor,* and the variety of its Species dependent meerly on
the various degrees, or more or lefs of that inhærent lufter.

*Art.* 4.
Of the *Pytha-
gorean* and *Sto-
ick.*

As for the *Pythagorean* and *Stoick* ; the *Former,* with inexcufable incogitancy, confounded the Tinctures of things with their Extrems, allowing
no real difference betwixt the *Superfice,* and the *Colour* it bears. *Pythagoras
Colorem effe extimam corporis fuperficiem cenfuit, hanc ob Canffam ; quod
Color Sectilem naturam habet, nontamen fit Corpus, aut Linea :* as *Plutarch
(de Placit. Philofoph.)* and out of him, *Bernhard. Cafius (de Mineral. lib.
2. cap. 2. Sect. 2. art.12.).* The *Later,* with unfatisfactory fubtility, (as
if, indeed, He meant rather to blanch over the 'Ακα-ταλε-ψία, or incom-
prehenfibility of the Subject, with ambiguous and Sophiftical Terms, than
confefs, or remove it.) makes Colour to be 'Επιπαιολα, a certain *Effloref-
cence,* arifing from a determinate Figuration of the Firft Matter; as we
have collected from the memorials of *Plutarch (lib. 1. de Placit. Philofoph.
cap. 15.)*

*Art.* 5.
Of the *Spagy-
rical Philofo-
phers.*

Laftly, the illuminated *Sons of Hermes,* who boaft to have, if not attained to the bottom of the myftery, yet out done the endeavours of all other
Sects of Philofophers, in profounding it ; confidently lead our curiofity to
their general Afylum, the three Univerfal Principles, *Sal, Sulphur* and *Mercury,* and tell us, that the Elemental Salts carry the mighty hand, or moft
potent Energy in the production of Colours. For, fuppofing three kinds
of Salt in all natural Concretions; the firft a Fixt and Terreftrial, the fecond
a Sal Nitre, allied to Sulphur, the Third a Volatile or Armoniac, referrible to
Mercury, and that all bodies receive degrees of *Perfpicuity,* or *Opacity,* refpondent to the degrees of *Volatility,* or *Terreftriety* in the Salts, that amafs them :
they thereupon deduce their various Colours, or vifible Gloffes, from the
various Commiftion of Volatile or Tralucent Salts, with Fixt or obfcure.

*Art.* 6.
The reafon of
the Authors
defert: n of all
thefe, and ele-
ction of *Demo-
critus* and *Epi-
curus* judg-
ment, touch-
ing the Gene-
ration of Co-
lours.

Now, notwithftanding all thefe *Sects* are as remote each from
other, as the Zenith from the Nadir, in their opinions touching the
Nature and Caufes of Colours, as to all other refpects ; yet do they generally Concur in this one particular, ειναι τα χρωμαζα συμφυη τοις σωμασιν,
*Colores effe Cohærentes corporibus,* that Colours are CONGENITE,
or COHÆRENT to bodies. Which being manifeftly repugnant to
reafon, as may be clearly evinced as well from the Arguments alledged by
*Plutarch (1. adverf. Colot.)* to that purpofe, as from the refult of our whole
fubfequent difcourfe, concerning this theorem : we need no other juftification of our Defertion of them, and Adhærence to that more verifimilous
Doctrine of *Democritus* and *Epicurus,* τω Νομω χρoιω ειναι, *Colorem Lege effe,* or more plainly in the words of *Epicurus,* τα χρωματα εν τοις
σωμασιν χ-λ ... τιναι τυξεις, η γεσεις, προς τ ω-ιν ; *Colores in corporibus gigni, juxta quofdam, refpectu vifus, ordines pofituff.* The Probability of which opinion, that we may with due ftrictnefs and æquanimity examine ; and enlarge what we formerly delivered, in our Origine of Qualities, touching the poffible Caufes of an inaffignable Variety of Colours :

We

*Art.* 7.
The Text of
*Epicurus,* fully
and faithfully
expounded.

We are briefly to advertife,

*Firſt*, That by the word, σωμασι, Bodies, we are not to underſtand *Atoms*, or *ſimple* bodies, for thoſe are generally præſumed to be devoyd of all Colour; but τα συγκριματα, *Concretions*, or *Compounds*.　　*Secondly*, that *Epicurus*, in this text, according to the litteral importance thereof, and the Expoſition of *Gaſſendus*, his moſt judicious and copious Interpreter, had this and no other meaning.　　That in the Extrems, or ſuperficies of all Concretions, there are ſuch certain Coordinations and Diſpoſitions of their component particles (which, according to our *Firſt Aſſumption* in the immediately præcedent Chapter, borrowed from the incomparable *Bullialdus*, are never contexed without more or leſs of Inæquality.) as that, upon the incidence of Light, they do and muſt exhibit ſome certain Colour, or other, reſpective to their determinate Reflection and Refraction, or Modification of the rayes thereof, and the poſition of the Eye, that receives them.　　That from theſe ſuperficial Extancies and and Cavities of bodies are emitted thoſe ſubſtantial Effluviaes, conſtituting the viſible Image; which ſtriking upon the primary Organ of Viſion, in a certain Order and Poſition of particles, cauſeth therein a ſenſation, or Perception of that particular Colour.　　But, that theſe Colours are not really Cohærent to thoſe ſuperficial particles, ſo as not to be actually ſeparated from them, upon the abſcedence of Light: and, conſequently that Colours have no Exiſtence in the Dark.　　Moreover, that the ſubſtance of Light, or the minute particles, of which its beams conſiſt, are neceſſarily to be ſuperadded to the ſuperficial particles of bodies, as the Complement, nay the Principal part of Colour: as may be inferred from theſe words of *Epicurus*, regiſtred by *Plutarch* (1. *adverſ. Colot.*) *Quinetiam hâc parte* (*luce, viz.*) *ſecluſâ, non video, qui dicere liceat, corpora quæ in tenebris in conſpicua ſunt, colorem habere.* Of which perſuaſion was alſo that admirable Mathematician, *Samius Ariſtarchus*; who poſitively affirmed (*apud Stobæum, in Ecl. Phyſ.* 19.) *Incidentem in ſubjectas res Lucem, Colorem eſſe; ideoque conſtituta in tenebris corpora colore prorſus deſtitui.* To which, doubtleſs *Virgil* ingeniouſly alluded in his

　　　　　　　　　　——*Ubi Cælum condidit Umbra
Jupiter*, *& rebus nox abſtulit atra Colorem.*

And *Lucretius* in his

　　　　*Qualis enim cæcis poterit Color eſſe tenebris,
　　　　Lumine qui mutatur in ipſo; propterea quod
　　　　Recta aut obliqua percuſſus luce refulget?* &c.

And, laſtly, that Light doth create and vary Colours, according to the various condition of the minute Faces, or ſides of the Particles in the ſuperfice, which receive and reflect the incident rayes thereof, in various Angles, toward the Eye.

Bb　　　　　　　　　　Sect.

## SECT. II.

*Art.* 1.
A PARADOX
That there are
no Colours in
the Dark.
HAving thus recited, explicated, and espoused the Conceptions of *Epicurus*, of the Creation of Colours, it behoves us to advance to the Examination of its Consistency with right reason, not only in its *General* capacity, but deduction and accommodation to *Particulars*.

But, First, to prævent the excess of your wonder, at that so Paradoxical assertion of his, *That there are no Colours in the dark*, or that all colours vanish upon the Amotion or defection of Light; we are to observe that it is one thing to be *Actually* Colorate, and another to be only *Potentially*, or to have a *Disposition* to exhibit this or that particular Colour, upon the access of the Producent, Light. For, as the several *Pipes* in an Organ, though in themselves all æqually *Insonorous*, or destitute of sound, have yet an equal Disposition, in respect of their Figuration, to yield a sound, upon the inflation of Wind from the Bellows; and as the seeds of *Tulips*, in Winter, are all equally *Exflorous*, or destitute of Flowers, but yet contain, in their seminal Virtues, a Capacity or Disposition to emit various coloured flowers, upon the access of fructifying heat and moysture, in the Spring : so likewise may all Bodies, though we allow them to be actually Excolor, in the Dark, yet retain a Capacity, whereby each one, upon the access and sollicitation of Light, may appear clad in this or that particular Colour, respective to the determinate Ordination and Position of its superficial particles.

*Art.* 2.
A familiar *Experiment*, attesting the Verity thereof.
To inculcate this yet farther, we desire you to take a yard of Scarlet Cloth, and having extended it in an uniform light, observe most exactly the Colour, which in all parts it bears. Then extend one half thereof in a primary light, *i. e.* the immediately incident, or direct rayes of the Sun; and the other in a secondary, or once reflected light : and then, though perhaps, through the præoccupation of your judgment, you may apprehend it to be all of one colour; yet if you engage a skilful Painter to pourtray it to the life, as it is then posited, He must represent the Directly illuminate half, with one Colour, viz. a bright and lightsome Red, and the Reflexly illuminate half, with another, i. e. with a Duskish or more obscure Red; or shamefully betray his ignorance of *Albert Durers* excellent Rules of shadowing, and fall much short of your Expectation. This done, gently move the extended Cloth through various degrees of Light and shadow : and you shall confess the Colour thereof to be varied upon each remove; respondent to the degree of Light striking thereupon. Afterward, fold the Cloth, as Boyes do paper for Lanterns, or lay it in waves or pleights of different magnitude; and you shall admire the variety of Colours apparent thereon : the Eminent and directly illustrate parts projecting a lively Carnation, the Lateral and averted yeilding an obscure sanguine, clouded with Murrey, and the Profound or unilluftrate putting on so perfect

sables,

sables, as no colour drawn on a picture can counterfeit it to the life, but the deadest Black. Your *Sense* thus satisfied, be pleased to exercise your *Reason* awhile with the same Example; and demand of your self, *Whether any one of all those different Colours can be really inhærent in the Cloth?* If you pitch upon the Scarlet, as the most likely and proper; then must you either confess that Colour not to be really inhærent, since it may, in less than a moment, be varied into sables, only by an interception of Light: or admit that all the other Colours exhibited, are æqually inhærent, which is more, we præsume, then you will easily allow. And, therefore, you may attain more of satisfaction, by concluding, that indeed no one of all those Colours is really so inhærent in the cloth, as to remain the same in the absence of Light; but, that the superficial particles of the Cloth have inhærent in them (*ratione Figuræ, Coordinationis & Positus*) such a *Disposition*, as that in one degree of Light it must present to the eye such a particular colour; in another degree, a second gradually different from that; in another, a third discriminate from both, until it arrive at perfect obscurity, or Black.

And, if your Assent hereto be obstructed by this DOUBT, *why that Cloth doth most constantly appear Red*, rather then Green, Blew, Willow, &c. you may easily expede it, by admitting, that the Reason consisteth only herein, that the Cloth is tincted in a certain Liquor, whose minute Particles are, by reason of their Figure, Ordination and Disposition, comparate or adapted to Refract and Reflect the incident rayes of Light, in such a manner, temperation, or modification, as must present to the eye, the species of such a Colour, viz. Scarlet, rather then a Green, Blew, Willow, or any other.    For, every man well knows, that in the Liquor, or Tincture, wherein the Cloth was dyed, there were several ingredients dissolved into minute particles; and that there is no one Hair, or rather no sensible part in the superfice thereof, whereunto Myriads of those dissolved particles do not constantly adhere, being agglutinated by those Fixative Salts, such as Sal Gemmæ, Alum, calcined Talk, Alablaster, Sal Armoniack, &c. wherewith Dyers use to graduate and engrain their Tinctures. And, therefore of pure necessity it must be, that, according to the determinate Figures and Contexture of those adhærent Granules, to the villous particles in the superfice of the Cloth, such a determinate Refraction and Reflection of the rayes of Light should be caused; and consequently such a determinate species of Colour, and no other, result therefrom.

*Art. 3.* The Constancy of all Artificial Tinctures, dependent on the constancy of *Disposition* in the superficial Particles of the Bodies that wear them.

Now, insomuch, as it is demonstrated by *Sense* that one and the same superfice doth shift it self into various Colours, according to its position in various degrees of Light and Shadow, and the various Angles, in which it reflecteth the incident rayes of Light, respective to the Eye of the Spectator; and justly inferrible from thence by *Reason*, that no one of those Colours can be said to be more really inhærent than other therein, all being equally produced by Light and Shadow gradually intermixt, and each one by a determinate Modification thereof: What can remain to interdict our total Explosion of that *Distinction* of Colours into *Real* or *Inhærent*, and *False*, or *only Apparent*, so much celebrated by the Schools? For, since it is the Genuine and Inseparable Propriety of Colours, in General,

*Art. 4.* That so generally magnified *Distinction* of Colors into *Inhærent*, and meerly *Apparent*; redargued of manifest *Contradiction*.

Bb 2    to

to be *Apparent*; to suppose that any Colour Apparent can be *False*, or less Real than other, is an open Contradiction, not to be dissembled by the most specious Sophistry; as *Des Cartes* hath well observed (*in Meteor. cap.* 8. *art.* 8.). Besides, as for those *Evanid* Colours, which they call *ἐμφαλικοὶ, meerly Apparent* ones, such as those in the Rainbow, Parheliaes, Paraselens, the trains of Peacocks, necks of Doves, Mallards, &c. we are not to account them Evanid, because they are not True: but, because the *Disposition* of those superficial particles in the Clouds, and Feathers, that is necessary to the Causation of them, is not *Constant*, but most easily mutable; in respect whereof those Colours are no more permanent in them, than those in the Scarlet cloth, upon the various position, extension, plication thereof. And Charity would not dispense, should we suppose any man so obnoxious to absurdity, as to admit, that the greater or less *Duration* of a thing doth alter the *Nature* of it. Grant we, for Example, that the particles of Water constituting the rorid Cloud, wherein the *Rainbow* shews it self, were so constant in that determinate position and mutuall coordination, as constantly to refract and reflect the incident beams of the Sun, in one and the same manner; and then we must also grant, that they would as constantly exhibite the same Species of Colours, as a Rainbow painted on a table: but, because they are not, and so cannot constantly refract and reflect the irradiating light, in one and the same manner; it is repugnant to reason, thereupon to conclude, that the *Instability* of the Colours doth detract from the *Verity*, or *Reality* of their Nature. For, it is only *Accidental*, or *Unessential* to them either to be varied, or totally disappear. So that, if you admit that Sea Green observed in the Rainbow, to be less True, than the Green of an Herb, because its Duration is scarce momentary in comparison of that in the Herb; you must also admit that Green in the Herb, which in a short progress of time degenerates into an obscure yellow, to be less true, than that of an Emrauld, because its Duration is scarce momentary, in comparison of that of the Emrauld.

*Art.* 5.
*The Emphatical, or Evanid Colours, created by Prisms; no less Real & Inherent, than the most Durable Tinctures.*

But, perhaps, Prejudice makes you yet inflexible, and therefore you'l farther urge; that the Difficulty doth cheifly concern those *Evanid* Colours, which are appinged on Bodies, reflecting light, by *Prisms* or *Triangular Glasses*, vulgarly called *Fools Paradises:* because these seem to have the *least* of *Reality*, among all other reputed *meerly Apparent.* And, in case you assault us with this your last Reserve; we shall not desert our station, for want of strength to maintain it. For, that those Colours are as *Real*, as any other the most Durable, is evident even from hence; that they have the very same *Materials* with all other, *i. e.* they are the substance of Light it self reflected from those objected Bodies, and (what happens not to those eyes, that speculate them without a Prism) twice refracted.

Experience demonstrates, that if a man look intently upon a polite Globe, in that part of it superfice, from which the incident Light is reflected, in direct lines toward his eye; He shall perceive it to appear clad in another Colour, than when He looks upon it from any other part of the Medium, toward which the Light is not reflected: and yet can He have no reason, why He should not account both those Different Colours to be *True*; the *Reflection* of light,

which

which varieth the Apparition according to the various Position of the eye, in several parts of the Medium, nothing diminishing their *Verity*. If so, why should not those Colours created by the Prism, be also reputed *Real*; the *Refraction* of Light, which exhibiteth other Colours in the objected Bodies, than appear in them without that Refraction, nothing diminishing their Reality?

By way of COROLLARY, let us here observe; that the Colours created by Light, reflected from objects on the Prism, and therein twice refracted, are *Geminated* on both sides thereof. For, insomuch as those Colours are not appinged but on the *Extremes* of the Object, or where the superfice is *unequal* (for if that be Plane and Smooth; it admits only an Uniform Colour, and the same that appears thereon, when beheld without the Prism): therefore are two Colours alwayes observed in that Extreme of the Object, which respecteth the *Base* of the Triangle in the Glass, and those are a *Vermillion* and a *Yellow*; and two other Colours in that extreme, which respecteth the *Top* of the Triangle, and those are a *Violet blew*, and a *Grass green*. And hence comes it, that if the Latitude of the Superfice be so small, as that the extremes approach each other sufficiently near; then are the two innermost Colours, the Yellow and Green connected in the middle of the Superfice, and all the four Colours constantly observe this order, beginning from the Base of the Triangle; a *Vermillion, Yellow, Green,* and *Violet*, beside the inassignable variety of other *intermediate* Colours, about the Borders and Commissures. We say, *Beginning from the Base of the Triangle*; because, which way soever you convert the Prism, whether upward or downward, to the right or to the left, yet still shall the four Colours distinguishably succeed each other in the same method, from the Base: however all the rayes of Light reflected from the object on the Prism, and trajected through it, are carried on in lines parallel to the Base, after their incidence on one side thereof, with the obliquity or inclination of near upon thirty degrees, and Refraction therein to an Angle of the same dimensions; that issuing forth on the other side, they are again Refracted in an Angle of near upon 30 degrees, and with the like obliquity, or inclination.

These Reasons equitably valued, it is purely Consequent, that no other *Difference* ought to be allowed between these *Emphatick*, or (as the Peripatetick.) *False* Colours, and the *Durable* or *True* ones, than only this; that the *Apparent* deduce their Creation, for the most part, from Light Refracted in *Diaphanous* Bodies, respectively Figurated, and Disposed, and sometimes from light only *reflected*: but, the *Inharent*, or *True* (as they call them) deduce theirs from Light variously *Reflexed* in *opace* bodies, whose superficial particles, or Extancies and Cavities are of this or that Figure, Ordination, and Disposition.

Not that we admit the *Durable* Colours, no more than the *Evanid*, to be *Formally* (as the Schools affirm) *Inharent* in Opace bodies, whose superficial Particles are determinately configurate and disposed to the production of this or that particular species of colour, and no other: but only *Materially*, or *Effectively*. For, the several species of Colours depend on the several *Manners*, in which the minute particles of

*Art. 6.*
COROLLARY.
The Reasons of *Emphatical* Colours, appinged on Bodies objected, by a *Prism*.

*Art. 7.*
The true *Difference* of *Emphatical* and *Durable* Colours, briefly stated.

*Art. 8.*
No Colour *Formally* inharent in objects; but only *Materially*, or *Effectively*: contrary to the constant Tenent of the Schools.

of Light strike upon and affect the *Retina Tunica*; and therefore are we to conceive, that opace Bodies, reflecting Light, do create Colours only by a certain *Modification* or *Temperation* of the reflected light, and respondent *Impression* thereof on the Sensory: no otherwise than as a Needle which though it contain not in it self the *Formal* Reason of *Pain*, doth yet *Materially*, or *Effectively* produce it, when thrust into the skin of an Animal; for, by reason of its Motion, Hardness, and Acuteness, it causeth a dolorous sensation in the part perforated.

*Art.9.*
The same farther vindicated from Difficulty, by the tempestive Recognition of some precedent *Assumptions* of the Atomists.

To diminish the Difficulty yet more, we are to recognize; that the *First Matter*, or Catholique Principles of all Material Natures, are absolutely devoyd of all *Sensible Qualities*; and that the Qualities of *Concretions*, such as *Colour, Sound, Odour, Sapor, Heat, Cold, Humidity, Siccity, Asperity, Smoothness, Hardness, Softness, &c.* are really nothing else but various MODIFICATIONS of the insensible particles of the First Matter, relative to the various Organs of the Senses. For, since the Organs of the Sight, Hearing, Tasting, Smelling, and Touching, have each a peculiar Contexture of the insensible particles that compose them; requisite it is, that in Concretions there should be various sorts of Atoms, some of such a special Magnitude, Figure and Motion, as that falling into the Eye, they may conveniently move or affect the Principal Sensory, and therein produce a sensation of themselves; and that either Grateful or Ingratefull, according as they are Commodious or Incommodious to the small Receptaries thereof (for the *Gratefulness* or *Ingratefulness* of Colours ariseth from the *Congruity* or *Incongruity* of the particles of the Visible Species, to the *Receptaries* or small Pores in the Retina Tunica): Some, in like manner, that may be convenient to the Organ of Hearing; Others to that of smelling, &c. So that, though Atoms of all sorts of Magnitude, Figure and Motion contexed into most minute Masses, arrive at all the Organs of Sense; yet may the Eye only be sensible of Colour, the Ear of Sound, the Nostrils of Odour, &c. Again, that Colour, Sound, Odour, and all other sensible Qualities, are varied according to the various situation, order, addition, detraction, transposition of Atoms; in the same manner as Words, whereof an almost infinite variety may be composed of no more then 24 Letters, by their various situation, order, addition, detraction, transposition; as we have more copiously discoursed, in our precedent Original of Qualities.

## SECT. III.

TO defcend to *Particulars*. It being more than probable, that the
various fpecies of Colours have their Origine from only the
various *Manners*, in which the incident particles of Light,
reflected from the exteriours of Objects, ftrike and affect the principal
fenfory; it cannot be improbable, that the fenfe of a *White* Colour is
caufed in the Optick Nerve, when fuch Atoms of light, or rayes confifting of them, ftrike upon the Retina Tunica, as come Directly from the
Lucid Fountain, the Sun, or pure Flame; or Reflexedly from a body,
whofe fuperficial particles are *Polite* and *Sphærical*, fuch as we have formerly conjectured in the fmalleft and hardly diftinguishable Bubbles of Froth,
and the minute particles of Snow.

*Art.* 1.
The Nativity
of *White*; of
the reafon of
its perception
by the fight.

And, as for the perception of its Contrary, *Black*; generally, though
fcarce warrantably reputed a Colour; we have very ground for
our conjecture, that it arifeth rather from a meer *Privation* of Light,
than any *Material Impreffion* on the fenfory. For, *Blackneſs* feems
identical, or coeffential with *Shadow*: and all of it that is pofitively perceptible, confifteth in its participation of Light, which alone caufeth it not to be abfolutely Invifible. And hence is it, that we have feveral *Degrees*, or gradual *Differences* of Black, comparative to the feveral degrees of fhadow, progreffing till we arrive at perfect *Darkneſs*:
and that we can behold nothing fo black, which may not admit of deeper and deeper blackneſs, according to its greater and greater receſs from
light, and nearer and nearer acceſs to abfolute Opacity. To reafon, therefore,
is it confonant that all Bodies, whofe natural Hew is *Black*, are compofed
of fuch infenfible particles, whofe furfaces are *fcabrous*, rough, or craggy,
and their Contexture fo *Rare*, or loofe, as that they rather *imbibe*, or
fwallow up the incident rayes of light, than reflect them outwardly toward the eye of the Spectator. Of this fort, the moft memorable,
yet difcovered, is the *Obfidian ftone*, fo much admired and celebrated
among the Romans; whofe fubftance being conflated of fcabrous and
loofely contexed Atoms, caufeth it to appear a perfect *Negro*, though held
in the Meridian Sun-fhine: becaufe the rayes invading it are for the
moft part, as it were abforpt and ftifled in the fmall and numerous Ca-
verns and Meanders varioufly interfperfed among its component particles.
Which common and illiterate eyes beholding, delude their curiofity with
this refuge; that it hath an Antipathy to Light, and doth therefore reflect
it converted into fhadows.

*Art.* 2.
*Black*, a meer
Privation of
Light.

The Generation of the Two *Extreme* and *Ground* Colours, White and
Black, being attained by this kind of inqueft into the Rolls of reafon;
the *Former* deriving it felf from Light, either immediately and in direct
lines profluent from its fountain, or by reflection from bodies, whofe
fuperficial particles are fphærical and polite; the *Later* from the Negation
of Light: there can be no great difficulty remaining concerning the Genealogy of all other INTERMEDIATE ones, fince they are but
the

*Art.* 3.
The Genealo-
gy of all *Inter-
mediate* Colors.

the off-spring of the Extreme, arising from the intermission of Light and shadow, in various proportions; or, more plainly, that the sense of them is caused in the Retina Tunica, according to the variety of Reflections and Refractions, that the incident Light suffers from the superficial particles of objects, in manner exactly analogous to that of the *Evanid* Colours, observed in sphærical Glasses, replete with Water, in Prisines interposed betwixt the object and eye, in angular Diamonds, Opalls, &c. For, even our sense demonstrates, that they are nothing, but certain Perturbations, or Modifications of Light, interspersed with Umbrelches, or small shadows.

**Art. 4.**
The Causes of the *Sympathy* & *Antipathy* of some Colours.

The Verisimility of this may be evinced from the *Sympathy* and *Antipathy* of these intermediate Colours, among themselves. For, the Reason, why *Yellow* holds a sympathy, or symbolical relation with *Vermillion* and *Green*, and *Green* with *Sky-colour* and *Yellow*, (as the experience of Painters testifieth, who educe a yellow Pigment out of Vermillion and Green, in due proportions commixt, upon their Palatts: and reciprocally, Green out of Yellow and Sky-colour, in unæqual but determinate quantities contempered) is no other but the *Affinity of their respective Causes*, or only gradually different manners of Light reflected and refracted; and intermixt with minute and singly imperceptible shadows. And, on the contrary, the Reason of the *Antipathy*, or Asymbolical relation betwixt a *saffron Yellow* and a *Cærule*, betwixt a *Green* and a *Rose colour*, into which a saffron yellow degenerates, and betwixt a *Yellow* and *Purple*, into which a Cærule degenerates: can be nothing else, but the *Dissimilitude or Remoteness of their respective Causes*; since things so remotely Discrepant, are incapable of Conciliation into a Third, or Neutral, or (rather) *Amphidectical* Nature, but by the mediation of something, that is participant of both. This the *Philosopher* glanced at in his; *Colores misceri videntur, quemadmodum soni; ita enim qui eximium quoddam proportionis genus servant, his Consonantiarum more, omnium suavissimi sunt, ceu purpureus & puniceus, &c. (de sens. & sensil. cap. 3.)*

**Art. 5.**
The intermission of small shadows, among the lines of Light; absolutely necessary to the Generation of any Intermediate Colour.

We say, *that all these Intermediate Colours emerge from the various intermission of Light, and small shadows*; because, to the production of each of them from reflected, or refracted Light, or both, the interposition of minute, and separately invisible *shadows*, is indispensably Necessary. Which may be evidenced even from hence, that Colors are not by Prisines appinged on bodies, but in their *Margines* or Extremes, there where is not only the general Commissure of Light and Shadows; but also an *Inæqualty* of superfice: which, by how much the more scabrous or rough, by so much the more are the Colours apparent thereon, ampliated in Latitude. For, since there is no superfice, however smooth and equal to the sense, devoid of many *Extancies* and *Cavities*; as we have more then once profestly declared: it is of necessity, that betwixt the confronting sides of the Extancies, reflecting the rays of light hither and thither, there should be intercedent small shadows, in the interjacent Cavities, from which no light is reflected. And hence is it, that in an object speculated through a Prism, the Cærule colour appears so much the more Dense and lively, by how much the nearer to the limbus, or Extreme of the Object it is appinged; because, in that place, is the greater proportion of small shadows: and *è contra*, so much more Dilute and Pale, by how much farther it recedeth from the

margin

Margin, insomuch that it degenerates, or dwindles at last into weak Sea-Green, or Willow, in its inmost part; because, in that place is the greater proportion of Light. Conformable to that rule of *Athanas. Kircher.* (*Art. Magn. Lucis & Umbræ. lib.* 1. *part.*2. *cap.* 1.) *Different autem & Umbra & Fulgores, majore & minore vel candore, prout vel Fonti lucis, aut tenebrarum propriores fuerint, vel à fonte longius recesserint, in quo luce & obscuritate summa sunt utraque. Unde patet, quanto Fulgores à luce magis recesserint, tanto plus Nigredinis: & quanto à tenebris magis recesserint Umbræ, diminuto nigrore, tanto plus albedinis acquirere: quæ omnia Visus judicare potest.* The same, proportionately, we conceive to hold good also in all Bodies, whose Colours are Genuine, or apparent to the naked Eye: chiefly because we may lawfully conceive, that every particle of every hair in a Scarlet, or Violet coloured Cloth, is consimilar in disposition to the particles in the extremes of an Object speculated through a Prism: and hold it purely Consequential thereupon, that light may arrive at the Eye from them, with the like Reflections and Intermistion with shadows, as from the extremes of the Reflectent Body, through the Glass, which advanceth its commixture with small shadows. And what we affirm of Scarlet and Violet, may also, with no less Congruity, be accommodated to Yellow and Sea-Green; allowing the same proportion and modification of Light and Shadows in them as in that part of the superfice of any other body, on which the Prism doth appinge them: and in like manner to all other Colorate objects, whose Tinctures bear any Affinity to either of these four specified, or arise from the Complexion of any two or more of them.

Art.6.

Two eminent PROBLEMS concerning the *Generation* and *Transposition* of the Vermillion and Cærule, appinged on Bodies by Prismes.

But here we are arrested by Two notable, and to our præcedent theory seemingly inconsistent PROBLEMS: which though of Difficulty enough to deserve the wealthy speculations of *Archimedes*, do yet require from us at least a plausible Solution, on the pænalty of no less than the loss of reputation, and the posting up a Writ of Bankrupt against our reason, by that austere Creditor, Curiosity.

( 1 ) How comes it, *that those two so discrepant and assymbolical Colours, created by a Prism, Vermillion and Cærule, arise from Causes so Cognate; the former only from the Commistion of a greater proportion of Light with a less of Shadows; the Later from a less proportion of Light with a greater of Shadows?*

( 2 ) Why, *when those two Colours Emphatical, Vermillion and Cærule are by a Prism intermediate, projected on a Wall or sheet of white paper beyond it, from the light of a Candle; if you put your eye in that place, on which either of the two Colours is appinged, so that another person, conveniently posited in the same room, may behold the same distinctly shining on the pupil of your eye; yet shall you plainly and distinctly perceive the other Colour in the Glass? For Example; if the Vermillion appear on your eye, you shall nevertheless clearly see a Cærule in the Glass: and transpositively, though your eye be manifestly and totally tincted with a Cærule, yet shall you see a Vermillion.*

C c                                    Touching

*Art. 7.*
The *Solution* of the *Former;* with a rational Conjecture of the Cause of the *Blew,* apparent in the Concave of the Heavens.

Touching the Former, we shall adventure to desume the Solution thereof meerly from the *Figure* of the Prisme, and determine the Reason on this only; that the Rayes of Light arriving at the *Base* of the Triangle, are trajected through it by a longer tract or way, than those arriving at or nearer to the *Top* thereof: and therefore, the Glass being in that part most crass, there must be more impervious particles obsistent to the Rayes of Light, each one whereof repercussing its raye back again into the medium from the Glass, causeth that the number of shadowes is multiplyed in that part of the object, which the Base of the Triangle directly respecteth; and consequently produceth a Cærule Tincture thereon. Such as that, not only by vulgar, but many transcendently learned Heads adscribed to the *Firmament*: which yet belongs rather to that vast (many have said infinite) *Space* betwixt it and our Terrestrial Globe, being caused by the rayes of the Cœlestial Lamps, from swarms of minute bodies interposed, thinly reflected toward our eyes: For, each of those impervious particles swarming in that immense space, must repercuse a ray of Light deradiated from above, and so by multiplying the number of shadows, make the Firmament (which otherwise, according to probability, would wear the mourning livery of Midnight) appear totally invested in an *Azure* mantle.

This, though meer Conjecture (and, indeed, the subject is too sublime to admit of other than conjecture, since St. *Paul* hath left us no observation concerning it, in his rapture up into the third Heaven, and the design of the *Ganzaes* is desperate) hath in it somewhat more of reason, then that confident conceipt of *Athanaf. Kircherus* (*Art. Magn. lucis & umbræ, lib. 1. part. 3. cap. 3. de Chromatismis rerum naturalium.*) *Medium inter utrumque Caruleum, proximum, viz. à nigro, seu tenebroso, colorem ad jucundissima illa Calorum spatia, inoffenso visu contemplanda, Natura providentissima mundo contulit, &c.* "that the "Providence of the Creator chose this Azure Tincture to invest the "Firmament withal, as the middle colour between the two Extreams, "White and Black, that so our sight might not, when we speculate "that universal Canopy, be either perstringed with the excessive lustre of "the one, nor terminated by the absolute opacity of the other. Because, if the natural Colour of the Firmament were *Azure,* as He præsumes; then would it, by reason of the vast Space betwixt it and our sight, and the repercussion of the greatest part of the rayes of Light, from our eye, by those Myriads of Myriads of Myriads of small bodies replenishing that intermediate Space, necessarily appear of some other colour: the experience of Sea-men assuring, that all Colours, (White and that of pure Flame, retaining to Whiteness, only excepted) lose themselves in long trajection through the medium, and that even Land, which is but few degrees removed from Opacity, appears to the first discovery like a blewish Cloud lying level to the Horizon. It being certain, therefore, that by how much the farther any Colour recedeth from Whiteness, by so much the less way it is visible (which the Græcian intimates in the word, λδυχος, *Albus,* ωξα το λδιοσω, *quod procul videatur.*) and that even the Earth, an Opace body, to Sea-men first Kenning it, at large distance, appears clad in a kind of obscure blewish Mantle: it cannot bee

dissonant

dissonant to reason to conceive, that the natural Colour of the Firmament cannot be Azure, since it so appears to us; and that it is rather Opace, because it appears Azure, when illustrate by the reflected Light of the Coelestial Luminaries.

Again, because the rayes of Light, incident on the *Top* of the Prism, are trajected through it by a shorter cut, or passage, than those incident on the Base; and so meet with fewer impervious and retundent particles, the Glass being in that part thinnest: therefore is the number of shadows much less in that part of the object, which respecteth the Cone or Top of the Triangle, than in that, which confronts the Base; and those few shadows which remain undiminisht, being commixt with a greater number of lines of light, are transformed into the species of a *Vermillion Red.* Such as that daily observed in the impure *Flame* of our Culinary Fires; which having many particles of Fuligenous Exhalations commixt with its pure luminous particles, that continuedly ascending, avert as many rayes of light from the eye of the Spectator, and so in some degree obnubilate it throughout: doth therefore put on the semblance of Redness. Or such as the *Sun* and *Moon*, commonly wear at their rising; when the minor part, though many of their rayes are refused, and averted from our sight, by the particles of dense vapours diffused through the spatious Medium.

However this may be disputed, yet is it warrantable to conceive, that the superficial Particles of all Bodies, clad in either of these Liveries, Vermillion and Cærule, may have in their Contexture obtained such a Disposition, as to reflect Light permixt with small shadows, in that definite Temperation, or Modification, in which it usually arrives at the eye, after its Trajection through a Prism; when it thereupon impresseth the sense of a Vermillion, or Cærule.

As for the Enodation of the *Later Difficulty*, it is comprehended in the Reasons of the Former. For, it being certain, that the Vermillion projected by a Prisme, doth consist of a greater proportion of Light mingled with a less of Shadows, and the Cærule, on the contrary, of a greater proportion of shadows interspersed among the lines of a less Light; and as certain, that the Vermillion appeareth on that side of the Prisme, where the Light is more copious, as therein meeting with fewer retundent impervious particles, in the substance of the Glass; and the Cærule in that part, where the Light is diminished, as meeting with more impervious particles, and being by them repercussed: it must inevitably follow thereupon, that, if an *opacous* body be posited within the bounds of this light, so that the light may fall on each side thereof, and as it were fringe it, a symptome quite contrary to the former shall evene, i. e. the Vermillion will appear on that side of the species, which is over against the Cærule, and the Cærule will be transposed to that side of the species, which confronteth the Vermillion. This is easily *Experimented* with a piece of narrow black Ribbon affixt longwise to either side of the Prisme. For, in that case, the light is bipartited into two Borders, or Fringes, the opace part veyled by the Ribbon on each side environed with light, and each border of light environed with two shadows; or, more plainly, between each border of shadows conterminate to each extreme of Light, trajected through the

*Art.* 8.
The *Solution* of the *Later.*

C c 2	unopacate

innopacate parts of the Glaſs : and, therefore, in the commiſſure of each of the two lights with each of the conterminous ſhadows, there muſt be Vermillion on one ſide, and Cærule on the other.

Now to drive this home to the head, the ſolution of the preſent Problem, the *Reaſon* why, when the light of a Candle is trajected through a Priſm, on a White paper or Wall, poſited at convenient diſtance beyond it, and there transformed into theſe two luminous Colours, Vermillion and Cærule, if you put your eye in that place of the Paper or Wall, whereon the Vermillion ſhines , you ſhall perceive only the Cærule in the Glaſs, and *è contra :* we ſay, the Reaſon of this alteration of ſite in the Colours ſeems to be only this, that the circumſtant Aer about the flame of the Candle being opacous, and ſo ſerving in ſtead of two Blacks to environ the borders of light, cauſeth that ſide of the Candle, which is ſeen through the thicker part of the Glaſs, to appear Blew ; and that which is ſeen through the thinner , to appear Red, according to the conſtant Phænomenon in Priſmes. But, if the ſpecies be beheld by Reflection from any illuſtrate and repercuſſing Body, ſuch as the paper, or wall, then muſt the ſeries or method of the borders of light and ſhadow be inverted, for the reaſon immediately præcedent, and conſequently, the ſituation of the Colours , emergent from their various contemperations, be alſo inverted.

*Art. 9.*
*The Reaſons,*
*why the Au-*
*thor proceeds*
*nor to inveſti-*
*gate the Cau-*
*ſes of Compound*
*Colours in*
*Particular.*

And thus have we, by the twilight of Rational Conjecture , given you a glimpſe of the abſtruſe Original of the *Extreme* and *Simple* Colours ; and ſhould now continue our Attempt to the diſcovery of the Reaſons of each of thoſe many COMPOUND ones, wherewith both Nature and Art ſo delightfully imbelliſh moſt of their peices : but, ſince they are as Generally, as rightly præſumed to be only the *multiplied removes of Light and Darkneſs,* i. e. to be educed from the various *Commixtures* of the *Extreme,* or *Simple,* or both ; and ſo it cannot require but a ſhort exerciſe of the Intellect to inveſtigate the determinate proportions of any two , or more of the Simple ones, neceſſary to the creation of any Compound Colour aſſigned (eſpecially when thoſe excellent Rules of that Modern Apelles , *Albertus Durerus,* præſcribed in his *Art of Limning* ; and the common Experience of Painters , in the Confection of their ſeveral Pigments, afford ſo clear a light toward the remove of their remaining obſcurity, and the ſingling out their particular Natures ): we cannot but ſuppoſe, that any greater ſuperſtructure on this *Foundation,* would be lookt upon rather as Ornamental and Superfluous , than *Neceſſary* to the entertainment of moderate Curioſity. Eſpecially when we deſign it only as a decent *Refuge,* for the ſhelter of ingenious Heads from the Whirlwind of *Admiration :* and not as a conſtant *Manſion* for *Belief.*

*Art. 10.*
*He confeſſeth*
*the Erection*
*of this whole*
*Diſcourſe, on*
*ſimple Conje-*
*ture: and enu-*
*merates the*
*Difficulties to*
*be ſubdued by*
*him, who*
*hopes to attain*
*an Apodiſtical*
*Knowledge of*
*the Eſſence &*
*Cauſes of Co-*
*lours.*

For, as we cautiouſly præmoniſhed, in the *Firſt Article,* the Foundation of it is not layed in the rock of abſolute Demonſtration, or deſumed *à Priori* ; but in the ſofter mould of meer *Conjecture,* and that no deeper than *à Poſteriori.* And this we judge expedient to profeſs , becauſe we would not leave it in the mercy of Cenſure to determine, whether or no we pretend to underſtand, What are the proper Figures and other eſſential Qualities of the inſenſible Particles of Light ; with what kind of Vibration, or Evolution they are deradiated from their Fountain ; What are the determinate Ordinations, Poſitions, and Figures of thoſe Reflectent and Refringent

particles

particles in the extreams of Bodies, Diaphanous and Opace, which modifie the Light into this or that species of Colour; What sort of Reflection or Refraction, whether simple or multiplyed, is required to the creation of this or that Colour; What are the præcise proportions of shadows, interwoven with Light, which disguise it into this or that colour. Besides, had we a clear and apodictical theory of all these niceties; yet would it be a superlative Difficulty for us to advance to the genuine Reasons, Why Light, in such a manner striking on the superfice of such a body, therein; suffering such a Reflection or Refraction, or both, and commix with such a proportion of shadows in the medium, should be transformed into a Vermillion, rather then a Blew, Green, or any other Colour. Again, were our Understanding arrived at this sublimity, yet would it come much short of the top of the mystery, and it might hazard a dangerous Vertigo in our brains to aspire to the Causes, Why by the appulse of Light so or so modified, there is caused in the Eye so fair and delightful a Sensation, as that of Vision; and why the sentient Faculty, or soul therein operating, becomes sensible not only of the particular stroak of the species, but also of the Colour of it.

For, where is that Oedipus, that can discover any *Analogy* betwixt the Retina Tunica, Optick Nerve, Brain, or Soul therein resident, and any one Colour? and yet no man can deny that there is some certain Analogy betwixt the *Species* and *Sensory*: since otherwise there could be no *Patibility* on the one part, nor *Agency* on the other.

We are not ignorant, that the aspiring Wit of *Des Cartes* hath made a towring flight at all these sublime Abstrusities, and boldly fastned the hooks of his Mechanick Principles upon them, thinking to stoop them down to the familiar view of our reason. But supposing that all Colours arise from *the various proportions of the process and circumvolutions of the particles of Light in bodies, respective to various Dispositions of their superficial particles, which accordingly more or less Accelerate, or Retard them*; as He hath copiously declared (*in Dioptric. cap. 1. & Meteor. cap.* 8.): and erecting this upon his corner stone, or grand Hypothesis, that *Light is nothing but an Appulse or Motion of the Æther*; or most subtile, and so most agile matter in the Universe; which is meerly *precarious*, and never to be conceded by any, who fears to ensnare himself in many inextricable Difficulties, Incongruities, and Contradictions, in the deducement of it through all the Phænomena of Light, Colours, and Vision: all that we can allow him, as to this particular, besides our thanks for his laborious Endeavours, is that close of Phaetons Epitaph, *Magnis tamen excidit ausis.*

*Art. 11.* *Des Cartes* attempt to dissolve the chief of those Difficulties, unsuccessful; because grounded on an unstable Hypothesis;

CHAP.

# CHAP. V.

## THE

# NATURE

### OF

# LIGHT.

*Art. 1.*
The *Clasp*, or
Ligament of
this, to the
præcedent
Chapter.

IN our three immediately præcedent Chapters, we have often mentioned the RAYES OF LIGHT, as the *Material Principle* both of all *Visible Species*, and *Colours*; and that we may not leave our Reader unsatisfied in any particular, the communication whereof seems necessary, or advantageous to His full comprehension of all our Conceptions relating to those Arguments, or any other of Affinity to them, that may hereafter occurr: we judge it our Duty, here to let him clearly know, What Notion we have of the Nature of that so admirably glorious and universally comfortable an Entitie, Light.

*Art. 2.*
The Authors
Notion of the
*Rays of Light.*

By the *Rayes of Light*, we understand, *certain most tenuious streams of Igneous Particles, in a continued fluor, and with ineffable pernicity succeding each other in direct lines, either immediately from their Lucid Fountain, or mediately from solid bodies reflecting them, towards the eye, and sensibly affecting the same.*

*Art. 3.*
A *Parallelism*
betwixt a
*stream of Water*
exsilient from
the Cock of a
Cistern, and a
*Ray of Light*
emanant from
its Lucid
Fountain.

This *Description* may receive somewhat more both of perspicuity and credit, if we consider the parallelism, or analogy, that each distinct Ray of Light holds to a stream of water, exsilient from the Cock of a Cistern, or tube of an Artificial Fountain. For, the reason why a stream of water issues from a tube in a kind of arch, and flows to some distance from its source

through

through the aer, is only this, that the particles of Water first exsilient, upon the remove of the stopple or obstacle, are so closely and contiguously pursued by other particles immediately following, and those again by others indesinently emanant, that they are impelled forward and driven on with such rapidity, as overcomes their natural propensity to direct descent, by reason of their Gravity, and carries them in a tense line from the vent so long as their impulse is superior to that of their Gravity; which encreasing more and more in each degree of distance, doth at length become victor over the force of the Motion, and præcipitate them downright. And as this gradual Tensity, or Rigidity of a stream of Water ariseth to it only from the Pressure or impulse of the Antecedent particles by the Consequent, in an uninterrupted succession: so may we conceeve, that a Ray of Light, or Wind (many of our Modern and most discovering Philosophers call a stream of Light, *Virgula Lucis*; and that by an unstrained Metaphor.) consisting of many rayes seemingly united, such as we observe shining in a room through some hole in the Window, or other inlet; doth therefore become in a manner *Tense*, or *Direct*, only because the particles first emanant from the Lucid Fountain are so urged and prest on by the subsequent, and those again by others, with equal pernicity, that they cannot deflect from a direct line, or obey the inclination of their Gravity, until some solid Body, interposed, cut off the fluor, by interrupting the succession, and then the Tensity, or Pressure ceasing, the Particles become incontiguous and disappear: as is observable, upon closing the inlet, through which a stream of Light is admitted into an otherwise opace room. For, immediately the successive supply of luminous particles being intercepted, the Antecedent droop, fail, and surrender that part of space, which they possess with splendour sufficient to affect the sense, to the horrid encroachment of Darkness.

This full *Comparison* præmised, we shall comply with opportunity, and here concisely observe

PRÆCONSI-
DERABLES.

(1) That *Aquilonius*, and most other *Opticomathematicians* do excellently distinguish Light into so many gradual Differences, as are the Reflections of which it is capable; denominating that Light, *Primary*, whereby a Body is immediately, or in direct lines from the Lucid Fountain, illustrated; that *Secondary*, which reflected from one solid body, illuminates another; that a *Third* Light, which illuminateth a body, after two Reflections from others: and so forward up to the *Centenary*, and *Millenary* light, if, at least, it be capable of so many reflections, from bodies most solid and polite.

*Art. 4.*
Light distinguisht into *Primary, Secondary*, &c.

(2) That Light at Second hand is more weak than at First; at Third than at Second; at Fourth than at Third, &c. or, that Light becomes so much Weaker, by how many more Reflections it hath suffered. Not (as is vulgarly concluded) that a *Reflex* ray is less Tense, or the successive pressure of its particles less violent or rapid, than those of a *Direct*; for, the motion of Light, however frequently reflected, is incomprehensibly swift: but, that every reflection doth much diminish it, some rayes being always diverted and scattered into other parts of the medium, by reason of the Asperity, or Inæquality of the particles in every superfice; and so there being no superfice that remits in a direct line the full number of rayes

*Art. 5.*
All Light Debilitated by Reflection; and why.

(some

(some have adventured to fay, fcarce half fo many as ) it received, and confequently the eye receiving fewer and fewer rayes fucceffively from every Reflectent, muft be more weakly affected and moved by the thin remainder. For, if all the rayes of the Sun directly incident on a Wall, were thence reflected on another wall fituate at a right angle ; the Second wall would be fully as luminous as the Firft ; and confequently, the Secondary light would be as ftrong and refplendent as the Primary : but, fince the fuperfice of the Firft Wall is unequal and fcabrous, it muft of neceffity come to pafs, that though many of the rayes incident thereon are from thence projected on the Second, yet as many are repercuffed into other regions of the Medium, fome upward, others downward ; fome to the right hand, others to the left, &c. according to the various faces, or fides of the fmall particles , with afperity contexed in the fuperfice of each ftone therein. So that one half , if not the major part of the directly incident rayes being diverted from the Second Wall, the Light thereon appearing muft be proportionately lefs ftrong and fulgent, than that , which illuminates the Firft. By the fame reafon, if the Second Wall by reverberation derive the Light to a Third , it muft likewife play the Publican , and remit but half fo many rayes, as it received from the Firft : and fo muft the Third tranfmit a thinner ftock of light to a Fourth , and a Fourth to a Fifth, &c.

*Art. 6*
An *Example*, fenfibly demonftrating the fame.

If this Example feem fcarce prægnant enough, let us defcend into a deep Pit, or with the Troglodites creep into the bowels of fome fubterraneous Cavern, and there our fenfe will demonftrate, that multiplied Reflections of Light gradually diminifh it even to abfolute infenfibility. For, the rayes of the Sun falling into the aperture of either Mine, or long Cave, are by oblique repercuffions from their fides conveyed inwards, and fo often bandied from fide to fide, that few or none attain to the bottom to diminifh the opacity thereof : every reflection remitting fome rayes, more or lefs, toward the mouth of the pit, or cave. And this ufhers in our Third obfervable.

*Art. 7.*
That light is in perpetual Motion; according to *Arift.*

(3) That *Ariftotles* affertion, *Lumen effe in continuo motu* , that Light is in perpetual motion, or reverberated to infinity ; is profound and orthodox. For, notwithstanding the illufion of our fenfe perfuades us , that all things in the aer about us, and within our houfes, are calm and unmoved : yet doth that better Criterion, our Reafon, affure that the Light diffufed through the aer is in perpetual inquietude, and confifteth of nothing elfe but a moft tenuious Contexture of innumerable rayes, fwarming from and to all regions unceffantly , fo long as the Lucidum ceafeth not to maintain the fucceffion of frefh rayes, that may be reflected from all obvious bodies. So that in what ever part of the medium the eye is pofited ; it fhall ever have fome object or other præfented: and particularly that, from whence fome rayes are more directly reflected into its Pupil. Not that we conceive the Light diffufed through the whole aer to be Continued, or United in all points, as are the parts of Water in the Sea : but, that, as a Spiders Web appears to be one entire and united body, though it confift of diftinct Filaments, varioufly intricate, and mutually decuffating each other ; fo alfo is Light, *Non unum quid Simplicifsimum, fed Compofitifsimum,* fome one thing not moft Simple or confifting of parts continuedly united, but moft Compound, or confifting of parts fo interwoven as to exclude all fenfible difcon-

difcontinuity; though our fenfe deprehend it to be *Incompofitifsimum*; becaufe the acies of the fight is too blunt to difcern the fingle rayes, which like moft flender Filaments with exquifite fubtilty interwoven into a vifible invifible Web, replenifh the whole Medium.

(4) That, though Light be ever *debilitated* by *Reflection*, yet is it many time *Corroborated* by *Refraction*; as that tranfmitted through Convex Glaffes, and Glafs Vials replete with limpid water: and then only debilitated, when it is Refracted by a Concave fuperficies of a pellucid body, or after refraction on a Plane fuperfice, is lookt upon obliquely. For as no reafon can be given for the Debilitation of Light by Reflection; but the Attenuation or Diminution of the number of its Rayes: fo can none be affigned for the Corroboration of it by Refraction in a Convex Glafs, or Vial filled with clear water; but the *multiplication* of its Rayes, in fome part of the Medium. Nor is there, on the contrary, why we fhould conceive Light to be made weaker by fome Refraction, unlefs in this refpect only; that if it had not fallen foul of a Refringent body, a greater number of rayes would have continued their direct progrefs in a clofer order, or more united ftream: and fo their Debility depends meerly on their Difgregation; not Diminution of Pernicity. Certainly, that Light which is corroborated by refraction in a Convex Glafs, would be yet more ftrong and energytical, if all thofe Rayes, that ftrike upon the obverted fide of the Glafs, were fo refracted, as to permeate and unite in the aer beyond the averted fide thereof: and thofe rayes which are trajected through the bottome of a Glafs Vial filled with water, arrive at the eye fo much the more Difgregate, by how much the more obliquely the eye is pofited; becaufe the water being in the bottom more copious, and fo containing more retundent particles, doth divert the greateft number of them into the ambient. And hence we infer, that if the beams of the Sun be conceded more weak in the Morn and Evening than at Noon, only becaufe of a greater Refraction by more vapours then interpofed; that effect muft chiefly arife from hence, that the Rayes come unto us obliquely, after their trajection through thofe fwarms of denfer vapours, and confequently more Diffipated, the major part of them being diverted into other regions of the Medium. Moreover, infomuch as all *Mafters in the Optiques* clearly demonftrate that the Image of an illuftrate object, fpeculated through water in the bottome of a veffel indiaphanous, doth appear lefs lively to thofe, that look on it obliquely, than to thofe that behold it in direct lines refpective to the tendency of the Light refracted by the Water; and that the fuperfice of every object hath fo much the fewer parts difcernable, by how much more obliquely it is fpeculated: therefore is it purely neceffary, that the Image of an object appear more *Contracted*, when fpeculated by a *Vertical* line, than when exhibited to the eye in a *direct*, and Irrefracted one. And this alfo we judge to be in fome part the Caufe, why the Sun when neareft to our Horizon, either Orient or Occident, appears in a Figure more Elliptical or Oval, than Sphærical: for then do we behold it *per lineam Verticalem.* We fay, *in part*; becaufe the fame Effect may alfo be induced by the Form of the Vaporous Sphære. However this may be controverted, yet moft certain it is, that the Lucid Image of the Sun is alwayes more *Vitiated*, when it arrives at our fight from an *Humble* pofition, than a fublime or Meridional: *Non quod pauciores quidem radÿ Directi mane, quàm meridie; fed reflexi tamen pauciores, quæ cum illis mifceantur, ipforumÿ Vim augeant.*

D d

*Art.8.*
Light, why Corroborated, in fome cafes, and Debilitated, in others, by Refraction.

COROLLARY.
Why the Figure of the Sun, both rifing & fetting, appears rather Elliptical, than Sphærical.

*augeant. Quia Directi supra liberam horizontis planitiem prætereant, nec redeant; cum sub meridiem in terram impacti non resilire regredique non valeant; as Gassendus, in Epist. ad Bullialdum, de Apparent. Magnitud. Solis Humilis & sublimis.* And this hath a near relation to our fifth observable.

*Art. 9.*
PARADOX.
That the proportion of Solary Rayes reflected by the superiour Aer, or Æther, toward the Earth, is so small, as not to be sensible.

(5) That the Body from which the rayes of a Lucid object more eminently the Sun, are repercussed so as to diminish the shadow round about it, seems not to be the Conterminous Aer, but rather some *Opacum* constitute beyond both it and the Aer. Not that we deny the necessary reflection of many of the Luminous rayes proceeding from the Sun, by those myriads of myriads of particles floating in the Atmosphere; and so the remission of them back again toward their source, and the consequent diminution of the shadow invironing the same: but that we conceive the proportion of rayes so diverted, to be so small, as to be much below the observation of our sense. For, He that is in the bottom of a deep Mine, hath his sight so little advantaged by the Aer illuminated by the meridian beams of the Sun, that though he can clearly behold the Starrs in the Firmament, immensely beyond that vast tract of Aer then illustrate; yet can he hardly perceive his own hand, or ought else about him, since all the rayes of Light, which affect his eyes, are only those few that have escaped repercussion upward, by those many oblique refractions in the sides of the Mine. Thus also in the night are we no whit relieved by the aer, or Æther surrunding our Horizon, or more properly, our Hemisphere beyond that region, to which the Cone of the Earths shadow extends: though the Sun doth as freely and copiously diffuse its light through all that vast Ocean of Aer, or Æther beyond the extent of the Earths shadow, at our Midnight, or when it is Vertical to the Antipodes, as at our noon when it is Vertical to us: which could not be, if any sensible proportion of light were reflected toward us by the particles of the Aer, or Æther, replenishing the subcælestial space. Hence comes it, that what Light remains to our Hemisphere in the night, ought to be referred, not to any Reflection of the Suns rayes from the sublime aer, or Æther; but to the Stars, or Moon, or both. And this is also no contemptible argument, that the Concave of the Firmament is *Opace*, and not azure, as most suppose.

*Art. 10.*
That every Lucid Body, as Lucid, doth emit its Rayes Spherically: but, as Visible, Pyramidally.

(6) That every Lucid Bodie is considerable in a double capacity; (1) *Qua Lucidum*, as shining with either native, or borrowed light, it illuminateth other bodies: (2) *Qua Visibile*, as it emits the visible Image of it self. In the *First* Respect, we may conceive it to be the *Center*, from which all its luminous Rayes are emitted by Diffusion *Spherical*, according to that establisht maxime of *Alhazen, Omne punctum luminosum radiare sphaeraliter*: in the *Second*, we may understand it to emit rayes in a diffusion *Pyramidal*, the base whereof is in it self, and cone in the eye of the Spectator. For, particularizing in the Sun, which being both a Lucid Body and a Visible object, falls under each acceptation; we must admit the Rayes thereof illuminating that vast ocean of Space circumscribed by the concave of the Heavens, to be deradiated from it sphærically, as so many lines drawn from one common Centre; because they are diffused throughout a region far greater than the Sun it self: and those rayes, that Constitute the Visible Images of it, stream from it in Cones or pyramids; because they are terminated in the pupil of the beholders eye, a body by almost infinite degrees

less

lefs than it felf. This is fully demonftrated by the Forms of Eclipfes, which no man can defcribe but by affuming the Sun as the Bafe, from whofe Extremes myriads of Rayes emanant, and in their progrefs circularly environing the Margin of the Earth, or Moon, pafs on beyond them till they end in a perfect Cone ; the Orbs of the Earth and Moon being by many degrees lefs in circumference, than that of the Sun. This confirms us, that thofe *Optico-mathematicians* are in the centre of truth, who teach, that the rayes of the Sun, and all other luminous Objects as they conftitute its vifible fpecies, are darted only Pyramidally ; infomuch as they are received in the eye of each Spectator, fo much lefs than the Sun, or other Luminary ; but that they progrefs in a fphærical Diffufion, in refpect of the circumambient Aer, in each point whereof the Luminary or Lucidum is Vifible. Since, fhould we allow the Concave of the Firmament to be as thickly fet with eyes, as Joves vigilant Pandars head was imagined by Poets ; we could not comprehend how the orb of the Sun could be difcernable by them all, unlefs by conceding this fphærical diffufion of Pyramids to all parts of the fame. And this doth as well illuftrate as confirm a former *Antiperipatetical Paradox* of ours, that the vifible Species of an object is neither total in the totall Space, nor total in every part thereof ; but the General Image is in the whole Medium, and the Partial or Particular Images, whofe Aggregate makes the General Image, in the fingular parts of the Medium : becaufe no fingular eye from any fingular part of the Medium, can perceive the whole of the object, but thofe parts only, which are directly obverted to that part of the Medium, in which the eye is pofited. Which affertion we inferred from hence, that not only the whole, but alfo every fenfible particle of an object doth emit certain moft fubtile rayes, conftituting the fpecies of it felf, in a fphærical diffufion, fo that the various particles emit various rayes, that varioufly decuffate and interfect each other, in all parts of the Medium : and as thefe rayes are emitted fphærically, *ex fe*, according to that maxime, *Omne Vifibile fui fpeciem effundere fphæraliter* ; fo do moft of them, *ex Accidente*, convene in their progrefs, and fo reciprocally interfect, as to fulfill the figure of a Pyramid. Whence it naturally follows, that becaufe fome Rayes muft convene, in all parts of the Medium, in this manner ; therefore are Pyramids of rayes made in all points of the Medium, from whence the object diffufing them is vifible. Notwithftanding this, we fhall fo farr comply with the Vulgar doctrine, as to allow ; that in refpect even of *one fingle eye*, in whatever part of the Medium pofited, the diffufion of rayes from an object may be affirmed to be *Spherical* : infomuch as no part in the object at confiderable diftance fingly difcernable, can be affigned, which is not lefs than the pupil of the Eye.

(7) That the Light diffufed through the Medium, is not feen by us : but that thing beyond the Medium from which fome rayes are ultimately reflected into the eye. For, if it chance that we perfuade our felves, that we perceive fomething in the Medium ; it is not pure Light it felf, but fome crafs fubftance, the fmall particles of Duft, Vapours, Smoak, or the like, which having received light from fome luminous fource, reflect the fame toward the eye.

*Art. 11.*
That Light is
invifible in
the pure me-
dium.

## Sect. II.

**Art. 1.**
The Neceſſity
of the Authors
confirmation
of the *Firſt*
*Præconſiderable*

NOw, of all theſe Præconſiderables only the *Firſt* can be judged *Præcarious*, by thoſe whoſe Feſtination or Inadvertency hath not given them leave to obſerve the Certitude thereof inſeparably connected to the evidence of all the others, by the linkes of genuine Conſequence. And therefore, that we may not be wanting to them, or our ſelves, in a matter of ſo much importance, as the full Confirmation of it by nervous and apodictical Reaſons, eſpecially when the Determination of that eminent and and long-lived Controverſie concerning the QUIDDITY or Entity of *Light, Whether it be an Accident, or Subſtance, a meer Quality, or a perfect Body?* ſeems the moſt proper and deſiderated ſubject of our præſent ſpeculations, and the whole Theory of all other ſenſible Qualities (as Vulgar Philoſophy calls them) is dependent on that one cardinal pin, ſince Light is the neareſt allied to ſpiritual natures of all others, and ſo the moſt likely to be Incorporeal: we muſt devote this ſhort Section to the perſpicuous Eviction of the CORPORIETY of Light.

**Art. 2.**
The CORPO-
RIETY of
Light, demon-
ſtrated by its
juſt *Attributes:*
*viz.*

Not to inſiſt upon the grave Authority either of *Empedocles*, who, as *Ariſtotle* (1. *de ſenſu & ſenſili: & de Gener. Animal.* 1. *cap.* 8.) teſtifieth, affirmed Light to be Ἀπορροιαν, *Effluxionem*, a material Emanation, and required certain proportionate Pores, or moſt ſlender paſſages in all Diaphanous bodies, for their tranſition; or *Plato*, who defined Colour, or Light diſguiſed, to be φλόγα ἀπορρέουσαν, *Effluentem quandam Flammulam*; or of *Democritus* and *Epicurus*, both which are well known to have been grand Patrons, if not the Authors of that opinion, that Light is corporeal: we judge it alone ſufficient to demonſtrate the Corporiety of Light, that the Attributes thereof are ſuch, as cannot juſtly be adſcribed to any but a Corporeal Entity.

**1.**
Locomotion.

Such are (1) *Locomotion*; for manifeſt it is, that ſome ſubſtance, though moſt tenuious, is deradiated from every Lucidum to the eye of the diſtant Spectator: nor is a Bullet ſent from the mouth of a full charged Cannon with the millionth part of ſuch velocity, as are the arrows ſhot from the bow of *Apollo*; ſince the rayes of the Sun are transformed from one end of the heavens to the other, in a far leſs diviſion of time, than a Cannon Bullet is flying to its mark.

**2.**
Reſilition.

(2) *Reſilition*; for the rayes of light are ſenſibly repercuſſed from all ſolid bodies, on which they are projected; and that with ſuch pernicity or rapid motion, as tranſcends, by inaſſignable exceſſes, the rebound of a Cannon Ball from a Rock of Adamant.

**3.**
Refraction.

(3) *Refraction*, for our ſenſe confirms, that Light is ever refracted by thoſe Bodies, which allow its rayes a paſſage, or through-fare, but not an abſolute free and direct one.

(4) *Coition*

( 4 ) *Coition*, or Union, or Corroboration, from bodies either reflecting, or transmitting many rayes to one common point of concurse, where they become so violent as to burn any thing applied.

( 5 ) *Disgregation* and Debilitation, from the didaction of its rayes reflected or trajected: so that those which before during their Union were so vigorous as to cause a conflagration, being one distracted become so languid as not to warm.

( 6 ) *Igniety*; since Light seems to be both the Subject, and Vehicle to Heat, and those speak incorrigibly, who call Light, Flame attenuated. Which we shall less doubt, if we consider the natural Parallelism betwixt *Flame* and *Water*, *Light* and a *Vapour*. For, as Water by Rarefaction, or Attenuation becomes a Vapour; so may we conceive Flame by Attenuation to become Light circumfused in the aer: and as a Vapour is nothing else but Water so rarefied into small discontinued particles, as that it doth scarce moisten the body on which it is impacted; so is Light nothing else but Flame so dilated by Rarefaction, that it doth hardly warm the body it toucheth. Lastly, as a Vapour how finely soever rarefied, is still substantially Water; because only by the Coition of its diffused particles it returns again to Water, as in all distillations: so must we account Light however rarefied, to be still substantially Flame; because only by the Coition, or Congregation of its dispersed rayes it is reducible into absolute Flame, as in all Burning-glasses.

These Attributes of Light considered, it is not easie for the most prævaricate judgment not to confess, that Light is a Corporeal substance, and the Rayes of it most tenuious streams of subtle Bodies: since it is impossible they should be deradiated from the Lucid Fountain with such ineffable pernicity, transmitted through the Diaphanum in a moment, impacted against solid bodies, repercussed, corroborated by Unition, debilitated by Disgregation, &c. without essential Corpulency.

Notwithstanding this apodictical evidence of the Corporiety of Light, the refractary *Peripatetick* will have it to be a meer *Quality*, and objects

( 1 ) That his master *Aristotle*, impugning the doctrine of *Democritus*, *Epicurus*, and others, who ascribed Materiality to Light, defined it to be meerly 'Ενεργεῖα *perspicui*, an *act of the Perspicuum*.

To this we answer, ( 1 ) That though *Aristotle* thought it sufficient barely to deny that light is 'Ασωμάτοι Σώματος. νδενός, *ullius corporis Effluxum*, and to affirm it to be *Energian perspicui, ut perspicuum*; yet will the judicious discover it to be rather an ambage to circumvent the incircumspect, than a demonstration to satisfie the curious. For, though *Philoponus* ( 2. *de Anim.* 7. ) willing to conceal or guild over his Masters error, interpreteth his *Perspicuum actu*, or illustrate Nature, and so Light to be a kind of Chord, which being continuedly interposed betwixt the object and the eye, causeth that the Colour thereof posited beyond the Medium, doth affect and move the eye to the act of intuition: yet hath He left the *Reason* and *Manner* of this supposed Act
of

of the Perspicuum on the eye, the chief thing necessary to satisfaction,
involved in so many and great Difficulties, as proclaim it to be absolutely inexplicable. (2) That albeit we deny not *Illumination* to be meerly *Accidental* to opace Bodies; yet therefore to allow the *Light*, wherewith they are illuminate, to be an Accident, and no Substance, is a manifest Alogie. And to affirm, that the Aer, Water, or any Diaphanous
body is the *subject of Inhasion* to Light, is evidently incongruous;
because every Medium is simply *Passive*, and remains unmoved while
the Light pervades it : and how can Light pervade it, if it be not
Corporeal ? or how can the rayes thereof conserve their Tensity and Directness in the Aer, while it is variously agitated by wind and other
causes, if they were not absolutely independent thereupon ? (3) What
*Aristotle* saith concerning the *Propagation* of the species of Light even
to infinity in all points of the *Medium*, besides its incomprehensibility, is absolutely inconsistent to the *Pernicity* of its motion, which is too rapid and momentany to proceed from a fresh Creation of Light in every
point of the medium : since the multitude of fresh productions successively
made, would require a far longer time for the transmission of the light of a
candle to the eye of a man at the distance of but one yard, than our sense demonstrates to be necessary to the transmission of the light of the Sun from
one end of Heaven to the other.

<table>
<tr><td>

*Art. 4.*
The Corporie-
ty of Light
imports not
the Coexi-
stence of two
Bodies in one
Place; contra-
ry to the *Peri-
patetick.*

</td><td>

(2) *That by allowing Light to be Corporeal, we incurr the absurdity
of admitting two Bodies into one and the same place.*

Which is soon solved by reflecting on what we have formerly and
frequently said, concerning *Inanity interspersed*, and observing what we
shall ( God willing ) say of those eminent Qualities, *Rarity and Perspi-
cuity :* from either of which it may be collected, how great a Multitude
of Pores are in every Rare and Perspicuous Body, which remain tenantable,
or unpossessed.

</td></tr>
<tr><td>

*Art. 5.*
Nor the moti-
on of a Body
to be *Instanta-
neous.*

</td><td>

(3) That from the Corporiety of Light it must follow, *that a Body may
be moved in an Instant.* But he hath not yet proved that the motion of
Light is *instantaneous :* and we have, that it is not, but only Momentany,
i. e. that Light is moved in a certain space of time, though imperceptible,
yet divisible, and not in one individual point, or Instant.

</td></tr>
<tr><td>

*Art. 6.*
The *Invisibili-
ty of Light* in
the limpid me-
dium, no Ar-
gument of its
*Immateriality :*
as the *Peripa-
tetick* pre-
sumes.

</td><td>

(4) *That the Rayes of Light are Invisible in pure Aer, and by conse-
quence Immaterial. Solut.* Their Invisibility doth not necessitate their
Immateriality ; for the Wind, which no man denies to be Corporeal, is
invisible : and as it sufficeth that we feel the Wind in its progress through
the aer, so also is it sufficient that we perceive Light, in the illumination of
Opace Bodies, on which it is impinged, and from which it is reflected. Be-
sides, whoso maketh his sense the measure of Corporiety, doth strain
it to a higher subtility, than the constitution of its Organs will bear,
and make many more spiritual Entities, than can be found in the Uni-
verse ; nay, He implicitely supposeth an Immaterial Being naturally
capable of Incorporation meerly by the Unition of its dispersed par-
ticles ; since many rayes of Light congregated into one stream become
visible.

</td></tr>
</table>

(5) *That*

(5) *That the Materiality of Light is repugnant to the Duration of the Sun; which could not have lasted so long, but must have, like a Tapour exhausted its whole stock of Luminous Matter, and winked out into perpetual night, long since, if all its Rayes were substantial Emanations,* according to our Assumption.

*Art.* 7.
The Corporiety of Light fully consistent with the Duration of the Sun: contrary to the *Peripateticks.*

But this Refuge may be battered with either of these two shots. 1 The superlative *Tenuity* of the Luminous particles continually emitted from the body of the Sun, is such as to prevent any sensible minoration of its orb, in many 1000 yeers. (2) If the Diametre of the Sun were minorated by 100000 miles less than it was observed in the days of *Ptolomy*; yet would not that so vast Decrement be sensible to our sight: since being in its Apogæum, in summer, it doth not appear one minute less in Diameter to the strictest astronomical observation, than in winter, in its Perigæum, and yet *Snellius, Bullialdus,* and *Gassendus,* three Astronomers of the highest form, assure us that it is about 300000 miles more remote from us, in its Apogæum, than Perigæum.

(6 and Lastly) *That if Light were Flame, then would all Light warm at least:* but there are many Lights actually Cold, such as that in the *Phospher-Mineralis,* or *Lapis Phenggites,* of whose admirable Faculty of imbibing, retaining and emitting a considerable light, the excellent *Fortunius Licetus* hath written a singular Tract, and *Athanas. Kircherus* a large chapter (*in Art. magn. Lucis & Umbræ lib.* part. 1. cap. 8.) in Gloworms, the scales and shells of some Fishes, among which the most eminent are those *Dactyli* mentioned by *Kircher* (*in libri jam citati part.* 1. cap. 6.) in these words, *sunt & Dactyli, ostreacei generis, qui vel manibus triti lumen veluti scintillas quasdam ex se spargunt: quemadmodum Melita, in Sicilia, Calabria, & Ligustici maris oris non sine admiratione à piscatoribus & nautis instructoribus observasse memini;* in Rotten Wood, &c. Ergo, &c.

*Art.* 8.
The insensibility of Heat in many Lucene Bodies, no valid Argument against the prætent Thesis, that Light is Flame Attenuated.

*Answer.* The Defect of actual Heat in these things, doth arise, in part from the abundant commistion of Gross and Viscid Humidity with those igneous Particles that are Collucent in them; but mostly from the exceeding *Rarety* of those Luminous Sparks; which being so thin and languid, as to disappear even at the approach of a Secondary Light, cannot be expected vigorous enough to infuse an actual warmth into the hand that toucheth them; especially when experience attesteth, that the Rayes of the Sun, after two Reflections, become so languid by Attenuation, as they can hardly affect the tenderest hand with any sensible Heat. And therefore, unless it can be evinced, that the disgregation of the parts of a Body, doth destroy the Corporiety of it; and that the simple Attenuation of Light doth make it to be no Light: we ask leave to retain our perfuasion, that the existence of many lights, which are devoyd of Heat, as to the perception of our sense, is no good Argument against the Igniety and Corporiety of Light.

CHAP.

# CHAP. VI.

## THE

# NATURE

## OF A

# SOUND.

### SECT. I.

Art. I.
An Elogy of
the sense of
Hearing: and
the Relation of
this and the
præcedent
Chapter.

IT was a hypochondriack conceit of *Plato*, that all our *Cognition* is but *Recognition*, and our acquired *Intellection* a meer *Reminiscence* of those primitive lessons the Soul had forgotten since her transmission from the sphere of supreme Intelligences, and Immersion into the Opacity of Flesh. For, *Proper Science* is proper only to *Omniscience*; and not to have knowledge by infusion or acquisition, is the attribute only of the *Essence of Wisdom*; and a priviledge due to none but the *Ancient of Dayes*, to have his knowledge derived beyond Antiquity: but *Man*, poor ignorant Thing, sent to School in the World, on the design of Sapience, must sweat in the exploration and pursuit of each single Verity; nor can he ever possess any science, in this dark region of life, but what he hath dearly purchased with his own anxious discovery, or holds by inhæritance from the charitable industry of his Fore-fathers. And, that the naked Mind of man, endowed only with a simple Capacity of Science, might by degrees adorn it self with the notions of whatever concerns his well-being either in this state of Mortality, or that future one of Immortality; hath the Bounty of his Creator furnished him with the Sense of HEARING: a sense particularly and eminently ordained for *Discipline*. For, though we sing Hymns to the *Eye*, for the *Invention*: yet must we acknowledge a sacrifice of gratitude due to the *Ear*, for the *Communication* and Diffusion of Arts and Sciences. *Quemadmodum aspectus ad vita dulcedinem, & commoda est magis necessarius;*

*ita*

*ita Auditus ad excipiendam artem, scientiam, & sapientiam est accommodatior: ille ad inventionem, hic ad communicationem aptior est;* saith that accurate and eloquent Anatomist, *Julius Cafferius Placentinus,* (*in premio ad libr. de fenf. organ.*). Thus much the antique Ægyptians intimated in their Hieroglyphick of Memory, the figure of a mans Ear; and the *Philofopher* exprest in his Character of the Hearing, *Auditus est fenfus Difcipline;* as alfo that Modern Ornament of Germany, *Sennertus* (*in hypomn. Phyf.*) in this memorable fentence; *Aures in homine quafi porta mentis funt, per quam illi communicantur, quæ doctrina & inftitutione de Deo, & alijs rebus neceffarijs traduntur, quæque nullo alio fenfu addifci poffunt.* Now, to bring you home to the fcope of this (not otherwife or unreafonable, or unneceffary) Elogy of the Hearing; fince the Relation betwixt the Sight and Hearing is fo great, as to the point of mans acquifition of Knowledge, as that the one can be no more juftly called the Difcoverer, than the other the Propagator of all Arts and Sciences: it is evident we have made no undecent Knot in the Clue of our Method, by immediately fubnecting this Enquiry into the Nature of a Sound, the adæquate and proper object of the Hearing, to our præcedent fpeculations of the Nature of Vifion, Colours, and Light.

Befides, as thefe two Senfes are Coufin-Germans, in their *Ufes* and *Ends:* fo likewife are they of near Alliance, in their *objects;* there being no fmall, nor obfcure *Analogy,* betwixt the nature and proprieties of a *Vifible Species,* and the nature and proprieties of an *Audible Species,* or Sound. For

*Art. 2.* The great Affinity betwixt Vifible and Audible fpecies; in their reprefentation of the fuperficial Conditions of objects.

( 1 ) As it is the property of *Light,* transfigured into Colours, to reprefent the different Conditions or Qualities of bodies in their fuperficial parts, according to the different Modification and Direction of its rayes, either fimply or frequently reflexed from them, through the Aer, to the Eye: fo is it the propriety of *Sounds* to reprefent the different Conditions or Qualities of Bodies, by the mediation of the Aer percuffed and broken by their violent fuperficial impaction, or collifion, and configurate into fwarms of fmall confimilar maffes, accommodable to the Ear. So that He fpeaks as Philofophically, who faith, that various founds are no more but the various Percuffions and impreft Motions of the Aer: as He that faith, Colours are no more then the various Immerfions of Light into the fuperficial particles of bodies and refpective Emerfions or Reflections from them, through the diaphanous medium to the Eye. Nor can we much diflike the conceipt of *Athanaf Kircher.* (*Mufurgiæ Univerfalis l. 9. part. 4. præluf.* 1. ) that if it were poffible for a man to fee thofe fubtle motions of the aer, caufed by the ftrings of an inftrument, harmonically playd upon (as we may the Circular Undulations, and Tremblings of water, raifed by a ftone thrown into it, in a river or ftanding lake ) the whole Tune would appear to him like a well drawn Picture, ingenioufly and regularly adumbrate with admirable variety of Colours, each one diftinctly reprefenting the particular Condition of that ftring or fonant Body, that created it.

( 2 ) As *Light* immediately fails and difappears upon the remove or eclipfe of its lucid fountain; as is manifeft by the fucceffion of darknefs in a room at night, when a candle is either removed out of it or extinguifhed; the

*Art. 3.* In the Caufe and manner of their Deftruction.

E e

the fucceffion of its rayes being intercepted: fo doth a *Sound* inftantly
perifh upon the Ceffation of the undulous motion of the Aer, which conduceth not only to the Creation, but Delation of it, as the principal, if
not the fole Vehicle. For, the fubfiftence of Sounds is not by way of
dependence upon the folid bodies, by which they were produced; according to the 7 *Propofit.* of *Merfennus* (*Harmon. lib.* 1. *pag.* 3.) *Soni non pendent à corpore, à quo primum producti funt:* but upon the Continuation
of the motion impreft upon the Aer, fo that the Duration of a Sound is
equal to the duration of the Agitation of the aer. And therefore *Bapt.
Porta, Cornelius Agrippa, Wecherus, Alexius,* and others of the fame tribe,
that fo highly pretend to *Phonocamptical Magick,* are worthy more than de-
rifion, for their infolent undertaking to Conferve a voice, or articulate found
of many fyllables, by including it in a long Canale of Lead, or other impervious matter; fo that upon unftopping the extreme of the Tube, after
many not only hours, but months, the voice fhall iffue out as quick and
diftinct as at the firft pronunciation, or infufurration into the cavity
thereof. Which (whether more impudent, or ignorant (for both Experience and the Nature of a Sound evidence the contrary) is difputable)
Rhodomantade is demonftrated to be abfolutely impoffible, by *Athanaf.
Kircher.* (*Mufurgia Univerfal. lib.* 9. & *Magia Echotectonica cap.* 1.) whether we remit the unfatisfied.

*Art.* 4.
In their *Acti-
nobolifm,* or
Diffufion, both
*Spherical* and
*Pyramidal.*

(3) As the Actinobolifm, or Deradiation of Light from the Luminary,
is *Spherical,* in refpect of the circumambient fpace illuminate by it: fo is the
Diffufion of a Sound in *excentral* lines from the fonorous body, through
the whole fpace, or medium within the fphere of its vertue; for, otherwife
a General, fpeaking in the midft of his Army, could not be heard in round.
Here is the only *difference* betwixt the Actinobolifm of Light and Sounds;
that the one is performed in time *imperceptible,* though not inftantaneous:
the other in moments *diftinguifhable,* which are more or lefs according to
the degrees of diftance betwixt the fonant and audient. Again, as the Deradiation of Light, confidered meerly as *Vifible,* not as *Lucidum,* is *Conical,*
or Pyramidal, in refpect to the Eye of the Spectator; as we have profeffedly demonftrated in the 10. *Article of the* 1. *Sect. of our Chapt. concerning the Nature of Light:* fo likewife doth every found make a Cone, or
Pyramid in the medium, whofe Bafe confifteth in the extreme of the body
producing the found, and cone in the ear of him that hears it; or as fome
Mathematicians, as *Blancanus* and *Merfennus,* whofe Bafe is in the Ear, and
Cone in fome one point of the fonorous fubject. Allowing only this
Difference; that the Cones or Pyramids of Vifible Species are more
*Geometrical, i.e.* more exactly conform to proportion Geometrical,
than thofe of Audible Species; which in regard as well of the grofsnefs
of their Particles, as lefs velocity of their motion, are eafily injured and
perturbed by Winds. And this, in truth, is the beft ground they have
to ftand upon, who opinion Sounds to be no more but fimple *Motions* of
the Aer.

*Art.* 5.
In their certi-
fying the fenfe
of the *Magni-
tude, Figure,*
and other
Qualities of
their Origi-
nals.

(4) As Vifible Species, fo do Sounds inform the Senfe, of the
Magnitude, Figure, and other Qualities of the Bodies, from which they
are emitted. For experience affureth, that Greater Bodies emit a Graver
Sound, than fmaller; that Concaves yeild a ftronger and more lafting
Sound, than Planes; that Hard things found more Acutely than Soft;

ftrings

strings diftended yeild a fharper found, than lax; Empty veffels than full, &c. Hence is it, that Goldfmiths, and Coyners diftinguifh good mony from bad, pure Gold from that largely allayed with Copper; and Metallifts judge of fimple and compound Metals, only by their Ring or found.    And we have heard of Vintners, who could exactly diftinguifh the Kinds and Goodnefs of Wines, only by the found of the Veffels that contained them: and therefore ufed to choofe them more by their Ear, than Palate.    But, what we here fay, that *Harder Bodies emit a found more Acute than fofter*; we defire may be underftood only of the *Plurality*, not Generality of Bodies.    For the examining *Merfennus*, having experimented the different founds of Metalls, tells us (*in præfat. ad Harmonic.*) that He found a Cylindre of Iron to be Unifone to another of fteel, equal in diametre and length; and both in acutenefs to tranfcend a Cylindre of Brafs of equal dimenfions, by a whole Diateffaron: nay more, that a Cylindre of Firr Wood yeilded, upon equal percuffion, a found more acute by a whole Ditone, than a Cylindre of Brafs, which yet yeilds a found more ftrong, lafting and grateful than any of the reft.    Each of which obfervations is fufficient to cut off the general intaile of that Canon, *Sonos eò acutiores, quo duriora fuerint corpora. Legendus eft Athanaf. Kircher. Art. Magn. Confoni & Diffoni lib. 1. appendice de Phonognomia.*

(5) As a Greater Light alwayes obumbrates a Lefs, fo a Great Sound alwayes drowns a Lefs: for it is manifeft, that the found of a Trumpet conjoyned to the low or fubmiffive voice of a man, makes it wholly unaudible, and the loud clamours of Mariners are fcarce heard in a tempeft. *Art. 6.* In the obfcuration of *Lefs* by *Greater.*

(6) As a too great Light offends alwayes, and often deftroyes the fight, as is eminently exemplified in the tyranny of *Dionyfius*, the Sicilian: fo, too great founds injure and lacerate the Hearing. For, many men have been ftrucken deaf for ever, by great Thunder-claps, and as many by the reports of grand Artillery. *Art. 7.* In their offence of the organs, when exceffive.

(7) As Light, meerly by the Condenfation of it rayes, produceth Heat in the aer: fo Sounds meerly by their Multiplication. For, it is obferved, that in all Battails, and chiefly in Naval fights, where many Cannons are frequently difcharged, the aer becomes foultry and hot; not fo much from the many fulphureous or igneous particles of the Gunpowder commixt with, as the violent concuffions, and almoft continued agitation of the Aer.  So that even in this particular, that Axiom, that *Motion is the Mother of Heat*, holds exactly found. *Art. 8.* In their production of Heat by *Multiplication.*

(8) The Effects of Audible Species, as well as of Vifible, are fubject to variation, according to the divers Condition of the Medium. For, as Flame, beheld through fmoak, feems to tremble: fo do founds, trajected through aer varioufly waved by Winds, rife and fall betwixt every Guft; as is obfervable moft eafily in the ringing of Bells, whether the wind be favourable, or adverfe. *Art. 9.* In their *Variability,* according to the various difpofition of the *Medium.*

(9) And what moft conduceth to our comprehenfion of the Nature of a Sound; For, as Light, fo is a Sound capable of Locomotion, Exfilition, Impaction, Refilition, Difgregation, Congregation; all which are the proper *Art. 10.* In their chief Attributes, of Locomotion, Exfilition, Impaction, Refilition, Difgregation, Congregation;

E e 2                                        and

and incommunicable Attributes of Corporiety. Only we muſt confeſs them diſcrepant in this, ( 1 ) That Sounds are delated from their Original not only in direct lines, but circular, elliptical, parabolical, and all others; for a ſound heard on the other ſide of a high Wall, comes not to the ear in a direct line through the Wall, as *Kircherus* contends ( *in Muſurgiæ Univerſal. lib.* 1.) with tædions arguments, but in an *Arch*, as the incomparable St. *Alban* hath firmly evinced (*in Cent. 3. Natural. Hiſt.* ): whereas Light conſtantly progreſs through the Medium to the Eye, in Direct lines, whether primary, reflex, or refracted. ( 2 ) A Sound is diffuſed through it ſphere of activity in a longer ſpace of time, by much, than Light, as is ſenſibly demonſtrated by this, that the flaſh of a Cannon arrives much ſooner at the Eye, than the report at the Ear : and the immediate Reaſon hereof is the leſs velocity of motion in the ſound, which conſiſting of groſſer particles than thoſe of Light, muſt be proportionately ſlower in its Delation. For, a Sound ſeems to be nought but the Aer, at leaſt the ſubtler or more æthereal part of aer, extrite and formed into many ſmall ( *Moleculæ* ) maſſes, or innumerable minute Contextures, exactly conſimilar in Figure, and capable of affecting the Organ of Hearing in one and the ſame manner : which configurated ſmall maſſes of aer fly off from bodies compulſed or knockt each againſt other, with ſome violence; and progreſs by Diffuſion in round. For, becauſe upon preſſure they mutually recede, each particle going off in that point where it finds the freeſt egreſs : therefore muſt ſome tend upward, others downward, ſome to the right, others to the left, ſome obliquely, others tranſverſly, &c. but all more ſlowly than the particles of Light, whoſe Tenuity being far greater, cauſeth them not to be ſubject to retardment by the like tumultuous Convolution. But, as the greater Corporiety of Sounds makes them *ſlower* in their *Diffuſion*; ſo doth it make them more *impetuous* and forcible in their *Impaction*, than the Species of Light : it being obvious to obſervation, that Violent Sounds, ſuch as great Thunders, Volleys of Cannon ſhot, the breaking of Granades, &c. uſually ſhake the largeſt Buildings, and ſhiver Glaſs windows at a mile diſtance and more. And yet are Sounds far eaſilier impeded, perturbed, and flatted, than the rayes of Light; every man knowing that no ſound can penetrate Glaſs, but in one caſe, or exigent of Nature, of which we ſhall particularly ſpeak, in the laſt Section of this Chapter : and ſince Sounds are repercuſſed more ſlowly; they are Diſgregated more hardly, and Congregated more faintly, than the rayes of Light. Laſtly, the Proportion of Retardation in the diffuſion of Sounds to the utmoſt of their ſphere of activity, is ſuch even from Winds; that as *Merſennus* hath computed, the diameter of the ſphere of a ſound, heard againſt the wind, is by almoſt a third part leſs than the diameter of the ſphere of the ſame ſound, aſſiſted by a favourable or ſecund Wind : but the Diameter of a Lucid Sphere is alwayes equal, which way ſoever and how violently ſoever the wind blows. ( 3 ) Bodies of narrow Dimenſions make a ſenſible reflection of Light, as is manifeſt from a Burning-glaſs of an inch diameter : but a Body of far greater dimenſions is required to the ſenſible Reflection of a Sound, i. e. to the production of an Eccho; though it is not to be doubted, but a ſound may be reflected from every Hard bodie on which it is impinged. This conſidered, we cannot but ſmile at the Credulity of many great *Ariſtoteleans*, who are perſuaded that an Echo is made

by

by the meer Repercuffion of the Sound from the particles of the Aer. For, notwithftanding we deny not, but the particles of the aer, within the fphere of the Sounds diffufion, encountring and arietating thofe particles of the found, may in fome fmall meafure repercufs them: yet we think it unfafe, therefore to admit this aereal Repercuffion to arife to Senfibility, or to be obfervable by the Creation of an Echo. And therefore we conceive, that whatever fenfible Reflection or Multiplication of a Sound, feems to proceed from the Aer, is not caufed really by the Aer, but fome Denfe and Hard Bodies, fuch as Rocks, Ædifices, Arches, &c. whofe Concavities reflect the particles of a Sound for the fame reafon, that Concaves Multiply Light.

---

## Sect. II.

THE *Congruities* of *Vifible* and *Audible* Species being fo many and Effential, and their *Incongruities*, or points of Difcrepancy fo few, and thofe altogether confifting in the meer Degrees of Velocity, and fome other Circumftances relating to the Medium: we have a fair and direct way opened to our Enquiry into the *Quiddity* or *Effence* of a Sound. Wherefore fince to conclude a parity of *Effence*, from a parity of *Attributes* and *Effects*, in any two Entities, is warrantable even by the ftricteft laws of Reafoning: we fhall adventure to affume a *Sound to be a Corporeal Ens.* Which before we farther confirm by Arguments, it behoveth us to lift that block of contrary *Authority* out of our Readers way; at which the credulity and incircumfpection of many have made them ftumble and hault ever after in their Opinions concerning this Subject.

*Art. 1.* The *Product* of the Præmiffet, concerning the points of Confent, & Diffent of Audible and Vifible Species: viz. That Sounds are Corporeal.

True it is, that *Pythagoras, Plato*, and *Ariftotle*, according to the Memorials of *Plutarch* (4. *Placit.* 20.) unanimoufly held a Sound to be *Incorporeal*, a meer *Accident*, or *Quality*, or *Intentional Species*; contrary to the doctrine of *Democritus, Epicurus*, and the *Stoicks*, who, as *Laertius* (*in lib.* 7.) exprefly records, affirmed it to be *Corporeal*, or a Material Efflux, the words of *Epicurus* being [ τ̄ φωνὴν εἶναι ῥεῦμα ἐκπεμπόμενον ἀπὸ τῶ̂ φωνούντων, ἢ ἠχούντων, ἢ ψοφούντων ] *Vocem feu Sonum, fluxum effe emiffum ex rebus aut loquentibus, aut fonantibus, aut quomodocunque ftrepitum edentibus.* But yet we conceive this repugnancy of Authority infufficient to infirm our Thefis of the CORPORIETY of Sounds; as well becaufe fimple Authority, though never fo reverend, is no demonftration, and fcarce a good argument, in points Phyfiological, where the appeal lies only to Reafon: as for this weighty confideration, that *Thefe* accepted a found in *Concreto*, i.e. for the fubftance of the Aer, or its moft tenuious particles, together with their proper Configuration; but *Thofe* in *Abftracto*, or only for the Figure impreft upon the fuperfice of the Aer, which they therefore inferred to be Incorporeal, that is, devoyd of Profundity. For, otherwife *Plato* (*apud Agellium*, *lib.* 5. *cap.* 15.) defines a found *Aeris validaque aeris percuffio*, a fmart and ftrong percuffion of the aer: and *Ariftotle* (2. *de Anim. cap.* 8.) calls it downright a *Motion of the Aer*; as the

*Art. 2.* An obftruction of præjudice, from the generally fuppofed repugnaut Authorities of fome of the Ancients; expeded.

Stoicks,

*Stoicks, Iŝtus aëris,* a stroke of the aer. So that the Difference seems occasioned only by their diverse Acceptation of the word Sound. This obstruction removed, we progress to the discharge of our province, *viz.* the Eviction of the Corporiety of a Sound.

*Art. 3.*
An Argument
of the Corpori-
ety of Sounds.
The *First* Argument of the Corporiety of a Sound, is (*Quod vim habet agendi, sive efficiendi aliquid*) that it is *Active* or *Effective.* For, the voice of a man violently emitted, or highly elevated by a kind of grating offends the vocal organs, and changes their sweetness or evenness into a hoarsness, and being long continued, leaves them misaffected with lassitude: as the experience of Hunters and Orators demonstrates.

Hither are we to referr *Lucretius* his

> *Præter radit enim vox fauces sæpe, facitque,*
> *Asperiora foras gradiens arteria clamor,* &c.

*Art. 4.*
A Second Ar-
gument.
The *Second* is desumed from its Capacity of *Repercussion,* or Resilition from solid bodies; which is the evident cause of our hearing one sound twice, or more often, according to the multiplicity of its Reflections: as in all *Echoes,* monophone or polyphone. Which *Aristotle* fitly compares not only to a *Ball* frequently rebounding, but also to *Light,* which Himself confesseth capable of reflections even to infinity: thereon concluding a sound subject to the same laws of Reflection with either. To which *Virgil* seems to allude in his

> *Saxa sonant, vocisque offensa resultat Imago.*

Intimating, that an Echo holds a perfect analogy with an Image reflected from a Mirrour. For, as beside that Image, which tends in a direct line from the Glass to the eye, innumerable others are so transferred from it into all points of the Medium, that divers other eyes variously posited therein shall behold the same general Image, each one receiving a particular Image: so likewise, beside that sound or voice, which arrives at your ear, innumerable others are so dispersed through all parts of the medium or sphere of diffusion, that if there were as many ears therein as the space could contain, each one would hear the same general sound or voice; and if it chance that any one particular voice be impinged against solid and lævigated or smooth bodies (for solids that are very Spungy or porous, suffer sounds to pass through them, and too scabrous or rough destroy them by dissipation) it may be repulsed in a direct line toward your ear, and you shall hear it again at second hand or Echoed.

COROLLA-
Ry.
*Art. 5.*
The Causes of
Concurrent E-
choes, where
the Audient is
equally (al-
most) distant
from the So-
nant and Re-
percutient.
Touching the *Reflection* of Sounds, we shall here, by way of *Corollary,* briefly observe. That in case you stand somewhat near to the smooth solid that reflecteth the sound, and the Creation of the sound be not very remote; then though an Echo thereof be made, yet shall not you hear it: because the *Direct* sound and the *Reflex* enter the ear so continently, i. e. the space of time betwixt their ingress is so imperceptible, that they seem but one intire sound. But, in this case, the sound becomes both stronger and longer; in respect of their Union.

And

And this comes to pass chiefly, when the Reflection is made from divers bodies at once ; as in all Arches, and Concamerated or vaulted rooms : in which for the most part, the sound or voyce loseth its Distinctness, and degenerates into a kind of long confused Bombe.

And hence, *viz.* the many Repercussions of a Sound from divers places together, or with so short intervals of time , as the sense cannot distinguish them ; is it, that the sound of *Concaves* percussed, lasteth much longer, than the sounds of bodies of any other figure whatever : especially when the Concave hangs at liberty, in the aer, so that its Tremulation be not hindred as are all Bells in Churches, and clocks. For, not only the External or ambient aer, but the Internal is agitated by those frequent Tremblings in the body of the Concave , and continually repercussed from side to side : and therefore, till the trembling ceaseth, the Bombination is continued.

*(right margin)* COROLLA-RY 2. *Art.6.* Why *Concaves* yeild the strongest and longest Sounds.

Again , if you stand far from the sonant bodie, and near to the Reflectent ; in this case also will the sound appear single, and coming only from the Reflectent : because both the Direct and Reflex sound invade the ear without any sensible difference in time ; and yet the Reflex sound as it is really the posterior, so doth it very much intend or increase the Direct, and consequently makes the impression observable only from it self.

*(right margin)* COROLLA-RY 3. *Art.7.* The reason of Concurrent Echoes , where the Audient is near the Reflectent, and remote from the sonant.

It is observable moreover , that by how much nearer the Ear is to the Anacamptick, or Reflectent ( yet at such distance , as is required to the discernment of the Direct voyce from the Reflex. ) by so much the fewer syllables of a word pronounced are Echoed : and *è contra*, by how much farther from the Reflectent (provided the distance exceed not the sphere of diffusion ) so many more syllables are repeated. The Reason being this, that the interval of time betwixt the Cessation of the Speaker, and the audition of the Reflex voice, is much less in the first case, and much greater in the later : and consequently , the less interval of time sufficeth to the Distinction of a fewer syllables, and the greater for more. This considered, we can no longer admire the distinct rehearsal of a whole Hexameter by some strong Echoes ; provided the voice pronouncing the verse be sufficiently strong to drive it to the Reflectent, and thence back again to the Ear, at large distance, such as is necessary to the allowance of time enough for the successive repercussion of each syllable : for otherwise the voice faileth by the way.

*(right margin)* COROLLA-RY 4. *Art.8.* Why Echoes *Monophon* rehearse so much the fewer syllables, by how much nearer the audient is to the Reflectent.

What hath been hitherto said, concerns only Echoes *Monophone* , that repeat the same syllable but once ; but there are Echoes *Polyphone*, such as repeat one and the same note, or syllable divers times over, and of them the Reason is far otherwise. For, the frequent rehearsal of the same syllable by an Echo , ariseth from the multitude of Reflectent Bodies, situate beyond each other in such order, that the nearer bodies referr it first, and the remoter successively : and sometimes from Bodies mutually Confronting each other, and alternately reflecting the same sound. Of this sort were those observed by *Lucretius*, in this Tristich.

*(right margin)* COROLLA-RY 5. *Art. 9.* The reason of *Polyphon* Echoes.

> *Sex etiam, aut septem loca vidi reddere voces,*
> *Unam cum jaceres ; ita colles collibus ipsis*
> *Verba repulsantes , iterabant dicta referre.*

Such

Such also was that prodigious one that entertained the Curiosity of *Gassendus* at *Pont Charenton* standing upon the river Seine, four miles from *Paris*. For in a square old ædifice of free-stone, uncovered at the top, and having a row of 5 Pillars on each side, as commonly our Churches, He heard a Monosyllable, which himself pronounced, clearly and orderly repeated by several Echoes, 17 times over; and when he uttered the Monosyllable in the Centre of the Ædifice, it was brought back to his ear 17 times from each extream (the area being somewhat oblong ) so distinctly, as He could easily numerate the repetitions on his fingers. If so *sileat Miracula Memphis*, let the *Ægyptian Pyramids* no longer boast their *Pentaphone* Echoes; nor the *Porticus Olympia* challenge the garland from the world for her *Heptaphone* Resonance, which is highly celebrated by the pens of *Plutarch* (*lib. 4. de placit. Philosoph. cap. 20.*) and *Pliny*, (*lib. 36. cap. 15.*). For, this at *Pont Charenton*, of which our Lord *St. Alban* was also an ear-witness, and not without some admiration, as Himself hath recorded (*in Centur. 3. Nat. Hist.*) hath no Rival, but that many tongued Echo in a Village called *Simoneta*, near *Millan* in *Italy*, which at some seasons, when the aer is serene, will iterate any Monosyllable, in which is no S. (which being but a kind of sibilation, or interior sound, few or no Echoes can reherse) 30 times over very distinctly, if credit be due to the testimony of *Blancanus* (*in Echometria, & in suo additione ad theorem. 20. de Echo polyphona.*)

**Art. 10.**
*A Third Argument of the Materiality of Sounds:*

A *Third* Argument of the materiality of a Sound, results to us from the *Pleasure* and *offence*, or *Gratefulness* and *Ingratefulness* of Sounds, as they are Concinnous, or Inconcinnous. For it is highly concordant to truth, that the suavity of a Sound proceeds from hence, that those minute Particles, which enter the ear and move the Auditory Nerve, are in their configuration so accommodate to the Receptaries, or Pores thereof, that they make a gentle, smooth or equal impression on the filaments, of which the Acoustick Nerve consisteth: and on the contrary, the Acerbity, or Harshness of a Sound, only from hence, that the minute particles invading the sensory, being asper or rough in their configuration, in a manner exulcerate, grate, or dilacerate the slender Filaments thereof.

**Art. 11.**
*The necessity of a certain Configuration in a Sound; inferred from the Distinction of one sound from another, by the Sense.*

That a certain *Configuration* of its minute particles, is essentially necessary to every Sound, may be concluded safely even from hence; that so great variety of Sounds, and chiefly of Words, or Letters, as well Vowels as Consonants, could not be so exactly distinguished by the Hearing, unless the sensory were variously, or in a peculiar manner percelled and affected by each: nor can that variety of *Affection* be made out, but by a variety of *Sigillation*, or Impression, dependent respectively on the various *Configuration* of those (*molecula*) small masses, that compose the sound.

**Art. 12.**
*The same confirmed by the Authority of Pythagoras, Plato, and Aristotle.*

To sweeten the harshness of this Assertion yet more; we alledge the unison Authority of no less than *Pythagoras* (whom all knowing men allow to have lighted the tapour to posterity, in the investigation of the Nature, and causes of proportions among Musical Sounds ) *Plato* and *Aristotle*, all which affirmed the same, if *Plutarch* be faithful (*in 4. de placit.*) while He introduceth them saying, τὸ σχῆμα, *Figuram*, *qua in aere, ejusque superficie fit certo ex ictu* ( *ᾗ σχουσαν πληγὴν* ) *evadere vocem,*

*vocem,* that the Figure made in the aer, and then it superfice, by some certain percussion, becomes a voice. And, that *Plutarch* hath done no more than justice to *Aristotle,* in this particular; is evident from his own words, (*in Problem.* 13. & 51.) where He expresly enquires, *Quare Vox, cum sit* Ἀηρ τις ἐχηματισμένος, *Aer quidam Figuratus, & qui dum transfertur, plerumque,* τὸ σχῆμα, *Figurum amittit, illam tamen dum a solido corpore repercutitur, incolumem servet?* " Why a voice, which is aer con-" figurate, and for the most part loseth its Figure, in its [long] transmissi-" on, doth yet conserve it intire and unimpaired, when repercussed from a " solid body, as in all Echoes?

*Art.* 13.
And by the
*Capacity* of the
most subtile
parts of the
Aer.

Nor can it be rightly denied, but that Flux of minute aereal Bodies, or most æthereal parts of the aer, which are excussed in round by two bodies arietating, are easily *Capable of Configuration:* when as much is subindicated even by those sensible Vortices, or Whirlings and Eddies of Winds, which are frequent in summer. Under this title fall those words of *Epicurus,* τῦτο δὲ ρεῦμα εἰς ὁμοιοσχήματα θρύπτεσθαι θρύσματα, &c. *Hunc vero fluxum in frustula consimilis Figuræ comminui:* the full sense whereof seems to be this. That when a Voyce is emitted from the mouth, or other sound from what body soever; the Contexture of the minute bodies effluent is so comprest, and confracted into smaller contextures, that of the Original are made swarms of Copies, or lesser masses exactly consimular in their Formation: and that those are instantly dispersed sphærically, or in round through the whole circumfused space, still conserving their similitude to the Original, or General voyce, or sound, till their arrival at the Eare; and so retaining the determinate signature of their Formation, are distinguisht accordingly by the sensory. By this it appears, that *Epicurus,* in this point, dissented inconciliably from *Democritus*; who conceived that all sounds were delated to the Ear by *Propagation,* i. e. that the sound being broken into myriads of small Fragments, each fragment did form the contiguous Aer into Contextures of the same Configuration with the Prototype, and those again formed the particles of aer next adjacent into the like, and so successively through all parts of the medium till they came home to the Organ of Hearing; not much unlike the dream of the *Aristoteleans,* concerning the Propagation of the species of Light in each point of the medium. Whereas the Conception of *Epicurus* is this, that the Primitive Configuration of the most tenuious particles of the Aer, by the percussion or Collision, is broken into many small masses; and each of those, at farther remove from the sonant into many smaller, and those again into smaller, all exactly respondent to the First in figure: after the same manner, as we observe a spark of Fire exsilient from a Firebrand, to be broken into a multitude of less sparks, and each of those shivered again into many less, until their exility makes them totally disappear.

*Art.* 14.
The Reason
and manner of
the *Diffusion* of
Sounds, explicated by a
congruous *Simile.*

This Reason and manner of the Diffusion of a Sound throughout so great a space of the medium, They may easily comprehend, who have observed the Sewers of Princes in *Italy* spout Orang-flower water, or other Fragrant Liquors, out of their mouths, with such dextrous violence, as to disperse it in a kinde of mist, through the aer of a spacious room, so that the aer contained therein becomes impregnate with the Odour, for the more noble entertainment of the sense. For the

　　　　　　　　　　　　Consent

Confent betwixt this *Exfufflation* of Water, and the fpherical Diffufion
of a Sound, is very manifeft, the greater Drops of water being in their trajection through the aer, broken, by reafon of the impulfe of the breath,
that difcharged them in diftrefs, into fwarms of lefs drops, and thofe a-
gain into lefs, fucceffively in the feveral degrees of remove, until they
attain fuch exiguity, as we obferve in the particles of a mift : and that
fmall proportion of Aer, emitted from the mouth of him that fpeaks, being difperfed into a denfe mift of voyces, replenifhing the whole fphere of
Diffufion.

*Art. 15.*
*The moft fub-*
*tle Particles of*
*the Aer onely,*
*the material of*
*Sounds.*

Here we are conftrained to a cautionary advertifement; that when we
fay, the *Aer is the Material of all voyces*, we do not mean all the Breath
expired from the Lungs, together with thofe Fuliginous Exhalations, that
the Denfation of the aer, in Cold weather, fubjects to the difcernment of
our fight; but onely the moft *fubtle part of the Aer* infpired, and modu-
lated in the Vocal Artery and other organs of fpeech : becaufe fuch onely
can be judged capable of Configuration. Nor can fo fmall a quantity of
pureft Aer be thought infufficient upon Difperfion to poffefs fo capacious
a fphere, as that of every ordinary voice; fo that of a whole Theatre
of Auditors, each one fhall diftinctly hear it : infomuch as onely a mouth-
ful of Water blown from a Fullers mouth, is fo diffufed as to irrigate the
aer replenifhing a room of confiderable amplitude. Efpecially, when the
Analogy holds quite through. For, as the Drops of Water are fo much
both larger and denfer, by how much neerer they are after exfufflation
to the mouth of the Fuller : fo alfo are the Vocal maffes of aer fo much
more large and denfe or agminous, by how much neerer they are to the
mouth of the Speaker; and *è contra.* Which alone is the reafon, why the
Voyce of an Orator in a Theatre is more ftrong and diftinct to thofe of his
Auditory, that fit neer at hand, than to thofe far off; provided the place
afford no *Concurrent Eccho*, for in that cafe, the Reflex voyce entering
the eare united with the Direct or Original, magnifies the impreffion on
the fenfory.

*PARADOX.*
*Art.16.*
*One and the*
*fame numeri-*
*cal voyce, nor*
*heard by two*
*men, nor both*
*ears of one*
*man.*

Now, infomuch as it is confentaneous to right reafon, to conceive, that
the Voice, at its firft Emiffion from the mouth, is one General Configuration of the moft tenuious particles of the Aer, with fome vehemency efflated from the vocal organs, after frequent collifions and tremulous repercuffions, and that this General voice, in its diffufion through the medium,
is contracted and difperfed into myriads of minute vocal configurations or
Particular voyces, fome of which invade the ears of one perfon, others
of another, &c. Hence is it a clear, though perhaps new and very paradoxical, truth, *That the fame numerical voyce of an Orator, is not*
*heard by any two of his Auditors, nay not by the 2 ears of any one;*
*but every man, and every Eare is affected with a diftinct voyce.* And
yet he incurs no Contradiction, that affirms the whole Auditory to receive the fame voyce. For, as all the water exfufflated into a mift from
the mouth of an Italian Sewer, or common Fuller, may be faid to be one
and the fame Water; though all the minute Drops, diffufed into feveral
parts of the aer, and irrigating the feveral parts of the Floor or cloth,
on which they are rained down, be not the fame drops : fo likewife may
we allow all the Aer efflated from the mouth of the fpeaker, to be one
and the fame Aer; though the Particular Voyces, delated to particular
Ears

Ears, are not the same Numerically. Besides, should we, with the major part of Scholers, admit a voice to be an Entity meerly *Intentional*, or simple *Quality*, or *Accident*, yet should we not detract one grain of weight from this our *Paradox* : since, to conceive any one Particular voice to be in divers places, or subjects, at once, is manifestly absurd.

Here opportunity would prompt us to insist upon the admitable *Conformation* of an *Articulate* Sound, and to enquire how each Vowel and Consonant is created by such and such motions of the Vocal Instruments : but the exceeding Difficulty countermands that inclination. For, though *Casserius, Placentinus, (in Anatom Sirmorin. Organ.)* & *Athanasius Kircherus (in lib. Anatomico de natura Sonis & Vocis, a cap. 10. ad finem libri.)* have attempted laudably in that abstruse theme : yet the Audit of their discoveries riseth no higher than this single rule, That the Vocal Artery and Lungs onely conduce to the Acuteness and Gravity of the Voice, as they discharge the inspired aer more Pressly, or Laxly ; and *Kircher (in cap.*10.) ingenuously confesseth, *At quomodo voces in gutture formentur, qua proportione elisionis aeris nascantur, tam obscurum est, quam voces hujusmodiclara sunt & manifesta auditui.* The difficulty, indeed, seems to consist chiefly in this, *How from the various motions of one single Organ, the Tongue* (the Author of *Distinction* in all *Articulate* sounds, though the Palate, Epiglottis, Uvula and Teeth are in their respective degrees of assistance inservient to the Elision of aer made by the Tongue) *and that two-leaf'd Door of the mouth, the Lips, such infinite variety of Letters and words doth most easily and almost insensibly result.* To solve this, the General answer is, that the wonder ought to be no greater, how one Tongue can suffice to the Articulation or Distinction of innumerable words, by its various Motions ; than that, how one Hand sufficeth to the Distinction of innumerable Characters. But, the Motions of the Hand requisite to Distinction of every Character, are observable by the sense : and those of the Tongue and Lipps requisite to the Formation of every word, together with the proportion of the Aers Elision in every Articulation, is deeply obscure : and therefore the Disparity being manifest, the *Problem* remains untoucht ; and our Admiration not so much as palliated.

*Art.* 17.
A PRO-
BLEM
not yet solved
by any Philo-
sopher : viz.
How such in-
finite variety
of Words is
formed onely
by the various
motions of the
Tongue and
Lips.

This Place might also admit another *Considerable*, as terrible to the most daring Curiosity as the Former ; and that is the *ineffable Pernicity*, whereby the Aer is exploded from the Lungs, that so it may attain the Form of a voice. For, to the Creation of a voice Consonous, or Unison to the sound of some one string on a Lute ; it is necessary, that the Aer be exploded by the Lungs, with the same Pernicity, as the other Aer is impelled by the string in each of its most rapid Vibrations, or alternate Recurses, after its smart percussion by the finger, or plectrum. But this Arcanum requires a *Galilæo* or *Mersennus*, at least, to its due speculation.

*Art.* 18.
A Second (also
yet unconjue-
red ) *Difficulty*,
viz. the deter-
minate *Perni-
city* of the Aers
motion, when
exploded
from the
Lungs, in
Speech.

The *observable* most proportionate to our Capacity, and Competent to our præsent Designation, is this ; *That no Sound is created without Motion* : and consequently, that the Thing Sonant, being endowed with solidity in some degree or Compactness sufficient to Resistence, ought either to be strook against another, that is solid and resistent ; as when a Hammer is strook upon an Anvil ; or against the Aer, in Flux and not much resisting, and that either by Pulsation of the Aer by a solid, as when the string of

*Art.* 19.
All Sounds
Created by
Motion, and
that either
when that in-
termediate Aer
is contracted
by two solids,
mutually resi-
sting ; or when
the aer is per-
cust by one
Solider when
a solid is per-
cust by the
Aer.

　　Lute

Lute percusseth the aer; or the Pulse of the solid by the Aer, violently agitated, as in all Pneumatick, or Wind instruments, where the stroke of the aer against the sides of the Concave causeth the Sound.

**Art. 20.**
*Rapidity of motion necessary to the Creation of a Sound, not in the First Case.*

In the *Former instance*, it is not necessary to the Creation of a Sound, that the Collision be made by a motion rapid; because the Resistence, on either part equal, causeth that when the Access or Appropinquation of one Solid to the other is Continent, the Aer interposed is Continently impelled and repelled reciprocally: and as the Aer becomes the more hardly distrest on each part, by how much neerer the two Solids approach each other; so proportionately is the motion more rapid. So that, by that time the two solids touch each other superficially, the motion is encreased to the highest rapidity, and the distrest Aer, no longer able to endure Compression, or to go and come alternately between the Solids, now contingent, breaks forth laterally in round, and is diffused in shivers through all parts of the medium, so that arriving at the Ear, it puts on the species of a Sound.

**Art. 21.**
*But, in the Second and Last.;*

But, in the *Second* and *Third* instances, it is necessary the motion of Collision be far more rapid, in order to the Creation of a Sound: because the Resistence, which is wanting on the part of the Aer, must be compensated by the frequent pulses and repulses of it, as when the Chord of an Instrument percust, doth very frequently impel the aer, by its Vibrations (the Greeks call them, χραδασμοι) or Reciprocations; or, as in Wind instruments, where the inflated Aer is, by quick reverberations from the sides of the Concave, very often impulst and repulst.

**Art. 22.**
*That all Sounds are of equal Velocity in the Delation.*

As for the Motion of the Aer, after its Formation into a Sound, from the Sonant to the Ear, therein is one particular worthy the wonder even of Scholars: and that is, *Whatever be the vehemence or remissness of the Collision, or force, by which the Aer is exagitated, yet is the Translation of the Sound, thence resulting, through the intermediate space to the term of it sphære, always equally swift.* For Experience demonstrates, that all Sounds small and great, excited in one and the same place, though they differ much. In the extent of their sphears of Audibility, are delated to that place in which they are heard, in equal time. This is easily observable in the reports of a Cannon and a Musquet, successively discharged at a mile distance. For, standing on a Tower, or other eminent place, and noting the moment, first when the Cannon is fired (the report and Flash being made both at the same instant) and numbring how many Pulses of your artery, or how many Seconds in a Watch denoting them, intercede betwixt your sight of the flame, and hearing the report, and then accounting how many Pulses, or Seconds intervene betwixt the flash and report of a Musquet: you shall finde the number of these equal to the number of those.

**Art. 23.**
*The Reason thereof.*

The *Reason* of this Æquivelocity of unequal Sounds, the *Stoicks* (*apud Plutarch.* 4. *placit.* 19. & *Laertium lib.* 7.) well insinuate, while they affirm, that the Aer percussed, in regard of its Continuity, is formed into many Rounds, such as those successively rising and moving on the surface of Water, upon striking or throwing a stone into it; which Circles made on the surface of Water by a small stone, move in the same tenor, and successively arrive at the margin of the River, or
Pool,

Pool, in as small time, as those caused by a great stone. And *Aristotle* (2. *de Anim. cap.* 8.) expressly declares his judgement, that the reason of the Delation of a Sound from the Sonant to the Audient, is the Continuity of the Aer : though *Simplicius* and *Alexander* differently interpret that Text, the one conceiving that he meant that a Sound was translated through the medium by reason of *sympathy* among the parts thereof; the other, by *Propagation* of the like Sound in all points of the medium successively, after the manner of species Visible, according to the dream of *Aristotle*. But all one it is to us, whether we conceive the motion of a Sound made by Propagation, or Undulous Promotion; as to our præsent scope : since either sufficeth to explicate the Cause, Why a Sound is longer before it arrive at the Eare, than a Visible species before it arrive at the Eye; because the Visible species is transmitted from the Object, neither by Propagation, nor Undulation, but Directly, and therefore is capable of no Retardment from the Medium.

As for the definite Velocity of Sounds, or determinate space of time, in which all Sounds are delated to the Extremes of their spheres; we conceive it to be *Rhodus* and *Saltus*, in the General, inassignable : in regard of the vast disparity in their several Extents, some sounds being scarce audible at the distance of 20 yards, and others cleer and distinct at as many, nay twice as many miles distance. But, if we assume this or that determinate Sound, and attain the præcise diametre of its sphere; it is no difficulty to commensurate its Velocity. For, *Mersennus* (*in reflexian. physicomath. cap.* 14. *& Proposit.* 39. *Ballistica* ) upon exact Experiment, found the Fragor of several Cannons discharge in the Court of the Bastile at *Paris*, to arrive at his eare, after the flashes, at such a rate, that the found pervaded 233½ Fathoms (each containing six feet *Paris* measure) in the space of every Second, or Sixtieth part of a minute : and thereupon rightly concluded, that the Report of a Cannon flyeth at the constant rate of neer upon 14000 Fathoms every minute, until it attain the extremes of it sphere. If this expedient for the measure of the Time wherein Sound is delated, seem either too costly or laborious; you have another most cheap and easie præscribed by the Lord *St. Alban* (*in Cent.* 3. *Nat. Hist.*) which is this. Let one man stand in a steeple, having a lighted taper with him, and some vail put before the flame thereof; and another, confæderate in the tryal, stand a mile off in the open field : then let him in the steeple strike the Bell with a weighty hammer, and in the same instant withdraw the vail; and so let him in the field account by his pulse what distance of time intervenes betwixt his sight of the Light, and hearing of the Sound. If the strokes of the Artery, which are subject to variation, for many causes, seem less certain; the Seconds in a minute watch (which are ἰσόχρονοι, æquitemporaneous) will be an exact measure of the interval, and so of the velocity of a Sound. *Plura vid. apud Mersennum lib.* 2. *Harmonic. proposit.* 40.

Another admirable secret there is in the Motion of Sound, which is, that no Winde can accelerate, or retard it, but it is delated from the Sonant to the Audient in equal time, whether the wind be high or gentle, secund or adverse. For, a Secund or favourable Wind is incomparably slower in motion than a sound, as appears by the Rack or drift of clouds, the undulation of Corn fields, the successive inclination of the tops of trees in woods, the

*Art.* 24.
To measure the Velocity of great Sounds.

*Art.* 25.
Sounds, not subject to Retardation, from adverse, nor Acceleration, from Secund Winds.

the rowling of waves at sea, &c.but an Adverse wind,though it may indeed disturb a sound, or weaken it by suppressing some of its particles (which is evident from hence, that all sounds attaining the eare against the wind, are not so clear and distinct, as when they are heard with the wind; as in Bells, whose noyse alternately riseth and falleth in contrary gusts) yet do all the particles that remain uninterrupted, permeate the medium with equal velocity. This may be soon Experimented either by Cannons, as *Mersennus*, or a candle and bell, as the Lord *Bacon*.

## SECT. III.

Art. 1.
That all Sounds, where the Aer is percussed by one solid, are created immediately by the Frequency, not the Velocity of motion; demonstrated.

THe Præmisses duly considered, it can seem no Paradox, *That a Sound is created in the Aer, not so much by the Velocity, as CREBRITY of motion:* and no unnatural Consequence thereupon, that the *Difference of an Acute and Grave Sound ariseth not from the greater and less swiftness or rapidity of the motion,* as *Aristotle* and most of his *Sectators* imagined; but *from the Frequency and Infrequency thereof, as Galilæo, Mersennus,* and *Gassendus.*

To secure this by plain *Demonstration*, take a Lute string in your hand, and having fastened one end thereof to some hook or pin in a wall, distend it gently; and then percussing it with your finger, you may perceive the Vibrations, or accurses and recurses alternately succeeding, but you shall hear no sound resulting from it: because, as every vibration of the string is performed in perceptible time, so doth the aer thereby percussed arrive at the eare with such sensible intervals betwixt each appulse,as that it leaves no impression therein remaining, which is not expunged and consolidated before the invasion of a second appulse. Then stretch the string somewhat streighter, so that the Vibrations thereof may become inobservable by the eye, in respect of their Frequency; and you shall hear a certain dull stridor, or kind of sibilation; because the Appulses of the aer, percussed by each Vibration,at the eare,will be almost Continent,so that the time interjected betwixt each stroke on the eare becomes imperceptible, and indistinguishable, nor can the first impression on the sensory be consolidated before a second renew it, &c. This done and observed, encrease the distension of the string yet more, and percussing it you shall perceive a clear sound to arise; because as the Vibrations,so are the percussions of the aer, and their Appulses to the Eare far more Continent, or more one,in regard the moments of Time intercedent betwixt the successive strokes,are more short and imperceptible.

Art. 2.
And likewise, where the Aer is the Percutient

And what we here say of the reason of a Sound resulting from a Lutestring, the same, in proportion, is to be conceived of all other Sounds created in Wind instruments, where the Aer is the Percutient. For,the breath easily and gently inflated into a Flute, Cornet, Trumpet, &c. yields no sound at all; onely because the pulses and repulses of the aer from the sides of the Concave are so infrequent, as to have the intervals of time distinguishable: and the aer likewise slowly emitted from the Lungs (the great Exemplar to all Pneumaticks) makes no voice, onely because it is not frequently

quently enough reverberated from the fides and annulary cartilages of the Vocal Artery, and consequently the Appulses of it to the eare being proportionately infrequent, cannot, by their Coition or Union into one stronger Appulse, make any sensible impression on the sensory. But the Aer then becomes sonant, when it is efflated with vehemency, in respect of its more frequent Appulses to the sensory, respondent to the more itterated pulses and repulses, or reverberations of the sides of the Vocal Artery. Thus also, when you draw your finger gently along a Table, or put a Hammer on an Anvil easily, you shall hear no sound; because the Repercussions of the Aer caused by that gentle motion, are so far asunder in time, as never to become Continent, or Conjoyned: and consequently, the Appulses of the percussions to the eare being alike infrequent, can never make a sensible impression on the Acoustick Nerve. And this we conceive more than sufficient evidence of the Verity of the First part of our Thesis; That a Sound is not generated in the Aer by the Velocity, but Crebrity of motion: unless in a remote dependence, as Velocity is the Cause of Crebrity.

As for the Remainder, viz. *That an Acute sound ariseth from more frequent, and a Grave Sound from less frequent percussions of the Aer :* the Certitude hereof may be easily concluded from this Experiment. Fasten a long Lute-string at one extreme on a hook nayled to a wall, and suspend a small weight at the other; then strike the string at convenient distance above the weight: and you shall observe the Swings, or Vibrations of it to be so slow, as that you may measure the time of each, by the systole and diastole of your Pulse, or the Seconds in a Minute Watch. Then wind up the Chord exactly to the half, the same weight continuing appended, and percuss it, as before: and you shall finde the Vibrations of it to be doubly swifter than the former, so that one Vibration shall be in time respondent to two Pulses. Again, abbreviate the Chord to half, and having percussed or abduced that half, which is now but a fourth part of the whole; you shall observe the Vibrations to be again doubled in Frequency, in respect of the Second, and quadrupled in respect of the First; so that now 4 Reciprocations shall be isochronical to one pulse. This effected, continue this determinate abbreviation of the Chord, by subdividing it into halfs successively, until the Reciprocations become so swift and frequent, as to be indistinguishable by the sense (though still you deprehend their Velocity and Crebrity to be encreased at a certain rate, i. e. duplicated upon each Dimidiation of the chord) when the Aer is so frequently percussed by it, as that it becomes Sonorous, or actually sonant. Then again Dimidiate the sonant remainder of the Chord, and upon percussion you shall observe the sound thereof to be more Acute by a whole *Octave,* than the Former: and thence you cannot but concede, that the Acuteness of this half of the sonant chord, above that of the whole sonant chord, is caused only by the doubly more frequent Percussions of the Aer, and proportionate strokes of the Sensory. And, because a Quadruplicate weight produceth the same Effect, being appended to the whole of the sonorous chord, as a simple weight doth in the half, as to the Duplication of the Celerity and Frequency of the Vibrations, in the same moments: hence is it, that if you encrease the weight, retaining the same Longitude of the Chord, by degrees, until you advance the sound thereof to an *Eighth*; it is manifest, that the Reciprocations of it are still doubly more swift and frequent, than those caused by the former weight. Moreover, what we affirm concerning the Half of the sonorous

Chord,

*Art. 3.*
That all Acute sounds arise from the *more,* and Grave from the *less* frequent percussions of the aer, demonstrated.

Chord, in respect of an Octave; holds true, in proportion also of the 2 thirdparts of the Chord, in respect of a *Fifth*, of the Dodrantal, or 3 quarters, in respect of a *Fourth*, and so of the rest of the musical Notes.

For, in a very long Chord, if you stop upon the third part of the half thereof, and percuss the Bessal, or two thirds of the half remaining at liberty: the proportion of its Reciprocations will not be Duple, but sesquialteral in respect to those of the whole length; i. e. 2 Vibrations of the Chord will not respond in time to one pulse of the Artery, nor 4. to 2. but 3. to 2. And, if you stop on the fourth part; then will the Reciprocations of the remainder be in proportion sesquitertial, i. e. 4 Vibrations shall be isochronical to 3 pulses. According to the same method, if you stop on the 5th. part of the Chord; the proportion of its Vibrations, to that of the former, will be sesquiquartal: if the 6th part, sesquiquintal; and so consequently of all other Notes. So that it seems easily determinable, by this scale, What is the proportion of the strokes inflicted on the Eare in every Acute sound, comparatively to those inflicted by every Grave: and this not onely in the sounds of a string, but all others of the like Original. To instance; when a Boy sings with a Man, and emits a note more Acute by an Eighth; it is to be conceived, that the Aer efflated from the Vocal Artery of the Boy, is doubly swifter in its motion, or doubly more frequent in its reverberations from the sides of the Wind-pipe, in respect of the double narrowness thereof, than that expired from the Vocal Artery of the man. And, hence we may occasionally advertise, that by how much the more Acutely any man would sing; by so much more streightly or narrowly must he Compress his Wind-pipe: that so the Aer may issue forth more distrest and streightned, having suffered the more Frequent reverberations from the sides and rings of the same.

**Art. 4.**
*The suavity of musical Consonances, deduced from the more frequent; and Insuavity of Dissonances from the less frequent Union of the Vibrations of strings, in their Termes.*

And this is that noble Fountain from which many of our modern *Theorical Musicians* have drawn the Reason of the *Suavity* of their CONSONANCES, and *Acerbity* or ingratefulness of their DISSONANCES: and that not without mature consideration. For, when two Sounds, synchronical in their creation, arrive at the eare in the same instant, and affect it with pleasure, or a kinde of sweetness; the Cause of that sweetness can be no other but this, that the percussions of the Aer generating those two Sounds, become so united, as to leave no sensible discrepancy, that might grate or exasperate the tender sensory: and on the other side, the reason of the Discord or Insuavity of two sounds, at once emitted, is onely this; that they are not united, so that the eare deprehends and dislikes their Discrepancy. Again, the several Degrees of this Suavity and Insuavity among musical sounds, cannot be deduced with equal probability from any other original, as from the variety of Coition, and Discrepancy of the Percussions creating the Sounds. To exemplifie in the Sounds resulting from strings; take two strings, equal in their materials, length, and thickness, and distended with equal weights, or force; and when you percuss them with one stroke, they will emit equal sounds, or that Consonance, which is called an *Unison:* which will be therefore grateful, because as the Vibrations of the strings, so will the strokes inflicted on the sensory, have the same proportion each to other, as one hath to one (the proportion of Equality) and consequently will be equal in number and time,

so

so as to affect the sensory most equally and Unitedly. But if you abbreviate one of the strings exactly to half; because (according to the præmisses) the sounds resulting from them, at once percust, must make an *Eighth*, or that Consonance, which the Greeks name διὰ πασῶν, and we a *Diapason*: therefore must that Eighth be eminently grateful also; insomuch as though after the Coalition of two strokes, one resulting from the shorter string be insociate, yet doth the immediately consequent stroke thereof perfectly unite with that of the longer string, and so the Unition is made Alternately, or at every other stroke; and therefore doth this Consonance invade the sense of all others, an Unison only excepted, most unitedly and equally, and consequently is the most pleasant and charming of all Consonances, after an Unison.    And when you make the proportion of the short string exactly Sequialteral to that of the long; because the sounds resulting from them, both at once percussed, make a *Fifth*, or *Diapente*: therefore will that Consonance also have a considerable degree of sweetness, though short of that of an Eight; insomuch as though two strokes pass insociated, yet doth the Union follow in every Third, and so the Unition is sufficiently frequent to please the sense, which is best delighted with that object, in which is the least difference of parts, according to that *fourth Præcogn. of Des Cartes* (in compend. Musicæ, pag. 6.) *Illud objectum facilius sensu percipitur, in quo est minor differentia partium.*    Again, if you make the proportion of the short string Sesquitertial to that of the long, because a Fourth, or Diatessaron, doth result from the percussion of them together, therefore will that Consonance be likewise competently grateful: in respect that after three insociated strokes, the Coition falls in every fourth. To Contract; the same holds in proportion exactly true also in Sesquiquartal and Sesquiquintal proportions, from which arise Thirds major and minor; and of superbiparting Thirds, and supertripartting Fifths, from which arise Sixths major and minor; and finally, in all Compound Consonances, such as Disdiapason, &c. For, always the Consonance is by so much more grateful, by how much more frequently the strokes unite in the Sensory: and *è contra*. Whence is generated the Dissonancy, or ingratefulness of Sounds, when ever the strokes either too rarely, or never unite: because, in those cases, the sense is held in a kind of lasting distraction, and unless a restitution of the distracted parts of the Sensory be made by some Coalitions, and those sufficiently frequent, (which are a kind of Balsam, to cure the gratings and dissolutions) the sensory must be mis-affected with a kind of Laceration, and undergo that dolour unwittingly. This the skilful Musician foreknowing, endeavours to prævent, by making a Diapason, or perfect Consonance tread upon the heels of a Dissonance, for varieties sake usually inserted into Tunes: thereby with advantage consolidating the ulceration of the sensory caused by the præcedent Discord, and making the Harmony the more grateful; as Health is most grateful immediately after sickness, and a Calme after a Tempest. And this is the reason, why an Eighth is by many reputed a more pleasing Consonance, than an Unison; viz. in respect of the Distraction, which succeeds alternately from the Dissociated strokes of one of two strings together percust: and not in respect of its Comprehension of all other Consonances, as *Des Cartes* seems to conclude (*in cap. 8. Compend Music.*)

G g    If

If this Genealogy of all Mufical Confonances feem either obfcure, or tædious; you may pleafe to accept it in Epitome, thus. The Vibrations of Chords are, according to moft exact obfervation reciprocally proportional

$$\text{to the} \begin{cases} \text{Length of} \\ \text{Weight at} \end{cases} \text{the ftring, having the fame} \begin{cases} \text{Weight.} \\ \text{Length.} \end{cases}$$

Whence many have concluded, that all Confonances in Mufick proceed from the fpeedier Union of thefe Vibrations in their Terms.

| The Terms of an | | are in proportion, as | therefore the fpace of | Vibrations, in the Graver Term, are juft equal to | Vibrations in the Acuter Term of an | |
|---|---|---|---|---|---|---|
| Eighth | | 2 to 1 | 1 | 2 | | Eighth. |
| Fifth | | 3 — 2 | 2 | 3 | | Fifth. |
| Fourth | | 4 — 3 | 3 | 4 | | Fourth. |
| Sixth major | | 5 — 3 | 3 | 5 | | Sixth major. |
| Third major | | 5 — 4 | 4 | 5 | | Third major. |
| Third minor | | 6 — 5 | 5 | 6 | | Third minor. |
| Sixth minor | | 8 — 5 | 5 | 8 | | Sixth minor. |

Hereupon our Harmonical Authors (whofe Pythagorean fouls feaft themfelves with the ravifhing, though filent Mufick of Numbers) for the moft part account an Eighth the Firft of Confonances, becaufe an Union is made before a fecond Vibration in the Graver Term; a Fifth the fecond Confonance, becaufe an Union is made before a third Vibration in the Graver Term, &c. according to the Scheme.

But this fo univerfally celebrated Melotherical Foundation hath been very lately fhook by that no lefs Erudite, than Noble Author of the *Animadverfions on Des Cartes Mufick Compendium, the Lord Vifcount Brouncker*; (whofe conftant Friendfhip, and learned Converfation, I muft profefs to have been one of the cheifeft Confolations of my life.) who having, upon profound, and equitable examination, found this great defect therein, that according to the former Derivation of all Mufical Confonances, a Third Major muft fucceed a Fourth and Sixth Major, and the proportion of 7 to 5 makes a Confonance as well, and before a Sixth minor; which is manifeftly repugnant to Experience: hath enriched the world with a new *Hypothefis* of his own happy invention, fufficiently extenfible to the full folution of all Mufical Phænomenes. According to which the Confonances arife (phyfically) from the Vibrations of Chords, not in refpect of their Union, but *Ratio-Harmonical Proportion*, as He is pleafed to call it: and this upon very good reafon, fince, the Vibrations being proportional to the Chords, and the Chords fo proportionally divided, it is of meer neceffity, that their Vibrations have the fame proportions. But of this, the Competent Enquirer may underftand more from his *Animadverfions*, &c.

And

And this speculation, touching the Nativity of Musical Consonances, harh engaged us to touch upon that Quicksand, from which none the most adventurous Curiosity hath ever yet returned with full resolution; and that is that eminent PROBLEM, *Quando sonus Harmonicus à nervo fieri incipiat ?* In what instant an Harmonical Sound, created by a Chord of an instrument percussed, or abduced from its directness, is begun ?

*Art. 7.*
PROBLEM ι.
In what in-
stant, an Har-
monical
Sound, result-
ing from a
Chord percuf-
sed, is begun.

For the clear understanding of this *Question*, we are first to advertise; that from the percussion of any Chord distended, there are made two different Sounds : *one* arising from the allision of the Aer betwixt the finger, or plectrum, and the Chord; which is so far from being Concinnous, that it frequently diminisheth the integrity or sweetness of the Musick, and alwayes makes a kind of Discord, where the unskilfull hand strikes too hard or foul; the other, from the Chord verberating the Aer in its Vibrations, which is the Concinnous, or Harmonical Sound, by the Græcians, for distinction sake, called φθόγγ@·, and in our language the *Twang*. And this is the subject of the præsent Enquiry. Secondly, we are to præmise this

*DIAGRAM.*

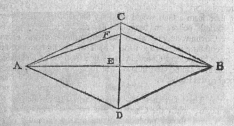

Wherein A B denotes the Chord, in a streight line, either perpendicularly, or horizontally distended; A B C the same Chord abduced, or impelled from the direct line to C; and F the same in the extreme or term of its spontaneous Flexion, after some certain recurses. And lastly, we are to state the Question, thus. Whether the Concinnous Sound begins from the First Excurse, which is made by the Chord from E to C, when it is impelled by the force of the percussion; or, when it returns, by spontaneous reflection, from C to E; or, when it hath past beyond E to D; or, in its whole Recurse from D, by E, to C?

(1) Some there are, who observing that, when a Chord is abduced from its direct line E to C, and returns it self from C to E, if a piece of wollen cloth, a mans finger, or ought else that may suppress its motion, be so set as to arrest it at E; then is no Harmonical sound created, either in its first Excurse from E to C, nor Recurse from C to E : have upon this Experience concluded, that the Concinnous Sound is begun in the first Recurse of the Chord from D to E; because they suppose, the Chord then to reverberate the Aer, which pursued it (*à tergo*) from C to D, and force it by contrary

violence

violence to fly back again from D to C by E : fo that the Aer at E, being on both fides diftreſt by that moving violently from C to E, on one hand, and that laftly impelled from D to E , on the other, muſt fuffer the higheſt Condenfation, of Compreſſion, or Percuſſion of all the other aer within the ſpace C D, and confequently be the original of the Sound.

( 2 ) *Others* have affirmed the original of the Sound to be from C to E, the ſpace of the firſt Recurſe : and their inducement thereto is this Experience. If a Chord of 30 perches length be with fufficient force extended , and then abduced from its line of direction to the diftance of 15 feet, more or leſs; it will yeild a kind of ftridor, or grave fibilation, in its fpontaneous Recurfe from C to E: which found would perhaps be Concinnous , if included in fome Inſtrument of ſufficient capacity. To which they add, that wands or rodds being ſwitcht in the aer, and Gun-ſhot in their flight , emitt a finging noyſe, though they are impelled only one way , and have no Recur- ſes, or doublings in the aer. But, to this it may be *Anfwered*, ( 1 ) That all theſe Bodies may more juſtly be conceived to yeild a ſound only in this reſpect ; that the inæqualities in their ſuperficies fo diftreſs the aer in their rapid Motion , and by frequent reciprocations in their finall cavities variouſly agitate the ſame , that it fuffers ſuch Circumvolutions as are tantamont to their Recurſes. ( 2 ) That no Bullet ſhot from a Gun would yeild any ſound at all , if it were ex- actly ſphærical, polite , and hard , and flew directly without that Voluta- tion , or Circumvolution, which the reſiſtence and circular returns of the aer conſtantly impreſs upon it. ( 3 ) That the Sibilation or Hiffing noiſe made by the long Chord, in its Recurſe from the 15 feet abduction, is not, nor ever can be Concinnous : and therefore the Experience is im- pertinent to this Problem.

( 3 ) A Third fort there is , who opinion the Harmonical Sound then to begin , when the Chord is firſt impelled from E to C; ſo that the Chord ſhould produce a Sound in the extremity or period of every Flexion, i. e. in C and D, at alternate Recurſes: and conſe- quently , that no fenfible Sound is produced in any part of the whole intermediate ſpace betwixt C D. And the Ground Theſe ſtand up- on , is the Experience of Cloth , which being violently ſhook in the aer, for the excuſſion of duſt, doth only then emit a ſmart ſound , or Rapp, when attaining the extremity of its Flexion , it percuſſeth the ſuperior aer , and is in the manner of Sails, fwelled up by the inferior aer. But, in this inſtance , and that confimilar one of Coach-whips, it is almoſt evident even to the eye , that the Rapp is made only by the Doub- ling of the Cloth, or Chord , at the end of their Flexion : and there- fore we are not convinced , that the Concinnous Sound is then be- gun, as theſe perſuade , in either C or D the period of each Flexion; eſpecially , when the Chord in C and D feems rather to quieſce , than move ; and fome quiet muſt intercede betwixt two contrary mo- tions of the ſame thing.

( 4 ) But , infomuch as all founds are cauſed by the Motion of the Aer; and the Sound alwayes is loudeſt , where the Motion of the Aer is moſt rapid ; and in the whole fonorous line, or ſpace betwixt C and E , the
motion

motion of the Aer intercluded is moſt ſwift, when the Chord returns
from C to E: therefore doth *Merſennus* (to whoſe judgment we moſt
incline, in this nicety) conclude; that the Harmonical ſound is begun
in the beginning of the firſt Recurſe of the Chord from C to E: and
that it is then of the ſame Acuteneſs, as are all the ſubſequent ſounds
made by the ſubſequent Recurſes; becauſe the reaſon of the Firſt Recurſe
ſeems to be the ſame with that of all the conſequent.

To this ſome have *objected*; that the ſound of the Firſt Recurſe is
too Expedite and ſhort, to be perceived by the Ear : ſince even the
Eye, incomparably more prompt in the diſcernment of viſibles, cannot
behold an object, whoſe Apparence, or Præſence exceeds not the Duration
of the foreſaid Recurſe of the Chord from the extreme of its flexion C to
E; which doth ſcarce endure the ⅙00 part of a minute. But this objection
is ſoon *diſſolved* by Experience, which teſtifieth, that if a quill, or other impediment be placed ſome ſmall ſpace beyond E towards D, ſo that the Chord
may complete its firſt Recurſe from C to E, without interruption : then will
a ſound be created, and ſuch as hath ſufficient Acuteneſs; though it be ſcarce
momentany in Duration, becauſe the frequency of its Recurſes is prævented.

Many other *Problems* there are, concerning the Reaſons of Sounds,
wherewith the inſatiate Curioſity of Naturaliſts hath entertained it ſelf,
in all ages : but, among them all we ſhall take cognizance of only
thoſe more eminent ones, which as they ſeem moſt irreconcilably repugnant to our Theory, when propoſed; ſo muſt they much confirm
and illuſtrate the dignity thereof, when clearly Diſſolved by us, without
the leaſt contradiction to, or apoſtacy from our Principles aſſumed. Since
the unſtrained Solution of the moſt difficult Phænomenaes, by the vertue of any Hypotheſis, is the beſt argument of its Verity and excellency above others, that fail in their Deduction to remote Particulars.

## PROBLEM 2.

*Whether may a Sound be created in a Vacuum, if any ſuch be in
Nature ?*

### SOLUT.

Art. 8.
That a Sound
may be crea-
ted in a Va-
cuum; contra-
ry to *Athanaſ.
Kircher in Art.
Magn. Conſoni
& Diſſoni lib.
1, cap. 6. Di-
greſſione.*

To ſolve this (by many accounted inexplicable) Ænigme, we need
only to have recurſe to our long ſince antecedent Diſtinction of a Vacuity
*Diſſeminate*, and *Coacervate*: for, that once entered our judgment, we
cannot indubitate that ingenious Experiment of *Gaſpar Berthius*, laureat Mathematician at *Rome* (frequently, and alwayes with honourable Attributes,
mentioned by Father *Kircher*, in ſundry of his Phyſicomathematical diſcourſes) which ſenſibly demonſtrateth the actual production of a Sound, in a
Diſſeminate Vacuity.

The *Experiment* is thus made. Having præpared a large Concave
and almoſt ſphærical Glaſs, æmulating the figure of a Cucurbite or
Cupping-glaſs, fix a ſmall Bell, ſuch as is uſual in ſtriking Watches of
the

the largest size, on one side of the concave thereof, and a moveable Hammer, or striker, at fit distance, on the other, so as the Hammer being elevated may fall upon the skirts of the Bell; and then lute or cœment on the Glass, firmly and closely (that all sensible insinuation of the ambient aer be prævented) to one extreme of a Glass Tube, of about an inch diametre in bore, and 8 or 10 feet in length. Then, reversing the Tube, pour into it a sufficient quantity of Quickfilver, or Water, to fill both it and the Head exactly. This done, stop the other extreme of the Tube with your finger, or other stopple accommodate to the orifice; and after gentle inversion, immerge the same to a foot depth in a Vessel of Water, and withdraw your stopple, that so much of the Quickfilver contained in the Head and Tube, as is superior in Gravity to the Cylindre of Aer, from the summity of the Atmosphere incumbent on the surface of the Water in the subjacent Vessel may fall down, leaving a considerable void Space in the superior part of the Tube. Lastly, apply a vigorous Loadstone to the outside of the Glass Head, in the part respecting the moveable extreme of the Hammer; that so, by its Magnetical Effluxions transmitted through the incontiguities or minute pores of the Glass, and fastned on to its *Ansulæ* or smal Holds, it may elevate the same: which upon the subduction of its Attrahent, or Elevator, will instantly relapse upon the Bell, and by that percussion produce a clear and shrill sound, not much weaker than that emitted from the same Bell and Hammer, in open aer.

Now, that there is a certain Vacuity in that space of the Head and Tube deserted by the delapsed Quickfilver, is sufficiently conspicuous even from hence; that the ambient Aer seems so excluded on all hands, that it cannot by its *Periosis* (to borrow *Platoes* word) or *Circumpulsion*, succeed into the room abandoned by the Quickfilver, and so redintegrate the solution of Continuity, as in all other motions.

And that this Vacuity is not Total, or Coacervate, but only Gradual or *Desseminate*, may be warrantably inferred from hence; (1) That Nature is uncapable of so great a wound, as a Coacervate Vacuity of such large dimensions, as we have argued in our Chapter of a *Vacuum Præternatural*, in the First Book: (2) That a Sound is produced therein, for since a Sound is an Affection of the Aer, or rather, the Aer is the Material Cause of a Sound, were there no aer in the Desert space, there could be no Sound. Wherefore, it is most probable, that in this so great distress ingenious Nature doth relieve herself by the insensible transmission of the most æthereal or subtile particles of the Circumpulsed Aer, through the small and even with a microscope invisible Pores of the Glass, into the Desert Space; which replenish it to such a degree, as to prævent a Total though not a Dispersed Vacuity therein: and though the Grosser Parts of the extremly comprest Aer cannot likewise permeate the same slender or narrow Inlets; yet is that no impediment to the Creation of a Sound therein, because the most tenuious and æthereal part of the aer, is not only a sufficient, but the sole material of a Sound, as we have more than intimated in the 15. *Art. 2. Sect. of the present Chapter.*

The

The only Difficulty remaining, therefore, is only this ; *Why the sound made in the disseminate Vacuity should through the Glass-head pass so easily and imperturbed, as to be heard by any in the circumstant space ; when common Experience certifieth, that the Report of a Cannon, at the distance of only a few yards, cannot be heard through a Glass window into a room void of all chinks or crannies ?*

Not need any man despair of expeding it. For, whoso considers the extraordinary and inscrutable wayes to which Nature frequently recurrs, in cases of extreme Necessity ; and that the Distress she undergoes in the introduction of this violent Vacuity (where her usual remedy the Peristaltick motion, or Circumpulsion of the Aer, is prævented by the interposition of a Solid) is much more urgent than that she is put to in the Compression of the ambient aer by the explosion of Canons (where the amplitude of uninterrupted space affords freedome of range to the motion imprest) we say, whoso well considers these things, cannot doubt, but that it is much easier to Nature to admit the trajection of the Sound produced in the Disseminate Vacuity, through the pores of the Glass-head, than the transmission of an External Sound into a close Chamber, through a Glass window, where is no Concavity for the Corroboration or Multiplication of the Sound, and consequently where the impulse is far less (respective to the quantity of the aer percussed) and the resistence as much greater.

## PROBLEM 3.

*Whence is it, that all Sounds seem somewhat more Acute, when heard far off ; and more Grave, near at hand : when the Contrary Effect is expected from their Causes, it being demonstrated, that the Gravity of a Sound ariseth (mediately, at least) from the Tardity, and Acuteness from the Velocity of the Motion, that createth it ; and many great Clerks have affirmed, that the motion of a Sound is less swift far off from, than near to its origine, according to that General Law of Motion, omnia corpora ab externo mota, tanto tardius moventur, quanto à suo principio remotiora fuerint ?*

*Art. 9.*
Why all
Sounds appear
more Acute, at
large, than at
small distance.

#### SOLUT.

No Sound is Really, but only *Apparently* more acute at great, then at small distance ; and the Cause of that semblance is meerly this : that every Sound, near its origine, in regard of the more vehement Commotion, and proportionate resistence of the Aer, dependent on its natural Elater, or Expansory Faculty, doth suffer some Obtusion, or Flatning ; which gradually diminishing in its progress or Delation through the remoter parts of the Medium, the Sound becomes more Clean, Even and Exile, and that *Exility* counterfeits a kind of *Acuteness.*

## PROBLEM 4.

*Why doth Cold Water, in its effusion from a Vessel, make a more full and acute noise, than Hot or Warm ?*

*Art. 10.*
Why Cold
water falling,
makes a fuller
noise, than
warm.

*SOLUT.*

*SOLUT.*

The substance of Cold Water, being more Dense and Compact, must be more weighty, and consequently more swift in its fall, and so the noise resulting from its impulsion of the aer, more sharp than that of Hot: which being rarefied by the fire, or made more lax in the contexture of its particles, looseth something of its former weight, and so hath a flower descent, and in respect of that flowness, produceth a weaker and flatter sound. And this is also the reason, why *Iron* hot yieldeth not so smart and full a sound, as when 'tis cold.

## PROBLEM 5.

*Art. 11.*
Why the
voice of a
Calf is more
Base than that
of an Ox, &c.

*Why is the Lowing of a Calf much more Deep, or Base, than that of an Oxe, Cow, or Bull, at their standard of growth: contrary to all other Animals, which have their voices more shrill and acute, when they are young, than when they are old?*

*SOLUT.*

The Cause of this singularity is found only in the peculiar Constitution of the *Larynx* of a Calf; which is in amplitude equal to, and in laxity and moysture much exceeds that of an Oxe, Cow, or Bull full grown; and so Age doth Contract and Harden, not ampliate the same, as in all other Animals: and it is well known that the wideness and laxity of the *Asper Artery*, is the cause of all Grave or Base Voyces.

## PROBLEM 6.

*Art. 12.*
Why a Disso-
nance in a
Base is more
deprehensible
by the ear,
than in a Tre-
ble voice.

*Why is a Dissonance more easily discovered by the ear, in a Barytonous, or Base Voyce, or Tone, than in an Oxytonous or Treble?*

*SOLUT.*

Because the Barytonous voyce is of a flow Motion, and the Oxytonous of a swift: and the sence doth ever deprehend that object whose apparence is more durable, more clearly and distinctly than that, whose apparence is only instantaneous, or less lasting.

CHAP.

# CHAP. VII.

## OF

# ODOVRS.

### Sect. I.

*Art. I.*
That the Cognition of the Nature of Odours is very difficult ; in respect of the *Imperfection* of the sense of *Smelling*, in man : and

Hoever is natively deprived of any one sense, saith *Aristotle (in Analyticis)* is much less capable of any Science, than He who hath all five Fingers on the left hand of his soul (to use the metaphor of *Casserius Placentinus, in præfat. ad lib. de sens. Organ* ) or all the Organs of the sensitive Faculty complete : and His reason is that General Canon, *Nihil est in intellectu, quod non prius fuerit in sensu* ; the senses being the Windows, through which the soul takes in her ideas of the nature of sensible Objects. If so, whoever hath any one sense less perfect than the others, can hardly attain the Knowledge of the nature of objects proper to that sense : and upon consequence, the Cognition of the Essence of an O-DOURE must be so much more difficult to acquire, than that of Visibles and Audibles, by how much less perfect the sense of SMELLING is in man, than the sight and Hearing. And, that Man, generally, is not endowed ( for, we may not, with our noble Country man Sir *Kenelme Digby* charge this imperfection altogether upon the Errors of our Diet ; because we yet want a Parallel for his *John of Liege*, who being bred savagely among wild beasts, in the Forrest of *Ardenna*, could wind his pursuers at as great distance, as Vultures do their prey, and after his Cicuration or reduction to conversation with men, retained so much of the former sagacity of his nose, that He could hunt out his absent friends by the smell of their footsteps, like our Blood-Hounds) we say, that man is not generally endowed with exquisiteness of smell ; needs no other eviction, but this : that He doth not deprehend or distinguish any but the stronger, or vehement sorts of Odours ; and those either very offensive, or very Grateful.

*Art. 2.*
The contrary
opinions of
*Philosophers,*
concerning it.
But, albeit this difficulty of acquiring the knowledge of the Essence and immediate Causes of Odours, hath its origine in the native Imperfection of our sense accommodate to the perception thereof: yet hath it received no small advance from the obscurity of our *Intellectuals*, the Errors of human judgement, and the common Effect thereof, the contrary Opinions of Philosophers. For, however they unanimously decree, that the proper object of smelling is an *Odour*; and the adæquate *sensory*, ordained for the apprehension of it, the *Mammillary Processes* of the brain, or two nervous productions derived to the basis of the nose : yet could they never agree about the chief subject of their dispute, the *Quiddity*, or *Form* of an Odour; or the *Commensuration* betwixt the same, and the odoratory Nerves, the theory whereof seems most necessary to the explanation of the Reason and Manner of its Perception and Distinction by them.

*Art. 3.*
Some determining an Odour
to be a *substance*
Thus, on one side of the schools, *Heraclitus*, cited by *Aristotle* (*de sensu & sensili, cap. 5.*) is positive, that the smell is not affected with only an Incorporeal Quality, or spiritual species; but that a certain subtle *substance* [κνισώδης αναθυμίασις] or Corporeal Exhalation, emitted from the odorous object, doth really and materially invade and affect the sensory.

(2) And *Epicurus* (*in Epist. ad Herodot. apud Diogen. Laertium, lib. 10.*) seconds him with somewhat a louder voice; *Existimandum est, Odorem non facturum ullam sui impressionem, nisi ab odora re usq; deferentur moleculæ seu Corpuscula quædam, ea ratione Commensurata ipsi olfactus sensorio, ut ipsum moveant afficiant ve; alia quidem perturbate ac discrepanter, ex quo odores Ingrati sunt; alia placide & accommodate, ex quo Jucundi sunt odores :* "men are to conceive, that an Odour could make no sensi-
"ble impression of it self, unless there were transferred from the odorous
"object certain substantial Effluxes, or minute Bodies, so Commensurate or
"Analogous to the peculiar Contexture of the Organ of smelling, as to
"be capable of affecting the same; and those either perturbdly and discor-
"dantly, whence some Odours are Ingrateful, or amicably and conveni-
"ently, and those Odours are Grateful.

(3) And *Galen*, attended on by most of the Æsculapian Tribe, sings the same tune, and in as high a key as either of the Former; saying, (*in lib. de instrum. olfact. cap. 2.*) *Id quod à rerum corporibus exhalat, Odoris substantia est :* though *Casserius Placentinus* (*de fabric. Nasi, Sect. 2. cap. 3.*) hath endeavoured to corrupt the genuine sense of those words, by converting *substantia* into *subjectum*, as if *Galen* intended only that the Exhalation from an odorous body was only the *subjectum inhæsionis*, and the Odour it self meerly the *Quality* inhærent therein. Contrary to the rules of Fidelity and Ingenuity; because incongruous both the Letter of the Text, and the Syntaxis thereof with his whole Enquiry.

(4) And the Lord St. *Alban*, though a modern, yet not unworthy to enter the Chorus with the noblest among the Ancients, though He had too frequently used his tongue to the Dialect of Immaterial
Qualities,

Qualities, and spiritual Images, in his discourses of the other senses; doth yet make a perfect unison with *Galen*, in this particular, delivering his judgement in most full and definite termes, thus: *Certain it is, that no smell issueth from a body, but with emission of some Corporeal substance; (Sylva sylvar. Cent.9.experim.834.)*

Now the moments of *Authority* being thus equal on both sides, our province is to determine the scales by the præpondium of *Reason,* i.e, with an even hand to examine the weight of the Arguments on which each of these contrary Opinions is grounded

On the other side, we hear the great Genius of Nature, as his Idolaters miscall him, *Aristotle*, and that most numerous of Sects, the *Peripatetick*, vehemently contending, that an Odour belongs to the classis of simple, or *Immaterial Qualities*; and that though it be wasted or transported on the wings of an Exhalation, from the Odorate body to the Sensory: yet is the sensory affected onely with the meer *Image*, or *Intentional species* thereof.

To begin with the *Later*, as the most Epidemical and generally entertained; we find the principal Base of it to be only that common Axiome, *Sensus non percipiunt substantias, sed tantum earum Accidentia,* that no sense is invaded and actuated into sensation by the Real or Material; but onely the intentional species of the Object: which being weak of it self, and by us frequently subverted in our præcedent Discourses; the whole superstructure thereon relying is already ruined, and they who will reædifie it, must lay a new foundation.

But, as to the *Former,* that an *Odour is a perfect substance,* by material impression on the Sensory causing a sensation of it self therein; this seems a Truth standing upon such firm feet of its own, that it contemns the crutches of sophistry. For

(1) No Academick can be so obstinate, as not to acknowledge, that there is a certain Effluvium, or Corporeal Exhalation from all odorous bodies, diffused and transmitted through the aer; as well because his own observation doth ascertain him, that all Aromatiques and other odorous bodies, in tract of a few years, confess a substantial Contabescence, or decay of Quantity; which makes our Druggists and Apothecaries conserve their parcels of Ambre Grise, Musk, Civit, and other rich Perfumes, in bladders, and those immured in Glasses, to prævent the exhaustion of them by spontaneous emanation : as for this, that the odour doth most commonly continue vigorous in the medium, a good while after the remove of the source, or body from which it was effused.    And *Aristotle* himself, after his peremptory Negative, *Odorem non esse* Ἀτορροὰν, *Effluxionem:* could not but let slip this Affirmative, ὅσον ἀπορρέἰ τῶν σωμάτων τότ' ἐλ τῶ ὀσφρεάτον, *quod effluit ex corporibus, ipsa est odorum substantia.*

(2) Common Experience confirms, that odours are vigorous and potent, not only in the production of sundry Affections in the brain, good or evil, according to their vehemency and Gratefulness or Noysomness, by the refocillation or pollution of the spirits; but also in the

Hh 2                                    Vellication

Vellication and frequently the Corrosion of tender investment of the
Nostrills. Thus much the reverend Oracle of *Cous* well observed in
28 *Aphorisme* 5 *Sect.* ; *Odoramentorum suffitus muliebria educit , &
ad alia plerumque utilis esset , nisi gravitatem capitis inferret :* and
*(8 de Compos. Galen* supports with his opinion and arguments , that pleasant Odours
*medic.secund.lo-* are a kinde of Nourishment of the spirits. Besides , *Plutarch* reports,
*ca, cap 4.)* that He observed Catts grow mad onely by the smell of certain odoriferous Unguents : and *Levinus Lemnius (de Natur. miracul )* hath a memorable story of certain Travellers , who passing through large fields of
Beans in the Flower, in Holland , become Phrantick meerly with the
strength of their smell. And all Physicians dayly finde , that good
smels , by a recreation of the languid spirits , speedily restore men
from swooning fits ; as evil scents often induce Vomitings , syncopes,
Vertigoes , and other suddain symptomes. Nay , scarce an Author,
who hath written of the Plague and its Causes , but abounds in relations of those accursed miscreants , who have kindled most mortal infections, by certain Veneficious practices, and Compositions of putrid and
noysom Odours : witness *Petrus Droetus (de pestilentia,cap.10.) Wierus
(de Veneficiis lib. 3.cap.37.) Horatius Augenius (lib.de peste,cap.3 ) Hercules Saxonia (de plica, cap.2.& 11.) Thomas Jordanus (de pestis phæno-
men.tr.1.cap 18.)* and *Sennertus,* out of *Nich. Polius in Hæmerologia Si-
lesiæ, (in lib. de peste, cap.2.)* Which prodigious Effects clearly proclaim the mighty energy of their Causes , and are manifestoes sufficient,
that Odours justly challenge to themselves those Attributes , which are
proper onely to *Corporiety :* nor can ought but downright ignorance expect them from the naked *Immaterial Qualities* , or imaginary *Images* of
the *Peripatetick.*

*Art. 8.*
That the Rea-
*son of an O-
dours affect-
ing the senso-
ry,consists on-
ly in a certain
Symbolisme be-
twixt the Fi-
gures and Con-
texture of its
Particles, and
the Figures
and Conex-
ture of the
Particles of
the Odoratory
Nerves.*

(3) The Manner of the Odours moving, or Affecting the Sensory can
never be explained, but by assuming a certain *Commensuration,* or Correspondency betwixt the Particles amassing the Odour , and the Contexture of the Olfactory Nerves, or Mammillary Processes of the brain
delated through the spongy bone. For (1) it is Canonical, that no
Immaterial can Operate upon a Material, *Physically* ; the inexplicable
activity of the Rational Soul upon the body by the mediation of the
spirits , and that of Angelical essences excepted. (2) Though an O-
dour, diffused through the aer , chance to touch upon the hands, cheeks,
lips, tongue, &c. yet doth it therein produce no sensation of it self, because the Particles of it hold no proportion to either the pores, or particles of which those parts are composed : but arriving at the organ of smelling , it cannot but instantly excite the Faculty therein resident to an actual sensation, or apprehension of it ; in regard of that correspondency in
Figure and Contexture , which the particles of it hold to the pores and
particles of the Odoratory Nerves. Certainly, as the Contexture of the
Odoratory Nerves is altogether different from that of the Tongue ; and so
the minute bodies of them, as well as the small spaces intercepted among
those minute bodies, in all points of their superficies not contingent,
are likewise of a dissimilar configuration from the particles and intercepted *vacuola* of the Tongue : so also is it necessary , that the small bodies,
which commove and affect the Contexture of the Odoratory Nerves , be
altogether dissimilar to those, which commove and affect the contexture of
the Tongue ; since, otherwise all objects would be in common , and the
Distinction of senses unnecessary. Now

Now (left we fhould feem to beg the Quæftion) that the fenfation is
effected in the Odoratory Nerves, only by the *Figures* of the particles of
an Odour ; and that the variety of Odours depends on the variety of im-
preffions made on the fenfory, refpective to their various figures and contextures : this is not obfcurely intimated in thofe formerly recited words
of *Epicurus, Molecularum, five Corpufculorum quædam perturbate ac dif-
crepanter, quædam verò placide ac leniter , feu accommodatè fe habere, ad
olfactus fenforium.* The fubftance whereof is this , that becaufe the particles and Contexture of fome Odours are fuch , that they ftrike the fenfory roughly and difcordantly to the contexture thereof ; therefore are
they *Ingrateful :* and on the contrary , becaufe other Odours have fuch
particles and fuch contextures, as being fmooth in Figure, ftrike the fenfory gently , evenly and concordantly to the contexture thereof ; therefore
are they *Grateful* and defiderable.   We might have introduced *Plato* himfelf, as lighting the tapor to us, in this particular ; infomuch as He faith (*in
Timæo*) that the fweet fort of Odours [χατιπραύνον ϗ ἢ πέφυκεν, ἀγαπη-
τῶς ἀποδιδòν] *de mulcere, & quâ inferitur, amicabiliter fe habere ,* doth
foftly ftroke, and caufe a certain blandifhment in the fenfory : but, that the
kinde of noyfom or ftinking Odours [τεαχύνον τέ ϗ βιαζόμενον] doth in
a manner Exafperate and wound it.   To this Incongruity or Difproportion betwixt offenfive fmells and the compofure of the Odoratory Nerves,
the profound *Fracaftorius* plainly alludeth , in his ; *proportionaliter autem
fe habent & odores , quorum ingratifsimus eft , qui Fætidus appellatur,
quique abominabili in faporibus refpondet ; nam & hic ex iis pariter refultat, quæ nullam habent digeftionem, nec rationem miftionis, fed confufionem
è multis fere ac diverfis , qualia fere funt Putrefcentia , in quibus diffoluta
miftione evaporatio diverforum contingit.* (*de fympath. & antipath. cap.
14.*) importing withal, that the reafon why the ftink of corrupting Car-
caffes is of all other moft noyfom , is becaufe the odours effuming from
them confift of heterogeneous or divers particles.   If you had rather
hear this in Verfe, be pleafed to liften to that Tetraftich of *Lucretius ;*

*Art.9.*
That the Di-
verfity of O-
dours depends
on the Diver-
fity of Impref-
fions made on
the fenfory,
refpondent to
the varicus Fi-
gures and Con-
texture of
their Parti-
cles.

> *Non fimile penetrare, putes, primordia formâ
> In nares hominum, cum tatrâ Cadavera torrent ;
> Et cum Scena Croco Cilici perfufa recens eft,
> Araq; Panchæos exhalat propter odores.*

Upon which we may juftly thus defcant.   As the hand touching a lock
of wool, is pleafed with the foftnefs of it ; but grafping a Nettle, is injured
by that phalanx of villous ftings, wherewith Nature hath guarded the leaves
thereof : fo are the Noftrills invaded with the odour of Saffron , delighted therewith, becaufe the particles of it are fmooth in figure , and of equal
contexture ; but invaded with the odour of a putrid Carcafe , they are
highly offended, becaufe the particles thereof are afper in figure
and of unequal contexture , and fo prick and dilacerate the tender fenfory.

Moreover , whereas there is fo great variety of individual Temperaments among men , and fome have the Contexture of their odoratory
Nerves exceeding diffimilar to that of others ; hence may we well derive
the Caufe of that fo much admired fecret ; *Why thofe Odours, which are*
*not*

*Art.10.*
Why fome
Perfons abhor
thofe fmells,
which are
grateful to
moft others.

*not onely grateful, but even highly cordial to some persons, are most odious and almost poysonous to others.* Infinite are the Examples recorded by Physicians, in this kinde; but none more memorable than that remembred by *Plutarch* (*lib.* 1. *adverf. Coloten.*) of *Berenice* and a certain Spartan woman, who meeting each other inftantly difliked and fainted, becaufe the one fmelt of Butter, the other of a certain fragrant Ointment. However, the rarity of the Accident will not permit us to pafs over the mention of a Lady of honor and eminent prudence, now living in *London*; who doth ufually fwoon at the fmell of a Rofe (the Queen of fweets:) and fometimes feafts her nofe with Affa fætida (the Devils Turd, as fome call it) than which no favour is generally held more abominable; and this out of no Affectation, for her wifdom and modefty exclude that prætence; nor to prevent Fitts of the Mother, for fhe never knew an Hyfterical paffion, but in others, in all her life, as fhe hath frequently protefted to me, who have ferved her as Phyfician many years.

*Art.* 11.
Why, among *Beafts,* fome *fpecies* are offended at thofe fcents in which others highly delight

Again, as this Affumption of the Corporeity of an Odour doth eafily folve the Sympathies and Antipathies obferved among men, to particular fmells; fo likewife doth it yield a plain and fatisfactory reafon, why fome Bruit Animals are pleafed with thofe Odours, which are extremely hateful to others. Why Doggs abhorr the fmell of Wine; and are fo much delighted with the ftink of Carrion, as they are loath to leave it behind them, and therefore tumble on it to perfume their skins therewith? Why a Cat fo much diflikes the fmell of Rue, that fhe will avoid a Moufe that is rubbd with the juice thereof, as *Africanus* (*in Geoponicis*)? Why Mice are poyfoned with the fcent of Rododaphne, or Oleander, commonly named Rofe-bay-tree; as *Apuleius*, and from him *Weckerus* (*de fecretis Animal.*)? Why Serpents are driven from Gardens by the fmell of Citrons as *Galen* affirms; when yet they folace themfelves with that of Savin, which our nofe condemns? Why Cocks cannot endure the breath of Garlick; which is foveraign incenfe to Turkeys, and pure Alchermes to their drooping yong ones? Why Moths are deftroyed by the fume of Hopps; which is Ambre Grife to Bees, as *Mouffet* (*de infectis*)? For the Caufe hereof wholly confifts in the Similitude or Diffimilitude betwixt the particular Contexture of the Senfory, and the Figures of the particles of the odour.

*Art.* 12.
The *Generation* and *Diffufion* of Odours, due only to *Heat.*

The Materiality of an Odour being thus firmly commonftrated; the next Confiderable is the *Generation*, and proxime *Efficient Caufe* thereof. And herein *Ariftotle* came neerer the truth, than in his conception of the Effence of it; for that Affertion of his, *Odorem gigni & moveri beneficio Caloris*, that Heat conduceth both to the Generation and Motion or Diffufion of an Odour, doth well deferve our affent. For, whether thofe minute Maffes, or fmall Concretions, that conftitute the body of an Odour, be contained chiefly in fome fulphurous fubftance, as the Diffolutions and Experiments of Chymiftry feem to conclude; or ambufcadoed in any other confiftence whatever: yet ftill is it manifeft, that they are deduced into act and fequeftred from thofe diffimilar or heterogeneous bodies of Earth and Water that furrund and opprefs them, and fo becoming more at liberty and united, they more vigoroufly affect the fenfe, and all this by the energy of *Heat*. Hence comes it, that all Fruits are fo much more Fragrant, by how much more Concocted and Maturated by the warmth of the Sun.

           That

That all Aromaticks grow in Hot Climats. That all smells are stronger in Summer, than Winter; as *Plutarch* observes (*lib. de Cauf. Natur. cap. 25.*) where he enquires, why in Frost wild beasts leave but a cold scent behind them, when they are hunted. That all odoriferous Druggs are Hot, and suffer a perpetual exhaustion or expence of their halituous substance: so that who so would conserve their Fragrancy, must embalm them in Oyl, or incorporate them with Gumms, or other substance not easily evaporable; according to the common practice of all Perfumers and Confectioners; or immure them in close conservatories, and that rather in great lumps, than small fragments, and in Cold rather than Hot rooms. Hence it is also, that all Botanicks hold it for an unquestionable Axiome, *Omnia Odorata esse calida*; so that some have undertaken to distinguish of the degrees of Heat in Plants and other Simples, meerly by the vehemence or languor of their Odour: and that *Aristotle* (*problem. sect. 12. quæst. 12.*) affirms that all Odorous seeds are Calefactive, because Heat is the Efficient of an Odour; to which *Galen* also subscribes (*4 de simpl. medicament. facul. cap. 22.*)

From the Nature & Efficient of Odours, we are conducted to their *Difference*, or *Distinct species*; which is an Argument involved not in the least Difficulties. For, since the imperfection of our sense of smelling is such, that it is affectable only with the more vehement sort of them, which are but few in comparison to those many, which the sagacity of most Bruit Animals makes familiar to their deprehension, and so we remain ignorant of the greatest part of them; and did we know them, yet should we be to seek for proper Appellatives to express their particular natures: to deliver an exact Table of all their Distinctions, is not only difficult, but impossible. Which Naturalists well understanding, have been forced to the cleanly shift of transferring the distinct names of *sapours* over to the specifical Differences of *Odours*; there being some manifest symbolism betwixt the two senses, and no obscure Analogy betwixt the Conditions of their objects: as *Aristotle* insinuates in his Affirmation, *Nullum corpus esse odoriferum, quod non pariter saporiferum existat (de sens & sensil. cap. 5.)* that all Odoriferous bodies are also saporiferous; and in his definition of an *olfactile*, or odorable object to be, *Quod sapidæ siccitatis diluendæ ac diffundendæ vim sortitur.* Well may we, therefore, content our selves with the Discrimination of those kinds of Odours, that fall under the Cognizance of our sense; and those are *Sweet*, *Sower*, *Austere*, *Acerb*, and *Fatt* or *Luscious*: as for *Putrid* or *Fætid* Odours, they have resemblance to *Bitter Sapours*, because as Bitter things are odious and distastful to the pallate, and no man swallows them without some horror and reluctancy, so likewise doth the Nose never admit rotten and cadaverous smells without loathing and offence. There is also another Difference of smells, whereof *one* kind is either pleasant or unpleasant by Accident, or upon Circumstance; as the smell of Meats and Drinks is pleasant to the Hungry, but offensive to the Full-gordged, and this sort is in common as well to Beasts, as Men: the *other* is pleasant, or unpleasant of their own Nature, as the smells of Herbs, Flowers, Perfumes, &c. which conduce neither to the Excitement, nor Abatement of Appetite, unless they be admixt to meats or drinks; to which *Stratis* alluded, when taxing *Uripides* he said, *Cum lens coquitur, unguenti nil infundito*, and this Difference is proper only to man. Lastly, Authors have divided Odours into *Natural*, and *Artificial*, or *Simple* and *Compound*; the Latter whereof our Luxury and Delicacy have enhanced to such immoderate

rates;

rares, that the Confection of them is become an Arte, and reduced to certain Dispensatories and set Præscripts, and that Lady is not *al-a-mode*, who hath not her Manuscript of Recipes for Perfumes, nay every street hath its *Myropolies* or shops of sweets, of all sorts.

**Art. 14.**
**The Medium of**
**Odours.**

Finally, the *Medium* inservient to Odoration, is either *Aer*, or *Water*: yet neither according to Essence, but *Infection*, or Imprægnation. That the Aer is a convenient Convoy, or Vehicle of an Odour, no man did ever doubt: and that water hath the like Capacity, or perodorable Faculty, though in an inferiour degree; we may, with *Aristotle* (*de histor. Animal.* 4. *cap.* 8.) conclude from the vulgar Experiment of betraying Fishes with perfumed Baites.

CHAP.

# CHAP. VIII.

## OF

# SAPOURS.

### Sect. I.

THE Nature of SAPOURS, the proper object of the Taste, *Aristotle* (*de sens. & sensil. cap. 4.*) concludes to be more easily Cognoscible, than that of Odours, Visibles, or the Objects of the other Senses; because as He præsumes, the sense of Tasting in Man, is more Exquisite, than his Smelling, Sight, &c. Whether his *Reason* be not præcarious, we need not determine: but it too nearly concerns us to affirm, that the extreme slenderness of his doctrine, touching the *Essence* and *Principles*, of Sapours as well in General as Particular; erected on that common imaginary base of Immaterial Qualities, hath given us just occasion to suspect the solidity of his *Inference* or *Conclusion*; and left us cause to account that sentence, much more Canonical, *That things most manifest to the Sense, often prove most obscure to the Understanding.* For, notwithstanding we have the demonstration of our sense, that, as He and all other Philosophers unanimously assert, the Object of the Tasting, in General, is το γευςον, *Gustabile:* yet doe his endeavours afford so dimme a light to our profounder inquisitions, as to leave us in the dark of insatisfaction, when We come to explore, What is the *Formal Reason* of a Sapour; What are the *Principles,* or *Material* and *Efficient* Cause thereof; and What *Relation* it bears unto, or *Manner* how it affects the Tongue, the prime and adæquate instrument of Tasting. Which that we may with due fulness and perspicuity declare, it behoveth us to invite your attention to a faithful Summary of His Speculations concerning that Subject.

*Art. 1.* From the superlative Acuteness of the sense of Tasting, *Aristotle* concludes the cognition of the Nature of Sapours to be more easily acquirable, than the nature of any other sensible object: but refutes himself by the many Errors of his own Theory, concerning the same.

I i                                                                    *Aristotle*

*Ariflotle*, from whose Text all the *Peripateticks* have not receded insomuch as in a title, as to the particular under debate, fixeth the original of a Sapour, in a certain Contemperation of three prime *Elemental Qualities*, viz. (1) *Terreflrious Siccity.* (2) *Aqueous Humidity.* (3) *Heat.* The two former as the *Material* Causes, the laft as the *Efficient*, to which, according to his cuftome, He configns the mafculine and determinative Energy, as in this, fo in all natural productions. The neceffity or the Concurrence of thefe three Firft Qualities to the Generation of a Sapour in any Concretion, He inferrs chiefly from hence; that Water, being in the purity or fimplicity of its effence, absolutely infipid, if percolated through *Siccum terreflre*, a-duft Earth, doth alwayes acquire a Sapidity, or Savourinefs, proportionate to the intenfe, or remifs aduftion of the terreftrious material diffolved by, and incorporated to it felf: as is commonly obfervable in Fountains, which become impregnate or tincted with the fapours of thofe veins of Earth, through whofe Meanders and ftreights they have fteered in their long fubterraneous voyages; and in all Lixivial decoctions, or Lees, which obtain a manifeft Saltnefs only by tranfcolation through Afhes, the Earthy and aduft reliques of compound bodies, diffolved by Fire. To which, He moreover addes, that becaufe the Contemperature may be various, according to greater or leffer proportion of either of the three ingredients; and the Aqueous *Humidum*, united to the Earthy *Siccum*, hath its confiftence fometimes participant of Craffitude, fometime of Tenuity: therefore are not all Sapours alike, but different according to the feveral Gradualities of their refpective and fpecifical Caufes. And thus much in the *General.*

To progrefs to the brief furvey of *Particulars*, it feems requifite that we obferve; that *Galen*, *Avicenna*, *Averrhoes*, and moft *Phyfitians* after them, have conceived this Theory of *Ariflotles* fo firm and impregnable, as they have thereon founded one of their pillars for the invention of Remedies, and advanced rules for the Conjectural inveftigation of the manifeft Faculties of Medicaments, by the Tafte: to that end conftituting *Eight Differences*, or Generical Diftinctions of Sapours, *viz.*

(1) *Acer*, which affects the mouth and chiefly the Tongue, with a certain acrimony and pungent ardor; fuch as is eminently confpicuous in *Pepper*, *Pellitory*, *Euphorbium*, *Caffea lignea*, *Winterian Bark*, &c. It arifeth from a Compofition of tenuious, dry and hot parts, and cannot fubfift in a fubject of any other conftitution.

(2) *Acid*, or *Sharp*, which likewife penetrateth and biteth the tongue, but with fome conftringency, and without any fenfe of heat: fuch as is deprehended in *Vinegre*, juice of *Limons*, *Citrons*, *Woodforrel*, *Berberies*, and in fome *Malacotones* and *Quinces*. It refults from a Concretion of fubtle and dry parts, either where the innate heat is refolved by fome degree of putrefaction, as in Vinegre: or where the innate heat is fo fmall as to be inferior to Cold, and that affociated with extreme ficcity; as in juice of Limons, &c.

(3) *Fat;*

(3) *Fat*, or *Luscious*, which sollicites the Gusts neither with heat, nor acrimony, but furrs and daubs the mouth with an unctuous lentor, or viscidity. Such is remarkable in *Oyle Olive*, *Oyle of sweet Almonds*, *Wallnuts*, in *Marrow*, *Butter*, and the *Fats* of Beasts, which have no rancidity, either acquired by antiquity, or natural, such as is perceivable in the Fat of *Lions*, *Wolves*, and *Tigers*: and in all *Mucilaginous Plants*, as in *Althæa* and *White Lilly* roots, &c. This hath its production from a thin aereal matter, temperate in heat and cold.

(4) *Salt*, which doth not much calefie, but with a sharp and penetring siccity bite the tongue, as is observed in the degustation of Common *Salt*, *Nitre*, and among Vegetables chiefly in *Rock Sampier*. This Sapour is also sensible in all *Chymical Salts*, extracted from Bodies by the sequestrating activity of Fire, cinefying their dry and terrestrious remains: nor is there any Compound in Nature, from which pyrotechny may not extract the Calx or proper Salt thereof, discernable by the taste. And therefore it is manifest, that all saltness subsisteth in a matter, whose principal ingredients, Heat and Siccity are equal.

(5) *Austere*, which being moderately adstringent, doth with some asperity coarctate the particles of the tongue, and therefore according to the judgment of the pallate, it seems dry and cooling. This is more properly called the *Crude* Sapour, as being peculiar to all Fruits during their immaturity, as is generally noted in the juice of unripe *Grapes*, green *Apricocks*, *Pears*, *Apples*, *Medlars*, *Porcellane*, &c. The substance wherein it consisteth, must be equally participant of Earth and Water, but where Cold hath the upper hand of Heat.

(6) *Sweet*, which being not offensive by the uneveness or exuperance of any Quality, affects the sense with suavity or delight. Such every man knows to be in *Sugar*, *Honey*, *Liquorice*, *Jujubes*, *Dates*, *Figgs*, and in most *Fruits* after their maturity: as also in *Manna*, and, in some degree, in *Milk*.

(7) *Bitter*, the Contrary to Sweet, which offending by the asperity and tenuity of its parts, doth in a manner corrade and divell the sensory. This superlatively discovers it self in *Aloes*, *Coloquyntida*, *Rhubarb*, *Wormwood*, the lesser *Centaury*, *Bitter Almonds*, and the *Galls* of Animals. The matter of it is crass and terrene, but adust by immoderate Heat, and hence that Galenical Axiome, *Omne amarum est calidum & siccum*.

(8) *Acerb*, or *Sower*, which bordereth upon the Austere or Pontick Sapour, being distinguishable from it, only by a greater ingratefulness to the sense; for it more constringeth and exasperateth all parts of the mouth, and so seems more exsiccative and refrigerative. It is prodigally perceived in the *rind* of *Pomegranates*, *Galls*, *Sumach*, *Cypress Nuts*, the *Bark* of *Oak*, the *Cups* of *Achorns*, &c. Its residence is alwayes in a Composition totally terrene and drye, whose languid heat is subdued to inactivity by the superior force of its antagonist, Cold.

To these some Modern Physitians ( to whom that Mystagogus or Priest of the Arabian Oracles, *Fernelius*, seems to have been the Coryphæus ) have superadded a ninth Sapour; ἀποϊόν, the *Fatuous*; which affecting the sense with no impression, is indeed no Sapour, but rather the Privation of all Sapidity. To this Heteroclite are commonly referred the several species of *Bread Corn, Gourds, Citrals, Cucumbers,* &c. Whose materials though crass, are not yet terrene, dry and adstrictive; but diluted with a plentiful portion of aqueous moisture, not exquisitely permixt, because of the small allowance of heat to their Composition.

**Art. 4**
An Examination and brief redargution of the same Doctrine

Now (to pass from the faithful Abridgment to the æquitable *Examen* of this Doctrine, of such sacred estimation in the Schools. ) though the Enquiries of most have steered this course, directed by the Chart of *Aristotle*, and attempted the deduction of all Sapours from *Primitive Qualities:* yet have they missed the Cape of truth. For, as *Scaliger* (in lib. de Plantis.) excellently argues, we may as safely derive Life, Sense, Increment, voluntary Motion, nay Risibility and Intellection ( actions flowing from Forms more noble and semi-divine) from Elements immediately, as Sapours from their First Qualities : unless it can be first evinced, that each Element hath some sapour actually inexistent; which but barely to suppose, is an absurdity gross enough to degrade the owner from the dignity of a Physiologist forever, and openly repugnant to the Fundaments of the *Aristotelean* Philosophy. To which argument of *Scaligers*, we shall superadd this weighty exception of our own; that according to the Hypothesis of First Elemental Qualities, it is absolutely impossible to Explicate the Causes of that so great Diversity of Tasts not only among Animals of different species, but Individuals of the same species; of which we shall discourse more expressly in opportunity.

**Art. 5.**
The postposition thereof to the more verisimilous Determination of the sons of *Hermes*, who adscribe all Sapours to Salt.

Wherefore we account it both more honourable and satisfactory, to incline rather to that laudable opinion of the Chymist, whose Flames have so farr enlightned our reason, as to shew, that the *Primary Cause* of Sapours doth consist in *Salt*; because all pyrotechnical Dissolutions seem to establish that Axiome, *Sal est primum Sapidum & Gustabile, & omnia quæ saporem habent, eam propter salem habent; ubicunque enim sapor deprehenditur, ibi sal est, & ubicunque sal, ibi sapor:* as the judicious *Sennertus* hath observed (*de Consensu Chymicorum cum Galenic. cap.* 11.) and *Lucius Grillus* hath copiously and solidly declared in that elaborate treatise of his, *de Sapore Amaro & Dulci*, to which we remit the farther Curious.

**Art. 6.**
But far more to that most profound and satisfactory Tenent of *Democritus* and *Plato*; which deduceth the Nativity of Sapours from the various Figures and contextures of the minute particles of Concretions.

But, if we would Anatomize the Heart of this Subject, and establish a more exact theory of the *First* Principles of a Sapour; we must consult the Oracles of *Democritus* and *Plato*, which tell us in short, that all *Sapours arise from the minute particles of Bodies, of such determinate Figures and Contextures, as being applied to the tongue, they naturally produce that Affection therein, which we call Gustation*, or Tasting. Of *Democritus* auctority, in this point, no man can justly doubt while *Aristotle* (*de sens. & sensil. cap.* 4.) avoucheth that He [ εἰ τὸ χημερὸν ἀνάγκη τὰς χυμὸς ] did referr Sapours to Figures: and *Theophrastus*, in a

more

more ample descant upon the text, affirms that He defined the particular sorts of Figures, which constitute the particular species of Sapours; in these words, *Rotundas esse, congruaque mole figuras, quæ Dulcem faciant; magna figura, quæ Acerbum; multangula minimæque orbiculari, quæ Acrem; angulata distorta, quæ Salsum; rotunda, lævi, distorta, quæ Amarum; tenui, rotunda, parva, quæ Pinguem.* And, what was *Platoes* persuasion, concerning the same Argument, Himself most perspicuously explains (*in Timæo*) where He in short ascribes the production of all Sapours [τραχυτησι τε, ἢ λιοτησιν] to *Asperity* and *Lævity*: and distinguishing all Sapours into two general orders, the First a *Pleasant* or Sweet sort, the other an *Unpleasant*, which runs up into several branches (for as it stands opposed to Sweet, it is either Bitter, or Salt, or Acid, or Acerb, or Acer, or Austere, &c.) He derives the *First* kind from hence, that the sapid object consists of particles so configurate, that effused upon the organ of Tasting, and entering the small pores or receptaries thereof, they become symbolical or correspondent to its small particles in figure and contexture, and so affect it gently, evenly, and concordantly; and the *Latter* from hence, that the sapid object is composed of such Particles, as have their Figures and Contexture so disproportionate and incommensurable to the pores and particles of the tongue, that invading it and entering its contexture, they exasperate, corrade and offend the same. And hence was it, that *Lucretius* seems to have borrowed his,

> *Ut facile agnoscas, è lævibus atque rotundis*
> *Esse ea, quæ sensus jucunde tangere possunt:*
> *At contra, quæ amara, atque aspera cunque videntur,*
> *Hæc magis hamatis inter se cumque teneri;*
> *Proptereaque solere vias rescindere nostris*
> *Sensibus, introituque suo perrumpere corpus.*

And this is the opinion to which we have espoused our constant assent, as well upon the obligation of those Reasons formerly alledged, in our *Original of Qualities*; as upon this important Consideration, that no other Hypothesis can afford a satisfactory Reason either of *manner* of the Sapours moving and affecting the sensory, or why there is such infinite *Variety* of Tasts not only among Animals of different Species, but even in individuals of the same Species, and particularly in men, among whom Millions are found, who delight in Wormwood, and abhorr Sugar; some that feast their Pallates with Aloes, others that think their mouths quite out of taste, unless they be ruminating the leaves of Tobacco; nay, we have known a Noble person of our own Nation, who had so singular a Pallate, that whenever He took a Purging Potion, would swallow it down by spoonfuls, as judging the pleasure too great to be shortned by a hasty draught, and when twas wholly exhausted, would wish himself a Ruminating Animal, that so He might taste it over and over, as if *Philoxenus* wish for a Cranes neck were too short to reach the height of so desireable a delight; and another, who would not be persuaded but the Forbidden Fruit was a Coloquyntida Apple, because he thought the taste of that the most Ambrosiack of all others.

*Art. 7.*
The *advantages* of this sentence, above all others touching the same subject.

But,

But, conceding with *Democritus* and *Plato*, that the Variety of Sapours is caused meerly by the Diversity of Impressions on the spongy
substance of the Tongue, respective to the various Figures and Contextures of the minute Particles of Bodies applied thereto, and by the
salivous moisture thereof so admitted into the pores, as sensibly to affect
it : we say, conceding this, we soon may solve this *Dissimilitude* of
Tastes, only by saying, that because the Contexture of the particles of
the tongue of one man, is different from that of the particles of another;
therefore doth one delight in the savor of one thing, the other of another:
every man being of necessity most pleased with the taste of that, whose
particles in figure and contexture are most symbolical or Correspondent to
the Figures and Contexture of the Particles of his tongue; and *è contra*.
To which we shall only add, that the Reason why to men in Feavers the
sweetest things seem bitter, is only this; that the Contexture of the Particles of the Tongue being altered, as well by the intense Heat of the Feaver,
as the infusion of a Bilious Humour into the pores thereof: those things,
whose Particles being formerly accommodate, appeared in the species of
sweetness, are now become asymbolical and inconvenient to the particles of
the tongue, and therefore appear Bitter.

*Art. 8.*
The *Objections*
of *Arist.* con-
cisely, though
solidly solved.

Nor is *Aristotles* reprehension of *Democritus*, of weight enough to Coun-
ter-encline our judgment; his chief *Objections* being rather Sophistical,
than Solid, and so no sooner urged than dissolved.

His *First* is of this importance; if the particles of Sapid Objects were
Figurate, according to *Democritus* Assumption, then would the sight,
as a Sense far more acute in perception, deprehend their various
Figures rather than the Taste: but the Sight doth not discern them;
*Ergo.*

Which is soon exped, by *Answering*, that it is not in the jurisdiction of one sense to judge of objects proper to another; nor is the quæstion
about the Figures, as they are in themselves, i. e. without relation to the
sense, but as they produce such a determinate Effect on the sensory,
of which the *Tasting* is the sole and proper Criterion. For Qualities are
to be reputed, not so much Absolute and constant Realities, as simple and
*Relative Apparencies*, whose Specification consisteth in a certain Modification of the First General Matter, respective to that distinct Affection they
introduce upon this or that particular Sense, when thereby actually deprehended.

His *Second* of this. Insomuch as there is a Contrariety among sensible
objects of all kinds; but none among Figures, according to that universally embraced Canon, *Figuris nihil esse Contrarium* : if the Diversity of
Sapours were derivative from the Diversity of Figures, then would there
be no Contrariety betwixt Sapours; but Sweet and Bitter are Contraries; *Ergo.*

Which is soon detected to subsist upon a Principle meerly precarious;
for we are yet ignorant of any reason, why we should not account an *Acute*
Figure the Contrary to an *Obtuse*, a *Gibbous* the opposite to a *Plane*, a *Smooth*
the Antagonist to a *Rough*; an *Angular* the *Antitheton* to a *Sphere*, &c.
His

His *Third*, and moſt conſiderable, of this. Becauſe the variety of Figures is infinite, at leaſt, inaſſignable; therefore would the variety of Sapours, if their diſtinct ſpecies were dependent on the diſtinct ſpecies of Figures, be æqually infinite: but all the obſervable Differences of Sapours
exceed not the number of Eight, at moſt; *Ergo.*

*Anſwer*; ſhould we allow *Ariſtotles* diſtinction of Sapours to be genuine: yet would it not follow, that therefore there are no more *Specifical
Subdiviſions* of each Genus; becauſe from the various commiſtions of
thoſe Eight Generical Differences one among another, an incomprehenſible variety of Diſtinct Sapours may be produced. Beſides, is not
that *Sweetneſs*, which the tongue perceives in *Hony*; manifeſtly different
from that of *Milk*? that of *Sugar* eaſily diſcernable from both? that
of *Canary* Sack different from that of *Malago*? that of an *Apple* diſtin-
guiſhable from that of a *Plumm*? that of *Fleſh* clearly diſtinct from all the
reſt? yet doth that Genus of *Sweet* comprehend them all. On the
other ſide, is the *Amaritude* of *Aloes, Coloquyntida, Rhubarb, Wormwood,*
&c. one and the ſame? or the *Acerbity* of *Cherries, Prunes, Medlars,* &c.
identical? no man, certainly, dares affirm it. Why therefore ſhould
we not write our names in the Catalogue of thoſe, who conceive as great
variety of *Taſtes*, as there is of *Sapid objects* in Nature. Or, ſince the
Experiments of Chymiſtry have made it probable, that all Sapours derive
themſelves from *Salts*, as from their *Primary Cauſe*; why may we not concede ſo many ſeveral ſorts of ſalts, and ſo many poſſible Commiſtions of
them, as may ſuffice to the production of an incomprehenſible variety of
Sapours?

And this gives us occaſion to obſerve, that Nature ſeems to have fur-
niſhed the Tonge with a certain *peculiar Moiſture*, chiefly to this end, that
it might have a General *Menſtruum*, or Diſſolvent of its own, for the eduction of thoſe *Salts* from hard and drye bodies, and the imbibition of them
into its ſpongy ſubſtance, that ſo it might deprehend and diſcern them.

*Art. 9.*
That the ſali-
tous *Humidity*
of the Tongue
ſerveth to the
*Diſſolution* and
*Imbibition* of
the *Salt*, in all
Guſtables.

CHAP,

# CHAP. IX.

## *Of Rarity, Density, Perspicuity, Opacity.*

### Sect. I.

**Art. 1.**
This *Chapters* right of *suc-* cession to the former.

Aving thus steered through the deepest Difficulties touching the proper objects of the other Senses, the Chart of Method directs us in our next course to profound the particular natures of all those *Qualities,* which belong to the apprehensive jurisdiction of the Sense of Touch-ing, either immediately, or relatively. But, before we weigh Anchor, that we may avoid the quicksands of too *General Apprehensions,* and draw a Map or Scheme of all the *Heads* of our intended Enquiries; that so we may præpare the mind of our Reader to accompany us the more easily and smoothly : it is requisite that we advertise,

**Art. 2.**
The Divers acceptation of the term, — *Touching.*

(1) That the Attribute of *Touching* is sometimes in *Common* to *all Bodies,* as well Inanimate, as Animate, when their superficies or extremes are Contingent; according to that Antithesis of *Lucretius,* *Tactus Corporibus cunctis, intactus Inani.* Sometimes in Common to *all Senses,* insomuch as all Sensation is a kind of Touching, it being necessary, that either the object it self immediately, or some substantial Emanation from it, be contingent to the Sensory; as we have apodictically declared in our præcedent considerations of Visible, Audible, Odorable, and Gustable Species. Sometimes (and in præsent) *Proper* to the *Sense of Touching in Animals* ; which, however it extend to the Perception of Objects, in number manifold, in nature various and frequently even repugnant (whereupon some Philosophers have contumaciously contended for a Plurality of Animal Touchings; others gone so high as to constitute as many distinct Powers of Touching, as there are [ διαφοραὶ, ἐ ϝναντιώσεις κτ̃ ϑ ἀφὴν ] Differences and Contrarieties of conditions in Tangibles) doth yet apprehend them all under one and the same common reason, and determinate qualification, after the same manner, as the sight discernes White, Black,

Black, Red, Green, &c, all *fub communi Coloris ratione*, in the common capacity of Colours.

*Art. 3.*
A pertinent
(though ſhort)
*Panegyrick* on
the *ſenſe* of
*Touching.*

And this is that fertile ſenſe, to whoſe proper incitement we owe our *Generation*; for, had not the Eternal Providence endowed the Organs official to the recruit of mankind, with a moſt exquiſite and delicate ſenſe of Touching, the titillation whereof tranſports a man beyond the ſeverity of his reaſon, and charmes him to the act of Carnality; doubtleſs, the Deluge had been ſpared; for the Firſt age had been the Laſt, and Humanity been loſt in the grave, as well as innocence in the fall of our firſt Parents. *Quis enim, per Deum immortalem, concubitum, rem adeo fœdam, ſolicitaret, amplexaretur, ei indulgeret ? quo Vultu Divinum illud Animal plenum rationis & conſilii, quem vocamus Hominem, obſcœnas mulierum partes, tot ſordibus conſpurcatus attrectaret, niſi incredibili voluptatis œſtro percita eſſent Genitalia ?* And let us but abate the temptation of this ſenſe, and libidinous invirement of it præambulous to the act of Congreſſion; and we ſhall ſoon confeſs that ſo magnified delight of ſenſuality, to be no other than what the nobleſt of Stoicks, *Marcus Antoninus* defined it, Ἐντερίου παράτριψις, ἢ μετά τίνος ſπαρμμῇ μυξαρίου ἔκκρισις, but the attrition of a baſe entrail, and the excretion of a little ſnivel, with a kind of convulſion, as *Hippocrates* deſcribes it, This is that *Fidus Achates*, or conſtant friend, that conſerves us in our firſt life, which we ſpend in the dark priſon of the womb; uſhers us into this, which our improvidence trifles away for the moſt part on the blandiſhments of ſenſual Appetite; and never forſakes us, till Death hath tranſlated us into an Eternal one. For when all our other unconſtant ſenſes periſh, this faithful one doth not abandon us, but at that moment, which determines our mortality. Whence *Ariſtotle* drew that prognoſtick (*de Anim. lib. 3. cap. 13.*) "that if any Animal be once deprived of the ſenſe of "Touching, death muſt immediately enſue; for neither is it poſſible "(ſaith He) that any living Creature ſhould want this ſenſe, nor to the be- "ing of it is it neceſſary that it have any other ſenſe beſide this. In a word, this is that perſuaſive ſenſe, and whoſe teſtimony the wary *Apoſtle* choſe to part with his infidelity, and to conclude the preſence of his revived Lord. That painful ſenſe, on the victory of whoſe torments the patient ſouls of Martyrs have aſcended above their faith. That Virtual and Medical ſenſe, by which the *Great Phyſician of diſeaſed nature*, was pleaſed to reſtore ſight to the blind, agility to the lame, hearing to the deaf; to extinguiſh the Feaver in *Peters* Mother-in-Law, ſtop the inveterate iſſue in his Hæmorhoidal Client; unlock the adamantine gates of death, and reſtore the widows ſon from the total privation, to the perfect habit of life.

*Art. 4.*
Some Tactile
Qualities, in
common to the
perception of
other ſenſes alſo.

( 2 ) That ſome Qualities are ſenſible to the Touch, which yet are common to the perception of other ſenſes alſo; for no ſcholler can be ignorant of that Diviſion of ſenſibles into *Common* and *Proper*; and that among the *Common* are reckoned *Motion*, *Quiet*, *Number*, *Figure*, and *Magnitude*, according to the liſt of *Ariſtotle* (*2 de Anim. cap.6.*)

*Art. 5.*
A *Scheme* of all Qualities, or Commonly, or Properly appertaining to the Sense of Touching, as they stand in their several Relations to, or Dependencies on the Universal Matter, Atoms; and so, of all the subsequent *Capital Arguments* to be treated of, in this Book.

(3 and principally) That the Qualities of Concretions, either Commonly or Properly appertaining to the sense of Touching, are to be considered in their several *Relations* to the Principles on which they depend. First, some result from the Universal matter, Atomes, in this respect, that they intercept Inanity, or space betwixt them; and of this original are *Rarity* and *Density*, with their Consequents, *Perspicuity* and *Opacity*. Secondly, Some depend on the Common Materials, in this respect, that they are endowed with their three essential Proprieties, Magnitude, Figure, Motion: and that either Singly, or Conjunctly. (1) Singly, and either from their Magnitude alone; of which order is the *Magnitude* or Quantity of any Concretion; and the Consequents thereof, *Subtility* and *Hebetude*: or from their Figure alone, of which sort is the *Figure* of every thing; and the Consequents thereof, *Smoothness* and *Asperity*, &c. or only from their Motive Virtue, of which kind is the *Motive Force* inhærent in all things in the General, and that which assisteth and perfecteth the same in most things, the *Habit* of Motion, and particularly *Gravity* and *Levity*. (2) Conjunctly, from them all; of which production are those commonly called the Four First Qualities, *Heat, Cold, Dryness, Moysture*; as also those which are deduced from them, as *Hardness, Softness, Flexility, Ductility*: and all others of which *Aristotle* so copiously (but scarce pertinently) treateth in his fourth book of Meteors: and lastly, those by vulgar Physiologists named *Occult Qualities*, which are also derivative from Atoms, in respect of their three essential Proprieties; and among these the most eminent and generally celebrated, is the *Attractive Virtue* of the Loadstone.

*Art. 6.*
The right of *Rarity* and *Density*, to the Priority of consideration.

Now on each of these we intend to bestowe particular speculation, allowing it the same order, which it holds in this scheme, which seems to be only a faithful Transsumpt of the method of Nature: and we shall begin at *Rarity* and *Density*. (1) Because nothing can be generated but of Atoms commixt, and that Commixture cannot be without more or less of the Inane space intercepted among their small masses; so that if much of the Inane space be intercepted among them, the Concretion must be Rare, if little, Dense, of meer necessity: (2) Because, the Four First reputed Qualities, Heat, Cold, Dryness, Moysture, are posterior to Rarity and Density, as appears by that of *Aristotle* (*physic. 8. cap. 16.*) where, according to the interpretation of *Pacius*, He intimates, that Heat and Cold, Hardness and Softness are certain kindes of Rarity and Density; and therefore we are to set forth from them, as the more Common in Nature, and consequently the more necessary to be known, *a Generalioribus enim, tanquam notioribus ad minus Generalia procedendum*, is the advice of *Arist.* (*physic. 1. cap. 2.*)

Sect.

## Sect. II.

COncerning the immediate *Causes* of Rarity and Density in Bodies, divers Conceptions are delivered by Philosophers. (1) *Some*, observing that Rare bodies generally are *less*, and Dense more Ponderous, and that the Division of a body into small parts, doth usually make it *less* swift in its descent through aer or water, than while it was intire; have thereupon determined the Reason of *Rarity* to consist *in the actual division of a body into many small parts :* and, on the contrary, that *of Density to consist in the Coadunation or Compaction of many small parts into one great continued mass.* But, These considered not, that Chrystal is not more rare, though less weighty (proportionately) than a Diamond: nor that the Velocity of bodies descending, doth not encrease in proportion to the difference of their several Densities; as their inadvertency made them præsume; there being sundry other Causes, besides the Density of a body, assignable to its greater Velocity of motion in descent, as the Heroical pen of *Galileo* hath clearly demonstrated (*in 1. Dialog. de motu.*) and our selves shall professedly evince in convenient place.

*Art. 1.* The Opinion of those Philosophers, who place the Reason of Rarity, *in the actual Division of a Body into small parts;* and the brief Refutation thereof.

(2) *Others*, indecently leaping from Physical to Metaphysical speculations, and imagining the substance of a body to be a thing really distinct from the Quantity thereof; have derived Rarity and Density *from the several proportions, which Quantity hath to its substance ;* as if in Rarefaction a Body did receive no mutation of Figure, but an Augmentation, and in Condensation a Diminution of its Quantity. But the excessive subtility, or rather absolute incomprehensibility of this Distinction, doth evidently confess it to be meerly Chimerical, as we have formerly intimated, in our discourse concerning the proper and genuine notions of *Corporiety and Inanity.*

*Art. 2.* A second Opinion, deriving Rarity and Density from the several proportions, which Quantity hath to its substance ; convicted of incomprehensibility, and so of unsatisfaction.

(3) A *Third* sort there are, who having detected the incompetency of the first opinion, and absolute unintelligibility of the Second; judiciously desume the more or less of Rarity in any body, from *the more or less of Vacuity intercepted among the parts thereof; and on the contrary, the more or less of Density from the greater or less exclusion of Inanity, by the reduction of the parts of a body to mutual Contingency.* And this is that opinion, which only hath subjugated our judgement, and which seems worthy our best patronage: in regard not only of its sufficiency to explicate all the various Apparences among bodies, resulting from their several Differences in Rarity and Density; but also of its exuperance of reason above the First, and of intelligibility above the second; it being the duety of a Philosopher, always to prefer Perspicuity to Obscurity, plain and genuine notions to such as are abstracted not farther from matter, than all possibility of Comprehension.

*Art. 3.* A Third, desuming the more and less of Rarity in Bodies, from the more and less of VACUITY intercepted among their particles; and the Advantages thereof above all others, concerning the same.

According

*Art. 4.*

*The Definitions of a Rare, and of a Dense Body; according to the assumption of a Vacuity Disseminate.*

According to this Hypothesis, therefore, of Vacuities interspersed (of which *Epicurus* seems to have been the Author) we understand, and dare define a *Rare* Body to be such, *as obtaining little of Matter, possesseth much of Place*; and on the contrary, a *Dense* one to be that, *which obtaining much of Matter, possesseth little of Place :* intending by *Place*, all that space circumscribed by the superfice of the Ambient, such as is the space included betwixt the sides, or in the concave of a vessel.

*Art. 5.*

*The Congruity of those Definitions, demonstrated.*

For, supposing any determinate space to be one while possessed by Aer alone, another while by Water alone; the Aer therein contained cannot be said to be Rare, but only because though it hath much less of matter, or substance, yet it takes up as much of space, or room as the Water : nor the Water to be Dense, but only because though it hath much more of matter, yet doth it take up no more of space, than the Aer. Whence it is purely Consequent, that if we conceive that Water to be rarified into Aer, and that Aer to be condensed into Water; the Aer made of the Water rarified, must replenish a vessel of capacity not only ten-fold, as *Aristotle* inconsiderately conjectured, but a hundred-fold greater, as *Mersennus* by statick experiments hath demonstrated : and transpositively, the Water made by the Aer condensed, must be received in a Vessel of capacity an hundred-fold less; when yet in that greater mass of Aer, there can be no more of Matter, or Quantity, than was in that smaller mass of Water, before its Rarefaction; nor in that smaller mass of Water less of Matter, or Quantity, than was in that greater mass of Aer, before its Condensation. Evident it is, therefore, that by those, contrary motions of *Rarifaction* and *Condensation*, a Body doth suffer no more than the meer *Mutation of its Figure*, or the *Diffusion* and *Contraction* of its parts : its Quantity admitting no Augmentation in the one, nor Diminution of the other.

*Art. 6.*

*That Labyrinth of Difficulties, wherein the thoughts of Physiologists have so long wandered; reduced to a point, the genuine state of the Question.*

This being Apodictical, the sole Difficulty that requires our Enodation, is only this; Whether a Rare Body possessing a greater space, than a Dense, proportionately to its Quantity, doth so possess all that space circumscribed by its superfice, as to replenish all and every the least particle thereof, not leaving any space or spaces, however exile, unreplenisht with some adæquate particle of its matter : Or whether there are not some small parts of space, intermixt among its diffused or mutually incontingent particles, in which no particles of its matter are included, and so there remain small Vacuola, or Empty spaces, such as we have formerly more than twice described, in our Chapter of a Disseminate Vacuity in Nature :

And this descends into another Doubt, whose clear solution is of so much importance, as richly to compensate our most anxious Enquirie; *viz.* Whether Rarity be caused from the interception of much Inanity, when the parts of a Body, formerly Adunate, are separated each from other (at least, in some points of their superficies) and so the Body become so much more Rare, by how much the more, or more ample empty spaces are intercepted among its incontingent particles : or Whether Density and Rarity depend on any other possible Causes besides this, i. e. without the intermistion of inane spaces among the particles of Bodies : And this we conceive to be the whole and nd true state of that Controversie, which hath so perplexed the minds of many the most eminent Philosophers in the world.

That the Rarity and Density of Bodies can arise from no other Cause immediately, but the more or less of Inanity intercepted among their particles; may be thus *Demonstrated*.

If in a Rare body there be admitted no *Vacuola*, or small empty spaces, but it be assumed, that the particles of Matter are adæquate both in Number and Dimensions to the particles of space, wherein it is contained; then must it necessarily follow, that in Condensation many particles of Matter must be reduced into one particle space, which before Condensation was adæquate onely to one particle of Matter: and, on the contrary, in Rarefaction, one and the same particle of matter must possess many of space, each whereof, before Rarefaction, was in dimensions fully respondent thereto. For Example; in Aer condensed into Water, an hundred particles of Aer must be reduced into one particle of space: and in Water rarified into Aer, one particle of the matter of Water must possess an hundred particles of space. Again, according to the Assumption of no Vacuity, since in a Vessel replete with Aer, the parts of Aer must be equal in number and dimensions to the parts of space, thereby circumscribed, none the least particle of space being admitted to be Inane; if you fill the same Vessel with Water, or Lead, or Gold, it must follow, that the parts of the matter of Aer, and the parts of the matter of Water, Lead, or Gold, shall be equal in number, because *Quæ sunt uni tertio æqualia, æqualia sunt etiam inter se:* and if so, needs must Aer be æqually Dense with Water, Lead, or Gold, which all men allow to be the most dense and compact body in Nature in regard it transcends all others in weight and difficulty of Solution, or Division;

(2) All bodies in the Universe must be equally Dense, or equally Rare;

(3) And so nothing can be capable of Condensation or Rarefaction. The least of which unconceaable *Absurdities*, (not to enumerate any others of those many that depend on the same Concession of an absolute Plenitude, or no Vacuity) is great enough to render those Heads, which have laboured to destroy the *Vacuola* of *Epicurus*, strongly suspected of Incogitancy, if not of stupidity.

Twere good manners in us to præsume, that no man can be so Facile, as to conceive, that *Aristotle* hath prevented these *Exceptions*, by that Distinction of his, *de Actu & Potentia* : but, because Præjudice may do much, we judge it expedient a while to insist upon the Examination of the *importance* and *congruity* thereof. He ratiocinates (4 *physic.cap 9.*) *that the matter of Contraries, E. G. of Heat and Cold, Rarity and Density is one and the same, so that as the same matter is one while Actually Hot another while Actually Cold, because it is both Hot and Cold Potentially : so is one and the same matter now Actually Rare, now Actually Dense, because it is both Rare and Dense Potentially.* But, in strictness of Logick, all that this Argument enforceth, is only that the same matter

is

*Art. 7.*
That *Rarity* and *Density* can have no other Causes immediate, but the more and *less of Inanity* intercepted among the particles of Concretions; DEMON-STRATED

*Art. 8.*
*Aristotles* Exceptions, against Disseminate Inaniey; neither important nor competent.

is Capable of Rarefaction and Condenſation; which no man ever diſputed. The Quæſtion is, Whether the ſame Matter, when Actually Rare, hath its parts diſſociated and diffuſed into a greater ſpace, than what they poſſeſſed while it was onely Potentially Rare, and that without the intermixture of Inanity among them? And all that can be collected from his diſcourſes touching that, is no more than this; that *as a matter or ſubſtance actually Hot, doth become more Hot, without the Emerſion, or Acceſſion of any new part, which was not actually Hot before : ſo likewiſe doth the ſame matter actually Extenſe, become more Extenſe, without the Emerſion, or Acceſſion of any new part, which was not actually Extenſe before.* But this Arrow was ſhot at random, not directly to the mark, nor hath it attained the Difficulty; For the Quæſtion again is not, Whether in Rarefaction, any part of the matter were not formerly Extenſe : but, Whether that matter, which was formerly Extenſe, can be made more Extenſe without the Diſſociation of its particles; and whether the particles of it can be actually Diſſociated, without the interception of Inanity among them? Beſides, His *Compariſon* is as incongruous, as his Argument is weak; for (1) His Aſſumption concerning Heat is not only Precarious, but falſe, as ſhall be demonſtrated, *in ſuo loco :* (2) were it true, yet doth that part of matter, which is actually Hot, remain indivulſe or indiſtracted; otherwiſe than a part of matter, which being actually Extenſe, becomes more Extenſe, and therefore the Analogy faileth.

In concluſion, to mend the matter, He recurrs to that ſimilitude of a *Circle, which though contracted into a leſs, hath yet none of its parts more incurvate than they were before :* But, alas the Quæſtion ſtill remains untoucht, and (that we may not ſtay to impeach him of indecorum, in making an indecent tranſition from a Phyſical to a Mathematical ſubject; contrary to his own Dialectical inſtitutes) his ſimilitude will bear no more of inference but only this, that a thing may be made more Denſe, which is Rare and Lax; which is impertinently diſputed, when all men concede it.

**Art. 9.**
The Hypotheſis of a certain Æthereal ſubſtance, to repleniſh the pores of Bodies, in Rarifaction; demonſtrated inſufficient, to ſolve the Difficulty, or demoliſh the Epicurean Theſis of ſmall Vacuities

The *Advocates* of *Ariſtotle* generally alleage in his Defence, that He ſuppoſed a certain *Æthereal,* or as ſome have called it, *Animal* ſubſtance, which inexiſtent in all Bodies, doth repleniſh their pores, and more eſpecially if their Contexture be Rare; and that when a Denſe Bodie is rarified, there are no ſmall Inane ſpaces intercepted among its Diſſociated particles, but that the ſpaces betwixt them are immediately poſſeſſed by that ſubtile Æthereal ſubſtance : and that when a Rare Body is Condenſed, that Æthereal ſubſtance, which did repleniſh its pores, is excluded.

But this *ſuppoſition,* though it come neerer to the Quæſtion, or center of the Difficulty, is yet far ſhort of *ſolving* it. For, take we (for Example) a Cubical foot of Aer, and inſomuch as the ſubſtance of the Aer is more groſs, or leſs exile, than the ſubſtance of the ſuppoſed Æther, therefore muſt it conſiſt of fewer particles, than the Æther : and upon conſequence, in the whole Cubical foot of Aer there are not more particles of Matter, the Aereal and Æthereal ones being conjoyned, than if it conſiſted only of Aereal particles. Now we enquire of *Ariſtotles* Champions

pions, Whether or no in that Cubical foot confifting of the Aggregate of both forts of particles, there are as many particles of Matter, as are in a Cubical foot of Water, Lead, or Gold? The *Affirmative* is more than they dare own; nor can they deny, but that the fpace poffeffed by one foot containeth as many fmall parts of fpace, refpondent to the particles of matter, as the other: and if fo, muft not there be in the Foot of Aer, many particles of fpace, which are poffeffed neither by the Aereal nor Æthereal particles, and are not thofe unpoffeffed particles of fpace abfolutely Empty? If you undertake the *Negative*, you infnare your felf in this Abfurdity, that the particles of a Cubical Foot of Aer and Æther conjoyned, muft be equal in number to the particles of a Cubical foot of Water, Lead, or Gold.

The Difficulty of underftanding the Formal and Immediate Reafon of Rarity and Denfity in Bodies, by that fo popularly applauded Hypothefis of an *Æthereal fubftance* (imagined to maintain an abfolute Plenitude, and fo a Continuity through the whole vaft Body of Nature) being thus evinced; let us a while confider, how eafily even the meaneft Capacity may comprehend the full Nature of thofe Primary and Eminent Affections, from the conceffion of *fmall Vacuities*. We have formerly explicated the matter, by the convenient fimilitude of an Heap of Corn, or Sand; which being lightly and gently poured into a Veffel, takes up more room then when preft down: and we fhall yet more facilitate the Conception thereof by another fimile, fomewhat more pregnant, becaufe more Analogous. When a *Fleece*, or *Lock of Wool* is deduced, or diftended, we fay, it is made more Rare; and when Compreffed, more Denfe: now the Rarity thereof confifteth only in this, that the Hairs, which were formerly more Confociate, United, or at clofer Order among themfelves, are Diffociated, Dif-united, or reduced to more open Order, and the fpaces betwixt them, become either more, or larger, in which no particle of Wool is contained: and on the contrary, the Denfity thereof confifteth onely in this, that the Particles or Hairs, which were before more Diffociated, or at open order, are by Compreffion brought to more Vicinity, or to clofer order, and the fpaces betwixt them become fewer and leffer. And thus are we to conceive, how the fame Matter, without Augmentation or Diminution of Quantity, may be now Rarified into Aer, and anon Condenfed into Water: for, inftead of the Hairs in the Fleece of Wool, we need only put the Particles of the matter, which in Rarifaction are Diffociated, in Condenfation Coadunated. And this Conception may be extended alfo to a Spunge, Flaxe, or any other Porous and Lax bodie; becaufe they are capable of Expanfion and Contraction onely in this refpect, that the fmall fpaces intercepted in the incontiguities or diftances of their particles, are now enlarged, now contracted. We confefs, this fimilitude is not adæquate in all points, there being this *Difference*, that when a Fleece of Wool is expanfed, the ambient Aer doth inftantly infinuate into the fmall fpaces intercepted betwixt the diffociated particles of it, and fo poffefs them; but, nothing of Aer, or Æther, or other fubftance whatever doth infinuate it felf into the fmall fpaces intercepted betwixt the diffociated particles of Aer, or Water, when either of them is Rarified: we fay, notwithftanding this Difparity,

*Art. 10.*
The *Facility* of underftanding the Reafons and Manner of Rarifaction and Condenfation, from the Conceffion of fmall Vacuities; illuftrated by a congruent Similitude.

Diſparity, yet doth it hold thus far good and quadrant, that as nothing of Wool poſſeſſeth thoſe ſpaces, which would therefore remain abſolutely Empty, in caſe the ſociable Aer did not inſtantly ſucceed in poſſeſſion of them; ſo, ſince the parts of the matter of Water are Expanſed or Diſſociated after the ſame manner, as are the Hairs of Wool, and after the ſame manner Contracted or United; and certain ſmall Loculaments are likewiſe intercepted betwixt the particles of that matter, in which nothing of Water can be contained, during the ſtate of Rarifaction, and which no other ſubſtance can be proved to poſſeſs; it muſt thence follow, that thoſe deſerted ſmall ſpaces, or Loculaments remain abſolutely Empty. And more than that, our ſimilitude is not concerned to impart.

**Art. 11.**
**PARADOX.**
That the Matter of a Body, when Rarified, doth poſſeſs no more of true Place, than when Condenſed, and the Conciliation thereof to the præpoſed Definitions of a Rare and of a Denſe Body.

But, that we may make ſome farther advantage thereof, we obſerve, that as when a Fleece of Wooll is expanſed, it is of a greater circumference, and ſo includes a greater Capacity therein, than when it is compreſſed; not that the ſingle Hairs thereof take up a greater ſpace in that capacity, for no Haire can poſſeſs more ſpace, than its proper bulk requires, but becauſe the inane ſpaces or Loculaments intercepted betwixt their diviſions are enlarged : exactly ſo, when the ſame Matter is now Rarified into Aer, anon Condenſed into Water, the Circumference thereof becomes greater and leſs, and the Capacity included in that circumference is augmented and diminiſhed accordingly; not that the ſingle Particles of the Matter poſſeſs a greater part of that capacity in the ſtate of Rarifaction, than in that of Condenſation, becauſe no particle can poſſeſs more of ſpace than what is adæquate to its dimenſions; but only becauſe the Inane ſpaces intercepted betwixt their diviſions are more ample in one caſe, than in the other. And hence it is purely conſequent, *that the matter of a Body Rarified can not be juſtly affirmed to poſſeſs more of true or proper Place, than the matter of the ſame body Condenſed*; though, when we ſpeak according to the cuſtomary Dialect of the Vulgar, we ſay, that a Body Rarified doth poſſeſs more of ſpace, than when Condenſed : inſomuch as under the terme Place is comprehended all that Capacity circumſcribed by the extremes or ſuperfice of a Body; and to the Matter, or Body it ſelf are attributed not onely the ſmall ſpaces poſſeſſed by the particles thereof, but alſo all thoſe inane ſpaces interjacent among them, juſt as by the word *City*, every man underſtands not only the dwelling Houſes, Churches, Caſtles, and other ædifices, but alſo all the ſtreets, Piazzaes, Church-yards, Gardens, and other void places contained within the Walls of it. And in this ſenſe onely are our præcedent *Definitions* of a *Rare*, and *Denſe* Body to be accepted.

**Art. 12.**
**PROBLEM.**
Whether Aer be capable of Condenſation to ſo high a rate as it is of Rarifaction : and the Apodictical ſolution thereof.

The Reaſons of Rarity and Denſity thus evidently Commonſtrated, the pleaſantneſs of Contemplation would invite us to advance to the examination of *the ſeveral Proportions of Gravity and Levity among Bodies, reſpective to their particular Differences in Denſity and Rarity; the ſeveral ways of Rarifying and Condenſing Aer and Water; and the means of attaining the certain weights of each, in the ſeveral rates, or degrees of their Rarifaction and Condenſation;*

*sation; according to the evidence of Aerostatick and Hydrostatick Experiments:* but in regard these things are not directly pertinent to our present scope and institution, and that *Galilæus* and *Mersennus* have enriched the World with excellent Disquisitions upon each of those sublime Theorems; we conceive ourselves more excusable for the Omission, than we should have been for the Consideration of them, in this place. However, we ask leave to make a short Excursion upon that PROBLEM, of so great importance to those, who exercise their Ingenuity in either *Hydraulick*, or *Pneumatick* Mechanicks: *viz.*

> *Whether may Aer be Rarified as much as Condensed; or whether it be capable of Rarifaction and Condensation to the same rate, or in the same proportion?*

That common Oracle, for the Solution of Problems of this abstruse nature, Experience hath assured, that Aer, may be Rarified to so great a height, in red-hot Æolipiles, or Hermetical Bellows, that the 70 part of Aer formerly contained therein, before rarifaction, will totally fill an Æolipile upon extreme Rarifaction thereof. For, *Mersennus*, using an Æolipile, which being Cold, would receive exactly 13 ounces, one Drachm and an half; and when Hot, would suck in only 13 ounces: found, that the whole quantity of Aer ignified, and replenishing the same Æolipile, when glowing Hot, being reduced to its natural state, did possess only the 70. part of the whole Capacity, which was due to the Drachm and half of Water. We say, upon *Extreme* Rarifaction; because this seems to be the highest rate, to which any Rarifaction can attain, in regard the Metal of the Æolipile can endure no more violence of the Fire, without Fusion.

As for the Tax, or Rate of its utmost *Condensation*; though many are persuaded, that Aer cannot be reduced, by Condensation, to more than a Third part of that Space, which it possesseth in its natural state; because they have observed, that Water infused into a Vessel of three Heminæ, doth not exceed two Heminæ, in regard of the Aer remaining within: yet certain it is, that Aer may be Condensed to a far higher proportion. For, Experience also confirms, that into the Chamber of a Wind-Gun (of usual Dimensions) Aer may be intruded, to the weight of a Drachm, or sixty Grains: and that in that Capacity, which contains only an ounce of Water, it may be so included, as that yet a greater proportion of Aer may be injected into it.    Now, therefore, insomuch as the Aer in *Mersennus* his Æolipile amounts to four Grains (at least) or sixe (at most) which number is ten times multiplied in sixty; and that the Concave of the Æolipile is to the Concave of the Pipe of the Wind-Gun, in proportion sesquialteral: by Computation it appears, that the Aer condensed in the Chamber of the Wind-Gun must be sufficient to fill the Æolipile ten times over, or the same Chamber 15 times over, if restored to its natural tenour. And hereupon we may safely Conclude, that Aer may be Compressed in a Wind-Gun, to such a rate, as to be contained in a space 15 times less, than what it possessed during its natural Laxity; and that by the force only of a Mans hand, ramming down the Embolus, or

Charging Iron : which Force being capable of Quadruplication, the Aër may be reduced into a space subquadruple to the former. If so, the rate of the possible Condensation of Aer, will not come much short of that of its extreme Rarefaction : at least, if a Quadruple Force be sufficient to a Quadruple Condensation ; and Aer be capable of a Quadruple Compression : both which are Difficulties not easily determinable.

## Sect. III.

Art. 1.
*The opportunity of the present speculation, concerning the Causes of Perspicuity and Opacity.*

PERSPICUITY and OPACITY we well know to be Qualities not præcisely conformable to the Laws of Rarity and Density ; yet, insomuch as it is for the most part found true (*cæteris paribus* ) that every Concretion is so much more Perspicuous, by how much the more Rare ; and *è contra,* so much the more Opace, by how much more Dense; and that the Reason of Perspicuity can hardly be understood, but by assuming certain small *Vacuities* in the Body interposed betwixt the object and the eye, such as may give free passage to the visible Species ; nor that of Opacity, but by conceding a certain *Corpulency* to the space or thing therein interposed, such as may terminate the sight : therefore cannot this place be judged incompetent, to the Consideration of their severall originals.

Art. 2.
*The true Notions of a Perspicuum and Opacum.*

By a *Perspicuum* [ τὸ διαφανὲς ] we suppose, that every man understands that Body , or Space, which though interposed betwixt the Eye and a Lucid, or Colorate Object, doth neverthelefs not hinder the Transition of the Visible species from it to the Eye : and by an *Opacum*; that which obstructing the passage of the Visible Species, terminates the sight in it self.

Art. 3.
*That every Concretion is so much the more Diaphanous, by how much the more & more ample Inane Spaces are intercepted among its particles ; cæteris paribus.*

We suppose also, that (according to our præcedent Theory) the Species Visible consist of certain *Corporeal Rayes* emitted from the Object , in direct lines toward the Eye ; and that where the Medium , or interjacent space is *free*, those Rayes are delated through it without impediment; but, where the space is præpossessed by any solid or Impervious substance, they are repercussed from it toward their Original, the Object. And hence we inferr , that because the total Freedom of their Transmission depends only upon the total Inanity of the Space intermediate ; and so the more or less of freedome trajective depends upon the more or less of Inanity in the Space intermediate : therefore must every Concretion be so much more Perspicuous, by how much the more , and more ample Inane Spaces it hath intercepted among its Component particles ; which permit the Rayes freely to continue on their progress home to the Eye.

This we affirm not *Universally* , but under the due limitation of a *Cæteris Paribus* , as we have formerly hinted. Because, notwithstanding a piece of Lawn is more or less Perspicuous , according as the Contexture of its Threads is more or less Rare ; and the Aer in like manner is more or less pellucid , according as it is perfused with more or fewer Vapours : yet do we not want Bodies , as *Paper, Sponges*, &c. Which though more then meanly Rare, are nevertheless Indiaphanous ; and on the contrary,

contrary, we see many Bodies, sufficiently Dense, as *Horn*, *Muscovy-glass*, *common glass*, &c. which are yet considerably Diaphanous.

Now, that you may clearly comprehend the *Cause* of this *Difference*, be pleased to hold your right hand before your eye, with your fingers somewhat distant each from other; and then looking at some object, you may behold it through the chinks or intervals of your fingers: this done, put your left hand also over your right, so as the fingers of it may be in the same position with the former; and then may you perceive the object, at least as many parts of it as before. But, if you dispose the fingers of your left hand so as to fill up the spaces or intervals betwixt those of your right; the object shall be wholly eclipsed. Thus also, if you look at an object through a Lawn, or Hair Sieve, and then put another Sieve over that, so as the holes or pores of the second be parallel to those of the first; you may as plainly discern it through both as one: but, if the tivils of the second sieve be objected to the pores of the first, then shall you perceive no part of the object, at least so much the fewer parts, by how much greater a number of pores in the first are confronted by threads in the second. And hence you cannot but acknowledge that the Liberty of inspection doth depend immediately and necessarily upon the Inanity of the pores; the Impediment of it upon the Bodies that hinder the trajection of the Rayes emitted from the Object: and yet that to *Diaphanity* is required *a certain orderly and alternate Position of the Pores and Bodies, or Particles.* This considered, it is manifest, that the Reason *why Glass, though much more Dense, is yet much more perspicuous than Paper,* is only this; that the Contexture of the small filaments, composing the substance of Paper, is so *confused,* as that the Pores that are open on one side or superfice thereof, are not continued through to the other, but variously intercepted with cross-running filaments: as is more sensible in the Cohtexture of a *Spunge,* whose holes are not continued quite thorow, but determined at half way, (some more, some less) so that frequently the bottome of one hole is the cover of another, as the Cells in a Hony-comb : but, Glass, in regard of the *uniform and regular Contexture* of its particles, which are ranged as it were in distinct ranks and files, with pores or intervals orderly and directly remaining betwixt them; hath its pores not so soon determined by particles oppositely disposed, but continued to a greater depth in its substance.

*Art. 4.* Why *Glass,* though much more Dense, is yet much more Diaphanous than Paper.

Though this make the whole matter sufficiently intelligible, yet may it receive a degree more of illustration, if we admit the same Conditions to be in the substance of Glass, that are in a *Mist,* or *Cloud*; through which we may behold and object, so long as the small passages or intervals betwixt the particles of the Vapours, through which the rayes of the visible species may be trajected, remain unobstructed: but yet perceive the same so much the more obscurely, by how much the more remote it is; because, in that case, more impervious particles are variously opposed to those small thorow-fares, that obstruct them, and so impede the progress of most of the rayes. For, thus also Glass, if thin, doth hinder the sight of an object very little, or nothing at all; but if very thick, it wholly terminates the progress of the species: and, by how much the thicker it is, by so much the more it obscures the object. And this, only because Glass, consisting of small solid Particles, or Granules, and insensible Pores alternately situate, hath many of its pores running on in direct lines through its sub-

*Art. 5.* Why the Diaphanity of Glass is gradually diminished, according to the various degrees of its Crassitude.

stance to some certain diftance, but fometimes thefe, fometimes thofe are obturated by fmall folid particles fuccedent, when at fuch a determinate *Crafsitude*, it becomes wholly opace.

*Art. 6.*
An Apoditti-
cal Confurati-
on of that po-
pular Error,
that *Glafs is to-
tally, or in every
particle, Dia-
phanous.*

And this gives us an opportunity to refute that vulgar *Error, That the fubftance of Glafs is totally Diaphanous , or that all and every Ray of the the Vifive Species is trajected through it, without impediment.* To demon-ftrate the contrary, therefore, we advife you to hold a piece of the fineft and thinneft Venice Glafs againft the Sun, with two fheets of white paper, one betwixt the Sun and the Glafs, the other betwixt the Glafs and your Eye : for, then fhall all the *Trajected* Rayes be received on the paper on this fide of the glafs, and the *Reflected* ones be received on that beyond it. Now, infomuch as that paper, which is betwixt your eye and the glafs, doth receive the *Trajected* rayes, with a certain apparence of many fmall *fhadows* intercepted among them ; and that paper beyond the glafs, doth receive the *Reflected* rayes with an apparence of many fmall *lights* : therefore we de-mand (1) from whence can that fpecies of fmall fhadows arife, if not from the Defect of thofe rayes, that are not tranfmitted through the Glafs , but averted from it ? (2) Whence comes it, that in neither paper the Bright-nefs or Splendour is fo great, as when no Glafs is interpofed betwixt them; if not from hence, that the reflected rayes are wanting to the neareft, the tra-jected ones to the fartheft? (3) Whence comes it that fome rayes are refle-cted, others trajected; if not from hence, that as a Lawn fieve tranfmits thofe rayes , which fall into its pores, and repercuffeth others that fall upon its threads: fo doth Glafs permit thofe rayes to pafs through, that fall into its pores ; and reverberate thofe, that ftrike upon its folid particles? And what we here fay of Glafs, holds true alfo (in proportion) of *Aer, Water, Horn, Vernifh, Mufcovy-glafs*, and all other *Diaphanous* Bodies.

CHAP.

# CHAP. X.

## OF

# MAGNITUDE, FIGURE:

### And their Consequents,

### SUBTILITY, HEBETUDE, SMOOTHNESSE,

### ASPERITY.

---

### Sect. I.

The MAGNITUDE and FIGURE of Concretions, in regard our Reason doth best derive them from the Two First Proprieties, or Essential Attributes of the Universal Matter, Atoms; are the Qualities which justly challenge our next Meditation. Concerning their Origination, therefore, we advertise

*Art. 1.* The Context-ture of this Chapter, with the præcedent

First, that although it be not necessary, that a Body made up of greater Atoms should therefore be greater, nor contrariwise, that a Body composed of lesser Atoms, should therefore be lesser; nor that a Body consisting of Atoms of this, or that determinate Figure, should constantly retain that Figure, without capacity of determination to any other: yet doth it seem universally true, that every Concretion therefore hath Magnitude, because its Material Principles, or Component Particles have their certain Magnitudes, or are essentially endowed with real Dimensions; and as true, that every Concretion is therefore determined to this or that particular Figure, because the Component Particles thereof are not immense, or devoyd of circumscription, but terminated by some Figure or other.

*Art. 2.* That the Mag-nitude of Con-cretions, ariseth from the Mag-nitude of their Material Prin-ciples.

Secondly, that the term *Magnitude* here used, is not to be accepted in a Comparative intention, or as it stands in opposition to Parvity; in which sense

*Art. 3.* The præsent intention of the term, *Magnitude.*

sense vulgar ears alwayes admit it : but a *Positive*, or as it is identical and importing the same thing with Quantity, or Extension. For, as every Atom, or that ultimate and indivisible portion of Matter, so called, is no Mathematical point, but possesseth its own simple Magnitude, or Quantity, without respect or comparison to Greater or Less. So must every Concretion be considered, as it stands possessed of its own compound Magnitude, or Quantity, without respect to any other Body, in comparison whereof it may be said to be Greater or Less. Because without the relative conception of any other Body, the Mind doth most clearly and distinctly apprehend the Magnitude of a Concretion by a Positive notion; insomuch it conceives it to have various parts, whereof none are included within other, but all situate in order, and each in its proper place : so that from thence must follow the Diffusion of them, and consequently the Extension of the whole consisting of them. And well known it is, that the Magnitude, or Quantity of a Body, is nothing but that kind of Extension, which amounts from the aggregate of the singular Extensions of its component particles: of which if any be conceived to be Detracted, or Apposed; so much is instantly understood to be Detracted from, or Apposed to the Extension of the whole Body. To this alludes that Distich of *Lucretius*,

> *Propterea, quia quæ decedunt Corpora quoique,*
> *Unde abeunt, minnunt ; quo venere, anguine donant.*

**Art. 4.**
*That the Quantity of a thing, is meerly the Matter of it.*

This duely perpended, no man need hereafter fear the drilling of his ears by those clamorous and confused litigations in the Schools, about the Formal reason of Quantity; for nothing can be more evident than this, that the Extension or Quantity of a thing is meerly *Modus Materiæ*, or ( rather ) the *Matter* it self composing that thing; insomuch as it consisteth not in a Point, but hath parts posited without parts, in respect whereof it is Diffuse: and purely consequent from thence, that every Body hath so much of Extension, as it hath of Matter, extension being the proper and inseparable Affection of Matter or Substance. Hence also may we detect and refute the extreme absurdity of those high-flying Wits, who imagine that a Body, when Rarified, though it hath no more of Matter, hath yet more of Quantity or Extension, than when Condensed: because from the præmises it is an apodictical verity, that the Extension attributed to a Body Rarified, is not an Extension of the Matter of it alone, but of the Matter and small Inane Spaces, intercepted among its dissociated particles, together; so that if you suppose the Extension of those small Vacuities to be excluded from the Aggregate, you cannot but confess, that the Matter hath no more of Extension in its parts Dissociated, than it had in the same parts Coadunated.

**Art. 5.**
*The Quantity of a thing neither augmented by its Rarefaction; nor diminished by its Condensation: contrary to the Aristotelians, who distinguish the Quantity of a Body, fr m its Substance.*

Moreover, this sufficiently instructs us to give a decisive Response to that so long debated Quæstion, *An per Rarifactionem acquiratur, per Condensationem deperdatur Quantitas ?* Whether the Quantity of a Body is Augmented in Rarifaction, and Diminished in Condensation, or no? For, as nothing of Matter is conceived to be added to a body, while it is Rarified; nothing of Matter detracted from it while Condensed: so is it uncimiable, at least unrefutable, that nothing of Quantity is acquired by Rarifaction

or

or amitted by Condenſation ; but only that thoſe empty ſpaces are admitted, or excluded, which being in a Rarified body conjoined to the ſmall ſpaces, that the particles of its matter poſſeſs, make it appear to be Greater, or to repleniſh a greater place, than before ; and in a Condenſed body, detracted from the ſmall ſpaces, that the particles of its matter do poſſeſs, make it appear Leſs, or to fill a leſs place than before.  If ſo, it may be cauſe of wonder even to the wiſeſt and moſt charitable Conſideration, that the Defendants of *Ariſtotles* doctrine of Quantity, have with ſo much labour and anxiety of mind betrayed themſelves into ſundry not only inextricable Difficulties, but open Repugnances; while on the one ſide they affirm, that as well Quantity as Matter, is Ingenerable and Incorruptible: and on the other admit, that the ſame Matter may be one while Extended to the occupation of all and every part of a greater ſpace ; and another while again ſo contracted, as to be wholly comprehended in the hundreth part of the former ſpace (as in the Condenſation of Aer into Water) than which no Contradiction can be or more open, or more irreconcileable.  And yet we ſee thoſe, who have eaſily ſwallowed it, and upon digeſtion become ſo tranſcendently exalted to ſublimities, as to imagine the Quantity of a thing to be abſolutely diſtinct from the matter, or ſubſtance of it : and thereupon to conclude, that Rarity and Denſity doe conſiſt only in the ſeveral proportions, which ſubſtance hath to Quantity.

Much more plauſible were their Explication, had they derived the Extenſion of a thing, meerly from *ſpace,* or *Place* ; becauſe, whenever any thing is ſaid to be Extenſe, the mind inſtantly layes hold of ſome determinate part of ſpace, referring the Extenſion of it ſimply and entirely to the Place, wherein it is, or may be contained, and which is exæquate to its Dimenſions : nor is it, indeed, eaſie to wean the Underſtanding from this habitual manner of Conception.  Whereof if we be urged to render a ſatisfactory Reaſon, we confeſs, we know no better than this ; that by the Law of Nature, every Body in the Univerſe is conſigned to its peculiar Place, i. e. ſuch a canton of ſpace, as is exactly reſpondent to its Dimenſions : ſo that whether a Body quieſce, or be moved, we alwayes underſtand the Place wherein it is Extenſe, to be one and the ſame, i. e. equal to its Dimenſions.

*Art. 6.*
The reaſon of Quantity, explicable alſo meerly from the notion of *Place.*

We ſay, *By the Law of Nature* ; becauſe, if we convert to the Omnipotence of its Author, and conſider that the Creator did not circumſcribe his own Energy by thoſe fundamental Conſtitutions, which his Wiſedom impoſed upon the Creature : we muſt wind up the nerves of our Mind to a higher key of Conception, and let our Reaſon learn of our Faith to admit the poſſibility of a Body exiſtent without Extenſion, and the Extenſion of a Body conſiſtent without the Body it ſelf ; as in the ſacred myſtery of our Saviours Apparition to his Apoſtles, after his Reſurrection [τῷ Θυρῶν κεκλεισμένων] the dores being ſhut.  Not that we can comprehend the *manner* of either, i. e. the Exiſtence of a Body without Extenſion, and of Extenſion without a Body ; for our narrow intellectuals, which cannot take the altitude of the ſmalleſt effect in Nature, muſt be confeſt an incompetent meaſure of ſupernaturals : but that, whoever allowes the power of God to have formed a Body out of no præexiſtent matter, cannot deny the ſame power to extend to the reduction of the ſame Body to nothing of matter again.  Which the moſt pious S. *Auguſt.* (*Epiſt.* 3. ) had

*Art. 7.*
The Exiſtence of a Body, without real Extenſion ; & of Extenſion without a Body : though impoſſible to *Nature,* yet eaſie to God.

had regard unto, in his excellent Adhortation, *Ut demus Deum aliquid posse, quod nos fateamur investigare non posse, & in quo tota ratio facti sit ipsa potentia facientis.*

**Art. 8.**
COROLLA
BY.
That the primary Cause, why Nature admits no Penetration of Dimensions, is rather the Solidity, than the Extension of a Body.

And here we have opportunity to observe, by way of Corollary, that insomuch as every Philosopher considers in a Body as well its *Solidity,* or Corpulency, as *Extension,* or Quantity ( though not as things really distinct, yet the same under a twofold acceptation ) we say, we therefore observe, that the primary Cause, why Nature cannot endure a *Penetration of Dimensions,* or that two Bodies cannot be admitted into the same place at once, seems not to be the Extension or Quantity of it, præcisely accepted, as the Disciples of *Aristotle* commonly conceive; but its Corpulency, or Solidity. For from this, rather than that, may be understood that *Opposition,* which is betwixt a *Vacuum* and a *Body*: and the *Renitency* which is in one Body against the admittance of another into it self.

**Art. 9**
The reasons of Quantity Continued and Discrete, or Magnitude and Multitude.

Concerning the *Continuity* of a Body we also observe; that a Body is to be reputed Continued, in respect its Parts are Copulate, Cohærent and Indivulse among themselves, so that notwithstanding they are in reality no more then mutually Contiguous, yet are their Commissures or conjoinings so exile, as not to be deprehensible by the sense. And thus may we understand Magnitude, or *Quantity Continued* (as the Schools phrase it ) to be distinguished from *Quantity Discrete,* or *Multitude,* only by this; that the parts of Magnitude may, indeed, be separated each from other, but are not actually separate : but the parts of Multitude are actually separate. Not that the parts of Multitude may not be mutually contingent (as many stones lying in one heap together) but that they do not reciprocally take hold of, or bind in each other, so as to make a sensible Continuity; and yet it is manifest, that an heap of Hairs dextrously twisted into small threads and woven into a close webb, makes a Continued body, though the Hairs do not penetrate each other, but are meerly contingent in their extremes. Thus mudd is likewise a Continued Body, though it be only a composition of Granules of Earth and Water reciprocally contingent; as appears by the separation of them, upon the easie evaporation of the watery particles by fire. Thus also, in a word, all Bodies, which are dissoluble by fire, or otherwise, have their parts only mutually contingent; all that the Dissolvent effects upon them, being only to divorce them from reciprocal Contact, and so destroy their apparent Continuity.

**Art. 10.**
That no Body is perfectly Continued, beside an Atome.

This considered, if any man enquire, what Body is so perfectly Continued, as not to consist of only Contingent, and consequently of separable particles; it is evident, that the whole vast stock, or Magazine of Nature can afford him none such, an *Atome* only excepted : and therefore of an Atome alone are we to understand that ænigmatical sentence of *Democritus,* recorded by *Aristotle* ( 7. *Metaphys.* 13. ) *Neque ex uno duo, neque ex duobus unum fieri posse*; because the *Insectility* of an Atome makes the emergency of two out of one, clearly impossible, and its *Solidity* interdicts the mutual penetration of two, necessary to their perfect Coalition into one. So that it is absolutely necessary, that all Atoms remain single and inconfused :
and

and yet this hinders not, but a Body, which is not actually divided into parts, may be said to be Continued; insomuch as it so appears to the sense, which cannot discern the several Commissures of its particles.

Again, forasmuch as *Aristotle* defines a *Continuum* to be that, *whose Parts are conjoyned by some Common mean, or Term*; it is requisite we observe how far forth his definition is consistent with right reason. We allow it to be true *Physically* so far forth, as there are no two parts assignable, which are conjoyned by some third intermediate part, either *sensible* (as in a magnitude of three feet, the two extreme feet are copulated together by the third intermediate) or *Insensible* (as in the magnitude of two feet, which are joyned together by some interjacent particle, so small as to evade the detection of sense) : But, if with Him we accept that Common Mean, or Terme, for a *Mathematical Point*, or individual (for He expresly affirms, that the parts of a Line are copulated by a Point, the parts of a Superfice, by a Line; the parts of a Body, by a Line, or Superfice ) 'tis plain, that our Conceptions must be inconsistent with Physical verity; because such *Insectiles*, or Individuals are not real, but only Imaginary, as we have copiously asserted in our Discourse concerning the *Impossible Division of a Continuum into parts infinitely subdivisible*. Besides, who can conceive that to be a *Cæment* or *Glew* to unite two parts into one Continued substance, which hath it self no parts designable either by sense or reason? Nor can any thing be rightly admitted to conjoyn two Bodies, unless it hath two sides, Extremes, or faces; one whereof may adhære to one of the two Bodies, the other to the other, so as to make a sensible Continuity.

*Art.* 11.
*Aristotle's* Definition of a *Continuum*, in what respect true, and what false.

Concerning the Quality of a Body called FIGURE, that which is chiefly worthy our præsent adversion, is onely this; that if Figure be considered *Physically*, it is nothing but *the superficies, or terminant Extreames of a Body*. We say, *Physically*, because Geometricians distinguish Figures into *Superficial*, or *Plane*, and *Profound*, or *Solid* : but the Physiologist knows no other Figure properly, but the *Superficial*; because, in strict truth, the Profound or Solid one seems to Him, to be rather the Magnitude, or Corpulency of a thing circumscribed or terminated by its Figure, than the Figure it self abstractedly intended. Nay, if we insist upon the rigour of verity, the Figure of a Body is really nothing but the *Body it self*; at least, the meer *Manner* of its Extreme parts, according to which our sense deprehends it to be smooth or rough, elated or depressed. This may be most fully evinced by only one Example, *viz.* the figure made upon Wax by the impression of a Seal. For, that Figure really is nothing but the very substance of the Wax, in some parts made more Eminent, in others more depress, or profound, according to the Reverse of its Type ingraven in some hard substance; and that without Adjection, or Detraction of any Entity whatever. And what we affirm of the Figure made in Wax by Sigillation, is of equal truth (proportionately) if accommodated to any other Figure whatever : nor doth it imply a Difference, whether the Figure be *Natural*, such as in Animals, Vegetables, Minerals; or *Artificial*, such as in Ædifices, Statues, Characters, &c.

*Art.* 12.
*Figure* (Physically considered) nothing but the *superficies, or terminant Extreams* of a Body.

M m                                        SECT.

## Sect. II.

THe Causes of Magnitude and Figure in Concretions, being thus investigated; it follows, that we explore their Effects, *i. e.* the *Qualities* which seem so immediately cohærent to the Magnitude and Figure of Bodies, as that reason cannot consigne them to more likely and probable Principles, than the two First Proprieties of the Universal Matter, Atoms.

*Art.* 2.
*Subtility* and *Hebetude,* how the Consequents of Magnitude.

The Consequents, therefore, of Magnitude, are SUBTILITY and its contrary, HEBETUDE. Not that the Emergency of a Great Body from Atoms the most Exile; or of a small body from great Atoms, is impossible; as we have formerly intimated: but, that a Body consisting of more Exile, or subtile Atoms, hath a greater subtility, or obtains a Faculty of penetrating the contexture of another body, by subingression into the pores, or inane spaces thereof; and a body consisting of grosser Atoms, must have more of Grossness or Hebetude, and so hath not the like Faculty of penetrating the Contextures of other bodies, by subingression into the inane spaces, or intervals betwixt their particles. This may be Exemplified in *Fire* and *Water*; *Wine and Oyle*; *Aqua Fortis and Milk,* &c.

*Art.* 3.
A considerable Exception of the Chymists (*viz.* that some Bodies are dissolved in liquors of grosser particles, which yet conserve their Continuity in liquors of most subtile and corrosive particles) prevented.

We are not now to learn the truth of that Chymical Canon, *Cuique fermè rei solvendæ, vel extrahendæ eligendum esse idoneum menstruum, quod ejus naturæ respondeat :* experience having frequently ascertained us, that Aqua Regis, which soon dissolves the most compact of bodies, Gold, will not at all impair Resine, Pitch, Wax, and many other Unctuous and Resinous Concretions; which yeild almost at first touch to the separatory faculty of oyle: that Mercurial Waters expeditely insinuate into the substance of Gold, dissolve the Continuity of its stiffly cohærent particles, and convert it from a most solid into an oyly substance; not so much by Corrosion, as symbolisme or Affinity of nature : that Salt, Nitre, and Sulphur, which being added to Sand, Flints, and many Metals, promote the solution, in a reverberatory fire; have yet no accelerating, but a retarding energy upon Turpentine, Balsome, Myrrh, &c. in the extraction of their Oyls, or Spirits: that all Waters, or Spirits extracted from Saline and Metalline natures are most convenient Menstruaes for the solution of Metals &Minerals; not so much in respect of their Corrosion, as similitude of pores and particles: and consequently that every Concretion requires to its dissolution some peculiar dissolvent, that holds some respondency or analogy to its contexture. But, yet have we no reason, therefore to abandon our *Assumption,* that the dissolution of one body, by the subingression or insinuation of the particles of another, must arise from the greater subtility of particles in the Dissolvent; until it be commonstrated to us, that a Body, whose Particles are less exile, can penetrate another Body, whose Pores are more exile, the contrary whereto is demonstrated to us by the frequent Experiments of Chymistry.

And,

And, therefore, the Reafon, Why *Oyle Olive* doth pervade fome Bo- dies, which yet are impenetrable even by *fpirit of Wine* (by *Raimundus Lullius*, and after him by *Libavius* and *Quercetan*, accounted the true Sul- phur and Mercury of Hermetical Philofophers, extracted from a Vegetable, for the folution of Gold into a Potable fubftance, and the Confection of the Great Elixir; and as General a Diffolvent, as that admired (but hard- ly underftood) Liquor Alkaheft of *Paracelfus*, if not the fame) can be no other but this: that in the fubftance of Oyle are fome Particles much more fubtile and penetrative, than any contained in the fubftance of Wine; though thofe fubtile particles are thinly interfperfed among a far greater number of Hamous, or Hooked particles, which retard their penetrati- on. Thus alfo in that affrighting and Atheift-converting Meteor, *Light- ning*, feem to be contained many particles much more exile and fearching than thofe of our Culinary Fires: becaufe it fometimes diffolves the hardeft of Metals in a moment, which preferves its integrity for fome hours in our fierceft reverberatory furnaces. Which *Lucretius* well expreffeth in this Tetraftich;

*Art.* 4.
Why *Oyle* dif- folutes the parts of fome Bodies, which remain invio- late in *Spirit of Wine*; and why *Lightning* is more penetra- tive; than *Fire*.

> *Dicere enim pofsis, cæleftem Fulminis ignem*
> *Subtilem magis, è parvis conftare Figuris;*
> *Atque ideo tranfire foramina, quæ nequit ignis*
> *Nofter hic è lignis ortus tædaque creatus.*

Secondly, the Qualities Confequent to Figure, are SMOOTH- NESSE, and its contrary, ASPERITY. Not that, if we appeal to the judgement of the fenfe, the fuperfice of a Body may not be fmooth, though it confift of angulous Atoms; or rough, though compofed of plain and polite Atoms: for, all Atoms, as well as their Fi- gures, are fo Exile, as that many of them that are angular, may cohære in- to a mafs, without any inequality in the fuperfice deprehenfible by the fenfe; and on the contrary, many of thofe that are plane and polite, may be convened and concreted into fuch maffes, as to make angles, edges, and o- ther inequalities fufficiently fenfible. But, that if we refer the matter part- ly to the judicature of Reafon, partly to the evidence of our fenfes in Ge- neral; we cannot but determine it to arife from the Figuration of Atoms alone. Firft, to the judicature of *Reafon*; for, as the mind admits no- thing to be perfectly continued, befides an Atom: fo can it admit nothing to be exquifitely fmooth, befides either the whole fuperfice of an Atom, if the fame be orbicular, oval, or of the like Figure; or fonf parts of it, if the fame be tetrahedical, hexahedrical, or of fome fuch poligone figure. Becaufe, look by what reafon the mind doth conclude the fuperfice of no Concretion in nature to be perfectly continued: by the fame reafon doth it conclude the fuperfice of every thing, feemingly moft equal and polite, to be varioufly interrupted with afperities, or eminent, and depreft particles; and this it refers immediately and folely to many fmall maffes of Atoms, in the Contexture coadunated, like as it referrs the interruptions in the fu- perfice of a piece of Lawne, or Cambrique, which to the eye and touch appears moft fmooth and united, to the fmall maffes of Filaments inter- woven in the webb. And here the Experiment of a *Microfcope* is opportune; for, when a man looks through it upon a fheet of the fineft and fmootheft Venice Paper, which feems to the naked eye, and moft exquifite touch, to be equal and terfe in all parts of it fuperfice; He fhall difcern it to be fo full

*Art.* 5.
*Smoothnefs* and *Afperity* in Con- cretions; the Confequents of *Figure* in their Material Principles.

of

of Eminences and Cavities, or small Hills and Valleys, as the most prægnant and præpared Imagination cannot suppose any thing more unequal and impolite. Secondly, to the *Evidence of our senses in General*; because, the very Affection of Pleasure or Pain, arising to the sensory from the contact of the sensible object, doth sufficiently demonstrate, that *smoothness* is a Quality resulting either from such Atoms, or such small masses of Atoms contexed, as are smooth and pleasant to the sense, by reason of their correspondence to the pores and particles of the Organ: and contrariwise, that *Asperity* is a Quality, resulting either from such single Atoms, or such most minute masses of Atoms concreted, as dilacerate, or exasperate the sense, by reason of their incongruity or Disproportion to the Contexture of the Organ; as we have, even to redundancy, Exemplified in the Grateful and Ungrateful Objects of each sense.

CHAP.

# CHAP. XI.

## OF THE

## Motive, Vertue, Habit, Gravity, and Levity

### OF

# CONCRETIONS.

SECT. I.

He Third Propriety of the Universal Matter, Atoms, is *Mobility*, or
*Gravity:* and from that fountain is
it that all Concretions derive their
*Virtue Motive*. For, though our
deceptable *sense* inform us, that the
minute Particles of Bodies are fixt
in the act of their Coadunation,
wedged up together, and as it were
fast bound to the peace by reciprocal concatenation and revinction:
yet, from the Dissolution of all
Compound natures, in process of
time, caused by the intestine Com-

*Art.* 1.
The *Motive*
*Virtue* of all
Concretions,
derived from
the essential
*Mobility* of A-
toms.

motions of their Elementary Principles, without the hostility of any External Contraries, may our more judicious *Reason* well infer, that Atoms are
never totally deprived of that their essential Faculty, Mobility; but are
uncessantly agitated thereby even in the centrals of Concretions, the most
solid and compact; some tending one way, others another, in a perpetual
attempt of Eruption, and when the Major part of them chance to affect
one and the same way of emancipation, then is their united force determined to one part of the Concretion, and motion likewise determined to
one region, respecting that Part. That same MOTIVE VIRTUE, therefore, wherewith every Compound Bodie is naturally endowed, must owe
its origine to the innate and co-essential Mobility of its component particles, being really the same thing with their Gravity, or *Impetus:* which
yet receives its determinate manner and degree from their mutual Combination. In respect whereof it necessarily comes to pass, that when Atoms,
mutually adhæring unto, and detaining each other, cannot obey the impulse

of

of their tendency singly, they are not moved with that pernicity, as if each were at abſolute liberty : but impeding and retarding each other in their progreſs, are carried with a ſlower motion, But that more or leſs ſlow, according to the rate or proportion of common Reſiſtence : becauſe always ſome of them are carryed to an oppoſite , others tranſverſly , others obliquely to a different region.

*Art. 2.*
Why the Motive Virtue of Concretions doth reſide principally in their ſpiritual Parts.

And hence is it, that becauſe Atoms are at moſt freedom of range in ſubtile and ſpiritual Concretions ; every degree of Denſity and Compactneſs cauſing a proportionate degree of Tardity in their ſpontaneous motions : therefore is the Motive Faculty not more generally , than rightly conceived, to reſide chiefly in the *ſpiritual*, or (as vulgar Philoſophy) *Æthereal* Parts of al Concretions.     And, whether the ſpirits of a thing are principally determined to move , thither do they not only themſelves contend, with great impetuoſity and ſpeed, but alſo carry along with them the more ſluggiſh ; or leſs moveable parts of the Concretion ; as is ſuperlatively manifeſt in the Voluntary motions of Animals.

*Art. 3.*
That the Deviation of Concretions from motion Directs and their Tardity in motion ariſe from the Deflexions and Repercuſſions of Atoms compoſing them

We need not here inſiſt upon the Redargution of that Blaſphemous and Abſurd (for the former Epithite always implies the later) dream of *Epicurus*, that Atoms were not only the Firſt *Matter* , but alſo the Firſt and ſole *Efficient* of all things ; and conſequently that all Motions, and ſo all Actions in the Univerſe are Cauſed meerly by the inhærent Mobility of them : becauſe we have expreſly refuted the ſame in our Treatiſe againſt *Atheiſm,* (*Chap. 2. Sect. 1. artic. ultim.*).     Eſpecially , ſince it is more opportune for us here to advertiſe ; that inſomuch as the motion of all Atoms is ſuppoſed to be of itſelf Direct , and moſt rapid ; therefore doth the Deviation , as well as the Tardity of Concretions ſeem to ariſe from the Deflection, Repercuſſion, or multiplied Repreſſion of the Atoms compoſing them. For, the Occurſation or meeting of two Atoms, may be in direct lines; ſo that among Atoms, either by ſingle percuſſion, or repercuſſion overcoming the firſt begun motion, as the aſſembly or Convention will bear , there may be cauſed ſome motions Direct, though more or leſs ſlow : and their Occurſations may be alſo according to Oblique angles , and ſo, by the ſame reaſon may enſue a motion, not only more or leſs ſlow, but alſo more or leſs Oblique. Moreover, if after one repercuſſion made to oblique angles , there chance to follow a ſecond, a third repercuſſion to angles equally oblique; then muſt the motion be purſued in obliquity multangular , according to the multiplicity of Repercuſſions : and if the Angles be very frequent and indiſtant , the motion becomes, at leaſt to appearance, to be of an uniform Curvity , and may therefore be termed a motion Circular , Elliptical , Helicoidal, or the like , according to the condition of its Deflection and Crookedneſs.

*Art. 4.*
Why the motion of all Concretions neceſſarily preſuppoſeth ſomething, that remains unmoved; or that, in reſpect of its ſlower motion, is equivalent to a thing unmoved.

Moreover we are to obſerve, that every Body, whether Simple or Compound, i. e. Atom or Concretion, from which a Repercuſſion is made, muſt either quieſce , or not be moved the ſame way , as is the repercuſt , or not with ſo ſwift a motion ; becauſe, otherwiſe there can be no mutual *Antilipta*, or *Reſiſtence* , nor the impingent body rebound from the repercutient.     And this is the only reaſon, why (excepting only the motion eſſential to Atoms.) the motion of all Concretions doth ever ſuppoſe ſomething
that

that remains Unmoved, or that, in respect of its less motion, is tantamount to a thing Unmoved: because, otherwise there could be no reciprocal Resistence, and so all motion might both begin and repair it self.

Having thus premised these few fundamental Laws of Motion in General, opportunity commands us to descend to the consideration of the FACULTY of Motion: insomuch as it seems not to be any thing distinct from that Motive Force, inhærent in all Concretions, which we have now both described, and deduced from its immediate origine, the Mobility of Atoms; and that it is well known to all Book-men, to appertain to the second species of Qualities, according to the method of *Aristotle.* To which we may add these lessons also, that it comprehends the Third species of Qualities, and obtains the First, or Habit, as its proper appendix. Know we, therefore, that the Faculty or Power of Motion doth therefore seem to be one and the same thing with the coëssential Mobility, now described; because every thing in Nature is judged to have just so much of Efficacy, or Activity, as it hath of Capacity to move either it self, or any other thing.

*Art.* 5.
What the *Active Faculty* of a thing is,

And hence is it, that in Nature there is no Faculty (properly) but what is *Active*; because, though the motions of things be really the same with their Actions: yet must all motion have its beginning only from the Movent, or Agent. Nor can it avail to the contrary, that all Philosophers have allowed a Passive Faculty to be inhærent in all Concretions; since, in the strict dialect of truth, that Passiveness is no other than a certain Impotency of Resistence, or the Privation of an Active Power, in defect whereof the subject is compelled to obey the Energy of another. If you suppose an obscure Contradiction in this our Assertion, and accordingly Object, that therefore there must be a Faculty of Resistence, in some proportion, and that that Resistence is Passive: we are provided of a satisfactory salvo, which is, that though the Active Virtue, which is in the Resistent, doth sometimes scarce discover it self, yet is it manifest, that there are very many things, which make resistence only by motion, which no man can deny to be an Active Faculty; as when we rowe against wind and tide, or strive with a Bowe in the drawing of it, for all these evidently oppose our force by contrary motion. And, as for other things, which seem to quiesce, and yet make some resistence; such we may conceive to make that resistence by a kinde of motion, which Physicians denominate a Tonick motion; like that of the Eye of an Animal, when by the Contraction of all its muscles at once it is held in one fixt position. Thus not only the whole Globe of the Earth, but all its parts are held unmoved, and first by mutual cohærence, and resist motions as they are parts of the whole: and thus also may all Concretions be conceived to be made Immote, not that the Principles of which they consist, are not in perpetual inquietude and motion; but, because their particles reciprocally wedge and implicate each other, and while some impede and oppose the motions of others, they all conspire to the Consistence of the whole. However the more Learned and Judicious shall further dispute this paradoxical Argument; yet dare we determine the Common Notion of a Faculty to be this, that there is inherent in every thing a Principle of Moving itself, or Acting, if not *Primary* (which the schools terme the *Forme*) yet *Secondary* at least, or profluent from the

Forme,

*Art.* 6.
That in Nature every Faculty is *Active*; none *Passive*.

Forme, being as it were the immediate *Instrument* thereof.

**Art. 7.**
*A Peripatetick Contradiction, affirming the Matter of all Bodies to be devoid of all Activity; and yet assuming some Faculties à tota substantia.*

And here we cannot conceal our wonder, that the *Peripatetick* hath not for so many ages together discovered himself to be intangled in a manifest Contradiction; while on one part He affirms, that there are certain Faculties flowing *à tota substantia*, from the whole substance of a thing, as if they were derived from the matter of Concretions: and on the other, concludes, as indisputable, that the Matter is absolutely devoid of all Activity, as if it were not certain, that the Faculties frequently perish, when yet not the whole and intire substance of the thing perisheth, but only the spiritual, or more tenuious parts thereof.

**Art. 8.**
*That the Faculties of Animals (the Ratiocination of man only excepted) are Identical with their spirits.*

Now, what more prægnant Argument than this can the most circumspect desire, in order to their Conviction, that the *Faculties* of an Animal (we exclude the Rational Faculty of man, from the sphere of our assertion) are *Identical* with the *Spirits* of it, i.e. the most subtile, most free, and most moveable or active part of its materials? For, though the spirits are by vulgar Philosophers conceived to be only the Primary Organ, or immediate Instrument, which the Faculty residing in one part, occasionally transmits into another: yet, to those Worthies, who have with impartial and profound scrutiny searched into the mystery, hath it appeared more consentaneous, that the spirits are of the same nature with the Faculty, and not only movent, but Instrument; nor can it stand with right reason to admit more than this, that as water in the streams is all one specifically with that in the fountain, so is the Faculty, keeping its court or chief residence in one part of the body, as it were the Fountain, or Original, from whence to all other parts, inservient to the same function, the diffusion of spirits is made, in certain exile rivolets, or (what more neerly attains the abstrusity) Rayes, like those emitted from the Sun, or other fountain of light. And, what we here say, of the Faculties of Animals, holds equal truth, also concerning those of Inanimate Concretions; allowing a difference of proportion.

**Art. 9.**
*The Reasons of the Coexistence of Various Faculties in one and the same Concretion.*

But here ariseth a considerble *Difficulty*, that at first view seems to threaten our Paradox with total ruine; and this it is: if the Faculties of Concretions be not distinct in essence from their spirits, or most agile particles; *how then can there be so many various Faculties coexistent in one and the same concretion*, as are dayly observed; for in an Apple, for example, there is one Faculty of affecting the sight, another of affecting the taste, a third affecting the smell. Concerning this, therefore, we give you this solution, that the coexistence of various Faculties in one Concretion, doth depend upon (1) *the variety of multiforme particles, of which the whole Concretion doth consist,* (2) *the variety of particles and special contexture of its divers parts,* (3) *the variety of External Faculties, to which it happens that they are applied.* To keep to our former Example, in an Apple, tis manifest, there are some particles, in which consisteth its faculty of affecting the smell, others in which consisteth its faculty of affecting the Tast; for, the Experiments of Chymistry demonstrate, that these different particles may be so sequestred each from other, as that the tast may be conserved, when the smell is lost,

and

and the smell conserved, when the taste is abolished. And in an Animal
it is no less evident, that the organ of one sense hath one peculiar kind
of contexture, the organ of another sense another: and finally, that
when we shall refer the Faculties of Odour and Sapour, which are in an
Apple, to the Faculties of smelling and tasting in Animals, they become
subject to a further discrimination. Since the same particles, which move
the smelling, shall create a sweet and grateful odour, in respect of one
Animal, and an offensive or stinking, in respect of another: and in like manner, those particles, which affect the Taste, shall yeild a most grateful and
desireable Sapour, to one Animal, and as odious and detestable a one to
another. Ought we, therefore, to account that Faculty of an Odour,
which is in an Apple, either *Single*, or *Multiplex*? If we would speak
strictly, it is *Single Absolutely: Respectively, Multiplex*. And thus, indeed, may we affirm, that in the General, or absolutely, an Apple is Odorous and Sapid: but Comparatively and in Special, that it is fragrant, or
foetid; sweet or bitter.

As for that Appendix of a Faculty, which not only Philosophers, but    *Art.10.*
the People also name a HABIT; Experience daily teacheth, that there   *Habit defined.*
are some Faculties, (in Animals especially) which by only frequency of
acting grow more prompt and fit to act: and upon consequence, that
that Hability or promptness for action, is nothing but a Facility of doing,
or repeating that action, which the same Faculty, by the same instruments,
hath frequently done before.

And, as to the *Reason* of this Facility; though it arise in some measure    *Art. 11.*
from the Power or Faculty it self, or the Spirits, as being accustomed   *That the Rea-*
to one certain motion: yet doth it chiefly depend upon the Disposition   *son of all Ha-*
of the *Organs*, or instruments which the Faculty makes use of in the   *bits in Ani-*
performance of its proper action. For, because the Organ is alwayes a   *mals, consist-*
Dissimilar or Compound Body, consisting of some parts that are crass and   *eth principal-*
rigid, we are to conceive it to be at first somewhat stubborn, and not   *ly in the con-*
easily flexible to such various motions, as the Faculty requires to its seve-   *formity and*
ral operations: and therefore, as when we would have a Wand to be every   *flexibility of*
way easily flexible, we are gently and frequently to bend it, that so the   *the Organs,*
tenour of its fibres running longwise through it, may be here and there   *which the re-*
and every where made more lax, without any sensible divulsion; so if we   *spective Facul-*
desire to have our hands expedite for the performance of all those difficult   *ty makes use*
motions that are necessary to the playing of a Lesson on the Lute, we must   *of, for the per-*
by degrees master that rigidity or clumsiness in the Nerves, Tendons, Mu-   *formance of*
scles and joints of our fingers, yea in the very skin and all other parts of our   *its proper*
hands. Thus also Infants, while they stammer, and strive again and again to   *Actions.*
pronounce a word clearly and distinctly, do no more than by degrees master
the stiffness and sluggishness of their tongues and other vocal organs, and so
make them more flexible and voluble: and when by assuefaction they have
made them easily flexible to all the motions required to the formation of
that idiome, then at length come they to speak it plainly and perfectly.
The same is also true, concerning the Brain, and those Organical parts
therein, that are inservient to the act of Imagination, and by the imagination
to the act of Discourse. For, though the Mind, when divorced from the
the body, can operate most readily, and knows no difficulty or impediment
in the act of Intellection; as being Immaterial, and so wanting no organs

N n    for

for the exercise of its reasoning Faculty : yet nevertheless, while it is adliged to the body and its material instruments, doth it remain subject to some impediment in the execution of its functions ; and because that impediment consisteth only in the less aptitude or inconformity of its proper organs, therefore the way to remove that impediment, is only by Assuefaction of it to study and ratiocination. And from this Assuefaction may the Mind be affirmed to acquire a certain Habit or Promptitude to perform its proper Actions ; insomuch as by reason of that Habit, it operates more freely and expeditely : but, yet, in stricter Logick, that Habit ariseth chiefly to its Organs ; as may be inferred only from hence, that the Organs are capable of increment and decrement, and to increase and decrease, is competent only to a thing that consisteth of parts ; such as is the Organ, not the Mind.

*Art. 12.*
*Habits, acqui-*
*rable by Bruits:*
*and common*
*not only to*
*Vegetables , but*
*also to some*
*Minerals.*

Nor is the acquisition of a Habit by assuefaction proper only to Man, but in common also to *all Living Creatures*, such especially as are used to the hand and government of Man, as Horses, Doggs, Hawks, and all prating and singing Birds. And where we affirmed, that some Faculties are capable of advancement to perfection by Habit ; we intended, that there are other Faculties which are *incapable* thereof, as chiefly the *Natural Faculties* in Animals, and such as are not subject to the regiment of the Will : though still we acknowledge that some of these there are , which upon change of temperament in their respective Organs, may acquire such a certain Habit , as may oppose the original inclination ; and of this sort the principal is the Nutrient Faculty , which may be accustomed even to Poison. Lastly, when we, said *Chiefly in Animals*; we were unwilling totally to exclude *Plants* ; because they also seem (at least Analogically ) to acquire a kind of Habit : as is evident from their constant retaining of any posture or incurvation , which the hand of the Gardiner hath imposed upon them , while they were tender and flexible; as also that they may by degrees be accustomed to forein soils, and (what is more admirable) if in their transplantation those parts of them, which at first respected the South or East, be converted to the North or West, they seldome thrive, never attain their due procerity. Nay, if the Experiments of some Physitians be true, *Minerals* also may be admitted to attain a Habit by assuefaction ; For *Baptista van Helmont*, (*in lib. de Magnetica Vulnerum curatione, & lib. de Pestis tumulo* ) reports that He hath found a Saphire become so much the more efficacious an Attractive of the pestilential Venome from the Vitals, by how much the more frequently it hath been circumduced about Carbuncles or Plague Sores ; as if Custome multiplied its Amuletary Virtue and taught it a more speedy way of conquest.

## SECT. II.

A Mong all Qualities of Concretions, that deduce themselves from the Mobility of Atoms, the most eminent is GRAVITY, or the motion of perpendicular Descent from Weight. Which, though most obvious to the observation of Sense, hath much of obscurity in its Nature; leading the Reason of Man into various and perplext Conceptions concerning its Causes: nor hath the judgment of any been yet so fortunate as to light upon a Demonstrative Theory concerning it, or fix upon such a determination as doth not lye open to the objection of some considerable Difficulty. So that it may well seem Ambition great enough for us, onely with due uprightness to examine the Verisimility of each opinion, touching the Formal Reason, or Essence of Gravity : that so we may direct younger Curiosities, in which they may, for the præsent, most safely acquiesce.

*Art.* 1.
Gravity, as to its Essence, or Formal Reason, very obscure.

*Epicurus*, indeed, well desumes the Gravity of all Concretions, immediately from the Gravity of Simple Bodies, or Atoms : insomuch as all things are found to have so much more of Weight, as they have of Atoms, or Matter, that composeth them; and *è contra*. Which reason the exact *Joh. Bapt. Balianus*, a Nobleman and Senatour of *Genoa*, seriously perpending; sets it down as a firm ground, *Gravitatem se habere ut Agens, Materiam vero, seu Materiale corpus, ut Passum ; & proinde gravia moveri juxta proportionem gravitatis ad materiam : & ubi sine impedimento naturaliter perpendiculari motu ferantur, moveri aqualitèr ; quia ubi plus est Gravitatis, plus ibi paritèr fit Materiæ, seu Materialis quantitatis ;* ( *de motu Gravium Solidorum & Liquidorum, lib.* 1. *cap.* 1. ). But, this being too General, and concerning rather the Cause of Comparative, than Absolute Gravity; leaves our Curiosity to a stricter search.

*Art.* 2.
The opinion of *Epicurus*, good as to the Cause of Comparative insufficient as to the Cause of Absolute Gravity.

The Grand Dictator of the Schools, *Aristotle*, taking it for granted [ *Unumquodque sensilium ita in suum locum ferri, ut ad speciem* ] that every corporeal Nature is by native tendency carried to its proper place, as to its particular Species ; confidently inferrs this doctrine : that Gravity and Levity are Qualities essentially inexistent in Concretions ( 4. *de Cælo, cap.* 2. ) and passionately reprehending *Democritus* and *Leucippus*, for affirming that there is no such thing in Nature, as Absolute Gravity, or Absolute Levity; concludes, that in Nature is something absolutely Heavy, which is Earth, and something Absolutely Light, which is Fire, ( *de Cælo, lib.* 4. *cap.* 4. ) But, neither of these Positions are more than Petitionary ; and so not worthy our assent : as the Context of our subsequent Discourse doth sufficiently convince.

*Art.* 3.
*Aristotles* opinion of Gravity, recited.

N n 2                    The

*Art. 4.*
*Copernicus the-*
*ory of Gravi-*
*ty, insatisfacto-*
*ry; and where-*
*in.*

The Third opinion worthy our memory, is that of *Copernicus*, who considering, that all Heavy Bodies, either projected Upwards by external violence, or dropt down from some eminent place, are observed to fall perpendicularly down upon the same part of the Earth, from which they were elevated, or at which they are aimed, and so that the Earth might be thence argued not to have any such Diurnal Vertigo, as His Systeme ascribes unto it, insomuch as then it could not but withdraw it self from Bodies falling down in direct lines, and receive them at their fall not in the same place, but some other more Westernly : we say, considering this, *Copernicus* determined Gravity to be, not any Internal Principle of tendency toward the middle, or Centre of the Universe ; but an innate propension in the parts of the Earth, separated from it, to reduce themselves in direct lines, or the nearest way, to their Whole, that so they may be conserved together with it, and dispose themselves into the most convenient, i.e. a sphærical figure, about the centre thereof. His words are these ; *Equi-*
*dem existimo, Gravitatem non aliud esse, quàm Appetentiam quandam na-*
*turalem, partibus inditam à Divinâ Providentiâ Opificis Universorum,*
*ut in unitatem integritatemque suam sese conferant, in formam Globi*
*coeuntes : quam Affectionem credibile est etiam Soli & Lunæ, cæterisque*
*Errantibus fulgoribus inesse, ut ejus efficacia in eâ, quâ se repræsentant,*
*rotunditate permaneant. (lib. 1. cap. 9.).* So that according to this *Copernican Assumption,* if any part of the Sun, Moon, or other Cœlestial Orb were divelled from them, it would, by the impulse of this natural tendency, soon return again in direct lines to its proper Orb, not to the Centre of the Universe. Which as *Kepler (in Epitom. Astronom. pag. 95.)* well advertiseth, is but a Point, i. e. Nothing, and destitute of all Appetibility ; and therefore ought not to be accounted the Term of tendency to all Heavy Bodies, but rather the Terrestrial Globe together with its proper Centre, yet not as a Centre, but as the *Middle* of its Whole, to which its Parts are carried by Cognation.

But, this opinion hath as weak a claim to our Assent, as either of the former ; as well because it cannot consist with the Encrease of Velocity in all Bodies descending perpendicularly, by how much nearer they approach the Earth, unless it can be demonstrated, that this encrease of Velocity in each degree of descent, ariseth only from the Encrease of Appetency of Union with the whole ( which neither *Copernicus* himself, nor any other for Him, hath yet dared to assent ): as in consideration of many other Defects, and some Absurdities, which, that wonder of the Mathematicks, *Ricciolus*, hath demonstratively convicted it of ( *in Almagisti novi parte posteriori, lib. 9. sect. 4. cap. 16. de Systemate terræ motæ.* ). Who, had He but as solidly determined all the Difficulties concerning the immediate Cause of this Affection in Bodies, called Gravity, as He hath refuted the Copernican Thesis of an Innate Appetency in the parts of the Earth to reunite themselves to the Whole : doubtless He had much encreased the obligations and gratitude of his Readers. But, making it his principal design to propugn the Physiology and Astronomy of the Ancients, especially such Tenents as are admitted by the Schools, and allowed of by the Doctors of *Rome*, as most concordant to the
litteral

litteral sense of Sacred Writ: He waved that Province, seeming to adhære to the common Doctrine of the *Stagirite*, formerly recited, and only occasionally to defend it.

Lastly, there are Others ( among whom *Kepler* and *Gassendus* deserve the richest Minervals ) who, neither admitting with *Aristotle*, that Gravity is any Quality essentially inhærent in Concretions; nor, with *Copernicus*, that it is an Appetency of Union, implanted originally in the parts of the Earth, by vertue whereof they carry themselves towards the Middle of the Terrestrial Globe: define it to be an *Imprest Motion, Caused immediately by a certain Magnetick Attraction of the Earth.*

*Art.5.*
The Determination of Kepler, Gassendus, &c. that Gravity is Caused meerly by the Attraction of the Earth: espoused by the Author.

And this opinion seems to carry the greatest weight of Reason; as may soon be manifest to any competent and equitable judgment, that shall exactly perpend the solid Arguments alledged by its Assertors: which for greater decorum, we shall now twist together into one continued thread, that so our Reader may wind them into one bottome, and then put them together into the ballance.

Insomuch as frequent and most accurate observation demonstrates, that the Motion of a Body downward doth encrease in the same proportion of Velocity, that the motion of the same Body, violently projected upward, doth decrease; therefore is it reasonable, nay necessary for us to conceive, that there are *Two distinct External Principles*, which mutually contend about the same subject, and execute their contrary forces upon the same Moveable. Now, of these two Antagonistical Forces, the one is *Evident*; the other *obscure*, and the argument of our instant Disquisition. *Manifest* it is, when a stone is thrown upward from the surface of the Earth into the Aer, that the External Principle of its motion *Upward*, is the Hand of Him, who projected it : But somewhat *obscure*, what is the External Principle of its motion *Downward*, when it again returns to the Earth. Nevertheless, this obscurity doth not imply a Nullity; i.e. it is high temerity to conclude that there is no External Cause of the stones Descent, because that External Cause is not equally manifest with that of its Ascent : unless any dare to affirm, that because He can perceive, when Iron is attracted to a Loadstone, no Externall Cause of that Attraction, therefore there can be none at all. Many, indeed, are the wayes, by which an External Cause may move a Body: and yet they all fall under the comprehension of only two Cardinal wayes, and those are *Impulsion*, and *Attraction*.

*Art. 6.*
The External Principle of the perpendicular Descent of a stone, projected up in the Aer; must be either Depellent, or Attrahent.

This præconsidered, it followes, that we cast about to finde some Cause, or Impellent, or Attrahent ( or rather two Causes, one Impellent, the other Attrahent, operating together ) to which we may impute the perpendicular motion of Bodies Descending. The *Impellent* Cause ( if any such there be ) of the perpendicular motion of a stone Descending, can be no other but the *Aer*, from above incumbent upon, and pressing it downward : because of any other External Cause of that effect, no argument can be given. For, should you suppose a sphere of Fire, or some other

*Art. 7.*
That the Resistence of the Superior Aer is the only Cause which gradually refracteth, and in fine wholly overcometh the Imprest Force, whereby a stone projected, is elevated upward,

Or vated upwar

or some other Æthereal Substance, to be immediately above the
convex Extreme of the sphere of Aer; which closely and with some
kind of pressure invironing the Aer, might compel all its parts to
flow together toward the Terraqueous Globe: yet could that super-ae-
real sphere, bounded and urged by the circumvolutions of the Cœlestial Orbs, do no more, than cause the Aer, being it self prest downward, to bear down upon the stone, and so depress it; and so the Aer
must still be at least the *Proxime* Cause impelling the stone downward.
Moreover, that the Aer alone may be the Impellent Cause of the
stones perpendicular Decidence from on high, even *Aristotle* Himself
seems to concede; insomuch as He is positive in his judgment, that when
a Heavy body projected upward is abandoned by its Motor, it is afterward moved only by the Aer, which being moved by the Projicient,
moves the next conterminous Aer, by which again the next neighbouring
Aer is likewise moved, and so successively forward untill the force of
the Imprest motion gradually decaying, the whole communicated motion
ceaseth, and a quiet succeeds. But, because *Aristotle* could not tell, what
Cause that is, which in every degree of the stones ascent opposing, at
length wholly overcomes the imprest force; unless it should be the
occurrent superiour Aer, which continually resisteth the inferior aer,
whereby the projected stone is promoted in its ascent: may not we
safely enough conclude, that the Aer from above incumbent upon
the projected stone, may by the same force depress it Downward,
wherewith it first resisted the motion of it Upward? Doubtless,
what force soever the Hand of a man, who projects a stone upward into the Aer, doth impress upon it, and the contiguous Aer;
yet still is it the superiour Aer, that both continually resisteth the
tendency of the stone upward, and at its several degrees of ascent
refracteth the force thereupon imprest by the hand of the Projicient, untill having totally overmastered the same, it so encreaseth
its conquering Depellent force, as that in the last degree of the
stones Descendent motion, the Depressive force of the Aer is become as great, as was the Elevating force of the Hand, in the beginning of its Ascendent motion. Suppose we, that a Diver should
from the bottome of the Sea throw a stone directly upward, with
the same force, as from the surface of the Earth up into the Aer;
and then demand, Why the stone doth not ascend to the same
height in the Water, as in the Aer. Is it not, think you, because the water doth more resist, and refract the Imprest force,
and so sooner overcome it, and then begins to impress its own con-
contrary Depressing Force thereupon, never discontinuing that impression, till it hath reduced the stone to the bottom of the Sea,
from whence it was projected? The Difference, therefore, betwixt
the Resistence of the Imprest force, by the Water, and that of
the Aer, consisteth only in Degrees, or more and less. And, though
the Renitency of the Aer may be thought very inconsiderable in comparison of that great Violence imprest upon a Cannon Bullet, shot
upward into the Aer: yet be pleased to consider, that it holds
some investigable proportion, with the Renitency of the Water.
Which proportion that we may understand, compare we not only
the very small Ascent of a stone, thrown upward from the bottome

of

of the Sea, to the large afcent of the fame ftone, with equal force, from the Earth, thrown up into the Aer; but alfo the almoft infenfible progrefs of a Bullet fhot from a Cannon tranfverfly through Water, with that vaft progrefs it is commonly obferved to make through the Aer: and we fhall foon be convinced, that as the Great Refiftence of the Water is the Caufe, why the Stone, or Bullet makes fo fmall a progrefs therein; fo is the fmall Refiftence of the Aer the Caufe why they both pervade fo great a fpace therein. And thus is it Demonftrable, that the *Refiftence of the fuperior Aer, is the External Agent, which conftantly refifteth, by degrees refracteth, and at length wholly overcomes the impreft Force, whereby Heavy Bodies are violently elevated up into the Aer.*

The Difficulty remaining, therefore, doth only concern the *Impellent Caufe of their Fall Down again;* or, whether the Aer, befides the force of Refiftence, hath alfo any *Depulfive Faculty,* which being impreft upon a ftone, bullet, or other ponderous body, at the top, or higheft point of its mountee, ferveth to turn the fame Downward, and afterward to continue its perpendicular defcent, till it arrive at and quiefce on the Earth. Which, indeed, feems well worthy our Doubt, becaufe it is obfervable, that Walls, Pavements, and the like folid and immote Bodies, though they ftrongly refift the motion of bodies impinged againft them; doe not yet imprefs any Contrary motion thereupon: the Rebound of a Ball or Bullet from a Wall, being the effect meerly of the fame force impreft upon it by the Racket, or Gun-powder fired, which firft moved it; as is evident even from hence, that the Refilition of them to greater or lefs diftance, is according to the more or lefs of the Force impreft upon them. Which thofe Gunners well underftand, who experiment the ftrength of their Powder, by the greatnefs of the bullets rebound from a Wall.

*Art.* 8.
That the Aer, diftracted by a ftone violently afcending, hath as well a *Depulfive,* as a *Refiftent* Faculty; arifing immediately from its *Elaterical,* or *Reftorative* motion.

And to folve this Difficulty, we muft *diftinguifh* betwixt Bodies, that are devoid of Motion, and which being diftracted, have no faculty of Reftitution, whereby to recollect their diffociated particles, and fo repair themfelves; of which fort are Walls, Pavements, &c: and fuch bodies that are actually in motion, and which by reafon of a natural Elater, or Spring of Reftitution, eafily and fpeedily redintegrate themfelves, and reftore their fevered parts to the fame contexture and tenour, which they held before their violent diftraction; to which claffis the Aer doth principally belong. Now, concerning the *Firft* fort, what we object of the non-impreffion of any Contrary motion upon Bodies impinged againft them, is moft certainly true: but not concerning the *Latter.* For, the Arm of a Tree, being inflected, doth not only refift the inflecting force, but with fuch a fpring return to its natural fite, as ferveth to impel any body of competent weight, that fhall oppofe its recurfe, to great diftance; as in the difcharge of an Arrow from a Bow. Thus alfo the Aer, though otherwife unmoved, may be fo diftracted by a Body violently pervading it, as that the parts thereof, urged by their own native Confluxibility (the Caufe of all Elaterical or Reftorative Motion) muft foon return to their natural tenour and fite, and not without a certain violence, and fo replenifh the

place

place formerly poſſeſt, but now deſerted by the body, that diſtracted them. That there is ſo powerful a *Reſtorative* faculty in the Aer, as we here aſſume, innumerable are the Experiments, thoſe eſpecially by Philoſophers uſually alledged againſt a Vacuum Coacervate, which atteſt. However, that you may the leſs doubt of its having ſome, and a conſiderable force of propelling bodies notwithſtanding it be *Fluid* in ſo high a degree· be pleaſed only to reflect your thoughts upon the great force of Winds; which tear up the deepeſt and firmeſt rooted Cedars from the ground, demoliſh mighty Caſtles, overſet the proudeſt Carracts, and rowle the whole Ocean up and down from ſhoar to ſhoar. Conſider the incredible violence, wherewith a Bullet is diſcharged from a Wind-Gun, through a firm plank of two or three inches thickneſs. Conſider that no effect is more admirable, than that a very ſmall quantity of Flame ſhould, with ſuch prodigious impetuoſity, drive a Bullet, ſo denſe and ponderous, from a Cannon, through the Gates of a City, and at very great diſtance: and yet the Flame of the Gunpowder is not leſs, but more Fluid than Aer. Who, without the certificate of Experience, could believe, that meerly by the force of ſo little Flame (a ſubſtance the moſt Fluid of any, that we know) not onely ſo weighty a Bullet ſhould be driven with ſuch pernicity forward through the aer to the diſtance of many furlongs; but alſo that ſo vaſt a weight, as a Cannon and its Carriage bear, ſhould at the ſame time be thereby driven backwards, or made to recoyle? What therefore will you ſay, if this could not come to paſs, without the concurrence of the Aer? For, it ſeems to be effected, when the Flame, at the inſtant of its Creation, ſeeking to poſſeſs a more ample room, or ſpace, doth convert its impetus, or violence as well upon the breech, or hinder part of the Canon, as upon the bullet lying before it in the bore or Cartridge; which diſcharged through the concave, is cloſely preſt upon by the purſuing flame: ſo that the flame immediately periſhing, and leaving a void ſpace, the Aer from the front or adverſe part inſtantly ruſheth into the bore, and that with ſuch impetuous pernicity, as it forceth the Cannon to give back, and yeilds a Fragor, or Report, as loud as Thunder; nay, by the Commotion of the vicine Aer, ſhakes even the largeſt ſtructures, and ſhatters Glaſs-windows, ſituate in the ſphere of its violence. And all meerly from the Elaterical Motion of the Aer, reſtoring its diſtracted parts to their natural tenour, or Laxity: ſo that you may be ſatisfied of its Capacity not only to reſiſt the Aſcent of a ſtone thrown upward; but alſo of *Depelling it downward*, *by an impreſt Motion.*

*Art. 9.*
That nevertheleſs, when a ſtone, projected on high in the Aer, is at the higheſt point of its mountee; no Cauſe can begin its Downward Motion, but the *Attractive Vertue of the Earth.*

Notwithſtanding our conqueſt of the main body of this Difficulty, about the Reſtorative Motion of the Aer, we are yet to encounter a formidable Reſerve, which conſiſts of theſe *Scruples.* When a ſtone is thrown upward, doth not the Aer in each degree of its aſcent, ſuffer a *Diſtraction* of its parts; and ſo is compelled by a *Perioſis*, or circular motion, to ſucceed into the place left below by the ſtone? Doth it not therefore impreſs rather an Elevatory, than a Depulſive Force thereupon, and ſo promote

the

the force impreſt upon it by the hand of Him, who projected it ? And muſt it not thence follow, that the firſt impreſt motion is ſo far from being decreaſed by the ſuppoſed Renitency of the ſuperior Aer, that it is rather increaſed and promoted by the Circulation thereof : and upon conſequence, that the ſtone is carried upward twice as ſwiftly, as it falls downward, ſince it is impelled upward by two forces, but falls down again only by a ſingle force ? True it is, that while a ſtone is falling down, the diſtra-ēted aer beneath ſeems to circulate into the place above deſerted thereby; but, in caſe a ſtone be held up on high in the Aer by a mans hand, or other ſupport, and that ſupport be withdrawn ſo gently, as to cauſe no conſiderable commotion in the Aer; in this caſe there ſeems to be no reaſon, why the Aer ſhould flow from above down upon it in the firſt moment of its de-lapſe. Beſides, when a ſtone projected upward, hath attained to the higheſt point of its aſcent, at which there ſeems to be a ſhort pauſe, or re-ſpite from motion, cauſed by the æquilibration of the two Contrary Forces, the Movent and Reſiſteut : why doth not the ſtone abſolutely *quieſce* in that place, there being in the Aer no Cauſe, which ſhould rather Depel it downward, then elevate it upward ?

Theſe conſiderations, we ingenuouſly confeſs, are potent, and put us to the exigent of exploring ſome other External Principle, beſide the motion of Reſtitution in the Aer; ſuch as may *Begin* the Downward motion of the ſtone, when gently dropt off from ſome convenient ſupporter, or when it is at the zenith or higheſt point of its aſcent, and and at the term of its Æquilibration overcome the Reſiſtence of the ſubjacent Aer, that ſo it may not only yeeld to the ſtone in the firſt moment of its Deſcent; but by ſucceſſive Circulations afterward promote and gradually accelerate its mo-tion once begun. Depellent Cauſe there can be none, and ſo there muſt be ſome *attrahent*, to begin the ſtones præcipitation : and that can be no other, but a *Certain peculiar Virtue of the whole Terreſtrial Globe, whereby it doth not onely retain all its Parts, while they are contigu-ous or united to it, but alſo retraēt them to it ſelf, when by any violence they have been avulſed and ſeparated.* And this Virtue may therefore be properly enough called *Magnetique.*

In Nature, nothing is whole and entire, in which there is not radi-cally implanted a certain ſelf-Conſervatory Power, whereby it may both contain its ſeveral parts in cohærence to it ſelf, and in ſome mea-ſure reſiſt the ſeparation or diſtraction of them; as all Philoſophers, upon the conviction of infinite Experiences, decree : and if ſo, it were a very partial Abſurdity to bereave the Terraqueous Globe, being a Body whole and entire, of the like conſervatory Faculty. And hence comes it, that if any Parts of the Earth be violently avelled from it; by this Conſervatory, (which muſt be *Attractive*) Virtue, it in ſome meaſure reſiſteth their avulſion, and after the ceſſation of the Avelling violence, retracteth them again; and this by inſenſible Emanations, or ſubtile threads, deradiated continually from its whole body, and hookt or faſtned to them : as a man retracts a Bird flown from his hand, by a line or thread tyed to its feet.

*Art. 10.*
*Argument, that the Terraque-ous Globe is endowed with a certain Attra-ēive Faculty, in order to the Detention and Retraction of all its Parts.*

*Art.* 11.
What are the
*Parts* of the
Terreſtrial
Globe.

By the *Parts of the Terreſtrial Globe* we intend not only the parts of Earth and Water (the liquid part of the Earth, and as Blood in an Animal) nor only all ſtones, Metalls, Minerals, Plants, Animals, and whatever Bodies derive their principles from them, ſuch as Rain, Dew, Snow, Hail, and all Meteors, Vapours, and Exhalations; nor only the Aer, wherewith the globe of Earth is circumveſted, as a Quince or Malacotone is periwigg'd about with a lanuginous or Hoary ſubſtance, (becauſe, if we abſtract from the ſurface of the Earth all vapours, expirations, fumes, and emanations of ſubtle bodies from water and other ſubſtances, which aſcend, deſcend, and everywhere float up and down in the Atmoſphere; nothing can remain about the ſame, but an Empty ſpace,) but alſo Fire it ſelf, which hath its original likewiſe from terreſtrial matter, as wood, oyl, fat, ſulphur, and other unctuous and combuſtible ſubſtance. Becauſe all theſe are Bodies, which as Parts of it ſelf the Earth containeth and holds together; not permitting any of them to be avelled from its orbe, but by ſome force that exceeds its retentive power: and when that avellent force ceaſeth, it ſuddainly retracts them again to it ſelf. And, inſomuch as two bodies cannot coexiſt in one and the ſame place at once; therefore comes it to paſs, that many bodies being at once retracted toward the Earth, the more terrene are brought neerer to the ſurface thereof, extruding and ſo ſucceeding into the rooms of the leſs terrene: whence the neerer adduced and Extruding Bodies are accounted *Heavy*, and the Extruded and farther removed, are accounted *Light*.

*Art.* 12.
A Second *Argument* that the
Earth is *Magnetical*.

Secondly, that the Earth is naturally endowed with a certain Magnetical Virtue, by which perpetually diffuſed in round, it containeth its parts in cohærence, and reduceth thoſe, which are ſeparated from it ſelf; after the ſame manner, as a Loadſtone holds its own parts together, and attracts Iron (which is alſo a Magnetique Production, as *Gilbert* (*de magnet. lib.* 1. *cap.* 16.) from the obſervation of Miners, and other ſolid reaſons, hath confirmed) to it ſelf, and retracts it after divulſion or ſeparation: we ſay, all this may be argued from hence, that the whole Globe of the Earth ſeems to be nothing but one *Grand Magnet*.

(1) Becauſe a Loadſtone, tornated into a ſphere, is (more than Analogically only) a Little Earth: being therefore nicknamed by *Gilbert* (*de magnet. lib.* 1. *cap.* 3.) Μικρόγη, *Terella*; inſomuch as the one, ſo alſo hath the other its Poles, its Axis, Æquator, Meridian, Paralels.

(2) Excepting only ſome parts, which have ſuffered an alteration and diminution, if not a total amiſſion of Virtue, in the Exteriors of the Earth; all parts thereof diſcover ſome magnetick imprægnation: ſome more vigorous and manifeſt, as the Loadſtone, and Iron; others more languid and obſcure, as White Clay, Bricks, &c.

Whereupon *Gilbert* erects his conjectural judgement, that the whole Globe Terreſtrial is compoſed of two General parts, the ſhell, and Kernel: the *Shell* not extending it ſelf many hundred fathoms deep (which is very ſmall comparatively to the vaſtneſs of its Diametre, amounting to 6872 miles, Italian meaſure) and all the reſt, or Kernel, being

one

one continued Loadstone substantially. (3) The Loadstone always converteth those parts of it self toward the Poles, which respected them in its mineral bed, or while it remaind united to the Earth. All which are no contemptible Arguments of our Thesis, that the whole Earth is endowed with a magnetique Faculty, in order to the Conservation of its Integrity.

Whether the Entrals of our Common Mother, and Nurse, the Earth, be, as *Gilbert* would persuade us, one Great Loadstone *substantially*; is not more impossible to prove, than impertinent to our present scope : it being sufficient to the verisimility of our assigned Cause of the perpendicular motion of Terrene Bodies, to conceive the Globe of the Earth to be a Loadstone only *Analogically*, i. e. that as the Loadstone doth perpetually emit certain invisible streams of exile particles, or Rays of subtile bodies, whereby to allect magnetical bodies to an union with it self; so likewise doth the Earth uncessantly emit certain invisible streams, or Rays of subtile bodies, wherewith to attract all its distracted and divorced Parts back again to an Union with it self, and there closely to detain them. And justifiable it is for us to affirm, that from the Terraqueous Orbe there is a continual Efflux, not only of Vapours, Exhalations, and such small bodies, of which all our Meteors are composed; nor only of such, as the general mass of Aer doth consist of : but also of other particles far more exile and insensible, nor less subtile than those, which deradiated from the Loadstone, in a moment permeate the most solid Marble, without the least diminution of their Virtue. — Because, as the Attractive Virtue of the Loadstone is sufficiently demonstrated by the Effect of it, the actual Attraction of Iron unto it : so is it lawful for us to conclude the Earth to be endowed with an Attractive Virtue also, meerly from the sensible Effect of that Vertue, the actual Attraction of stones, and all other bodies to it self; especially since no other Conception of the Nature of that Affection, which the world calls Gravity, can be brought to a cleer consistence with that notable Apparence, the gradual Encrease of Velocity in each degree of a bodies perpendicular fall.

*Art.* 13.
A Parallelisme betwixt the Attraction of Iron by a Loadstone, and the Attraction of Terrene bodies by the Earth.

Besides, the *Analogy* may be farther deduced from hence; that as the Virtue of the Loadstone is diffused in round, or spherically, and upon consequence, its Effluvia, or Rays are so much the more rare, by how much the farther they are transmitted from their source or original; and so being less united, become less vigorous in their attraction, and at large distance, i. e. such as exceeds the sphere of their Energy, are languid and of no force at all : so doth the Terrestrial Globe diffuse its Attractive Virtue in round, and upon consequence, its Effluvia, or Rays become so much the more rare or dispersed, by how much farther they are transmitted from their fountain; and so being less united, cannot attract a stone or other terrene body at excessive distance, such as the *Supralunary* and *Ultramundane* spaces. Which that we may assert with more perspicuity, let us suppose a stone to be placed in those Imaginary spaces which are the outside of the World, and in which God, had He so pleased,

*Art.* 14.
That as the sphere of the Loadstones Allective Virtue is limited; so is that of the Earths magnetism.

might have created more worlds; and then examine, whether it be more reasonable, that that stone should rather move toward this our Earth, than remain absolutely immote in that part of the Ultramundan spaces wherein we suppose it posited. If you conceive, that it would tend toward the Earth; imagine not only the Earth, but also the whole machine of the world to be Annihilated, and that all those vast spaces, which the Universe now possesseth, were as absolutely Inane, as they were before the Creation: and then at least, because there could be no Centre, and all spaces must be alike indifferent, you will admit, that the stone would remain fixt in the same place, as having no Affectation, or Tendency to this part of those spaces, which the Earth now possesseth. Imagine the World to be then again restored, and the Earth to be restituate in the place as before its adnihilation; and then can you conceive that the stone would spontaneously tend toward it ? If you suppose the *Affirmative*; you will be reduced to inextricable difficulties, not to grant the Earth to affect the stone, and upon consequence, to transmit to it some certain Virtue, consisting in the substantial Emanations, not any simple and immaterial Quality, whereby to give it notice of its being restored to its pristine situation and condition. For, how otherwise can you suppose the stone should take cognizance of, and be moved toward the Earth. Now, this being so, what can follow, but that stones, and all other Bodies accounted Heavy, must tend toward the Earth, only because they are *Attracted* to it, by rays or streams of Corporeal Emanations from it to them transmitted ? Go to then, let us farther imagine, that some certain space in the Atmosphere, were, by Power supernatural, made so Empty, as that nothing could arrive thereat either from the Earth, or any other Orbe : can you then conceive, that a stone placed in that Inanity, would have any Tendency toward the Earth, or Affectation to be united to its Centre ? Doubtless, no more, than if it were posited in the Extramundan spaces : because, having nothing of Communication therewith, or any other part of the Universe, the case would be all one with the stone, as if there were no Earth, no World, no Centre. Wherefore, since we observe a stone from the greatest heighth, to which any natural force can elevate the same, to tend in a direct or perpendicular line to the Earth ; what can be more rational than for us to conceive, that the Cause of that Tendency in the stone is onely this, that it hath some communication with the Earth ; and that not by any naked or Immaterial Quality, but some certain Corporeal, though most subtile Emanations from the Earth ? Especially, since the Aer incumbent upon the stone, is not sufficient to Begin its motion of Descent.

*Art.* 15.
An *Objection* of the Disproportion between the great Bulk of a large stone and the Exility of the supposed magnetique Rays of the Earth: *Solved by three weighty Reasons.*

If you shall yet withhold your Assent from this Opinion, which we have thus long endeavoured to defend ; we conjecture the Remora to be chiefly this : that it seems improbable, so great a Bulk, as that of a very large stone, and that with such pernicity, should be attracted by such slender means, as our supposed magnetick Emanations : and therefore think it our duty to satisfie you concerning this Doubt. We Answer (1) That a very great quantity of Iron (proportionately) is easily and nimbly rusht into the arms of a Loadstone meerly by Rays of most subtile particles, such as can be discovered no way, but by their Effect. (2) That stones, and other massy Concretions have no such great ineptitude, or Resistence to

motion,

motion, as is commonly præsumed. For, if a stone of an hundred pound weight be suspended in the Aer, by a small wier, or chord: how small a force is required to the moving of it hither? Why therefore should a greater force be required to the Attraction of it downward. (3) When you lift up a stone or other body from the Earth, you cannot but observe that it makes some Resistence to your Hand, more or less according to the bulk thereof; which Resistence ariseth from hence, that those many magnetique lines, deradiated to, and fastned upon it, by their several Deflexions and Decussations, hold it as it were fast chained down to the Earth, so that unless a greater force intervene, such as may master the Earth Retentive power, and break off the magnetique lines, it could never be avelled and amoved from the Earth. And hence is it, that by how much the greater force is imprest upon a stone, at its projection upward; by so many more degrees of excess doth that imprest force transcend the force of the Retentive Magnetique lines, and consequently to so much a greater Altitude is the stone mounted up in the Aer: and è *contra*. Which is also the Reason, why the Imprest Force, being most vigorous in the first degree of the stones ascent, doth carry it the most vehemently in the beginning; because it is not then Refracted: but afterward the stone moves slower and slower, because in every degree of ascention, it looseth a degree of the Imprest Force, until at length the same be so diminished, as to come to an Æquipondium with the Contrary force of the magnetique Rays of the Earth detracting it Downward.

Lastly, from hence is it, that the perpendicular Delapse of most Bodies, though of far different weights, is observed to be Æquivelox; contrary to that Axiome of *Aristotle* (2. *de Cælo*, *text* 46) *quo majus fuerit corpus, eo velocius fertur,* and (*text* 77.) *parvum terræ particulum, si elevatu dimittatur, ferri deorsum, quo major fuerit, velocius moveri;* upon which the Aristoteleans have grounded this erroneous Rule, *Velocitates gravium descendentium habere inter se eandem proportionem, quàm gravitates ipsorum,* that the Velocities of Heavy bodies falling downward have the same proportion one to another as their Gravities have.

*Art.* 16.
The Reason of the Æquivelocity of Bodies, of different weights, in their perpendicular Descent: with sundry unquestionable Authorities to confirm the Hoti thereof.

And the Reason of this Æquivelocity of Unequal weights, seems to be this; that of two Bullets, the one of only an ounce, the other of an hundred pounds weight, dropt from the battlements of an high tower, at the same instant, though the Greater Bullet be attracted by more magnetique lines deradiated from the Earth, yet hath it more particles to be attracted, than the Lesser: so that there being a certain Commensuration betwixt the Force Attractive, and the quantity of Matter Attracted; on either part the Force must be such, as sufficeth to the performance of the motion of either in the same space of time; and consequently, both the Bullets must descend with equal Velocity, and arrive at the surface of the Earth in one and the same moment. All which that Lynceus, *Galilæo* well understood, when (in the Person of *Salviatus*) desiring to calculate the time, in which a Bullet might be falling from the concave of the Moon to our Earth; and *Sagredus* had said thus to Him, *Sumamus igitur globum determinati ponderis,*

*ponderis, & quidem illum ipsum, cujus descentionis ex Luna tempus metiri volumus :* He positively answered Him, *Id vero nihil interest, &c.* It makes no difference whatever the weight of the Bullet be, because if four Bullets, the one of one pound, the second of ten, the third of an hundred, the fourth of a thousand pounds weight, be let fall together from the altitude of an hundred cubits, they shall all perform their perpendicular motions in the same proportion of time, and attain the Earth in the same moment. (*Dialog. 2. de systemat. cosmico, pagina Latina* 1 4.) The same also exactly consists with the frequent Experiments of *Joh. Baptista Balianus,* who (*in lib. 1. de motu Gravium, pag.* 4.) saith thus ; *Inter alia dum anno millesimo sexcentesimo undecimo, per paucos menses, ex patriæ legis præscripto, Præfectum Arcis savonæ agerem, ex militaribus observationibus quæ occurrebant, illud maxime deprehendi, ferreos, & lapideos tormentorum bellicorum globulos, & sic corpora gravia, seu ejusdem, seu diversæ, speciei, in inæquali satis Mole, & gravitate, per idem spatium æquali tempore & motu, naturaliter descendere ; idque ita uniformiter, ut repititis experimentis mihi plane constiterit, duos ex prædictis globis, vel ferreos ambos, vel alterum lapideum, alterum plumbeum, eodem plane momento temporis dimissos sibi, per spatium quinquaginta pedum, etiam-si unus esset libræ unius tantum, alter quinquaginta, in indivisibili temporis momento, subjectum solum ferire, ut unus tantum amborum ictus sensu perciperetur.*

To this Certificate we might subscribe the concurrent testimonies of *Nich. Cabæus* (*in meteor lib. 1. text. 11. quæst.* 5, & 6.) of *Arriga* (*Disputatione* 4. *de Generatione, Sectione* 5, *subsectione* 3) of *Gassendus* (*de motu impresso à motore translato, Epist.* 1.) : but we think it better, to refer our Reader to the touchstone of his own easie and cheap *Experiment,* as the most certain way of conviction.

**Art. 17.**
*That the whole Terrestrial Globe is devoyd of Gravity : and that in the universe is no Highest, nor Lowest place.*

Moreover, insomuch as the Terrestrial Globe, considered in its whole, hath no need of any Direct or Perpendicular motion, whereby to tend to its proper place in the Universe, because it never receedeth from its proper place therein : but the Parts of it only have need of a Direct or Perpendicular motion, whereby they may be reduced to their proper place, the whole Earth, from which they are frequently separated : therefore must it have been unnecessary for the Creator to have endowed the whole Terrestrial Globe with Gravity or any Force, whereby it might be directly carryed to a place, out of which it should be constitute ; and sufficient only to endowe it in the whole with such an Attractive Virtue, whereby it might retain its parts in adhærence to it self, and retract them to an union, when violently distracted from it. For, that Motion Direct or Perpendicular, which the Vulgar ascribes to Gravity, is *Motus Unitivus,* a Motion Unitive or Congregative of all the Parts of the Earth ; as may be argued from hence, that it is the same in the Antipodes, as in our Hemisphere, and from all points of the Earths circumference conspires to one and the same common Centre. But, though this motion is Congregative of all Parts of the Earth related or brought back to an union

with

with the great body or Globe thereof; yet is it not Congregative of the whole Globe to any thing else, as if the Globe of the Earth were to be united to the Moon, or any other Orbe in the World. Nor can it be affirmed, that Gravity, or this Virtue to motion Direct, is conceded to the Terraqueous Orbe, to the end it should, at the Creation, carry it self to that place, which is Lowest in the Universe; or being there posited, constantly retain it self therein: since in the Universe is neither Highest, nor Lowest place, but only Respectively to the site of an Animal, and chiefly of Man, whose Head is accounted the Highest, and Feet the Lowest part; in the same manner as there is no Right, nor Left side in Nature, but comparatively to the site of the parts in mans body, and in reference to the Heavens. For, those Lateralities are not determined by any general and certain standard in Nature: but variously assigned according to our Imagination. The Hebrews, Chaldeans, and Persians, confronting the Sun at his arising in the East; place the Right side of the world in the South: as likewise did all the Roman Soothsayers, when they took their Auguries. The Philosopher takes that to be the East, from whence the Heavens begin their Circumgyration: and so assigns also the right hand to the South. The Astronomer, regarding chiefly the South and Meridian Sun, accounts that the Dextrous part of Heaven, which respecteth his right hand, and thats the West. And Poets, differing from all the rest, turn their faces to the West, and so assign the term of Right to the North: for otherwise *Ovid* must be guilty of a gross mistake in that verse, *Usque dua dextrá zona, totidemque sinistrá.* Hence is it, that as the East cannot be the Right side of the World, unless to Him, who faceth the North: so is the Vertical point of the world not to be accounted the Highest part of the Universe, but onely as it respecteth the Head of a man standing on any part of the Earth; because, if the same man travail to the Antipodes, that which was before the Highest, will then be the Lowest part of the World. This considered, we must præfer that solid opinion of *Plato*, that in the World there is an Extreme, and a Middle Place, but no Highest and Lowest; to that meerly petitionary one of *Aristotle*, that all Bodies tend toward the Centre of the Earth, as to the Lowest place in the Universe.

How, saith the offended *Peripatetick*, the *meerly Petitionary opinion of Aristotle*? Why, do not all men admit that to be the Lowest part of the World, which is the Middle or Centre thereof? And is not that the Centre of the Earth?

*Art.* 18.
That the Centre of the Universe is not the Lowest part thereof: nor the Centre of the Earth, the Centre of the World.

And our *Reply* is, that, indeed, we can admit *Neither*. (1) Because, should we allow the World to have a Middle, or Centre; yet is there no necessity, that therefore we should concede the Centre to be the Lowest place in the World; no more than that the Navil, or Central part of a man should therefore be the Lowest part. For, to speak like men, who have not enslaved their reason to præjudice; what is opposed to the *Middle*, is not suprem, but *Extreme*: and Highest and Lowest are opposite points in the same Extreme. So likewise in the Terrestrial Globe, whose middle part we account not the Lowest, but the contrary point in the sphear: since, otherwise we must grant the Earth to have a double *Infinity*, one in regard of its Centre, the other in respect of the extreme points

points of its Diametre, according to which the Antipodes are Lowest to us, and we Lowest to them.

(2) Who dares prætend to demonstrate, that there is an Extreme in the Universe; or if there be, to determine where and what it is: and upon consequence, whether the Universe hath any Centre, and where that Centre is? Tis more than *Galilæo* durst, as appears by that his modest confession; *Nescimus quidem ubi sit Universi centrum, neque an sit: quodque, si maxime detur, aliud nihil est, nisi punctum imaginarium, adeoque nihilum, omni facultate destitutum.* (*systemat Cosmici dialog.1 pag.22* ) Besides, we see it to be, and upon very good grounds, disputed amongst the most Curious and Learned wits of the world, whether the Fixt stars are moved about the Earth, or the Earth by a Diurnal motion upon its own axis? Whether the Fixt stars be all in one and the same concave superfice: or rather (as the Planets, which notwithstanding the deluded sight, are demonstrated not to be in one, but different spheres) some farther from, some neerer to the Earth, dispersed in the immense space? For, from hence, that the Distance betwixt them and us is so vast, that our sight not discerning the large spaces intercepted betwixt them in their several orbes; they all appeare at the same distance, all in the same circumference, wose Centre must be there, where the Eye turning it self about, doth behold them: so that in whatsoever part of the immense space of the World, whether in the Moon, Sun, or any other Orb, you shall imagine your self to be placed; still you must, according to the evidence of your sight, judge the World to be spherical, and that you stand in the very centre of that Circumference, in which you conceive all the Fixt stars to be constitute.

Truly, it is worthy the admiration of a wise man, to observe, that the very Planets are admitted by the *Aristoteleans* to have certain motions Excentrique, i.e. to be moved in such Gyres, as have not their Centres in the Earth, but in places immensly distant from it: and yet that the same Persons should so openly Contradict themselves, as to account that the Centre of the Earth is that common Centre of the world, about which all the Cœlestial orbes ar circumduced. These Difficulties perpended, we cannot infallibly determine, whether or no Earthy Bodies, when descending in direct lines to the Earth, are carried toward the Centre of the World: and though they should be carried toward the Centre of the World, yet doth that seem to be only by Accident, as it is also by Accident, that they are carried toward the Centre of the Earth; in which as being a meer imaginary Point, they can neither be received nor attain quiet. For, *per se*, they are carried toward the Earth, as to their Whole, or Principle; and having once attained thereto, so acquiese on the surface of it, as they no more seek to pass on from thence to its Centre, than an Infant received into his Nurses armes or lap, cares to sink farther into her Entrals: and meerly *per Accidens* is it, that they are directed toward the Centre of the Earth; because tending in the neerest cut, or shortest line to the place of their quiet, they must be directed toward the Centre, since if we suppose that direct line to be continued, it must pass through the Centre of the Earth. And thus have we left no stone unsubverted, in all *Aristotles* Theory of Gravity, which is, *that Weight is a Quality essentially inherent in all terrene Concritions, whereby they spontaneously tend toward the Centre of the Terrestrial Globe, as to the Common Centre, or Lowest place in the Vniverse.* The whole Remainder of our præsent Assumption, therefore, concerns our farther Confirmation of that opinion touching the Essence of Gravity, which we have espoused; which is, that *it is the meer Effect of the Magnetique Attraction of the Earth.*

Let

Let us therefore once more refume our Argument *à Simili*, confidering
the Analogy betwixt the Attraction of Iron by a Loadftone, and that of
Terrene Concretions by the Earth ; not only as to the Manner of their refpective Attractions, but cheifly as to the parity of Reafons in our judgements upon their fenfible Effects.   When a man holds a plate of Iron of 6
or 7 ounces weight, in his hand, with a vigorous Loadftone placed at convenient diftance, underneath his hand ; and finds the weight of the Iron to
be encreafed from ounces to pounds : If *Ariftotle* on one fide fhould tell
him, that that great weight is a Quality effentially inhærent in the Iron, and
*Kepler* or *Gilbert*, on the other, affirm to him, that that weight is a quality
meerly *Adventitious*, or impreft upon it by the Attractive influence of
the Loadftone fubjacent ; 'tis eafie to determine, to which of thofe fo contrary judgements he would incline his affent.   If fo, well may we conceive
the Gravity of a ftone, or other terrene body, to belong not fo much to the
Body it felf, as to the Attraction of that Grand Magnet, the Terraqueous
Globe lying underneath it.  For, fuppofing that a Loadftone were, unknown
to you, placed underneath your hand, when you lifted up a peice of Iron
from the earth ; though it might be pardonable for you to conclude, that
the great weight, which you would obferve therein, was a Quality effentially inhærent in the Iron, when yet in truth it was only External and At-
tractitious ; becaufe you were ignorant of the Loadftone fubjacent ; yet, if
after you were informed that the Loadftone was placed underneath your
hand, you fhould perfevere in the fame opinion, the greateft Candor imaginable could not but condemn you of inexcufable pertinacity in an Error.  Thus
alfo your ignorance of the Earths being one Great Loadftone may excufe
your adhærence to the erroneous pofition of *Ariftotle*, concerning the
formal Reafon of Gravity ; but, when you fhall be convinced, that the
Terreftrial Globe is naturally endowed with a certain Attractive or Magnetique Virtue, in order to the retention of all its parts in cohærence to it felf,
and retraction of them when by violence diftructed from it, and that gravity is nothing but the effect of that virtue ; you can have no Plea left for the
palliation of your obftinacy, in cafe you recant not your former perfuafion.

*Art.* 19.
A Fourth Argu-
ment, that Gra-
vity is only
*Attraction.*

Nor ought it to impede your Conviction, that a far greater Gravity, or
ftronger Attractive Force is impreft upon a piece of Iron by a Loadftone,
than by the earth, infomuch as a Loadftone fufpended, at convenient diftance,
in the aer, doth eafily elevate a proportionate mafs of Iron from the earth ;
becaufe this gradual Difparity proceeds only from hence, that the Attractive Vertue is much more Collected or United in the Loadftone, and fo is
fo much more intenfe and vigorous (according to its Dimenfions) than in the
Earth, in which it is more diffufed ; nor doth it difcover how great it is in
the fingle or divided parts, but in the Whole of the Earth.  Thus, if you
lay but one Grain of falt upon your tongue, it fhall affect the fame with
more faltnefs, than a Gallon of Sea-water : not that there is lefs of falt in
that great quantity of Sea Water, but that the falt is therein more diffufed.

*Art.* 20.
Why a greater
Gravity, or
ftronger At-
tractive force
is impreft up-
on a piece of
Iron by a
Loadftone,
than by the
Earth.

But to lay afide the Loadftone and its Correlative, Iron, and come to
our tafte and *Incomparative Argument* ; fince the Velocity of the motion
of a ftone falling downward, is gradually augmented, and by the acceffion

*Art.* 21.
A Fifth Argu-
ment, almoft
Apodical ;
that Gravity is
the Effect of
the Earths At-
traction.

P p                                         of

of new degrees of Gravity, grows greater and greater in each degree of its Defcent; and that Augmentation, or Acceffion of Gravity, and fo of Velocity, feems not fo reafonably adfcriptive to any other caufe, as to this, that it is the Attraction of the Earth encreafing in each degree of the ftones Appropinquation to the Earth, by reafon of the greater Denfity or Union of its Magnetique Rayes: What can be more Rational, than that the *Firft degree of Gravity*, belonging to a ftone not yet moved, fhould arife to it from the fame Attraction of the Earth? When, doubtlefs, it is one and the fame Gravity that caufeth both thofe Effects, the fame *in Specie*, though not *in Gradu*: And no Quality can be better intended, or augmented, than by an Acceffion of more Degrees of force from the fame Quality,

---

## Sect. III.

*Art.* I.
*Levity, nothing but lefs Gravity.*

Aftly, as concerning LEVITY, which is vulgarly reputed the Contrary to Gravity, and by *Ariftotle* defined to be a Quality inharrent in fome Bodies, whereby they fpontaneoufly tend upward, we underftand it to be nothing but a *lefs Gravity*: and fo that Gravity and Levity are Qualities of Concretions, not Pofitive, or Abfolute, but meerly *Comparative*, or *Refpective*. For, the fame Body may be fayd to be Heavy, in refpect to another that is Lighter; and Light, in refpect to another that is Heavier. For Example, let us compare a Stone, Water, Oyle, and Fire (which we have formerly annumerated to Terrene Concretions) one to another; to the end that our Affertion may be both illuftrated and confirmed at once. Water, we fee, being poured into a veffel, immediately defcends to the bottom thereof, and if permitted to fettle, doth foon acquiefce: but, upon the dropping of a Stone into the fame veffel, as the Stone defcends, the Water afcends proportionately to give it room at the bottom. And Oyle, infufed into a veffel alone, doth likewife inftantly defcend, and remains quiet at the bottom thereof: but, if Water be poured thereupon, the oyle foon afcends, and floats on the furface of the Water. If the Veffel be repleat only with Aer, the Aer quiefceth therein: but when you pour oyle into it, the Aer inftantly afcends, and refignes to the oyle. Laftly, thus Fire would be immediately incumbent upon the furface of the Earth, and there acquiefce; but that the Aer, being circumftant about the fuperfice of the Terreftrial Globe, and the more weighty body of the two, doth extrude it thence by depreffure, and fo impell it upwards, to make room for it felf beneath. And thus are all thefe Bodies Heavy and Light, *Comparatively* or *Refpectively*. The Heavieft of them all is the Stone, as being the moft ftrongly attracted by the Earth: or, is the leaft Light among them all, as being the leaft abduced from the Earth. And, Water, which is Light, in comparifon of the Stone, is yet Heavy in comparifon of Oyle:

and

And Oyle, though Light in comparison of Water, is yet Heavy in comparison of the Aer. And Aer, though Light comparatively to Oyle, is yet Heavy in respect of Fire; which is the Lighteſt of them all, becauſe it is the moſt elevated from the Earth: Or is the leaſt Heavy among them all, becauſe it is the leaſt attracted by the Earth.

This conſidered, we cannot but ſmile at their Credulity, who can admit *Ariſtotles* dream of a peculiar *Sphere of Fire*; and thereupon contend, that Fire ſpontaneouſly aſcends in queſt of its ſphere: When it is manifeſt, that Fire doth not mount up upon the wings of any native Tendency, or of that Imaginary Faculty, call'd Levity; but is driven upward by the impulſion of the Aer. Who is there dares affirm, that oyl, when pour'd forth of a veſſel by ſome expert Diver, in the bottome of the Sea, doth aſcend to the top of the water, in queſt of a Sphere of Oyle? or that Water, elevated to the brim of a veſſel upon the injection of ſand into the ſame veſſel, doth aſcend ſpontaneouſly, and in purſuit of a Sphere of Water? Or that Aer, deſcending into a mine, doth ſpontaneouſly deſcend in queſt of an Aereal Sphere? Or, that Fire it ſelf, when it ſtoops down to catch hold of ſome uncteous and eaſily inflammable ſubſtance, as is often noted, doth ſtill obey its eſſential Levity, in order to its reunition to its proper Sphere? And yet for all this, the world is full of thoſe (ſo epidemick is the contagion of Præjudice) who dare affirm that ridiculous and groſsly abſurd Figment of the Aſcenſion of Fire to an Igneous Sphere conſtitute we know not where belowe the Moon.

*Art. 2.*
*Ariſtotle's* Sphere of Fire, extinguiſht.

But we are yet to prove, that *Fire is impelled upward by the Aer.* Conſider therefore, that Fire will not burn in a chymny, if all the doors, windows and chinks of the room be ſo cloſely ſhut, as that no ſupply of freſh Aer can be admitted into it : and the Reaſon is plainly this, that unleſs there be a ſource of freſh Aer to ſucceed into the place of that, which impels the Fire upward in the chimny, there can be no Continuation of the impuls or elevation of the Fire, and ſo the Fire muſt be extinguiſhed; but when a liberty of ingreſſion is left to the External Aer, then is the Internal Aer cloſely purſued by a freſh ſupply, and ſo the motion continued. Conſider alſo that Fire always burns the clearer and ſooner, if the fewel be laid hollow in a grate of Iron, or upon andyrons; than if it be impoſed flatly upon the bare hearth: becauſe, in the former caſe, the ambient Aer doth more eaſily and fluently inſinuate it ſelf underneath the Fire, and as it impells the flame upward, fan and blow the coals, like a pair of bellows. And this gave the Chymiſt the hint for the invention of his *Wind-Furnace*, which needs no other bellows but that conſtant ſtream of Aer, which flows in beneath the fewel, and ventilates the coals moſt ſtrongly. And then Conclude, with *Copernicus (lib. 1. cap. 8.) Ignem nihil aliud eſſe, quàm huncit terrenum ſeu famam ardentem; cujus proprium eſt, extendere quà invaſerit: motum autem Extenſivum eſſe à centro ad circumferentiam; ſed terreſtrem illum habitum,*

*Art. 3.*
That Fire doth not Aſcend ſpontaneouſly, but *Violently*; i.e. is impell'd upward by the Aer.

P p 2                                                    *bitum, litum, seu fumum rapi in sublime, & extrudi suum extra locum, ideoque statim languescere tanquam confessâ causâ violentiæ, qua terrestri materia illata fuit: quapropter Levitatem non dari, aut non esse Connaturalem hisce corporibus.* Conclude also, with Us; that in the Earth indeed, there are Direct Motions Upward and Downward: but those Motions are proper only to the *Parts* ( as Gravity and Levity are likewise proper only to the Parts) not to the Whole, or Globe of the Earth.

CHAP.

# CHAP. XII.

# HEAT and COLD.

### Sect. I.

He Genealogy of those sensible Qualities of Concretions which arise from *either* of the three Essential Proprieties of Atoms, in its *Single* capacity, thus far extending it self, here begins that other of those, which result from any *Two*, or *All* of the same Proprieties, in their several *Combinations*, or *Associations*.

*Art.* 1.
The Connection of this to the immediately prece-dent Chapter.

Of this order, the *First* are *Heat*, *Cold*, *Humidity*, *Siccity*; which though the Schools, building on the fundamentals of their Dictator, *Aristotle*, derive immediately and solely from the 4 First Qualities of the vulgar *Elements*, Fire, Aer, Water, Earth; yet, because those reputed Elements are but several Compositions of the Universal matter, and so must desume their respective Qualities from the consociated Proprieties of the same ; and because the original of no one of those Qualities can be so intelligibly made out from any other Principles : therefore doth our reason oblige us, to deduce them only from the *Magnitude*, *Figure*, and *Motion* of *atoms*.

*Art.* 2.
Why the Author deduceth the 4 *First Qualities*, not from the 4 vul-gar *Elements*; but from the 3 *Proprieties* of *Atoms*.

Concerning the First of this Quaternary, H E A T ; we well know, that it is commonly conceived and defined by that relation , it bears to the sense of touching in Animals; or, as it is the Efficient of that passion, or Acute Pain, as *Plato* (in *Timæo*) calls it, which Fire, or immoderate Heat impresseth upon the skin, or other organ of touching ; yet, forasmuch as this Effect, which it causeth in the sensient part of an Animal, is only special and Relative ; therefore ought we to understand its Nature, from some *General* and *Absolute* Effect, upon which that Special and Relative one depends, and that is the *Penetration*, *Discussion* and *Dissolution* of Concretions.

*Art.* 3.
The Nature of *Heat is to be* conceived from its *General Effect* ; *viz.* the *Penetration*, *Discussion*, and *Dissolution* of *Bodies* con-crete.

To

*Art. 4.*
*Heat defined*
*as no Immate-*
*rial, but a*
*Substantial*
*Quality.*

To come therefore to the Determination of its Essence, by the explanation of its Original; by *Heat*, as from our præcedent Disquisition of the Origine of Qualities in General may be præsumed, we do not understand any Aristotelean, *i.e.* naked or Immaterial Quality, altogether abstract from matter: but *certain Particles of matter, or Atoms, which being essentially endowed with such a determinate Magnitude, such a certain Figure, and such a particular Motion, are comparated to insinuate themselves into Concrete Bodies, to penetrate them, dissociate their parts, and dissolve their Contexture*; or, to produce all thus mutations in them, which are commonly adscribed to Heat, or Fire. Not that we gainsay, but Heat may be considered *Abstractly*, or as it is a certain peculiar *Manner*, without which a substance cannot calefie; in which sense *Anaximenes (apud Plutarch. de Frigore primigen.)* may be allowed to have spoken tollerably, when he said, *Neither Heat, nor Cold is substantial*, but affirm only, that it is not any thing abstracted from, and independent upon matter (as most have incircumspectly apprehended) or ought else, in Reality, but *Atoms themselves*, the substantial Principles as of all Concretions; so of all their Faculties or Qualities, and to which, as all Motion, so all Action ought to be imputed.

*Art. 5.*
*Why such*
*Atoms, as are*
*comparated to*
*produce Heat,*
*are to be*
*Named the*
*Atoms of Heat:*
*and such Con-*
*cretions, as*
*harbour them,*
*are to be cal-*
*led Hot, either*
*Actually, or*
*Potentially.*

And albeit these Atoms, from which we derive this noble and most eminent Quality, Heat, be not Hot essentially; yet do they deserve the name of the *Atoms of Heat*, or *Calorifick Atoms*, insomuch as they have a capacity or power to Create Heat, *i.e.* cause that Effect, which consisteth in Subingression, Discussion, Exsolution. Likewise, those Bodies which contain such Atoms, and may emit them from themselves; ought also to be accounted Hot, insomuch as that by the emission of their Calorifick Atoms they are empowered to produce Heat in other bodies: and when they do Actually emit them, *i.e.* give their Calorifick Atoms liberty to pursue their own native Motions, after exsilition; then may they be said to be *Actually Hot*, or Formally Hot, as the Schools phrase it; but, while they contain them within themselves, and hinder their exsilition, they are Hot only *Potentially*. To the *First* of these Differences, we are to refer Fire: To the *Second*, not only all those things, which Physicians call Calefactive Medicaments, such as Wine, Euphorbium, Peper, &c. but also all such as are capable of ignition, combustion, incalescence and the immission of Heat into other bodies objected, such as Wood, Resine, Wax, Oyle, &c. For, all such may be conceived to contain igneous or Calorifick Atoms, which during their revinction or imprisonment in Concretions, cannot pursue their motion, and so not produce Heat; but immediately upon the obtaining of their liberty, or emption, they manifest their nature in the production of heat.

*Art. 6.*
*The 3 necessa-*
*ry Proprieties*
*of the Atoms*
*of Heat.*

Now, if we enquire What kind of Atoms these Calorifick ones are, and upon what their power of producing Heat depends; *Democritus, Epicurus,* and all the tribe of *Atomists* unanimously tell us, that they are *Exile in Magnitude, Spherical in Figure, most Swift in Motion*. And this upon very good reason. For, (1.) That they must be most *Exile in bulk*, is inferrible even from hence, that no Concretion can be so compact and solid, in which they will not find some pores or small inlets, whereat to insinuate themselvss into the Centrals of it, and penetrate thorow its substance; though perhaps not in so great a number, as is required, to the total dissolu-

tion

tion of its Contexture, as in the Adamant, which as Naturalists affirm, no Fire can demolish or diſſolve. (2.) That they ought to be *Spherical in Figure*, is probable, yea neceſſary from hence ; that of all others they are moſt Agile, and evolve themſelves *quoquoverſùm*, on all parts of the Concretion, into which they are admitted. And Geometry teacheth, that no figure is ſo eaſily moved, as a Sphere, whether naturally, or violently. Firſt, *Naturally* ; becauſe, by how much neerer to a Sphere the figure of any ſolid body approacheth, by ſo much the more ſpeedily doth it deſcend, as is obſerved of globular ſtones in Water : and a round ſtone rowles it ſelf farther and ſwifter downe hill, than a plane or angular one. Secondly, *Violently*, becauſe a globular ſtone may be projected much farther, than one of any other figure. This is alſo evident in the Motion of Volutation ; ſo that the line of direction to the Centre of the World (if any ſuch there be) conſiſting in the axis of the Globe, the motion of it is moſt hardly refracted and arreſted. For, there are 3 points, thorowe which the direct imaginary line, in which alone a Globe can quieſce, muſt paſs, *viz.* the *Centre of the World*, the *Centre of Gravity* in the Globe, and the *point of Contact* : and if either of theſe 3 be without, or beſide the line of quiet, a Globe once moved ſhall never reſt, but be continually moved, until all the 3 points be in the line of direction. Furthermore, how eaſie it is to impel a Globe, is demonſtrable meerly from hence, that being poſited upon a perfect plane, it can touch the ſame but only in one point ; and ſo relying upon that point, may moſt eaſily be deturbed from that ſlender ſupport ; but in all other Figures the reaſon of innixion or Relying, is quite contrary. Laſtly, as a ſphere doth moſt eaſily admit an impreſt motion ; ſo doth it longeſt retain the ſame, moſt violently preſs upon other occurring bodies, and moſt equally diſpence its conceived force ; as hath been profoundly demonſtrated by *Magnenus* (*in theoricæ militaris lib.* 1. *theorem.* 4. & 5 ) (3) And that they muſt be alſo *ſuperlatively ſwift in motion*, may be argued not only *à poſteriori*, from the impetuous diſcuſſion and ſeparation of the particles of bodies by them, and their unceſſant æſtuation among themſelves arietating each other : but alſo *à Priori*, becauſe, being ſpherical, they are moſt mobile. Thus much, at leaſt in importance, we have from *Philoponus* (*in* 1 *phyſic.*) where he ſaith, *Sphæricus Atomos, tanquàm facillimè mobiles, eſſe Caloris, igniſque cauſſas ; quatenus enim ſunt facilè mobiles, dividunt, ſubeuntque velocius : id quippe igni proprium eſt, & dividere, & moveri facilè poſſe.* And albeit *Plato* would not have the Atoms of Fire to be ſpherical, but *Pyramidal* ; becauſe having moſt exile points, ſlender angles, and acute ſides, they might be more accommodate for Penetration or ſubingreſſion : yet, to the *Diviſion* or *Cutting* of bodies, He requires τῶν τε μορίων Σμικρότητα, ϗ τ̃ φορᾶς τάχ̃ι. the Exiguity of particles, and celerity of Motion. So that the Patrons of Atoms præſuming the Calorifick Atoms to be extreamly Exile, *i.e* as ſmall as *Plato* ſuppoſeth the points and angles of his Pyramids to be : we do not perceive any conſiderable difference betwixt their opinion and his. But before we take off our pen from this ſubject, we are to advertiſe ; that indeed all Atoms, of their own nature, are inexcogitably ſwift ; and ſo that our aſſertion of the ſuperlative Velocity of Calorifick Atoms, doth appertain only to Atoms as they are in Concretions, where their native Velocity and Agility is retarded and diminiſhed by reciprocal cohærence and revinction. And, therefore, ſeeing that all Atoms, agitated by their eſſential mobility, are in perpetual attempt to extricate themſelves from Concretions, that ſo they may attain their primitive

mitive freedom of motion; that none can so soon extricate and disengage themselves; as those that are spherical; because such cannot be impeded by the small hooks, or angles of others. *Cum enim sphæra omnibus angulis careat, nihil hamati, aut retinentis offendet, facilè permeabit, & quoquover- sus ad naturæ penetrabit instituta, dividet instar cunei, & (quod nulli alteri figuræ contingit) contactu puncti labefaciens planum, statim amplo sinu sibi viam facit, cum nihil habet angulosi, quo possit detineri; quod ejus acti- vitati necessarium fuit:* saith *Magnenus* (*de Atom. lib.2 cap.3.*) As also, that we speak the Dialect of *Democritus,* when we call these Calorifick Atoms, sometimes the Atoms of *Heat,* sometimes the Atoms of *Fire,* in- discriminately; because Heat and Fire know none but a Gradual Diffe- rence; at least, because Heat, in a General sense, implies all degrees, and Fire, in a Special, the highest degree of Heat; *Aristotle* himself (1 *Mete- or.*3) excellently defining Heat to be nothing else, but *Caloris Hyperbole,* the *Excess* of Heat.

*Art. 7.*
That the Atoms of Heat are capable of *Expedition* or deliverance from Concre- tions. Two *wayes;* viz. by *Evocation* and *Motion.*

The Proprieties, or requisite Conditions of these Calorifick Atoms, be- ing thus explored; our next Enquiry must be concerning the Manner of their *Emancipation,* or *Expedition* from the fetters of Concretions. We observe, therefore, that the Atoms of Fire, imprisoned in Concretions, have *Two* ways of attaining liberty. (1.) By *Evocation,* or the Assistance of other Atoms of the same nature; when such invading and insinuating themselves into the centrals of a body, do so dissociate its particles, as that dissolving the impediments or chains of the igneous Atoms therein con- tained, they not only give them an opportunity, but in a manner sollicite them to extricate themselves. And by this way do the Atoms of Fire, in- cluded in Wood, Wax, Turpentine, Oyle, and all other Inflammable Con- cretions, extricate themselves, when they are set on fire; the sparks or flame, wherewith they are accensed, penetrating their contexture, and re- moving the remoraes, which detained and impeded their internal Atoms of Fire, and exciting them to Emption : Which thereupon issue forth in swarms, and with the violence of their exsilition drive before them, in the apparence of fuliginous Exhalations or smoak, those dissimilar parti- cles, which suppress and incarcerated them, during the integrity of the Concretion. (2.) By *Motion,* or *Concussion* ; and that either Intestine, or External. First, *Intestine*; when, after many evolutions, the igneous Atoms, included in a Body, do of themselves dissociate and discuss those heterogeneous masses, wherein they were imprisoned: Which they chiefly effect, when after some of them have by spontaneous motion attained their freedom, if any thing be circumstant, which hath the power of repelling them, as cold; for, in that case, returning again into the centrals of the body, from whence they came; and so associating with their fellows, pro- mote the discussion of the remaining impediments, and concur to a gene- ral Emption. From this Motion ariseth that Heat, or Fire, which is vul- garly ascribed to the *Antiperistasis,* or *Circumobsistence of Cold* ; as, for Example, when a heap of new Corn, or Mowe of green Hay, being kept too close, during the time of its fermentation, or sweating (as our Husband- men call it) sets it self on fire : the cold of the ambient aer, repelling the Atoms of Fire(which otherwise would expire insensibly)back again into t ; and so causing them to unite to their fellows : and upon that consociation, they suddainly engage in a general cumbustion, and dissolving all impedi- ments, acquire their liberty. Hence also proceed all those Heats, which are
                                                              observed

obferved in Fermentation, Putrifaction and all other inteftine Commotions and Mutations of Bodies.

Hither likewife would we refer that fo generally believed Phænomenon, the *Warmnefs of Fountains*, Cellars, Mines, and all fubterraneous Foffes, *in Winter :* but that we conceive it not only fuperfluous, but alfo of evil confequence in Phyfiology, to confign a Caufe, where we have good reafon to doubt the verity of the Effect. For, if we ftrictly examine the ground of that common Affertion, we fhall find it to confift only in a mifinformation of our fenfe ; *i.e.* though Springs, Wells, Caves, and all fubterraneous places are really as Cold in Winter, as Summer ; yet do we apprehend them to be warm : becaufe we fuppofe that we bring the organs of the fenfe of Touching alike difpofed in Winter and Summer ; not confidering that the fame thing doth appear Cold to a hot, and warm to a Cold hand, nor obferving, that oyle will be conglaciated, in Winter, in fubterraneous Cells, which yet appear warm to thofe, who enter them, but not in Summer, when yet they appear Cold. Secondly, by Motion *External*, when a Sawe grows Hot, by continuall affriction againft wood, or ftone ; or when fire is kindled by the long and hard affriction of 2 dry fticks, &c. This is manifeft even from hence ; that unlefs the bodies agitated, or rubbed againft each other, are fuch as contain igneous Atoms in them ; no motion, however lafting and violent, can excite the leaft degree of Heat in them. For, Water agitated moft continently and violently, never conceives the leaft warmth : becaufe it is wholly deftitute of Calorifick Atoms. Laftly, as for the Heat, excited in a body, upon the Motion of its *Whole*, whether it be moved by it felf, or fome External movent ; of this fort is that Heat, of which motion is commonly affirmed to be the fole Caufe : as when an Animal grows hot with running, &c. and a Bullet acquires heat in flying, &c. And thus much concerning the manner of Emancipation of our Calorifick Atoms.

The next thing confiderable, is their peculiar *Seminarie* or *Confervatory*; concerning which it may be obferved, that the Atoms of Fire cannot, in regard of their extreme Exility, fphærical Figure, and velocity of motion, be in any but an *Unctuous* and vifcous matter, fuch whofe other Atoms are more hamous, and reciprocally cohærent, than to be diffociated eafily by the inteftine motions of the Calorifick Atoms ; fo that fome greater force is required to the diffolution of that unctuoufnefs and tenacity, whereby they mutually cohære. And hereupon we may fafely conclude, that an Unctuous fubftance is as it were the chief, nay the fole *Matrix* or *Seminary* of Fire or Heat ; and that fuch Bodies only, as are capable of incalefcence and inflammation, muft contain fomewhat of Fatnefs and unctuofity in them. Sometimes, we confefs, it is obferved, that Concretions, which have no fuch Unctuofity at all in them, as Water, are Hot, but yet we cannot allow them to be properly faid to wax Hot, but to be made Hot ; becaufe the principle of that their Heat is not Internal to them, but *External* or *Afcititious*. For inftance ; when Fire is put under a veffel of Water, the fmall bodies, or particles of Fire by degrees infinuate themfelves thorowe the pores of the veffel into the fubftance of the Water, and diffufe themfelves throughout the fame ; though not fo totally, at firft, as not to leave, the major part of the particles of the Water untoucht : to which

*Art. 8.*
An *Unctuous* matter, the chief *Seminary* of the Atoms of Heat ; and why.

Q q                                                              other

other igneous Atoms succeffively admix themfelves, as the water grows hotter and hotter. And evident it is, how fmall a time the Water doth keep its acquired heat, when once removed from the fire: becaufe, the Atoms of Heat being meerly Adventitious to it, they fpontaneoufly defert it one after another, and leave it, as they found it, Cold: only this Alteration, they caufe therein, that they diminifh the Quantity thereof, infomuch as fucceffively afcending into the aer, they carry along with them the more tenuious and moveable particles of the Water, in the apparence of vapours, which are nothing but Water Diffufed, or Rarefied.

*Art. 9.*
*Among Unctu-*
*ous Concreti-*
*ons, why fome*
*are more eafi-*
*ly inflammable*
*than others.*

But, if what we affirm, that only Unctuous Bodies are Inflammable, be generally true; whence comes it, *that amongft Unctuous and Pinguous Concretions, fome more eafily take fire, than others?* The *Caufe*, certainly is this, that the Atoms of Fire, incarcerated, in fome Concretions, are not fo deeply immerft in, nor fo oppreft and overwhelmed with other Heterogenous particles of matter, as in others: and fo acquire the liberty of Eruption much more eafily. Thus dry Wood is fooner kindled, than Green; becaufe, in the green, the Aqueous moyfture, furrounding and oppreffing the Atoms of Fire therein contained, is firft to be difcuffed and attenuated into vapours: but, in the Dry, time, by the mediation of the warmth in the ambient aer, hath already abfumed that luxuriant moyfture, fo that none but the oleaginous, or unctuous part, wherein the Atoms of Fire have their principal refidence, remains to be difcuffed; which done, the Atoms of Fire inftantly iffue forth in fwarms, and difcover themfelves in flame. Thus fpirit of Wine is fo much the fooner inflammable, by how much the more pure and defæcated it is; becaufe the igneous Atoms therein concluded, are delivered from the greater part of that Phlegme, or aqueous humidity, wherewith they were formerly furrounded and oppreft. On the contrary, a ftone is not made Combuftible without great difficulty; becaufe the fubftance of it is fo compact, as that the Unctuous humidity is long in difcuffion. We fay, a Stone, not a Peble, or Arenaceous one, becaufe fuch is deftitute of all Unctuofity, and fo of all igneous particles: but, a Limeftone, fuch as is capable of reduction to a Calx: or a Flint out of which by concuffion againft fteel, are excuffed many fmall fragments, plentifully fraught with Atoms of Fire.

*Art. 10.*
*A CONSE-*
*CTARY.*
*That Rarefa-*
*ction is the*
*proper Effect*
*of Heat.*

The Nature and Origine of Heat being thus fully explicated, according to the moft verifimilous Principles of *Democritus*, *Epicurus*, and their Sectators; it follows, that we progrefs to thofe *Porifmata* or *Confectaries*, which from thence refult to our obfervation; and the *Solution* of fome moft confiderable *Problems*, retaining to the fame Argument, fuch efpecially as have hitherto eluded the folutive capacity of any other Hypothefis, but what we have here afferted.

Infomuch, therefore, as the Atoms of Heat, which are always incarcerated in an Unctuous Matter, doe, upon the acquifition of their liberty, iffue forth with violence, and infinuating themfelves

into

into Bodies, which they meet withal, and totally pervading them, diffociate their particles, and diffolve their Compage or Contexture: Hence is it manifeft, that *Rarefaction*, or *Dilatation* is upon good reafon accounted the proper Effect of Heat; fince thofe parts of a body, which are Conjoyned, cannot be Disjoyned, but they muft inftantly poffefs a greater part of fpace (underftand us in that ftrict fenfe, which we kept our felves to, in our Difcourfe of Rarefaction and Condenfation) than before. Hence comes it, that Water in boyling, feems fo to be encreafed, that what, when cold, filled fcarce half the Caldron, in ebullition cannot be contained in the whole, but fwells over the brim thereof. Hence is it alfo, that all bodies attenuated into Fume, are diffufed into fpace an hundred, nay fometimes a thoufand degrees larger than what they poffeffed before.

From this Confectary we arrive at fome *Problems*, which ftand directly in our way to another; and the *Firft* is that Vulgar one, *Why the bottom of a Caldron, wherein Water, or any other Liquor is boyling, is but moderately warm, at moft not fo hot, as to burn a mans hand applyed thereto?*

*Art.* 11.
PROBLEM 1.
Why the bottom of a Caldron, wherein Water is boyling, may be touched by the hand of a man, without burning it:
Sol.

The Caufe of this culinary Wonder (fo our Houfewifes account it) feems to be this; when the Atoms of Heat, paffing through the pores in the bottom of the Caldron into the water, do afcend through it, they elevate and carry along with them fome particles thereof: and at the fame time, other particles of Water, next adjacent to them, fink down, and inftantly flowe into the places deferted by the former, which afcended, and infinuate themfelves into the now laxarated pores in the bottom of the caldron. And though thefe are foon repelled upwards by other Atoms of Fire afcending thorowe the pores of the Veffel, and carried upwards, as the former, yet are there other particles of Water, which finking down, infinuate alfo into the open pores of the veffel, and by their conflux or downward motion, much refract the violence of the fubingredient Atoms of Fire: and fo, by this viciffitude of Heat and Moyfture, it comes to pafs, that the Heat cannot be diffufed throughout the bottom of the Caldron, the Humidity (which falls into the pores of it in the fame proportion, as the Heat paffeth thorow them) hindering the poffeffion of all its empty fpaces by the invading Atoms of Fire. Nor doth it availe to the contrary, that the Water which infinuates into the pores of the veffel, is made Hot, and fo muft calefie the fame, in fome proportion, as well as the Fire underneath it; becaufe boyling Water poured into a cold Caldron, doth more than warm it: For, thofe particles of Water, which fucceffively enter into the void fpaces of the veffel, are fuch as have not yet been penetrated *per minimas*, by the Atoms of Fire. For, all the cold, formerly entered into the water, is not at once difcuffed, though the Water be in boyling; the Ebullition arifing only from the cohærence of the calefied with the uncalefied particles of the Water. And from the fame Caufe is it, that a fheet of the thinneft Venice Paper, if fo folded upward in its Margines, as to hold Oyle infufed into it, and laid upon a gridiron over burning coals; doth endure the fire without inflammation for a good fpace:

Q q 2                                                        Which

Which some Cooks observing, use to fry Bacon upon a sheet of Paper only.

*Art.* 12.
PROBLEM 2.
Why Lime becomes ardent upon the affusion of Water.
Sol.

Secondly, *Why doth Lime acquire an Heat and great Ebullition upon the affusion of Water? since, if our præcedent Assertion be true, the Heat included in the Lime ought to be suppress so much the more, by how much the more Aqueous Humidity is admixt unto it.*

This Difficulty is discussed by Answering; that the Aqueous Humidity of the Lime-stone is indeed wholly evaporated by fire in its calcination; but yet the Pingous, or Unctuous for the most part remains, so that its Atoms of Fire lye still blended and incarcerated therein: and when those expede themselves, and by degrees expire into the ambient aer, if they be impeded and repelled by water affused, they recoyle upon the grumous masses of the Lime, and by the Circumobsistence of the Humidity, become more congregated; and so upon the uniting of their forces make way for the Exsilition of the other Atoms of Fire, which otherwise could not have attained their liberty but slowly and by succession one after another. So that all the Atoms of Fire contained in the Lime, issuing forth together, they break through the water, calefie it, and make it bubble or boyle up; the calefied parts thereof being yet cohærent to the uncalefied.

*Art.* 13.
PROBLEM 3.
Why the Heat of *Lime* burning is more vehement, than the Heat of any *Flame* whatever.
Sol.

The Third Problem is, *Why the Heat of Lime, kindled by Water is more intense than that of any Flame whatever?*

Answer, that forasmuch as Flame is nothing but Fire Rarefied, or as it were an Explication, or Diffusion of those Atoms of Fire, which were lately ambuscadoed in some Unctuous matter; and that all Fire is so much more intense or vehement, by how much more Dense it is, *i.e.* by how much the more congregated the Atoms which constitute it are: therefore is the Heat of Lime unslaking more vehement than that of any Flame, in regard the smallest grains of Lime contain in them many Atoms of Fire, which are not so diffused or disgregated in a moment, as those in Flame. So that a mans hand being waved to and fro in Flame, is invaded by incomrably fewer particles of Fire, than when it is dipt into, or waved through water at the unslaking of Lime thereby; the small granes of Lime adhæring unto, and insinuating into the pores of the hand, the many Atoms of fire invelloped in them, incontinently explicate themselves, violently penetrate and dilacerate the skin, and other sentient parts, and so produce that Pungent and Acute pain, which is felt in all Ambustions. From the same Reason also is it, that a glowing Coale burns more vehemently than Flame: and the Coals of more solid wood, as Juniper, Cedar, Guaiacum, Ebony, Oke, &c. more vehemently than those of Looser wood, such as Willow, Elder, Pine tree, &c. The like Disproportion is observable also in the Flames of divers Fewels; for in the flame of Juniper are contained far more Igneous Atoms, than in that of Willow: and consequently they burn so much more vehemently. True it is, that spirit of Wine enflamed, is so much more Ardent, by how much more refined and cohobated: yet this proceedeth from another Cause; *viz.* that the Atoms of Fire issuing from spirit of Wine of the first Extraction, have much of the Phletegme, or Aqueous moysture of the Wine intermixt among them; and so cannot be alleaged

as

as an Example that impugne's our Reason of the Different Heats of several Flames.

The *Fourth*, is that Vulgar Quære,*Why boyling Oyle doth scald more dangerously, than boyling Water?*

To which it is easily Answered; that Oyle, being of an Unctuous and Tenacious consistence, and so having its particles more firmly cohærent, than Water, doth not permit the Atoms of Fire entered into it, so easily to transpire : so that being more agminous, or swarming in oyl, they must invade, and dilacerate the hand of a man, immersed into it, both more thickly and deeply, than those more Dispersed ones contained in boyling Water. Which is also the Reason, why Oyle made fervent is much longer in cooling, than Water: and may be extended to the Solution of the

*Art. 14.*
PROBLEM 4:
Why boyling *Oyle* scalds more vehemently, then boyling *Water*. Sol.

Fifth Problem, *viz. Wherefore do Metals, especially Gold, when melted, or made glowing hot, burn more violently, than the Fire that melteth, or heateth them; especially, since no Atoms of Fire can justly be affirmed to be lodged in them, as in their proper seminary, and so not to be educed from them, upon their Liquation, or Ignition.*

*Art. 15.*
PROBLEM 5:
Why *Metals*, melted or made red hot, burn more violent than the Fire, that melteth or heateth them; Sol.

For, the Heat, wherewith they procure Ambustion, being not domestick, but only Adventitious to them from the Fire, wherein they are melted, or made red hot ; the reason why they burn so extreamly, must be this, that they are exceedingly Compact in substance, and so their particles being more tenacious or reciprocally cohærent, then those of wood, oyle, or any other body whatever, they more firmly keep together the Atoms of fire immitted into them : insomuch that a man cannot touch them with his finger, but instantly it is in all points invaded with whole swarms of igneous Atoms, and most fiercely compunged and dilacerated. And, as for the *Derasion of the skin* from any part of an Animal, immersed into melted metal ; this ariseth partly from the total dissolution of the tenour of the skin by the dense, and on every side compungent Atoms of Fire ; partly from the Compression and Resistence of the parts of the Metal, now made Fluid, which are both so great, that upon the withdrawing of the member immersed into the metal, the part which is immediately prest upon by the particles thereof, is detained behind, and that's the skin. Hence also is it no longer a Problem, *Why red hot Iron sets any Combustible matter on Fire;* for it is evident, that it cannot inflame by its own substance, but by the Atoms of Fire immitted into, and for a while reteined in its Pores. And this brings us to a

Second CONSECTARY, *viz.* That as the Degrees of Heat are various (Physicians, indeed, allow only 4, and Physiologists but double that number ; the Former, in order to the more convenient reduction of their Art to certain and established principles ; the Latter, meerly in conformity to the Dictates of *Aristotle* : but Neither upon absolute necessity, since it is reasonable for any man to augment their number even above number, at pleasure) So also must the *Degrees of fire be various*. For, since Fire, even according to *Aristotle* is only the Excess of Heat, or Heat encreased to that height, as to Burn, or Enflame a thing ; if we begin at the gentle Meteor called

*Art. 16.*
CONSECTARY 2.
That, as the degrees of Heat, so those of fire are innumerably various.

called *Ignis Fatuus* (which lighting upon a mans hand, and a good while adhereing thereto, doth hardly warm it) or at the fire of the purest spirit of Wine enflamed (which also is very languid, for it is frequent among the Irish, for a Cure of their Endemious Fluxes of the belly, to swallow down small balls of Cotton, steept in spirit of wine, and set on fire, and that many times with good success.) We say, if we begin from either of these weak Fires, and run through all the intermediate ones, to that of melted Gold, which all men acknowledge to be the Highest: we shall soon be convinced, that the Degrees of Fire are so various, as to arise even to innumerability.

**Art. 17.**
*That to the Calefaction, Combustion, or Inflammation of a body by fire, is required a certain space of time; and that the space is greater or less, according to the paucity, or abundance of the igneous Atoms invading the body objected; and more or less of aptitude in the contexture thereof to admit them.*

Most true it is, in the General, that every Fire is so much the more intense, by how much more numerous, or agminous the Atoms of Fire are, that make it: yet, if we regard only the Effect, there must be allowed a convenient space of time, for the requisite motion of those Atoms, and a supply of fresh ones successively to invade and penetrate the thing to be burned or enflamed. For, since the Igneous Atoms, exsilient from their involucrum, or seminary, and invading the extrems of a body objected to them, are subject to easy Repercussion, or (rather) Resilition from it; therefore, to the Calefaction, Adustion, or Inflammation of a body, it is not sufficient, that the body be only moved along by, or over the Fire: but it must be held neer, or in it, so long as till the first invading Igneous Atoms, which otherwise would recoyle from it, be impelled on, and driven into the pores of the same, by streams of other Igneous Atoms contiguously, succeeding and pressing upon them. And, however the space of time, be almost in assignably short, in which the finger of a man, touching a glowing Coale, or melted metal, is burned; because, the Atoms of Fire are therein exceeding Dense and Agminous, and so penetrate the skin, in all points: yet nevertheless common observation assures, that in the General a certain space of time is necessary to the Effect of Calefaction or Ambustion; and that so much the Longer, by how much the Fewer, or more Disgregated the Igneous Atoms are, either in the Body Calefying, or the Aer conterminous thereto. And this (as formerly) to the end, that the Motion of the Igneous Atoms first assaulting the object may be continued, and a supply of fresh ones, promoting and impelling the former, be afforded from the Focus, or Seminary. Hence is it, that a mans hand may be frequently Waved to and fro in Flame, without burning; because the Atoms of Fire, which invade it, are repercussed, and not by a continued aflux of others driven foreward into its pores, the motion of his hand preventing the Continuity of their Fluor: but, if his hand be held still in the flame, though but a very short time, it must be burned; because the first invading Atoms of Fire are impelled on by others, and those again by others, in a continent fluor, so that their Motion is continued, and a constant supply maintained. Hence comes it also, that no Metal can be molten only by a Flash, or transient touch of the Fire (for, we are not yet fully satisfied of the verity of that vulgar tradition, of the instantaneous melting of money in a purse, or of a sword blade in its sheath, by Lightning: and if we were, yet could we assign that prodigious Effect to some more propable Cause, *viz.* the impetuosity of the motion, and the exceeding Coarctation of those Atoms of Fire, of which that peculiar species of Lightning doth consist) but it must be so long held in, or over the Fire, as until the Igneous Atoms have totally pervaded its contexture, and

disso-

dissociated all its particles: and therefore, so much the longer stay in the fire doth every Metal require to its Fusion; by how much the more Compact and Tenacious its particles are.

As the Degrees of Fire are various, as to the more and less of Vehemency, respective to the more and less Density, or Congregation of the Igneous Atoms: So likewise is there a considerably variety among Flames, as to the more and less of *Duration.* Concerning the *Causes,* therefore of this Variety, in the General, we briefly observe; that Flame hath its Greater or Less Duration, respective to the

*Art. 18.*
*Flame more or less Durable, for various respects.*

(1.) *Various Materials, or Bodies inflammable.* For, such Bodies, as have a greater Aversion to inflammation, being commixt with others, that are easily inflammable, make their flame less Durable; as Bay Salt, dissolved in spirit of Wine, shortens the duration of its flame, by almost a third part, as the *Lord Bacon* affirms upon exact experiment (*Nat. Hist. cent. 4.*) and contrariwise, such as approach neerer to an affinity with fire, *i.e.* have much of Unctuousness, and plenty of igneous Atoms concealed therein, yield the most lasting Flames; as Oyle and Spirit of Wine commixt in due proportions; and spirit of Salt, to a tenth part, commixt with Oyle Olive, makes it burn twice as long in a Lamp, as Oyle alone, from whence some Chymists have promised to make Eternal Lamps with an Oyle extracted from common Salt, and the stone Amianthus.

(2.) *The more or less easie Attraction of its Pabulum, or Nourishment.* For, Lamps, in which the Flame draweth the oyle from a greater distance, always burn much longer, than Candles, or Tapers, where the circumference of the fewel is but small; and the broader the surface of the Oyle, or Wax, wherein the Wiek is immersed, so much the longer doth the flame thereof endure; not only in regard of the greater Quantity of Nourishment, but of its slower Calefaction, and so of its longer Resistence to the absumptive faculty of the flame. Since it is observed, that the Coolness of the Nourishment, doth make it more slowly consumable: as in Candles floating in water. This was experimented in that service of our *quondam* English Court, called *All night*; which was a large Cake of Wax, with the Wiek, set in the middest: so that the flame, being fed with nourishment less heated before hand, as coming far off, must of necessity last much longer, than any Wax Taper of a small circumference.

(3) *Various Conditions of the same Materials.* For, Old and Hard Candles, whether of Wax, or Tallowe, maintain flame much longer than New, or soft. Which good Houswives knowing, use no Candles under a year old, and such as have, for greater induration, been laid a good while in Bran, or Flower. And, from the same reason is it, that Wax, as being more firm and hard, admixt to Tallowe and made up into Candles, causeth them to be more lasting, then if they were præpared of Tallowe alone.

(4.) *Different Conditions, and Tempers of the ambient Aer.* For, the Quiet and Closeness of the Aer, wherein a Taper burneth, much conduceth to the prolongation of its flame: and contrariwise, the Agitation
thereof,

thereof, by winds, or fanning, conduceth as much to the shortning of it: insomuch as the motion of flame makes it more greedily attract, and more speedily devour its sustenance. Thus a Candle lasteth much longer in a Lanthorne, than at large in a spacious roome. Which also might be assigned as one Cause of the long Duration of those subterranean Lamps, such as have been found (if credit be due to the tradition of *Bapt. Porta,* (*lib.* 12 *Magia natural. cap. ultim.*) *Hermolaus Barbarus* (*in lib.* 5. *Dio cap.* 11.) *and Cedrenus Histor. Compend.*) All which most confidently avouch it, upon authentique testimonies,) in the Urns of many Noble Romans, many hundreds of years after their Funerals. Here should our Reader bid us stand, and deliver him our positive judgement, upon this stupendious Rarity, which hath been uged by some Laureat Antiquaries, as a cheif Argument of the transcendency of the Ancients Knowledge as in all Arts, so in the admirable secrets of Pyrotechny, above that of Later Ages; as we durst not be so uncharitable, to quæstion the Veracity of either the Inventors, or Reporters of it: so should we not be so uncivil, as not to releive his Curiosity, at least with a short story, that may light Him towards farther satisfaction. A certain *Chymist* there was, not many years since, who having decocted Litharge of Gold, Tartar, Cinnaber, and Calx vive, in spirit of Vinegre, until the Vinegre was wholly evaporated; closely covering and luting up the earthen vessel, wherein the Decoction was made, buried it deeply in a dry Earth, for 7 moneths together (in order to more speedy maturation, expected from the Antiperistasis of Cold) came at length to observe what became of his Composition: and opening the vessel, observed a certain bright Flame to issue from thence, and that so vehement, as it fired the hair of his eyebrowes and head. Now, having furnished our Reader with this faithful Narrative; we leave it to his owne determination: Whether it be not more probable, that those Coruscations, or Flashes of Light, perceived to issue from Vials of Earth, found in the demolisht sepulchres of the Great *Olybius,* and some eminent Romans, at the instant of their breaking up by the spade, or pickaxe; did proceed rather from some such Chymical Mixture, as this of our Chymist (who acquired Light by the hazard of Blindness) which is of that nature as to be in a moment kindled, and yield a shortlived flame, upon the intromission of Aer into the vessel, wherein it is contained; than from any Fewel, that is so slowly Absumable by Fire, as to maintain a constant Flame, for many hundred years together, without extinction, and that in so small a vial, as the Fume must needs recoyle and soon suffocate the Flame. But we return from our Digression, and directly pursue our embost Argument. It much importeth the greater and less Continuance of Flame, whether the Aer be *Warm,* or *Cold, Dry* or *Moist.* For *Cold* Aer irritateth flame, by Circumobsistence, and causeth it burn more fiercely, and so less durably; as is manifest from hence, that Fire scorcheth in frosty weather: but *Warme* Aer, by making flame more calm and gentle, and so more sparing of its nourishment, much helpeth the Continuance of it. If *Moist,* because it impedeth the motion of the igneous Atoms, and so in some degree quencheth flame, at least, makes it burn more dimly and dully; it must of necessity advance the Duration of flame: and contrariwise, *Drie* Aer, meerly as drie, produceth Contrary Effect, though not in the same proportion; nay so little, that some Naturalists have concluded the Driness of Aer to be only indifferent, as to the Duration of Flame.

And

And now we are arrived at our Third and Last CONSECTARY, That the *immediate and genuine Effect of Heat, is Disgregation,* or *Separation:* and that it is only by Accident that Heat doth Congregate Homogeneous natures. To argue by the most familiar way of Instance; when Heat hath dissolved a piece of Ice, consisting of water, earth, and perhaps of gravel and many small Festucous bodies commixt; the Earth, Sand and other Terrene parts sink downe and convene together at the bottom, the water returns to its native fluidity, and possesseth the middle region of the Continent, and the strawes swim on the surface of the water: not that it is essential to the Heat so to dispose them; but essential to them, being dissociated and so at liberty, each to take it proper place, according to the several degrees of their Gravity. Thus also, when a Mass of various Metals is melted by Fire, each metal, indeed, takes it proper region in the Crucible, or fusory vessel: but yet the Congregation of the Homogeneous particles of each particular Metal, is not immediately caused, but only occasioned, *i. e.* Accidentally brought to pass by the Disgregation or præcedent separation of the particles of the whole Heterogeneous Concretion, by heat. Again, the Energy of every Cause in Nature ceaseth, upon the production of its perfect Effect; but the Effect of Heat ceaseth not, when the Homogenieties of the mass of Ice, or Metal, are Congregated, but continues the same after, as before, *i. e.* to Dissolve the compage of the Metal, or Ice, and Dissociate all the particles thereof: for, so long as the Heat is continued, so long do the Ice and Metal remain Dissolved and Fluid. This considered, what shall we say to *Aristotle,* who makes it the Essential Attribute of Heat, *Congregare Homogenea,* to Congregate Homogeneous Bodies. Truly, rather then openly convict so great a Votary to truth of so palpable an Error; we should gladly become his Compurgator, and palliate his mistake with an indulgent comment; that in his Definition of Heat, to be a Quality genuinely Congregative of Homogeneous natures, He had his eye, not upon the General Effect of Heat (which He could not but observe, to Disgregate the particles of all things, aswel Homogeneous, as Heterogeneous.) but upon some special Effect of it upon some particular Concretions, such as are Compounded of parts of Divers natures, as Wood and all Combustible bodies Concerning which, indeed, His Assertion is thus far justifiable, that the whole Bodie is so dissolved by fire, as that the Dissimilar parts of it are perfectly sequestred each from other, and every one attains it proper place; the Aereal part ascending and associating with the Aer, the Aqueous evaporating, the Igneous discovering themselves in Flame, and the earthy remaining behind, in the forme of Ashes. But alas! this favourable Conjecture cannot excuse, nor gild over his Incogitancy; for, the Congregation of the Homogenous particles of a Body, dissolved by Fire, in the place most convenient to their particular Nature, ariseth immediately from their own Tendency thither, or (that we may speak more like our selves, *i. e.* the Disciples of *Epicurus*) from their respective proportions of Gravity, the more Heavy extruding and so impelling upward the less heavy: and only Accidentally from Heat, or as it hath dissolved the cæment, and so the Continuity of the Concretion, wherein they were confusedly and promiscuously blended together. So that Truth will not dispense with our Connivence at so dangerous a Lapse, though in one of Her choicest Favorites; chiefly, because it hath already deluded

*Art.* 19.
CONSECTA-
RY 3.
That the immediate and genuine Effect of Heat, is the Disgregation of all bodies, as well Homogeneous, as heterogeneous and that the Congregation of Homogeneous Natures, is only an Accidental Effect of Heats contrary to *Aristotle.*

so many of Her seekers, under the glorious title of a Fundamental Axiome: but strictly enjoynes Us, to Conclude; that Heat, *per se*, or of its own nature, is alwayes a *Disgregative Quality*; and that it is of of meer Accident, that upon the sequestration of Heterogeneities, Homogeneous Natures are associated, rather than, *è contra*, that it is of meer Accident, that while Heat Congregates Homogeneous, it should Disgregate Heterogeneous Natures, as *Aristotle* most inconsiderately affirmed and taught.

---

## Sect. II.

*Art.* 1.
*The Link connecting this Section to the former.*

AS in the Course, so in the Discourse of Nature, having done with the principle of Life, Heat, we must immediately come to the principle of Death, COLD: whose Essence we cannot seasonably explain, before we have proved, that it hath an Essence; since many have hotly, though with but cold Arguments, contended, that it hath none at all, but is a meer Privation, or Nothing.

*Art.* 2.
*That Cold is* no *Privation of Heat; but* a *Real* and *Positive Quality;* demonstrated.

That Cold, therefore, is a *Real Ens*, and hath a *Positive Nature* of its own, may be thus demonstrated. (1.) Such are the proper Effects of Cold, as cannot, without open absurdity, be ascribed to a simple Privation; since a Privation is incapable of Action: for, Cold compingeth all Bodies, that are capable of its efficacy, and congealeth Water into Ice, which is more than ever any man durst assigne to a privation. And, when a man thrusts his hand into cold Water, the Cold He then feels, cannot be sayd to be a meer privation of the Heat of his hand; since, his hand remains as Hot, if not hotter than before; the Calorifick Atoms of his hand being more united, by the circumobsistence of the Cold. (2.) All Heat doth Concentre and unite it self, upon the Antiperistasis of Cold; not from fear of a privation, because Heat is destitute of a sense of its owne being, and so of fear to lose that being; and if not, yet Nothing can have no Contrariety, nor Activity: but, from Repulsion, as we have formerly delivered. (3.) Though many bodies are observed to become Cold, upon the absence, or Expiration of Heat: yet is it the intromission of the Quality contrary to Heat, that makes them so; for, if External Cold be not introduced into their pores, they cannot be so properly sayd, *Frigescere*, to wax Cold, as *Decalescere*, to wax less Hot. Thus a stone, which is not Hot, nor Cold, unless by Accident, being admoved to the fire, is made Hot; and removed from the fire, you cannot (unless the ambient Aer intromit its Cold into it) so justly say, that it growes Cold, as that it grows Less hot, or returnes to its native state of indifferency. (4.) When Water (vulgarly, though untruely præsumed to be naturally or essentially cold) is congealed into Ice by the Cold of the aer, it would be most shamefully absurd, to affirm, that the Cold of the Ice ariseth meerly from the Absence of Heat in the water; because it is the essential part of the supposition, that the Water had no Heat before. (5.) Privation knowes no Degrees; for the Word imports the totall Destitution, or Absence of somewhat
formerly

formerly had, otherwife, in rigid truth, it can be no Privation (and therefore our common Diftinction of a Partial, and Total privation, hath lived thus long meerly upon indulgence and tolleration.): but Cold hath its various Degrees, for Water is colder to the touch than Earth, Ice than Water, &c. therefore Cold is no Privative, but *a Pofitive Quality.*

The Reality of Cold being thus clearly evicted, we may, with more advantage undertake the confideration of its *Formality*, and explore the roots of thofe Attributes commonly imputed thereunto.

*Art. 3.*
That the adæquate *Notion* of Cold, ought to be defumed from its *General Effect*, viz. the *Congregation* and *Compaction* of bodies.

Firft, therefore, we obferve; that though Cold be Scholaftically defined by that paffion caufed in the organs of the fenfe of touching, upon the contact of a Cold object; yet doth not that fpecial Notion fufficiently exprefs its Nature: becaufe there is a more *General Effect* by which it falls under our cognizance; and that is the *Congregation* and *Compaction* of the parts of bodies. For, fince Cold is the Antagonift to Heat, whofe proper vertue it is, to Difcufs and Difgregate; therefore muft the proper and immediate virtue of Cold be, to *Congregate* and *Compinge*: and confequently, ought we to form to our felves a notion of the Effence of Cold, according to that general Effect, rather than that fpecial one produced in the fenfe of Touching, which doth adumbrate only a Relative part of it.

Secondly, that by *Cold*, we underftand not any *Immaterial Quality*, as *Ariftotle* and the *Schools* after him; but a *Subftantial* one, *i.e. certain particles of Matter, or Atoms whofe determinate Magnitude and Figure adapt or empower them to congregate and compinge bodies, or to produce all thofe Effects obferved to arife immediately from Cold.* And, as the *Atoms*, which are comparated to the Caufation of fuch Effects, may rightly be termed, the *Atoms of Cold*, or *Frigorifick Atoms*: fo may thofe *Concretions*, which harbour fuch Atoms, and are capable of Emitting them, be named *Cold Concretions*; either *Actually*, as Froft, fnowe, the North-wind, &c. or *Potentially*, as Nitre, Hemlock, Night-fhade, and all other fimples afwel Medical, as Toxical or Poyfonous, whofe Alterative Virtue confifteth cheifly in Cold.

*Art. 4.*
Cold, no Immaterial; but a *Subftantial* Quality.

Now, as for the determinate *Figure* of Frigorifick Atoms; our enquiries can hope for but fmall light from the almoft confumed vaper of Antiquity: For, though *Philoponus* (in 1 *phyfic.*) & *Magnenus* (*de Atomis, difput.* 2. *cap.*3.) confidently deliver, that *Democritus* affigned a *Cubical* Figure to the Atoms of Cold; and endeavour to juftifie that affignation, by fundry Mathematical reafons: yet *Ariftotle*, a man afwell acquainted with the doctrines of his Predeceffors, as either of thofe, exprefly affirms, that nor *Democritus*, nor *Leucippus*, nor *Epicurus* determined the Atoms of Cold to any particular Figure at all; for, His words are thefe (3 *de cælo, cap.* 4.) ὐκ ἐτὶ ἐπιδιώριστι, *Nihilpende determinarunt.* So, that rather than remain altogether in the dark, we muft ftrike fire out of that learned Conjecture of our Mafter *Gaffendus*; and taking our indication from the rule of Contrariety, infer, that the Atoms of Heat being fpherical, thofe of Cold, in all reafon, muft be *Tetrahedical*, or Pyramidal, confifting of 4 fides, or equilateral Triangles. To make the reafonablenefs of this fuppofition duly evident, let us

*Art. 5.*
*Gaffendus* conjectural Affignation of a *Tetrahedical* Figure to the Atoms of cold; afferted by fundry weighty confiderations.

consider (1.) That as Heat hath its origine from Atoms most exile in magnitude, spherical in figure, and so most swift of motion: so must its Contrary, Cold, be derived from principles of Contrary proprieties, *viz.* Atoms not so exile in magnitude, of a Figure most opposite to a sphere, and so of most slow motion. (2.) That none but *Tetrahedical* Atoms can justly challenge to themselves these proprieties, that are requisite to the Essensification of Cold. For (1.) If we regard their *Magnitude*, a Tetrahedical Atom may be Greater than a Spherical, by its whole Angles: because a Sphere may be circumscribed within a Tetrahedon. (2.) If the *Figure* it self, none is more opposed to a Sphere, than a Tetrahedon: because it is Angular, and farthest recedeth from infinity, or (rather) innumerability of small insensible sides, which a profound Geometrician may speculate in a Sphere. (3.) If their *Mobility*; no body can be more unapt for motion, than a Tetrahedical one: for, what vulgar Mathematicians impute to a Cube, *viz.* that it challengeth the palme from all other Figures, for Ineptitude to motion, doth indeed more rightfully belong to a Tetrahedon; as will soon appear to any equitable consideration, upon the perpension of the reasons alleagable on both parts. But here we are to signifie, that this ineptitude to motion proper to Tetrahedical Atoms, is not meant of Atoms at liberty, and injoying freedom of motion, in the Inane space; since, in that state all Atoms are præsumed to be of equal velocity: but only of Atoms wanting that liberty, such as are included in Concretions, and by intestine evolutions continually attempt Emancipation and Exsilition. (4.) It cannot impugne, at least, not stagger the reasonableness of this conjectural Assignation of a Tetrahedical figure to the Atoms of Cold, that *Plato* (*in Timæo*) definitely adscribeth a *Pyramidal Figure* to Fire, not to the Aer, *i.e.* to the Atoms of Heat, not to those of Cold: because, if any shall thereupon conceive, that a Pyramid is most capable of penetrating the skin of a man, and consequently of producing therein the sense of Heat, rather than Cold; He may be soon converted by considering a passage in our former section of this Chapter, that the Atoms of Heat may, though spherical, as well in respect of their extreme Exility (which the point of no Pyramid can exceed) as of the velocity of their Motion, prick as sharply, and penetrate as deeply, as the Angles of the smallest Pyramid imaginable. To which may be conjoyned, that the Atoms of Cold, according to our supposition, are also capable of Pungency and Penetration; and consequently that a kind of Adustion is also assignable to great Cold; according to that expression of *Virgil* (1 *Georg.*) *Boccæ penetrabile frigus adurit*. For, in fervent Frosts (to use the same Epithite, as the sweet-tongued *Ovid*, in the same case) when our hands are, as the English phrase is, Benumm'd with Cold; if we hold them to the fire, we instantly feel a sharp and pricking pain in them. Which ariseth from hence, that the Atoms of Heat, while by their agility and constant supplies they are dispelling those of Cold, which had entred and possessed the pores of our hands, do variously commove and invert them; they are hastily driven forth, and in their contention and egress, cut and dilacerate the flesh and skin, as well with their small points, as edges lying betwixt their points, and so produce an acute and pungitive pain. Whereupon the sage *Sennertus* (*de Atomis*) grounds his advice, that in extreme cold weather, when our hands are so stupified, as that an Extinction of their vital heat may be feared; we either immerse them into cold water, or rub them in snow, that the Atoms of Cold, which have wedged each other

into

into the pores, may be gently and gradually called forth, before we hold them to the fire: and this, least not only grievous pain be caused, but a Gangrene ensue, from the totall dissolution of the Contexture of our hands by the violent intrusion of the Cold Atoms, when they are forcibly impelled and agitated by the igneous; as the sad experience of many in *Ruscia*, *Groenland*, the *Alps*, and other Regions obnoxious to the tyranny of Cold, hath taught. Concerning this, *Helmont* also was in the right, when He said, *Mechanicè namque videmus, membrum fere congelatum sub nive recalescere, & à syderatione præservari; quod alias aer mox totalitèr congelare pergeret, vel si repentè ad ignem sit delatum, moritur propter extremi alterius festinam actionem &c. (in cap. de Aere articul 8.)* (5.) Nor doth it hinder, that *Philoponus* and *Magnenus* affirm, that the Atoms of Cold ought to be *Cubical*, in respect of the eminent aptitude of that figure, for Constipation and Compingency, the General Effects of Cold: because, a Pyramid also hath its plane sides, or faces, which empower it to perform as much as a Cube, in that respect; and if common Salt be Constrictive, only because, being Hexahedrical in form, it hath square plane sides, as a Cube; certainly Alum must be more Constrictive, because being Octahedrical in form, it hath triangular plane sides, as a Pyramid. Besides, it is manifest, that these plane sides must so much the more press upon and wedge in the particles of a body, by how much more of the body, or greater number of its particles they touch: and that by how much more they are entangled by their Angles, so much more hardly are they Expeded, and so remain cohærent so much more pertinaciously. Hence comes it, that all Concretions consisting, for the most part, of such figurated Atoms, are *Adstrictive Effectually:* for, interposing their particles amongst those of other bodies, that are Fluid; they make their Consistence more Compact and somewhat Rigid, as in Ice, Snow, Haile, Hoar-frost, &c.

The Consignation of a Tetrahedical Figure to Frigorifick Atoms appearing thus eminently verisimilous; to the full Explanation of the Nature of Cold, it remains only, that we decide that notable *Controversy*, which so much perplexed many of the Ancients: *viz. Whether Cold be an Elementary Quality;* or (more plainly) *Whether or no the Principality of Cold belongs to any one of the four vulgar Elements; and so whether Aer, or Water, or Earth may not be conceived to be Primum Frigidum, as rightfully as Fire is sayd to be Primum Calidum?* Especially, since it is well known, that the *Stoicks* imputed the principality of Cold to the *Aer*; *Empedocles* to *Water*, to whom *Aristotle* plainly assented, though He sometimes forgot himself, and affirmed that no Humor is without Heat (as *in 5, de Generat. animal cap. 2.*); and *Plutarch* to *Earth*, as we have learned from Himself *(lib. de frigore primigenio.)* *Art. 6.* Cold, not *Essential* to Earth, Water, nor Aer.

To determine this Antique Dispute, therefore, we first observe; that it arose chiefly from a *Petitionary* Principle. For it appears, that all Philosophers, who engaged therein, took it for granted, that the Quality of Heat was eminently inhærent in Fire, the chief of the 4 Principal or Elementary substances; and thereupon inferred, that the Contrary Quality, Cold, ought in like manner to have its principal residence in one of the other 3: when, introth, they ought first to have proved, that there was such a thing as an Element of Fire in the Universe; which is more than any Logick
can

can hope, since the Sphere of Fire, which they suppofed to poffefs all that vaft space between the convex of the Sphere of Aer, and the concave of that of the Moon, is a meer Chimæra, as we have formerly intimated, and *Helmont* hath clearly commonftrated (*in cap. de Aere.*) And Secondly we affirm, that as the Higheft degree of Heat is not juftly attributary to any one Body more than other, or by way of fingular eminency (for, the Sphere of Fire failing, what other can be fubftituted in the room thereof ?) but to fundry fpecial Bodies, which are capable of Exciting or Conceiving Heat, in the fuperlative degree : fo likewife, though we fhould concede, that there are 3 Principal Bodies in Nature, namely Aer, Water, Earth, in each whereof the Quality of Cold is fenfibly harboured ; yet *is there no one of them, of its own nature more principally Cold than other, or which of it felf containeth Cold in the higheft degree ; but fome fpecial Bodies there are, compofed of them, which are capable of Exciting and Conceiving Cold, in an eminent manner.* But, in Generals is no Demonftration ; and therefore we muft advance to *Particulars*, and verify our Affertion, in each of the Three fuppofed Elements apart.

For the *Earth* : forafmuch as our fenfe certifieth, that it is even Torrified with Heat, in fome places, and Congealed with Cold in others, according to the temperature of the ambient Aer in divers climats, or as the Aer, being calefied by the Sun, or frigified by froft, doth varioufly affect it, in it fuperficial or Exterior parts ; and fo it cannot be difcerned, that its External parts are endowed with one of thefe oppofite Qualities more than the other : and fince we cannot but obferve, that there are many great and durable fubterraneous Fires burning in, and many fervid and fulphlureous Exlations frequently emitted, and more Hot Springs of Mineral Waters perpetually iffuing from its Interior parts, or bowels ; and fo it is of neceffity, that vaft feminaries of Igneous Atoms be included in the Entrals thereof : We fay, confidering thefe things, we cannot deny, but that the Earth doth contain as many Particles of Heat, or Calorifick Atoms, both without and within, as it doth of feeds of Cold, or Frigorifick Atoms, if not more ; and upon confequence, that it cannot be *Primum Frigidum*, as *Plutarch* and all his *Sectators* have dreamt. What then ; fhall we conclude Antithetically, and conceive that the Globe of the Earth is therefore Effentially rather Hot, than Cold ? Truely, No ; becaufe experience demonftrateth, that the Earth doth belch forth Cold Exhalations, and congealing blafts, as well as Hot Fumes, and more frequently : witnefs the Northwind, which is fo cold, that it refrigerates the Aer even in the middft of Summer, when the rivers are exhaufted by the fervor of the Sun ; to which *Elihu*, one of *Jobs* forry Comforters, feems to have alluded, when He faid, That *Cold cometh out of the North, and the Whirlwind out of the South.* All, therefore, we dare determine in this difficult argument (the decifion whereof doth chiefly depend upon Experiments of vaft labour and cofts) is only thus much ; that the Earth, which is now Hot, now Cold, in its extreme or fuperficial parts, may, as to its Internal or profound parts, be as reafonably accounted to contain various feminaries of Heat, as of Cold : and that the principal feeds of Cold, or fuch, as chiefly confift of Frigorifick Atoms, do convene into *Halinitre*, and other Concretions of natures retaining thereto. And our Reafon is, that *Halinitre* is no fooner diffolved in Water, than it congealeth the fame into perfect Ice, and ftrongly refrigerates all bodies, that it toucheth ; infomuch that we may not only conclude,

conclude, that of all Concretions in Nature, at leaſt that we have diſcovered, none is ſo plentifully fraught with the Atoms, or ſeeds of Cold, as
Halinitre; but alſo adventure to anſwer that Problem propoſed to *Job,
Out of whoſe womb came the Ice, and the Hoary Froſt of heaven, who hath
gendred it ?* by ſaying, that all our Freezing and extreme Cold winds
ſeem to be only copious Exhalations of Halinitre diſſolved in the bowels of
the Earth; or conſiſting of ſuch Frigorifick Atoms, as compoſe Halinitre;
and this becauſe of the identity of their Effects, for the Tramontane Wind
(the coldeſt of all winds, as *Fabricius Paduanus,* in his exquiſite Book
*de Ventis,* copiouſly proveth) which the Italians call *Chirocco,* can pretend to no natural Effect, in which Halinitre may not juſtly rival it. Long
might we dwell upon this not more rare than delightful ſubject: but, beſides that it deſerves a profeſt Diſquiſition, apart by it ſelf, our ſpeculations are limited, and may not, without indecency, either digreſs from their
proper Theme, or tranſgreſs the ſtrict Laws of Method. May it ſuffice,
therefore, in præſent, that we have made it juſtifiable to conceive that the
Earth containeth many ſuch Particles, or Atoms (whether ſuch as pertain
to the Compoſition of Halinitre, or of any other kind whatever) upon
the Exſilition of which the body containing them may be ſaid to become
Cold, or paſs from Potential to Actual Cold: and upon the inſinuation of
which into Aer, Water, Earth, Stones, Wood, Fleſh, or any other terrene Concretion whatever, Cold is introduced into them, and they may be
ſaid to be Frigefied, or made Cold.

Secondly, as for *Water;* that the prætext thereof to the prærogative of
Eſſential Frigidity is alſo fraudulent, and inconſiſtent with the *Magna
Charta* of right Reaſon, may be diſcovered from theſe conſiderations.
1. When Water is frozen, the Ice always begins in it ſuperfice, or upper
parts, where the Aer immediately toucheth it: but, if it were Cold of its
own Nature, as is generally præſumed, upon the auctority of *Ariſtotle,*
the Ice ought to begin in parts fartheſt ſituate from the Aer, that is in the
middle, or bottom, rather than at the top; at leaſt, it would not be more
ſlowly conglaciated in the middle and bottom, than at the top. (2.) In all
Froſts, the Cold of Water is encreaſed; which could not be, if it were the
principal ſeat of Cold. For, how could the Aer which according to the
vulgar ſuppoſition, that Water is the ſubject of inhæſion to extreme Cold,
if leſs cold, infuſe into water a greater cold, than what it had before of its
owne? or, how could Nitre, diſſolved in water, ſo much augment the
Cold thereof, as to convert it into Ice, even in the heat of ſummer, or by the
fires ſide; as is experimented in Artificial conglaciations: if Nitre were not
endowed with greater cold than Water? (3.) If Water be formally in-
gravidated with the ſeeds of Cold; why is not the ſea, why are not all Rivers, nay, all Lakes and ſtanding Pools (in which the excuſe of continual
motion is prævented) conſtantly congealed, and bound up in ribbs of Ice?
Whence comes it, that Water doth conſtantly remain Fluid, unleſs in
great froſts only, when the Atoms of Cold, wafted on the wings of the
North-wind, and plentifully ſtrawed on the waters, doe inſinuate themſelves among its particles, and introduce a Rigidity upon them? Certainly,
it is not conform to the Laws of Nature, that any Body, much leſs ſo eminent and uſeful a one as Water, ſhould for the moſt part remain alienated
from its owne native conſtitution, and be reduced to it again only at ſome
times, after long intervals, and then only for a day or two. (4.) Were
Cold

Cold essentially competent to Water, it could not so easily, as is observed, admit the Contrary Quality, Heat, nor in so high a degree, without the destruction of its primitive form. For, no subject can be changed from the Extreme of one Quality inhærent, to the extreme of a contrary, without the total alteration of that Contexture of its particles, upon which the inhærent quality depended; which done, it remains no longer the same: but Water still remains the same, *i. e.* a Humid Fluid substance, both at the time of, and after its Calefaction by fire, as before. And, therefore, that common saying, that *Water heated doth reduce it self to its native Cold,* though it be tollerable in the mouth of the people; yet He that would speak as a Philosopher, ought to change it into this, that *Water, after calefaction, returns to its primitive state of Indifferency to either Heat, or Cold:* for, though after its remove from the fire, it gradually loseth the Heat acquired from thence, the Igneous Atoms spontaneously ascending and abandoning it one after another; yet would it never reduce it self to the least degree of cold, but is *reduced* to cold by Atoms of Cold from the circumstant Aer immitted into its pores. What then; shall we hence conclude, that Water is Essentially *Hot?* Neither; because then it could not so easily admit, nor so long retain the Contrary Quality, Cold, for Hot springs are never congelated. Wherein therefore can we acquiesce? Truly, only in this determination; that *Water is Essentially Moist, and Fluid: but neither Hot, nor Cold, unless by Accident, or Acquisition, i.e.* it is made Hot, upon the introduction of Calorifick, and Cold, upon the introduction of Frigorifick Atoms; contrary to the tenent of *Empedocles,* and *Aristotle.*

Lastly, as for the *Aer:* insomuch as it is sometimes Hot, sometimes Cold, according to the temperature of the Climate, season of the year, præsence or absence of the Sun, and diversity of Winds: we can have no warrant from reason, to conceive it to be the natural Mother of Cold, more than of Heat; but rather that it is indifferently comparated to admit either Quality, according to divers Imprægnation. Whoever, therefore, shall argue, that because in the Dogg daies, when the perpendicular rayes of the Sun parch up the languishing inhabitants of the Earth in some positions of its sphere, if the North-wind arise, it immediately mitigates the fervor of the Aer, and brings a cool relief upon its wings; therefore the Aer is Naturally Cold: may as justly infer, that the Aer is Naturally Hot; because, in the dead of Winter, when the face of the Earth becomes hoary and rigid with frost, if the South-wind blowe, it soon mitigates the frigidity of the Aer, and dissolves those fetters of Ice, wherewith all things were bound up. Wherefore, it is best for us to Conclude, that the *Essential Quality* of the Aer, is *Fluidity*; but as for *Heat* and *Cold*, they are Qualities meerly *Accidental* or *Adventitious* thereto; or, that it is made Hot, or Cold, upon the commixture of Calorifick, or Frigorifick Atoms. So that where the Aer is constantly imprægnate with Atoms of Heat, as under the Torrid Zone, there is it constantly Hot, or Warme at least: where it is Alternately perfused with Calorifick and Frigorifick Atoms, as under the Temperate Zones; there is it Alternately Hot and Cold: and where it is constantly pervaded by Frigorifick Atoms, as under the North Pole; there is it constantly Cold.

*Art.7.*
But to some *special Concretions, for the most part, consisting of Frigorifick Atoms.*

To put a period, therefore, to this Dispute; seeing the Quality of *Cold* is not Essentially inhærent in Earth, Water, or Aer, the Three Principal
Bodies

Bodies of Nature; where shall we investigate its *Genuine Matrix*, or proper *subject of inhæsion*? Certainly, in the nature of some *Special Bodies*, or a particular species of *Atoms* (of which sort are those whereof Salnitre is for the most part composed)which being introduced intoEarth,Water,Aer, or any other mixt Bodie, impregnate them with cold.

But, haply, you may say, that though this be true, yet doth it not totally solve the doubt,since it is yet demandable,*Whether any one,and which of those Three Elements is highly Opposite to the Fourth*, viz. *Fire?* We Answer, that forasmuch as that Bodie is to be accounted the most Opposite to Fire, which most destroyes it : therefore is *Water* the chief Antagonist to Fire, because it soonest Extinguisheth it. Neverthele*s* there is no necessity, that therefore Water must be Cold in as high a degree, as Fire is Hot : for, Water doth not extinguish Fire, as it is Cold (since boyling water doth as soon put out fire, as Cold)but as it is *Humid*, i.e. *as it enters the pores of the enflamed body, and hinders the Motion and Diffusion of the Atoms of Fire.* Which may be confirmed from hence (1.) That Oyle, which no man conceives to be Cold, if poured on in great quantity,doth also extinguish fire,by suffocation, which is nothing but a hindering the Motion of the igneous Atoms : (2.)That in case the Atoms of Fire issue from the accensed matter, with such pernicity and vehemence, and reciprocal arietations, and in such swarms, as that they repel the water assused, and permit it not to enter the pores of the fewel (as constantly happens in Wild-fire,where the ingredients are Unctuous,and consist of very tenacious particles.) in that case, Water is so far from extinguishing the flame, that it makes it more impetuous and raging. However,we shall acknowledg thus much, that if the Principality of Cold must be adscribed to one of theThree vulgar Elements;the *Aer* doubtless,hath the best titlethereunto:because,being the most Lax and Porous bodie of the Three it doth most easily admit,and most plentifully harbour the seeds of Cold ; and being also subtile and Fluid, it doth most easily immit, or carry them along with it self into the pores of other bodies, and so not only Infrigidate,but some times Congeal,and Conglaciate them;in case they be of such Contextures and such particles, as are susceptible of Congelation and Conglaciation.

The Fable of the *Satyr* and *Wayfering man*,who blew hot and cold,though in the mouth of every School-boy,is yet scarce understood by theirMasters; nay, the greatest Philosophers have found the reason of that Contrariety of Effects from one and the same Cause,to be highly problematical.Wherefore since we are fallen upon the cause of the Frigidity in the Aer; and the Frigidity of our Breath doth materially depend thereon:opportunity invites Us, to solve that Problem,which though both *Aristotle(sect 3.prob.7.& Anaximenes (apud Plutarch. de frigore primigenio)* have strongly attempted ; yet have they left it to the conquest of *Epicurus* principles : *viz. Why doth the breath of a man warme,when efflated with the mouth wide open, and cool,when efflated with the mouth contracted?* To omit the opinions of others, therefore,we conceive the cause hereof to be only this;that albeit the Breath doth consist of aer,for the most part fraught withCalorifick Atoms,emitted from the lungs and vital organs,yet hath it many Frigorifick ones also interspersed among its particles : which being of greater bulk,than the Calorifick, and so capable of a stronger impuls,are by the force of efflation transmitted to greater distance from the mouth ; because, the Calorifick Atoms commixt with

Sf                                                                              the

*Art.8.*
*Water,the chief Antagonist to Fire;not in respect of its Accidental Frigidity,but Essential Humidity : and that the Aer hath a juster title to the Principality of Cold, than either Water, or Earth.*

*Art.9.*
PROBLEM: *Why the breath of a man doth Warme , when expired with the mouth wide open ; & Cool, when efflated with the mouth contracted.*

the breath, in regard of their exility, are no sooner dischaged from the mouth, than they instantly disperse in round. Wence it comes, that if the breath be expired in a large stream, or with the mouth wide open; because the circuit of the stream of breath is large, and so the Hot Atoms emitted are not so soon dispersed : therefore doth the stream feel warme to the hand objected there, and so much the more warme, by how much neerer the hand is held to the mouth; the Calorifick Atoms being less and less Dissipated in each degree of remove. But, in case the breath be emitted with contracted lipps; because then the compass of the stream is small, and the force of Efflation greater : therefore are the Calorifick Atoms soon Disgregated, and the Frigorifick only ramain commixt with the Aer, which affects the objected hand with Cold: and by how much farther (in the limits of the power of Efflation) the hand is held from the mouth, by so much colder doth the breath appear, and *è contra.* That Calorifick Atoms are subject to more and more Dispersion, as the stream of a Fluid substance, to which they are commixt, is greater and greater in circuit, may be confirmed from hence; that if we poure hot Water, from on high, in frosty weather, we shall observe a fume to issue and ascend from the stream all along, and that so much the more plentifully, by how much greater the stream is. Thus we use to cool Burnt wine, or Broth, by frequent refunding it from vessel to vessel, or infunding it into broad and shallow vessels; that so the Atoms of Heat may be the sooner dispersed : for, by how much larger the superfice of the liquor is made, by so much more of liberty for Exsilition is given to the Atoms of Heat contained therein, and as much of Insinuation to the Atoms of Cold in company with the circumstant Aer. Thus also we cool our faces in the heat of Summer, with fanning the aer towards us : the Hot Atoms being thereby dissipated, and the Cold impelled deeper into the pores of the skin : which also is the reason, why all Winds appear so much the Colder, by how much stronger they blowe; as *De s Cartes* hath well observed in these words : *Ventus vehementior majoris frigiditatis perceptionem, quàm aer tranquillus, in corpore nostro excitat; quod aer quietus tantùm exteriorem nostram cutim, quæ interioribus nostris carnibus frigidior est, contingat : ventus verò, vehementiùs in corpus nostrum actus, etiam in penetralia ejus adigatur, cumque illa sint cute calidiora, idcirco etiam majorem frigiditatem ab ejus contactu percipiunt.*

Art. 10.
Three CON-
SECTARIES
from the pre-
mises.

In our precedent Article, touching the necessary assignation of a Tetrahedical Figure to the Atoms of Cold, we remember, we said, that in respect of their several sides, or plane faces, they were most apt to Compinge, or bind in the particles of all Concretions, into which they are intromitted; and from thence we shal take the hint of inferring Three noble CONSECTARIES.

(1.) That Ice, Snow, Hail, Hoarfrost, and all Congelations, are made meerly by the intromission of Frigorifick Atoms among the particles of Fluid bodies : for, being once insinuated and commixt among them, in sufficient plenty, they alter their fluid and lax consistence into a rigid and compact, *i.e.* they Congeal them.

(2.) That the Horror, or Trembling sometimes observed in the members of Animals; as also that Rigor, or Shaking, in the beginning of most putrid Fevers, and generally when the Fits of Intermittent fevers invade, are chiefly caused by Frigorifick Atoms. For, when the Spherical Atoms

of

of Heat, which swarm in and vivifie the bodies of Animals, are not moved *quaquaversum* in the members with such freedom, velocity, and directness excentrically, as they ought; because, meeting and contesting with those less Agile Atoms of Cold, which have entred the body, upon its chilling, their proper motion is thereby impeded: they are strongly repelled, and made to recoyle towards the Central parts of the bodie, in avoydance of their Adversary, the Cold ones; and in that tumultuous retreat, or introcession, they vellicate the fibres of the membranous and nervous parts, and so cause a kind of vibration or contraction, which if only of the skin, makes that symptome, which Physicians call a *Horror*; but if of the Muscles in the Habit of the bodie, makes that more vehement Concussion, which they call a *Rigor*. Either of which doth so long endure, as till the Atoms of Heat, being more strong by Concentration and Union, have re-encountered and expelled them. That it is of the Nature of Hot Atoms, when invaded by a greater number of Cold ones, to recoyle from them, and concentre themselves in the middle of the body, that contains them; is demonstrable from the Experiment of Frozen Wines: wherein the spirits concentre, and preserve themselves free from Congelation in the middle of the frozen Phlegm, so that they may be seen to remain fluid and of the colour of an Amethyst: as *Helmont* hath well declared, in his *History of the Nativity of Tartar in Wines.*

(3.) That the Death of all Animals, is caused immediately by the Atoms of Cold; which insinuating themselves in great swarms into the body, and not expelled again from thence by the overpowered Atoms of Heat; they wholly impede and suppress those motions of them, wherein Vitality consisteth: So that the Calorifick ones being no longer able to calefy the principal seat of life, the Vital flame is soon extinguished, and the whole Body resigned to the tyranny of Cold. Which is therefore well accounted to be the grand and profest Enemy of Life.

# CHAP. XIII.

### OF

### *Fluidity, Stability, Humidity, Siccity.*

### Sect. I.

*Art. 1.*
Why *Fluidity*
and *Firmness*
are here con-
fidered before
*Humidity* and
*Siccity.*

Ere our very Method muft be fomewhat Paradoxical, and the Genealogy we fhall afford of thofe Two vulgarly accounted Paffive Qualities,
*Humidity* and *Siccity*, very much
different from that univerfally embraced in the Schools. For, fhould
we tread in the fteps of *Ariftotle*,
as moft, who have travelled in this
fubject, have conftantly done ; we
muft have fubnected our Difquifition into the Nature and Origine of
Moifture and Drynefs, immediate-
ly to that of Heat and Cold, as the other pair of Firft Elemental Qualities,
and *è diametro* oppofite to them. But,having obferved, that thofe 2 Terms,
*Moift* and *Dry*, are not, according to the fevere and præcife Dialect of
truth, rightly accommodable to all thofe things, which are genuinely im-
ported by thofe Greek Words, ὑγρὸν and ξηρὸν, according to the defini-
ons of *Ariftotle*, and confequently that we could not avoid the danger of lofing ourfelves in a perpetual *Æquivocation* of Terms, unlefs we
committed our thoughts wholly to the conduct of Nature Herfelf, progreffing from the more to the lefs General Qualities, and at each ftep explicating their diftinct dependencies : we thereupon inferred, that we
ought to præmife the Confideration of *Fluidity* and *Firmnefs*, which are
more General, to that of *Humidity* and *Siccity*, as lefs General Qualities, and which feem to be one degree more removed from Catholick
Principles.

*Art. 2.*
The Latin
Terms, *Humi-
dum* and *Sic-
cum*, too nar-
row to com-
prehend the
full fenfe of
*Ariftotle*,
ὑγρὸν & ξηρὸν.

That thofe 2 Terms fo frequent in the mouth of *Ariftotle*, ὑγρὸν ἢ
ξηρὸν, are more ample in fignification, than *Humidum* and *Siccum*, by
which His Latin Interpreters and Commentators commonly explicate
them ; is manifeft even from hence, that under the word ὑγρὸν is comprehended not only, in General, whatever is ρυτὸν, Fluid and *Liquid*,
but

but also, in special, that matter or body, whereby a thing is moistned, when immersed into, or perfused with the same: and likewise, under the contrary term ξηρὸν, is comprehended as well, in General, whatever is πυκνὸν, *Compact* or *Firme* and *Solid*, as in special, that matter or body, which being applyed to a thing, is not capable of Humectating or Madefying the same, and which is therefore called also ξεραῦνον, *Aridum*. Now, this duely perpended doth at first sight detest the Æquivocation of the Latin Terms, and direct us to this præcise determination; that whatever is *Fluid*, is not *Humid*; nor whatever is *Dry*, *Compact* or *Firme*; but that a *Humid* body properly is that, whereby another body, being perfused, is moistned [ἕρεκτὸς] or madefied [διαίνεται]: and, on the contrary, that a Dry or Arid body is that, which is not capable of Humectating, or madefying another body, to which it is applied.

Again, forasmuch as *Aristotle* positively defines τὸ ὑγρὸν, *id quid facile, terminum admittens, proprie tamen non terminatur*, that which being destitute of self-termination, is yet easily terminated by another substance; tis evident, that this His Definition is competent not only to a *Humid* thing, in special, but also to a *Fluid*, in General: such as are not only Water, Oyle, every Liquor, yea and Metal or other Concretion, actually fused or melted; but also the Aer, Flame, Smoke, Dust, and whatever is of such a nature, as that being admitted into any vessel or other continent of whatever figure; or however terminated in it superfice, doth easily accomodate it self thereunto, put on the same figure, and confess termination by the same limits or boundaries; and this, because it cannot terminate it self, as being naturally comparated only to Diffusion. On the other side, since He defines τὸ ξηρὸν, *quod facile terminatum proprio termino, terminatur ægre alieno*; to be that which is easily terminated by its owne superfice, and hardly terminated by another; it is also manifest, that this Definition is not peculiar only to a *Dry* or *Arid* substance, but in common also to a *Firme* or *Solid* one: such as not only Earth, Wood, Stones, &c. but also Ice, Metal unmolten, Pitch, Resine, Wax, and the like Concreted juices, and (in a word) all bodies, which have their parts so consistent and mutually cohærent, as that they are not naturally comparated to Diffusion, but conserve themselves in their own superfice, and require compression, dilatation, section, detrition, or some other violent means, to accommodate them to termination, by the superfice of another body. And, certainly, if what is præcisely signified by the Terme ὑγρὸς, be no more than what is meant by the Latin substitute thereof, *Humidum*; then might the Aer be justly said to be Humid; which is so far in its owne nature from being endowed with the faculty of Humectating bodies, that its genuine virtue is to exsiccate all things suspended therein; nay even Fire it self might be allowed the same Attribute, together with Smoke, Dust, and the like Fluid substances, which exsiccate all bodies perfused with moisture. On the advers part, if what is præcisely intended by the Terme ξηρὸς, were fully expressible by the Latin, *Siccum*, or *Aridum*; then, doubtless, might Wax, Resine, and all Concreted juices be accounted actually Dry, nay Ice it self, which is only Liquor congealed, could not be excluded the Categorie of Arid substances. These Considerations premised, though we might here enquire, Whether *Aristotle* spake like Himself, when He confined Fluidity (and that according to his owne definition) to only 2 Elements, Water and Aer; when yet the Element of Fire, which He placed

above

*Art. 3.*
*Aristotles* Definition of a *Humid* substance, not præcise enough; but, in common also to a *Fluid*, and his Definition of a *Dry*, accommodable to a *Firme*.

above the Aereal region, must be transcendently Fluid(else how could it be so easily terminated by the Concave of the Lunar Sphere, on one part, and the Convex of the Aereal, on the other?) And whether His Antithesis or Counter assertion, *viz,* that the 2 Firme Elements are Fire and Earth, be not a downright Absurdity: yet shall we not insist upon the detection of either of those two Errors, because they are obvious to every mans notice; but only Conclude, that though every Humid body be Fluid, and every Arid or Dry body be Firm; yet will not the Conversion hold, since every Fluid is not Humid, nor every Firme, Dry; and upon natural consequence, that Humidity is a species of Fluidity, and Siccity a Species of Firmity; and also that it is our duety to speculate the Reasons of each accordingly beginning at the Generals.

*Art. 4.*
Fluidity defined.

*FLUIDITY* we conceive to be a Quality, arising meerly from hence, that the Atoms, or insensible particles, of which a fluid Concretion doth consist, are smooth in superfice, and reciprocally contiguous in some points, though dissociate or incontiguous in others; so that many inane spaces (smaller and greater according to the several magnitudes of the particles, which intercept them) being intersperfed among them, they are, upon the motion of the mass or body, which they compose, most easily moveable, rowling one upon another, and in a continued fluor, or stream diffusing themselves, till they are arrested by some firm body, to whose superfice they exactly accommodate themselves.

*Art. 5.*
Wherein the Formal Reason thereof doth consist.

That the Essence of Fluidity doth consist only in these Two conditions, the smoothness of insensible particles, and interruption of small inane spaces among them, where their extrems are incontiguous; may be even sensibly demonstrated in an heap or measure of *Corne.* Which is apt for Diffusion, or Fluid, only because the Grains, of which it doth consist, are superficially smooth and hard, and have myriads of inane spaces intercepted among them, by reason of the incontiguities of their extrems, in various points: so that, whenever the heap is moved, or effused from one vessel into another, the Grains mutually rowling each upon other, diffuse themselves in one continued stream, and immediately upon their reception into the concave of the vessel, the Aggregate or mass of them becomes exactly accommodate to the figure, or internal superfice of the same. And, forasmuch as the different magnitudes of composing particles, do not necessitate a difference of formal qualities; but only variety of Figures, contexture and motion: well may we conceive the same reasons to essence the Fluidity of Water also; because betwixt an heap of Corne, and an heap or mass of Water, the Difference is only this, that the Grains, which compose the one, are of sensible magnitude, and so have sensible empty spaces interposed among them; but the Granules, or particles, which compose the other, are of insensible magnitude, or incomparably more exile, and so have the inane spaces intercepted among them, incomparably less. For, that Water doth consist of small Grains, or smooth particles, is conspicuous even from hence; that Water is capable of conversion into Fume, or Vapour, only by Rarefaction, and Fume again reducible into Water, meerly by Condensation; and the reason why Fume becoms visible, is only this, that the least visible part of fume is a Collection or Assembly of many thousand of those singly-invisible particles, which constitute the Water, from whence the fume ascends; as may be ascertained from hence, that to the

com-

composition of one single drop of Water, many myriads of myriads of insensible particles must be convened and united. So that Water contained in a Caldron, set on the fire and seething, doth differ from the Fume exhaled from it, only in this respect; that the one is Water Condensed, the other Rarified: or, that Water is made Fume, when its particles are violently dissociated, and the aer variously intercepted among them; and Fume is returned to Water, when the same particles are reduced to their natural close order, and the intercepted aer again excluded. Again, that the Fluidity of Water depends on the same Cause (proportionately) is that of an heap of Corne, may, according to the Lawe of Similitude, be justified by the parallel capacity of Water to the same Effects, *viz*. Diffusion, Division, and Accommodation to the figure of the Recipient, or Terminant: For, the result hereof is, that it hath no Continuity or mutual Cohærence of its particles, which should hinder their easy Dissociation. Nor is it a valid Argument to the contrary, that Water appears to be a *Continued* body, but an heap of Corne, a Discontinued; for, that is only according to *Apparence*, caused from hence, that by how much smaller the component particles of a Concretion are, by so much smaller must the inane spaces be, which are intercepted among them, where they are incontiguous, and upon consequence, so much the less interrupted, or more continued must the mass or Aggregate appear: as may be most familiarly understood, if we compare an heap of Corne, with one of the finest *Calis* sand; that with an heap of the most volatile or impalpable Powder, that the Chymist or Apothecary can make; and so gradually less and less in the dimensions of Granules, till we arrive at the smallest imaginable. So that we cannot wonder, that the substance of Water should be apprehended by the dull sense, as wholly Continued, though really it be only less interrupted than an heap of sand: when the Grains, whereof Water is amassed, are incomparably smaller, than those of the finest sand, and intercept among them inane spaces incomparably smaller such as are by many degrees belowe the discernment of the acutest sight, though advantaged by the best Microscope.

*Art.6.*
The same farther illustrated by the twofold Fluidity of Metals; and the peculiar reason of each.

If this Argument reach not the height either of the Difficulty it self, or your Expectation and Curiosity concerning it; be pleased to imp the Wings of it with the feathers of another, of the same importance, but more perspicuity. It is well known, especially to Chymists and Refiners, that every metall is capable of a *twofold Fluidity*: *one*, in the forme of an impalpable or volatile *Powder*; the other, of a *Liquor*, whose fluor is continued, according to the judgement of sense. For, when a Metal is Calcined by Præcipitation, *i.e.* by Corrosive and Mercurial Waters, specifically appropriate to its nature; being thereby reduced into small Grains, it becomes Fluid, after the manner of sand, and therefore may as conveniently be used in Hour-glasses, for Chronometry, or the measure of time: but, because each of those visible Grains is made up of millions of other more exile and invisible Granules or particles, which are the component principles or matter of the Metal; hence it is, that if we put them all together in a Crucible, and melt them in a reverberatory fire, whose igneous Atoms invade, penetrate and subdivide each Granule into the smallest particles (to which the Corrosive Virtue of the *Aqua fortis* could not extend) then will the whole mass put on another kind of Fluidity, such as that of Water, Oyle, and all other Liquors. Now, the *Reason* of the *Former* Fluidity is manifestly the same with that of *Corne* and *Sand*, newly explicated:

explicated : and that of the *Latter*, the fame as of *Water*, *i.e.* the Granules
of the Calcined powder, being diffolved into others of dimenfions incomparably fmaller, do intercept among themfelves, or betwixt their fuperficies,
where thofe are incontiguous, innumerable multitudes of Inane fpaces, but
thofe incomparably lefs than before their ultimate fubtiliation ; and confequently (as hath been faid) make the Metal diffolved to be deprehended by the fenfe, as one entire and continued fubftance.  To Conclude,
therefore ; we can difcover no Reafon againft us, of bulk fufficient to ob-
ftruct the current of our Conception, that the Fluidity of *Fire, Flame,*
*Aer*, and all *Liquid* fubftances whatever, cannot well be deduced from
any other Caufe, but what we have here affigned to Water and Metals
diffolved : efpecially when we confider, that it is equally confentaneous to
conceive, that every other Fluid or Liquid body is compofed alfo of certain fpecially-configurate Granules, or imperceptible particles; which being only contiguous in fome points of their fuperficies, not reciprocally Cohærent cannot but intercept various inane fpaces betwixt them; and be therefore eafily emovable, diffociable, externally terminable, and capable of making
the body apparently Continuate, as Water it felf.

*Art. 7.*
Firmnefs de-
fined :

And, as for the other General Quality, FIRMNESS, or STABILITY ;
fince Contraries muft have Contrary Caufes, and that the folidity of Atoms
is the fundament of all folidity and firmnefs in Concretions: well may we
underftand it to be radicated in this, that the infenfible particles, of which
a Firme Concretion is compofed (whether they be of one or diverfe
forts, *i.e.* fimilar or diffimilar in magnitude and figure) do fo reciprocally comprefs and adhære unto each other, as that being uncapable
of rowling upon each others fuperfice, both in refpect of the ineptitude of their figures thereunto, and the want of competent inane
fpaces among them, they generally become uncapable (without extream violence) of Emotion, Diffociation, Diffufion, and fo of
Termination by any other fuperfice, but what themfelves conftitute.

*Art. 8.*
And derived
from either of
3 *Caufes.*

If it be farther Enquired, Whence this reciprocal Compreffion, Indif-
fociability, and Immobility of infenfible particles in a Firme, Concretion
doth immediately proceed; we can derive it from Three fufficient Caufes.
(1.) *The many fmall [ Hamuli, Uncinulivè ] Hooks or Clawes* by which
Atoms of unequal fuperficies are adapted to implicate each other, by mutual cohærence : and that fo clofely, as that all Inanity is excluded from
betwixt their commiffures or joynings ; and this is the principal and moft
frequent Caufe of ftability. (2.) *The Introduction and preffure of Ex-*
*traneous Atoms,* which invading a Concretion, and wedging in both themfelves, and the inteftine ones together, and that cheifly by obverting
their plane fides or fuperficies thereunto; caufe a general Compreffion
and Cohæfion of all the particles of the mafs.  And by this way doth
froft congeal Water and all Humid Subftances; for, fince the Atoms
of Cold are tetrahedical, and thofe of Water octahedical, as is moft
reafonably conjecturable; thofe of Cold infinuating themfelves into the
fubftance of Water, by obverfion of their plane fides to them, they
arreft the rowling particles thereof, and fo not permitting them to be
moved as before, impede their fluidity, and make the whole mafs Rigid
and Hard, or Firme.  Hither alfo may we moft congruoufly referr
the

the Coagulation of milk, upon the injection of Rennet, Vinegre, juice of Limons, and the like Acid things. For, the Hamous and invisicating Atoms, whereof the Acid is mostly composed, meeting with the Ramons and Grosser particles of the milk, which constitute the Caseous and Butyrous parts thereof; instantly fasten upon them with their hooks, connect them, and so impeding their fluiditie, change their lax and moveable contexture into a close and immoveable or Firme: while the more exile and smooth particles of the milk, whereof the serum or whey is composed, escape those Entanglings and conserve their native Fluidity. This may be confirmed from hence; that whenever the Cheese, or Butter made of the Coagulation, is held to the fire, they recover their former Fluidity: because the tenacious particles of the Acid are disentangled and interrupted by the sphærical and superlatively agile Atoms of fire. (3.) *The Exclusion of introduced Atoms,* such as by their exility, roundness and motion, did, during their admission, interturbe the mutual Cohæsion and Quiet of domestique ones, which compose a Concretion. Thus, in the decalescence of melted metals, and Glass, when the Atoms of fire, which had dissociated the particles thereof and made them Fluid, do abandon the metal, and so cease to agitate and dissociate the particles thereof: then do the domestique Atoms returne to a closer order, mutually implicate each other, and so make the whole mass Compact and Firme, as before. Thus also when the Atoms of Water, Wine, or any other dissolvent, which had insinuated into the body of Salt, Alume, Nitre, or other Concretion retaining to the same tribe; and dissolving the continuity of its particles, metamorphosed it from a solid into a fluid body, so that the sight apprehends it to be one simple and uniforme substance with the Liquor: we say, when these dissociating Atoms are evaporated by heat, the particles of the Salt instantly fall together again, become readunated, and so make up the mass compact and solid, as before, such as no man, but an eye-witness of the Experiment, could persuade himself to have been so lately diffused, concorporated, and lost in the fluid body of Water.

## SECT. II.

*Art.* 1.
*Humidity de-*
*fined.*

BY the light of the Præmises, it appears a most perspicuous truth, that HUMIDITY is only a certain Species of Fluidity. For, whoever would frame to himself a proper and adæquate Notion of an *Humor,* or *Humid* substance; must conceive it to be *such a Fluid or Fluxile body, which being induced upon, or applied unto any thing, that is Compact, doth adhare to the same* (per minimas particulas) *and madify or Humectate so much thereof as it toucheth.* Such, therefore, is Water, such is Wine, such is Oyle, such are all those Liquors, which no sooner touch any body not Fluid, but either they leave many of their particles adhærent only to the superfice thereof (and this, because the most seemingly polite superfice is full of Eminences and Cavities, as we have frequently asserted) and so moisten it; or, penetrating through the whole contexture thereof, totally Humectate or wett the same. But, such is not Aer, such is not any Metal fused, such is not Quick-silver, nor any of those

T t                                                    Fluors,

Fluors, which though they be applied unto, and fubingrefs into the pores of a Compact body, doe yet leave none of their particles adhærent to either the fuperficial or internal parts thereof; but, without diminution of their own quantity, run off clearly, and fo leave the touched or pervaded body, unmodified, or unhumectate, as they found it.

**Art. 2.**
**Siccity defined.**

On the other fide, it is likewife manifeft, that SICCITY or ARIDITY, is only a certain fpecies of Firmnefs, or ftability: becaufe a Dry or Arid fubftance is conceived to be Firm or Compact, only infomuch as it is void of all moifture. Of this fort, according to vulgar conception, may we account all Stones, Sand, Afhes, all Metals, and whatever is of fo firme a conftitution, as contains nothing of Humidity, either in it fuperfice, or entrals, which can be extracted from it, or, if extracted, is not capable of moiftning any other body: but, not Plants nor Animals, nor Minerals, nor any other Concretion; which, though apparently dry to the fenfe, doth yet contain fome moifture within it, and fuch as being educed, is capable of humectating another body.

**Art. 3.**
**Siccity, rather**
**Comparative,**
**than Abfolute.**

We fay, *According to Vulgar Conception*; becaufe, not *Abfolutely*: for, though Siccity be oppofed to Humidity, not as an Habit, to which any Act can be juftly attributed, but as a meer Privation (for, to be Dry, is nothing elfe but to want moifture; yet, becaufe a Moiftned body may contain more or lefs of Humidity, therefore may it be faid to be more or lefs Dry Comparatively, and a body that is imbued with lefs moifture, be faid to be dryer than one imbued with more. Thus Green Wood, or fuch as hath imbibed extraneous moifture, is commonly faid to grow more and more dry by degrees, as it is more and more Dehumectated; and then at length to be perfectly dry, when all the Aqueous moifture, as well natural as imbibed, is confumed, though then alfo it contain a certain unctuous moifture, which Philofophers call the *Humidum Primigenium*: but, this only Comparatively, or in refpect to its former ftate, when it was imbued with a greater proportion of Humidity.

**Art. 4.**
**All moifture**
**either Aqueous**
**or Oleaginous.**

For the illuftration of this, we are to obferve, that there are *Two* forts of Moifture, wherewith compact bodies are ufually humectated; the one, *Aqueous* and *Lean*; the other, *Oleaginous* and *Fat*. The *Firft* is eafily diffoluble and evaporable by heat, but not inflammable: the *other*, though it eafily admit heat, and is as eafily inflammable, in regard of the many igneous Atoms contained therein; is not eafily exfoluble, nor attenuable into fume, in regard of the Tenacious cohærence of its particles. To the Firft kind may be referred that moifture in Concretions, which Chymifts extracting, call the *Mercury* of Vegetables: becaufe, though it moiftens as Water, and is as incapable of inflammation, yet is it much more volatile or evaporable. And, to either or both forts, though in a diverfe refpect belongs that, which they call *Aqua Vitæ*, or the fpirits of a Vegetable, fuch as fpirit of Wine: becaufe though it doth moiften as Water, yet is it far more eafily diffoluble and evaporable by heat, and as inflammable as oyle. And thus much we learn in the School of Senfe, that fuch bodies as are humectate with the Aqueous and Lean moifture, are eafily capable of Exficcation: but fuch as are humectate with the Unctuous and Fat, very hardly: Why? becaufe the Atoms, of which

the

the Aqueous doth confift, are more lævigated or fmooth in their fuperfice, and fo having no hooks, or clawes, whereby to cohære among
themfelves, or adhære to the concretion, are foon difgregated; but
thofe, which compofe the Oleaginous, being entangled as well among themfelves, as with the particles of the body, to which they are admixt, by
their Hamous angles, are not to be expeded and difengaged, without great
and long agitation; and after many unfuccefsful attempts of evolution. Thus
Wood is fooner reduced to Afhes, than a ftone: becaufe that is compacted
by much of Aqueous Humidity; this by much of Unctuous. For the fame
reafon is it likewife, that a clodd of Earth, or peice of Cloth, which hath imbibed Water, is far more eafily reficcated, than that Earth or Cloth, which
hath been dippt in oyle, or melted fat. And this gives us fomewhat more
than a meer Hint toward the clear Solution of Two PROBLEMS, frequently occurring, but rarely examined.

The one is, *Why pure or fimple Water cannot wafh out fpots of Oyle, or*
*Fat from a Cloth, or filk Garment : which yet Water, wherein Afhes have*        PROBLEM 1.
*been boyled, or foap diffolved, eafily doth?* For, the *Caufe* hereof moft        Why pure wa-
probably is this; that though Water of it felf cannot penetrate the unctu-        ter cannot
ous body of oyle, nor diffociate its tenacioufly cohærent particles, and con-        wafh out oyle
fequently not incorporate the oyle to it felf, fo as to carry it off in its        from a Cloth;
fluid arms, when it is expreffed or wrung out from the cloth: yet, when        which yet wait is imprægnated with Salt, fuch as is abundantly contained in Afhes, and        ter, wherein
from them extracted in decoction; the falt with the fharp angles and        Afhes have
points of its infenfible particles, penetrating, pervading, cutting and di-        been decocted,
viding the oyle, *in minimas particulas,* the Water following the particles        or foap dif-
of falt at the heels, incorporates the oyle into it felf, and fo being wrung        folved, eafily
out from the cloth again, brings the fame wholly off together with it felf.        doth? Solut.
Which doubtlefs, was in fome part underftood by the Inventor of foap;
which being compounded of Water, Salt and Oyle moft perfectly commixt, is the moft general Abfterfive for the cleanfing of Cloathes pol-
luted with oyle, greafe, turpentine, fweat and the like unctuous natures:
for, the particles of oyle ambufcadoed in the foap, encountring thofe
oyly or pinguous particles, which adhære to the hairs and filaments of
Cloth and ftain it, become eafily united to them, and bring them off together with themfelves, when they are diffolved and fet afloat in the Water
by the incifive and diffociating particles of the Salt; which alfo is brought
off at the fame time by the Water, which ferveth only as a common vehicle
to all the reft.

The other, *Why ftains of Ink are not Delible, with Water, though de-*
*cocted to a Lixirium, or Lee, with Afhes, or commixt with foap : but*        PROBLEM 2
*with fome Acid juice, fuch as of Limons, Oranges, Crabbs, Vinegre, &c.*        Why ftains of
For, the Reafon hereof feems to be only this; that the Vitriol, or Cope-        ink are not to
rofe, which ftrikes the black in the Decoction of Galls, Sumach, or other        be taken out
Adftringent Ingredients, being Acid, and fo confifting of particles con-        of cloth, but
generous in figure and other proprieties to thofe which conftitute the        with fome
Acid juices: whenever the fpot of Ink is throughly moyftned with an        Acid Liquor?
acid liquor, the vitriol is foon united thereto, and fo educed together with        Solut.
it upon expreffion, the union arifing (*propter ὁμοίωσιν*) from the *Simili-*
*tude* of their two natures. For, there always is the moft eafy and
perfect *union,* where is a *Similitude* of Effences, or formal proprieties;

as

as is notably experimented in the eduction of Cold from a mans hands or other benummed parts by rubbing them with ſnow; in the evocation of fire by fire; in the extraction of ſome Venoms from the central to the outward parts of the body, by the application of other Venoms to the skin (which is the principal cauſe, why ſome Poyſons are the Antidotes to others); the alliciency and evacuation of Choler by Rhubarb, &c.

Laſtly, in this place, we might pertinently inſiſt upon the Cauſes and Manner of Corroſion and Diſſolution of Metals and other Compact and Firme bodies, by *Aqua Fortis, Aqua Regis,* and other Chymical Waters; the Exſolubility of Salt, Alume, Nitre, Vitriol, Sugar and other Salin concreted juices, by Water; the Exhalability or Evaporability of Humid and Humectating ſubſtances, and other uſeful ſpeculations of the like obſcure nature: but, each of theſe deſerves a more exact and prolix Diſquiſition, than the time conſigned to our præſent province will afford; and what we have already ſaid, ſufficiently diſchargeth our debt to the Title of this Chapter.

CHAP.

# CHAP. XIV.

## Softneſs, Hardneſs, Flexility, Tractility, Ductility, &c.

### Sect. I.

He two Firſt of this Rank, of Secundarie Qualities HARDNESS and SOFTNESS, being ſo neer of Extraction and Semblance, that many have confounded them with Firmneſs and Fluidity, in a General and looſer acceptation (for, ſo *Virgil* gives the Epithete of *Soft* to Water, & *Lucretius* to Aer, Vapors, Clouds, &c.) becauſe a *Firme* bodie, or ſuch whoſe parts are reciprocally cohærent, and ſuperfice more than only apparently continued, as Wax, may be *Soft*; and on the other ſide, a *Fluid* body, or ſuch whoſe particles are not reciprocally cohærent, nor ſuperfice really continued, as ſand, may be *Hard*; therefore ought we to begin our examination of the nature of *Hardneſs* and *Softneſs*, and their Conſequents, *Flexility, Tractility, Ductility, &c.* where that of Firmneſs and Fluidity ends; that ſo we may, by explicating their *Cognation*, when mentioned in a *general* ſenſe, manifeſt their *Differences*, when conſidered in a *Special* and præciſe, and ſo prevent the otherwiſe imminent danger of æquivocation.

*Art. 1.*
The *Illation* of the Chapter.

To come, therefore, without farther circumambage, to the diſquiſition of the proper nature of each of theſe Qualities, according to the method of their production; conforming our conceptions to thoſe of *Ariſtotle*, who (4. *Meteor.* 4.) defines *Durum* to be, *Quod ex ſuperficie in ſeipſum non cedit*; and *Molle*, to be *Quod ex ſuperficie in ſeipſum cedit*; and referring both to the cognizance of the ſenſe of Touching, we underſtand a HARD body to be ſuch, *whoſe particles are ſo firmely coadunated among themſelves, and ſuperfice is ſo continued, as that being preſt by the finger, it doth not yeeld thereto, nor hath it ſuperfice at all indented or depreſſed thereby*; ſuch is a ſtone; and on the contrary, a SOFT one to be ſuch, *as doth*

*Art.2.*
*Hard* and *Soft*, defined.

*doth yield to the pressure of the finger in the superfice, and that by retrocession or giving back of the superficial particles, immediately prest by the finger,* versus profundum, *towards it profound or internal;* such as Wax, the Flesh of Animals, Clay, &c.

*Art. 3.*
The Difference betwixt a Soft and Fluid.

For, the chief Difference betwixt a *Fluid,* and a *Soft* body, accepted in a Philosophical or præcise, not a Poetical or random sense, consisteth only in this; that the *Fluid,* when prest upon, doth yield to the body pressing, not by indentment or incavation of it superfice, *i.e.* the retrocession of it superficial particles, which are immediately urged by the depriment, toward its middle or profound ones, which are farther from it; but by rising upwards in round and equally on all sides, as much as it is deprest in the superfice: and a *Soft* doth yield to the body pressing, only by retrocession of it superficial inwards toward it central particles, so that they remain during, and sometimes long after the depression, more or less lower than any other part of the superfice. Which being considered, *Aristotles* judgement, that *Softness is incompetent to Water,* must be indisputable: because tis evident to sense, that Water, being deprest in the superfice doth not recede towards its interior or profound parts, as is the property of all soft things to doe; but riseth up in round equally on all sides of the body pressing, and so keeps it superfice equally and level as before.

*Art. 4.*
Solidity of Atoms, the Fundament of Hardness and Inanity, intercepted among them, the fundament of Softness, in all Concretions.

As for the *Fundamental Cause* of *Hardness* observed in Concretions; it must be the chief essential propriety of Atoms *Solidity*: and upon consequence, the *Original* of its Contrary, *Softness* must be *Inanity.* For, among Concretions, every one is more and more Hard, or less and less soft, according as it more and more approacheth to the solidity of an Atom, which knowes nothing of softness: and on the other side, every thing is more and more soft, or less and less hard, according as it more and more approacheth the nature of Inanity, which knowes nothing of Hardness. Not that the Inane space is therefore capable of the Attribute of *Soft,* as if it had a superfice, and such as could recede inwards upon pression: but, that every Concretion is alwayes so much the more soft, *i.e.* the less hard, by how the more it yields in the superfice upon pressure; and this only in respect of the more of Inanity, or the Inane space intercepted among the solid particles, whereof it is composed. It need not be accounted Repetition, that we here resume what we have formerly entrusted to the memory of our Reader; *viz.* that touching the deduction of these two Qualities, Hardness and Softness, the provident Atomist hath wonn the Garland from all other Sects of Philosophers: for, supposing the Catholike materials of Nature to be Atoms, *i. e.* Solid or inflexible and exsoluble Bodies, he is furnished with a most sufficient, nay a necessary Reason, not only for the Hardness or Inflexibility, but also for the Softness or Flexibility of all Concretions; insomuch as it is of the essence of his Hypothesis, that every compound nature derives its Hardness only from the Solidity of its materials, and softness only from the Inane space intercepted among its component particles; in respect whereof each of those particles is moveable, and so the whole Aggregate or mass of them becomes flexible, or devoid of rigidity in all its parts, and consequently yeelding in that part, which is pressed. But, no other Hypothesis excogitable is fruitful enough to afford a satisfactory, nay not so much as a meerly plausible solution of

this

this eminent and fundamental Difficulty; for, thofe who affume the univerfal matter to be voyd of Hardnefs, and fo infinitely exfoluble, *i.e.* not to be Atoms, though they may, indeed, affign a fufficient reafon, why fome Concretions are foft; yet fhall they ever want one to anfwer him, who demands, why other Concretions are Hard; becaufe themfelves have exempted Atoms, from whofe folidity all Hardnefs arifeth to Concretions.

And this moft eafily detecteth the grofs and unpardonable incogitancy of *Ariftotle*, when He determined the Hardnefs and Softnefs of Concretions to be *Abfolute* Qualities; for, fince Atoms alone are abfolutely void of all Softnefs, and the Inane fpace alone abfolutely void of all Hardnefs; and all Concretions are made up of Atoms: nothing is more manifeft, than that Hardnefs and Softnefs, as attributary to Concretions, are Qualities meerly *Comparative*, or more præcifely, that Softnefs is a Degree of Hardnefs; and confequently, that there are various Degrees of Hardnefs, according to which Concretions may be faid to be more or lefs Hard, and fuch as are hard, in refpect of one, may be yet foft in refpect of another, that is more hard, or lefs foft.

*Art. 5.*
Hardnefs and Softnefs, no Abfolute, but meerly *Comparative* Quali. ries; as adfcriptive to Concretions: contrary to *Ariftotle.*

As for the præcife *Manner*, how the feveral Degrees of Hardnefs and Softnefs refult from Atoms and Inanity commixt; we need not much infift thereupon; fince the production of each degree may be eafily and fully comprehended, from our præcedent explanation of the Caufes of Fluidity and Firmnefs. For, though Softnefs be obfervable in bodies endowed with Firmnefs, or Influxibility; yet becaufe the degrees of Firmnefs are alfo various, and proceed from the more or lefs Arrefting or Impeding of Fluidity, and fo that the thing confift of Atoms more or lefs Coarctated, moveable among themfelves, and diffociable each from other (from whence alone doth the yeeldingnefs of it in the fuperfice arife): therefore is it neceffary, that in Firme things the fame is the caufe of Softnefs, which in Fluid things is the caufe of Fluidity. Nor is the Difference betwixt their productions other than this, that to *Softnefs*, fpecially and ftrictly accepted, are required Atoms fomewhat *Hooked*, and fo Retentive each of other, as not to be wholly diffociated, or to permit a manifeft abruption or breach of continuity, upon preffure: but, to ftrict Fluidity it is not requifite, that the Atoms be at all Hamous, or reciprocally retentive.

*Art. 6.*
Softnefs in Firme things, deduced from the fame caufe, as Fluidity in Fluid ones.

Infomuch, therefore, as there is fome certain Compactnefs (more or lefs) even in all Soft Concretions; from thence it may be eafily inferred, that the *General* reafon of the *Mollification* of *Hard* bodies, doth confift in this; that their infenfible particles be in fome degree diffociated, *i.e.* fo feparated each from other, in many points, as that more and larger inane fpaces be intercepted among them, than while they were clofely coadunated: and on the contrary, that the *General* reafon of the *Induration* of *Soft* bodies, doth confift only in this; that their infenfible particles, before in fome degree diffociated, be reduced to a clofer order, or higher degree of Compactnefs, and fo moft of the inane fpaces intercepted, be excluded from among them. To this the doubting *Merfennus* fully fubfcribes (*in lib. 2. Harmonicor. propofit. ultima*) where deducing the caufes of Hardnefs, Rigidity, and the like qualities from the Atoms of *Democritus* and *Epicurus*, he plainly faith; *Duritiem fieri*

*Art. 7.*
The *General* Reafon of the *Mollification* of *Hard*, and *Induration* of *Soft* bodies.

*fieri ab Atomis ramosis, quæ suis hamatis implicationibus perexigua spa-*
*tia relinquunt inania, per quæ nequeant ingredi corpuscula caloris, &c.*
Nay, such is the urgencie of this truth, that *Aristotle* Himself seems to
confess it, in these words: *quæ humoris absentia concrescunt & durantur, ea*
*liquefacere humor potest; nisi adeo sese (particulæ nimirum) collegerint coierint-*
*que, ut minora partibus aquæ foramina sint relicta: id quod fictili accidit,*
*&c.* (4. *Meteorum. cap.*8.)  And we need seek no farther than a ball of
wool, for the *Exemplification* of both; for, that being so relaxed, as that
the hairs touch each other more rarely, or in fewer points, and thereupon
more of the ambient Aer be intercepted among them, instantly becomes
soft: and then being so compressed, that the hairs touch each other more
frequently, or in more points, and the aer be thereupon again excluded from
among them, it as soon becomes hard.

*Art.* 8.
The *special*
manners of the
Mollification
of Hard: and
Induration of
Soft bodies.

But if we wind up our curiosity one note higher, and enquire the *Spe-*
*cial Manner* of Mollifying Hard bodies; we shall find it to rest upon either *Heat,* or *Moisture.* Upon *Heat,* when the Atoms of fire, subingressing
into the pores of a Hard Concretion doe so commove and exagitate
the insensible particles thereof, that they become incontiguous in more
points, than before, and so the whole mass being made more lax and rare,
upon the interception of many new inane spaces among its particles, puts
on a capacity of yeelding to any thing that presseth it, and of receding
from it superfice toward its interiors, according to the property of softness.  Thus Iron made red hot, is mollefied, and hard Wax liquefied
by heat.  Upon *Moisture,* when the particles of an Humor so insinuate
themselves among the closely cohærent particles of a Hard body, that
dissociating them in some measure, they intermix among them, and so
(themselves being sufficiently yeelding upon pressure) cause the bodie
to become yeelding and recessive from it superfice inwards.    Thus
Leather is softned by lying in Water, or Oyle; and Clay assumes so
much the more of softness, by how much the more of water it hath
imbibed.

On the other side, if we pursue the *Induration* of *Soft* bodies up to its
*Special* Manner, we shall secure it either in *Cold,* or *Siccity.* In *Cold,*
whether we understand it to be a simple expulsion of Calorifick Atoms,
lately contained in the bodie; as in the growing hard of Metals after fusion: or the introduction of Frigorifick Atoms into the bodie, naturally
void of them; as in the induration of Water into Ice. In *Siccity,* whether
we conceive it to be a meer expulsion of the particles of moisture from a
Concretion; as when Earth is baked into Bricks: or a superinduction of
drie particles upon a moist concretion; as in the composition of Pills, which
for the most part consist of Drie Powders and Syrupe, or some other viscid
moisture.

*Art.* 9.
PROBLEM.
Why Iron is
Hardned, by
being immer-
sed red-hot
into Cold Wa-
ter; and its
SOLUTION.

But here we feel a strong Remora, or *Doubt*; How it comes about,
*that Iron made glowing hot, and immediately plunged into cold Water, ac-*
*quires a greater degree of hardness, than is had before?*  And to remove
it, we *Answer*, that the particles of the Water subingress into the amplified pores of the Iron, and are not again excluded from thence, though
the particles thereof returne to their former close order, and reciprocally implicate each other, as before in candescence; but, remaining
imprisoned

imprifoned in the fmall incontiguities, or inane fpaces, which otherwife would have been empty, make the body of the iron fomewhat more folid or hard than otherwife it would have been: That this is a fufficient Caufe of that Effect; may be warrantably inferred from hence; that if the fame feafoned iron be afterwards brought to the fire again, and therein made red hot, fo that the contexture of its particles be relaxed, and the particles of Water, which poffefs the inane fpaces betwixt them, be evaporated; there doth it refume its former Softnefs; and this our Smiths call *Nealing* of Iron.

To fteer on, therefore, the fame courfe of Difquifition we have begun; forafmuch as Softnefs is defined by the *Facility*, and Hardnefs by the *Difficulty* of bodies *yielding* in the fuperfice: the only Confiderable remaining to our full explanation of the formal Reafon of each of thefe two Qualities, is, *How the yielding of a Soft body in the Superfice is effected*; for, that being once explicated, the rule of Contraries will eafily teach us, *Wherein the Refiftence of a Hard doth immediately confift*. And this requires no tædious indagation, for from the Præmifes it may eafily be collected; that a foft body doth then yeild, when its particles immediately preffed in the fuperfice, do fink down and fubingrefs into the pores immediately beneath them, and then prefs down the next fubjacent particles into pores immediately beneath them; and thofe likewife prefs down the next inferior rank of particles into void fpaces below them; and thofe again prefs down others fucceffively until (the number of pores or void fpaces fucceffively in each fubingreffion decreafing) there be no more room to receive the laft preffed particles,and then the fubingreffion ceafeth. If this feem not fufficient to make the yeildingnefs of Soft bodies clearly intelligible; we muft remit our Reader to our præcedent Difcourfe concerning the incapacity of Aer to be Condenfed or Compreffed, in a Wind-gun, beyond a certain proportion, or determinate rate. Farther, becaufe a foft body cannot be fqueezed, unlefs it reft upon or againft fomething that is hard, at leaft, lefs foft than it felfe; fo that, though the lower fuperfice thereof, relying upon the fupport,is fo bounded, that it hath no liberty of fpace, whether to recede *Verfus profundum*; yet hath it full liberty of fpace *Verfus latera*: therefore comes it to pafs, that the fubingreffion of particles into pores, and the Compreffion of others, is made not only *Verfus profundum*, in that part of the foft body, which directly confronteth the hard, whereupon it refteth; but alfo *Verfus latera*, toward the fides, or circumambient. And that after a various manner, according to the various Contextures of foft bodies in the fuperfice.

For, if the fuperfice (*i. e.* the outward part) of a foft body, be of a more Compact and tenacious Contexture, than the interior mafs or fubftance; as is the fkin of an Animal, compared to the fubjacent flefh, and a bladder in refpect of the oyle therein contained: in that cafe, the compreffion of the particles is, indeed, propagated by fucceffion to fome diftance as well toward the bottom, as the fides, to which the fuperior particles being preffed directly downward, and there refifted, deflect; yet not to that diftance, as where the fuperfice is of the fame Contexture with the interior mafs, as in Wax and Clay, in both which, the Compreffion, and fo the yeilding may be propagated quite thorow,

*Art.10.*
The Formal Reafons of Softnefs and Hardnefs.

*Art.11.*
The ground of Ariftotles Diftinction betwixt Formalitia and Preffilia.

U u                                                                      or

or from the ſuperior to the inferior ſuperfice, where it immediately reſteth upon the hard body, all the intermediate particles ſtarting toward the
ſides, as being preſſed above and reſiſted belowe. And hereupon, doubtleſs, was it that *Ariſtotle* properly called thoſe ſoft bodies, whoſe ſuperfice
is either of a weaker, or of the ſame contexture with their internal ſubſtance,
πλαςά, *Formatilia*; inſomuch as when a Seal or other Solid body doth
preſs them, they ſuffer ſuch a Diffraction or Solution of Continuity in
their ſuperficial parts, as that the diſſociated particles are not able to
reſtore themſelves to their former ſituation and mutual cohæſion, but
retain the figure of the body which preſſed them: and, on the contrary,
ſuch as have the contexture of their ſuperfice more firm and tenacious than
that of their internal maſs, πιεςά, *Preſſilia*; inſomuch as upon preſſure
they ſuffer not ſo great a Diffraction or Solution of Continuity in their
ſuperficial parts, but that they ſtill have ſome mutual cohærence, and ſo
are able to reſtore themſelves to their former ſituation, upon the remove
of the body that preſſed them.

**Art. 12.**
Two *Axioms*,
concerning &
illuſtrating the
nature of
Softneſs.

For the illuſtration of this, it is obſervable   (1) That to the yielding
of every ſoft body, when preſſed, it is neceſſary, that it have *freedom
of ſpace on its ſides*: becauſe, if the lateral particles, when preſſed by
the intermediate ones, have not room whether to recede, they cannot
yield at all; and ſo the Compreſſion muſt be very ſmall. This may
moſt ſenſibly be Exemplified in a tube filled with Water; for, if you
attempt to compreſs the Water therein contained, with a Rammer ſo
exactly adapted to the bore of the tube, as that no ſpaces be left betwixt it and the ſides thereof, whereat the water may riſe upward, you
ſhall make but a very ſmall and almoſt inſenſible progreſs therein.
(2) *That no ſuperfice of what contexture ſoever, can be depreſſed* verſus
*profundum, or be any way dilated, but it muſt ſuffer ſome Diffraction or
Solution of Continuity*, more or leſs. For, inſomuch as each particle of
the ſuperfice doth poſſeſs a peculiar part of ſpace proportionate to its
dimenſions; and though upon the Dilatation of the ſuperfice, *i. e.* the
remove of its particles to a more lax order, greater ſpaces are intercepted among them, yet are not the particles multiplied in number, nor
magnified in dimenſions, and ſo cannot poſſeſs more or greater ſpaces
than before: therefore is it neceſſary, that the ſuperfice be variouſly crackt,
and the continuity thereof infringed in many places. The Neceſſity hereof doth farther evidence it ſelf in the Flexion of a Twig, Cane, or other
[ΰμμπύλον] *Flexile* body; for, when a Twigg is bended, as the Concave
ſuperfice becomes Contracted and Corrugated, the particles thereof
being not able to penetrate each other, nor crowd themſelves into fewer places: So at the ſame time, is the Convex Dilated, and ſuffers
many ſmall breaches or cracks, the particles thereof being uncapable
either to multiply themſelves, or poſſeſs more ſpaces, than before.
The ſame likewiſe is eaſily intelligible in a *Tractile* body, ſuch as (*Ariſtotle* names Ἑλκλόν) a Nerve, or Luteſtring: for allbeit the interruption of Continuity be not ſo manifeſt to the ſenſe in a Tractile
as in a Flexile body: yet may we obſerve, that when a *Tractile* body
is extended or drawn out in length, it is extenuated or diminiſhed in
thickneſs. And, what, think you, becomes of thoſe interior particles,
which compoſe its Craſſitude or thickneſs? Certainly, they muſt come
                                                     forth

forth into the fuperfice, that fo they may interpofe themfelves among the
Diffociated particles thereof, poffefs the void fpaces left betwixt them,
and with their fmall clawes or hooks on each hand cohæring to them,
make the fuperfice apparently continued. Would you obferve the
Interruption of Continuity among the fuperficial particles of a Tractile
body, and the iffuing forth and intermiftion of interior particles among
them; be pleafed to paint over a Luteftring with fome oyled Colour,
and afterward vernifh it over with oyle of Turpentine: then ftrain
it hard upon the Lute, and you fhall plainly perceive the fuperfice
of it to crack and become full of fmall clefts or chinks, and new particles (not tincted with the colour) to iffue forth from the entralls of the
ftring, and interpofe themfelves among thofe fmall breaches. Laftly, the
fame is alfo difcoverable by the fight in a *Ductile* body [Ἐλατὸν] fuch
as every Metal; for, no metal, when preffed or hammerd, is dilated or
expanded on all fides, for any other reafon but this, that it is as much
attenuated in thicknefs, and the particles in the fuperfice are fo diffociated, as that the interior particles rife up, poffefs the deferted fpaces, and cohære to the difcontinued exterior particles, as may be more plainly difcerned if the fuperfice of the Metal be tincted with fome colour.

## Sect. II.

From the Præmifes, whereupon we therefore infifted fomewhat the
longer, it is manifeft, that FLEXILITY, TRACTILITY,
DUCTILITY, and other Qualities of the fame Claffis, are
all the Confequents of Softnefs: as the Contrary to them all RIGI-
DITY, is the Confequent of Hardnefs; infomuch as whoever would
frame to himfelf an exact notion of a Rigid body, meerly as a Rigid, muft compofe it of the Attributes, inflexile, intractile, indu-
ctile.

*Art.1.*
*Flexility, Tra-
ctility, Ductili-
ty, &c. de-
rived from
Softnefs: and
Rigidity from
Hardnefs.*

Nor doth any thing remain to our clear underftanding of the nature of FLEXILITY, but the Solution of that great Difficulty,
*Cur flexilia, poftquam inflexa fuerint, in priftinum ftatum refiliant?*
Why a flexile body, fuch as a Bowe of wood, Steel, Whalebone,
&c. doth, after flexion, fpring back again into its natural figure and fituation?

*Art.2.*
PROBLEM.
What is the
Caufe of the
motion of Re-
ftoration in
Flexiles? and
the SOLUT.

The Reafon of this Faculty of *Reftitution*, we conceive (with the immortal *Gaffendus*) to be this; that the Recurfe or Refilition of a flexile
body is a certain Reflex motion, which is continued with a Direct motion: as we fhall have opportunity profeffedly to demonftrate, in our fubfequent Enquiry into the nature of Motion. In the mean while, it
may fuffice to ftay the ftomach of Curiofity, that we evidence the caufe
of it to be the fame with that of the Rebound of a ball, impelled by a
racket, from a Wall: for, as the force, which makes the ball rebound
from the wall, is the very fame which firft impelled it againft the Wall;
fo is the force, which reflecteth a bowe, after bending, the very fame
which bended it. To *Exemplifie*; when a man layes a ftaff tranfverfly

Uu 2                  upon

upon a beam, and ſtrikes the end that is toward him, downward; the end
that is from him, muſt riſe, as much upward: as well becauſe of the reſiſtence of the beam (which here performs the office of an *Hypomochlion,*
or middle Fulciment) as of the continuity and compactneſs of the ſtaff it
ſelf; and ſo the ſame cauſe, the hand of the man, which impelled the one
extreme of the ſtaff downward, is alſo the cauſe of the riſing of its other
extreme upward. Again, let the ſtaff have liberty of play between two
beams, the one above, the other beneath it; and upon the Depulſion of
one end, the other ſhall riſe up, and be impinged againſt the upper beam,
and from thence rebound back again upon the lower, and thence again to
the higher, and thence again to the lower, and ſo alternately be reflected
from one to the other, till the force of reſiſtence in the 2 beams hath wholly
overcome that of the firſt percuſſion or impulſe: yet ſtill doth the laſt
Rebound, no leſs than the firſt, owe it ſelf to the ſame Cauſe, which impreſſed the firſt motion upon the ſtaff, which was the hand of the man,
who impelled it. To approach one degree neerer; ſet up a ſtaff perpendicularly in ſome hole in the floore or pavement, ſo that it may have ſome
liberty of motion to each hand: and then, if you impel or inflect the upper extreme to the right hand, the part of the lower extreme, which re-
ſpecteth the upper part of the right ſide of the hole, will preſs upon the
ſame, and the other ſide of the lower extream, where it toucheth the loweſt part of the left ſide of the hole, ſhall be at the ſame time impinged likewiſe againſt the left ſide; and that ſo forcibly, that it ſhall rebound from
thence to the oppoſite ſide, and at the ſame time, the upper part, which
you inflected, ſhall rebound from the right to the left: and thus ſhall the
ſtaff be agitated from ſide to ſide, by alternate reſilitions, till the reſiſtence
of the hole hath wholly overcome the force thereupon impreſt, by your
hand. This laid down, we infer, that the cauſe of Returne in the ſtaff, is
the ſame with that of the Self-reſtorative motion in bodies Flexile; for,
that you may be able to inflect one end of the ſtaff, it is neceſſary, that ſome
part of it be held faſt in your hand, ſome hole, chink, or other hold,
that ſo you may diſtinguiſh the *Hypomochlion,* or point of Reſt, from the
part inflected.

*Art.* 3.
Two Obſtacles
are expeded.

Nor is it ought available to the contrary, to *Object* (1) that the ſtaff
is not bent with one ſingle ſtroke, but a continent preſſion: becauſe a
Continent preſſion is nought elſe but a continent Repetition of ſtrokes;
and that is the laſt ſtroke, immediately upon which the laſt and non-im-
peded Reflexion doth enſue. 2. that our Example of the Reſilition
of a ſtaff is incongruous, there being a conſiderable Rigidity therein,
but none in Flexile bodies: for, though there be no perfect or Abſolute
Rigidity in Flexile ſubſtances, yet is there a ſufficient Firmneſs, which
is a degree of Rigidity; and by how much greater that is, by ſo much
the greater force of impulſe is required to the inflexion, and conſequently ſo much ſtronger is the Reflection. So that while the bottome of
the ſtaff, and its Hypomochlion alternately performe their offices, the
one reflecting this, the other the contrary way, ſo many more Alternate Reflexions, or Excurſes and Recurſes are made, by how much
greater the Rigidity of the ſtaff, and firme fixation in its hold, are;
and *è contra.* And, ſince the Reflection, which is made from the
firmely fixt part, is as it were the *Fundamental,* or *General* Reflexion;
innumerable *Special* or *Particular* Reflexions, exactly like the General,

are

are made *in singulis partibus* : infomuch as the parts of the Concave fuperfice are fo compreffed, in order, one after another, from the Deflected Extrem to the Fixt, that fuffering mutual refiftence, they are compelled to ftart back in the fame order, one after another ; and the parts of
the Convex fuperfice, from the Fixt Extreme to the Deflected, are fo
retracted in order one after another, that they return in order to their natural fite ; and fome parts thus confpiring with others, reduce the whole inflected bodie to its natural fituation and figure.

Finally, becaufe every Reflex Motion is alwaies ( though, perhaps,
not fenfibly ) weaker, than the Direct ; therefore is it, that in every Deflexion, both to the Concave fuperfice, fome particles fubingrefs to the
interiors of the Flexile bodie, which cannot returne forth again to the
fuperfice ; and to the Convex, other particles egrefs to the fuperfice,
which cannot returne in again to the interiors : Whereupon it comes to
pafs, that by how much the longer the Inflexion is continued, or how
much the more frequently repeated ; by fo much the more Contracted
is the Concave fuperfice made, and fo continues, and fo much more
Deduced or Dilated is the Convex fuperfice made, and fo continues,
and confequently both the Inflexion and Reduction become as fo much
the weaker, fo as much the fmaller. Nay, where the Deflexion is fo
great, as that fome parts of either fuperfice are wholly Diftracted and
Diffociated, and fo can no longer maintain that mutual cohærence and
continuity, which is neceffary to the feries of Reflexion and Retraction :
there doth no Reduction at all followe, after Inflexion, at moft only fo
much, as is made by the parts, which yet remain cohærent, in which alfo
we muft allowe the diftinction of Concavity and Convexity. Thus, when
a Twigg is broken half off in the middle, by overmuch bending ; it makes
no more Reflexion, than what depends only upon the half which is un-
broken.

*Art. 4.*
Why *Flexile*
bodies grow
weak, by over-
much, and
over frequent
Bending.

As for T R A C T I L I T Y likewife, all the obfcurity which
remains upon its nature, depends upon this Difficulty ; *Cur Nervus
diftentus, & è fuo fitu diftractus toties hinc inde redeat ?* Why doth a
Tractile bodie, fuch as a Nerve or Luteftring, when diftended, and
abduced from the line of direction to either fide, not only reduce it
felf from that obliquity to directnefs ; but recurr beyond it, and then
returns toward the place of its firft abduction, and thence back again
to and beyond the line of direction, and fo makes many excurfes and
recurfes ?

*Art. 5.*
The Reafon
of the frequent
*Vibrations*, or
*Diadroms* of
Luteftrings, &
other *Tractile*
Bodies ; de-
clared to be
the fame with
that of the
*Reftorative Mo-
tion* of *Flexiles*:
and Demon-
ftrated.

And this may be foon folved, by *Anfwering* ; that the Caufe of this
Tremulation or Vibrations of a Tractile thing, diftended and percuffed,
or abduced, feems to be the fame with that of the Reflexion of a Flexile,
newly rendred. For (1.) A chord diftended, is nothing but a Flexile
body ; and fo much the more apt for Reflection, by how much more
it is Diftended : becaufe Tenfion is a kind of Rigidity. (2.) A
chord diftended hath the reafon not only of one fimple Flexile bodie,
but alfo of two conjoyned ; infomuch as it hath 2 Extrems, in each
of which we may diftinguifh the Hypomochlion, or fixt part, from the
Reflectent ; and in the middle, or that part, which is percuffed or abduced by the plectrum or finger, there are as it were 2 other Extrems

tremes conjoyned, which being naturally reluctant each to other, cauſe the reciprocal Reduction each of other. (3.) As a Twigg, after inflexion, doth return beyond the middle, or line of directneſs, and goes and comes frequently, till it hath overcome the firſt impreſſed motion, and recovered its natural ſite  becauſe after the firſt Reflexion is made, a ſecond ſucceeds, for the ſame reaſon, as the firſt, a third for the ſame reaſon as the ſecond, and ſo a fourth, fifth, &c. ſucceſſively: So alſo, is it neceſſary, that many Vibrations, or Excurſes and Recurſes be alternately made, by a Chord diſtended and percuſſed; becauſe the ſame cauſe remains to the ſecond, third, fourth, &c. which was to the firſt. *Lege Merſennum, Harmonicor. lib. 3. Propoſ. 22. Corollario de Aſomis.*

This may be fairly demonſtrated in this Chord A. B. vertically diſtended, by a weight appenſed. For, being elevated to the point C. falling from thence, it will make its firſt diadrome to I. not to L. becauſe of the reſiſtence of the Aer: and thence by new force returning over the center B. it will make it ſecond diadrome, not quite home to N. becauſe of the re-

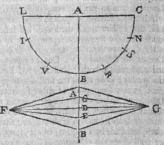

ſiſtence of the Aer, but only to S. and thence relapſing, it will make its third diadrome no higher then V. and thence back again, its fourth to R. and at length, its diadroms ſucceſſively diminiſhing, it reſteth at the centre B.   And thus you ſee how the force or impetus, whereby it is moved, is by ſenſible degrees and proportionately diminiſhed: and that it is impoſſible, it ſhould make any two Diadroms Æquiſpatial, during the whole time of its motion.  For, if we concede two diadroms to be equal in ſpace; we muſt find them to be produced by an equal impetus.  Therefore, if the Chord recurring from C. ſhould on the other ſide aſcend as high as L. it would of neceſſity thence returning make its ſecond Diadrome to C. where it began its firſt, and thence recur to L. again, and thence to C. and ſo the motion would be perpetual.  Leſt, therefore, that Abſurdity be admitted in nature, it is neceſſary that the impetus be proportionately diminiſhed, that ſo the Chord may after various Vibrations arrive at the centre or terme of its motion. You ſee alſo, that the Natural impetus, by whoſe ſwindge or rapt, the weight appenſed at the lower extreme of the Chord, is carried to the Centre, is the Cauſe of all its Tranſcurſions or diadroms: and that the Reſiſtence of the conſtipated or compreſſed Aer, is the cauſe of the Diminution of them.

But

But here comes the P R O B L E M (such a one as put e-
ven *Mersennus* Himselfe to the *Eruditis Physicomathematicis discu-
tiendum relinquo; Harmonicor. lib.* 2. *proposit.* 29.) and that is,
*Cur Diadromus Chordæ maximus eodem tempore conficit totum spaci-
um, quo minimus, aut reliqui singuli diadromi intermedii illud confici-
ant?* Whence is it, that all the Excurses and Recurses, or diadroms
of a Chord, either Vertically, or horizontally distended, and abduced from the line of Direction; are Isochronical, or Æquitemporaneous, though not Æquispacial: as also are All the Vibrations of a Flexile body, fixt at one extream, and deflected at the
other.

*Art.* 6.
PROBLEM.
Why the Vi-
brations, or
Diadroms of
a Chord di-
stended and
percussed, are
Æquitempora-
neous, though
not Æquispaci-
al: and the
SOLUT.

This stupendious Phænomenon may be thus Demonstrated.  Let F.
G. (in the second diagram) be the Chord horizontally distended; which,
being distracted from its direct situation, F. G. to A. makes its several
Diadroms, A.B. B.C. C.E. and E.D.  Now we say, that All these
Diadroms, though greatly disproportionate in point of space, are yet
exactly proportionate in point of Time, *i. e.* the first Diadrom, A. B.
doth measure its whole space, in the same proportion of time, as doth
the second Diadrom, B. C. or the third, C. E. or the fourth E. D. For,
since the Violence or impetus, whereby the Chord is abduced from the
line F. G. to the point A. is so much the greater, by how much the
longer the line of the Epidrom is, the Chord must pervade it space so
much the more speedily, by how much the space is greater, compared
to that of the subsequent ones : it necessarily followes, that all the subsequent Diadroms must be Æquidiurnal, because look how much is detracted from the Longitude, Magnitude, and Impetus of the subsequent
Diadroms exactly so much accedeth to the Brevity of the space, which
they are to percurr; and so the longitude of the posterior Epidrom becomes inverted in proportion to the Time, and its Brevity of space com-
pensateth the decay of that Impetus, which was in the Prior Diadrom.
For Example; Let the Chord, which makes an hundred Diadroms, pervade a foot space, in its first Diadrom, and the hundredth part of a foot,
at its last, or hundredth Diadrom: we affirm, that the first Diadrom must
be an hundred times swifter than the Last; which is an hundred times
slower, as being to the same proportion less violent, and that which
immediately præcedeth the Quiet of the Cord, in the Direct line,
F. G.

More plainly; the First Diadrom, A. B. as it is the Greatest, so is it
the most Violent ; and as it is the most Violent, so must the Velocity,
whereby it pervades the whole space betwixt A.B. be also the Greatest :
and the Second Diadrom, B. C. how much it comes short, in violence
of tension, and Celerity of motion, of the First, so much doth it come
short of the Magnitude also thereof ; so that though the space of the former,
A. B. be much larger than that of the second, B.C. yet doe they both pervade their several spaces in the same proportion of Time, because, as the second Diadrom, B. C. hath less of violence and of Celerity, than the first,
A. B. so hath it just so much less of space to pervade, and so the Diminution

nution of fpace Compenfateth the Diminution of Violence and Celerity. Wherefore, the Reafon of the Third Diadrom being the fame to the Second, as that of the Second to the Firft; and of the Fourth to the Third, as that of the Third to the Second: it is manifeft and neceffary, that all the Diadroms be i Æquidiurnal, though not Æquifpatial; which is what we Affumed.

*Art. 7.*
*PROBLEM.*
Why doth a
Chord of a
duple length,
perform its di-
adroms in a
proportion, of
time duple, to
a Chord of a
fingle length;
both being
difitended by
equal force; &
yet, if the
Chord of the
duple length
be diftended
by a duple forc
or weight, it
doth not per-
form its Dia-
droms, in a
proportion of
time duple to
that of the o-
ther; but on-
ly if the Force
or weight di-
ftending it, be
quadruple to
the Firft fup-
pofed: and its
SOLUT.

But yet the Lees of the P R O B L E M remain behind; for it is worthy farther Enquiry: *Why a Chord of a Duple length, v. g. of 4 foot, doth performe its Diadroms in a Duple proportion of Time, to a Chord of a fingle length, v. g. of 2 foot; when both are diftended by equal Force, or Weight: and yet, if the Chord of 4 foot be diftended by doubly as great a Force or Weight as that of only 2 foot, it doth not performe its Diadroms with Velocity Duple thereunto; but only if the force of its Diftenfion be Quadruple to the force firft fuppofed?*

And to exhauft them, though fomewhat rough and crabbed, we A N S W E R, As in a *Penfile* bodie, or Chord vertically diftended by a weight, the time of each fingle Excurfe, is equal to that time', in which the fame weight would, if permitted, be falling from fuch an Altitude, as is commeafurable by the diametre of the Circle, whereof Arches are defcribed by the Excurfes of the Penfile body abduced from the perpendicular: So in a *Tenfile* body, fuch as a Chord ftrained upon a Lute, All the times, in which a part of the Chord accepted exactly in the middle, excurreth from one fide, are equal to one time, in which one of its Extrems, if cut off, would directly pervade the whole length, and come into the place of the other, toward which the force, being ftill the fame behind, would draw it. For, the fame Force, certainly, is alwaies able to produce the fame Effect: and if the lateral fpaces of the Diadroms doe continually decreafe; the Velocity of the motion muft alfo continually decreafe. And the caufe of that continual Decrement, can be no other but the Force Drawing or diftending the Chord, which continually refracteth the contrary Force, by the plectrum or finger impreffed thereupon. Now, fince All the Excurfes of a Chord, of whatever length, are exæquated to one and the fame direct Trajection thereof, as we faid even now; in the *Former* Cafe, the Trajection cannot but be performed in a duple proportion of Time, as a Duple proportion of Space is affumed to be trajected or pervaded, by the fame Motive or Attractive Force: but in the *Latter* not, becaufe Three Equal things being fuppofed, *viz.* Time, Space, and the Weight or Attractive Force, it is of pure neceffity, that the fame fpace remaining, look how much of Time is diminifhed, fo much is the motive Force encreafed, and what is the proportion of fpace to Time, the fame is the proportion of the Motive Force to Space. And hence comes it, that the proportion of fpace to Time being as that of 2 to 1; the Motive Force muft have to fpace the proportion of 4 to 2: and confequently to Time, not as 2 to 1, but as 4 to 1.

Laftly,

Lastly, as for DUCTILITY, little remains Additional to what we have formerly said, concerning the Formal Reason thereof, but the Solution of that notable PROBLEM, about the admirably vast *Extensibility* of that King not only of Metals, but of the whole Earth, *Gold.* And, indeed, since we have it upon the testimony of our Experience, that one Ounce of pure Gold may be, by Malleation, extended to such an amplitude, as to cover ten Acres of Land; and that one Grain thereof may be Wier drawne into a thread of such incomparable fineness, as to commensurate 400 foot; and consequently, that one Ounce of Gold is capable of deduction into a thread, whose length may fusill the measure of two hundred and thirty thousand, and four hundred feet, of six inches apiece: we say, this being avouched by those Mechaniques, who deale in Beating of Gold into Leaves, and Drawing it out into Wier, it seems well worthy our Enquiry, upon what Cause this stupendious Prærogative of Gold doth chiefly depend. In a word, therefore, we conceive this superlative EXTENSIBILITY of Gold, to be warrantably referrible to a *Threefold Cause*, *viz.* the unparalleld Compactness of it substance, the great Tenuity of its Component particles, and the Multitude of small Hooks or Clawes, whereby those particles reciprocally implicate each other, and maintain the Continuity of the whole Mass. For (1) the *exceeding Compactness of its Contexture* doth afford parts sufficient to so great Extension, *i. e.* such an abundance of them, as upon the Decrement of the Mass in Profundity, may rise up into the superfice and enlarge the Latitude, or Longitude: (2) The *Tenuity of its component particles* maketh the mass capable of Diminution in profundity, and so of Augmentation in superfice, even to an incredible proportion: and (3) The *Multitude of small Hooks*, whereby those Exile particles reciprocally cohære, sufficeth to the constant Continuity; for, while the mass is suffering under the Hammer, no sooner can the stroke thereof dissociate one particle from its neighbour, but instantly it layes hold of and fastneth upon another, and as firmely cohæreth thereunto, as to its former hold: So that the mutual Cohæsion is maintained even above the highest degree of Extension or Attenuation, which any imaginable Art can promise. Nay, so sufficient a Cause of incredible Ductility doth this last seem to be, that *Mersennus* regarded no other: as may be collected from these his words: *Sunt autem Corpora maximè Ductilia, quæ habent Atomos undique Hamatas, ut Aurum; cujus Atomi non ita possunt evolui, ut sese deserant in inferioribus, aut superioribus partibus, quin laterales succedant, quibus usque ad insignem tenuitatem perveniant;* (*Harmon. lib. 3. propos. 22. Corollario de Atomis.*) This apprehended, the Chymist needs not longer to perplex himself about the Cause of the *Incorruptibility*, and *incapacity of Volatilization* in Gold: and if his so promising Art can attain to the investment of any Metal with these Proprieties; let other men dispute, whether it be Gold or no, for our parts, we oblige our selves so to accept it.

Now, that we may run through all other Secondary Qualities, in this one Course, we farther observe; that to the prædominion of Softness, men ought to refer SECTILITY, such as is seen in wood Cut transversly: and FISSILITY, such as in wood cleft along the Grain. For, whatever

X x

whatever is [τὸ τμπτὸν] *Sectile*, muſt in ſome ſort return to the nature of *Flexility*; ſeeing that the parts of it, which are immediately preſſed upon by the edge of the Axe, Knife, or other Cutting inſtrument, muſt recede inwardly, *i.e.* from the ſuperſice to the profundity of the Maſs, and the Lateral parts, at the ſame time, give back on each hand, for otherwiſe there could be no yeilding, and ſo no cutting; and in like manner, whatever is ·[τὸ χιϛὸ] *Fiſſile*, muſt have ſo much of Flexility alſo, as that, when the parts of it, in the place, upon which the Force is firſt diſcharged, begin to be diſſociated, a certain Compreſſion muſt run along ſucceſſively to all the other parts, which are afterwards to be diſſociated. But, though a *Fiſſion*, or Cleaving may be made without any Deperdition of Subſtance, or exceſſion of parts from the body cleft, thoſe parts, which were coadunated *Sec. Longitudinem*, being only ſeparated *Sec. Longitudinem*: yet is that impoſſible in any *Section* whatever, though made by the acuteſt edge imaginable; becauſe, look how much of the body doth commenſurate the bredth of the edge of the Cutting inſtrument, ſo much, at leaſt, is beaten off and deſtracted from the body, betwixt the ſides of the inciſion. And thus much concerning the Conſequents of *Softneſs.*

**Art. 10.**
*Tractility and Friability, the Conſequents of Hardneſs.*

  As for thoſe of *Hardneſs*; they are T R A C T I L I T Y and F R I A B I L I T Y. For, whatever is [τὸ κατακλον] *Fractile*, capable of fraction into pieces, as a Flint and moſt other ſtones, muſt have ſo much of *Rigidity*, ( the chief propriety of Hardneſs ) as may ſuffice to hinder the yeilding of it ſuperſice, upon preſſure or percuſſion; and conſequently all ſubingreſſion of ſuperior particles into the ſmall vacuities intercepted among the inferior ones; and ſo to cauſe, that the ſuperſice is firſt diffracted, and ſucceſſively all the ſubjacent particles diſſociated, quite thorow to the contrary ſuperſice, the inferior particles being ſtill pulſed by the Superior [κατὰ τὸ ζυνεχὲς] by reaſon of their Continuity. So that the fragments into which the body is ſhattered, are greater or leſs, either according to the diverſe contexture thereof in divers parts, in reſpect whereof ſome parts may be contexed more Compactly and Firmly, and others again more Laxly and Weakly: or according to ſituation, in reſpect whereof thoſe parts, which are neerer to the Circumference, flie off more eaſily than thoſe, which are more remote. In like manner, whatever is properly [τὸ θραυϛὸν] *Friabile, Brittle*, as Marble, Glaſs, Earthern Veſſels, &c. muſt alſo have ſo much of Rigidity, as to make it uncapable of Flexion, Traction, Diduction, or Extenſion, by any means whatever: ſo that upon any forcible preſſion, or percuſſion, the whole maſs or ſubſtance of it is ſhivered into duſt, or broken into greater fragments, which are eaſily ſubject to be Crumbled into duſt afterward. Now, that a Hard or Rigid bodie being percuſſed, or preſſed, with force ſufficient, in one Extreme or Superſice, the percuſſion or preſſure may be propagated from part to part ſucceſſively, till it arrive at and be determined in the other extreme; may be evinced by ſundry moſt eaſie Experiments, ſome whereof are recited by the *Lord St. Alban* (*in Sylva ſylvarum Cent.* 1.) But this one will ſerve the turne. When an Oyſter, or Tortois ſhell is let fall from a ſufficient altitude, upon a ſtone, it is uſually ſhattered into many peices; and that for no other Reaſon but this, that the lower ſide, whether Convex or Concave, being

being vehemently impinged againft the ftone, the particles thereof immediately knockt by the ftone, as vehemently give back, and in their quick Retroceffion impell the particles fituate immediately above them; whereupon thofe impelled particles with the fame violence impell others next in order above them, until the percuffion being propagated from part to part fucceffively quite home to the upper fuperfice, it comes to pafs, that each percuffed part giving back, the whole fhell is fhattered into fmall Fragments.

All which may feem but a genuine Paraphrafe upon the Text of *Mersennus*. (*Harmonicor. lib. 2. propof. 43.*) *Duritiei verò proprietas appellatur Rigiditas; quæ fit ab Atomis ita fibi invicem cohærentibus, ut Deflexionem impediant: quod contingit in Corporibus, quæ conftant Atomis Cubicis, octuedris & tetruedie, ex quibus refultat perfecta fuperficiecularum inter fe cohæfio; hinc fit ut Rigida Corpora Fructilia fint, non autem Sectilia, & ictu impacto tota in frufta diffiliant. Qui adum prædictæ fuperficiunculæ fe invicem premunt, quæ funt ex una parte, dimoventur ab iis, quæ ex alia; adeo ut unico impetu externo Corpori impreffo, Contufio fentiatur per totum, & partium eodem, momento fit feparatio.*

Art. 11.
Ruptility, the
Confequent
partly of
Softnefs, partly
of Hardnefs.

There yet remains a Quality, which is the Ofspring neither of Softnefs alone, nor Hardnefs alone; but ought to be referred partly to the one, partly to the other: and that is RUPTILITY. For, not only fuch Bodies, as challenge the Attribute of *Softnefs*, are fubject to *Ruption*, when they are diftreffed beyond the tenour of their Contexture, either by too much *Inflexion*, as a Bow over bent; or too much *Diftention*, as Leather or Parchment over ftrained; or too much *Malleation*, as a plate of Lead, Iron, or other Metal over hammerd: but fuch alfo as claim the title of *Hardnefs*, and that in an eminent proportion, as Marble; for, a Pillar of Marble, if long and flender, and laid tranfverfly or horizontally, fo as to reft only upon its two extrems, is eafily broken afunder by its own Weight. For, as Soft bodies, when rackt or deduced beyond the rate of mutual Cohærence among their parts, muft yeeld to the External Force, which diftreffeth them, and fo fuffer total difcontinuity: fo Hard ones, when the Internal Force, or their owne Weight, is too great to be refifted by their Compactnefs, as in the example of a long Marble Pillar, not fupported in the middle; then muft they likewife yeeld to that fuperior force, and break afunder.

Art. 12.
PROBLEM,
Why Chords
diftended, are
more apt to
break neer the
Ends, than in
the middle?
and its
SOLUT.

And here the *Archer* and *Mufician*, put in, for a Solution of that PROBLEM, which fo frequently troubles them; *viz. Cur Chordæ facilius circa Extrema, quam circa Medium frangantur, cùm vi vel pondere, five horizontaliter, five verticaliter trahuntur?* Why Bowftrings, Luteftrings, and other Chords, though of uniforme Contexture throughout, and equally diftended in all parts, do yet ufually break afunder, not in the middle, or neer it, but at one End, where they are faftned?

The *Caufe*, certainly, muft be this; that the Weight or drawing force doth alwayes firft act upon the parts of the ftring, which are neereft to it, and fucceffively upon thofe, which are fartheft off, *i. e.* in the Middle: fo that the ftring fuffering the greateft ftrefs neer the Extrems, is more subject to break there, than in any other part. Wherefore, whenever a Bowstring breaks in or neer the middle; it may safely be concluded, that the string was weakest in that place. To which we may add this also, that Experienced Archers, to prævent the frequent breaking of their strings, and the danger of breaking the Bow thereby; injoyn their String-makers, to add a Link of Flax, or Twist more at the Ends of each string, than in any other parts of it : and that they call the *Forcing* , because Experience hath taught them, that the Force of the Bow is most violently discharged upon those parts of the string, which are neerest to the Horns.

CHAP.

# CHAP. XV.

# OCCULT QUALITIES
## made MANIFEST.

### SECT. I.

Aving thus long entertained it self with the most probable Reasons of the several wayes and means, whereby Compound Bodies exhibite their several Attributes and Proprieties to the judicature of the Sensitive Faculties in Animals, and principally in Man, the Rule, Perfection and grand Exemplar of all the rest; tis high time for our Curiosity to turn a new leaf, and sedulously addresse it self to the speculation of Another Order, or Classis of Qualities, such as are vulgarly distinguished from all those, which have hitherto been the subject of our Disquisitions, by the unhappy and discouraging Epithite, OCCULT. Wherein we use the scarce perfect Dialect of the Schools; who too boldly præsuming, that all those Qualities of Concretions, which belong to the jurisdiction of the senses, are dependent upon Known Causes, and deprehended by Known Faculties, have therefore termed them *Manifest*: and as incircumspectly concluding, that all those Proprieties of Bodies, which fall not under the Cognizance of either of the Senses, are derived from obscure and undiscoverable Causes, and perceived by Unknown Faculties; have accordingly determined them to be *Immanifest* or *Occult*. Not that we dare be guilty of such unpardonable Vanity and Arrogance, as not most willingly to confess, that *to Ourselves all the Operations of Nature are meer Secrets*; that in all her ample catalogue of Qualities, we have not met with so much as one, which is not really Immanifest and Abstruse, when we convert our thoughts either upon its Genuine and Proxime Causes, or upon the Reason and Manner of its perception by that Sense, whose proper Object it is: and consequently, that as the *Sensibility* of a thing doth noe way præsuppose its *Intelligibility*, but that many things, which are most obvious and open to the Sense, as to their *Effects*, may yet be remote and in the dark to the

*Art. 1.*
That the *Insensibility* of Qualities doth not import their *Unintelligibility*; contrary to the presumption of the *Aristotelean*.

the Understanding, as to their *Causes*: so on the Contrary, doth not the *Insensibility* of a thing necessitate, nay, nor aggravate the *Unintelligibility* thereof, but that many things, which are above the sphere of the Senses, may yet be as much within the reach of our Reason, as the most sensible whatever. Which being præcogitated, as, when we look back upon our præcedent Discourses, touching the Originals and Perception of Sensible Qualities, we have just ground to fear, that they have not attained the happy shoar of verity, but remain upon the wide and fluctuating ocean of meer Verisimility : . So also, when we look forward upon our immediately subsequent Disquisitions into the Causes of many Insensible Qualities, are we not destitute of good reason to hope, that though we herein attempt the consignation of Consentaneous and Probable Causes to sundry of those Effects, which Schollars commonly content themselves only to Admire, and without farther exercise of their Intellectuals, to leave wrapt up in the Chaos of *Sympathies* and *Antipathies* ; yet will not the Ingenious misunderstand us, or conceive that we esteem or propose those Reasons as *Oraculous* or *Apodictical*, or create an expectation of the Discovery of such Originals, whereupon those Rarer Operations and Magnalia of Nature do proximely and genuinely depend. However, some may think it expedient for us to profess, that as in our former Enquiries, so in *this*, our Designe is only to explain sundry admired Effects, by such Reasons, as may appear not altogether Remote and Incongruous, but *Consentaneous* and *Affine* to Truth ; that so no mans judgement may be impeached by embracing them for most Probable, untill the (in that respect, too slow) wheel of Time shall have brought up some more worthy Explorator, who shall wholly withdrawe that thick Curtain of obscurity, which yet hangs betwixt Natures Laboratory and Us, and enrich the Commonweal of Letters, by the discovery of the Real Verity. And this we must enterprize, by continuing our progress in the allmost obliterated Tract, that *Epicurus* and *Democritus* so long since chalk'd forth ; not by treading in the beaten road of *Aristotle* and his *Sectators*, who (for ought we have learned) were They, who first founded that ill contrived Sanctuary of Ignorance, called *OCCULT QUALITIES.*

*Art. 2.*<br/>
Upon what grounds, and by whom, the Sanctuary of Occult Qualities was erected.

For, generally setting up their rest in the Commission of Elements, and their supposed Immateriall Qualities ; and being not able ever to explicate any Insensible Propriety, from those narrow and barren Principles : they thought it a sufficient Salvo for their Ignorance, simply to affirme all such Proprieties to be *Occult* ; and without due reflection upon the Invalidity of their Fundamentals, they blushed not to charge Nature Herself with too much Closeness and Obscurity, in that point, as if she intended that all Qualities, that are *Insensible*, should also be *Inexplicable.*

*Art. 3.*<br/>
Occult Qualities and profest Ignorance, all one.

The ingenious *Sanchez*, among many Sceptical Arguments of the Uncertainty of Sciences, seasonably urgeth this one, as very considerable, against Physiologists, *that when any Natural Problem, such* as that of the Attraction of Iron by a Loadstone, of straws by Amber, &c. is objected to them ; instead of setting their Curiosity on work to
In-

to investigate the Causes thereof, they lay it in a deep sleep, with that infatuating opium of Ignote Qualities: and yet expect that men should believe them to know all that is to be known, and to have spoken like Oracles concerning that Theorem; though at the same instant, they do as much as confess, that indeed they know nothing at all of its Nature and Causes. For, what difference is there, whether we say, that such a thing is Occult; or that we know nothing of it?

Nor is it a Course either less dishonorable to the Professors, or dangerous to the Students of Philosophy, to refer such Effects, upon which men commonly look with the eye only of Wonder, to Secret *Sympathies* and *Antipathies*: forasmuch as those Windy Terms are no less a Refuge for the Idle and Ignorant, than that of Occult Proprieties, it being the very same in importance, whether we have recourse to the One, or to the other. For, no sooner doe we betake ourselves to *Either*, but we openly confess, that, all our Learning is at a stand, and our Reason wholly vanquisht, and beaten out of the field by the Difficulty proposed. We deny not, that most, if not All of those Admired Effects of Nature, which even the Gravest Heads have too long thought sufficient Excuses of their Despair of Cognition, do arise from some *Sympathy*, or *Antipathy* betwixt the Agent and Patient: but yet for all that, have we no reason to concede, that Nature doth institute or Cause that sympathy or Antipathy, or the Effect resulting from either, by any other Lawes, or Means, but what she hath ordained and constantly useth, to the production of all other Common and familiar Effects. We acknowledge also, that *Sympathy* is a certain *Consent*, and *Antipathy* a certain *Dissent* betwixt Two Natures, from one, or both of which there usually ariseth some such Effect, as may seem to deserve our limited Admiration: but is it therefore reasonable for us to infer, that those Natures are not subject unto, nor regulated by the General and Ordinary Rules of Action and Passion, whereto Nature hath firmly obliged Herself in the rest of Her Operations?

*Art. 4.*
*The Refuge of Sympathies and Antipathies, equally obstructive to the advance of Natural Science, with that of Ignote Proprieties.*

To lance and cleanse this Cacoethical Ulcer, to the bottom, Consider we, that the *General Laws* of Nature, whereby she produceth All Effects, by the Action of one and Passion of another thing, as may be collected from sundry of our præcedent Dissertations, are these: (1.) That every Effect must have its Cause; (2) That no Cause can act but by Motion; (3) That Nothing can act upon a Distant subject, or upon such whereunto it is not actually Præsent, either by it self, or by some instrument, and that either Conjunct, or Transmitted; and consequently, that no body can move another, but by contact Mediate, or Immediate, *i. e.* by the mediation of some continued Organ, and that a Corporeal one too, or by it self alone. Which considered, it will be very hard not to allowe it necessary, that when two things are said either to *Attract* and *Embrace* one the other by mutual *Sympathy*, or to *Repell* and *Avoid* one the other, by mutual *Antipathy*; this is performed by the same wayes and means, whereby we observe one Body to Attract and hold fast another, or one Body to Repell and Avoid conjunction with another, in all Sensible and Mechanique Operations. This small Difference only allowed,
that

*Art. 5.*
*That all Attraction, referred to Secret Sympathy; and all Repulsion, ascribed to secret Antipathy, betwixt the Agent and Patient, is effected by Corporeal Instruments, and such as resemble those, whereby one body Attracteth, or Repelleth another, in sensible and mechanique operations.*

that in Gross and *Mechanique* operations, the Attraction, or Repulsion is performed by *Sensible* Instruments: but, in those finer performances of Nature, called *Sympathies* and *Antipathies*, the Attraction or Repulsion is made by Subtle and *Insensible*. The means used in every common and Sensible Attraction and Complection of one Bodie by another, every man observes to be Hooks, Lines, or some such intermediate Instrument continued from the Attrahent to the Attracted; and in every Repulsion or Disjunction of one Bodie from another, there is used some Pole, Lever, or other Organ intercedent, or somewhat exploded or discharged from the Impellent to the Impulsed. Why therefore should we not conceive, that in every Curious and Insensible Attraction of one bodie by another, Nature makes use of certain slender Hooks, Lines, Chains, or the like intercedent Instruments, continued from the Attrahent to the Attracted, and likewise that in every Secret Repulsion or Sejunction, she useth certain small Goads, Poles, Levers, or the like protruding Instruments, continued from the Repellent to the Repulsed bodie? Because, albeit those Her Instruments be invisible and imperceptible; yet are we not therefore to conclude, that there are none such at all. We every day behold Spiders letting themselves down from high roofs, and as nimbly winding themselves up again at pleasure, by such slender threads of their own occasionall and extempory spinning, as tis not every common eye that can discern them. Nay, in a Mask at Court, we have seen a whole Chorus of Gods descend into the theatre, as from the clouds, only by Wires and other lines, so fine and slender, as that all the light of the tapers burning therein was not sufficient to discover them to the sight of the Spectators: and vast aud ponderous Scenes so suddenly and dextrously shifted, by the almost inobservable motions of Skrews, Elevators, Pulleys, and the like Archimedean Engines and Devices, that the common Beholders, judging only by the Apparence, or (rather) Non-apparence, have thought those great machines to have been Automatous, or to have moved themselves, and at last to vanish into nothing. And shall we not then allowe the incomparably more Curious Mechaniques of Natures, the Exemplar of Art, to be wrought by Instruments of Subtility incomparably greater: and that many of those small Engines, whereby she usually moves and sustains bodies of considerable bulk and weight, are Corporeal, though by incomputable excesses below the perception of our acutest sense? Certainly, for us to affirm, that nothing Material is emitted from the Loadstone to Iron, which by continuity may Attract it; only because our sense doth deprehend nothing intercedent betwixt them: is an Argument of equal weight with that of the Blind man, who denied the Being of Light and Colours, because He could perceive none. In a word, if there be any validity in what we have so plainly asserted, and frequently inculcated, touching the *Hebetude* or Grossness of our *Senses*, on one part, and the great *Exility* of all *Aporrxa*'s or Effuxes streaming from Bodies, on the other; and if that Oracle, Reason, be to be heard, which so long since persuaded *Hippocrates*, and many other, Secretaries of Nature, that most, if not All Bodies are [ διάπνοα κ ξύμπνοα ] *Perspirable* and *Conspirable*, *i.e.* that they continually emit insensible Effluvia's from themselves to others: We say, if there be any weight in all this, men cannot think it unreasonable in us to conceive, that those Admired Effects, which they commonly

monly afcribe to Hidden Sympathies and Antipathies, are brought about
by the fame ways and means, which Nature and Art ufe in the Caufation
of the like Ordinary and Senfible Effects; and that the Inftruments of
*Natural* Attraction, Compleétence, Repulfion, Sejunction are *Corporeal,*
and hold a neer Analogie to thofe of *Artificial*; only thefe are *Grofs* and
*Perceptible,* thofe *Subtile* and *Imperceptible.*

Notwithftanding the perfpicuity of thefe Arguments, we fhall not
fupererogate, to heighten the luftre of fo defirable a Truth, by the vernifh of a convenient and prægnant *Simile,* or two.  If we attentively obferve a *Chamæleon* catching Gnats and other fmall Flyes in the Aer, for
his food; we fhall fee him dart out a long and flender tongue, with a
fmall recurvation at the tip, and birdlimed with a certain tenacious and
invifcating moifture, wherewith, in a trice, laying hold of a Fly, at fome
diftance from his mouth, he conveys the fame into it with fuch cleanly
fpeed, as exceeds the Legerdemane of our cunningft Juglers, and may
have been the cheif occafion of that popular Error, *that he lives meerly upon Aer.*  And when we fee a peice of Amber, Jet, hard Wax,
or other Electrique, after fufficient friction, to attract ftraws, fhavings
of wood, quils, and other feftucous bodies of the fame lightnefs, objected within the orbe of their Alliciency; and that with a cleanly and quick
motion:  Why fhould we not conceive, that this Electricity or Attraction may hold a very neer Analogy to that attraction of Gnats, by
the exferted and nimbly retracted tongue of a Chamæleon.   For
(1) it is not improbable, that the Attraction of all Electriques is performed by the mediation of fwarms of fubtle Emanations, or Continued
Rayes of exile particles, comparative to fo many Chamæleons Tongues;
which through the whole Sphere of their Virtue, in various points mutually interfecting, or decuffating, and more efpecially toward their
Extreams, doe not only infinuate themfelves into the pores of thofe
fmall and light feftucous bodies occurrent, but lay hold upon feveral infenfible Afperities in their fuperfices, and then returning (by way
of *Retraction*) back to their Original or Source, bring them along in
their twined arms, and fo long hold them faft in their Complicate embraces, as the warmth and radial Diffufion, excited by affriction,
lafteth.    (2) All the Difparity, that can be objected, feems to
confift onely in the Manner of their Return, or Retraction; the
Tongue of the Chamæleon being both darted forth, and retracted
by help of certain Mufcles, wherewith Nature, by a peculiar providence, hath accommodated that otherwife Helplefs Animal: but,
Electriques are deftitute of any fuch organs, either for the Exfer-
tion, or Reduction of their Rayes.  And this is not fo great, but
it may be folved, by fuppofing, that as if the Chamæleons Tongue
were drawn forth at length by a mans hand, and not extruded by
the inftruments of Voluntary Motion, it would again Contract and
Reduce it felf fpontaneoufly, after the fame manner as Nerves and
Luteftrings retract and curle up themfelves, after violent Diftenfion: fo may the Rayes, which ftream from an Electrique, being abduced from their fountains, not fpontaneoufly, but by the force of
præcedent Affriction, be conceived to Reduce and Retract themfelves, after the manner of Sinews and Luteftrings violently extended.

*Art.6.*
The Means of
*Attractions*
fympathetical,
explicated by
a convenient
*Simile.*

Yy                              (3) That

(3) That ſuch tenacious Rayes are abduced from Amber and other Electriques, is eaſily convincible (beſides the experiment of their Attraction of convenient objects) from hence ; that all Electriques are *Unctuous* and *Pinguous* Concretions, and that in no mean degree : and manifeſt it is, that a viſcid and unctuous Bodie is no ſooner Warmed by rubbing, but there riſe out of it certain ſmall *Lines* or *Threads*, which adhære to a mans finger that toucheth it, and ſuch as may, by gentle abduction of the finger, be prolonged to conſiderable diſtance. But, however this may be controverted, and the Way of all Electrique Attractions variouſly explicated, according to the various Conceptions of men ; the Itch of Phancy being ſooneſt allayed by the liberty of ones ſingular Conjecture, in ſuch curious Theorems : yet ſtill is it firme and indubitable, that though the Attraction of ſtraws by Amber, be in ſome ſort Admirable, yet is it not *Miraculous*, as is implied in that opinion, which would have it to be by ſome *Immaterial* (i e. *Supernatural*) Virtue ; and that it is effected by ſome *Corporeal*, though both impalpable and inviſible Organs continued from the Attrahent to the Attracted.

*Art. 7.*
*The Means of Abaction and Repulſion Antipathetical, explicated likewiſe by ſundry ſimilitudes.*

On the Other ſide, as for the *Abaction*, or *Repulſion* of one thing by another, in reſpect whereunto Vulgar Philoſophers have thought and taught, that the Abacted or Repulſed doth (if an Animal) voluntarily (if Inanimate) ſpontaneouſly Flie from and avoid Conjunction with the Abacting, or Repellent, by reaſon of ſome hidden *Enmity* or *Antipathy* betwixt their Forms : though the Reaſons and Manner of ſuch *Fugation*, ſo far forth as concerns Animals, may be collected from our former Diſcourſes of the Gratefulneſs and Offenſiveneſs of Senſible Objects ; yet ſhall we here farther illuſtrate the ſame by certain *Analogies* and *Similitudes.* When a *Nettle* is objected to a mans Hand, why doth He withdraw it from the ſame ? Not upon the account of any Antipathy in his hand to the Nettle ; becauſe being bruiſed, or withered, no Childe but will boldly handle it : but, becauſe the Nettle is pallizado'd with millions of ſmall ſtings, or prickles, which like ſo many Darts, wounding the the skin, cauſe a pain therein, and ſo the man, for avoidance of harm, catcheth his hand from it, as an injurious object. Why likewiſe doth the *Noſe* abominate and avoid *ſtinking Odours*, whenever they are brought neer it ? Is it not becauſe ſuch Foetid and Offenſive Odours conſiſt, for the moſt part, of ſuch ſharp and pungent Particles, as holding no Correſpondence to the pores and contexture of the Odoratory Nerves, are no ſooner admitted, but they in a manner ſcratch, wound and dilacerate the Senſory ? And may we not conceive thoſe diſproportionate Particles of the ungrateful Odour to be as ſo many ſmall *Lancets* or *Darts*, which offer the ſame injury to the Mammillary Proceſſes of the brain, that the Prickles of a Nettle offer to the skin ? Certainly, as the Nettle ſtrikes its Darts into the skin, and not into the Nayles of a mans hand ; becauſe thoſe are of too cloſe and firm a Contexture to admit them : ſo doth an offenſive Odour immit its painted and angular Particles into the tender ſmelling Nerves, and not into the skin, becauſe its Contexture is more Compact, than to be capable of Puncture or Dilaceration thereby. Laſtly, Why doth the *Eye* abhor and turne from *Ugly and Odious Objects* ? Is it not only becauſe the Viſible Species emitted from ſuch Bodies, doth conſiſt of Particles of ſuch Configurations and Contexture, as carry no proportion to the particles and contexture of the Optique Nerves, but striking upon the *Retina Tunica*, instantly wound and exasperate the slender and tender filaments thereof, and so cause the Eye, for fear of farther injury, to close, or avert it self? And are not those Acute and Disproportionate Particles, composing the visible Species, worthily resemblable to so many small Prickles or Lancets, which though too subtile to wound the Skin, Nostrils, or other parts of the body, whose Composure is less delicate, do yet instantly mis-affect and pain the Optique Nerves, whose singular Contexture doth appropriate to them the Capacity of being sensible of that compunction? Now, putting all these Considerations into the scale together, and ponderating them with an equal hand, we shall find their weight amount to no less than this : that *as every Sympathy is displayd by certain Corporeal, though Invisible Organs, comparated to Attraction and Amplectence; so is every Antipathy, by the like invisible Organs, comparated to Repulsion and Sejunction*; which is what we Assumed.

Hence may we, without much difficulty, extract more than a Conjectural judgement, *What are the First and General Causes of all Love and Hatred.* For, look what kind of Motions, whether Grateful or Ungrateful, are by the Species impressed upon the Nerves peculiarly inservient to that sense, by which the Object is apprehended; the very same are continued quite home to the Brain, and therein accordingly move and affect the Common Sensory : so as that, according to the *Pleasure* or *Offence* of the Perception, there is instantly excited an Affection either of *Prosecution* of the thing, by whose species that pleasant motion was Caused, and that is the Hint and Ground of *Loving* and *Desiring* it; or of *Aversation* from it, and that is the Ground of *Hating* and *Declining* it.

*Art. 8.*
The First and General Causes of all Love and Hatred betwixt Animals.

Nay, the same may be well admitted also for the Cause, *Why things Alike in their Natures, love and delight in the Society each of other*; and on the contrary *Why Unlike Natures abhor and avoid each other.* For, as those which are *Consimilar* in their Temperaments, affect each other with *Congenerous* and *Grateful* Emanations : So doe those of *Dissimilar* mis-affect each other with *Discordant* and *Ungrateful.* And therefore it is no longer a wonder, that men Love, or Dislike each other commonly at first interview, though they scarce know why : nor can we longer withold our Assent to that unmarkable Opinion of *Plato*, that *Similitude of Temperaments and so of Inclinations, is not only the Cement, but Basis also of Amity and Friendship.*

*Art. 9.*
Why things Alike in their natures, love and delight in the Society each of other; and why Unlike natures abhor and avoid each other.

SECT. II.

*Art.* 1.
The *Scheme* of
Qualities (re-
puted) occult.

FRom this *General* Difquifition into the Reafons of All Sympathy, and Antipathy, to which moft of thofe Proprieties, which by Philofophers are celebrated as ftupendious and Abfcondite, are ufually referred; we muft advance to the Confideration of *Particular* inftances, that by the Solution of Singulars, we may afford the greater relief to mens Curiofity, and have fo many Opportunities of examining the Verifimility of our former Thefis, *that all fuch Effects, the knowledge of whofe caufes is generally defpaired of, are produced by Subftantial and Explicable Means.* And, in order hereunto, we fhall, according to the method of the no lefs Acute than Judicious *Fracaftorius* (de *Sympath. & Antipath. Rerum*) Diftinguifh All Occult Qualities into *Generall,* and *fpecial;* fubdividing the *Generall* into (1) *the Confpiration of the Parts of the Univerfe,* and (2) *the Influx of Celeftial upon Sublunary Bodies:* and the *Speciall* into fuch as Concern (1) *Inanimates,* (2) *Infenfibles,* (3) *Senfibles.*

*Art.* 2.
Natures Avoid-
ance of Vacuity,
imputed to the
ſyzygia or
Confpiration
of all parts of
the Univerſe;
no *Occult Qua-
lity.*

To the FIRST GENERAL ORDER, *viz.* the Confpiration and Harmony of all Parts of the Univerfe, Philofophers unanimoufly afcribe the *Avoidance of Vacuity;* whereupon many are the Secrets, that are prefumed to enfue, as the Afcention of Heavy, Defcent of Light Bodies, the Sejunction of Congenerous and Sociable Natures, the Conjunction and Union of Difcordant and Unfociable, and the like Irregular and Præpofterous Effects. But, as for all thefe Secrets, we have long fince declared them to be no Secrets, but the moft ordinary and manifeft operations of Nature. For, in our Examination and Solution of all the Appareuces in the late famous Experiment of introducing a Vacuum in a Tube, by Water or Quick-filver, invented by *Torricellius;* we have at large proved, that Nature doth not abhor any but Senfible, or Coacervate Emptinefs: nor that neither *per fe,* or upon the neceffity of an abfolute Plenitude of all places in the Univerfe; but by *Accident* only, and that either in refpect of the natural *Confluxibility* of the parts of Fluid Bodies, fuch as Aer and Water, which caufeth them with great velocity to flow into the parts of Space deferted by a body paffing thorow them; or of the Repugnancie of admitting two bodies into one and the fame place, at the fame time, their *Solidity* prohibiting the penetration of ones dimenfions by the other. Wherefore, let no man henceforth account the Confpiration of the Parts of the Univerfe, to be an *Occult Quality;* or fo much ftand amazed at all or any of thofe *Phænomena,* which arife from Natures Averfion from Vacuity Senfible; as if they had fome Extraordinary Lawes and Conftitutions particularly ordained for their production, and belonged to fome higher Oeconomy than that, according to which fhe regulates her Common Active and Paffive Principles.

To

To the SECOND, *viz*, the Influx of Cælestial upon Sublunary Bodies, innumerable are the Effects, which the Fraud of some, the Admiration of many, and the Credulity of most have confidently imputed: and therefore it cannot be expected, we should, in this place, so much as Enumerate the one Half, much less insist upon them All. Sufficient it is, to the Acquitance of our præsent Debt, that we select the most considerable among them, and such as seem Capital and Comprehensive of all the rest. As for the *Power and Influence of the Stars*, of which Astrologers talk such wonders, and with such pride and oftentation; truly, we have Reason to assure us, that our Cognation and Subjection to those radiant Bodies, is not so great, as that not only All the Actions, Fortunes, and Accidents of Particular men, but even the Warres, Peace, Mutations, Subversions of whole Empires, Nations, States, and Provinces should depend upon their Smiles or Frowns: as if All Occurrents on the theatre of our Lower Orb, were but the orderly and necessary Effects of the Præscriptions and Consignations of the Superior Orbs; or as if there were no Providence Divine, no Liberty of Mans Will.

*Art. 3.*
The *power and influence of Cœlestial bodies*, upon men, suppoſed by *Judicial Astrologers* inconſiſtent with *Providence Divine*, and the *Liberty* of mans will.

(2) As for the *Reciprocation*, or *Afflux and Reflux of the Sea*, so generally fathered upon the Influx and Motion of the *Moon*, which doth herself suffer the like Ebbs and Floods of her borrowed Light, 'tis well known, how *Seleucus* of old, and *Galilæus* of late, have more fully and roundly deduced it from the motion ascribed to the *Earth*. And though we should allow this great Phænomenon to depend upon the several *Aspects* or *Phases* of the Moon, yet is there no necessity to drive us to the subterfuge of any *Occult* and *Immaterial Influence* from her waxing and waning Light: since the System of *Des Cartes* (*in Princip. Philosoph. part.4. page 220.*) doth much more satisfactorily make it out, from the *Elliptical Figure* of the Sphere, wherein the Moon moves; as will soon appear to the Examiner.

*Art. 4.*
The *Afflux and Reflux of the ſea*, I deriva-ive from any *immaterial Influence* of the Moon.

(3) As for the *Diurnall Expansion, and Conversion of the Heliotrope toward the Sun*; though great notice hath been taken thereof by the Ancients, and most of our Modern Advancers of the Vanities of Natural Magick (who will have every Plant to retain to some one of the Planets, by some secret Cognation, and peculiar sympathie.) have laboured to heighten it to the degree of a Wonder: yet can we not conceive the Effect to be so singular, nor that any such Solemne Reason need be assigned thereunto. For, every mans observation may certifie him, that all Marygolds, Tulippa's, Pimpernell, Wartwoort, Mallow Flowers, and indeed most other Flowers, so long as they are in their Vigour and Pride, use to Open and Dilate toward noon, and somewhat Close and recontract themselves after Sun set. And the *Cause* (surely) is only the Warmth of the Suns Rayes, which discussing the Cold and Moisture of the præcedent Night (whereby the Leaves were loaden towards the bottom, or in the bowle of the Flower, and so made to rise more upright and conjoyn their tops) and somewhat Exsiccating the Flower, make the pedeftalls of its leaves more flaccid, so that they seem to expand and unfold themselves, and incline more outwards, meerly by reason of their want of strength to sustain themselves in an erect and concentrical posture: for always the hotter the Day, the greater is the Expansion. Likewise, as

*Art. 5.*
The Cauſes of the diurnal *Expanſion*, and *Converſion* of the *Heliotrope* and other Flowers,

as for the Flowers *Converſion* to, or *Confronting* the Sun in all its progreſs above the horizon, wherein our Darkſom Authors of Magick Natural, principally place the Magnale; the Cauſe thereof is ſo far from being more obſcure than, that it is the very ſame with that of its Expanſion. For, as the Sun running his race from Eaſt to Weſt, doth every moment vary the points of his Rayes vertical incidence upon the ſtalk which ſupports the Flower, and upon the leaves thereof; ſo muſt the whole Flower incline its head and wheel about accordingly: thoſe parts of the ſtalk upon which the rayes are more perpendicular, and ſo the heat more intenſe, becoming more dry and flaccid, and ſo leſs able to ſupport the burthen of the Flower, than thoſe, which ſuffer only from the oblique, reflected and weaker beams. Notwithſtanding this Solution, if any Champion of ſecret *Magnetiſm* ſhall yet defend this *Circulation* to be a *Propriety* of the Heliotrope, to which no other Flower can prætend; and that this Solar Plant diſcovers it Amours to the Sun, by not only diſcloſing its rejoycing head and boſom at the præſence, and wrapping them up again in the mantle of its owne diſconſolate and languiſhing leaves, during the abſence of its Lover, but alſo by facing him all day long: leſt He ſhould inſult, upon an apprehenſion, that our theory is at a loſs, we ſhall tell him, in a word; that that *Propriety*, which he ſuppoſeth, muſt conſiſt only in ſuch a peculiar Contexture and Diſpoſition of the particles, which compoſe its Leaves, as makes them more fit to receive, and be moved, and their ſpiritual and moſt ſubtle parts to be in a manner Circulated by the Rayes of the Sun, than the Leaves of any other Flower whatever. As in the Organ of Smelling, there is a certain Peculiar Contexture of its inſenſible Component Particles, which renders it alone capable of being moved and affected by Odours, that have no influence nor activity at all upon the Eye, Eare, or other Organ of Senſe.

*Art. 6.*
Why Garden Claver hideth it ſtalk, in the heat of the day.

(4) Great things have been ſpoken alſo of the *Garden Claver*, which *hareth its boſom, and hideth the upper part of its ſtalk, whenever the Sun ſhines hot and bright upon it:* but, this (doubtleſs) hath the ſame Cauſe, as the Former, the Hiding of the ſtalk being nothing but an over-expanſion of the Leaves, which by reaſon of the violent ardour of the Sun, grow more faint and flaccid, and ſo leſs able to ſupport themſelves.

*Art. 7.*
Why the Houſe Cock uſually Crowes ſoon after midnight; and at break of day.

(5) A Fifth Secret, found in the Catalogue of Cæleſtial Influxes, is the *Crowing of the Houſe-Cock,* at certain and periodical times of night and day, and more eſpecially ſoon after midnight, and about day break: for, moſt eſteem it an Occult Propriety, and all our *Crollians* and ſuch as promote the dreams of *Signatures* and *Sydereal Analogies*, reckon the Cock a chief *Solar* Animal, for this reaſon alone; as if his Phanſy received ſome magnetique touches and impreſſions from the Sun, which made him proclaime his Advent into our Hemiſphere, and like a faithful Watch or Clock, meaſure out the ſeverall ſtages in its race. Great enquiry alſo hath been made after the *Cauſe* hereof, in all ages, and various Conceptions entertained concerning it. Some with lofty and Rhetorical Diſcourſes endevouring to perſuade, that Nature intended this Οχιον αενiε, (as *Plutarch* calls it ) or *Gallicinium*, as an Alarme to rouſe up ſluggiſh man from the dull armes of ſleep, and ſummon him to the early Contemplation of her Works; as *Pliny* ( *Natural. Hiſtor. lib.* 10. *cap.* 21. )

Others

Others afcribing it to a Defire of Venery in this Animal, arifing from the turgefcence and ftimulation of his fperm, at certain periods; as *Erafmus*, who is therefore worthily and fufficiently derided by *Scaliger* (*Exercit.* 239) Others affigning it to an Appetite of Aliment, invading and exciting after determinate intervalls; as *Cardan.* And others alleaging we (nor themfelves) know not what peculiar influence of the Sun, caufing a fuddain mutation, or Evocation of the Spirits and blood of the Cock, which were Concentred by fleep; as *Cælius Rhodiginus* (*lib.* 16. *Antiq. Lection. cap.* 13.) But, All thefe Great Clerks feem to have grafpt the ear, and catched at fhadowes. For (1) it may be doubted, that all Cocks, in one and fome meridian, doe not Crow at the fame times of night or day; and that no Cock doth obferve fet and punctual times of Crowing; both which are præfumed: and whoever fhall think it worth the lofs of a nights fleep, as we have done, to obferve the Crowing of fundry Cocks in fome Country Village, where the Houfes ftand fcatteringly and far afunder, fo that the Cocks cannot awake each other; will, perhaps, more than doubt of either. (2) It is, as Natural, fo Familiar to the Cock, fo often as his Imagination is moved by a copious and frefh afflux of Spirits to his Brain, to rowze up himfelf, clapp his wings, and found his trumpet as well at noon, after noon, and at other times of day and night, upon feveral occafions; as when he hath efcaped fome late danger, obtained a victory, found fome treafury of grain, compreffed his miftrefs, and the like; as if his joy were not complete, till he had communicated the ridings thereof to his Wives and Neighbours, by the elevation of his gladfome and triumphat voice. (3) May we not allowe the Cock to have his fet times of Sleeping and Waking, as well as all other Living Creatures, that live *fuo jure*, and according to the Aphorifms of their Specifical Conftitutions, and regiment of their proper Archæa's; and likewife moft Men, who live healthfully and orderly, keeping to conftant hours for labour, meat, reft and fleep? (4) What need is there that we fhould have recourfe to fuch a far-fetcht (and never brought home) Caufe, as that of a Secret Commerce, and peculiar Sympathy betwixt this Fowl and the Sun in the other Hemifphere; when we have a more probable and manifeft one, neerer hand; *viz. The fuddain invafion of the Cock, by encreafed Cold foon after midnight?* For, when the Sun hath made fome fenfible advance in the lower world, beyond the Nadir point or midnight circle, and hafteneth toward our Eaft; He moves and drives along before him into our horizon, the (formerly) quiet and cold Aer of the Night: which invading the Cock, difturbs him from his reft, during which his Heat is retired inward, and awakens him on the fuddain: fo that rowzing up himfelf, exciting his courage, and diffufing his Spirits again into his members, to oppofe that Cold, and perhaps alfo to prevent his falling from the perch; he ftands up, clappeth his wings againft his fides, and chants a cheerfull Pæan to himfelf and Rooftfellowes, celebrating his fafety and conqueft with the loud mufick of his throat.

Art. 8.
Why Shell-
fish growe fat,
in the Full of
the moon, and
lean again at
the New.

(6) A sixth notable Secret, appertaining to the same Classis, is that of the *Encrease of the Substance of Shell Fish, of the Brains in Coneys, and of the Marrow in the bones of most Land Animalls, as the moon approacheth her Full; and the Decrease of them again, as her Light decreaseth toward her New.* But, laying aside all Lunar Magnetism, Immaterial Influxes, and the like Toyes put into Great Words; we take it, the Phænomenon may be well enough solved, by referring it meerly to the *Moons great Humidity*; at least, if those vast Duskish spots, apparent in her Orb, be her moist Element, carrying some analogy to our Seas, as the most and best of our Modern Astronomers have believed, and upon grounds almost demonstrative, and wholly irrefutable. For, insomuch as the Rayes of the Sun, in greater abundance falling upon the face of the Moon, toward and at her Full, than in her Wane, are accordingly more abundantly reflected from thence upon our Terraqueous Globe, bringing along with them no sparing Tincture of the Moons Moisture; so that the Light which is Reflected from the Oceans in the moon, being more moist than warm, must needs be more Prolifical, Generative, and prædisposed to the Nutrition of Animals: and that in the New of the Moon no such plentiful Abduction of her moisture can be expected, because fewer of the Suns Rayes are, at that time, Reflected from her Orb to ours; why should it be thought so strange, that either Aquatile, or Terrestrial Animals should be nourished more plentifully at the Full, than New of the Moon? Especially since it is no præcarious, nor novell Assertion, that the Light coming from the Moon, is tincted with Humidity, as being reflected from the Watery as well as solid parts of her Orb; Experience having frequently demonstrated, that the Calorifick Rayes not only of the Sun, but even of our terrestrial and culinary Fires, being trajected through various Liquors, and other Catoptricall bodies, or reflected from them, doe imbibe and carry off much of their Virtues, and become thereby imprægnate, so as to be prædisposed to the production of sundry noble Effects, such specially as relate to the Alteration, Germination, Pullulation, and Generation of Vegetables and Animals, both Aquatile, and Terrestrial. Nevertheless, in case this Cause assigned seem somewhat Remote and obscure, we shall alleage *Another*, sufficiently verisimilous to ease men of their wonder, at the Fullness of the Shell Fish in the Full moon, and their Leaness in the New; and that is the *Encrease of the Tides of the Sea,* which ascending higher upon the shoars, at the Full moon, and washing down more of Mudd, Slime and Saltness from thence, afford greater plenty of Aliment to all Shell Fish: which delight in, and thrive best upon such kind of food, and are observed therefore to frequent foul and slimy shoars, and yet neerer and neerer to land, as the Tides rise higher and higher, and again remove farther and farther off, as the tides sink lower and lower.

Art. 9.
Why the
Selenites re-
sembles the
Moon in all
her several As-
pects.

(7) To this Classis also belongs the Famous *Selenites*, or *Moon-Gemme*, a certain præcious stone, found only in *Arabia,* as *Dioscorides* (*lib.* 5. *cap.* 110.) delivers: whose rare and singular Faculty is this, *that it represents the Moon in all her several Dresses of Light, or Apparences,* encreasing its Lustre exactly as she encreaseth hers, and proportionately losing it, if the Relations be true, which have been made thereof by Authors of the highest form for Credit, namely *Pliny* (*lib.* 36. *cap.* 10.) S. *Augustine*

*guflin (de Civit. D. lib. 21. cap. 5.) Zanardus (de Univers. Element. quaff.*
*53.) Nichol. Caufsinus (lib. 11. Symbol.5.) oh. Daniel Mylius (Ba-*
*filica Chymic.lib. 5. cap.28.)* and many modern *Mineralogifts.* Now, for
the *Reafon* of this Rarity, in all liklihood, it muft be if not the very fame,
yet Coufin German to that of the former. Becaufe, it is very probable, that fome certain portion of a thin, fluid and fubtle matter (we may
conceive it to be Hydrargycal, or relating to Quickfilver, fince all the
forenamed Authors defcribe the ftone to be White and Candent of Colour;) wherein the Luftre of the ftone doth moftly confift, doth fuffer
fome Alteration, according to the more and lefs of the Lunar Light incident upon it; and is refpectively Circulated through the loofer or lefs
compacted parts of the ftone, after the fame manner as the more fubtle
and fpiritual parts of fome Flowers are Circulated by the rayes of the
Sun; the particular Configuration and Contexture of its infenfible particles being fuch, as difpofe to that Circulation, upon the influx of the
Moons Light.

In the Inventory of S P E C I A L Sympathies and Antipathies, the
Firft Divifion Concerns INANIMATE Natures, and among fuch the
firft place belongs to the *Attraction of Iron by the Loadftone,* the fecond
to the *Attraction of Straws* and other fmall and light bodies by *Amber* and
othe *Electrique:* but fuch is the fingular Excellency of the *Former,* that
it not only deferves, but challengeth a fingular Chapter to its Difquifition;
and the *Reafon* of the *other* we have plainly, though fuccinctly explicated,
in the precædent Section, the Confideration of the Wayes and Inftruments of all Attraction Natural, in the General, impelling us upon the
Anticipation thereof.

*Art. 10.*
Why the Con-
fideration of
the *Attraction*
*of Iron by a*
*Loadftone,* is
here omitted.

In the *Third,* we are to examine the fecret *Amity of Gold and Quick-*
*filver, of Brafs and Silver;* which is fo manifeft, that whenever Gold is
diffolved in Chryfulca or *Aqua Regis,* and the Spirit or Diffolution of
Quickfilver fuperadded thereto, the fubtle Effluvia ftreaming from the
particles of the Gold, will inftantly lay hold of, and at diftance attract and
firmly embrace the particles of the Quickfilver, into which the Diffolving
liquor hath fubtiliated it; and in like manner, when Brafs and Silver are
diffolved in the fame *Aqua Fortis,* their particles are obferved to unite even
to concorporation, though the Spirits iffuing from them, are not potent
enough to perform an Attraction, while the Metals remain entire and in
the mafs. Thefe Effects we conceive may well be referred to the *Cor-*
*refpondency* or Compoffibility betwixt the *Figures* of the infenfible particles, of which the Emiffions from the Gold, and Brafs confift, and thofe
of the *pores, inequalities,* and *faftnings* in the fuperficies of the Granules of
the Diffolved Quickfilver, and Silver: but what thofe Figures are on each
part, is above our hopes of determination; nor can we afford the Curious
any other light for Conjecture in this true Abftrufity, but what himfelf may
perceive to arife to him by Reflection from the Reafons, we fhall hereafter
give, for the Attraction of Iron by a Loadftone. In the mean while, we
præfent Him, for Diverfion of his Scrutiny, with a fhort and opportune
COROLLARY.

*Art. 11.*
The fecret
Amities of
Gold and
*Quickfilver,* of
*Brafs* and *Sil-*
*ver,* unriddled.

*Art.* 12.
A COROL-
LARY.
Why the Gra-
nules of Gold
and Silver,
though much
more pond-
rous then
those of the
*Aqua Regis*
and *Aqua For-
tis,* wherein
they are dif-
folved, are yet
held up, and
kept floating
by them.

Delightful it is, and indeed Admirable to behold the Granules of Gold and Silver, though much more ponderous than those of the *Aqua Regis,* and *Aqua Fortis,* to be notwithstanding held up, and constantly kept in a floating and elevated posture by them. And yet, in all likelihood the *Salt* dissolved in those Corrosive Waters, must be the Sole Cause of that strange Effect. For, the Salts which are plentifully dissolved in those Liquors, by a kind of mutual Cohæsion of their insensible particles supporting each other from the bottom to the top of the Glass, or other containing vessel; doe sustain and bear up the Granules of the Metals which they have Corroded and Embraced. And this seems the more probable from hence; that if common Water, impregnate with a few dropps of Oyle of Tartar (that Great instrument of Separation) be superinfused upon those Tinctures, the Granules of the dissolved Metals suddainly disengage themselves from the arms of the Corroding Salts, and sink to the bottom: the fresh Water yet farther dissolving those Salts, and giving them fuller Fluidity; so that becoming more Attenuate, they lose their mutual Cohæsion, and so their power of supporting; and tis well known, that Salt water will beare up such bodies, as will hardly swim in fresh. And this we take to be the General Reason of all sorts of Præcipitation, practised either by Chymists, or common Refiners of Metals: the Oyle of Tartar thereto conducing no otherwise, than meerly as it serves to the farther Attenuation of the Salt Armoniack and other Corrosive Salts formerly dissolved in the strong Waters.

*Art.* 13.
The Cause of
the Attraction
of a Less Flame
by a Greater.

(4) To the *Fourth,* we assign the *Attraction of a Less Flame by a Greater;* according to the erroneous Dialect of the People: for, really it is rather the *Extension of a Greater Flame to the Fewel of a Less.* For, the heat of a Greater Flame being proportionately more intense and diffusive, extends it self to the pabulum or nourishment of the less, where the same is situate within the Sphere of its power: and thence it comes to pass, that the Greater burning more strongly, by reason of that addition or augmentation of its fewel, doth more and more dilate it self that way, till at length it becomes wholly united to the Less. Which unexamining heads not understanding, have imputed to a certain Attractive faculty in the Greater Flame, depending upon the Identity of the two Natures, or more præcisely, the same Nature in two Divisions; and many have rackt their brains to erect subtle Discourses thereupon, as if they wanted other Opportunities to exercise their Learning, and entertain their Curiosity.

*Art.* 14.
The Cause of
the Involation
of flame to
Naphtha, at
distance.

(5) To the *Fifth* belongs the supposed *Attraction of Flame by Naphtha* of *Babylon,* at distance; which is also improperly accounted an *Attraction:* for the Flame of its own accord flyeth to, and layeth hold of the Naphtha; and the Cause of that *Involation* is only this. From the body of the Naphtha there is emitted in round a certain fat and unctuous, and so soon inflammable Halitus, or steam, which being extended to the borders of some flame posited at convenient distance, and thereby kindled in the extreme of its Sphere, becomes enflamed all along the Rayes, and they burning, soon bring home the flame to the body of the Naphtha, from which they are emitted, in a continued fluor.

(6) Next

(6) Next to this, Philosophers usually place the *Attraction of Water* Art. 15.
*by a Spunge*; wherein they are as much mistaken as in either of the two Of the Ascensi-
last. For, the Ascention of Water into the pores of a spunge, so placed on of Water in-
as to touch only the superfice of it, comes not from any Appetite of At-to the pores of
traction, or Suction inhærent in the Spunge, as is generally præsumed a *Spunge.*
and affirmed; but onely from the *Depression*, *or downward impulse*
*of the water by the swelling and sensibly dilating spunge*; and the manner
of that series of motions is thus. The skirts or lowest parts of the spunge,
touching the superfice of the Water, immediately imbibe some parts of
it into its pores, and becoming thereby dilated and tumid, press down the
subjacent Water to such a proportion as responds to the quantity of their
owne expansion; so that as they are more and more dilated by the admission of more and more parts of Water into their Cells or Receptaries, it
must be, that the Water being more and more depressed toward the bottom, must rise higher and higher on the sides of the Spunge, and insinuate
it self into other and other pores successively, till the whole spunge be
filled. Manifest it is by Experience, that if Water or any other Liquor,
when it is though never so gently pressed in the superfice, find any the
smallest *Chinks* in the body pressing it, it doth instantly rise up in round,
and insinuate it self into those pores or Chinks, the sides thereof in a
manner sustaining it, and so præventing its relapse or efflux. This we
cannot but observe, when we dip the nose of our Pen into ink; the
small *Cleft* or slit in the lowest part of the Quill, assisting the Assent
of the ink into the hollow thereof, and carrying up so much of it, as the
mutual Coherence of its parts will permit: for, if we dipp the point of a
Pen, which hath no slit, into a standish, we shall observe no such plentiful Assent of ink; there being no support or fastnings for it on each
side of the nose, and so no obstacles to its relapse and sudden efflux.
And, as for the Reason, Why *Water Ascends, when it meets with any*
*body, that is Dry, Filamentous or Fibrous, and full of pores or Chinks, such*
*as a Spunge, Cloth, Pen, &c.* it may be most fully explained by the Instance of a *Syphon*, or Pump.

Take a Pipe of Lead, of the figure of a Carpenters Squire, whose Art. 16.
one arme is longer then the other (such our Wine Coopers exhaust The same il-
their Buts of Wine withal) and immerse the shortest into a Cistern of lustrated by
Water, so as it may come very neer the bottom, and yet the longer the example
arme rest upon the margin of the Cistern, in a dependent or declining of a *Syphon.*
posture, then with your mouth suck forth the Aer contained in the
cavity of the pipe: and you shall observe the Water quickly to follow
on the heels of the Aer, and flow in full stream out of the Cistern
through the pipe, without ceasing till all the Water, that covers the
shortest arme of the pipe, and so hinders the ingress of the aer into its orifice, be exhausted. Of this the *Cause* is only, that as your Cheeks are
inflated and distended by the Aer, which upon exsuction comes rushing
into your mouth, doe strongly move and impell the ambient aer; so
doth that, receding, move and impell the neighbouring aer, and that
again moves and impels the next, till the impulse be propagated to the
surface of the Water in the Cistern: and the Water being thus depressed
in the superfice, riseth up into the Cavity of the pipe, which the extracted
Aer had newly deserted and left unpossessed; nor doth it thenceforth cease

Zz 2 to

to ascend and flow in a continued stream through the pipe, until all be exhausted. Because, how much of Water flows through the pipe, exactly so much of Aer is, by impulsion, Circulated into the place thereof; the last round of aer wanting any other place to receive it, but what it provides for its self in the Cistern, by depressing the water yet remaining therein: and thus the Circulation once begun, is continued, till all the Water hath past through the pipe.

*Art.17.*
*The reason of the Percolation of Liquors, by a cloth whose one end lieth in the liquor, and other hangs over the brim of the vessel, that contains it.*

Upon the same Cause, or some other so like it, as 'tis no easie matter to discriminate them, doth that kind of *Percolation* of Liquors, and especially of *Aqua Calcis*, depend, which is made by a long piece of *Woollen Cloth*, whose one end lies in the Liquor, and other hangs over the brim of the vessel that contains it. For, the Liquor gently ascends and creeps along the filaments of the Cloth, because, being though but very lightly prest in it superfice by the same, it doth proportionately ascend in round, so deliver it self from that pressure; and by that motion impelling the incumbent Aer upwards, it causeth the same to Circulate and depress the surface of the Liquor, and so makes it rise by insensible degrees higher and higher along the hairs and threads of the Cloth, till at length it arrive at the highest part thereof resting upon the margin of the vessel; and thence it slides down the decline or propendent half of the Cloth, and falls down into the Recipient, by dropps. And this Motion is Continued till all the Liquor hath passed the Percolatory, leaving the fæces adhærent to the fibres of the same: each drop impelling the Ambient Aer, and driving it in round, or by a *Periosis*, upon the surface of the Water, so long as any remains in the vessel. And this, we conceive, may suffice to any mans Comprehension of the Reason of the Repletion of a Spunge, by Water *Ascending* (not Attracted) into its Cavities or Pores.

*Art.18.*
*The reason of the Consent of two Lutestrings, that are Æquison.*

(7) Another eminent Secret of Sympathy, belonging to the same Division, is that *Consent betwixt two Lutestrings, that are Æquisone*: (for *Unisone* is hardly proper), which is thus experimented. Take 2 Lutes, or Vials, and their treble, mean, or base strings being tuned to an Equality of Sounds, lay one of them upon a table, with the strings upward, with a small short straw equilibrated upon the Æquison string: and then strike the Æquison string of the other instrument, and you shall observe, both by the leaping off of the straw, and the visible trembling of the string, whereon it was imposed, that it shall participate of the motions of the string of the other instrument percussed; all the other Dissonous strings, as wholly unconcerned in the motion imprest, remaining unmoved. The like also will be, if the Diapason or Eighth to that string be percussed, either in the same Lute or Vial, or other lying by: but, in none of these, the Consent is discernable by any report of sound, but meerly by motion. And yet the Cause of this Sympathy is not so very obscure, but the dullest Pythagorean might soon have discovered it to be only this; that the percussed string doth suffer a certain number of Diadroms, or Vibrations, and impress the like determinate motions upon the Aer: which lighting upon another string of equal Contexture and Extension with the former percussed, doth impress the same motions thereupon, and impell and repell it so correspondently, as to make it suffer an equal number

of

CHAP.XV. *Occult Qualities made Manifest.* 357

of Diadroms. Nor doth the Aer hinder it in its feveral Reciprocations or alternate excurfes and recurfes ; becaufe the percuffed ftring makes all its alternate excurfes and recurfes, at and in the fame time, as the untoucht ftring doth, and fo impels the Aer alternately to the contrary fide thereof. But, that agitated Aer which falls upon a ftring of a different degree of extenfion, and fo neceffarily of a different tone, though it imprefs various infenfible ftrokes thereupon, yet are thofe impreffed ftrokes fuch as mutually check and oppofe each other, *i.e.* the Excurfes hinder the Recurfes: and therefore the ftring remains unmoved, at leaft as to the fenfe. Likewife, the Confent of another ftring, which makes that Confonance, which Muficians call a *Diapafon* or *Eighth*, to that which is percuffed by the hand, arifeth only from hence, that the Excurfes and Recurfes of the ftring percuffed by the hand, do not at all clafh with, nor perturb and confound the Excurfes and Recurfes of the ftring moved immediately only by the Aer, but are Coincident and Synchronical to them, and obferve the fame periods; and fo both agree in their certain and frequent intervals : more particularly, in an Eight, every fingle Diadrom of the longer and more lax ftring, is coincident to every fecond, fourth, fixth, &c. Diadrom of the fhorter or more tenfe ftring. Nay farther, if the two ftrings be Confonous though but in the lefs perfect Confonance of a *Fifth*; yet fhall the fympathy hold, and manifeft it felf (which is not commonly obferved) by the tremulation of the untouched ftring, that is tuned to a Fifth: becaufe their Diadroms are not wholly confufed, each fingle diadrom of the longer or lower ftring, being coincident to every third, fixth, ninth, &c. diadrom of the fhorter or more tenfe ftring. But if the two ftrings be *Diffonous*, the fympathy fails, becaufe the Excurfes and Recurfes agree not in any of their Intervals or Periods, but perturb and confound each other ; as may be more fully underftood from our præcedent Difcourfe of the *Reafon of Confonances and Diffonances Mufical.*

*Art. 19.*
The reafon of the *Diffent betwixt Entestrings of fheeps Guts,* and thofe of *Woolfs.*

(8) Nor is it the Inæquality of Tenfion, difparity of Longitude and Magnitude, or Non-coincidence of the Vibrations in their feveral periods, that alone make Two ftrings Difcordant ; for, if we admit the common tradition of Naturalifts, where *an Inftrument is ftrung with fome ftrings made of Sheeps, and others of Woolfs Guts intermixed, the beft hand in the World fhall never make it yeeld a perfect Confonance, much lefs play an harmonious tune thereupon.* And the *Caufe*, doubtlefs, is no other than this ; that the ftrings made of a Woolfs Guts are of a *different Contexture* from thofe made of a Sheeps ; fo that however equally both are ftrained and adjufted, yet ftill fhall the Aer be unequally percuffed and impelled by them, and confequently the founds created by one fort, confound and drown the foundsrefulting from the other. To leave you in the lefs uncertainty concerning this, it is commonly obferved, that from one and the fame ftring, when it is not of an Uniforme Contexture throughout, but more clofe, even, and firme in fome parts than in others (all fuch our Muficians call *Falfe* ftrings) there doe alwayes refult various and unequal founds : the clofe, even and firm parts yeelding a fmart and equal found, the lax and uneven yeelding a dull, flat and harfh ; which two different founds at the fame time created, confound and drown each other ; and confequently where fuch a ftring is playd upon in Confort, it difturbs the whole Concent or Harmony. It is further obferved alfo, that the Mufick of an Harp
doth

doth infect the mufick of a Lute, and other fofter and milder inftruments with a kind of Afperity and Indiftinction of Notes: which Afperity feems to arife from a certain kind of Tremor, peculiar only to the Chords of that Inftrument. The like alfo hath been reported of other fcarce Confortive Inftruments, fuch as the Virginalls and Lute, the Welfh Harp and Irifh, &c.

But you'll Object, perhaps, that the Difcordance of Woolves and Sheeps Gutlings feemeth to arife rather from fome Formal *Enmity*, or inhærent *Antipathy* betwixt the Woolf and Sheep: becaufe it hath been affirmed by many of the Ancients, and queftioned by very few of the Moderns, *that a Drum bottomed with a Woolfs skin, and headed with a Sheeps, will yeeld fcarce any found at all; nay more, that a Wolfs skin will in fhort time prey upon and confume a Sheeps skin,* if they be layed neer together. And againft this we need no other Defenfe than a downright appeal to *Experience,* whether both thofe Traditions deferve not to be lifted among Popular Errors; and as well the Promoters, as Authors of them to be exiled the fociety of Philofophers: thefe as Traitors to truth by the plotting of manifeft falfehoods; thofe as Ideots, for beleiving and admiring fuch fopperies, as fmell of nothing but the Fable; and lye open to the contradiction of an eafy and cheap Experiment.

*Art.* 20.
The tradition of the Confuming of all Feathers of Foul, by thofe of the Eagle; exploded.

(9) Nor can we put a greater value upon the *Devouring of all other Birds Feathers by thofe of the Eagle commixt with them*; though the Author of *Trinum Magicum* hath bin pleafed to tell us a very formall and confident ftory thereof: becaufe we have no Reafon to convince us, that the Eagle preys upon other Fowls, out of an Antipathy or Hatred, but rather out of Love and Convenience of Aliment; and though there were an Enmity betwixt the Eagle and all his feathered fubjects, during life, yet is there no neceffity that Enmity fhould furvive in the fcattered peices of his Carcafs, efpecially in the Feathers (that are but one degree on this fide Excrements) which is præfumed to confift cheifly in the Forme; fince thofe Proprieties which are Formal, in Animals, muft of neceffity vanifh upon the deftruction of the Forme, from whence they refult. Thus Glow-worms project no luftre after death; and the Torpedo, which ftupefies at diftance, while alive, produceth no fuch effect though topically applied, after death: for there are many Actions of Senfible Creatures, that are mixt, and depend upon their vital form, as well as that of miftion: and though they feem to retain unto the Body, doe yet immediately depart upon its Diffunion.

*Art.* 21.
Why fome certain *Plants* befriend, and advance the growth and fruitfulnefs of others, that are their neighbours.

In the SECOND Divifion of *Special* Occult Qualities, *viz.* fuch as are imputed to *Vegetables,* the Firft that expects our Confideration, is the fo frequently mentioned and generally conceded *Sympathy, or mutually beneficial Friendfhip betwixt fome certain Plants,* as betwixt *Rew*, and the *Figg-tree,* the *Rofe* and *Garlick,* the *Wild Poppy* and *Wheat*; all which are obferved to delight and flourifh moft in the neighbourhood of each other, and our skilful Gardners ufe to advance the growth and fructification of the one, by planting its favourite neer it. Concerning this, therefore, we advertife; that men are miftaken not only in the Caufe, but

Denomination

Denomination also of this Effect: suppoſing a ſecret Friendſhip where is none, and imputing that to a certain Cognation, or Sympathy, which ſeems to proceed from a manifeſt Diſſimilitude and Antipathy betwixt Divers Natures. For, wherever two Plants are ſet together, whereof the one, as being of a far Different, if not quite Contrary Nature, and ſo requiring a different kind of nouriſhment, doth ſubſtract and aſſimilate to its ſelf ſuch a juice of the earth, as would otherwiſe flow to the other, and deprave its nouriſhment, and conſequently give an evil tincture to its Fruit and Flowers: in this caſe, Both Plants are reciprocally the remote Cauſe of the Proſperity each of other. And thus Rew, growing neer the roots of the Figg-tree, and attracting to its ſelf the Rank and Bitter moiſture of the earth, as moſt agreeable to its owne nature; leaveth the Milder and Sweeter for the aliment of the Fig tree, and by that means both aſſiſteth the procerity of the Tree, and Meliorateth the Fruit thereof. Thus alſo Garlick, ſet neer to a Roſe tree, by conſuming the Fœtid juice of the ground, and leaving the more Odorate and benigne to paſs into the roots of the Roſe tree; doth both farther the Growth and Germination thereof, and encreaſe the Sweetneſs of it Flowers. But, as for the Amity betwixt the *Wild Poppy* and Wheat, we ſhould refer it to another Cauſe, *viz.* the Qualification of the ground by the tincture of the Wheat, ſo as to præpare it for the Generation and growth of the Wild Poppy; not by ſubſtraction of Diſagreeing moiſture, but by Enriching the Soyle, or imprægnating it with a ferility, determinate to the production of ſome ſorts of weeds, and chiefly of that. For, moſt certain it is, that there are certain Cornflowers, which ſeldom or never ſpring up but amongſt Corn, and will hardly thrive, though carefully and ſeaſonably ſet in other places: ſuch are the Blew-bottle, a kind of yellow ſingle Marygold, and the Wild-Poppy.

(2) This diſcovered, we need not ſearch far after the Reaſons of thoſe *Antipathies*, which are reported to be between the *Vine* and *Cole-woort*, the *Oke* and *Olive*, the *Brake* and *Reed*, *Hemlock* and *Rew*, the Shrub called our *Ladies Seal* (a certain Species of Bryony) and the *Cole-woort*, &c. which are preſumed to be ſo odious each to other, from ſome ſecret Contrariety of their reſpective Forms, that if any two of them, that are Enemies, be ſet neer together, one or both will die. For, the truth is, all Plants, that are great Deprædators of the moiſture of the earth, defraud others that grow neer them, of their requiſite nouriſhment, and ſo by degrees impoveriſhing, at length deſtroy them. So the Colewoort, is an enemy not only to the Vine, but any other Plant dwelling neer it; becauſe it is a very ſucculent and rank Plant, and ſo exhauſts the fatteſt and moſt prolifical juice of the ground. And if it be true, that the Vine will avoid the Society of the Colewoort, by Averting its trunck and branches from it, this may well be only in reſpect of its finding leſs nouriſhment on that ſide: for, as the *Lord St. Alban* hath well obſerved, though the root continue ſtill in the ſame place and poſition, yet will the Trunk alwayes bend to that ſide, on which it nouriſheth moſt. So likewiſe the Oke and Olive, being large trees of many roots, and great ſpenders of moiſture, doe never thrive well together: becauſe, the ſtronger in Attraction of juice, deceives and ſtarves the weaker. Thus Hemlock is a dangerous neighbour to Rew; becauſe, being the Ranker Plant of the two, and living upon the like juice, it defrauds it of ſufficient

ſuſtenance,

*Art.* 22.
Why ſome Plants thrive not in the ſociety of ſome others.

ſuſtenance, and makes it pine away for penury. And the like of the reſt.

*Art. 23.*
The Reaſon of the great Friendſhip betwixt the *Male and Female* Palm-trees.

(3) But what ſhall we think of that ſemiconjugall *Alliance betwixt the Male and Female Palme trees,* which is ſo ſtrong and manifeſt, that the Femal, which otherwiſe would languiſh, as if ſhe had the Green ſickneſs, and continue barren ; is obſerved to proſper, and load her fruitful boughs, with braces of Dates ; when ſhe enjoys the Society of the Male : nay, to extend her arms to meet his embraces, as if his maſculine influence were neceſſary not only to her impregnation, and the maturity of her numerous iſſue ; but even to her own health and welfare ? Why, truly, we cannot better expound this dark Riddle of Nature, than by having recourſe to ſome *Corporeal Emanations,* deradiated from the male, which is the ſtronger and more ſpriteful plant, to the Female, which is the weaker, and wants an Acceſſion of heat and ſpirits. For, far enough from inprobable it is, that ſuch Emanation may contain much of the Males *Seminal* and *fructifying virtue* ; and it hath been avouched by frequent Experiments, that the bloſſoms and Flowers of the Male being dried and poudered, and inſperſed upon the branches of the Female, are no leſs effectual to her Comfort and Fertility, than the Vicinity of the Male himſelf. We are told, indeed, by *Herodotus,* and from his own ſtrict obſervation ; that the Male Palm produceth yearly a Dwarfiſh ſort of Dates, which being uncapable of maturity and perfection, men uſe therefore to gather early, and bind them on the loaden branches of the Female : that there corrupting, and breeding a kind of ſmall volant Inſect, reſembling our Gnats (which the Natives call *Pſene,* though *Theophraſtus* ſeems to appropriate that name only to thoſe Flyes, that are a ſpontaneous production out of the immature fruit of the Wilde Figg tree, ſuffering putrefaction) that they may advance the Growth and Maturity of her fruit ; not by any ſecret influence, but the manifeſt Voracity of thoſe Inſects, which continually preying upon the ripening fruit, both open the tops of them, and ſo make way for the rayes of the Sun to enter more freely and deeply into their ſubſtance, and ſuck out moſt of the luxuriant crude and watery juice, leaving the Alimentary and Unctuous to the more eaſie digeſtion and aſſimilation of the formerly overcharged Seminal Virtue of the Plant. This, we confeſs, is nice and plauſible, but not totally ſatisfactory ; becauſe it extends only to the Reaſon of the Males remote Aſſiſtance of the Female, in the maturation of her Fruit ; leaving us ſtill to enquire, Why ſhe herſelf remains in a ſteril and pining condition, unleſs ſhe enjoys the Society and invigorating irradiation of the Male ; and why ſhe inclines her amorous boughs toward his, as if meer Neighbourhood were a kind of Divorce, and nothing leſs than abſolute Union could ſatisfie her Affection. And what we have here ſaid, of the Sympathy betwixt the Male and Female Palms, will not loſe a grain of its Veriſimility, when our Reader ſhall pleaſe to accommodate it to the Explanation of the Cauſe of the like Amity betwixt the *Fig tree,* and *Capriſicus* or *Wild Fig tree:* of which *Pliny (lib. 15. cap. 19.)* relates the very ſame ſtory, as *Herodotus* doth of the Palms.

(4) This

(4) This puts us in mind of the great Sympathy betwixt *Vine* and *Wine*, expressed from its Grapes, and immured in Hoggheads, though at the distance of many miles. For, it seems most convenient, that it is from the like *Diffusion of subtile Emanations*, imbued with the *Seminal tincture* of the Vine, that Wines stored up in deep Cellars, in the same Country where they grew (for, in *England*, whither all wines are transported over sea, no such Effect hath been observed: the Remove being too large to admit any such Transmission of influence from the transmarine Vineyards to our Cellars) become sick, turbid, and musty in the Cask, at the same time the Vines Flower and Bud forth: and again recover their former Clearness and Spirit, so soon as that season is past. And, that this Conjecture may seem to smell the less of Phansy, we desire you to consider, through what large tracts of Aer even the *Odours* (Exhalations much less Subtile and Diffusive, than those we conceive emitted from Vines to Wines) of many Aromaticks are usually diffused, in serene weather ; especially in respect of such Persons, and Bruit Animals, as are exquisite in their sense of smelling. Hath it not been observed, that the Flowers of Oranges have transmitted their odours perfect and strong, from great Gardens, to the nostrils of Mariners, many leagues off at Sea: nay, so far, that some Sailers have discovered land by the smell of them, when their longest Perspectives could not reach it ? Doe not we frequently observe, that Ravens will scent a Carcass, at many miles distance; and fly directly to it by the Chart of a favourable wind ? Nay, are not there good Historians that assure us, that Eagles in *Italy*, have sometime received an invitation by the nose, to come and feast on the dead bodies of men, in *Africa* ?

Here, since we are occasionally fallen upon the large Diffusion of some Odours, especially to sage and unpræpossessed Noses ; we shall take the advantage of that Hint, to advertise you of a *Vulgar Error*, viz. *that Waters distilled of Orange Flowers and Roses, become wholly Inodorous, and Phlegmatick, at the time of the Blooming and Pride of those Flowers upon their trees*. For, really those distilled Waters are not in themselves, during the season of the Flowers, from which they were extracted, less fragrant than at other times : but, because in the season of those Flowers, they diffuse their odours so plentifully through the Aer, and præpossess the nostrils, as that the odours of the Waters, being somewhat less quick and strong, are less perceived, than at other times, when the Aer is not imbued with the stronger and fresher odours, nor the olfactory Nerves præoccupied. And this may be inferred from hence ; that when the season of those Flowers is past, and the smelling organ unoccupied ; the Waters smell as fragrant as ever. For, as to the *Assuefaction* of the sense of smelling, to particular odours, good or bad, we need not say much of that : since Experience doth daily confirme, that the sense is scarce moved and affected by the same odour, though closely præsented, after Custom hath once strongly imbued it with the same.

## SECT. III.

**Art. 1.**
Why this
*Section* confi-
ders only some
few select Oc-
cult Proprie-
ties, among
those many
imputed to
*Animals.*

IN the THIRD and last Division of *Special* Occult Qualities, or such as are vulgarly imputed to *Sensible* Creatures; the Pens of Schollars have been so profuse, that should we but recount, and with all possible succinctness, enquire into the Verity and Causes of but the one Half of them; our Discourses would take up more sheets of Paper, than are allowed to the Longest Chancery Bill: wherefore, as in the former, so in this, we shall select and examine only a Few of them, but such as are most in vogue, and whose Reasons, if judiciously accommodated, suffice to the Solution of the Rest.

**Art. 2.**
The supposed
Antipathy of a
*Sheep* to a
*Woolf,* solved.

(1) The *Antipathy of a Sheep to a Woolf,* is the common argument of wonder; and nothing is more frequent, than to hear men ascribe it to a provident Instinct, or hæreditary and invincible Hatred, that a Lamb, which never saw a Woolf before, and so could not retain the impression of any harme done or attempted by him, should be invaded with horror and trembling, at first interview, and run from him: nay, some have magnified the secret so far, as to affirme the Antipathy to be Equall on both sides. Concerning this, therefore, we observe; that the Enmity is not Reciprocal: For, He that can be persuaded, that the Woolf hates the Sheep, only because he worries and preys upon him, and not rather, that the Woolf loves the sheep, because it is a weak and helpless Animal, and its flesh is both pleasant and convenient food for him: we shall not despair to persuade Him, that Himself also hates a sheep, because he finds his pallate and stomach delighted and relieved with Mutton. Nor is the Enmity on the sheeps side Invincible; for, ourselves have seen a Lamb brought, by Custom, to so great familiarity with a Woolf, that He would play with him, and bleat, as after the Dam, when the Woolf hath removed out of the room: and the like Kindness have we very lately observed betwixt a Lamb and Lyon of the Lord Generall *Cromwells,* kept at *Sion* house, and afterward publikely shewed in *London.* Again, the Fear, which surpriseth the Lamb at first sight of a Woolf, seems not to arise from any Hereditary Impression derived from the Dam, or Sire, or Both; as well because all Inbredd or traduced Antipathies are invincible, as that none of the Progenitors of the Lamb, for many Ages, ever saw or received any impression of injury from a Woolf, here with us in *England.* Besides, in case they had, and though it be indisputable, that some Beasts are afraid of men, and other Beasts, meerly from the memory of some Harme received from some man, or Beast of the same species; the Idea of him, that did the Harme, remaining impressed upon the table of the Memory, and being freshly brought again to the Phansy, whenever the sense brings in the like species: yet is it not likely, that the same Idea should be propagated by Generation to the issue, after so many hundred removes; and traduced from one Individual to the whole species, throughout the world.

The

The Cause, therefore, why All Sheep generally are ſtartled and offended at ſight of a Woolf, ſeems to be only this ; that when the Woolf converts his eyes upon a ſheep, as a pleaſing and inviting object, and that whereupon Appetite hath wholly engaged his Imagination ; he inſtantly darts forth from his brain certain ſtreams of ſubtle Effluvia's, which being part of thoſe Spirits, whereof his newly formed Idea of dilaniating and devouring the ſheep, is compoſed, ſerve as Forerunners or Meſſengers of deſtruction to the ſheep ; and being tranſmitted to his Common Senſory, through his optick nerves, moſt highly miſaffect the ſame, and ſo cauſe the ſheep to fear, and endeavour the præſervation of his life, by flight.

This receives ſufficient Confirmation from hence ; that not only ſuch Averſions, as ariſe from the Contrariety of Conſtitutions in ſeveral Animals, are commonly obſerved to produce thoſe Effects of Fear, Trembling and flight from the objects, from which offenſive impreſſions are derived, by the mediation of diſagreeing Spirits or Emanations : but even the ſeeing them in a paſſion of Anger, or Fury, doth ſuddainly cauſe the like. For, violent Paſſions ever alter the Spirits, and Characterize them with the idea at that time moſt prævalent in the Imagination of the Paſſionate ; ſo that thoſe ſpirits iſſuing from the body of the Animal, in that height of Paſſion, and inſinuating themſelves into the brain of the other Animal contrarily diſpoſed, muſt of neceſſity highly diſguſt and offend it. Which is the moſt likely Reaſon that hath hitherto been given, Why *Bees* ſeldom ſting men of a mild and peaceful diſpoſition : but will by no means endure, nor be reconciled to others of a froward, cholerick, and waſpiſh nature. The ſame alſo may ſerve to anſwer that common Quære, Why ſome *Bold and Confident* perſons, having tuned their ſpirits to the higheſt key of Anger and Indignation, have *daunted* not only fierce *Maſtiffs*, but even *Lyons*, *Panthers*, and other Wild and ravenous Beaſts, meerly by their threatning looks, and put them to flight by the Artillery of their ſcornful Eyes. And this Key, wherewith we have unlockt the ſecret betwixt the Lamb and Woolf, will alſo open thoſe like Antipathies ſuppoſed to be betwixt the *Dove* and *Falcon*, the *Chicken* and *Kite*, and all other weak Animals, and ſuch as uſe to make them their prey.

*Art. 3.* Why Bees uſually invade froward and cholerick Perſons : and why bold and confident men have ſometimes daunted and put to flight, Lyons and other ravenous Wild Beaſts.

( 2 ) It is worthy a ſerious Remark, that *ſundry Animalls bear a kind of implacable Hatred to the Perſons of ſuch men, as are delighted or converſant in the Deſtruction of thoſe of the ſame ſpecies with them :* as we daily ſee, that ſwine are highly offended and angry at Butchers : that Dogs bark at and purſue Glovers, that deal moſt in Dog skins, and Beadles that are imployed in killing of Dogs, in time of the plague, to prævent the diffuſion of Contagion, and encreaſe of Putrefaction, by their means ; that Vermin will avoid the trapps and gins of Warrenners, wherein any of their owne kind hath been taken and deſtroyed, &c. As for theſe Antipathies, or ſtrong Averſions, tis manifeſt, that they ariſe not from any Specifical Inſtinct, or Character of Providence impreſſed upon their reſpective Natures, or Eſſential Forms, but only from the Activity of the præſent object upon the ſenſe. For the Blood commonly adhæring to the cloths of the Butcher, and Dogg-killer, and likewiſe to the traps and gins, wherein Vergin have been caught and deſtroyed, doth emit ſuch odours, as invading the Senſory of

*Art. 4.* Why divers Animals Hate ſuch men, as are uſed to deſtroy thoſe of their owe ſpecies : and why Vermin avoid ſuch Gins and Traps, wherein others of their kind have been caught and deſtroyed.

any

any Animal of the same species, excite a kind of Horror in the like Animal that smells them; and so cause it to abhor and avoid all such persons and places, for fear of the like harm and internecion, as their fellowes have suffered from them. Now, that which makes these odours insinuate themselves with such ease and familiarity into the Sensories of animals of the same species, is the similitude and Uniformity of their Specifical Constitutions, which yet the rough hand of Corruption seems not totally to have obliterated in the long since extravenated blood and spirits, but to have left some Vestigia or Remains of the Canine nature in the Doggs blood, of the Porcine in the Swines, &c. And, that which makes them so horridly *Odious*, is the great Alienation of the blood from its genuine temper and conditions. For, the smell of the Carcass, or blood of any Animal, having once suffered the Depravation of Corruption; is always most hateful and dangerous to others of the same Species: and it hath been observed, that the most pernicious Infections and Plagues have been such, as took their Original from the Corruption of Humane Bodies; which indeed, is the best reason that hath been yet given, why the Plague so often attends long and bloody Sieges, and is commonly the second to the Sword. We conceive, the same to be also the ground of that Axiom of the *Lord St. Alban* (*Nat. Hist. cent.* 10.) *Generally, that which is Dead, or Corrupted, or Excerned, hath Antipathy with the same thing, when it is Alive, and when it is found; and with those parts which do excern: as a Carcass of Man is most infectious and odious to man, a Carrion of an Horse to an Horse, &c. Parulent matter of Wounds and Ulcers, Carbuncles, Pocks, Scabbs, Leprousy, to Sound flesh. And the Excrements of every species to that Creature, that excerneth them. But the Excrements are less Pernicious, than the Corruptions.*

*Art.* 5.
The Cause of the fresh Cruentation of the Carcass of a murthered man, at the præsence and touch of the Homicide.

(3) The *Cruentation* (and, according to some reports, *the opening of the Eyes*) *of the Carcass of a murthered man, at the præsence and touch of the Homicide*; is, in truth, the noblest of Antipathies: and scarce any Writer of the Secrets or Miracles of Nature, hath omitted the Consideration thereof. This Life in Death, Revenge of the Grave, or loud language of silent Corruption, many Venerable and Christian Philosophers have accounted wholly *Miraculous* or *Supernatural*; as ordained and effected by the just judgement of God, for the detection and punishment of the inhumane Assassine. And, lest we should seem too forward, to expunge, from the mind of any man, the belief of that opinion, which to some may be a more powerful Argument, than the express Command of God, to deterr them from committing so horrid and execrable a Crime as Murder: we shall so far concurr with them, as to conceive this Effect to be *Divine* only in the *Institution*, but meerly *Natural* in the *Production*, or *Immediate* Causes. Because the Apparence seems not to transcend the Capacity of Natural Means, and the whole Syndrome and Series of it Causes may be thus explained. It is an Opinion highly Consentaneous, that in every vehement Passion there is formed a certain Idea as well of the Object, whereupon the Imagination is most intent, as of the Good or Evil connected unto, and expected from that Object; and that this Idea is as it were impressed, by a kind of inexplicable Sigillation; upon the Spirits, at the same instant the Mind determineth to Will the præsent Prosecution, or Avoidance of the object: So that, by the mediation of the Spirits (those Angels of
the

the Mind) the same Idea is transmitted to the Blood, and through the Arteries diffused into all parts of the body, as well as into the Nerves and Muscles, which are inservient to such Voluntary Motions, as are requisite to the execution of the Decrees and Mandats of the Will, concerning the Prosecution, or Avoidance of the Object. This being so, we may conceive, that the Phansy of the Person assaulted by an Assassine, having formed an Idea of Hatred, Opposition, and Revenge, and the same being Characterized upon the Spirits, and by them diffused through the blood; though the blood become much less Fluid in the veins after death, by reason the vital influence and Pulsifick Faculty of the Heart, which Animated and Circulated it, is extinct: yet, because at the præsence of the Murderer, there issue from the pores of his body such subtile Emanations, as are Consimilar to those, which were emitted from him, at the time He strove with, overcame, and killed the Patient; and those Emanations entering the Dead Body, doe cause a fresh Commotion in the blood remaining yet somewhat Fluid in its veins, and as it were renew the former Colluctation or Duell betwixt the yet wholly uncondensed Spirits of the slain, and those of the Homicide: therefore is it, that the Blood, suffering an Estuation, flows up and down in the veins, to seek some vent, or salley-port; and finding none so open as in that part, wherein the wound was made, it issues forth from thence. And, where the Murthered Person is destroyed by strangulation, suffocation, or the like unbloody Death, so that there is no manifest Solution of Continuity in the skin, or other Exterior parts of the body; in that case, it hath been observed, that the Carcass bleeds at the Mouth, or Nose, or both; and this only because in all vehement strivings, and especially in Colluctation for life, the Spirits and Blood flow most plentifully into the Arteries and Veins of the Head, as is visible by the great Redness of the Eyes and face of every man that Fights; and where the blood fixeth in most plenty, there will be the greatest tumult, æstuation and commotion, when it is fermented, agitated, and again set afloat, by the Discordant Effluvia's emitted from the body of the neer approaching or touching Murtherer and consequently, there must the vessels suffer the greatest stress, distension, and disruption, or apertion of their orifices.

( 4 )  And this magnale of the (as it were) Reanimation of the vindictive blood in the veins of a Dead body, by the Magick of those Hostile and Fermenting Aporrhæa's, transmitted from the body of Him, who violently extinguished its former life, ushers in Another, no less prodigious, not less celebrated by Naturalists: and that is the suddain *Disanimation of the Blood in Living Bodies, by the meer præsence of the Basilisk, Catablepa, and Diginus*; Serpents of a Nature so transcendently Venemous, that, according to popular Tradition, and the several relations of *Dioscorides*, *Galen*, *Pliny*, *Solinus*, *Ælian*, *Avicen*, and most other Authors, who have treated of the Proprieties of Animals and Venoms, they are Destructive beyond themselves, i.e. *they either kill by intuition, or Hiss out the flames of life by their Deleterious Expirations*. If Natural Historians have herein escaped that itch of Fiction, to which they are so generally subject, when they come to handle Rarities; and that Nature hath produced any such Species, whose optical Emissions, or Pectoral Expirations are fatal and pernicious

*Art. 6.*
How the *Basilisk* doth empoyson and destroy, at distance.

nicious to all, or moſt other Living Creatures; neither of which ſeems to be above Controverſie: the *Cauſe* of this ſtupendious Effect muſt conſiſt only in this, that thoſe Rayes which are emitted from the Eyes, or that Halitus expired from the Lungs (for, their Hiſſing is far more loud and vehement than that of any others) of theſe Serpents, are Deleterious in the ſuperlative degree, *i. e.* of ſuch *Subtlety* and *Vehemence*, that they no ſooner invade an Animal, but they as it were in a moment alter and ſubvert the requiſite temper of that ſpiritual ſubſtance, wherein its life doth proximly and principally depend, and ſo render it thenceforth wholly unfit to performe the Actions of Life. But, as for thoſe other Traditions (1) of the Baſiliſks deſtroying a man by prior Aſpect alone (2) of its Identity with the Cockatrice, which hath no real exiſtence in Nature, and is only an Hieroglyphical Fiction, or Symbolical Invention of the old Ægyptians (3) of its Production from the Egg of an old decrepite Cock; and (4) of its being an Animal with wings, legs, a long and ſpiral Taile, and a Criſt or Comb on the head, like that of a Cock, as it is vulgarly deſcribed and painted, and repræſented in thoſe artificial contrivances made of the skin of a Thornback, by Impoſtors: we may juſtly refer them partly to abſolute Impoſſibilities, partly to vain and ridiculous Follies; as the induſtrious *Aldrovand*, and ingenious Doctor *Brown* have done before us.

(5) The Rarity of the Baſiliſk, coming not much behind that of the Phenix (for, we have not heard of more than four or five, in the ſpace 2000 years) may, we confeſs, ſomewhat excuſe the Credulity of thoſe, who have ſo eaſily ſwallowed the Figment of it poyſoning a man by Priority of Aſpect alone; becauſe to the Refutation of it by Experiment, it is requiſite that the Opponent live at the ſame time, and in the ſame Country, with that King of Venoms. But, we doe not ſee, what extenuating plea can remain to thoſe ſoft and flexible minds, that ſo readily aſſent to that common Tradition, that *the ſight of a Woolf affects the Spectator with abſolute Dumbneſs, or very great Hoarſneſs, at leaſt:* when there are few Countries, but have Woolves enough to give any Enquirer the opportunity of Experiment; and Few of thoſe, who have encountred Woolves very often, and that in woods and deſerts, have been heard to complain of any Symptome or Miſ-affection thereupon. Which is evidence ſufficient, that either the Antipathy of man to a Woolf was the Dream of ſome vain and Romantique Phanſy; or, that men have deluded themſelves, by the heedleſs Conſignation of the Effect to a remote and unconcerned Cauſe, blindly aſcribing that to ſome ſpeciſical Hoſtility betwixt the inſenſible Emanations tranſmitted from the Eyes of the Woolf, and the temperament of the Tongue and other organs of ſpeech in man, which, in truth, belongs only to the Paſſion of Fear, wherewith any puſillanimous or cowardly Perſon may be ſtrongly ſurprized, at the ſuddain and unexpected ſight of a Woolf. For, manifeſt it is (1) that whoever fears not a Woolf, ſhall never find any ſuch Palſy in his tongue, or Aſperity in his throat and vocal Artery, at the ſight of him: as the daily Experience of ſuch, in *Ireland* and other Countreys, frequently infeſted with Woolves, as delight in Hunting them, doth demonſtrate. And (2) that whoever Fears, ſhall find in himſelf the ſame ſymptome of obmuteſcence, or difficulty of Vociferation,

ration, whether he fees the Woolf firft, or the Woolf him; fuddain filence being ever the Affociate,(or (rather) Confequent of great and fuddain Fear. The Aphonia, therefore, or Defect of voice, which hath fometimes, though very rarely, been obferved to invade men, upon the Confp tion of Woolves; is not the genuine Effect of any fecret and radicated Antipathy, or Fafcinating Virtue in the fubtle Aporrhæa's emitted from the eyes, lungs, or bodie of the Woolf: but only of their own *Fear* and *Terror*, arifing from a ftrong apprehenfion of Danger; the fuddain and impetuous Concentration of the Spirits, toward the Heart, by reafon of the violent Terror, at that time, caufing a Defection of fpirits, and confequently a kind of Relaxation in the Mufcles of the Tongue, and Nerves infervient to the vocal inftruments: So that the infpired Aer cannot be Efflated with that force and celerity, as is neceffary to the loudnefs and diftinct articulation of the voice.

(6) Nor is it the Eye alone, that the Folly of men hath made obnoxious to Antipathies, but the Ear alfo hath it fhare of wonderful Effects; for, there go folemn ftories of inveterate and fpecifical Enmities betwixt the *Lyon and Cock, Elephant and Swine*, and He hath read little, who hath not more than once met with fundry relations, that *the Crowing of the Cock is more terrible than death, to the fierceft Lyon, and the Grunting of a Swine fo odious to an Elephant, that it puts him into an Agony of Horror, Trembling, and Cold fweat.* Which notwithftanding, may well be called to the barre of Experiment, and many worthy Authors have more then queftioned, among whom, *Camerarius* (*in Symbol.*) exprefly affures us, that in his time, one of the Duke of *Bavaria*'s Lyons, breaking into a yard adjacent to his Den, and there finding a flock of Poultry, was fo far from being afraid of the Cock, or his Crowing, that he devoured him and his troop of Hens together. And as for the *other* Antipathy; ourfelves have feen an Elephant feed and fleep quietly in the fame ftable, with a Sow and her whole litter of Piggs. However, left fome fhould plead the power of *Cuftom*, in both thefe cafes, and object, that that Lyon and Elephant had been, by *Affuefaction*, brought to endure the naturally hateful Noifes of the Cocks Crowing, and the Swines Grunting; to eradicate the belief of the fuppofed Occult Antipathies, we fay: that fuch may be the Difcrepancy or Difproportion betwixt the Figures and Contextures of thofe fubtile particles, that compofe thofe Harfh Sounds, and the Contexture of the organs of Hearing in the Lyon and Elephant, as that they exafperate them, and fo highly offend thofe Animals. For, thus we fuffer a kind of fhort Horror, and our Teeth are fet on edge, by thofe harfh and vehement founds, made by fcraping of trenchers, filing the teeth of faws, fqueaking of doors, and the like: only becaufe thofe founds grate and exafperate the Auditory Nerves, which communicate the harfh impreffion to the Nerves of the Teeth, and caufe a ftridor therein.

(7) But if we pafs from thefe Imaginary, to *Real* Antipathies, and defire not to mifimploy our Underftanding, in the queft of *Dihoties* for fuch things, of whofe *Hori* the more fober and judicious part of Schollars juftly doubt; let us come to the wonderful Venome of the TARANTULA, a certain Phalangium or fmal Spider frequent in *Italy*, but moft in and about *Tarentum* in *Apulia*; which hath this ftrange Propriety, that being

*Art.8.*
The *Antipathies* of a *Lyon* and *Cock* of an *Elephant* and *Swine* meerly Fabulous.

*Art.9.*
Why a man intoxicated by the venome of a *Tarantula*, falleth into violent fits of *Dancing*; and cannot be cured by any other means, but *Mufick*.

ing communicated to the bodie of man, by biting, it makes him Dance most violently, at the same time, every year, till He be perfectly cured thereby, being invincible by any other Antidote but Musick. An Effect so truly admirable, and singular, that the Discovery of its abstruse Causes, and the manner of their operation, cannot but be most opportune and grateful to the Curious; who, we presume, would gladly knowe,

*Why such as are empoysoned by the biting of a Tarantula, fall into violent Fits of Dancing, and cannot be Cured by any other Remedies, but the Harmonious Straines of Musick alone?*

## SOLUTION.

How great the power of Musick is, as to the excitement, exaltation, and compescence or mitigation of the Passions of the Mind of Man; and wherein the Cause of that Harmonical Magick doth consist: would be a Digression, and perhaps somewhat superfluous for us here to enquire. And, therefore, cutting off all Collateral Curiosities, we shall confine our present scrutiny to the limits of our owne Profession; endeavouring only to explain the Reasons, why Musick hath so strong and generous an Energy, as certainly to cure the Bodie of a man, intoxicated with the Venome of the Tarantula, which eludes and despises the opposition of all other Alexipharmacal Medicaments. Forasmuch, therefore, as the strings of a Lute, Vial, or other Musical Instrument, do alwayes move and impell the Aer, after the same manner as themselves are moved and impelled, and by this proportionate misture of Sounds create an Harmony delightful not only to the Eare, but to that Harmonious Essence, the soul, which Animates the Eare; hence comes it, that by the musical Harmony, that is made by the Musicians playing to the person infected with the Tarantisme, the Aer, by reason of the various and yet proportionate motions of the strings, is harmonically moved and agitated, and carying those various motions of the harmony impressed upon it self, into the Eare, and so affecting the Phantastical Faculty with those pleasant motions, doth in like manner affect and move the spirits in the brain: and the spirits having received those impressions, and diffused into the Nerves, Muscles and Fibres of the whole body, and there meeting with a certain thin, acrimonious and pricking Humor, which is the chief fewel and vehicle of the Venome derived from the Tarantula; they attenuate and agitate the same, by a way very like that of Fermentation, and disperse it with a quick motion through all the parts. And this Humor being thus set afloat, and estuated, together with the venome, or seeds of the Poyson, which are contained therein, must needs affect all the Musculousand Nervous parts, upon which it toucheth, with a kind of Itch, or gentle and therefore pleasant vellication, or (rather) Titillation: So that the Patient feeling this universal Itch, or Tickling, can be no longer at ease and quiet, but is compelled thereby to dance and move all the members of his body with all agility and violence possible. This Dancing causeth a Commotion of all the Humors in his body; that Commotion augments the present Heat thereof; that Heat causeth a Relaxation and Apertion of the pores of the skin; and thereupon ensues a liberal and universal sweat;

and

and together with that sweat, the venome is dispersed and expelled. But, where the Venome is so deeply settled, and as it were radicated in the solid substance of the parts, as that one or two, or three Fits of Dancing and Sweating are not sufficient to the total Eradication and Expulsion thereof; in that deplorable case, the Patient becomes freshly intoxicated, and relapseth into his dancing paroxisms, at the same periodical season, every year, without omission, till his many and profuse Annual sweats have freed him from all Reliques of the Poyson.

Most true it is, that Divers Tarantiacal persons are affected with divers Musical Instruments, and divers Tunes and Ayrs; but this is to be imputed to the Diversity of Complexions and Temperaments either of the Tarantula's, which envenome them, or of the Persons themselves. For, such as are Melancholy of themselves, or intoxicated by the poyson of the duller and more sluggish sort of Tarantula's, are ever Affected and Sympathize rather with the musick of Drums, Trumpets, Sackbuts, and other loud and strong sounding instruments, than with that of Lutes, Vials, Violins, and other soft and gentle ones. For, since Melancholy is a thick, heavy and viscid Humor, and the Spirits alwaies follow the Disposition of the Humor praedominant; to the Concitation and Dissipation thereof, a greater force of motion is required. And this, doubtless, was the Reason, why a certain Girl of *Tarentum*, being there bitten by a Tarantula, and affected with the stupendious symptome of Tarantism, could never be excited to dance by any sounds, but those of Guns, Alarms beaten upon Drums, Charges and Triumphs sounded in Trumpets, and other military musick; the heavy and viscid venome, meeting with a body of a Cold and Phlegmatick Complexion; and so requiring very strong Commotions of the Aer and Spirits, to its Estuation and Dissipation. And, on the Contrary, Cholerick and Sanguine Complexions, are, by reason of the Subtility of their Spirits, and greater Fluidity of their Humors, soonest Cured by the Harmony of Lutes, Harps, Vials, Virginals, Guitarrs, Tiorba's, and other stringed Instruments.

*Art. 10.* Why Divers Tarantiacal Persons are affected and cured with Divers Tunes, and the musick of divers Instruments.

But, that which deserves our highest Admiration, is this; that *this Venome of the Tarantula doth produce the same Effect in the body of man, which it doth in that of the Tarantula it self, wherein it is generated:* as if there were some secret Cognation and Similitude betwixt the Nature of that venemous Spider, and that of Mankinde. For, as the Poyson, being infused into any part of mans body, and set a work by Musick, doth, by a continual vellication or Titillation of the Muscles and Membranes thereof, incite the intoxicated person to dance: So likewise, while it remains in its own womb and proper Conservatory, the body of the Tarantula being once set a work by Musick, doth it incite the Tarantula to dance, and caper, as is commonly observed by the Italians, and at large related by *Athan. Kirsherus (in opere Magnetico)* and some others of unquestionable veracity, who would admit no testimony in this particular, but what they received from their own exact observations. Among the sundry Narrations of Experiments in this kind, *Kircher* entertains his Reader chiefly with this one, as the most exact and commemorable. 'A certain Italian Duchess (sayes He) to the end she 'might be fully satisfied of the truth of this prodigy of nature, of which 'she had so often heard, and as often doubted, commanded that a Tarantula

*Art. 11.* That the venome of the Tarantula doth produce the same effect in the body of a man; as it doth in that of the Tarantula it self; and why.

Bbb                                    should

'should be brought into the Hall, or Refectory of a Colledge of Jesuits,
'all the Fathers being præsent; and there set upon a small chipp of wood,
'that floated in a dish of water. Then she gave order, that an Excellent
'Harper should stand by, and play over several of his best composed
'Tunes. The Tarantula, for a good while, seemed wholly unconcerned
'in the musick, discovering no motions of tripudiation in himself; but
'at length, when the Harper had hit upon some certain Notes Strains,
'and Ayres, such as held some proportion to the Humor and Specifical
'Venome of the Spider, the now enchanted Insect began to detect its sym-
'pathy to Musick; and natural inclination to dancing, not only by the
'frequent lifting up his feet, and nimble agitation of his whole body, but
'even most exactly observing time and measures, according to the Harmo-
'nical Numbers exprest in the Tune: and as the Musician plaid more slow-
'ly or swiftly, so did the little beast dance more slowly or nimbly; not
'moving a foot, after the Tune was ended.     But, this which then appeared so rare to the Dutchess and other Spectators, they soon after heard
to be very common to the Musicians of Tarentum, who being hired, with
an annual pension paid out of the Publique purse, to cure the meaner sort of the people, when any is bitten by a Tarantula; that they may
not miss of healing the Patient, and put themselves to the pains of playing long: they first enquire of the Patient, in what house, what field,
or place he was bitten, of what colour and bigness the Tarantula was,
that bit him. Being satisfied of these particulars, they forthwith go to
the place described, and there looking among the several species of Ta-
rantulas, as they are busie in weaving their Cobweb nets, for the en-
snaring of Flyes; they search for such a one as the Patient hath described,
and having once found the like, they instantly fall to their instruments;
and play over whole sets of Lessons one after another, till they light upon such a one, as holding some proportion to the Specifical temperament and venemous Humor of that Tarantula, inciteth him to dance.
And both exceeding delightful and strange it is to behold the great
variety of Humors among many Tarantula's together; one while this
sort, another while that exactly sympathizing with the Harmonious motions of the strings and aer. When the Musicians have thus informed
themselves of the particular Genius and Humor of that species of Tarantula's, by one of which the Patient was envenomed; they return home,
and set him a dancing almost at first touch of their instruments, playing over again and again those Tunes, whose Correspondency to the
poyson, that lieth ambuscado'd in the centrals of his bodie, they
had formerly experimented: and they seldom or never fail of the
Cure, where they are certain what Notes and Tunes are most accommodate to the Genius of the Spider, that hath intoxicated the Patient.

Nor is it at all inconsistent with Reason, that the Tarantula it self
should suffer the same strange Effect from the Charms of Musick, as
the man doth whom its Venome hath intoxicated: for seeing that
the Humor, which supplies the office of Blood in this Insect, is exceeding viscous, and impraegnate with subtle and hot spirits, and so becomes a
subject very convenient to receive the Motions impressed upon it, by the
most subtle parts of the Aer; whereof the Sounds are composed: it seems
almost necessary, that being astuated and set afloat, by the motions of the
blood                                                                              aer,

aer, which are Harmonical; it should cause the like Vibrissations in the nervous parts of the Tarantula, as the hand of the Musician hath caused in the Consonous strings of the instrument; the strings caused in the Aer, and the Aer caused in the spirits of the Animal: and consequently, that the Animal should suffer a kind of Itch, or gentle vellication in all its nerves, and muscles, and to ease it self of that troublesom Affection, move all its members, not only with great agility, but variety of motions correspondent to those of the Harmony impressed upon its spiritual substance; especially where the Harmony is proportionate to the specifical (and perhaps, individual) Constitution of the same.

That the vital Humor of these and most other Spiders, is both *viscous*, and a *subject capable of Sounds*, as we here assume, may be inferred from the relation of *Peter Martyr* (*in Histor. sua India Occidental*) that in the *West Indies* there is a certain species of Phalangiums, or Venenate Spiders, whose poyson, being expressed, is so exceedingly viscid and tenacious, that the Natives use to draw and spin it out into long threads, and twist those threads into Treble strings for their instruments of Musick: as also from our own ocular testimony, whenever we press a Spider to death.

*Art. 12.*<br>That the Venom of the Tarantula is lodged in a *viscous Humor*, and such as is capable of *Sounds*.

And (what is of greatest moment to our præsent Disquisition) that the *Venome of the Tarantula, by reason of the Acrimony, or Mordacity of its Spiritual and hot particles, causeth an uncessent Titillation, or Itching joyned with great heat, in the nervous and musculous parts of mans body,* when it is in æstuation and commotion therein, may be collected from the agreeing relations of all persons, who have known the misery of Tarantisme: every one complaining of an insufferable Itch in all parts of his body, during the paroxisme, and finding a remission of the same immediately after profuse sweating. For your farther Confirmation herein, be pleased to hear Father *Kircher* tell you a memorable and pertinent story. 'A certain Cappucine (saith He) of the Monastery belonging to that Order, in *Tarentum*, 'being bitten by a Tarantula, and by his (in that point, too severe) Superi- 'ors forbidden to have recourse either to Baths, or Dancing, for the cure 'of his infection, as means that might seem too light and inconsistent with 'the gravity and rigid rules of his Profession; was so miserably and beyond 'all patience tormented with an itching and burning in both the interior and 'and exterior parts of his body, that rest and quiet were things he had long 'since been a stranger to; and hoping to find some ease and allay of his 'restless pains by bathing in cold water, he, one night, privily conveyed 'himself out of the Covent, and leaped into an Arm of the Sea, that em- 'braced the town. Where, indeed, he met with a perfect cure of all his 'torments and grievances; being instantly drowned: leaving his Brethren 'to lament their own great loss, as well as the Sadness of his Fate; and his 'Superiors to repent the cruelty of that Superstition, which had denied him 'the use of those innocent Remedies, Musick and Dancing, which the 'happy experience of many thousands had præscribed.

*Art. 13.*<br>That it causeth an uncessent Itching, and Titillation in the Nervous and Musculous parts of mans body, when infused into it, and fermenting in it.

Lastly, as it is not every Harmonical Ayre that suits with the Genius of every Tarantula, but every particular species holds a secret Correspondence to some particular sorts of Instruments, Tunes, and Strains composed of such and such Notes: So likewise is it not the Musick of every instrument, nor every modulation of sounds that move and

excite

excite every person infected with this kind of poyson; but every Tarantiacal Patient requires such and such particular Harmonious Tunes,
Strains, and Notes as are proportionate to that Diathesis, or Disposition, which results from the Commixture and Confermentation of his
owne Humors, and the Venome infused into his body. Which is the Reason, why some dance to no musick but that of Drums, Trumpets and o-
ther loud and martial instruments; and others again are easily charmed to
Levolta's by the mild and gentle Consonances of Lutes and Tiorba's. And
if the Patient, being of a *hot and bilious* Complexion, be intoxicated by
the venome of a Tarantula of the like Cholerick temperament; upon the
æstuation and confermentation of those two consimilar Humors, the Patient shall become *Feverish,* insatiately *thirsty, restless,* and *furiously maniacal*:
but, where a *Melancholy* Tarantula hath empoysoned a man of the like
dull and sluggish Constitution; in that case, He shall be infested with
great and inexpugnable *Drowsiness, Stupidity, Spontaneous Lassitude,* love
of *Solitude,* unseasonable and affected *Silence,* and the like Symptoms
contrary to the former, and shall be relieved only by grave and solemne
tunes; the Accidents supervening upon this kind of intoxication, always
following and betraying the capacity of the prædominant Humor, and re-
sponding to that Harmony, which hath the most of proportion to the Genius of the Poyson.

*Art.* 14.
The cause of
the Annual Re-
cidivation of
the Tarantism,
till it be per-
fectly cured.

And as for the *Annual Relapses* of Patients, into their Tarantiacal Fits;
the *Cause* thereof must be only this, that the Reliques of the Poyson causing a fresh Commotion and Fermentation of the most susceptible Humors
of the body, and especially of the Serous and Bilious part of the blood(for,
most persons thus affected; have their Paroxysms in the hottest season of
the year) and imbuing them with exceeding great Acrimony and Mordacity; diffuse themselves through the Arteries and Veins into all parts of
the body; and fixing more especially on the thin membranes, that invest
the muscles, so oppress, prick and vellicate them, as that the infected shall
know no rest nor ease, till he hath danced and sweat; to the dissipation and expulsion of all those sharp and pungent particles, that were diffused into the
Habit of his body.

*Art.* 15.
A Conjecture,
what kind of
Tunes, Strains
and Notes seem
most accom-
modate to the
Cure of Taran-
tiacal Persons
in the General.

But, what *particular Sounds, and Notes, and Strains, and Ayres,* are
Accommodate to the Venome of this or that particular Tarantula; we
leave to the determination of the long experienced Musicians of Tarentum
only thus much we may say, in the General; that by how much the
more frequent Diminutions of Notes into halfs and quarters (which is called Division) and the more frequent permission of Sharps and Flats, in a
Tone charged with frequent Semitones, the Tune containeth; by so much
the more grateful will the same be to all Tarantulized Persons; because,
from the Celerity of the motions it comes, that the Dormant Venome is
more nimbly agitated, and so must sollicite them to dance the more spritely
and vehemently. Hence is it, that the Musicians of *Italy,* such especially
who profess the certain and speedy Cure of the Tarantisme, for the most
part, enrich and adorne their strains with various Divisions of Notes; and
that mostly in the *Phrygian* Tone, because it consisteth of frequent Semitones.

(8) What

( 8 ) What we have here faid, concerning the Magick of Harmonious Sounds both upon the Tarantula it felf, and thofe unhappy men, whom its Fafcinating venome hath Tarantulized; as it doth wholly take off the Incredibility of thofe Relations, which fome Natural Magicians have fet down, of the *Incantation of Serpents, by a wand of the Cornus, or Dog tree*: So doth it alfo give us no obfcure light into the dark Caufe of that Effect, which among the Ignorant and Superftitious hath ever paffed for meerly præftigious and Diabolical. For, it being certain, that all Serpents are moft highly offended at the fmell, and influx of thofe invifible Emanations proceeding from the Cornus, by reafon of fome great Difproportion or Incompoffibility, betwixt thofe fubtile Effluvia's, and the temperament of the Vital and Spiritual Subftance of Serpents: infomuch that, in a moment, they become ftrongly intoxicated thereby: Why fhould it feem impoffible, that He, who underftands this invincible Enmity, and how to manage a wand or rod of the Cornus with cunning and dexterity; having firft intoxicated a Serpent by the touch thereof, fhould, during that fit, make him obferve and readily conforme to all the various motions of that wand: So as that the unlearned Spectators perceiving the Serpent to approach the Enchanter, as he moves the wand neerer to himfelf; to retreat from him, as he puts the wand from him; to turne round, as the wand is moved round; to dance, as that is waved to and fro; and lye ftill, as in a trance, when that is held ftill over him; and all this while knowing nothing, that the fimple virtue of the wand is the Caufe of all thofe mimical motions and geftures of the Serpent: they are eafily deluded into a belief, that the whole fcene is fupernatural, and the main Energy radiated in thofe words, or Charms, which the Impoftor, with great Ceremony and gravity of afpect, mutters forth, the better to difguife his Legerdemane, and diffemble Nature in the Colours of a Miracle.

*Art.* 16. The Reafon of the Incantation of Serpents, by a rod of the Cornus.

And, as in this, fo in all other Magical Practices, thofe Bombaft Words, nonefenfe Spells, exotique Characters, and Fanatick Ceremonies, ufed by all Præftigiators and Enchanters, have no Virtue or Efficacy at all (that little only excepted, which may confift meerly in the founds, and tones in which they are pronounced, in refpect whereof the care may be pleafed or difpleafed) as to the Caufation of the Effect intended; nor doe they import any thing, more than the Circumvention of the Spectators judgement, and exaltation of his Imagination, upon whom they prætend to work the miracle. Which confidered, it will be an argument not only of Chriftianifm, but of found judgement in any man, to conclude; that excepting only fome few particulars, in which God hath been pleafed to permit the Devil to exercife his Præftigiatory power (and yet, whofo fhall confider the infinite Goodnefs of God, will not eafily be induced to beleive, that He hath permitted any fuch at all,) all thofe Volumes of Stories of Fafcinations, Incantations, Transformations, Sympathies of men and beafts with Magical Telefms, Gamahues or Waxen Images and the like myfterious Nothings, are meer Fables, execrable Romances. So Epidemical, we confefs, hath the Contagion of fuch Impoftures been, that among the People, when any Perfon waxeth macilent, and pines away, we hear of nothing but Evil Neighbours, Witchcraft, Charms, Statues of Wax, and the like venefical fopperies; and inftantly fome poor decrepite old woman is fufpected, and

*Art.* 17. DIGRESSION. That the Words, Spells, Charafters, &c. ufed by Magicians, are of no virtue or Efficacy at all, as to the Effect intended; unlefs in a remote intereft, or as they exalt the imagination of Him, upon whom they pretend to work the miracle.

and perhaps accused of malice and Diabolical stratagems against the life of
that person : who all the while lieth languishing, of some Common Disease, and the learned Physician no sooner examines the case, but he finds
the sick mans Consumption to proceed from some inveterate malady of the
bodie, as Ulcer of the Lungs, Hectique Fever, Debility of the Stomack,
Liver, or other common Concocting part, or from long and deep Grief
of mind. In like manner, when the Husband-man observes his field to
become barren, his chattel to cast their yong, or die, his corn to be blast-
ed, his fruits to fall immaturely, or the like sinister Accidents : nothing
is more usual with him, than to charge those misfortunes upon the Magical Imprecations of some offended Neighbour, whom the multitude supposeth to be a Cunning man, or Conjurer.  And yet, were the Philosopher consulted about those Disasters, he would soon discover them to
be the ordinary and genuine Effects of Natural Causes, and refer each
Contingent to its proper original.  True it is likewise, that many of those
Sorcerers, whom the vulgar call *White Witches*, in respect of the good
they prætend to do, frequently præscribe certain Amulets, or Periapts, for
the prævention or cure of some diseases : and in this case, if the Amulet or
Periapt, be composed of such Natural Ingredients, as are endowed with
Qualities repugnant to the Disease, or its germane Causes, we are not to
deny their efficacy.  But, as for those superstitious Invocations of Angels and Spirits, Salamons Characters, Tetragrammatons, Spells, Circles, and the like vain and ridiculous Magical Rites and Ceremonies, used
by the Sorcerer, at the time of the Composition or Application of those
Amulets or Periapts, they are of no power, or virtue at all, and signifie
nothing but the Delusion of the Ignorant.  Again, we grant, that the
Imagination and Confidence of the sick Person, being by such means exalted, may conduce very much to his Recovery ; for, it is no secret, that the
minds of Languishing men are, for the most part, erected, and their drooping
spirits as it were Re-inforced, by the good opinion they have entertained of
the Physician, and the Confidence they place in his præscripts : but, yet
are we not therefore to allow any Direct and Natural Efficacy to that superstitious præparation, and Ceremonious administration of Remedies,
which are alwaies observed by such Impostors, as prætend to Extraordinary skill, and some supernatural way, in the Cure of Diseases, and seem to
affect and glory in the detestable repute of Magicians.  And what we say of
the Cure of Diseases, by Periapts, Amulets, and the like, we desire should
be understood also of Magical Philtres, or Love-procuring Potions, of the
Ligature of the point, on the Wedding night, to cause Impotency in new
married men toward their Brides (a thing very frequent in *Zant* and *Gasco-
ny*) and the like effects : because each of these hath other Causes, than those
remote and unconcerned Nugaments præscribed by those Cheaters, and
all the influence and power they can have upon the persons, to whom they
are præscribed, consisteth only in the præpossession of their Phancy, and
the strength of persuasion to Hope, or Fear.

*Art.* 18.
The Reason of
the Fascination
of Infants, by
old women.
(9) There is, besides, a certain sort of *Fascination Natural*, about which
no small adoe is kept in the world, and most Nurses, when they observe
their Infants not to thrive, or fall into Cachexies, languishing conditions, Convulsions, or the like, instantly crie out, that some *envious
Beldam hath overlooked them.*  Concerning this secret therefore, in
which Imagination (on the Infants part) hath no interest at all, we say,
                                                                    tha

that if there be any thing of truth, as to matter of Fact, the Fascinating activity of the old malicious Crone must consist only in this: that she doth evibrate or dart forth from her brain, certain malignant Spirits, or rayes, which entering the tender body of the Infant, do infect the purer spirits, and so the blood in its Arteries, and assimilating the same to their depraved and maligne nature, corrupt all the Aliment of the body, and alienate the parts from their genuine and requisite temperament.   Not that those Malignant Emissions can arrive at, and infect an Infant that is absent, as is vulgarly conceived; but that the malicious old woman must be præsent, and look (with an oblique or wift look) and breath upon the Child, whose health she envies, nay, conjure up her Imagination to that height of malice, as to imbue her spirits with the evil Miasme or Inquinament of those vitious and corrupt Humors, wherewith her half-rotten Carcass is well stored; and to assist the Contention of her optique Nerves and Muscles, that so those Spirits may be ejaculated with great force.   For, that an old woman though as highly malignant in her Nature and Malice, as can be supposed, should be able to infect and envenome an Infant at great distance; is not to admitted by any, but such as have ignorance enough to excuse their perswasion of the highest Impossibility imaginable.   But, that she may, in some measure, contribute to the indisposition of an Infant, at whom she shoots her maligne Eye-beams, neer at hand; may receive much of credit from the Pollution of a Lookinglass by the adspect of a Menstruous woman; and from the Contagion of Blear Eyes, Coughing, Oscitation or Gaping, Pissing and the like: all which are observed to be somewhat infectious to the standers by.

(10) You may call it *Fascination* also, if you please, when the *Torpedo doth benumb or stupifie the hand of the Fisherman*.   For, as the Maleficiation of Infants is the Effect only of certain malign or ill conditioned Emanations transmitted to them from the brain of some malevolent and half venemous Ruines of a woman: so likewise must the stupefaction of the hand of the Fisherman, be the Effect of certain Stupefactive Emanations, either immediately, or by the mediation of a staff or other continued body, transmitted thereunto from the offended Fish; which Emanations, by a Faculty holding some neer Analogy to that of *Opium Hyosciamus*, and other strong *Narcoticks* or stupefactive Medicaments, do in a moment *Dull* and *Fix* the Spirits in the part, that they invade, and so make it Heavy, Senseless, and unfit for voluntary motion.

*Art.* 19.
The Reason of the stupefaction of a mans hand by a *Torpedo*.

(11) But, how shall we get free of that Difficulty, wherein so many high-going Wits have been Gravell'd; *the sudden arrest of a ship, under sail, by the small Fish Echineis*, thereupon general called a *Remora*?   We cannot expede our selves from it, by having recourse to any Fixing Emanations transmitted from the Fish to the ship; because the Motion thereof is not voluntary, but from External Impulse; nor hath the ship any spirits, or other Active principles of motion, that can be supposed capable of Alteration by any influx whatever.   Nor by alleaging any motion, contrary to that of the tide, winds, and oares, impressed upon the ship by the Remora; because, whatsoever kind of Impulse or Force can be imagined impressible upon it thereby: yet can it never be sufficient to impede and suppress the so violent motion thereof; insomuch as the Remora, neither adhæring to any rock, shelf, or other place more firme than the water, but only to the ship it self;

*Art.* 20.
That ships are not Arrested in their course, by the Fish called a *Remora*: but by the Contrary impulse of some special Current in the Sea.

self, muſt want that fixation & Firmitude, that is inevitably neceſſary, whenever any thing doth ſtop, or move another thing of greater weight then it ſelf. What then? ſhall we impeach of unfaithfulneſs all thoſe Authentick Hiſtorians, who have recorded the ſuddain and prodigious Arreſts of the ſhips of *Periander Antigonus*, and *Caius Caligula*, in the middeſt of their Courſes, though therein advantaged by the Conſpiring impulſes of Sails and Oares? Not ſo neither; becauſe many other veſſels, as well before as ſince, have been ſtopped in the like manner: and there is in nature Another Cauſe, incomparably more potent, and ſo more likely to have arreſted them, than that ſoft, ſmall and weak Fiſh Echineis; and that is the Contrary motion of the ſea, which our Mariners (who alſo have been often troubled with the experiments of its Retropellent Force) call the *Current*; which is alwayes moſt ſtrong and cumberſome in narrow and aufractuous Chanels. Which being ſcarce known to the Sea-men of thoſe times, when Navigation and Hydrography were yet in their infancy, and few Pilots ſo expert, as to diſcriminate the ſeveral Re-encounters, or Contrary Drifts of Waters in one and the ſame Creek or Arme of the Sea; when they found any veſſel ſuddenly retarded and impeded in its courſe, they never conceived that Remoration to ariſe from ſome Contrary Current of Waters in that place, but from ſome Impediment in the bottome or keel of the veſſel it ſelf. And as they ſearched there for it, if it hapned twice or thrice, that they found ſome ſmall Fiſh, ſuch as the *Concha Veneris*, or any other not much unlike a *Snail*, adhæring to the lower part of the Rudder, or Keel; they inſtantly, and without any examination at all, whether ſo weak a cauſe might not be inſufficient to ſo great an Effect, imputed the Remoration of their veſſel thereunto. Hiſtorians, indeed, tell us, that the Admiral Galley, which carried the Emperour *Caligula*, in his laſt voyage to *Rome*, was unexpectedly Arreſted, in the middeſt of all his numerous Fleet; and that an Echineis was found ſticking to the bottom thereof: but they forgot to tell us, whether or no there were any other Fiſhes of the ſame kind affixed to any other of the Galleys, that kept on their courſe; and we have good reaſon to conjecture, that there were, becauſe very few ſhips are brought into Havens and Docks to be carined, but have many ſmall fiſhes, reſembling Snails, adhæring to their bottoms, as ourſelves have more than once obſerved in *Holland*. Beſides, ſince, at *Caligula's* putting forth from *Aſtura*, an Iſland Port, and ſteering his courſe for *Antium*, his Galley, as is the cuſtome of Admirals, kept up in the middle Chanell; why might it not be encountred and oppoſed by ſome ſpecial current, or violent ſtream, in that place, ſo ſtreitly pent in on both ſides by the ſituation of certain Rocks and Shelves, as that its greateſt force was in one certain part of the Chanell, and ſo not extenſible to the other Galleys of his Navy, that were rowed neerer to the ſhoars, and ſo rode upon free water? For, thus ſhips are now adayes often Arreſted by ſpecial Currents, in the *Fretum Sicilienſe*, whoſe Chanels are rocky, aufractuous, and vorticous, or obnoxious to frequent Eddies and ſtrong Whirlepools; and neer *Gaditanum* you may every day behold the Contrary Drifts of ſhips by the Contrary Currents in the ſame Arme of the Sea; ſome veſſels being carried toward the ſhoars, whether the ſea runs out, while others ride toward the Chanel, where the ſea runs in.

(12) So unlimited is the Credulity of man, that some have gone further yet from the bounds of Reason, and imagined a *Second* wonderful Faculty in the *Remora*, viz. the *Præsagition of violent Death, or some eminent Disaster, to the chief person in the ship, which it arresteth.* For, *Pliny* (*lib. 9. cap. 25. & lib. 23. cap. 1.*) will needs have it a Prodigy portending the murder of *Caligula*, which ensued shortly after his arrival at *Rome* from *Astura :* and that by the like arresting of the ship of *Perianders* Ambassadors sent to obtain an edict for the Castration of all Noble youths, Nature did declare her high detestation of that Course so destructive to the way of Generation, that she had instituted for the Conservation of her noblest species. But, every man knows, how easie it is to make any sinister Accident the Omen of a tragical Event, after it hath happened: and that *Plinies* Remark upon the inhuman Embassie, and succeeding Infortune of *Perianders* Messengers, would better beseem the ranging pen or tongue of an *Orator*, than the strict one of a *Philosopher*.

*Art. 21.*
*That the Echineis, or Remora is not Ominous.*

(13) Here, we should open and survey the whole Theatre of *Venoms* or *Poisons*, on one hand; and that of *Antidotes*, or *Counterpoisons*, on the other: those operating to the *Destruction*, these to the *Muniment* and *Conservation* of *Life*; and both by such *Qualities* and wayes, as are generally both by Physiologists and Physitians, præsumed to be *Occult*, or beyond the investigation of Reason, and of which all that is known, is learned in the common School of Experience. But, worthily to examine the Nature of each particular Poison, among those many found in the lists of Animals, Vegetables, Minerals, and explicate the Propriety, by which its proper Antidote or Alexipharmacon doth encounter, oppose, conquer and expel it: must of necessity enlarge this Section into a Volume, besides the expence of more time, than what we have consigned to our whole Work. And, therefore, we hope our Reader will not conceive his expectation wholly frustrated, nor Curiosity altogether defrauded; though we now entertain Him only with the *General* Reasons, Why Poisons are Hostile and Destructive, why Counterpoisons friendly and Conservative of Life.

*Art. 22.*
*Why this place admits not of more than a General inquest into the Faculties of Poisons and Counterpoisons.*

*Gwoinus* (*de Venen. lib. 2. cap. 24.*) we well remember, defines *Venenum*, Poison, to be [ *quod in corpus ingressum, vim infert, Naturæ illamque vincit* ] *That which being admitted into the body, offers violence to Nature, and conquers it.* And, according to this Definition, by *Poisons* we understand not only such things, as bear a pernicious Enmity in particular to the temperament of the Heart, or that substance, wherein the Vital Faculty may be conceived principally and immediately to consist: but *all such as are hostile and destructive to the temperament of the Brain, or any other Noble and Principal Organ of the body, so as by altering the requisite Constitution thereof, they subvert the œconomy and ruine the frame of Nature, wherein the Disposition of the parts, to perform the Actions of Life, is radicated.*

*Art. 23.*
*Poisons defined.*

And that, wherein this *Deleterious* or *Pernicious Faculty* doth consist, we conceive to be a certain *Substance*, which being communicated or infused into any part of the body, though in very small quantity, doth, by reason of the exceeding *Subtility* and violent *Mobility* or

*Art. 24.*
*Wherein the Deleterious Faculty of Poison doth consist.*

C c c    *Agility*

*Agility* of the insensible particles, of which it is composed, most easily and expeditely transfuse or disperse it self through the whole body, consociate it self to the spirits, and invading the Heart, Brain, or other Principal Organ, so alter the requisite Disposition or temperament and habit thereof, as to make it thenceforth wholly uncapable of performing the Functions or Actions of life, to which it was destined and framed; and by that means introduceth extreme Destruction.

*Art.* 25.
*Counterpoisons*
*Defined.*

Likewise, by *Alexipharmacal Medicaments*, or *Counterpoisons*, we understand, not such things, as have only a propitious and benign Friendship particularly for the temperament of the Brain, Heart, or other Noble Organ in the body, and are therefore accounted specifically Auxiliant and Corroborative thereunto, in the Expulsion of ought, that is noxious and offensive unto it; because, in that sense, all Cardiacal, Cephalical, and Specifically Corroborative Medicaments would be Alexiterial, and every peculiar Venome would not require its proper Antivenome, both which are contradicted by Experience: But, *such things as are endowed with Faculties è diametro and directly Contrapugnant to Poisons, meerly as Poisons*; For, divers things that are absolute Poisons of themselves, and would destroy, if taken alone by themselves, do yet become powerful Præservatives and Antidotes against other poisons, and afford suddain and certain relief to nature, when taken to oppose them. Thus *Aconite*, than which scarce any venome is more speedy and mortal in its operation upon a sound body, doth yet prove a præsent remedy to one bitten by a *Scorpion*, if drank in Wine: as *Pliny* hath observed (*lib.* 27. *cap.* 2.)

*Art.* 26.
*Wherein their*
*Salutiferous*
*Virtue doth*
*consist.*

And that, wherein this *Salutiferous Virtue of Antidotes* doth consist, we conceive likewise to be a certain *substance*, which being received into the body, though in small quantitie, doth with expedition diffuse it self throughout the same: and encountering the venome formerly admitted, and then operating, refract its energy, prævent its further violence, extinguish its operation, and at length either totally subdue, or totally educe it. For, All Alexipharmacal Remedies do not bring relief to nature, assaulted and oppressed by Poison, by one and the same way or manner of operation; some working by way of *Repulsion*, others by way of *Abduction*, others by way of *Opposition* and downright *Conquest*, when they are taken *Inwardly*: some by *Retraction*, others by *Extinction*, where they are applied *Externally*.

*Art.* 27.
*How Triacle*
*cureth the ve-*
*nome of Vipers*

Thus *Triacle*, whose Basis or master ingredient is the *Flesh* of *Vipers*, doth cure a man empoisoned by the *Biting* of a *Viper*; only because, in respect of Consimilarity or Similitude of substance, it uniteth it self to the Venome of the Viper, which had before taken possession of and diffused it self throughout the body, and afterwards educeth the same together with it self, when it is expelled by sweating, procured by divers Cardiacal and Hidrotical, or Sudorifick Medicaments commixt in the same Composition: no otherwise than as Soap, whose principal Ingredient is oil, doth therefore take off oily and greasie spots from Clothes; because, uniting it self unto a Cognate or Consimilar substance, the Oil or Fat adhæring to the Cloth, and so assisting its Dilution and Concorporation with the Water,

in

in which it self is diffolved; it carrieth the fame away together with
it felf in the water, when that is expreffed or wrung out by the hand
of the Laundrefs. More plainly, As oyle is therefore commixed with
Afhes, or Salt, in the compofition of Soap, to the end it may not ftain
the Cloth anew, to which it is applyed, but being confufed with the oil
or Fat, wherewith the cloth was formerly ftained, Abduce or carry off
the fame together with it felf in the water, which is the Vehicle to both:
fo likewife is the Flefh of Vipers therefore commixt with fo many
Alexiterial Simples as concur to the Confection of Triacle, to the end
it may by them be hindred from envenoming the body a new, but
yet at the fame time be fo commixt with the Venome already diffufed
through the body, as that when thofe Alexiterial Medicaments are by
Sweat or otherwife educed from the body, carrying along with them
the Venome of the Vipers flefh; to which they are individually confociated, they may alfo abduce or carry away that venome of the Vipers
tooth, which was formerly diffufed through the body. And this, we
moreover conceive, may be the General Reafon not only of the Evacuation of Venomes by Sweat, where the Antidote works by Union and
Abduction; but alfo of the *Evacuation* of fuperfluous Humours by *Elective
Catharticks*, or Purging Medicaments, that fpecifically educe this, or that
Humor: for, it may be as lawfully faid, that *Like may be cured by Like, or
Unlike by Unlike*; as that oil may be abfterged by its Like, viz. the oil in
Soap, and by fomething that is Unlike, viz. the Salt, or Water carrying
the oil individually commixt with it.

Thus alfo doth the body of a *Scorpion*, being bruifed and layed warm to
the part, which it hath lately wounded and envenomed, fuddainly Retract,
and fo hinder the further Diffufion of the Poifon that it had immitted into the body; only becaufe the Nervous and Fibrous parts of the Scorpi-
ons body bruifed, by a motion of Vermiculation recontracting themfelves,
as Chords too much extended, and fo retracting the Venome that yet remains
adhærent to them: do at the fame time Extract that Confimilar Venome,
that was infufed into the wound. The fame alfo may be conceived of the
Cure of the venome of a *Spider*, by the body of the Spider contufed, and
applyed to the part envenomed: and of the Cure of the Biting of a *Mad Dog*,
by the Liver of the fame Dog, in like manner Contufed and impofed.

*Art. 28.*
How the body
of a *Scorpion*,
bruifed and
laid warm up-
on the part,
which it hath
lately wound-
ed and enve-
nomed; doth
cure the fame,

Nor is it by way of Union and Abduction alone, that fome Poyfons
become Antidotes againft others; but alfo by that of direct *Contrariety*,
*Colluctation* and *Conqueft*: for, there being great Diverfity of Venoms,
fome muft be Contrapugnant to others; and whenever any two, whofe
Natures and Proprieties are Contrary one to the other, meet together,
they muft inftantly encounter and combate each other, and at laft the
Activity of the Weaker fubmit to that of the ftronger, while Nature
acting the part of a third Combatant, obferves the advantage, and coming in with all her forces to the affiftance of her Enemies Enemie, com-
pletes the Victory, and delivers Her felf from the danger. Befides,
we have the teftimony of Experience, that Divers men have fortified
their bodies againft the affault and fury of fome Poifons, by a gradual Affuefaction of them to others, as *Mithridates*, and the *Attick* old Woman, &c.

*Art. 29.*
That fome
Poifons are
Antidotes a-
gainft others
by way of di-
rect *Contrariety*

Hence

Art. 30.
Why sundry
particular
men, and some
whole-Nations
have fed upon
Poisonous Ani-
mals and
Plants, with-
out harm.

Hence we remember Another considerable *Secret* concerning Poisons, much disputed of in the School of Physitians; viz. *Whence comes it, that not only sundry Particular Persons, but even Whole Nations have fedd upon venemous Animals and Plants, without the least of harm, nay with this benefit, that they have thereby so familiarized Poisons to their own Nature, as that they needed no other Praservative against the danger of the strongest Poison, but that Venenate one of their own Temperament?* Whereto, we Answer, in a word, that that Tyrant, *Custome,* alone challengeth the honour of this wonder; such men having, by sensible degrees, or slow advance from lesser to greater Doses of Poisons, so changed the temperament and habit of their bodies, that the wildest Venoms degenerated into wholesome Aliments, and Poisons were no more Poisons to them, than to the Animals themselves, which Generate and contain them. Which duely considered, we have little reason to doubt the verity of *Galens* relation (*de theriaca ad Pison.*) of the Marsi, and Ægyptians, whose ordinary Diet was Serpents: or of the like in *Pliny* (*lib. 6. cap. 29.*) concerning the Psyllæ, Tintyritæ, and Candei, who were all ophiophagi, or Serpent-Eaters: or of *Theophrastus* his story (*lib. 9. de histor. animal. cap. 18.*) of certain Shepherds in Thrace, who made their grand Sallads of white Hellebor: or of *Avicens* (*lib. 4. sen. 6. tract. 1. cap. 6.*) of a certain Wench, who living upon no other Viands but Toads, Serpents, and other the strongest poisons, and mostly upon that of Napellus, became of a Nature so prodigiously virulent, that she outpoisoned the Basilisk, kissed several Princes to death, and to all those unhappy Lovers, whom her rare beauty had invited to her bed, her Embraces proved as fatal, as those of *Jupiter* armed with his thunder, are feigned to have been to semele: or of *Jul. Cæs. Scaligers* (*Exercit.* 175.) concerning the Kings son of *Cambaia,* who being educated with divers sorts of poisons from his infancy, had his temperament thereby made so inhumane and transcendently Deleterious, that He destroyed Flyes only with his breath, killed several women with his first nights Courtship, and pistolled his Enemies with his Spittle; like the serpent *Ptyas,* that quickly resolves a man into his originary Dust, only by Inspuition, as *Galen* reports (*de theriaca ad Pison. cap. 8.*)

Art. 31.
The Armary
Unguent, and
Sympathetick
Powder, im-
pugned.

The Rear of this Division of Secrets concerning Animals, belongs to the ARMARIE or MAGNETICK UNGUENT, and its Cousin German, the SYMPATHETICK POWDER, or Roman Vitriol calcined; both which are in high esteem with many, especially with the Disciples of *Paracelsus, Crollius, Goclenius,* and *Helmont,* all which have laboured hard to assert their Virtue in the Cure of Wounds, at great distance, either the Unguent, or Powder being applyed only to the weapon, wherewith the wound was made, or to some piece of Wood, Linnen, or other thing, to which any of the blood, or purulent matter issuing from the wound, doth adhære. Concerning those, therefore, we say, in short; (1) That notwithstanding the stories of wounds supposed to have been cured by Hoplochrism, both with the Unguent and Vitriol, are innumerable; yet is not that a sufficient Argument to convince a circumspect and wary judgment, that either of them is impowered with such a rare and admirable Virtue, as their admirers præsume: because many of those stories may be Fabulous; and were the several Instances or Experiments of their Unsuccessfulness summed up and alledged to the contrary, they would, doubtless, by incomparable excesses overweigh those of their successfulness, and soon

counter-

counter-incline the minds of men to a suspicion at least of Error, if not of Imposture in their Inventors and Patrons. (2) Though the Examples of their success were many more than those of their Failing; yet still would it be less reasonable for us to flye to such remote, obscure, imaginary Faculties, as do not only transcend the capacity of our Understanding, but openly contradict that no less manifest than general Axiome, *Nihil agere in rem distantem :* than to have recourse to a proxime, manifest, and real Agent, such as daily producing the like and greater Effects by its own single power, may justly challenge the whole honour of that Sanative Energy to it self, which the fraud of some, and incircumspection of others have unduly ascribed to the Unguent, or Sympathetick Powder : We mean, the *Vital* (if you please, you may call it, the Animal, or Vegetative) *Faculty* it self; which rightly performing the office of Nutrition, doth by the continual apposition of the Balsam of the Blood, to the extremes of the small Veins, and to the Fibres in the wound, repair the lost flesh, consolidate the Disunited parts, and at length induce a Cicatrice thereupon. For, common Experience demonstrateth, that in men of temperate Diet and euchymical bodies, very deep and large wounds are many times soon healed of themselves; i. e. meerly by the goodness of Nature it self, which being vigorous, and of our own provision furnished with convenient means, wholesom and assimilable Blood, doth every moment freshly apply it to the part that hath suffered solution of Continuity, and thereby redintegrate the same: especially when those Impurities generated by putrefaction in the wound, which might otherwise be impediments to Natures work of Assimilation and Consolidation, are removed by the Detersive and Adstrictive Faculty of the Salt in the Urine, wherewith the wound is daily to be washed, according to the præscript of our Sympathetical Chirons. Nor is this more than what Dogs commonly do, when by licking their wounds clean, and moistning them with the saltish Humidity of their tongues; they easily and speedily prove their own Chirurgeons. (3) The Basis or Foundation of Hoplochrism is meerly Imaginary and Ridiculous; for, the Assertors thereof generally dream of a certain *Anima Mundi,* or Common Soul in the World , which being diffused through all parts of the Universe, doth constantly transferr the Vulnerary Virtue of the Unguent, & Vitriol, from the Extravenated blood adhæring to the weapon or cloth, to the wound, at any distance whatever, and imbuing it therewith, strongly assist Nature in the Consolidation of the Disunion. But, insomuch as this *Anima Mundi,* according to their own wild supposition, ought to be præsent to all other wounds in the world, no less than to that, from which the blood, whereunto the Unguent, or Vitriol is applied, was derived : therefore would it cure all other wounds, as well as that particular one; since it interveneth betwixt that wound and the Unguent or Vitriol, by no more special reason, than betwixt them and all other wounds; unless it can be proved, that some other special thing is transmitted to that particular wound from the Unguent, and that by local motion through all points of the intermediate spaces successively; which they will by no arguments be induced to concede.

This Verdict, I præsume, was little expected from *Me,* who have, not many years past, publickly declared my self to be of a *Contrary* judgment; written professly in Defence of the cure of wounds, at distance, by the Magnetick, or Sympathetick Magick of the Weapon-Salve; and Powder of Calcined Vitriol; and excogitated such *Reasons* of my own, to support and explicate

Art. 32.
The Authors
*Retraction* of
his quondam
Defence of the
Magnetick
Cure of
Wounds,
made in his
*Prolegomena* to
*Helmonts* Book
of that subject
and title.

explicate the so generally conceded and admired Efficacy of Both, as seemed to afford greater satisfaction to the Curious, in that point, than the Romantique *Anima Mundi* of the Fraternity of the Rosy-Cross, the Analogical Magnetism of *Helmont*, or, indeed, than any other whatever formerly invented and alledged. And, therefore, to take off my Reader from all admiration thereat, it is necessary for me here to profess, that the frequent Experiments I have, since that time, made, of the downright Inefficacy and Unsuccessfulness as well of the Armary Unguent, as Sympathetick Powder, even in small, shallow, and in dangerous Wounds; my discovery of the lightness and invalidity of my own and other mens Reasons, adferred to justifie their imputed Virtues, and abstruse wayes of operation; and the greater Probability of their opinion, who charge the Sanation of wounds, in such cases, upon the sole benignity and Consolidative Energy of *Nature it self:* these Arguments, I say, have now fully convinced me of, and wholly *Converted* me from that my former Error. And glad I am of this fair opportunity, to let the world know of my *Recantation*: having ever thought my self strictly obliged, to præfer the interest of *Truth*, infinitely above that of *Opinion*, how plausible and splendid soever, and by whomsoever conceived and asserted, to believe, that Constancy to any unjustifiable Conception, after clear Conviction, is the most shameful Pertinacity, a sin against the very Light of Nature, and never to be pardoned in a profest Votary of Candor and Ingenuity; and to endeavour the Eradication of any Unsound and Spurious Tenent, with so much more of readiness and sedulity, by how much more the unhappy influence of my Pen, or Tongue hath, at any time, contributed to the Growth and Authority thereof.

CHAP.

# CHAP. XVI.

## THE

# PHÆNOMENA

### OF THE

# LOADSTONE

### *EXPLICATED.*

#### SECT. I.

Hose Wit had the best edge, and came nearest the slitting of the hair; His, who said, that the LOAD-STONE is the *real Janus*, because of its Two opposite Faces, or Poles, one whereof confronteth the North, the other the South: or His, who called it the *Egg and Epitome of the Terrestrial Globe*; because as the Egg contains the Idæa of the whole and every part of its Protoplast or Generant, so doth the Loadstone comprehend the Idæa of the whole and every part of the Earth, and inherit all its Proprieties, being Generated thereby, at least therein: or His, Who named it *The Nest of Wonders*; because, as a Nest of Boxes, it includes many admirable Secrets, one within another, insomuch, that no man can well understand the mystical platform of its Nature, till he hath opened and speculated them all one after another: or His, who affirmed it to be the *Antitype of the Poets Hydra*; because, no sooner hath the Sword of Reason cut off one Head, or Capital Difficulty, but Two new ones spring up in the place of it, nor ought any man to hope the total and absolute Conquest thereof, but by Cauterizing the veins of every *Difficulty*, i.e. leaving not so much as the seeds of a Scruple,
but

*Art. I.*
The *Nature* and *Obscurity* of the Subject, hinted by certain *Metaphorical Cognomina*, agreeable thereunto, though in divers relations.

but folving all its various Phænomenaes to the full : or His, who thought it fufficient, with *Ariftotle*, to call it [ ἡ λίθος ] *The ftone*, that fingularity importing its tranfcendent Dignity : we freely leave to the judgment of our *Reader*.

*Art. 2.*
Why the Au-
thor infifteth
not upon the
(1) feveral
*Appellations.*
(2) *Inventor*
of the Load-
ftone, (3) *In-
vention* of the
*Pixis Nautica.*

And, as for fundry other Enquiries, that do not in any direct or oblique intereft concern the Inveftigation of the Caufes of All, or Any of thofe admirable Proprieties obferved in the Loadftone ; fuch as that of the various *Appellations* given it by feveral Philofophers of old, by feveral Nations, at this day, together with the proper Original, Etymology and Reafon of each : Whether it was firft *Difcovered* by the Shepherd *Magnes*, on Mount Ida ; as *Pliny* (*lib. 36. cap. 26.*) reports out of the records of *Nicander* : Whether its *Attractive Virtue* was known not only to *Hippocrates* and other Senior Philofophers of *Greece*, but alfo to the Primitive Hebrews, and Ægyptians ; as *Gilbert* conjectureth (*de Magnet. lib. 1. cap. 2.*) : Whether the Knowledge of its *Verticity, or Polary Virtue* cannot be derived higher than the top of the four laft Centuries, and ought to be afcribed to a French man, together with the honour of the Invention of the *Pixis Nautica*, or Navigators Compafs, about the year of Chrift, M. CC. as *Gaffendus* would perfuade, out of one *Guyotus Provinceus*, an old French Poet, who not long after, writ a Panegyrick in Verfe upon the Excellency and fundry ufes of the fame ; or to *John Goia* (alias *Gira*) of *Salerna*, who lived not till almoft an hundred years after the faid *Guyotus* had divulged his Poem, as *Blancanus* (*in Chronolog. Mathemat. Secul. 2.*) contends : Whether the Nations inhabiting the *Sinne* had the ufe of the Mariners Compafs, before the Europeans ; or whether they learned it of the European fhips, that firft advanced beyond the Cape of Good-hope, and coafted the Mare Rubrum, and begun Commerce with them : All thefe things, as being not only not eafie to determine, but alfo fcarce pertinent to our præfent fcope, we refer to our Readers own enquiry, in *Gilbert, Cabeus, Kircher*, and other Authors, who promife him all poffible fatisfaction therein.

*Art. 3.*
The *Virtues* of
the Loadftone,
in General ;
*Two*, the *Attra-
ctive*, and *Di-
rective.*

To come, therefore, directly to the profecution of our main defign, we obferve, that the VIRTUES of the Loadftone are, in General *Two*, one *whereby it attracteth Iron to it felf*, the other *whereby it directeth both it felf and Iron, which it hath impregnated by contact or influence, to the Poles of the Earth :* the *Firft* is called *Alliciency*, the *Other* its *Verticity* or *Polarity*. Concerning the Caufe of its *Alliciency*, or the reafon of the Attraction of Iron by the Loadftone, or ( if you would have us fpeak in the fenfe and dialect of Dr. *Gilbert*) the *Coition* of Iron and a Loadftone, various opinions have been conceived and afferted as well by Modern as Ancient Philofophers. Among thofe of the *Ancients*, that which beft deferves our commemoration and confideration, is the opinion of *Epicurus* : who, left He might feem fcarcely fufficiently confcious of the great difficulty of the fubject, excogitated a Two-fold Theory for its Explication and Solution ; the Former of which we may eafily collect from the Commentary of *Lucretius* thereupon ; the Latter from the Difpute of *Galen* (*lib. 1. de Natur. Facult.*) againft it. For,

*Art. 4*
*Epicurus* his
firft Theory,
of the Caufe
and Manner of
the Attraction
of Iron by a
Loadftone ;
according to
the Expofition
of *Lucretius.*

*Lucretius*, profeffing to explain the Reafon and Manner of the Attraction of Iron by the Loadftone, according to the Principles and judgment of
*Epicurus*

*Epicurus*, founds his Difcourfe upon thefe Four Pillars, or *Præconfiderables*; (1) *That all Concretions do continually emit fubtile Effluvia's, or Aporrhæa's: (2) That the contexture of no Concretion is fo compact, as not to have many fmall Vacuities, or infenfible Pores, varioufly intercepted among its folid and component particles: (3) That the Effluvia's ftreaming from Concretions, are not equally Congruous or Accommodate to all Bodies they meet with in the fphere of their Diffufion: (4) That the fmall Pores, or infenfible Inanities intercepted among the particles of Concretions, are not all of one and the fame Circumfcription, or Figure; and fo not indifferently accommodable or proportionate to all forts of Effluvia's iffuing from other bodies, but only to fuch, as are fymmetrical or Correfpondent to them in Figure and Magnitude.* And then *He* proceeds to erect this fuperftructure thereupon.

' The Attractive Virtue of the Loadftone, being determinate only to ' Iron and Steel (which is Purified Iron) feems to confift in this; that ' both from the Loadftone and Iron there perpetually iffue forth continued ' ftreams of infenfible particles, or bodies, which more or lefs, according ' to their number and force of diffufion, commove and impel the am- ' bient Aer: and becaufe the ftreams which flow from the Loadftone ' are both more numerous and more potent, than thofe which are emit- ' ted from the Iron; therefore is the ambient Aer alwayes more ftrongly ' difcuffed and impelled about the Loadftone, than about the Iron; and ' fo there are many more Inane Spaces therein created about the Load- ' ftone, than about the Iron. That forafmuch as, when the Iron is ' placed within the fphere of the Aer Difcuffed by the Effluxions of ' the Loadftone, there cannot but be much of Inanity intercepted (un- ' derftand infenfible Inanity) betwixt it and the Loadftone; thence it ' comes, that the Aporrhæaes of the Iron tend more freely or uninter- ' ruptedly toward that part, which faceth the Loadftone, and fo are carried ' quite home unto it: and becaufe they cannot tend thither in fuch ' fwarms, and with fuch freedome, but they muft impell the Particles ' of the Iron that are yet cohærent together; therefore muft they alfo ' move and impel the whole mafs of Iron, confifting of thofe recipro- ' cally Cohærent Particles, and fo carry it quite home to the Load- ' ftone. That, when a Loadftone Attracteth Iron, not only through ' the Aer, but alfo through divers compact and firm bodies, and par- ' ticularly through Marble; we are to conceive that there are more ' and more capacious Inanities made in that part of fuch interpofed bo- ' dies, which refpecteth the Loadftone, than in that part of them, which ' confronteth the Iron. That the reafon, why other things, as Straw, ' Wood, Gold, &c. being fituate within the fphere of the Aer Dif- ' cuffed by the Effluxes of the Loadftone, do not in like manner emit ' their fubtile particles in fuch numerous and potent ftreams, as carrying ' along their Cohærent Particles with them, fhould move and im- ' pel their whole maffes to a Conjunction with it: is only this, ' that the Particles emitted from the Iron are alone Commenfurable ' to the Inane Spaces in the Loadftone. That, becaufe Iron tendeth ' to the Loadftone indifcriminately, i. e. either upward or downward, ' tranfverfly or obliquely, according to the region of its Application; ' this indifferency could not be, but in refpect of the introduced ' Vacuities, into which the particles (otherwife prolabent only downward)

' are

'are carried without Distinction of region. And, lastly, that the mo-
'tion of the Iron towards the Loadstone, is assisted and promoted by
'the Aer, by reason of its continual Motion and Agitation; and first
'by the *Exterior Aer*, which being alwayes most urgent on that part,
'where it is most Copious, cannot but impel the Iron toward that part
'where it is less Copious, or more full of Inanities, i. e. toward the
'Loadstone: and afterward by the *Interior Aer*, which being likewise
'alwayes commoved and agitated, cannot but cause the stronger motion
'toward that part, where the Space is rendred more Inane. And this
we conceive to be the summary of *Lucretius* Exposition of *Epicurus*
Opinion touching the Reason of the Loadstones *Iron-attractive* Faculty.

Art. 5.
His other So-
lution of the
same, accord-
ing to the
Commentary
of *Galen.*
And *Galen* (*in loco citato*) impugning the Magnetick Theory of
*Epicurus*, first makes a contracted, but plain recital thereof, in these
words : *A lapide quidem Herculeo ferrum, à succino verò palens at-
trahi, &c. quippe effluentes Atomos ex lapide illo ita figuris congruere
cum illis, quæ ex ferro effluunt; ut in amplexus facile veniant; quam-
obrem impactas utrinque (nempe in ipsa tum lapidis, quam ferri
corpora concreta) & resilientes deinde in medium, circumplicari invicem, & ferrum simul pertrahi, &c.* Wherein, besides his usu-
all fidelity in the Recitation even of such opinions of other men,
as he thought good to endeavour to refute, we have good reason to believe, that *Galen* came as near as possible to the true
and genuine sense of *Epicurus* : forasmuch as those Four Præconsiderables alledged by *Lucretius* for the support of his exposition of
the Cause and Manner of the Coition of the Loadstone and Iron,
may be with equal Congruity accommodated also to this latter
Epicurean Solution of the same problem, according to this præsent
interpretation and abridgement of *Galen.* For, according to the
tenour thereof, both the Loadstone and Iron are præsumed to
consist of particles exactly alike in configuration, and to have the
like Inane Spaces, or insensible pores intercepted among those
particles : and this upon no slender ground, seeing that the Loadstone and Iron are perfect Twinns, being both generated not onely in the same Matrix, but of the same Materials, one the same
Mineral Vein of the Earth. And, therefore, it is the more probable, that the particles or Atoms issuing in continued streams from
the Loadstone, and invading Iron situate within the Orb of their
activity, should easily and deeply insinuate themselves into the
pores of the Iron; and there meeting with streams of other A-
toms so exactly consimilar to themselves, engage them to reciprocal Cohærence, and being partly repercussed or rebounded from
thence toward their Source, abduce those Atoms along with them,
to which they cohære, and by the impulse of other cohærent particles, abduce also the whole and entire mass : especially since it
is part of the supposition, that the Atoms transmitted from the
Iron to the Loadstone, do reciprocally move, engage, and compel the particles thereof, after the same manner ; it being almost
necessary that the Atoms on both sides, in good part rebounding or
resilient, toward their sources, and mutually implicated, should flow together into the medium, and so doing, that the whole bodies or masses
of

of the iron and Loadstone should be brought to a Conjunction in the Medium, because of the Cohæsion of both sorts of the flowing Atoms, with those, of which the whole masses are contexed. For, notwithstanding it be vulgarly apprehended and affirmed, that the Iron doth come to the Loadstone, rather then the Loadstone to the Iron; that the streams of Atoms emanant from the Loadstone, are both more numerous and much more potent; and found by Experiment that pieces of Iron do not only meet Loadstones half way, but come quite home to them, where the Loadstones are either much greater and weightier, or so held fast in a mans hand, or otherwise, as that they cannot exercise their reciprocal tendency : yet, as *Gilbert* speaks (*de Magnet. lib. 2. cap. 4.*) *Mutuis viribus fit Concursus ad unitionem*, the Coition is not from one single Attraction, but from a *Double*, συναιτελέχχα, or *Conactus*. And, as for the reason, why other things do not apply themselves to the Loadstone, as well as Iron; it may be said, that the streams of Atoms flowing from the Loadstone, and encountring those that are emitted from other bodies, do either pass uninterruptedly along by them, or are not, in respect of their Dissimilitude in Figures, so implicated or Complected with them, as in their resilition to flow together and concurr in the medium.

*Art. 6.*
*Galens three Grand Objection against the same, briefly Answered.*

And then He attempts the subversion thereof, by the opposition of some Arguments, and especially of these *Three Queries*. (1) *How such minute and insensible bodies, as those of which the Magnetick Aporrhæa's are supposed to consist, can be able to Attract [ βαρέιδη ὑπὸς νσιαν ] so great a weight as that of a mass of Iron ?* Whereto it may be Answered, in behalf of *Epicurus*, that the Magnetick Effluxes are not supposed to be so potent, as to draw any mass of Iron of what weight soever, but only such a one, whose bulk or weight carrieth some proportion to the force of the Attrahent, or Loadstone. Again, He might have considered, that the motions of the Grossest and Heaviest Animals are performed by their spirits, that are bodies as exile and imperceptible as the Magnetick Effluviaes : that Winds, which also consist of insensible particles, do usually overturn trees and vast ædifices, by the impetuosity of their impulses : and that subterraneous Vapours are frequently the Causes of Earthquakes. And, as for the reason, How the Magnetick Aporrhæa's can *Deduce, Apprehend,* and *Detain* a mass of Iron; He might have remembred, that the Atoms of the Magnet are conceived to have certain small *Hooks*, or *Clawes*, by which they may lay hold upon the *Ansulæ*, or Fastnings in the Iron; to have a violent *Motion*, which is the Cause both of their Impaction against, and Resilition from the Iron, and to have a perpetual *Supply* of the like Atoms continually streaming from the same fountain, by which they are assisted in their Retraction, whereupon the Attraction may ensue, and that so much the more forcible, by how much nearer the Iron is præsented, in regard of the more copious Efflux, or Density of the Magnetical rayes. (2) *How comes it, That a piece, or ring of Iron, being it self Attracted by a Loadstone, and on one part adharent unto it, should at the same time attract and suspend another ring on the contrary part; that second ring likewise attract and suspend a third,*

*that*

*that third a fourth, that fourth a fifth, &c.* To this we may apply that Responſe of *Epicurus*, which *Galen* himſelf commemorates; *An dicemus, effluentium ex lapide particularum nonnullas quidem, ubi ferro occurſaverint, reſilire; & has ipſas eſſe, per quas ferrum ſuſpendi contingat? nonnullas vero illud ſubeuntes, per inanes meatulos tranſire quam ocyſſimè, & conſequenter impactas in aliud ferrum proximum, cum illud nequeant ſubingredi, tametſi prius penetraverint, hinc reſilientes verſus prius, complexus alios prioribus ſimiles efficere?* For, herein is nothing ſo incongruous, as *Galen* conceives; it being not improbable, that ſome of the Magnetical Atoms, falling upon a piece of Iron ſhould be impinged againſt the ſolid particles thereof, and others of them, at the ſame time, penetrate the ſmall inanities or pores betwixt thoſe ſolid particles; after the ſame manner, as we have formerly aſſerted the particles of Light to be partly Reflected from the ſolid parts, and partly Trajected through the Pores of Glaſs and other Diaphanous bodies: nor that ſome of thoſe Magnetick Rayes, which paſs through the pores of the firſt Iron, ſhould invade a ſecond Iron poſited beyond it, and be impinged likewiſe againſt the ſolid particles of that, and ſo reflected toward their original, while ſome others pervading the Inanities of the ſecond, ſhould attract a third piece of Iron, and ſo conſequently a fourth, a fifth, and ſometimes more. And, certainly, in this caſe it is of no ſmall advantage to *Epicurus*, that the Force of the Magnetick Attraction is ſo *Debilitated* by degrees, as that in the ſecond iron it becomes weaker than in the firſt, in the third than in the ſecond, in the fourth than the third, &c. until at length it be totally evirate and decayed: becauſe, upon the ſecond there cannot fall as many rayes, as did upon the firſt, nor upon the third, as upon the ſecond, &c. as we have at large explicated, in our diſcourſe of the Cauſes of the Debilitation of Light. It may be further added alſo, in defence of *Epicurus*; that the Atoms of the Loadſtone, penetrating the ſubſtance of Iron, do ſo exſtimulate the Atoms thereof, that the Iron inſtantly ſuffering an Alteration of the poſition of all its component particles, doth in a ſort compoſe it ſelf according to their mode, and put on the nature of the Loadſtone it ſelf: and therefore it can be no ſuch wonder, that one iron Magnetified ſhould operate upon another iron, as the Magnet did upon it.

*Art. 7.*
The inſatisfaction of the *Ancients* Theory neceſſitates the Author to recur to the *Speculations* and *Obſervations* of the *Moderns*, concerning the Attraction of Iron by a Manner; and the Reduction of them all to a few Capital *obſervables. viz.*

But, all this, we confeſs, though it conferr ſomewhat of ſtrength and plainneſs to the opinion of *Epicurus*, cannot yet be extended ſo farr, as to equal the length of our Curioſity, concerning the Reaſon of the Coition of the Loadſtone and Iron; and therefore it imports us to ſuperadd thereunto ſo much of the *Speculations* and *Obſervations* of our Modern Magnetarian Authors, *Gilbert, Cabeus, Kircher, Grandamicus, &c.* (who have with more profound ſcrutiny ſearched into, and happier induſtry diſcovered much of the myſtery) as may ſerve to the enlargement at leaſt, if not the full meaſure of our ſatisfaction. And, in order hereunto, to the end Perſpicuity and Succinctneſs may walk hand in hand together through our whole enſuing Diſcourſe; we are to compoſe it of ſundry OBSER-VABLES: ſuch as may not only conduct our Diſquiſitions through all the dark and ſerpentine wayes of Magnetiſm, and acquaint us with the ſeveral Laws of Magnetick Energy; but alſo, like the links of a Chain, ſuſtain each other, by a continued ſeries of mutual Dependency and Connexion.

The

The FIRST OBSERVABLE is, *that as well the Loadstone, as its be-*
*loved Mistress, Iron, seems to be endowed with a Faculty, that holds some*
*Analogy to the sense of Animals* ; and that principally in respect of *At-*
*traction.* For (1) as an Animal, having its sensory invaded and affected
by the species of a grateful object, doth instantly desire, and is accordingly carried, by the instruments of Voluntary motion, to the same : so
likewise so soon as a lesser or weaker Loadstone, or piece of Iron, is invaded and percelled with the species of a greater or more potent one,
it is not only invited, but rapt on toward the same, by a kind of nimble
Appetite, or impetuous tendency.

(2) As sensible objects do not diffuse their species of Colour, Odour,
Sound, &c. to an Animal at any distance whatever, but have the spheres of
their Diffusion or transmission limitted : so neither doth the Loadstone,
nor Iron transmit their Species or Emanations each to other, at any distance
whatever, but only through a determinate interval of space, beyond which
they remain wholly insensible each of others virtue.

(3) As a sensible object, that is convenient and grateful, doth by its
species immitted into the sensory of an Animal, convert, dispose, and attract the Soul of the Animal ; and its soul being thus converted disposed
and attracted toward that object, doth by its Virtue or Power, carry the
body, though gross and ponderous, along to the same : exactly so doth
the Loadstone seem, by its species transfused, to convert, dispose and attract towards it the (as it were) soul, or spiritual substance of Iron ; which
doth instantly by its power or vertue, move and carry the whole mass, or
grosser parts of it along to an union with the same. Certainly, it would
not easily be believed, that a thing so exile and tenuious, as is the Sentient Soul of an Animal (which is only *Flos substantiæ,* the purer and sub-
tler part of its matter) should be sufficiently potent to move and from
place to place transfer so ponderous and unweildy a mass, as that of the
Body ; unless our sense did demonstrate it unto us, and therefore, why
should we not believe, that in Iron there is somewhat, which though it be
not perfectly a Soul, is yet in some respects Analogous to a Soul, that
doth though most exile and tenuious in substance, move, and transferr the
rest of the mass of Iron, though ponderous, gross and of it self very unfit for
motion ? All the Difficulty, therefore, which remains, being only about the
Manner, How the Sentient Soul of an Animal is affected by and attracted
toward a Grateful Object, let us conceive, that the sensible species, being
it self Corporeal, and a certain Contexture of small particles effluxed from
the object, such as do gently and pleasantly commove and affect the Organ
of Sense, being once immitted into the Sensory, doth instantly move the
part of the Soul, (which is also Corporeal, and a certain Contexture of
small particles) inhærent or resident in that Organ, and evolving the
particles of the Soul converted (perchance) another way, and turning them
about toward that part, from whence themselves are derived, i. e. toward
the object, it doth impress a kind of impulse upon them, and so determine
and attract the soul, and consequently the whole Animal, toward the object. For, admitting this Conception, we may complete the Parallelism
intended, thus ; as the particles of a sensible species, transmitted from
a grateful object, and subingressing through the organ into the contexture
of the Soul, or Sentient part thereof, do so sollicite it, as that it becomes
converted

converted toward, and is carried unto that particular object, not without a *certain* impulse of appetite: so do the particles of the Magnetical species, subingressing into the Soul of the Iron, so evolve its insensible particles, and turn them toward the Loadstone, as being thus sollicited, it conceives a certain appetite or impetus toward the same, and which is more, forthwith resalutes it, by diffusing the like species toward it. For, as if the Iron were before asleep and unactive, it is awakened and excited by this exstimulation of the Magnetical Species; and being as it were admonished, what is the propriety of its nature, it sets it self nimbly to work, and owns the Cognation. But, by what other way soever it shall be explicated, How an Animal is affected by, and rapt toward a sensible object: by the same way may it still be conceived, how Iron is affected by, and rapt toward a Loadstone. For, albeit as to divers other things, there be no Analogy betwixt the Nature and Conditions of an Animal, and those of Iron: yet cannot that Disparity destroy the Analogy betwixt them in point of *Alliciency* or *Attraction*, here supposed. Which well considered, *Scaliger* had no reason to charge *Thales Milesine* with ridiculous Madness, for conceding the Loadstone and Iron to have *Souls:* as Dr. *Gilbert* (*lib. 2. de Magnet. cap. 4.*) hath observed before us.

*Art. 9.*
That the Loadstone & Iron interchangeably operate each upon other, by the mediation of certain Corporeal Species, transmitted in Rayes: and the Analogy of the Magnetick, and Luminous Rayes.

The SECOND; that forasmuch as betwixt the Loadstone and its Paramour, Iron, there is observed not only an Attraction, or mutual Accession, or Coition, but also a firm Cohæsion of each to other, like two Friends closely entwined in each others arms; and that this Cohæsion supposeth reciprocal Revinction, which cannot consist without some certain corporeal Instruments, that hold some resemblance to Lines and Hooks: hence is it warrantable for us to conceive, *that the species diffused from the Loadstone to the Iron, and from the Iron to the Loadstone, are transmitted by way of Radiation, and that every Ray is Tense and Direct in its progress through the intermediate space, like a small thread or wire extended, and this because it consisteth of Myriads of small particles, or Atoms flowing in a continued stream, so that the precedent particles are still urged and protruded forward, in a direct line, by the consequent;* after the same manner as the rayes of *Light* flowing from a Lucid body, the Cause of whose Direction must be their Continued Fluor, as we have formerly Demonstrated, at large. We may further conceive, that as the rayes of Light do pass through a Perspicuous body; so do the Magnetical rayes pass thorow the body of Iron. That as among all the Lucid rayes incident upon a Perspicuous body, whose side obverted to the Luminary is of a Devex figure, only one ray, *viz.* that which falls upon the middle point or centre, is directly trajected; and all the rest are inclined or refracted toward that Direct one, in their progress through the aer beyond the Diaphanous body: so is only one of the Magnetick rayes, incident upon Iron, directly trajected through the same, and all the others are refracted or deflected toward that one direct. Only here is the Disparity, that from the Diaphanous body to the Luminary no rayes are interchangeably transmitted: but from the Iron to the Loadstone there are; and of these also, in their permeation thorow the Loadstone, only one is direct, and all the rest deflected toward that one. That forasmuch as these Magnetick rayes, being hence and thence refracted, and accordingly passing thorow the pores of the body of the Iron, on one side, and those of the Loadstone, on the other; do variously intersect each other at certain Angles, and in respect of those angles, become like so many Arms embowed, or Chords inflected; and so perstringe the solid particles interjacent among the pores: thence doth it come to pass, that the whole masses or bodies being thus, on this side and that interchangeably perstringed, there ensues the mutual Adduction of the one to the other, or of the less or weaker to the greater or stronger; and consequently the Cohæsion of the one to the other, the Devinction being, as the Adduction, reciprocal. We need not advertise, that the Magnetick rayes are so much stronger and tenser than the Luminous; by how much they are more Subtile and Agile: being such as that in a moment they pass thorow a very great mass of Marble, which the rayes of Light cannot doe. Nor that the Magnetique rayes do not attract Marble, though they do attract Iron posited beyond it; nor strawes, or other lighter things interposed: because, except the Loadstone and Iron, no other bodies whatever do reciprocally emit and effect each other with their rayes; nor have they that Disposition of their Pores or passages, which is necessary to the determinate Refraction of the Magnetique rayes, and to the constriction of their solid particles thereby.

The THIRD; the Magnetique Species being diffused by Deradiation Excentrical, and the Attraction of the Loadstone (of a Spherical figure) being therefore *Circumradious*, or from all points of the circumference of its sphere of Energy: it will be requisite that we allow it to have (1) a *Centre*, as that which is on all sides Corroborated by all the circumstant parts; (2) an *Axis*, as that to which the virtues of all the circumjacent Fibres are contributed; (3) the *Diametre of an Æquator*, which lying in the middle of all its Fibres, may also contain the strongest virtue of them all. For, having conceded this Geometrical Distinction of parts to a *Terrella*, or Spherical Magnet; we shall reap this advantage thereby, that we shall easily comprehend and describe the several reasons of Laws and Experiments Magnetical. To particularize; insomuch as the Magnetique Rayes are diffused from the Centre of the Loadstone to all points of it superfice, and beyond it to the bounds of their Orb of Activity; that ray, which passeth through either of its Poles, doth attract only by the force of the Axis; and that, which passeth through the Æquator, draws only by the force of the Diametre of the Æquator; and the other rayes, which like Meridians, pass through the other parts, draw by a Compound or Complicated force, insomuch as they are alwayes intermediate betwixt one ray, which proceeds directly from the Axis, and is parallel to the Æquator, and another which comes directly from the Diametre of the Æquator, and is parallel to the Axis. And, because the Æquator is æquidistant from either Pole; thence is it, that an Iron Obelus, or Needle, being præsented thereunto, shall be drawn parallel to the Axis, and in a direct line to the Diametre of the Æquator: because all the rayes expiring from the Axis, as they are the longest and strongest of all others, so are they also on each hand Equal, and equally attractive of the Extremes of the Needle; so that when it cannot incline to one Pole more than to the other, as being æquilibrated by two equal rivals, it must consist in the middle betwixt them both. Again, if the Needle be præsented to any part of the Terrella, beyond the Æquator, toward either Pole, in this case, because the ray issuing from the Diametre of the Æquator doth then display its virtue to the height, and that ray which is derived

from

*Art.* 10.
That every Loadstone, in respect of the Circumradiation of its Magnetical Aporrhæa's ought to be allowed the supposition of a Centre, Axis, and Diametre of an Æquator, and the Advantages thence accrewing.

from the Axis, is not of so much power as another longer one passing through, or near to the Æquator : therefore shall the extreme of the Needle, toward the nearest Pole, feeling that stronger virtue, be somewhat inclined ; as if affecting to be conformed to that ray, which is direct to the Diametre of the Æquator ; and it shall be always inclined so much the more, by how much longer that ray is, and the other, profluent from the Axis, the shorter.    Lastly, because in approaching very near to the Pole, the one ray becomes very long, the other very short ( comparatively ) ; and so the Needle must be now almost right to the Æquator : thence comes it, that at the very Pole, that Extreme of the Needle, which regards it, shall cohære to the Pole, and so the Needle shall be disposed in the same line with the Axis it self.

*Art.* II.
The Reason of that admirable *Biforin,* or *Janus-like* Faculty of Magneticks : and why the Poles of a Loadstone are incapable, but those of a Needle easily capable of *Transplantation* from one Extreme to the contrary.

The FOURTH ; the Loadstone being of such singular Contexture, and so admirably comparated by Nature, as that while it remains whole, the one half of its particles have a certain Polary respect, or manner of Conversion to one part, and the other half to the opposite part ; and when it is cut in two at the Æquator, each segment, which formerly had all its particles converted one and the same way, doth in a moment alter their respect, and convert the one half of them to one part, and the other to the Contrary part : therefore doth a Needle ( invigorated ) though all its particles were before indiscriminately and confusedly posited, likewise in a moment obtain a Conversion of one half of its particles to one part, and of the other half to the contrary part ; and this either from its long situation above the earth, or affriction to a Loadstone, or to another Needle strongly Magnetified.    And this is that prodigious Propriety of Magnetical Bodies, which *Cabeus* calls *Facultatem Duarum facierum,* a Faculty of Two Faces ; and *Kircher* [ δίμορφος ] *Biformem Facultatem :* though they differ beyond reconciliation in their reasons, or Explications of it.    But, though this Janus Quality be in common as well to Iron, as to the Loadstone it self ; to the former, onely by infusion, to the latter by essence : yet are we to allow this Difference, that the Poles of the Loadstone are never to be changed from one extreme to the other ; but those of a Needle are easily capable of transplantation, so that the Cuspis, which now is strongly affected to the North, may in a minute be alienated and inspired with a respect to the South, onely by a præposterous Affriction of it to the Loadstone.    And hence comes it, that as the North pole of one Loadstone doth not attract or unite with the North pole of another Loadstone ; so doth not the North Cuspis of a Needle conform it self to the North pole of a Loadstone ; provided it be only præsented, not applyed, or affricted upon it.    For, from the last Touch or Affriction of the Loadstone, the Cuspis of a Needle acquireth a Verticity è diametro opposite to its former :    in case it be rubbed upon a contrary pole, or upon the same pole with a contrary wipe or Ductus.    Hence also is it, that if you fill a Quill with the Filings or Powder of a Loadstone, and offer it to either of the Poles of a whole Loadstone ; it shall remain altogether insensible of its influence, and acquire no Verticity at all : because all the Granules of the Powder, intruded into the quill, have their poles confused, some respecting this, others that, others a
quite

quite contrary region. But, if you exchange the Filings of Loadstone for the Filings of Steel, and offer either of the extrems of the quill to either Pole of a Loadstone; it shall instantly own the Magnetique influx, and be imbued with the Polary Virtue, or Directive Faculty thereof : and this, because all the Granules of the Steel powder, wanting determinate poles of their own, are indifferently disposed to admit and retain the virtue of either Pole of the Loadstone, in any part.

If this be true, you'l ask us, *How it comes about, that the Northern Pole of one Loadstone doth not only not Attract, but nimbly Repel or Avert the Northern Pole of another Loadstone, if they be brought within the orb of their power ?*

And we *Answer*; that the Aversion is not really from the Repulsion of one North Pole by the other, but from the *Attraction* of the South Pole, which is felt and owned at that distance : but, because the South Pole cannot be detorted toward the North, but the North Pole of the other Loadstone must receed and veer from it; therefore doth that conversion seem, indeed, to be a kind of Fugation, which really is only an Attraction. The same is to be understood of the Austrine Pole of one Loadstone, in respect of the Austrine Pole of another; and also of either Cuspis of a Needle excited as well in respect of another Needle invigorated, as of a Loadstone. The same also of a Loadstone dissected according to its Axis, when the Divisions or Segments being never so little dissociated, doe not attract each other respectively to their former situation; but the Austral part of the one segment is wheeled about to the Boreal part of the other : and so of the other Poles : the contrary whereunto alwayes happens, when a Loadstone is dissected according to the Æquinoctial.

*Art. 12.*
An Objection, of the Aversion or Repulsion of the North Pole of one Loadstone, or Needle, by the North Pole of Another: *prævented.*

And from this one Fountain flow these Three Magnetique *Axioms.* (1) *Contraria Contrarijs sunt amica; similia similibus Inimica:* i. e. Magnetical Poles of the same Aspect and Apellation, are alwayes Enemies, and decline both commerce and conjunction each with other; and Poles of a Contrary respect and denomination, are alwayes Friends, and affect and embrace each other. For, to all Magneticks this is singular; that those parts, which are friends each to other, ever regard opposite regions, and convert to contrary points; but those, which are Enemies, regard the same region, and convert to the same point : because Friendly parts may constitute the same Axis; but Adverse cannot.

*Art. 13.*
Three principal *Magnetique Axioms,* deduced from the same Fountain

(2) *Quæ eadem sunt uni tertio, non sunt eadem inter sese;* i. e. Two Poles of the same respect and name, are both Friends to a Third pole of the Contrary respect and name : but yet they are Enemies and irreconcileable among themselves. And hence comes it, that a third Pole, being offered to either of two friendly Poles, cannot be a common friend, but a necessary Enemie to either. For, those Poles, which are Friends, are of a contrary respect, one Septentrional, the other Meridional : to which a Third cannot approach, unless it be a Meridional, that shall be an Enemy to the Meridional, or a

E e e                                                                 Septen-

Septentrional; that shall be an Enemy to a Septentrional : because Poles of the same Aspect, cannot compose the same Axis, but those of a Contrary doe. And this starts up another singularity of Magnetiques; that there can be no more than Two Twinds : insomuch as more than Two cannot compose the same Axis, in the same part.

(3) *Virtus ex eadem fonte petita, inimica & noxia ; ex Contrarys fontibus, amica & jucunda.* For, if you imbue the Heads of two Needles with the virtue of the same Pole, their Heads shall reciprocally turn away each from other, and mutually destroy each others verticity : but, if you imbue them with the virtue of Contrary poles, they shall unite and mutually conserve each others verticity. Likewise, if a long Needle be applyed, in the middle, to either pole of a Loadstone, and then be cut off in the place of the late Contact; the New Extremes (formerly united in the middle) shall instantly display Contrary Virtues, and reciprocally avoid each other.

*Art.* 14.
A DIGRES-
SION to the
Iron Tomb of
*Mahomet.*

And here, our Oath of Allegiance to Truth, whereby we are obliged to serve Her upon all occasions, will excuse our Digression, if we step a little aside to the so famous Sepulchre of that greatest of Impostors, *Mahomet*, and observe how egregiously false that common report is, concerning the suspension of his Iron Tomb in the Aer, by the equal Virtues of two Loadstones, the one fixt above in the arched roof, the other beneath in the floor of his Temple at *Medina Talnabi* in *Arabia.* If we consult the Relations of Travellers concerning it, we shall not only not meet with any, who affirms it upon any other grounds, but the Tongue of Popular Fame, and tradition of the multitude : but also with some, that expresly Contradict it ; for, as *Vossius* tells us, both *Gabriel Sionita*, and *Johannes Hesronita*, two learned *Maronites*, who journied to *Medina* on purpose to satisfie themselves and others in that point, positively deliver, that the Tomb of Mahomet is made of White Marble, and stands upon the ground in the East end of that Mosque.

*Les Voyages Fameux Du Sieur Vincent
Le Blanc Marseillois,* p. 21. l. 1, c. 4

*Quant à la ville de Medine, quelques-uns ont donné à entendre que le Sepulchre de Mahomet estoit là, ou à la Meque, tout de fer & suspendu en l' air par le moyen de quelques pierres d' aymant : Mais c' est une chose tres fausse, estant bien certain, comme ie l' ay appris sur le lieu mesme, que ce faux Prophete mourut & fut enterré à Medine, où l' on voit encore son sepulchre fort frequenté de pelerins Mahometans de tous les quartiers du monde, comme est le Sepulchre de Jerusalem de tous les Chrestiens. Ce Sepulchre est de marbre blanc ; avec les tombeaux de Ebnbeker, Ali, Omar, & Otman Califs, successeurs de Mahomet, chachun ayant au pres de soy les livres de sa vie & de sa Secte, qui sont fort divers ; &c.*

And

And, if we confult our own Reafon, confidering the fetled and unalterable Laws of Magnetical Attraction; we fhall foon be confirmed not onely of the monftrous Falfity, but abfolute Impoffibility of the Effect. For, fhould we grant it to be in the power of humane induftry, to place an Iron fo præcifely in the neutral point of theMedium betwixt two Loadftones, equally attracting it, the one upward, the other downward; as that the Gravity of the Iron, and downward Attraction of the Inferiour Loadftone might not exceed, nor be exceeded by the upward Attraction of the Superiour Loadftone, and fo the Iron fhould remain, without any vifible fupport, Æquilibrated betwixt them, in the Aer: yet could not that pofition of the Iron be of any Duration; becaufe, upon the leaft mutation of the temper of the Iron, or motion of it by the waving of the Aer from high winds, and divers other caufes, the Æquilibration muft ceafe, and the Iron immediately determine it felf to the Victor, or ftrongeft Attractor. But, fince what is here fuppofed, is wholly repugnant to the Experience of all, who have or fhall attempt fo to æquilibrate an Iron in the Aer betwixt two Loadftones, as that it fhall not feel the Attractive Virtue of one more ftrong than that of the other: we need not long ftudy what to think of the fufpenfion of Mahomets Iron Cheft.

Nor is it lefs impoffible, that an Iron fhould be held up, at diftance, in the Aer, by the Virtue of a Loadftone placed above it: infomuch as that force, which at firft is fufficient to overcome the refiftence of the Irons Gravity, and elevate it from the ground, muft, as the Iron approacheth nearer, be ftill more potent to attract it; and fo that cannot oppofe the Attractive Energy of the Loadftone, in the middle of it fphere, which was forced to fubmit and conform unto it, in the Extremes. This we may foon experiment, with a Needle by a thread chained to a table, and elevated perpendicularly in the aer, by the pole of a Loadftone: for, the Needle will nimbly fpring up to meet the Loadftone, fo farr as the thread will give it fcope; and if the thread be cut off, it inftantly quits the medium, and unites it felf to its Attractor, from whofe embraces it was before violently detained. Hereupon as we may affure our felves, that *Dinocrates*, that famous Architect, who, as *Pliny* relates (*lib.* 34. *cap.* 14.) began to Arch the Temple of *Arfinoe* in Alexandria, with Loadftones; that fo Her Iron Statue might remain Pendulous in the aer, to excite wonder and Veneration in the Spectators; but was interrupted in the middle of his Work both by his own death, and that of *Ptolomy*, Arfinoes Brother, who expired not long before him; died moft opportunely in refpect of his Reputation, becaufe He muft have failed of the chief Defign, though he had lived to finifh his ftructure: fo alfo can it be no longer doubted, that *Ruffinus* his ftory, of the Iron Chariot in the Temple of *Serapis*, and *Bedas* of the Iron Horfe of *Bellerophon*, fuftained by Loadftones fo cunningly pofited, as that their Virtues concurr and become adjufted in one determinate point; are meer Fables, and fit to be told by none but doating old women in the chimney corner.

*Art. 15.*
*That the Magnetique Vigour, or Perfection both of Loadstones and Iron, doth consist in either their Native Purity and Uniformity of Substance; or their Artificial Politeness.*

The FIFTH; *As one Loadstone is stronger in its Attractive Virtue than another, though of the same. nay, perhaps, much greater bulk and weight: so is some Iron more disposed than other, both to admit and conform to the Attraction of a Loadstone, and, after invigoration, to attract and impregnate other Iron.* As for the *Vigour* and *Perfection* of a Loadstone; it consisteth both in its Native Purity, and Artificial Politeness. (1) In its *Native Purity*; for, if no Dross or Heterogeneous substance be admixt to the Magnetick Vein in the earth, from which a Magnet is extracted; then is that Loadstone superlatively potent and energetical in Attraction: and among Loadstones of this sincere and homogeneous Constitution, there are found no degrees of Comparison, but what the Difference of their several Bulks doth necessarily create. But, in case any Heterogeneous matter be commixt with the Magnetick seeds or particles of a Loadstone, at its Concretion; as it for the most part falls out : then must the Attractive Energy of that stone be weaker, according to the proportion of that spurious matter admixed thereunto. This may be confirmed from hence; that some very small Loadstones are more potent than very Great ones; of which sort shall we account that of which *Mersennus* (*de Magnete*) affirms, that weighing but 7 Gr. in all, it would nimbly attract and elevate a mass of Iron 17 times higher than it self: and from hence, that some stones that were dull and languid before, after the secretion of their Drossy and Impure parts, become very active and potent. Thus, when any Heterogeneous substance hath been, like a Cortex or shell circumobduced about a Loadstone, in its concretion; if the same be pared or filed away, and the remaining Kernel be polished; its Virtue shall be augmented to a very great proportion. (2) *In its Artificial Terseness or Politeness*; for, by how much smoother a Loadstone is, in it superfice, with so many the more rayes of Virtue, both Attrahent and Amplectent or Connectent, doth it touch Iron oblated unto it, and *è contra.* Likewise, as for the more or less *prædisposition of Iron*, both to receive the Attractive influence of a Loadstone, and, after excitement to attract other iron; this also consisteth either in its more or less of *Native Purity*, or of *Acquired Politeness*: because, how much the nearer it comes to the pure nature of Steel, by so many the more parts hath it both Unitive unto the Loadstone, and susceptive of its rayes; and by how much more smooth and equal it superfice is made, by so many more are the parts, by which it doth touch and adhære unto the Loadstone; and consequently imbibe so much the more of its Virtue, and *è contra.*

And this introduceth

*Art. 16*
*That the Arming of a Magnet with polished Steel, doth highly Corroborate, but as much diminish the sphere of its Attractive Virtue.*

The SIXTH OBSERVABLE; *That a Loadstone, being Armed or Capp't with steel, is thereby so much Corroborated, that it will take up a farr greater weight of Iron or Steel, than while it remained naked or unarmed.* For, *Mersennus* had a Loadstone, which, (as himself avoucheth) being naked, could elevate no more than half an ounce of Iron; but when he had armed it with pure and polisht steel, it would easily suspend 320 times a greater weight, i. e. ten pounds of Iron: a proportion not credible, but upon the certificate of Experiment. Now, the *Cause* of this admirable Corroboration of the Loadstones Attractive Virtue, by a plate of polisht Steel, can be no

other

other than this, that the Loadstone being of such a rough contexture, as that in respect of the particles of some heterogeneous matter concorporated unto it, it is uncapable of that exquisite smoothness in the surface, which may be obtained by steel; therefore can it not touch Iron so exquisitely, or in so many points, as Steel may: and consequently not invade it with so many Direct and united rayes. But, Steel being of a more simple substance, and close contexture, may in all its substance be imbued with the Magnetique Virtue: and being polisht, touch an Iron, to which it is admoved, with more parts, and invade it with more dense and united rayes. For, those indirect rayes, which otherwise the Loadstone would diffuse scatteringly through the Medium, in respect of the various inequalities of it superfice, and multitude of small pores intercepted among its particles; the Steel doth recollect, unite and transmit to the Iron admoved, and thereby more strongly embrace and detain it. We say, *To Iron Admoved*; For, though the *Retentive* Virtue of a Loadstone Armed with Steel, be by many degrees stronger; yet is its *Attractive* Virtue by some degrees weaker than that of an unarmed Loadstone: i. e. it doth not diffuse its Attractive virtue half so farr, and a sheet of the finest Venice paper interposed betwixt an Armed stone and Iron, doth impede its Attraction; a manifest argument, that the Fortification is determined only to contact. This we confess *Mersennus* flatly denies, and upon his own observation : but till our Reader shall meet with such a stone, as *Mersennius* used, we advise him not to desert the common Experience of the impediment of the Attraction of Iron by an Armed Loadstone, by paper interposed, since *Grandamicus*, whose chief business was the exact observation of all Magnetique Apparences, expresly faith; *vix fit adhesio ferri ad lapidem armatum, si vel Charta, vel aliud tenuissimum Corpus interponatur.* It hath, moreover, observed, that if a Magnet be perforated along its Axis, and a rod of polisht Steel, exactly accommodated to the perforation, be thrust thorow it; its orb of Attraction shall be much enlarged, and its Energy fortified to an incredible rate. *Consule Jacob. Grandamicum, in Nova Demonstrat. Immobilitatis Terræ, ex Magneticis, cap. 5. Sect. 1. pag. 99.*

Having layed down these *sixe Observables*, which are of such Capital concernment, as that there is no Effect or Phænomenon of Attraction Magnetical, that may not conveniently be referred to one, or more of them; and consigned a probable Reason to each : the onely memorable Difficulty that remains, concerning the Attractive Virtue of Magnetiques, is, *Why a small or weak Loadstone doth snatch away an Iron from a Great or more potent one ?* But, as the incomparable *Kircher* hath subtely observed, a small or weak Loadstone doth remove a Needle from a Great and Potent one, while it self remains within the sphere of the Great or strong ones activity : because the virtue of the small or weak stone, is Corroborated by the Accession of that of the Great or strong. Which is demonstrable from hence, that if the Needle be so long, that its extremes reach beyond the orb of the Great Loadstones activity; then cannot a less or less potent one remove it away and elevate it : and in case one of the extremes be somewhat too near to either Pole of
the

*Art. 17.* Why a *smaller or weaker* Loadstone, doth snatch away a Needle from a *Greater, or more Potent* one; while the small or weak one is held within the sphere of the great or stronger ones *Activity*: and not otherwise the Great Loadstone, then is the Lefs stone much lefs able to fub-ftract the Needle than in the former cafe; becaufe fo, the Virtue of the Great Loadftone is augmented by the Addition of that of the Lefs.

*Art.* 18.
COROLLA-
RY.
Of the Abdu-
ction of Iron
from the Earth
by a Lead-
ftone.

And hence, by way of COROLLARY, we obferve, that the Abduction of a piece of Iron from the Earth by a Loadftone, is fo farr from being a good Argument againft the Earths being Magnetique, or one vaft Loadftone; that it rather makes for it : becaufe the Loadftone being applied to the Iron, and operating within the fphere of the Earths Virtue, is fo Corroborated thereby, that it abduceth the Iron from it, by the fame reafon, that a Lefs Loadftone fnatcheth a Needle from a Great one. And thus much concerning the *Attractive* Faculty of the Loadftone; both according to the moft confi-derable Doctrine of the Ancients, and the more exact Theory of the Moderns.

---

## SECT. II.

*Art.* 1.
The Method,
and Contents of
the Section.

TO enquire the Reafon, therefore, of the other General Proprie-priety of Magnetiques, their DIRECTION, or Converfion of their Poles to North and South; is all the remainder of our præ-fent Defign : which that we may accomplifh with as much plainnefs and brevity, as the quality of the Argument will admit of; we fhall obferve the fame advantageous Method of Difquifition as we have done in the former, touching the Caufes and Wayes of Magnetique Attraction, reducing all the obfervations of the Moderns, of the *Di-rection, Declination,* and *Inclination* of the Loadftone, and other Mag-netical bodies, to certain Heads, and difpofing them according to their or-der of fubalternate dependency.

*Art.* 2.
Affinity of the
Loadftone
and Iron.

The FIRST OBSERVABLE is; that the *Loadftone and Iron are Twinns in their Generation,* and of *fo great Affinity in their Na-tures,* that Dr. *Gilbert* might juftly fay, *that a Loadftone is Iron Crude, and Iron a Loadftone excocted* : For they are for the moft part found lodg-ed together in the fame fubterraneons bed; as the experience of all fuch as are converfant about Iron Mines in Germany, Italy, France, England, and moft other Countries, doth every day demon-ftrate.

And that is the moft probable Caufe, that can be given, why Loadftones generally are fo much the more Vigorous and perfect, by how much deeper in the Veins of Iron Mines they are digged. There is, indeed, a report diffufed not only among the People, but alfo fome of the higheft form of Learned Writers, and chiefly derived from the authority of *Strabo*; that in the Weftern Ocean are certain vaft *Magne-tick Rocks,* which drawing Ships that fail near them (by reafon of the Iron pinns, wherewith their ribbs and plancks are faftned, and held together)

gether) with irrefiftible violence and impetuofity, fplit them in pieces, or extracting the Iron pinns, carry them like arrowes flying to a Butt, through the aer : But, the light of Navigation hath long fince difcovered this ftory to be as highly Romantique, as the Enchanted Caftles of our Knights Errant, or the moft abfurd of Sir *John Mandevils* Fables; and herein we may fay of *Strabo*, as *Lucian* of the Indian Hiftory of *Ctefias* the Cnidian, Phyfician to *Artaxerxes* King of Perfia, *fcripfit de ijs, quæ nec ipfe vidit unquam, neque ex ullius fermone audivit.*

The SECOND; *That the Loadftone feems not only to have all the Conditions of the Terreftrial Globe, but also to imitate the positional respects thereof, conforming it self exactly unto it.* For, as the Terraqueous Globe hath *Two Poles*, by which it owns a refpect to the Poles of the Heavens, the one *Boreal*, the other *Auftral* : fo likewife hath the Loadftone two contrary Poles, alwayes difcoverable in the oppofite parts or extremes thereof, efpecially if it be turned into a fphere. And, as the Globe of the Earth hath an *Æquator*, *Parallels* and *Meridians*; fo hath the Loadftone : as may be demonftrated to the eye, by applying a fmall Steel Needle thereunto; for, at either of its Poles, the Needle fhall be erected perpendicularly, and lye in the fame line with its Axis; but at any of the intermediate Spaces, or Parallels, it fhall be neither plainly erected, nor plainly lye along, but obferve an oblique fituation, and more or lefs oblique, according to the variety of the Parallels; and at the middle interftice, or Æquator, it fhall difpofe it felf in conformity to the ductus of the Meridian, and fix in a pofition parallel to the Axis of the Loadftone.  That a Loadftone doth accommodate it felf exactly to the Earth, as a Needle doth accommodate it felf to the Loadftone; is evinced from this eafie Experiment. If you fufpend a Loadftone (whofe Poles you have formerly difcovered, and noted with the Characters, *N. S.*) in calme aer, or fet it floating at liberty in a veffel of Quickfilver, or a fmall Skiff of Cork fwimming upon Water, that fo it may freely perform the office of its nature; you fhall obferve it continually to move it felf from fide to fide, and fuffer alternate Vibrations or accefles and recefles, till it hath fo difpofed it felf according to the Meridian, as that one of its Poles, viz. that marked with *N.* fhall repair to the North, and the other, upon which *S.* is infcribed, to the South.  Nor that only, but, forafmuch as *England* is fituate near the North of the Earth, and fo hath the North pole fomewhat demerfed or depreffed below the horizon, nearer than the South Pole of the Earth: therefore doth not the Loadftone keep up both its Poles in a level or perfectly horizontal pofition, but depreffeth that pole which affects the N, fomewhat below the plane of the horizon, as much as it can, directing the fame to the *N.* pole of the Earth. Farther, being it is commonly obferved, that this Depreffion (fome call it the DECLINATION, others the INCLINATION) of the *N.* pole of the Loadftone, or point of an excited Needle, is fo much the greater, by how much nearer the ftone or needle is brought to the Boreal part of the Earth; fo much lefs, by how much nearer to the Æquator: therefore may we conclude

*Art. 3.*
The *Loadftone*
conforms it
felf, in all re-
fpects, to the
*Terreftrial*
*Globe*; as a
*Needle* con-
forms it felf to
the *Loadftone*

clude, that a Loadstone, being removed, in the same position of freedome, from the Æquator by degrees to each of the Earths poles, would more and more depress or decline its Boreal pole, by how much it should come nearer and nearer to the Boreal pole of the Earth; and on the otherside of the Æquator, more and more decline its Austral pole to the Austral pole of the Earth, by how much nearer it did approach the same; nor could it lye with both poles above the horizon at once, in any part of the Earth, but upon the Æquator, and at either of the Poles of the Earth, the Axis of the stone would make one with the Axis of the Earth.

*Art. 4.*
*Iron obtains a Verticity, not only from the Loadstone, by Affriction, or Aspiration; but also from the Earth it self: and that according to the laws of Position.*

The THIRD; That Iron acquireth a Verticity not only from the touch or affriction of a Loadstone, but also from its meer situation in, upon, or above the Earth, in conformity to the poles thereof. For, all Iron barrs, that have long remained in Windows, Grates, &c. in a position polary, or North and South; if you suspend them in æquilibrio by lines in the aer, so as they may move themselves freely, according to the inclination of their Virtue received from the Earth, will make several diadroms hither and thither, and rest not untill they have converted to the North that extreme, which in their former diuturne position regarded the North, and that to the South, which formerly respected the South : and having recovered this their Cognation, they shall fixe in a Meridional posture as exactly as the Loadstone it self, or a Magnetified Needle.

To experiment this, the most easie way is to offer, at convenient distance, a Magnetick Dial, or Marriners Compass, to the extrems of an Iron barr, that hath long layn *N* and *S* : for, then may you soon observe the Needle or Versory freely equilibrated therein to be drawn in that point, which respecteth the North, by that extreme of the barr, which is Australized, and, on the contrary, the South point of the Needle to be drawn by that extreme of the barr, which is Borealized. This Vertical imprægnation of Iron meerly by the Earth, is also evidenced from hence; that Iron barrs made red hot, and then set to cool in a Meridional position, do acquire the like polary Cognation, and being either at liberty of conversion suspended by small Chords in the aer, or set floating in small boats of Cork, or applyed to the Needle of a Pixis Nautica, immediately discover the same.

This being most manifest, why may not our Marriners, in defect of a Loadstone, make a Needle or Fly for their Chard, of simple Iron alone; since, if it hath layn in a Meridional situation above the earth, or been extinguished according to the same lawes of position, it will bear and demonstrate as strong an affection to the poles of the Earth, as a Needle invigorated by a Loadstone, nor shall the Depression or Declination of the one, in each degree of remove from the Æquator toward either pole, be less or greater than that of the other.

The

The FOURTH; that infomuch as both the Loadstone and Iron *Art. 5.* have fo neer a cognation to the Earth, and conformity of fituation to One and the fame Nature, the parts of it: nothing, certainly, can feeme more confentaneous, in common to than that they both hold one and the fame nature in common the Earth, with the Earth, at leaft with the Internall parts, or Kernell there- *Loadftone and* of; but yet with this difference, that Iron, being a part of the *Iron.* Earth very much altered from its originall conftitution by the activity of its feminall principle, cannot therefore fo eafily manifeft its extraction, or prove it felf to be the genuine production and part thereof, without præcedent Repurgation, and Excitation, or frefh Animation from the Effluviums of the Earth; but a Loadftone, having not undergon the like mutations from concoction, and fo remaining nearer allied to the Earth, doth retain a more lively tincture of its polary faculty, and by the evidence of fpontaneous Direction demonftrate its Verticity to be purely native, and it felf by confequence, to be onely a divided part, or legitimate iffue of the Earth. Further, from hence, that the Loadftone and the Terreftriall Globe have both one and the fame power, though in different proportions, of impregnating Iron with a polary affection, impreffing one and the fame faculty thereupon; it is juftly inferrible, that the Loadftone, not onely in refpect of other Conditions wherein it refembleth the Earth, but alfo, and in chief of this noble Efficacy of invigorating and renovating the magneticall quality of Iron, may well be accounted (as the Father of Magnetique Philofophy, *Dr. Gilbert* hath named it) Μικρόγη, *Terrella*, the Globe of Earth in epitome, and that the Earth it felf may be reputed *Ingens Magnes*, a Great Loadftone. Though, in truth, the Earth may challenge the title of a Great Magnet by another right, though fomewhat lefs evident, and that is its Attraction of all terrene bodies in direct lines to it felf (as we have formerly made moft verifimilous, in our Chapt. of Gravity and Levity) by the fame way and inftruments, as the Loadftone attracteth Iron. And though it cannot be denied, that the Cortex of the Terreftriall Globe, which may be many miles thick, is varioufly interfperfed with waters, Vapours, exhalations, ftones, metalls, metalline juices, and divers other diffimilar and unmagneticall bodies: yet notwithftanding may we juftly conceive, that the *Nucleus* Kernell or interior part of the Earth is a fubftance wholly Magneticall, and that many Veins or branches thereof, being derived unto the exterior parts, are thofe very fubterraneous Veins from which by effoffion Loadftones are extracted. Efpecially fince nature doth invite us to this conception by certain clear evidences not onely in Iron, which may be digged out of moft places in the Earth, but alfo in moft Argillous and Arenaceous Concretions; all which are found to be endowed with a certain, though obfcure? Polary inclination, as appears in Bricks and Tiles, that have a long time enjoyed a meridionall fituation, regarding the N. with one extreme, and the S. with the other, or been made red hot and afterward cooled

Fff North

north and south, or perpendicularly erected, as hath been said of Iron barrs.

*Art. 6.*
The Earth, im-
prægnating
Iron with a
Polary Affe-
ction, doth
cauſe therein
a *Locall Immu-*
*tation* of its
inſenſible par-
ticles.

The FIFTH; It being then moſt certain, that Iron obtaines a magneticall Verticity, or faculty of ſelf-direction to the poles of the earth, meerly either from its long ſituation, or refrigeration after ignition, in a poſition reſpective thereunto: we may be almoſt as certain, that this Affection ariſeth to the Iron from no other but a *Locall immutation, or change of poſition of its inſenſible particles,* ſolely and immediately cauſed by the magneticall Aporrhæa's of the Earth invading and pervading it. When we obſerve the Fire by ſenſible degrees embowing or incurvating a peice of wood, held neer it, how can we better ſatisfy our ſelves concerning the cauſe and manner of that ſenſible alteration of the figure of the wood, then by conceiving, that its inſenſible particles are all of them ſo commoved by the Atoms of Fire immitted into it ſubſtance, as that ſome of them are conſociated which were formerly at diſtance, and others diſſociated, which were formerly contingent, all being inverted and ſo changing their priſtine ſituation, and obtaining a new poſition, or locall direction, much different from their former? And, when we obſerve a rod of Iron, freſhly infected with the Polary virtue of the Earth, to put on a certain ſpontaneous inclination in its extremes, and convert it ſelf exactly according to the meridian, and with a kind of humble homage ſalute that pole of its late inſpirer, from whence it received the ſtrongeſt influence: how can we more reaſonably explain the reaſon of that effect, than by conceiving, that upon the immiſion of the Earths magneticall Rayes into the ſubſtance of the Iron, the inſenſible particles thereof are ſo commoved, diſtructed, inverted, and turned about, as that they all are diſpoſed into a new poſture, and acquire a new locall reſpect or Direction, according to which they become as it were reinnimated with a tendency, not the ſame way, but another much different, and (when the cognation of their extremes are varied by an inverted ignition and refrigeration) quite contrary to that, whither they tended before this mutation of their poſition and reſpect? This Conjecture may ſeem ſomewhat the more happy from hence; that a barr of Iron, when made red hot, doth acquire this Polary Direction in a very few minutes of time: but being kept cold, it requires many years ſituation North and South, to its imprægnation with the like virtue; a ſufficient manifeſt, that the particles of the Iron being, by the ſubingreſſion of the Atoms of Fire among them, reduced to a greater laxity of contexture, are more eaſily commoved and inverted by, and more expeditely conforme themſelves unto the diſpoſition of the magnetique influence of the Earth. When a red hot barr of Iron is cooled, not in a meridian poſition to the poles of the Earth, but tranſverſly or equinoctionally; why doth it not contract to it ſelf the like verticall diſpoſition? doubtleſs, the beſt reaſon that can be given for it, is this; that the inſenſible particles of it are

not

not converted, nor their situation varied so much in the one position of the whole mass, as in the other: the magnetically Rayes of the Earth invading the substance of the Iron in indirect and so less potent lines. Likewise, if the same barr of Iron, after it hath imbibed a Verticity, be again heated and cool'd in a contrary position, what reason can be assigned to the change of the Southern Verticity into a Northern, and its Northern into a Southern, by the contrary obversion of its ends: unless this, that the particles of the Iron doe thereby suffer a fresh conversion, and quite contrary disposition, no otherwise than those of a piece of wood, when it is incurvated by the fire according as this or that side is obverted thereunto?

The SIXTH, forasmuch as Iron doth derive the same Verticity or Direction from its Affriction against a Loadstone; as it doth from the magnetically influence of the Earth, when posited respectively to its poles: it appears necessary, that it doth suffer the same Locall Immutation of its insensible particles, from the efficacy of the magnetically rayes of the Loadstone, as from those of the Earth; especially since we cannot comprehend, how a Body should acquire a strong propension or tendency to a new place, without some generall Immutation, and that a Locall one too, of all its component particles. The strength of this our conception consisteth chiefly in this; that after a rod or needle of Iron hath contracted a sprightly Verticity from a Loadstone, by being rubbed thereupon from the middle toward the ends, it doth instantly lose it again, if it be rubbed upon the same, or any other Loadstone, the opposite way, or from either end toward the middle. For, how can it be imagined, that a right-hand stroak of a knife upon a Loadstone should destroy that polary Faculty, which it had obtained from a left-hand stroak upon the same; unless from hence, that the insensible particles of the blade of the knife, were turned one way by the former affriction, and reduced again to their former naturall situation by the latter? It seems to be the same, in proportion, as when the ears of Corn in a field are blown toward the South by the North wind, and suddainly blown from the South toward the North by the South wind. Nor doth Iron, after its excitement retain any of the magnetically Atoms immitted into it either from the Earth, or a Magnet; but, suffers only an immutation of its insensible particles, which sufficeth to its polary respect a long time after: for, a Needle is no whit heavier after its invigoration by a Loadstone, than before, as *Mersennus* and *Gassendus* together experimented, in such a *Zygostata* or Ballance, wherewith Jewellers are to weigh Pearles and Diamonds; which is so exact, that the ninety-sixth part above four thousand of a grain, will turn it either way.

*Art. 7.*
The Loadstone doth the same.

The SEVENTH, that the Virtue immitted into Iron, either from the Earth it self, or a Loadstone, is no simple, or immateriall Quality, as both *Gilbert* and *Grandamicus* earnestly contend; but a certain *Corporeal Efflux*, or Fluor, consisting of insensible bodies, or particles, which

*Art. 8.*
The Magnetique Virtue, a Corporeal Efflux.

which introduce upon the particles of Iron the same Disposition, and Local respect, as themselves have.

For (1) That an Immutation is caused in the particles of Iron, as well by the influence, or Magnetical rayes of the Loadstone (which doth also invigorate Iron, at some distance, though not so powerfully, as by immediate contact, or affriction) as of those transmitted from the Earth; we have already declared to be not only verisimilous, but absolutely necessary: & that nothing should yet be derived unto the Iron from them; as the Instrument of that Immutation; is openly repugnant to the Fundamental Laws of all Physical activity, since nothing can act upon a distant subject but by some Instrument, either continued or transmitted.

(2) What is immitted into the Iron from the Earth and Loadstone, cannot be any naked Quality, or Accident without substance; because, what wants substance, must also want all Activity.

(3) The Materiality of the Magnetique Virtue is inferrible likewise from hence, that it decayes in progress of time (as all Odours do) and is irreparably destroyed by fire, in a few minutes, and is capable of Rarity and Density (whence it is more potent near at hand, than at the extremes of it (sphere)all which are the proper and incommunicable Attributes of Corporiety.

(4) Insomuch as it changeth the particles of Iron, that have Figure and Situation; therefore must it self consist of particles also, and such as are in figure and situation consimilar to those of Iron: no less being assumable from the Effect even now mentioned, *viz.* the Ablation of that Verticity, by a right hand draught of a Needle upon a Loadstone, which it lately acquired from it, by a left hand one. Nor, indeed, doth the Loadstone seem to act upon Iron, otherwise than as a Comb doth upon wool or hair; for as a Comb being drawn through Wool, one way, doth convert and dispose the hairs thereof accordingly, and drawn præposterously or the contrary way, doth invert & præposter the former ductus of the hairs: so do the Magnetical Rayes invading and pervading the substance of Iron, one way, dispose all the insensible particles thereof according to their own ductus, toward the same way; and immitted into it the quite contrary way, they reduce the particles to their native situation and local respect; and so the formerly imprinted Verticity comes to be wholly obliterated.

*Art. 9.*
*Contrary Objections, & their Solutions.*

OBJECTED, we confess it may be; that the Incorporiety, or Immateriality of the Loadstones Virtue seems inferrible from hence, that it most expeditely penetrateth and passeth through many bodies of eminent solidity, and especial Marble: (2) That it is (Soullike) total in the total Loadstone, and total in every part thereof: seeing that into how many sensible pieces soever a Loadstone is broken or cut, yet still doth the Virtue remain entire in every one of those pieces, and there instantly spring up in each single fragment, two contrary Poles, an Axis, Æquator, Meridians and Parallels.

But,

But, as to the *subtility of Particles and Pores* in Concretions, our Book is even surcharged with discourses upon that subject, in the Generall: so that notwithstanding the first objection, we may adhære to our former Conception, that the particles flowing from the Earth and Loadstone, are of such superlative Tenuity, as without impediment to penetrate and permeate the most compact and solid Concretions, and specially Marble, whose small pores may be more accommodate to the figures of the magnetick Atoms, and so more fit for their transmission, than those of divers other bodies much inferior to it in compactness and solidity. And being we have the oath of our sense, that the Atoms of Fire doe instantly find out many inlets or pores in the body of Marble, by which they insinuate themselves into its centrall parts, and so not only calefie the whole mass or substance thereof, but reduce it suddainly into a brittle Calx: why should we not concede, that the Magnetick Atoms may likewise find out convenient inlets or pores in the same, and by them nimbly pervade the whole mass; and that with so much more of ease and expedition, by how much more subtile and active they are, than those of Fire? True it is, that we can discerne no such Particles flowing from magneticks, no such Pores in Marble, but how great the Dulness or Grosness of our senses is, comparatively to the ineffable subtility of many of Natures Instruments, by which she bringeth admirable Effects to pass, we need not here rehearse. (2) As for the other Argument desumed from the *Frustulation* of a Loadstone, we Answer; that the single Virtues of the single fragments, are nothing else but so many Parts of the Totall Virtue: nor being taken singularly, are they equally potent with the whole, only they are like the Totall, because in the whole Loadstone they follow the ductus or tract of its Fibres, that run parallel each to other, and conjoyn their forces with that Fibre, which being in the middle, stands for the Axis to all the rest. But, in each Fragment, they follow the same ductus or Grain of the Fibres, and one Fibre must still be in the middle: which becomes an Axis, and that to which all the circumstant ones confer and unite their forces.

The EIGHTH, that the Magnetick Virtue, both existent in the Loadstone, and transfused into Iron, seems by a lively Analogy, to resemble the *Vegetative Faculty* or *soul* of a *Plant*; not only in respect of the Corroboration of the force of its median Fibre, or Axis, by the conference of the forces of all the circumstant ones thereupon, as the centrall parts of a Plant are corroborated by the circumambient: but also, and principally, in respect of the *situation, Ductus, or Grain of its Fibres*; which run *meridionally*, as those in Plants perpendicularly, or upward from the roots to the tops of the spriggs. For, as in the Incision or Engraffing of the shoot of one tree, into the trunck or stock of another, the Gardiner must observe to insert the lower extreme of the shoot, into a cleft in the upper extreme of the stock, as that from whence the nutritive sap and vegetative influence are to be derived unto it; because, if the shoot were inverted, and its upper extreme inserted into the stock, it would necessarily wither and die, as being in that præposterous position made uncapable of the influx of the Alimentary juice and vitall Faculty, both which come from the root upward to the branches, and

*Art. 10.*
*A Parallelisme of the Magnetique Virtue, and the Vegetative Faculty of Plants.*

and cannot defcend again from them to the root: exactly fo, when we would difpofe a Loadftone in conformity of fituation to the Earth, from which it hath been cut off, or to another Loadftone, a quondam part of it felf; 'tis not every way of Appofition, that will be convenient, but only that, when it is difpofed in a direct line, refpondent to the fame Ductus or fituation of its Fibres, according to which it was continued to the Earth, before its feparation. Nor is this meer Conjecture, but a truth as firme as the Earth it felf, and as plain as fenfe can make it; it being conftantly obferved, that what fituation a Loadftone had in its Matrix, or minerall bed, the very fame it fhall ftrongly affect, and ftrictly obferve ever after, at leaft, while it is a Loadftone, i. e. un-till time or Fire have deftroyed its Verticity. And, as for the Ufe thereof; it is fo fruitfull, as to yield us the moft probable Reafon in Generall, for fundry the moft obfcure among all Magneticall Ap-parences.

*Art. 11.*
*Why Poles of the fame re-fpect & name, are Enemies: and thofe of a Contrary re-fpect & name, Friends.*

(1) Forafmuch as the Loadftone ever affects its native fituation, and that its Northern part did, while it remained in its matrix, ad-hære to the Southern parts of the fame magnetique vein, that lay more North, and its Southern part did adhære to the Northern part of the magnetick vein, that lay more South: therefore is it, that the North pole of a Loadftone doth never affect an union with theNorth pole of the earth, nor its South pole direct to the South pole of the Earth: but quite contrary, its North pole converts to the South, and its South to the North. So that whenever you obferve a Load-ftone, freely fwimming in a boate of Cork, to convert or decline one of its poles to the North of the Earth; you may affure your felf, that that is the South pole of the Loadftone: and è contra.

*Art. 12.*
*When a Mag-net is diffected in-to two pieces, why the Bo-reall part of the one half, de-clines Conjun-ction with the Boreall part of the other; and the Auftrall of one with the Auftrall of the other.*

(2) From the fame and no other Caufe is it alfo, that when a Mag-net is diffected or broken into two pieces, and fo two new poles created in each piece; the Boreall pole of the one half fhall never admit Coi-tion with the Boreall pole of the other, nor the Auftrall extreme of the one fragment affect conjunction with the Auftrall extreme of the other: but contrariwife, the Auftrall end fhall feptentrionate, and the feptentrionall Auftralize. The fame alfo happens, whenever any two Loadftones are applied each to other; the Caufe being Generall, viz. the Native *Ductus* or Grain of the Magnetique Fibres: which is inverted, whenever the Boreall part of a Loadftone is applied to the Boreall part of the Earth, or of another Loadftone; or the Meridionall part of a Loadftone be converted to the meridionall part of the Earth of another Loadftone; as the Ductus of the Fibres in a fhoot of a Plant is inverted, when the upper extreme thereof is inferted into the upper part of a ftock. This confidered, when we obferve the Ani-mated Needle of a Mariners Compafs, freely converting it felf round, upon the pin, whereon it is æquilibrated; that end, which direct-eth to the North pole of the Earth, muft be the South point of the Needle, and viceverfally, that muft be the North cufpis of the Needle, which confronteth the South of the Earth. And; when præfent a Loadftone to a magnetified Verfory, that part of the Loadftone muft be the North pole, to which the South cufpis of the Needle comes;

and

and that, to which the North point of the Needle approaches, muſt be the South of the Loadſtone. The ſame alſo may be concluded, of the extremes of Irons, when a Loadſtone is applied unto them; for, that part of an Iron barr, which laied meridionally, hath reſpected the North, muſt have been ſpirited by the Southern influence of the Earth; and è contra: and among our Fire Irons, the upper end muſt have imbibed the Northern influence of the Earth, and the Lower the Southern; contrary to the aſſertion of ſome of our Magneticall Phi-loſophers.

The NINTH; the Analogy of the Earth to the Loadſtone, and other magnetically inſpired bodies, being ſo great, and the Cauſe there-of ſo little obſcure; it may ſeem a juſtifiable inference, *That the Ter-rieſtriall Globe doth inwardly conſiſt of certain continued Fibres, run-ning along from North to South*, or from South to North, in one uninterrupted ductus: and conſequently, that ſince the middle Fibre is as it were the Axis, whoſe oppoſite extremes make the two Poles, in caſe the whole Earth could be divided into two or more great parts, there would inſtantly reſult in every part or diviſion, a ſpecial Axis, two ſpeciall Poles, a ſpeciall Æquator, and all other conditions as formerly in the whole Globe; ſo that the ſeptentrionall part of one piece would conjoin it ſelf to the Auſtrine part of another, and the ſeptentrionall parts reciprocally avert themſelves each from other, as the parts of a Loadſtone. And this we may underſtand to be that migh-ty and ſo long enquired Cauſe, *why all the parts of the Terreſtriall Globe do ſo firmly cohære, and conſerve the primitive Figure*; the Cohæſion, Attractive Virtue, conſtant Direction, and ſpontaneous Verticity of all its genuine parts, all whoſe Southern Fibres doe mag-netically, or individually conforme and conjoyn themſelves to the Nor-thern, and their Northern to the Southern, being the neceſſary Cauſes of that Firmneſs, and conſtancy of Figure. Impoſſible, we con-feſs, it is, to obtain any ocular Experiment of this conſtitution of the Earths internall Fibres; the very Cortex of the Earth extending ſome miles in profundity: but yet we deſume a reaſonable Conjecture there-of, as well from the great ſimilitude of effects wrought by the Earth and other Magneticks, as the Experience of Miners, who frequently obſerve, and conſtantly affirme, that the Veins of ſubterraneous Rocks, from whoſe chinks they dig Iron oare, doe allwayes tend from South to North; and that the Veins or eminent Rocks, which make the Giant Mountains upon the face of the Earth, have generally the ſame Direction. And though there are ſome Rowes or Tracts of Moun-tains, that run from Eaſt to Weſt, or are of oblique ſituation; yet are there alwayes ſome conſiderable intercifures among them, from South to North: ſo that that can be no ſufficient argument, that the interior Fibres of the Earth, which are truely and entirely magneticall, and ſub-jacent under thoſe Mountainous rocks, doe not lye in a meridionall po-ſition, or conforme to the Axis of the Earth.

*Art. 13.*
The Fibres of the Earth ex-tend from Pole to Pole; and that may be the Cauſe of the firme Cohæ-ſion of all its Parts, conſpi-ring to con-ſerve its Sphe-ricall Figure.

The TENTH; that ſince the obſervations of Miners aſcertain us, that the Ranges or Tracts of Rocks, in the Cortex or acceſſible part of the Terreſtriall Globe, do for the moſt obſerve a præciſely Meri-dionall

*Art. 14.*
Reaſon of Mag-neticall *Varia-tion*, in divers climates and places.

dionall situation, and tend from South to North, and sometimes (i.e. in some places) deflect toward the East and West, with less and greater obliquity; and that our Reason may from thence, and the similitude of the Earth and Loadstone, naturally extract a Conjecture, that the Fibres of the Earths Kernell or inaccessible parts, though for the most they tend præcisely from the South to the North; may yet in many places more and less Deflect toward the East and West: we need no longer perplex our minds with enquiring, *Why all Magnetiques, and especially the Versory or Needle of the Sea-mans Compass, being horizontally æquilibrated, doe in some places point directly to the North and South, and in others deflect toward the East and West, with more and less of obliquity;* which Navigators call (for distinction of it from the Depression, or Inclination, formerly explicated) the VARIATION of the Loadstone, or Needle. From the Mariners Tables (though they are full of discord, as to the degrees of the Needles Deflection or Variation from the true Meridian, in severall parts of the Earth) we learn, that the Needle doth exactly conforme it self to the Axis of its great Inspirer, the Earth, without any sensible deflection at all, in the Iland Corvus, one of the Azores, in the Iland of the Trinity, in the promontory of the Needles, neer the Cape of Goodhope, in the Fretum Herculeum, Syllæum, the Thracian Bosphorus, the Iland Malta, at Vienna, and divers other places. But in others, and particularly in England, it declines somewhat toward the East, yet with considerable diversity, so that in some countries its Variation exceeds not 1. 2. or 3. degrees at most, and in others it amounts to no less than 40, or 50. Again there are other meridians, in which the Declination of the Compass is toward the West, as frequently upon the Orientall coast of the Northern America; on the Occidentall coast of Nova Zembla, and Goa; the Eastern side of Africa; in our Mediterrane, at Naples, and sundry other places. Nay, oftentimes in the same Meridian, and in various degrees of Latitude, it hath been observed, that the Needle doth not vary at all, and vary both Eastward and Westward; for, though in the Iland Corvus the Declination be insensible, where the Latitude is of about 40 degrees; yet on this side of it, in the Latitude of 20 degrees the Declination amounts to 12 degrees Eastward: and beyond it, in the Latitude of 46 degrees the Declination toward the West, ariseth to 8 degrees; and farther off, in the Latitude of 55 the Westward Declination equalls 74 degrees. So also, in the Iland Elba, at one promontory, the Needle deviates toward the East only 5 degrees; at another promontory, 8; and at a third, as high as 20. which being duely perpended, doth soon detect the unadvisedness and incircumspection of Those, who have referred the Declination of the Magnet to the Deviation of the Asterisme, Ursa Minor, or Pole of the Ecliptick from the poles of the World; and attempted to explain it by imagining some certain Magnetick Rocks, which being situate on the East side of the Artick Pole of the Earth, constitute a speciall Magnetick Pole, or that whereunto the Versory Needle is generally deflected. Much more happy than this, was the invention of *Dr. Gilbert*; who supposing that the Magnetique Virtue of the Earth was more powerfully impressed upon the Needle from the Extant or Eminent parts thereof, and especially in great Continents: makes out the cause of the Magnets indirection,

"direction, or Variation, thus. If the Needle be placed in the middle
"betwixt two vast Continents, as in the Azores, which have Europe
"to the East, and America to the West, it suffers no sensible Distra-
"ction to either part: but, if it be brought nearer to the Continent of
"Europe and Asia, it must be invited and deflected toward the East;
"and nearer to the Continent of America, it shall deviate as much to-
"ward the West. For the same Cause also, upon the Western coast of
"Africa the Declination is toward the East; and on the Orientall, to-
"ward the West: and betwixt them both, as at the Cape of Good-hope
"none at all. And yet this subtle Theory of *Dr. Gilbert* is more then
suspected of Imperfection: For, since that, on the Western coast of A-
merica, and of Goa, the Declination of the Needle is Westward; and
not onely on the Orientall side of the Meridionall America, and chiefly
about the streights of Megellan, but also on the Orientall side of the
Septentrionall America, as at Virginia, the Declination teaseth not to be,
in the same manner, toward the East, absolutely contrary to *His* Hypothesis: therefore hath the incomparable Father, *Kircher*, to his own
immortall honour, and our greater satisfaction, advised us, to leave the
Attraction of adjacent Continents, and have recourse onely to the divers
Positions of the interior Magneticall Fibres of the Earth, over which
the Magnet, or Needle stands; considering that they have their situation
sometimes exquisitely Meridionall, sometimes more and less oblique, and
tend in some places in longer, in others in shorter tracts. For, it is no
difficult conception, the Virtue of the Earth is impressed upon the Needle
from the magneticall Fibres and Veins, that are nearest, i. e. directly subjacent thereunto; and disposed thereby into a situation respective to the
Ductus of those perpendicularly subjacent Fibres: so that whatever be the
Direction of the Needle; i. e. either without all Declination, or with
some, more or less, in one part toward the East, in another toward the contrary pole of the heavens; still may we suppose it to be exactly respondent to the Ductus, or Direction of the Fibres of the Earth, that perpendicularly lye underneath it. Nor is this meerly Petitionary, or excogitated onely for the solution of this grand Magneticall Problem, as
the Former of *Gilbert* seems to have been; but founded upon a Parallel
*Experiment*: for, if you place severall Barrs of Iron excited, upon the
ground, so that one may lye exactly according to the Meridian, and all the
rest in severall degrees of obliquity, untill you come almost to make an
Æquinoctionall line with one; and then gently and at requisite distance, move an invigorated Needle, equilibrated upon a pin, over them;
you shall observe the Direction of it to be varied to more and less
obliquity from the Meridian Barr, respectively to the situation of each
of the other Barrs, over which it is directly held. Now, if you suppose the Magnetique Fibres of the Earth to have the same Virtue upon
the Needle, as, if not much more than the subjacent Iron Barrs have:
you have attained the bottome of the Mystery, and that one of the
greatest in Nature.

*Art.* 15.
The *Decrement* of Magneticall *Variation*, in one and the same place, in divers years.

The ELEVENTH and last, that as the Conversion of the inspired Needle is not exactly meridionall in all places of the Earth, but siding more or less toward the East, in some Topicall meridians, and toward the West, in others: so also *is not the Declination thereof, though in one and the same place, constant to the same degree, at all times, but admits considerable Variation, and that in a few years.* For, *Mr. Burrows,* in the year 1580, making an exact observation of the quantity of the Needles Declination toward the East, at Limus, near London, found it to amount to no less than 11. degrees 15 minutes: and afterward, in the year 1622. *Mr. Gunter,* at the same place, observed it to be diminished to onely 6. degrees, and 13 minutes: and *Gellebrand,* in Anno Dom. 1634. in the same place, found it to come yet lower, and not to exceed 4 degrees 6 minutes: So that, in the meridian of London, as our Noble Countryman, Sir. *Kenelm Digby* hath well remarked, the Declination of the Needle Eastward hath been more Diminished in the latter years than in the former. The like Decrease of the Variation of the Needle hath been taken notice of also in France, at Paris by *Mercennus,* and at Aix, by *Gassendus.* And therefore we may præsume, if the Needles continue, in the same manner, and at the same rate, to lessen their Declination, that within a very few years, with us here in England, and other adjacent Countries, they will have no Declination at all toward the East, and perhaps wheele about toward the West, and every year more and more approach the contrary point of the Æquator.

*Art.* 16.
The *Cause* thereof not yet known.

Now, as for the *Cause* of this truely stupendious Effect of Magneticks; *Grandamicus,* indeed, thinks it best solved, by charging it onely upon the Errors of observation, not upon any Mutation of the Axis of the Earth, which would of necessity vary all Cælestiall observations, no less than Magneticall ones: enforcing this His opinion from hence, that the best of Astronomers are frequently not onely subject to, but guilty of great Errors, in their operations to find out the true Generall Meridian Line, of the Altitude of the Sun, of the point of the Heavens that is verticall to this or that place, where they use their instruments, &c; the certain knowledge of all these particulars being absolutely requisite to make a true compute of the Degrees of the Needles Variation. But, the Observators nominated being all eminent Mathematicians, well understanding the severall Causes, that might betray them into incertitude, and aswell how to prævent or avoyd them all; and each one setting about the work, with all possible care and circumspection: and it being very improbable, that they all should fall into one and the same delusion: the Ingenious, we hope, will excuse us, if we incriminate *Grandamicus Himself,* with much of temerity, and somewhat of injustice, in this detracting judgement of His, and assent to their more candid and reasonable one, who referr this sensible Declination of Declination in the Magnet, to some certain indigenary Cause, or Disposition proper to those Places and Countries, where such observations were made. But, what *indigenary and particular Disposition* that is, which should thus vary the Magneticall Variation, in the intervall of a few years; is a Problem indeed, and such as seems reserved for the exposion of Elias. *Kircher* and *Gassendus,*

*dus*, we acknowledge, have attempted moſt laudably, in ſuppoſing the Magneticall Fibres, that lye more diſtant from the Axis of the Earth, or neerer to the ſuperſice thereof, not to be ſo firmely cohærent each to other, but that they may be emoved, evolved, and ſeparated, by ſome ſubterraneous Cauſe or other, and ſo exchange their more oblique, for a leſs oblique, and at length for an abſolutely direct or truely meridionall ſituation; as the Fibres of the Muſcles of Animalls are obſerved ſometimes to ſuffer a certain Revulſion, or change of ſituation, under the skin, for ſeverall Cauſes: and that this Locomotion and Decrement of obliquity of the ſuperficiall magnetick Fibres of the Earth, may be the ſole Cauſe of the like Decrement of obliquity, or Declination of the Needle, in one and the ſame place, in divers years. But, foraſmuch as this Suppoſition is irreconcileable to our *Ninth obſervable* præcedent, touching the Cauſe of the firme Cohæſion of the parts of the Earth, and the Conſtancy of its Sphæricall Figure, from thence reſulting; and that neither *Kircher* nor *Gaſſendus* tells us, what ſubterraneous Cauſe that ſhould be, which might emove and tranſlate the Magneticall Fibres of the Cortex of the Earth, from a more to a leſs indirect ſituation (which in juſtice they both ought to have done:) we ſhall onely applaud the ingenuity of their Conjecture, and return to our former judgement, That the true Cauſe of the Decrement of the Magneticall Variation is yet in the bottome of *Democritus* Pit; and He, who ſhall be ſo happy to extract it from thence, ſhall have our vote, to have his ſtatue ſet on the right hand of that of *Gilbert*, in the Vatican.

There yet remains a Difficulty, which being left unreſolved, is of importance enough to make the intelligent and wary Reader ſomewhat coſtive in his Aſſent even to the chiefeſt and moſt Fundamentall of our Præcedent obſervables, concerning the Reaſon of Magneticall Verticity. And that is, *That ſome Loadſtones have more than Two Poles;* ſuch as that Tripolar one of *Furnerius*, of which both *Kircher* and *Gaſſendus* make ſingular mention.

*Art.* 17.
No Magnet hath more than *Two Legitimate* Poles: and the reaſons of *Illegitimate* ones.

Concerning this, therefore, we ſay; that in every Loadſtone there are two, and but *two true* and *Legitimate* Poles: and that all others apparent in them, either at the Æquator, or betwixt it and either of the Genuine Poles, are ſpurious or Illegitimate; ariſing either from ſome *Node* or Knot growing laterally on to a Magnee (ſuch as is commonly obſerved to interrupt the direct progreſs of the Fibres, or Grain of Trees, and of ſtones) or from an *irregular* and *horned Figure* of the ſtone it ſelf, in reſpect of either of which the Magnetick Virtue cannot be commodiouſly united at the two Genuine and directly oppoſite Poles, but is diſtracted obliquely to that Prominent Node, or Horn-like Protuberancy. For, if either the Node or horns of a Loadſtone, which cauſe it to have more than two Poles, be artificially cut off, and the remainder of the ſtone be poliſhed, a Needle, or the Filings of ſteel, thereunto applied, ſhall never be perpendicular erected at any part thereof, but onely at the *Artick* and Antarctick points; nor ſhall

the

the ftone difpofe it felf otherwife than conformably to the Meridian, both which are the moft certain Difcoverers of the true Poles of a Loadftone. Thofe Illegitimate Poles, therefore which fometimes (though very rarely) are found in a Loadftone, are as it were the oblique and Præternaturall parts of it, obtaining the reafon of Poles only by Accident. Which yet hinders not, but that many times, from the imperfection of the ftone, it may come to pafs, that the two Legitimate Poles of the fame Loadftone, though exactly polifhed, and reduced to a perfect Sphere, may not exift in the Extremes of its Diametre : for, unlefes the Magnet be Uniforme in fubftance and Virtue, the Poles thereof cannot be directly oppofite each to other.

*Art. 18.*
*The Conclu-*
*fion, Apologe-*
*ticall; and an*
*Advertife-*
*ment, that the*
*Attractive and*
*Directive Acti-*
*ons of Magne-*
*tiques, arife*
*from one & the*
*fame Faculty,*
*and that they*
*were diftin-*
*guifhed onely*
*διδασκαλιας*
*χάριν, for con-*
*venience of*
*Doctrine.*

And thus, in a naturall Method, and w$^{i}$th as much fuccinctnefs, as the copious fubject would beare (according to our engagement) have we enquired into the Caufes of the Two Generall Faculties of the Loadftone, the *Attractive* and *Directive*, with the moft confiderable Phænomena's arifing from either, or both of them. Wherein, if we have been fo happy, as to afford but the leaft of fatisfaction to others; we fhall account it no fmall content to Ourfelves, and think our ftudies thereby more than fufficiently compenfated. If not, we fhall yet confolate ourfelves with this; that we are not the Firft, who have fallen fhort of the Readers Expectation, in the Difcuffion of this fingularly Abftrufe Argument: which is a thing fo highly Admirable, that *Aphrodifæus (initio Problem.)* affirmed the Nature thereof to be underftood only by *Him,* that created it; and *Galen (de thetica ad Pifon.)* termed the *Attractive* Virtue thereof wholly *Divine.* To which we fhall add alfo this; that the Hypothefis, of the continued *Ductus* of the Magnetick Fibres of the Earth, efpecially of the Kernell, or Interior. fubftance thereof, from the South to the North Pole (upon which we have erected the folutions of fundry great Magneticall Apparences) is fubject to much lefs of Improbability, than that of *Gilbert* and *Grandamicus,* that the Magnetique Virtue is a *fimple,* or *Immateriall Quality;* than that of *De's Cartes,* that the Magnetique Aporrhæa's confift of ftreated or Screw'd Atoms, paffing through the Earth, by contrary and diverfly figurated infenfible pores, iffuing forth at either pole, and wheeling about interchangeably to the oppofite pole; than that of *Sr. Kenelm Digby,* that the Magnetique ftreams glide along from either Pole and Hemifphere of the Earth, by Attraction to the Æquator; or, in truth, than any other hitherto excogitated and divulged.

But, before we put an end to this Chapter; 'tis requifite to advertife you of a Confiderable, omitted in the beginning of it; which is, that though we affumed the Virtue Magnetick to be (in Generall) *Twofold, Attractive* and *Directive;* yet is that Diftinction to be admitted, not in an Abfolute, but *Refpective* intention, or only (καξ' ἐπίνοιαν) in order to our more diftinct Comprehenfion of the immediate, and particular Reafons of fundry refpective Magneticall Effects, which otherwife muft have wanted the advantage of order in their confideration. For, we are fully convinced of the truth of that Affertion of *Grandamicus* (*Nova Demonftrat. Immobilit. Terra, cap. 5. Sect. 2.*) that the *Attraction* and *Direction,* or *Alliciency* and *Polarity* of Magneticks, are caufed by *one* and the *fame Faculty:* which being conferred upon them, by the infinite Wifdome and Goodnefs of the Creator, in order to the Confervation

of

of the Earth, and all its genuine parts, in that position in the Universe, and that disposition among themselves, in which they are best supported, and most conveniently performe Actions conforme and proper to their Nature; may be yet termed *Attractive*, insomuch as it *Unites* Magneticall Bodies, violently separated; and *Directive*, insomuch as it *Disposeth* them in a due and commodious situation. And so, notwithstanding the Actions and Motions of Magnetiques seem exceeding Various, and in some cases, plainly Contrary; yet are they to be deduced from one simple principle, one and the same Generall Virtue, and they all may be conveniently explicated by the same Common Reason.

BOOK

# The Fourth Book.

## CHAP. I.

### OF

# GENERATION

## AND

# CORUPTION.

#### Sect. I.

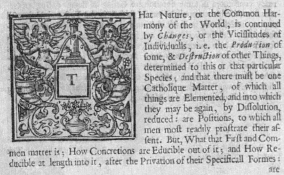

THat Nature, or the Common Harmony of the World, is continued
by *Changes*, or the Vicissitudes of
Individualls, i. e. the *Production* of
some, & *Destruction* of other Things,
determined to this or that particular
Species; and that there must be one
Catholique Matter, of which all
things are Elemented, and into which
they may be again, by Dissolution,
reduced: are Positions, to which all
men most readily prostrate their assent. But, What that First and Com-
mon matter is; How Concretions are Educible out of it; and How Reducible at length into it, after the Privation of their Specificall Formes:
are

are Quæſtions, whoſe Beginnings are more eaſily known, than their
ends. However, foraſmuch as we have endeavoured, in our immediately foregoing Book, to determine the *Firſt* of them, together with the
poſſible Emergency of all Qualities (whereof either our ſenſe, or Reaſon
can afford us any meaſure of cognizance) and the Reaſons of the Perception of them by Animals, from Atoms, ſo and ſo Configurated, and
ſo and ſo Diſpoſed in Commiſſion: it now neerly concerns us, to attempt the moſt hopefull Deciſion of the *other Two* that ſo we may not
ſeem to have thus long diſcourſed of the Principles, and Affections of
Compound Bodies, while we remained wholly ignorant of the moſt probable wayes both of their *Origination* from thoſe Principles, and of their
*Reverſion* into them again, when they have loſt the right of their former Denominations, and ſuffered to the utmoſt of their Diviſibility.

*Art. 1.*
The proper
Notions of *Ge-
neration & Cor-
ruption.*

By the terme, GENERATION, we ought præciſely to under-
ſtand *that Act of Nature, whereby ſhe produceth a Thing de novo, or gives
Being to a Thing, in ſome certain Genus of Bodies Concrete:* and conſequently, by its Contrary, CORRUPTION, *that whereby ſhe
Diſſolves a Thing, ſo that thenceforth it ceaſeth to be what it was.* For,
when Fire, a ſtone, a Plant, an Animal, or whatever is referrible to
any one determinate kind of Bodies Compound, is firſt produced, or
made, and begins to be ſo, or ſo Denominated; it is truely ſaid to be
Generated: and contrariwiſe, when a Thing periſheth, and loſeth the
right of its former Denomination; it is as truely ſaid to be Corrupted.
And this is that which *Ariſtotle* (1. *de Generat 2.*) frequently call's *Gene-
ratio* απλη & τελεια, Generation *Simple* and *Perfect*, ſo to prævent that
Confuſion of Generation with *Alteration*, into which many of his Præ-
deceſſors had often fallen, to their own and their Diſciples no little diſ-
quiet. For, "Ετεροιωσις, *Alteration* can be accounted a Generation only improperly, or *ſecundum quid*, foraſmuch as by Alteration a Body
is not produced *de novo*, but onely acquires ſome new Quality, or ſome
Accidentary Denomination: and Philoſophers accordingly define it to be
*Progreſſionem Corporis ex una qualitate in aliam,* a Progreſſion of a Body
from one Quality to another, as when water is changed from cold
to hot by fire. Again, every Mutation requires a ſubject to be Altered, and that ſubject muſt be ſomething Compound, complete, and already conſtituted in ſome determinate Genus of Beings: But, of Generation ſtrictly accepted the onely ſubject is the Firſt and Univerſall
matter, which being in it ſelf deſtitute of all Form *Ariſtole* doth therefore ſubtly call *ſimpliciter Non-ens,* ſimply, or determinately Nothing;
foraſmuch as he frequently inculteth, that Generation is made [εξ απλως
μη οντ©] ex *Non ente ſimpliciter.* Becauſe had He ommitted that adverb *ſimpliciter*, his Reader might juſtly have underſtood *Non ens abſolute* Nothing Abſolutely; and ſo have accuſed him of openly contra-
dicting his own Fundamentall Axiome, *Ex nihilo nihil fieri,* that nothing
can be made or generated of Nothing.

This

This being præmifed, to prævent the danger of Æquivocation, we observe First, with *Ariftotle* ( 3. *de cœlo* 1.) that among the Ancient Philofophers, *fome* held, that Nothing is Generated, nothing Corrupted; as *Parmenides* and *Meliffus*: *Others* again, that. All things are Generated and Corrupted; as *Hefiod* and *Heraclitus*. Secondly, that of Thofe, who admitted Generation, and confequently Corruption, *fome* conceived, that Generation is made by the Accefs of a Form to Matter; and that that Form is a certain New fubftance, abfolutely diftinct from that of the Matter, and together with it conftituting the Compofitum, or whole refulting from the Commiftion of Matter and Form: of which fect *Ariftotle* Himfelf deferves to be in the Chair, becaufe in order to his Affertion of this Opinion, He fuppofeth a Threefold fubftance, the *Matter*, *Form*, and *Compofitum* arifing from their Commiftion. But, *Others* though they concede, that Generation, indeed, confifteth in the Acceffion of a Form to Matter; yet will they not allow that Form acceding, to be fubftantiall, but onely a certain *Accident* or Modification of the Matter it felf: fo that according to their theory, in Generation there fuperveneth upon Matter fome certain Quality, of fuch a Condition, as that by reafon thereof a Thing obtain's a certain Being in Nature, and acquireth fome determinate Denomination, refpective to that Genus of Bodies, to which its Nature doth referre it. And in the Catalogue of Philofophers of this perfuafion, *Ariftotle* nominateth as Principalls, *Empedocles*, *Anaxagoras*, *Democritus*, and *Leucippus*; all which He fharply taxeth of Confounding Generation with Alteration, and of inferring, that afwell Generation as Corruption arifeth, not from the *Tranfmutation* of Principles, but onely from their [ Συγκρισις & διακρισις ] *Concretion* and *Secretion*: which is not only inconfiftent, but contrapugnant to His own great Hypothefis, that the Four Elements, or Catholique Principles of Generation, are fo Tranfmutable, both *fecundum fubftantiam* (at leaft, according to the Comments of all his modern Expofitors) & *fecundum Qualitates*, as to their fubftance and Qualities, as that from their Commiftion, Alteration, and Corruption, a certain New and diftinct fubftance doth arife, which is the Form of the Thing fo produced. For, having fuppofed for a Groundwork, that the Four Elements are not the Firft Principles; it could not ftand with his advantage, not to have affumed alfo, that the Elements may be fo Tranfmuted, as that the more Generall and Common Matter doth ftill remaine: and that the fame, upon the perdition of the Elementary Forms, may put on a New Forme, that is fubftantiall; and that very thing, by which the refulting or Generated Body is fpecified, and entituled to fuch a Denomination. But, as for *Empedocles*, and the reft enumerated ( to whom we may add alfo *Epicurus*) 'tis well known that notwithftanding they all admitted the Four Vulgar Elements, as readily as *Ariftotle* Himfelf, yet would they by no means hear of their *Tranfmutability* either as to *fubftance*, or *Qualities*: unanimoufly decreeing, that in their Commiftion each of them is divided into particles moft minute, which yet retain the very fame fubftance and qualities, that they had before, as that every particle of Fire doth ftill retain the fubftance and quality of Fire, namely Heat; and that every particle of Water doth like-

*Art.* 3.
Various pi-
nions of the
Ancient Pi-
lofophers,
touching the
reafon of Ge-
neration: and
the principall
authors of
pacti.

likewiſe conſtantly conſerve the ſubſtance, and quality of Water, *viz;*
Moiſture, and ſo of the other two: ſo that it is moſt evident, They
would have, that in Generation there is onely a [ Σύγκρισις ] *Concre-*
*tion* of the inſenſible particles of the the 4 Elements, but no Tranſmu-
tation of any one of them, either with the *Perdition* of their own, or
the *Adoption* of a new ſubſtantiall Forme, both which are præſumed by
*Ariſtotle.*

<table>
<tr>
<td>

*Art* 4.
The two great
opinions of
the ſame Phi-
loſophers con-
cerning the
manner of the
*Commiſtion* of
of the Com-
mon Princi-
ples, in Gene-
ration, faith-
fully & briefly
ſtated.

</td>
<td>

But this great Difficulty, about the Generation of Things from the
Commiſtion of the General Principles, ſoon loſeth it ſelf in a Greater,
which concerns the *Manner and Condition of their Commiſtion*, and
whoſe conſideration will beſt inſtruct us aſwell what is the main Dif-
ference among Philoſophers, touching this moſt weighty Theorem, as
what opinion can beſt deſerve our Approbation and Aſſent. Concer-
ning this, therefore, we find two neceſſary Remarks. (1) That there
are Two different Kinds of *Commiſtion*, whereof the one is, by *Ariſtotle*
(*de Generat.* 1. *cap.* 10.) termed Σύνθεσις, *Compoſition*, and by others,
παράθεσις, *Appoſition:* the *other* is called, in the Dialect of the
*Stoicks*, Σύγχυσις *Confuſion*, and in that of *Galen*, κρίσις, *Coalition*,
or *Temperation.* The *Former* is when thoſe things, whether Ele-
ments, or others, that are mixed together, do not interchangeably
penetrate each others parts, ſo as to be conjoyned *per minima*; but
either themſelves in the whole, or their parts, onely touch each other
ſuperficially: as in the Commiſtion of the Grains of wheat, Barly,
Rye and other Corn. The *Latter*, when the things commixed, are
ſo ſeemingly united, and concorporated, as that they may be concei-
ved mutually and totally to pervade and penetrate each other, *per*
*minimas partes*, ſo as that there is no one inſenſible particle of the
whole mixture, which hath not a ſhare of every ingredient; as when
Wine and Water (that we may uſe the Example, aſwell as Concep-
tion of *Ariſtotle*) are infuſed together into the ſame veſſel. Now the
*Stoicks* and *Ariſtotle* are equally earneſt to have this *Latter* way, or
manner of Commiſtion, *viz.* Σύγχυσις, *Confuſion*, to be that, accor-
ding to which the Elements or Principles of Bodies are commix't in
Generation: But *Empedocles, Anaxagoras, Democritus, Epicurus,*
with all their Sectators, allow none but the *Former*, or παράθεσις, *Ap-*
*poſition*; with very ſolid arguments (among which the eaſy ſepara-
bility of Wine from Water, either by a ſponge, or Cup of Ivie, is
not the leaſt ) aſſerting, that the Σύγχυσις of Elements, as alſo of all
other things, is really a meer Σύνθεσις, Compoſition of their ſmall par-
ticles, though apparently, or according to the judgement of ſenſe, it
may paſs for a Σύγχυσις, or Confuſion.

</td>
</tr>
</table>

(2) That, when either the Elements themſelves, or any other Bodies more
Concrete, as Water and Wine, are mixed together; they may reciprocally divide, diſſect, and reſolve each other into either very ſmall and
*inſenſible* [*molecula*] *maſſes*, which yet are each of them compoſed of multitudes of Atoms concreted; or moſt exſile particles, i. e. *Atoms* themſelves:
and where the reſolution is only into inſenſible Maſſes, there may the Commiſtion be accounted *Perfect*; but, where the parts of each ingredient
are

are so far resolved, as to be reduced quite down to the first Matter, Atoms, there is the Commistion *most Perfect*.

Now, upon this *Distinction* depends the whole Controversy betwixt *Aristotle* and the *Stoicks*, on one part, and the *Atomists*, on the other, about the Manner of the Commistion of the Common Principles in Generation: *Those* vehemently contending for their totall *Concorporation*, or Unition *per minimas partes*, so that every the most minute particle in the whole mistum, must be of the very same nature with the whole; *These* strongly asserting, that no Mistion of Elements, or Temperation of Principles, goes further than a meer *Apposition*, or *superficiall Contingency* of their several particles, so that the particles of each ingredient must still retain the very same nature they had before commistion, howbeit they may seem to be totally Concorporated, or Confused, in regard they are reduced to such Exility, as that each single one escapes the discernment of the sense.

These two so highly repugnant Opinions being thus rightly stated, it follows, that we uprightly perpend the Verisimility of each, that so we may confer our Assent upon the more ponderous. If we look no further than the *Common Notion*, or what every man understands by the Terme, *Mistion*; it is most evident, that the things commixed ought to *Remain* in the *Mistum*; for if they do not remain, but Perish, both according to substance and Qualities, as *Aristotle* and the *Stoicks* hold, then is it no Mistion but a *Destruction*: and since the propriety of this Notion cannot be solved by any other reason, but that of the *Atomists*, that the particles of things are in commistion onely apposed each to other, without amission of their proper natures; what Consequence can be more naturall and clear than this, that that their opinion is most worthy our Assent and Assertion? (2) Though *Chrysippus* attempts to conserve the integrity of this Common Notion, by a subtlety, saying, That the most minute particles of things mixed, do so remain entire both as to substance and Qualities, as that they reciprocally penetrate each other, and become mutually Coextended; and that thence it comes to pass, that in the whole *Mistum* there is none the smallest particle, which is not mixed, or which doth not partake aswell of the substance, as Qualities of every ingredient therein: yet doth He not onely fall short of his designe, but also further entangle himself, and subvert other more manifest Notions. For, from that his Position it necessary follows. (1) That two Bodies are at once in one and the same place, both mutually penetrating each others dimensions, or without reciprocall expulsion (2) That a pint of Water, and a pint of Wine commixed, must not fill a quart, but that both are no greater than one, i. e. be both contained in a pint together: forasmuch as it supposeth, that the particles of one have no other Ubi, but what is posse'st by the particles of the other. (3) That a very small Body may be Coextensive, or Coæquate to a very great one; as that a spoonfull of Water may be Coæquate to a But of Wine: since it supposeth, that, both being commix't, there is no part of space in the vessel including them, which doth not contain somewhat of the Water as well

*Art.* 5.
That of *Aristotle* and the *Stoicks*, refuted: and *Chrysippus* subterfuge, convicted of 3 Absurdities

as of the Wine. Now, all thefe things being manifeftly Repugnant, and yet naturally Confequent upon *Chryfippus* Pofition: it is no lefs repugnant, that the particles of things commixt fhould remain, by mutuall Penetration, and Coextenfion.

*Art. 6.*
*Ariftotles two-
fold Evafion
of the Incon-
gruities at-
tending the
pofition of the
Remanence of
things commi-
xed, notwith-
ftanding their
fuppofed recipro-
cal Tranfubftan-
tiation: found
likewife meer-
ly Sophifticall.*

(3) Nor, indeed, hath *Ariftotle* Himfelf been more happy than *Chryfippus*, in his invention of a way, to remove or palliate the grofs repugnancy of his opinion, to the proper importance of the term, Commiftion; as may eafily be evinced by a fhort adduction of it to the teft of reafon. That He might defend his Doctrine of the Remanence of things commixed, notwithftanding their reciprocall Tranfubftantiation; and at the fame time avoid thofe fundry manifeft Ἀσυςαζἃ, or Incongruities, to which that doctrine is fubject: He excogitated *Two fophifticall* fubterfuges. The one, that when two divers things are commixed, in very unequall proportions, fo as the one is very much prævalent o're the other (as when one fingle drop of Wine is inftilled into ten thoufand Gallons of water) in that cafe there is no Miftion, in ftrick acceptation; but an abfolute Exfolution and Tranfmutation of the fpecies of the weaker into that of the ftronger, (of the fpecies of the Wine, into that of the Water.) The Other, that when the things commixed are fo exactly equall in quantity or Virtues, as that one is not the leaft prævalent over the other; or when the one prævails upon the other but little: in both thefe cafes, though each put on the nature of the other, by reciprocall tranfmutation, or that which is a little inferior be altered from its own nature into that of the Superior; yet is not that Tranfmutation of both, a Generation of either, or the tranfmutation of the one, a Generation of the other, but onely of fome *Third* thing, which is *middle* betwixt, and *common* to both.

But, there is neither of thefe, which may not be called a *fnare*, more juftly than a fubterfuge. For, as to the *Firft*; were *He* living, and in the Schools, we fhould onely demand of him, if after the inftillation of one fingle drop of Wine into 10000 Gallons of Water, a fecond drop fhould be fuperinfufed, and after that a third, a fourth, and fo more and more fucceffively, till the mafs of Water were augmented to ten, a hundred, a thoufandfold: *of what Nature would the whole mixture of Wine and Water be?* He, doubtlefs, would *Anfwer* Us, that the whole would ftill be Water, though to one meafure of Water 10000 meafures of Wine were fuperaffufed drop after drop; fince, according to His own theory, it allwayes muft remain meer and fimple Water (otherwife the firft drop of Wine could not be tranfpecificated, or be converted into the nature of the Water) into which even the very laft drop of Wine was infufed: or elfe He muft teach us when, i. e. from what particular drop of Wine inftilled, the whole Aggregate or Mafs of both liquors began to put off the nature of Water, and on that of Wine. And, who is fo dull either by nature, or præjudice, as not to apprehend, that the Reafon is the fame for one, as for the other; for ten thoufand thoufand Gallons, as for one fingle Drop of Wine? Now this being *Abfurd*, as far beyond palliation, as pardon; is it not much better for Us to fay, that if one *drop of* Wine be infufed into fo large a quantity of Water, it is divided into very exile particles, each whereof doth
ftill

still retain the nature of Wine, but so commixed and adhæring to the incommensurably more dense and numerous particles of the Water, as that they seem to vanish, though really they still subsist the very same, as before commistion? That Two drops being infused into the same Water, the particles therof becoming doubly more numerous, would be contingent and cohærent to more particles of the Water? That, if ten, a hundred, a thousand, ten thousand, a hundred thousand, &c. Drops of Wine be successively superaffused into the same Water, the particles of the Wine would at length amount not only to an equall, but a greater number than those of the Water; and consequently so prævail over them, as to change their Virtue, and subdue them into the Apparence of Wine?

And as to the *Other*, we might very lawfully Except against it, as altogether *Unintelligible* (for, who can understand, How the Inferior Mistile can be transmuted into the Nature of the superior, and yet not be the very same thing with it?) but, least we appear all severity, we shall wave that cavill, and insist onely upon the most important part of the Assertion. *Aristotle* saith, *That from the Commistion of two divers things, a certain Third thing is Generated, or Produced, which is of a Nature Median betwixt, and Common to Both those things commixed.* Now, Whether is it His meaning, that the Resulting middle and Common thing doth participate of the Extremes of Each mistile; or, that it ariseth from the Destruction of both Mistiles? For the Text will endure no third interpretation. If the *Latter*, then do not either of the things mixed Remain, and so there can be no Mistion: expressly contrary to His own Assumption, and the tenour of that Common Notion, for the præservation whereof He excogitated and designed this Subterfuge. If the *Former*, as seems most genuinely inferrible from the Adjectives, *Medium* and *Commune*, then our Enquiry is, How, and in what respect, that Middle and Common thing comes to be participant of the Extremes of each Mistile? In the Wine (that we may retain his own Instance) there was Matter, there was Forme, there were Qualities; and likewise in the Water: shall we therefore conceive, that the Middle and Common thing produced, is participant of all, i. e. Matter, Forme, and Qualities of Both the Mistiles, or onely of those of one of them?

(1) For the *Matter*, He cannot deny, that the Mistum containes the whole Matter of Both; because neither the Matter of the one, nor of the other can be destroyed. And since the Matter of each hath Parts, the smallest of which is Extense or Quantitative, and so must possess a proportionate part of space in the Continent; therefore we demand, whether are the Parts of the Matter of the Wine existent in the very same places, with the Parts of the matter of Water, or in distinct places by themselves? If He should say, as the supposition implies, that the parts of Both do exist in one and the same place; He would ruine himself upon that Impossibility of the Coexistence of Two Bodies in one place: and if that they are in distinct places; then must it follow, that they onely touch each other superficially, and so are not mixed by mutual Penetration and Coextension (as He affirmed) but by meer *Apposition*, or *Composition*. (2) As to the *Forms*; *Aristotle* cannot but admit, that
the

the Forms of both Wine and Water do survive their Commistion, and exist in the Mistum; or Middle and Common thing resulting from them; because, otherwise, there would be a plain Corruption, not a simple Alteration of the things mixed, and consequently Mistion ought to be defined rather *Mistilium Corruptorum*, than *Alteratorum Unio*: Besides, if the Formes perish, the Emergent Form must be absolutely *New*, and so not participant of the Form of each Mistile. But, if He reply, that Both Forms are United and coexistent in the whole matter of the Mistum; then must every the smallest particle of the matter of each have both the Form proper to it self, and the Form of the other also, and so the, whose matter must have two whole distinct Forms at once: which is an Absurdity infinitely below the concession of *Aristotles* subtility, and whether or no his Sectators will defend it, we leave to themselves. To elude this Dilemma, He, indeed, hath determined, that the Form of the Mistum is one only, and that neither of the Præexistent Forms, in Act, but both in *Power*. But, alas! this is a poor shift for so great a Philosopher; for if the præexistent Forms of both Mistiles be not Actually in the Mistum, then are not the Mistiles onely Altered, but wholly *Corrupted*: nor can it enter into the thoughts of any sober man, How the Resulting Form should contain the Præexistent ones, in *Power*. For, if the Resulting Form is capable of being changed again into the præexistent ones, from which it did result; as when Wine and Water commixed, are again separated: that argues of necessity, that the Forme of the Mistum is not a New Forme (as He assumes) but one Composed of the two præexistent ones commixed.

( 3 ) And lastly, as for the *Qualities*; neither ought *Aristotle* to deny the *Remanence* of them: for, since in them consisteth the chief Capacity or Power of recovering the last Forms; if they perish, how can they be inservient to the recovery of the Forms? Necessary it is, therefore, that the Qualities of things commix't be onely interchangably *Refracted*, not Abolished. And thus have we demonstrated, that *Aristotle*, aswell as the *Stoicks*, engulfed himself in an Ocean of bottomless *Difficulties*, and irreconcilable *Incongruities*; while He sought to propugne that unreasonable Opinion, of the Mutuall *Confusion*, and *Transmutation* of the things commixed in Generation. For a Collateral *Remark*, be pleased to reflect upon this great Example, when you would enforce, *How heavy a burthen lye's upon those shoulders, which take upon them to support an Error: and how weak the Armes of the most Giant wits are found when they strive to bear up against the stream of Truth.*

*Art. 7.*
*That the Formē of things, arising in Generation, are no New substances, nor distinct from their matter: contrary to the Aristoteleans.*

Having detected the sundry Difficulties, that wait upon the Doctrine of *Aristotle*, touching the *Origination*, or *Emergency* of a *Form*, in a thing Generated from divers things commix't; let us proceed to Another Article of the same Chapter and enquire whether there be not also a very remarkable Difficulty inseparable from his Doctrine of the *Essence* of that Forme; that so at length we may the better determine, *Whether the Forme of a thing Generated from Elements,*

*or*

*or other more compound Bodies commix't, be a substance* (as *Aristotle contends*) *or onely some certain Quality, or Accident* (as *Democritus* and *Epicurus* assert.) But, first, we are to advertize, that from this Discourse of ours, against the substantiality of Forms Generated, we exempt the *Rationall Soul of Man*; for, that being an Essence separable from the Body, and subsisting entire and complete after separation (as we intend, if God shall be pleased to grant us health, and the world vacation from publique cares, to demonstrate at large, in a singular Treatise) may therefore be most justly termed a substance, or Form substantiall: as intending onely to examine the reasonableness of that opinion, by the Schools imputed to their Master *Aristotle*, that *the Forms of things are substantiall, and wholly distinct from Matter*. The Quæstion (and indeed a very Great one) is, *Wherein that substance, or Form, which Aristotle* affirm's to arise, *de novo, in Generation, lay hid before Generation?* His *sectators* unanimously tell us, that it was contained in the Matter, not in *Act*, but onely in *Power*, or *Capacity*: and we demand again, if it were not Actually contained in the Matter, how could it be Actually educed from thence? They reply, that it is educed out of the Matter onely by the Power of the Agent. But, this is a shamefull Desertion of the Quæstion, which is not about the Power of the Agent; but, How the Form of a thing, which themselves assume to be a substance, i. e. a reall and self-subsisting Entity, and so clearly Distinct from the Matter of the Mistum, can yet be Educed out of that very Matter? When they say, that the Form is concealed in the Power of the Matter; if they would but permit us to understand the Form to be a certain portion of the Matter, and as it were the Flower, or purer part thereof, which should afterward, in Generation, be attenuated, refined, sequestred from the grosser mass; and then be again conjoyned to the same, and as it were Animate it: then, indeed, might the Eduction of a Form, as a reall and substantiall Being, be easily conceived, and assented to. But, this they expresly prohibite, lest they should incur a double Contradiction: the one, in conceding the Matter to be Corruptible; the other, in allowing the Form to be indistinct from Matter. Forasmuch, therefore, as they protest against that Interpretation of the Text; and yet are peremptory, that the very substance of the Form educed, was before eduction potentially comprehended in the very substance of the Matter: they give us the trouble of still pressing them to explain How, or after what manner, the substance of the Form was Potentially contained in that of the Matter? And here they fly to their accustom'd refuge, an obscure Distinction, saying, that the Power of the Matter, in respect to the Form, is Twofold: (1) *Eductive*, forasmuch as the Form may be, by virtue of the Agent, educed out of it; (2) *Receptive*, forasmuch as it receives that same Form educed. And so they conclude, that the Matter doth contain the Form in both these Powers, or double Capacity. But, this will not blunt the edge of Curiosity. For, as to the First, *viz.* the *Eductive* Power; 'tis manifest, that to contain a thing by an Eductive Power, imports no more, nor less than this, to have Actually in it self that, which is capable of eduction

from

from it. Thus a Purse, wherein ten pieces of money are actually contained, may well be said to contain them by an Eductive power; because He that hath the purse, may at his pleasure Educe them from thence: but, if the Purse did not actually contain them, He that wanted money, might starve before He could prove, that they were contained therein by an Eductive power.

And therefore we may set up our rest in this Conclusion, that as a piece of Gold cannot be educed out of an Empty Purse: so doth not Ἄμορφος, or Exforme Matter (so themselves determine it to be) contain a Form, by an Eductive Power.

As to the Other member of the Distinction, the *Receptive* Power; tis also manifest, that to contain a thing by a Receptive Power, is no other than to be in a condition of Receiving it: but, this Capability, or Power Receptive comes much short of being sufficient, that any thing should be actually educed from that, which hath onely such a power of entertaining it; since otherwise the prodigall need not fear the exhaustion of all the money in his purse, because it is capable of more, when that's gone. Which being most grosly Absur'd, it cannot be less Absurd to conceive, that the Form of a thing may be educed from the matter thereof, because it is contained therein by a Receptive Power. Indeed, if they would allow the Form to be, not a substance, but a certain Quality, species, or modification of a substance or Matter; then might we understand how it might be contained in the Power of the Matter; because the sense would be no more than this, that the Matter is capable of being so changed and disposed, as to be put into such a Mode, or Form: by the same reason, as the species, or Image of Mercury may be said to be contained in the power of a piece of wood, or be educed out of it, insomuch as the wood is capable of being formed into the statue of Mercury, by the hands of the statuary.

But, while they make the species or Image of Mercury, to be a New substance, absolutely distinct from the wood, which is the substance, or Matter of that Image; and in Generall discriminate the Figure, or Forme of a thing, from the substance of the thing it self: we are to be excused, if we do not at all understand them, in more than this, that they endeavour to assert what themselves do not, nor cannot understand.

*Art.* 8.
*That the Form of a thing, is only a certain Quality, or determinate Modification of its Matter.*

But, as for the *other* Philosophers, formerly nominated; if you please to convert your attention to the summary of their theory concerning the same Argument, we doubt not but in the conclusion you will concur with us in this judgement, that They speak (at least) both much more intelligibly and satisfactorily. They deny not, that Generation is indeed, determined to a substance, because the the thing produced or generated, is a substance. Nor that in generation there alwayes ariseth a Forme, by which the thing generated is specified, because Generation supposeth specification, and specification imports a Forme. Not, again, that that Form is really a substance, *i. e.* a certain

most

moſt tenuious, moſt ſpiritual, and ſo moſt active part of the Body, ſuch as we have often hinted the ſoul of a Plant or Brute Animal to be.    But the points which they declare againſt, as manifeſtly unreaſonable, are theſe Two: (1) That ſuch a Forme is a New ſubſtance, or formerly not Exiſtent; becauſe it is unavoidably neceſſary, that that moſt tenuious, moſt ſpiritual, and moſt active portion of the matter ſhould be ſomewhere præexiſtent, before it was copulated to the groſſer and leſs active part of the maſs, and affected it with ſuch a particular mode, as ſpecifies the miſtum: (2) That that which is properly called the Forme of a thing, is ought elſe but a certain Quality, or determinate Manner of the ſubſtances exiſting, or ſpecial Modification of the matter thereof.    For, it being unanimouſly decreed by them All, that every thing is generated from an Aggeries of Matter, or Material Principles, coaleſcing in a certain Order and Poſition: they therefore determine, that the thing generated, or Concreted, is nothing but the very material Principles themſelves, as convened and coaleſced in this or that determinate Order and Poſition, and ſo exhibited to the cognizance of our ſenſes, under this or that determinate Forme, Species, or Quality.    And leſt we ſhould delude our ſelves, by a groſs apprehenſion, that the tenuious and more agile part of the body is only confuſedly blended together with the groſs and leſs agile part, *Empedocles and Anaxagoras* tell us præciſely, that the Forme of the whole, or Quality by which the Body is made ſuch as it is, doth yet reſult from as well the order and ſituation of the tenuious parts among themſelves, and of the groſſer among themſelves, as of the tenuious and groſſer conjunctively, or one among another.    And this they illuſtrate by the ſimilitude of an Houſe.    For, as an Houſe is nothing but Timber, Stones, Morter, and other materials, according to ſuch or ſuch a reaſon and order diſpoſed and contexed together, and exhibiting this or that Forme; and as there is nothing in it, which before the ſtructure thereof was not found in the wood, quarry, river, and other places, and which after its demolition (whereby its Forme periſheth) doth not ſtill exiſt in ſome place or other: ſo is a Horſe (for example) nothing elſe but thoſe material Principles, or exile Bodies, of which after a certain manner connected among themſelves it is compoſed, both with this determinate Conformation of Members, and this interior Faculty of Vegetation, and in a word, with this particular Forme, Quality, Species, or Condition, which denominates it a Horſe; when yet the Principles of which both its Groſſer members are coadunated, and its tenuious and ſpiritual ſubſtance, the ſoul, is contexed, were formerly exiſtent in his progenitors, in graſs, in Water, Aer, and other Concretions; and the Form alſo, ſo ſoon as the Compoſitum is diſſolved, vaniſheth, as well the tenuious as groſſer particles returning again to aer, water, earth, or other Bodies, as they were before their Concretion, or Determination to that particular ſpecies of things, by Generation.

But, *Democritus, Epicurus,* and *Leucippus* are ſomewhat more full and perſpicuous in their Solution of this Problem, declaring (1) That, when a Thing is Generated, multitudes of Atoms are congregated, commixed, compoſed, diſpoſed, & complicated after ſuch a determinate manner, as that from thence doth neceſſarily reſult a body of ſuch a particular ſpecies, apparence, and conſequently of ſuch a reſpective denomination.  (2) That in ſuch a Body there is no ſubſtance, which was not præexiſtent, it being impoſſible that New Atoms (which only conſtitute Corporeal ſubſtance) ſhould be created: but only that ſuch a certain Diſpoſition and Configuration of the Atoms, eternally præexiſtent, is made, from which ſuch a Form ariſeth, which

*Art. 9.*
An abſtract of the theory of the Atomiſts, touching the ſame.

is nothing really diſtinct from, but is the very Atoms themſelves , as they are thus, and no otherwiſe ordered and compoſed. (3) That the Forme of a thing,conſidered abſtractly or by it ſelf,is therefore onely a meer Quality, Accident, or Event,of which the Atoms,which compoſe that Body or ſubſtance, are naturally capable , when thus conſociated and mutually related : whether we underſtand it to be the Forme of the whole Compoſitum , or of that moſt ſubtile and active part of the ſubſtance commonly called the Soul, or ſpecifical Forme (V. G. of an Horſe) the ſame being (not a New, or freſhly created ſubſtance, as *Ariſtotle,*and the *Schools* upon his Authority conceive,but) only a certain Contexture of the moſt ſubtile and moveable Atoms in the compoſition. (4)That out of the infinite ſtock of the Univerſal and FirſtMatter,unceſſantly moving in the infinite ſpace,when ſuch Conſimular Atoms meet together , as are reciprocally proportionate or reſpondent,and mutually implicate each other by their ſmall Hooks and Faſtnings; then are generated certain very ſmall Bodies, or *maſſes,* ſuch as being much below the diſcernment of the ſenſe, may be accounted *Semina Rerum,* the ſeeds of things : differing from the *Homœomerical Principles* of *Anaxageras* in this,that though very hardly, yet at laſt they may be diſſolved, and reduced to the ſingle Atoms,of which at firſt they were compoſed, whereas the *Homœomera* of *Anaxagoras* are *Irreſoluble,*and *Firſt Principles.* (5) That theſe *Moleculæ,* Firſt Maſſes , or ſmalleſt Concretions of Atoms, are the Proxime and Immediate Principles of Fire , Water , Aer, and of other things more ſimple, ſuch as the Chymiſts conceive their Three Catholique Principles, Sal, Sulphur,and Mercury to be : from which afterward congregated and commit t into greater maſſes, artiſ various kinds of Bodies, reſpectively to the various manners of their commiſtion , diſpoſition , and concretion : as Animals, Vegetables, Minerals. (6) That from the Diſſolution of Bodies compoſed of divers ſorts of ſuch Firſt Maſſes of Atoms,ſuch as Animals, Plants, Minerals, and each of their ſeveral ſpecies , divers Bodies of more ſimple Compoſitions may be Generated , according as the ſmall maſſes or Complications of Atoms, ſeparated , by diſſolution , from them, ſhall be more or leſs Conſimilar, and convene again in this or that order and poſition,or particular ſpecies ; as when from wood diſſolved by Fire, are generated Fire,Smoke,Flame,Soot,and Aſhes. And this is the Summary of the Atomiſts Doctrine concerning the eſſence of Forms : which that we may conveniently illuſtrate, let us a while inſiſt upon that moſt opportune inſtance of the Generation of thoſe divers things, *Fire, Flame,Smoke Soot, Aſhes,* and *ſalt,* from the Diſſolution of *Wood.*

*Art.* 10
An illuſtration
thereof , by a
pragnant and
opportune in-
ſtance, viz. the
Generation of
Fire , Flame,
Fume, Soot,
Aſhes and Salt,
from Wood dil-
ſolved by Fire.

Let us conceive (1) That Wood is a Compound Body, made up of various Moleculæ, or ſmall maſſes of Atoms : (2) That thoſe ſmall maſſes of Atoms are ſuch, as that being congregated, commixt , and according to ſuch a determinate manner diſpoſed , they muſt in the whole compoſition, retain the ſpecies or Forme of Wood ; but being diſſociated , ſeparated, and after another manner again connexed and diſpoſed , they muſt exhibite other leſs compound Forms , or ſpecies of Bodies : (3) That in the Concretion there are exiſtent multitudes of ſpherical, moſt exile , and moſt agile Atoms, ſuch as, when they are expeded from the fetters of the groſſer maſs, and flye away together in great numbers, and conſociated, are comparated to make and exhibite the ſpecies of Fire : (4) That of theſe Igneous particles is generated *Flame.* Whoſe *Clarity* &*Splendor* ariſeth from the Abjection of other diſſimilar and impure parts, formerly commixt with the Igneous particles. Whoſe tendency *Vpwards,*and ſucceeding *Diſapparence* ariſe both from the force and pernicity of the Igneous particles in their exſſiliti-
on,

on, and the pressure or urgency of the ambient Aer. Whose gradual *Attenuation*, and *conicall Figure* arise from hence, that the Igneous particles, in respect of their roundness, exility, and superlative mobility, evolving and expeding themselves from the Concretion the soonest of all others contained therein, and in swarms diffusing themselves through the environing aer, on all sides, do create a Light, which is by degrees so exhausted, in regard of the speedy avolition of the igneous Atoms composing it, that it dwindles or consumes away to a cone or sharp point, which is also much more rare then the basis, where the igneous particles are most dense and agminous. Whose *Dilatation* from its base to some degrees, and *Tremulation* or *Undulation* arise from the copious, but indirect emption of the igneous particles, disengaging themselves from the grosser parts of the mixture. Whose *Obnubilation* by some smoke commixt with it, is caused by the many Fuliginous particles, that the Igneous ones carry off with them, as they flye away. Whose faculty of *Pungency*, *Penetration*, and *Dissolution* of most bodies objected, consisteth in the transcendent subtility of the Igneous particles, and in the pernicity of their motion, as we have largely declared in our præcedent Discourse of the Nature of Heat. (5) That the *Fume*, or smoke issuing from wood in combustion, together with Flame, is much more simple than the wood it self, but yet compounded of divers particles, some whereof are Watery, others Earthy, others Salt, others Fuliginous, as appears by the adhærence of the soot to the Chimny, by the præcipation of the earthy fæces of soot to the bottom of a vessel of Water, and the extraction of Salt from thence by a dissolution of soot in warm water, and the Denigration of things thereby. (6) And lastly, that what we have conceived of Flame and Smoke, may be equally reasonable, if applied also to the remaining *Ashes* of wood burned, they being likewise composed of various particles or small masses both of *Salt* and *Earth*; and the particles of Earth being again composed of Mud and Sand, or such as that of which Glass is made. And when we have perpended the verisimility of these Conceptions, we shall be fully convinced; that Wood is a thing composed of divers sorts of small bodies, or minute masses of Atoms; and that the Form thereof doth consist in the Congeties, Concretion, complexion, and determinate Disposition of them all; as also that the Fire, or Flame issuing from it in combustion, is a thing likewise consisting of various sorts of particles contained in the Wood, and which being separated, and again consociated (according to the Consimilarity or likeness of their natures) and concreted among themselves, obtain another Disposition, and Forme, and so exhibite the species of a New body.

## Sect.  II.

*Art.* I.
That in Cor-
ruption, no
*substance* perish-
eth; but only
that determi-
nate *Modifica-*
*tion* of sub-
stance, or Mat-
ter, which spe-
cified the
thing.

FRom Generation (as in the Method of Nature, so in our disquisitions concerning Her) we pass to CORRUPTION; which is no more but the Dissolution of the Forme; i. e. the determinate Modification of the matter of a thing, so that it is thereby totally devested of the right of its former Denomeration.     For, since it is most certain, that in Generation, there doth arise no such New *substantial* Forme, as *Aristotle* dreamt of, and most men have ever since disquieted their heads withal : it can be no less certain, that neither in *Corruption* can any such Form, as ever was *substanti-al*, perish or be annihilated.    Which verily that we may most commodi-ously enforce, resuming our late *Instance* of the Generation of Fire, Flame Smok, &c. from the combustion of wood, we shall to our præcedent re-marks there thereupon, superad this observation; that when wood perish-eth by Fire, and so is resolved into divers other Bodies, it is not resolved in-to any other, but those very same things, which were really præexistent and contained therein; and consequently, that nothing thereof perisheth, but only that determinate Connexion and situation of its parts, or that special manner of their existence, (you may call it Forme, Quality, Species, Acci-dent, or Event) in respect whereof it was wood, and was so denominated. A strange Assertion you'l say, that there is really existent in wood, Fire, that there is Flame, that there is Salt, that there are all those divers things into which it is resoluble by corruption.   And yet the Truth much transcends the strangeness of it; the difficulty, at which you are startled, consisting only in Name, not in the Thing it self.   For, if by Fire you understand burning Coales or Flame actually ardent and lucent; and if by Salt you conceive a Body sapid, really and sensibly corrading the tongue : then, in-deed, we shall confess that there is no such Fire, nor Flame, no such Salt ex-isting actually in wood : But, if you by the names of Fire, and Salt, under-stand (as the tenour of our Dissectation, both directeth and obligeth you to understand) the seeds, or small masses, or first Concretions of Fire and Salt, such which are so exile, as that each of them singly accepted is very much beneath the perception and discernment of the most acute of senses; but yet when multitudes of them are sequestred from the whole mass, and are again congregated and freshly complicated together, the seeds of Fire by themselves, those of Salt by themselves; then do these actually burn and shine, and those actually make a Sapour, sharply affecting and corrading the tongue : we see no reason, why you should wonder at our tenent, that both Fire and Salt, *viz.* that very Fire which burns and shines in the wood, that very Salt which may be extracted from the Ashes thereof, were præexistent in the wood.   Certainly, you cannot but admit as highly con-sentaneous to reason; that in a vapour to what rate soever attenuated, there are contained the seeds of Water, or the first concretions of Aqueous A-toms; which though singly existent they are wholly imperceptible, yet nevertheless are they really particles of water : for as much as they want only the convention and coalition of many of them together, to the discovery

very of their nature in sensible masses; for of many of them condensed are made very small drops of water, of those drops assembled together arise greater drops, of those rain is generated from that rain arise whole streams, and many of those streams meeting together swell into great and impetuous torrents. And if this be so easily, why should that be so hardly admittible?

But to desert this Example, and address to another so competent and illustrious, that it takes off all obscurity as well as difficulty from our conception; it is well known, that silver is capable of such exact permistion with Gold, as that though there be but one single ounce of Silver admixt by confusion to 1000 ounces of Gold: yet in the whole mass there shall be no sensible part, wherein somewhat of that small proportion of silver is not contained. Now, you cannot expect that each single molecula, or seed of silver should appear to the sense, so as to distinguish it self, by its proper colour from the small masses of Gold: because each molecula of silver is surrounded with, and immersed among 1000 particles or small masses of Gold. Nor can you believe, that the silver is wholly unsilvered, or Changed into Gold; as *Aristotle* affirmed, that a drop of Wine, infused into a great quantity of Water, is changed into Water: because the skilful Metallist will soon contradict you in that, by an ocular demonstration. For, by Aqua Fortis poured upon the whole mass, He will so separate the silver from that so excessive proportion of Gold, as that there shall not be left inhærent therein so much as one the smallest particle thereof; and in the superfice you may plainly discern multitudes of very small holes, (like punctures in wax, made by the point of the smallest needle) in which the moleculæ or small masses of the silver were resident, before its sequestration from Gold. Why therefore, according to the same reason, should it not be equally probable, that the seeds, or particles of Fire are so scatteringly diffused through the substance of wood, as that being surrounded and overwhelmed with myriads of particles of other sorts, they cannot therefore put on the apparence proper to their nature, and discover themselves to be what really they are, until being by the force of the external fire invading and dissolving the compage of the wood, set at liberty, and disengaged from their former oppression, they issue forth in swarms, and by their coemergency and consimilarity in bulk figure and motion being again congregated, they display themselves to the sense in the illustrious Forme of Fire and Flame, and proportionately diminish the quantity of the wood; which thereupon is first reduced to Coals, and afterward, the separation and avolation of more and more particles successively being continued, to Ashes, which containing no more igneous particles, can maintain the combustion no longer.

*Art.* 2.
Enforcement of the same Thesis, by an illustrious *Example.*

The like may be said also of the Salt, diffusedly concealed in Wood. For, insomuch as each single particle of Salt ambuscadoed therein, is blended among, and as it were immured by myriads of other particles: it is impossible they should exhibite themselves in their genuine Forme, while they remain in that state of separation or singular existence; which they must do, till the compage of the whole mass or Concretion be dissolved. And would you be, beyond all pretext of doubt, convinced, that they yet retain their proper nature, amidst such multitudes of other particles; be pleased only to make this easie Experiment. Take two pieces of the same Wood of equal weight, and steep one in water, for two or three days; and keep
the

*Art.* 3.
An *Experiment* demonstrating that the *Salt* of *Ashes* was præexistent in Wood; and not produced, but only educed by Fire.

the other from all moyſture; then by fire reduce each of them apart to Aſhes, and by Water affuſed thereunto, and boyled to a lee, extract the Salt from the Aſhes of each: this done, you ſhall find the Aſhes of the drie piece to have yeelded a quantity of Salt proportionate to its bulk, but thoſe of the wet one very little, or none at all. And the Reaſon is only this, that the water in which the one piece was macerated, hath exhauſted moſt part, if not all of the Salt, that was contained therein. Now this Example we alledge to prævent your falling upon that vulgar conceit, that the Salt of Aſhes is produced only by the Exuſtion of the Wood: ſince, according to that ſuppoſition, the macerated piece of wood would yeeld as much of Salt, as the Drie. This conſidered, it remains a firm and illuſtrious truth, that all the particles of the Fire, Salt, Smoke, &c. educible from wood, were really præexiſtent therein, though ſo variouſly commixt one among another, as that notwithſtanding each of them conſtantly retained its proper nature entire, yet could they not diſcover themſelves, in their own colours, proprieties, and ſpecies, till many of each ſort were diſ-engaged from the Concretion at once, and aſſembled together again.

*Art.* 4.
The true ſenſe
of three Gene-
ral *Axioms,* de-
duced from the
precedent do-
ctrine of the
*Atemiſts.*

Now ſuch are the Advantages of this Theory above that of *Ariſtotle,* that beſides the full ſuffragation of it to the Common Notions of Generation and Corruption, of ſubſtance, Forme, &c. it aſſiſts us in the expoſition of Three General Axiomes, which though drawn into rules by *Ariſtotle* himſelf, are partly inconſiſtent with, partly unintelligible from his doctrine.

The Firſt is, *ſi aliquid corrumpitur ultimum abire in primam Materiam,* That when any thing is corrupted, it is at laſt reduced to the Firſt matter: which doth expreſly contradict His grand theſis, that the Forme of a thing is a ſubſtance, which begins to be in Generation, and ceaſeth to be, or is annihilated in Corruption; for, had He ſpoken conformably thereto, He muſt have ſaid, that when the Compoſitum is diſſolved by Corruption, it is partly reduced to matter, partly to Nothing. But, if the Form be not ſubſtantial, and that what is Corrupted, is compoſed of no other ſubſtantial parts, but thoſe which are material; as we have aſſumed: then, indeed, doth the Axiome hold good, and we may with good reaſon ſay, that when any thing is Corrupted, it is reduced to matter, or the material parts, of which it was compoſed, as wood diſſolved by fire, is reduced to Fire, Smoke, Soot, Aſhes, &c. of which it did conſiſt. And foraſmuch as by that Adverb, *Ultimum,* Finally, He gives us the occaſion of Enquiring, *An in Corruptione detur reſolutio aduſque materiam Primam?* Whether or no in Corruption there be a Reſolution even to the Firſt matter? we cannot but obſerve, that the manner of that ultimate reſolution may be much more eaſily comprehended, according to our aſſumption, than according to His own. Becauſe Our Firſt matter is Atoms, and the ſecond matter certain ſmall maſſes of Atoms, or the firſt Concretions, which we therefore, obſerving the phraſe of *Epicurus* and *Lucretius,* call *Semina Rerum,* the ſeeds of Things, ſuch as thoſe whereof Fire, ſilver, Gold, and the like Concretions are compoſed: and ſo, if the Reſolution proceed to extremity, i. e. to Atoms, or inexſoluble particles (as in ſome caſes it doth) then may it well be ſaid, that the reſolution is made to the Firſt Matter; but if it go no farther then thoſe ſmall maſſes of Atoms (as moſt commonly it doth not) then can we juſtly ſay no more, than that the reſolution is made only to the ſecond matter.

The

The Second is, *Corruptionem Unius esse Generationem alterius*, that the Corruption of one thing is the Generation of another, which cannot confist with truth, if understood in any other sense but that of our supposition. For, since, Corruption is nothing else but a separation and exsolution of the parts, of which a thing was composed: we may conceive, how those parts so separated and exsolved, may be variously convened and commixt again afterward, as to constitute New Concretions, & put on other new Forms. Not that they were not formerly existent, as to all their substantial parts : but only that they were not formerly existent in a state of separation from others, nor coadunated again in the same compage, and after the same manner.

The Third, *Id quod semel Corruptum est, non posse idem numero naturæ viribus restitui*, that what is once Corrupted, cannot by Natures power be again restored numerically the same : which is to be understood in this sense. As a Watch, or other Artificial machine, composed of many several parts, may be taken in pieces, and easily recomposed again into the very same numerical Engine, both as to matter and Forme, the Artificer recollecting the divided parts thereof, and so disposing them, as that each possesseth the same place and position, as before its dissolution : so likewise might the same Natural Compositum, V. G. a piece of Wood, be, after the separation and exsolution of all its component parts, again recomposed numerically the very same, both as to matter and Forme, in case all those dissolved parts could be recollected, reunited, and each of them restored to its former place and position. But, though all the various parts thereof remain, yet are they so scattered abroad into so many and so various places, and commixt (perchance) with so many several things, that there is no Natural Power that can recollect and restore them to the same places and positions, which they held before their disunion and dissolution. And, therefore, if any man shall say, that such or such a thing, dissolved by Corruption, is capable of being restored again the same *in specie*; we ought to understand him no otherwise than thus : that some of the parts of that thing may so return, as that being conjoyned to others, not numerically the same, but like unto those, to which they were formerly conjoyned, they may make up a body exactly like the former, *in specie*, or of the same Denomination; as when the Carcase of an Horse is corrupted, some parts thereof are converted into Earth, some of that Earth is converted into Grass, some of that Grass eaten by another Horse, is again converted into Seed, whereof a third Horse is generated. And thus are we to conceive the endless Circulation of Forms.

As for the Principal CAUSES of Corruption, (omitting the consideration of such as are External, or invading from without, in respect they are innumerable; and of that Internal one also, the intestine war of Elements in every Concretion, of which *Aristotle* hath such large discourses, and the Schools much larger) the theory of *Epicurus* instructs us, that they are only Two. The *First* and *Grand* one is the *Intermistion of Vacuity among the solid particles of bodies*; in respect whereof all Concretions are so much more easily Exsoluble, or subject to Corruption, by how much more of Vacuity they have intercepted among the solid particles, that compose them : according to that Distich of *Lucretius*.

*Art. 5.*
The General Intestine Causes of Corruption, chiefly Two : (1) the *interception of Inanity among the solid particles of Bodies*; (2) The *mutual Gravity and inseparable Mobility of Atoms.*

Et

*Et quam quæque magis cohibet res intus Inane,*
*Tum magis his rebus penitus tentata labascit.*

The *other* is the *Ingenite Gravity*, or *natural and inamissible propensity of Atoms to Motion* which always inciteth them to intestine commotions and continual attempts of exsilition. So that where their Connexions and complications are but lax, and easily exsoluble, as in all Animals, all Plants, and some Metals, there do they sooner and more easily expede themselves, and so in short time totally dissolve the Concretions, which they composed. But, where they are bound to a more lasting peace, by more close compaction, and reciprocal complications, as in Gold and Adamants; there their inhærent propensity to motion is so suppreſt, as that they cannot disengage themselves each from other, without great difficulty, and after many hundred yeers continual attempts of evolution, convolution and exsilition. Which is the true Reason both why Gold is the least Corruptible of all things yet known; and why it is not wholly Incorruptible, but obnoxious to spontaneous Dissolution, though after perhaps a million of yeers, when after innumerable myriads of convolutions, the Atoms which compose it, have successively attained their liberty, and flye off one after another, till the whole of that so closely compacted substance be dissolved.

*Art. 6.*
The Generall *Manners,* or Ways of Generation and Corruption.

From the Causes, our thoughts are now at length arrived at the MANNERS, or Ways of Generation and Corruption; and find them to be of Two sorts, *General* and *Special.* Concerning the *General,* we observe, that according to the doctrine of *Epicurus,* (whose great præheminence in point of Verisimility and Concordance throughout, hath made us præfer it to that of *Aristotle,* which we have amply convicted of manifest Incomprehensibility, and self-contradiction) things are generated either immediately of Atoms themselves convened together and concreted, and resolved again immediately into Atoms; or immediately of præexistent Concretions, and resolved immediately into them again. Of the way how the *Former* is effected, we have said enough, in the second chapter of our Discourse against Atheism. As to the *Latter,* be pleased to understand, in a word, that all Generation is caused by either (1) *A simple Transposition of parts of the same numerical matter,* Or (2) *an Abjection of some parts of the old, or præexistent matter,* or (3) *An Accession of new parts.* For, howbeit all these three General ways of Generation are mostly so concurrent and commixt, as that one is hardly found without the association of the other two: yet when we consider each of them in special, and would determine which of them is prædominant over the others, in the generation of this, or that particular species of things: it will be necessary, that we allow this Discrimination. First, therefore, those things are said to be generated [κατα μεταθεσιν] by a meet *Transposition* of parts, which are observed to be spontaneous in their Production; as Frogs engendred only of mud or slime, Worms from putrid Chees, &c. because from the very self-same præexistent matter, only by a various transposition of its parts, & succeeding reduction of them to such, or such a determinate order & situation, something is generated, of a nature absolutely new or quite different from what that matter formerly had. And hither also are we to refer those *Transmutations of Elements,* of which *Aristotle* and the *Schools* have such frequent and high discourses: because, when Aer is conceived to be changed into Water, or Water
ter

ter transformed into Aer ; all the myſterie of thoſe reciprocal metamor-
phoſes amounts to no more, than a meer putting of the parts of the ſame
common and indifferent matter into different modes , and the interception
of more or leſs of Inanity among them, as we have frequently demonſtrated. Secondly, ſuch things are conceived to be generated [κατα προσθεσιν]
by *Addition* or Acceſſion, which are not ſpontaneous in their original, but of
ſeminal production , and ſpecificated by the univocal virtue of their ſeeds :
becauſe in Propagation, rightly accepted, a very ſmall quantity of ſeed, pervading a greater maſs of matter, doth ferment, coagulate , and ſucceſſively
appoſe more and more parts thereof to itſelf, and conform the ſame into
the ſpecies of that thing, from which it was derived, and impregnated with
the idea of the whole and every part thereof. And this Difference includes
not only all *Augmentation,* which is a kind of Aggeneration, and conſiſteth
only in the Appoſition of new matter or ſubſtance, and that in a greater proportion than what is decayed or exhauſted: but alſo every *Compoſition*
whatever , ſuch as is the *Inſition* or *Inoculation* of Plants. Thirdly, ſuch
things are ſaid to be generated [κατ αναιρεσιν] by *Detraction* which ariſe
from the Diſſolution of others, and ſubſiſt only by Excretion or Separation ; as Fire, Smoke, &c. are derived from the Diſſolution of wood , and
other combuſtible ſubſtances, to which they were formerly commixt ; and
Wax from the ſeparation of Hony, together with which it was blended in
the Combs. And , as for the Contrary, *Corruption,* 'tis eaſie to deduce it
from the contrary ways of diſpoſing matter.

And here again the incircumſpection of *Ariſtotle* manifeſtly diſcovers it
ſelf ; who multiplies the General ways of Generation, to a ſuperfluous number : expreſly teaching, that every ſimple Generation ariſeth from (1) either
*Transfiguration,* as when a ſtatue is made of molten metal ; or (2) *Addi-*
*tion,* as when Vegetables or Animals are Augmented ; or (3) *Ablation,* as
when a ſtatue is hewn out of Marble , all ſuch parts being cut off and abje-
cted, as were ſuperfluous to the perfection of the Figure deſigned ; or (4)
*Compoſition* , as in the ſtructure of a houſe of various materials compoſed,
according to the rules of Architecture ; or (5) *Alteration,* when a thing is
changed as to matter , as when Aſhes are produced out of wood combuſt.
When notwithſtanding, had not his accuſtomed diligence been laid aſleep,
or judgement perverted, he muſt ſoon have perceived , that his Transfiguration, Addition, and Ablation are really the very ſame with the Tranſpoſition, Adjection, and Detraction of our *Epicurus* ; and that Compoſition
is neceſſarily referrible to Addition, and Alteration to Tranſpoſition.

*Art. 7.*
Inadvertency
of *Ariſtotle* in
making *Five*
General Modes
of Generation

Concerning the *Special* modes, or ways of Generation , we need advertiſe you of only two Conſiderables, (1) That each of the three General
ways, newly mentioned, is ſo fruitful in poſſible variety , as that the ſpecial
ſubordinate ones, whereof it is comprehenſive, are (it not infinite, yet) abſolutely innumerable, ineffable, incomprehenſible. For, if the Letters of
our Alphabet, which are but 24 in number, may be ſo variouſly compoſed,
as to make ſuch a vaſt diverſity of words, which cannot be enumerated by
fewer then 39 cyphers, *viz.*

*Art. 8.*
The *ſpecial*
Manners of
Generation,
innumerable ;
and why.

29523279903960414084761860964352000000000.

*(Tantum Elementa queunt, permutato ordine ſolo)*

Kkk                                What

What Arithmetician can compute the several special ways of composition, whereof that incomprehensible variety of Figures which (as we have
frequently assumed) Atoms may bear, is easily capable?

*Art. 9.*
All sorts of A-
toms, not indif-
ferently com-
petent to the
Constitution
of all sorts of
thing.

(2) That, as the Image of Mercury cannot be carved out of every stone,
or every piece of wood; nor words fit for reading, or pronunciation arise
from every commistion of Letters: so, in Natural Concretions is it impossible, that all things should be made of all sorts of Atoms, or that all A-
toms should be equally accommodate to the constitution of every species of
Concretions. For, though Atoms of the same figure and magnitude may,
by their various transposition, adjection, ablation, compose things of various
forms or natures: yet are they not all indifferently disposed to the composition of all things, nor can they be connected after one and the same manner, in divers things. Because, to the composition of every thing in specie,
is required such a special disposition in the Atoms, which compose it, as that
they must appose to themselves such other Atoms, as are congruous and
suitable to them, and as it were refuse the society and combination of others
that are not. And hence is it, that in the Dissolution of every Concretion,
the consimilar or like Atoms always consociate together, and expede themselves from the Dissimilar and incongruous.

CHAP.

# CHAP. II.

# OF

# MOTION.

### Sect. I.

Ertainly, the Great *Galileo* did moſt judiciouſly and like himſelf, to lay the foundation of his incomparable Enquiry into the moſt recondite myſteries of Nature, in the Conſideratin of the Nature of MOTION, and ſevere Examination (that we may not ſay, ſubverſion) of *Ariſtotles* Doctrine concerning it. Becauſe, Motion being the Heart, or rather the Vital Faculty of Nature, without which the Univerſe were yet but a meer Chaos; muſt alſo be the nobleſt part of *Phyſiology*: and conſequently, the ſpeculation thereof muſt be the moſt advantageous Introduction to the Anatomy of all other parts in the vaſt and ſymmetrical Body of this All, or Adſpectable World. Again, if Motion and Quiet be the principal modes of Bodies Exiſting, as *Des Cartes (in princip.philoſoph. part.2. ſect.27.)* ſeems ſtrongly to aſſert; if Generation, Corruption, Augmentation, Diminution, Alteration, be only certain ſpecies, or more properly the Effects of Motion, as our immediately præcedent Chapter cleerly imports; and that we can have no other Cognizance of the conditions or qualities of ſenſible objects, but what reſults from our perception of the Impulſes made upon the organs of our ſenſes, by their ſpecies thither tranſmitted: aſſuredly, the Phyſiologiſt is highly concerned to make the contemplation of *Motion*, its *Cauſes, Kinds,* and *Univerſal Laws,* the *Firſt* link in the chain of all his Natural Theorems. And, truly, this we our ſelves had not endeavoured, had not our firm reſolution to avoid that ungrateful prolixity, which muſt ariſe from the frequent Repetitions of the ſame Notions, in the ſolution of various natural Apparences; and our deſign of inſenſibly præparing the minde of our Reader, with the gradual inſinuation of all both Cauſes and Effects of ſpecial motions, as they ſtood in relation to this or that particular ſenſible object, and principally to Viſibles, and the Gravitation of Bodies: not only inclined,

*Art. 1.*
Why the Nature of *Motion*, which deſerved to have been the ſubject of the firſt ſpeculation, was reſerved to be the Argument of the *Laſt,* in this Phyſiology.

but by a necessity of Method almost constrained us, to make that the *Hem*, or *Fringe*, which otherwise ought to have been the *First Thread* in this rawe and loosely contexed Web of our *Philosophy*.

Nor, indeed,can we yet prævent all Repetitions; for, our præsent Theorem being *Physicomathematical*, and such as must borrow some light, by way of Reflection,from sundry observables, occasionally diffused upon several of our Discourses præcedent: we need not despair of a Dispensation for our Recognition of a few remarkable passages, directly relating thereunto, and especially of these *Three Epicurean Postulates*, or Principles.

*Art. 2.*
An *Epicurean* Principle, of fundamental concern to motion.

The FIRST; *that the Adam or Radical and Primary Cause of all motion competent to Concretions, is the inherent Gravity of their Materials, Atoms : whether the Concretion be moved spontaneously, or violently, i.e. by it self, or another.* The *Reason* of its spontaneous or self-motion may be thus conceived. While Atoms are, by their own inamissible propensity to motion, variously agitated and tumultuous in any Concretion; if those which are more moveable and agile then the rest, so conspire together in the course of their tendency, as to discharge their united forces upon one and the same quarter of the body containing them, and so attempt to disengage themselves toward that region:then do they propel the whole body toward the same region,transferring the rest of their less active associates along with them. It being highly consentaneous, that motion may be expressed first in the singular Atoms themselves, then in the smallest masses, or insensible Combinations of Atoms; and successively in greater and greater, till the sensible parts of bodies, and at length the whole bodies themselves participate the motion, and undergo manifest agitation : as *Lucretius* (*in lib. 2.*) hath with lively Arguments asserted.

*Art. 3.*
*Aristotles* Position, that the *first Principle of motion,is the very Forme of the thing moved;* absolutely incomprehensible :unless the Form of a thing be conceived to be a certain tenuous Contexture of most subtile and most active Atoms.

And this, certainly, hath far a stronger claim to our assent, than that fundamental Position of *Aristotle*; that *the First Principle of motion in any thing, is the very Forme of the thing moved*. For, unless He shall give us leave, by the word Forme, to understand a certain tenuous Contexture of most subtile and most active Atoms, which being diffused through the body or mass consisting of other less subtile, and in respect of their greater compaction together, or more close reciprocal revinction, less active Atoms; doth, by the impression of its force or Virtue motive, upon the whole, or any sensible part thereof, become the Principle of motion to the whole body : we say, unless he shall be pleased to allow us this interpretation, we shall take the liberty to affirm, that it is absolutely incomprehensible. For, that the Forme of a thing, accepted according to His notion of a Forme, should be the Proto-cause or Principle of its motion; is unconceivable; since, according to the tenour of *Aristotles* doctrine, the Forme must be educed out of the Matter, or power of the Matter, that constituteth or amasseth that thing : and consequently, the Forme must owe as well its very Entity or Being, as all its Attributes onely to the matter it self; which yet He describes to be something (rather, nothing) meerly Passive, and devoid of all activity or Power whatever. How, therefore, can it appear other than a downright *Contradiction*, to any man,whose intellect is not eclipsed, by reason of some great disorder of its proper Organ; that that Matter, which in it self hath no Power or Faculty of Moving, should nevertheless be able to impress a Faculty of motion, and potent Activity, upon the Form, supposed

Pofed to be abfolutely diftinct from matter? Doubtleſs, the Forme doth not derive that Motive Virtue from the *Qualities* inhærent in the matter: foraſmuch as thoſe Qualities, as even the Ariſtoteleans themſelves furiouſly contend, are but the meer Reſults of the Power of the matter. Nor from the *Efficient*; becauſe They account the Efficient to be a Cauſe meerly External, and to transfuſe nothing of it ſelf into the thing Generated; but only to diſplay its Efficiency, or (to ſpeak in their own dialect) to execute its Cauſality upon the matter. Again, it being neceſſary, that all that Virtue of Moving, which is in the Efficient, ſhould depend ſolely and wholly upon its Forme; and that Forme alſo ought, by equal reaſon, to be educed out of the matter: They loſe themſelves in a round of Petitions, and ſtill reduce themſelves to the ſame Difficulty, *How it is poſſible, that the matter ſhould give that Faculty of Motion to the Forme, which it ſelf never had.*

The SECOND; *that in General there is no other but Local motion.* Wherein that we may plainly and briefly inſtruct you, how far *Epicurus* differs from *Ariſtotle, Plato,* and ſome other Philoſophers; give us leave to commemorate unto you.

*Art.* 4.
A ſecond general certain Fundamental, concerning motion: and the ſtate of the Difference betwixt Epicurus, Ariſtotle, and Plato, touching the ſame.

(1) That *Ariſtotle* putting a difference betwixt [κίνησιν, & μεταβολὴν] Motion and Mutation, is not ſufficiently conſtant in his doctrine: ſometimes making Mutation to be the Genus, and Motion only a certain ſpecies thereof; and ſometimes, by inverſion of the tables, making Motion the Genus, and mutation a ſpecies thereof. For, (*in* 5. *phyſic. cap.* 2.) ſtating Mutation betwixt two Terms, *à quo, & ad quem,* the from whence and to what; He aſſigns unto 4 diſtinct Modes, or Manners; the firſt, *à ſubjecto in ſubjectum*; the ſecond, *ex non ſubjecto in non ſubjectum*; the third, *ex non ſubjecto in ſubjectum*; the fourth *ex ſubjecto in non ſubjectum:* and thereupon infers, as of pure neceſſity, that ſince nothing can be changed according to the ſecond mode, therefore muſt mutation according to the third, be Generation; according to the fourth, be Corruption; and according to the firſt, be Motion, which is always either from Quantity to Quantity, or from Quality to Quality, or from Place to Place. Whereas, in another place (*viz.* 3. *Phyſic.* 1.) He poſitively teacheth, that Motion is a certain Act, to which that paſſeth, which is in Power; and ſo makes the ſpecies thereof to be not only thoſe motions, whoſe terms on either ſide are Poſitive, or (in his own phraſe) Contrary, as are thoſe which concern Quantity, Quality, Place: but thoſe alſo, whoſe each term is Privative, as are thoſe which concern ſubſtance. And hereupon He ſeems to have grounded that memorable Diviſion of Motion (*lib. de prædicam. cap. de motu.*) into ſix ſpecies, *viz.* Generation, Corruption, Accretion, Diminution, Alteration, and Lation or Loco-motion: whereof the firſt two are according to ſubſtance; the ſecond two, according to Quantity; the fifth, according to Quality; and the Laſt, according to Place.

(2) That *Plato* ſeems conſtantly to accept Mutation for the Genus, and motion for one ſpecies thereof: ſubdividing motion into two ſpecies, Lation and Alteration. Foraſmuch as in one place (*viz. in Polit*) He terms the Converſions of the Cœleſtial bodies, Mutations: and in another (*in Phæd.*) he takes Alteration for mutation; ſaying moſt eloquently in the perſon of *Socrates* (*in theæt.*) *Illudne moveri appellas, dum quidpiam locum*
*è loco*

*è loco mutat, aut in eodem convertitur?* Tho. *Equidem.* Socrat. *illa ergo
una fit species motus. At, cum in eodem quidem perstat, sed senescit tamen,
aut ex albo fit nigrum, ex molli durum, aut altereratione quapiam alterum e-
vadit: an non videri alium motus speciem necesse est?* Tho. *mihi quidem
videtur.* Socrat. *Necessarium id igitur, duas, inquam, esse motus species,
Alterationem, & Lationem, Circulationemve?* &c.

(3) That *other Philosophers*, insisting in the steps of *Plato* constitute only two kinds of Motion, only in this they differ from Him, that what He calls [φοράν, ἡ περιφορεάν], Latin, or *Circumlation*, They call [μετάβασιν, ἡ μεταβατικὴν κίνησιν] *Transition*, or *motion Transitive*: and what He names [ἀλλοίωσιν] *Alteration*, They denominate [μεταβολὴν] *Mutation*, or [μεταβλητικὴν κίνησιν] *Motion Mutative*; as *Empiricus* (2. *adverf. physic.*) hath judiciously observed.

(4) That *Epicurus* (as the same *Empiricus*, in the same place, attesteth) is chief of those Physiologists, who accounted the Motion of Transition as the Genus, and Mutation or Alteration as only the species thereof. And this upon irrefragable Reason. Forasmuch as Alteration is nothing else but the consequent of Local motion, whereby Atoms, or the insensible particles of Concretions usually accede, decede, concur, complicate, and change their former positions, so as to render the sensible parts or whole of them other than they formerly were. Which being considered, we are only to advertise farther, that the Argument of our præsent Enquiry, is not Motion as it is proper to Atoms, as they either concur to the first constitution of a body, or are disgregated at the dissolution thereof; in which respect it may comprehend Generation and Corruption: nor as they concur to the Augmentation of a body already constituted, or flye off from it, and by their decedence Diminish it, in which respect it may comprehend Accretion and Diminution: nor as they are variously transported, and so conduce to affect the same body with divers Qualities; in which respect it may include Alteration. Because concerning Motion under all these Terms and relations, we have sufficiently discoursed already, in places to which those considerations did genuinly refer themselves. But, our subject is *Motion as proper to a body Concrete, which sensibly changes the Place of its whole, or some sensible part.* For, herein motion plainly distinguisheth it self from mutation, that in *motion* the whole Body, V. G. of a man, or some sensible part thereof, as his hand or foot is translated from one place to another: but in *Mutation* only the insensible particles of a body, or any part thereof, change their positions and places, though the whole, or sensible parts thereof remain quiet.

*Art.* 5.
*Epicurus's De-
finition of mo-
tion, to be the
Remove of a bo-
dy from place to
place; much
more intelligi-
ble and proper,
than Aristotles,
that is the Act
of an Entity in
power, as it is
such.*

The THIRD; that *Motion or Loco motion* (for, the common Notion, which every man conceives, so soon as he hears the word motion pronounced, unites them) *is much more intelligibly and properly defined by Epicurus,* to be [μετάβασις ἀπὸ τὸ αὐτε, εἰς τοπον] *the migration of a body from place to place:* than by *Aristotle,* to be *Actus entis potestat, quatenus est tale.* For as nothing can be more manifest than the one; so nothing can be more obscure than the other.

And yet if your curiosity be great enough to furnish you with patience, while we endeavour to pick out the meaning of *Aristotle,* in that his ænigmatical

matical Definition ; we advise you to reflect upon the whole syntax of
those conceptions, from whence He seems to have deduced it. Know,
therefore, that He conceived, that there are some things, which always possess, and inamissibly retain the perfection due to their nature, [εντελεχεια
φυσιν] *Perfecti-habitione*, or (as his Expositors commonly render it) *Actu
solum*, in Act only : and others again, which are not indeed, without some
perfection, but such as they are capable of losing, and may at the same time
acquire another ; so that they may be said to be [εντελεχεια η δυναμει]
both in *Act* and *Power together*. For, He admits nothing to be meerly in
Power ; because He would not allow, either that matter can exist without
Forme ; or that any thing in nature can be altogether without some perfection. Now, those things, which are only in Act, must, according to His
opinion, be no other but the Cœlestial Bodies ; insomuch as they alone
seem constantly and inamissibly to possess their Forme, nor can their substance or matter be conceived, to have a Capacity of receiving any other
Forme whatever. But, those which are both in Act and Power at once,
are all sublunary Bodies ; insomuch as their substance, or matter so stands
possest of some one Forme in Act ; as that it still remains in a Capacity of
being devested of that Forme, and invested with a new one ; and the whole
Compositum so hath its certain Quantity, certain Quality, certain Place,
and whatever other (if there be any other) perfection requisite to its particular nature, as that it may notwithstanding be totally deprived thereof,
and obtain another. Know also, that He useth the word, Εντελεχεια
sometimes for the perfection already acquired ; sometimes for the very
manner of its acquisition, in which sense it is a certain Action, and so comes
to be called [ενεργεια] an Energy. This being præsupposed, He infers,
that Motion is [Εντελεχεια] an Act, according to the posterior mode :
understanding it to be as it were the Way, or manner, whereby the perfection is acquired, or the Acquisition it self : which is also a certain perfection, but competent to an Entity, or moveable, not as it hath a perfection,
which it loseth ; but as it hath a Power to that, which it receiveth. And
hence is it, that He resolved to define Motion to be *the Act of an Entity in
Power, as it is such.*

*Art. 6.*
*Empericus* his
*Objections
against that
Definition of
Epicurus : and
the full Solu-
tion of each.*

Which notwithstanding all the light this our most favourable Descant,
or any other can cast upon it, is yet much inferior in Perspicuity to that
most natural and familiar one of *Epicurus* ; that Motion is *the migration or
Remove of a body from one place to another.* Nevertheless, to verifie that
unhappy proverb, that no Truth can be made so plain, as not to be impugned ; *Empericus* (2. *adverf. physic.*) hath charged it with sundry Imperfe-
ctions. As

(1) That it doth not comprehend the motion of a Globe, or wheel circumvolved upon its Axis ; forasmuch as a wheel, when circumgyrated upon its Axe, is sensibly moved, but not removed from one place to another.
But to this we may readily *Answer* ; that though the whole wheel be not
removed out of its whole place, yet are the Parts of it sensibly transferred
from place to place ; the superior descending to inferior, while the inferior
ascend to superior places, the right hand parts succeeding into the places of
the left, as fast as the left succeed into those of the right, and all parts successively shifting their particular places. And upon this distinction of
Place into *Total* and *Partial* ; was it that some Philosophers have Defined
motion

motion to be *Migrationem de loco in locum, vel totius corporis, vel partis ipsius;* or, as *Chrysippus* and *Apollodorus* (*apud Stobæum, in Ecl. phys.*) *Mutationem secundum locum, aut ex toto, aut ex parte.* Nay, even *Plato* Himself seems to have had an eye upon the same Difference, when He said, that Local motion was conjunctly *Lation,* or *Circum-lation.*

(2) That likewise the point of that arme of a Compass, which is fixed in the Centre, while the other is moved round, in the description of a Circle; is moved, but not removed out of its place : as is also the Hinge of a door, while the door is opened or shut. But, this Objection must soon yeeld to the same Response, as the former : since tis manifest, that the parts of the point of the Compass, and Hinge change their Partial places.

(3) That there is a certain sort (He adds, Admirable) of motion, to which the importance of Epicurus Definition doth not extend; which is thus made. Let a man, in a ship under sail, walk, with a staff in his hand, from the forecastle to the poup of the ship; and with just so much speed, as the ship is carried forward : so that in the same space of time, as the ship is moved a yard forward, the man and the staff in his hand may be moved a yard backward. This done (saith He) doubtless there must be a motion both of the man and his staff; and yet neither of them shall be moved into new place, either as to their whole, or their parts : because both must remain in the same parts of the Aer, and Water, or in the same perpendicular line extended from the mans head to the bottom of the Sea; or, what is the same thing, they shall still possess the same Immoveable space. But, this so admirable Difficulty lies open to a double solution: for it may be *Answered.* (1) That in this case, the Thighs, Leggs, and feet of the man walking upon the deck of the ship, must be alternately moved into new places; because, as often as each of his feet is referred from the Anterior to the Posterior part thereof, it must be moved twice as swiftly, as the ship is moved from the Posterior toward the Anterior : since it is absolutely necessary, that the double velocity of one foot should compensate that space of time, in which the other foot resteth, while the ship is constantly carried forward in one uniform tenour of motion. And, therefore, his feet may be conceived to be alternately moved from place to place; after the same manner, as a man, sitting on a wooden, or standing Horse, doth move his leggs alternately forward and backward : the trunck and upper part of his body remaining unmoved, or still keeping the same Centre of Gravity. (2) That the Trunck of his body also must be moved from place to place; and also his head, and the staff in his hand : because, at every step, all of them must be somewhat elevated, and again depressed, or let down. For, in progression, the feet of a man cannot be alternately moved forward, but at every time the one foot is set plainly upon the ground, the trunck and so the head and arms, must sink a little downward; in regard of the Distension of the muscles of that thigh and leg : and again when the other leg is advanced, and the leg upon which the whole body resteth the while, is elevated upon the toes, to cast the body forward; the trunck, head and shoulders are lifted a little upward; in respect of the bodies inclining to a new Centre of Gravity.

Gravity. For, it is moſt true, what *Galilæo* hath moſt ſubtly Demonſtrated, that *a man goes, becauſe he falls :* ſince he could not advance forward, while he kept his body æquilibrated upon the ſame Centre of Gravity ; but falling forward at each ſtep, he ſuſtains himſelf with the fixing another foot upon a new Centre of Gravity.

(4) That if we ſuppoſe an Individual , or ſmalleſt thing to be turned round in the ſame place ; there will be motion, but no change of place, either as to the whole, or any part thereof. And we Demand , whether by that Individual He means *minimum mathematicum*, or *Phyſicum ?* If *Mathematical*, the ſuppoſition is not to be admitted : becauſe , what is meerly Imaginary is not capable of motion. But, if *Phyſical ;* then admitting the ſuppoſition, we *Anſwer ;* that the reaſon of the motion of an Individual moved round in the ſame place, is the ſame with that of the motion of a Globe or wheel upon its Axis. For , ſuch a body is not ſaid to be Individual, or ſmalleſt, becauſe it hath no magnitude or parts deſignable by the minde ; but becauſe there is no force in nature , that can divide and reſolve it into thoſe parts : and therefore , ſince it is not a meer point, but contains parts ſuperior, inferior, &c. the whole cannot be moved , but ſome parts muſt ſucceed into the places deſerted by others; and conſequently there muſt be Loco-motion. Though this alſo be of the number of ſuch Events, as can hardly be effected by the power of Nature ; foraſmuch as ſuch a phyſical Individual being either permitted to its own liberty, would move ſpontaneouſly in a direct line, not a circular ; or impulſed by another, could not be ſo exactly circumvolved in a Circle , as not to deflect ſomewhat, more or leſs, to one ſide or other. And thus have we Reſolved all the Difficulties, by *Empericus*, objected to the Definition of Motion, given by *Epicurus*.

But yet we have not aſcertained our Reader, *that there is ſuch a thing as Motion in the World :* and therefore, that we may not ſeem to be meerly Petitionary, in begging that at the hand of another mans charitable Belief, which the ſtock of our own Reaſon is rich enough to afford us : we ſhall briefly touch upon that Quæſtion , *An ſit Motus ,* Whether there be any Motion in Nature : Eſpecially, foraſmuch as it is very well known , that among the Ancients there was a notable Controverſie concerning it. For, ſome, as *Heraclitus, Cratylus, Homer, Empedocles* and *Protagoras* (as *Plato* [*in theat.*] notes at large) affirmed, that All things in the univerſe are in perpetual *Motion :* and others, of which number *Parmenides, Meliſſus* and *Zeno* were the Principal, (as *Ariſtotle* (1.*phyſic.*) particularly records) Argued, on the contrary, that All things are in perpetual *Quiet ,* or that there is no motion at all.

*Art. 7. That there is motion ; contrary to the Sophiſms of Parmenides, Meliſſus, Zeno, Diodorus, and the Scepticks.*

Now as to the *Former ;* our Quarrel againſt them is not ſo great, as that of *Ariſtotle* was : foraſmuch as it carries the face of very great probability that They intended no more than this ; that All ſublunary Bodies are in perpetual *Mutation* of their *Inſenſible* Particles , not *Loco-motion* of their *ſenſible* Parts, or Whole ; or, more plainly, that all Concretions unceſſantly ſuffer thoſe irrequiet Agitations, or inteſtine Commotions of their inſenſible particles, from which thoſe ſenſible Changes, Alteration, Augmentation, Diminution, Generation, and Corruption, are by ſlow and inſenſible degrees introduced upon them. And thus even *Ariſtotle* Himſelf inter-

L l l          prets

prets their opinion; saying (*in 8.phys. 3.*) they held, that All things are moved [ἀλλὰ λανθάνειν τῦτο τὴν ἐμπειρίας αἰσθήσεων] *verum id latere experientiam sensuum,* that that motion falls not under the observation of the senses. Which is no more, than what *Epicurus*, or any man else, imbued with his excellent principles, might have asserted.

And as for the *Latter* Sect; neither doth our Choler boyl up against them, to that height, as did *Sextus Empericus* his, when (*in 2.adverf.physic.*) He could not be content to nickname them [Στατιώται] the *standers*; but so far obeys the impulse of his passion, as to fly out into opprobrious language, and brand them with the ignominious character of [Ἄφυσικοί] *Unnatural* Philosophers. And our *Reasons,* why we look not upon them with so oblique and indignatory an eye, as the Vulgar use to do; are these.

(1) Experience doth so clearly Demonstrate, that there is motion; as that no man can deny it, but he must, at the same instant, manifestly refute himself with the motion of his tongue. And such is the constant verity of *Epicurus* his Logical Canon, concerning the Certitude of our senses, as to the information of our mind; as that every Philosopher, nay every man ought to allow them to be judges in cases of sensible Objects : and consequently to conclude, with *Aristotle; ad mentis imbecillitatem debet referri, si quis arbitretur omnia quiescere, & dimisso sensu, rationem requirat.* And, certainly, whoso seriously impugnes, what the evidence of sense confirms; is so easie an Adversary, as to deserve our smiles, rather than our Anger.

(2) Divers have apprehended, that those Philosophers, who seemed to impugn the being of Motion, did not oppose it in a serious, but purely Paradoxical humor, and an ambition of shewing themselves so transcendently acute, as to be able to indubitate Truths even of the most manifest Certitude. Nor are They, indeed, to be understood in that gross sense, which is so generally passant among Vulgar Authors; forasmuch as it is much more probable, that *Parmenides* and *Melissus*, when they laid down for a maxime *Esse omnia unum Ens immobile,* so intended *Nature*, or the All of things, as that they held it, or at least some certain Divine Virtue constantly diffused through, and animating the vast mass of the Universe, to be *God,* or the *Supreme Being*; whose propriety it is to be *Immoveable,* as being Ubiquitary and All in All. And, that *Zeno* himself, the Prince of Antimotists, had some such meaning; may be naturally collected, as well from the Contents of that Book, commonly ascribed to *Aristotle*, concerning *Xenophanes, Zeno* and *Gorgias :* as from those very Arguments He alleadged against motion; the importance of them all declaring, that his supposition was, there could be no motion, if as well motion it self, as Place and Time did consist of *Insectiles,* or *Indivisibles.* Likewise, as for *Diodorus,* so fervently addicted to the Eristick, or Contentious Sect; manifest it is, that his grand scope in his whole Discourse against motion, was only to evince, that a good Wit could not want Arguments wherewith to invade and stagger the belief of a thing, than which nothing can be more certain. Lastly, as for the *Pyrrhoneans,* or *Scepticks*; the design of all their stratagems against motion, seems to have been only this innocent one : to insinuate that no knowledge is exempted from Doubts; and that the mind of

man

man is obnoxious to so great infirmity, as to be able to raise such clouds of Dubitation, which its own dim light is not sufficient ever after to dispel again.

(3) But, granting them all to have been in Earnest, and to have aimed at the shafts of their Wit point blanck at the destruction of Motion; yet if we examine the sharpness of their best Arguments, we shall soon finde them not half so formidable, as most have, through incircumspection, conceived them. As for that Giant Difficulty urged against motion, by *Zeno* which a long time wore the reputation of *Invincible*; please you but to reflect upon our Chapter of a *Vacuum Natural*, you may there meet with a full *Dissolution* of it. If that be too great a trouble to you; we dare un-undertake, your belief shall not miscarry, though you adventure it barely upon the Refutation of *Zeno*, by *Diogenes* the *Cynick*: who hearing him somewhat proudly object the same in the schools, only rose up and walked; as wisely conceiving that to be a sufficient, as well as the most ready Demonstration of the Contrary. As for the other Goliah Objection, excogitated and urged by *Diodorus*; it runs thus: *Si quidquam movetur, aut in quo loco est, movetur, aut in quo non est: at neque in quo est, in eo enim manet, si in ipso est; neque in quo non est, ubi enim quid nam non est, ibi neque agere, neque pati quicquam potest: quam obrem quicquam non movetur.* "If any thing be moved, it must be moved either in that place, wherein it "is, or in that wherein it is not: but not in that place wherein it is, be- "cause if it be there, it remains there; nor in that wherein it is not, because "nothing can either act or suffer there where it is not: therefore nothing "is moved. For, thus *Empiricus (2. advers.physic. 3. pyrrhon. hypotyp. 8. & ibidem lib. 2. cap. 22.)* often præsents it; among other things seasonably commemorating, how pleasantly *Diodorus* was therefore derided by *Herophilus* the Physician. When *Diodorus* came to him, to entreat him to set his shoulder, that was out of joynt; *Herophilus* bad him be of good courage, since it was impossible his shoulder should be dislocated: for, saith He, either it was moved in the place, wherein it was, or that wherein it was not; but in neither: and therefore it was not dislocated. Which *Diodorus* hearing, became conscious of his own sophisme, and entreated him to lay aside his subtleties and mirth, and address himself to his speedy cure. But to return to the Difficulty proposed; we observe, that it was impertinently done of *Diodorus* to make this Interrogation; Either in the place, wherein it is, or in that, wherein it is not: unless perhaps He meant the Common place of a thing, such as is a Hall, from one end whereof a man may walk to the other. In which case, it may be Answered, that the man walking, is moved in the place, wherein he is; for he is moved in the Hall, though not in the same part of the Hall. But, the Quæstion is not of the Common, but *Proper* place of a thing; and therefore the Interrogation ought to be, if any thing be moved, it must be moved either *from* this to that, or from this to another place: not *In* the place, wherein it is; or wherein it is not; since according to the true Notion of motion, we understand it to be the passing of a thing *from* one place to another. And consequently, the *Answer* is; that a body moved, is moved neither in the place, where it is; nor in that, where it is not: but from one place, wherein it was, through a place which it passeth, or pervadeth, to a third place, where yet it is not. Perhaps, you'l yet reply, as a thing passeth through a place, is it not in a place? And we shall rejoyn, that that very Quæry

doth detect the sophisme; for, since the word *Esse*, to *Be*, is, according to
common signification, convenient as well to things Permanent, as Successive or Fluent; and according to a peculiarly accommodate signification,
competent only to things Permanent: it is understood in the former sense,
when the Quæstion is, *Either where it is, or where it is not ?* and in the latter, when the subsumption is, *But neither where it is, nor where it is not :*
according to which reason, you Doubt, Whether a thing *Be*, while it is moving. Which considered, when it is Enquired, whether a moveable be
moved in the place, where it is, or in that, wherein it is not : we are to Distinguish thus; it is moved in the place, wherein it is *Transiently*, and moved in the place wherein it is not *Permanently*. And, to your Quæstion,
Whether a thing be not in a place, when it passeth through a place ? We
Answer likewise, that it is in a place Transiently, not Permanently. Nor
ought this Language to sound strange; since nothing ought to be conceived to be in any other manner, than what the Nature thereof doth præscribe : and such is the Nature of Motion, that it should be conceived to
be [μετάβασις] a *Passing* through, not [μετεμψ̀γη] a *Permansion*, or stay-
ing in a place. Lastly, as for the Arguments of the *Scepticks*; they are
all grounded upon the same Difficulties as those of *Zeno* and *Diodorus :*
and therefore must submit to the same Resolutions.

## SECT. II.

*Art. I.*
*Aristotles Defi-
nitions of Na-
tural and Vio-
lent motion,
incompetent :
and more a-
dæquate ones
substituted in
the room of
them.*

BEing thus præpared with Considerations of the most Genuine Notion, most adæquate Definition, and Primary Cause of Motion in all
Concretions; and an infallible assurance, that there is such a thing as Motion in the world: the next degree to which our Enquiry is to advance, is
the more General and Principal K I N D S thereof; among which, the
First we meet with, is that common Distinction of motion into *Natural*
and *Violent*.

A *Natural* motion, saith *Aristotle* (8. *physic.*4.) is that, whose Principle is *Internal*; and a *Violent*, that, whose Principle is *External* : so that,
accordingly, that Body may be said to be moved Naturally, which is moved by it self; and that Violently, which is moved by another. But, for
as much as *Aristotle* himself doth much amuse us, while he ever and anon
affirms, that one body may be moved by another, and yet not be moved
violently; and that a motion may be said to be Natural or Violent, in
more than one respect; and that some more easie and familiar Notion is
to be accommodated to each of those Contrary Terms, Natural and Violent : therefore is it much more convenient for us, to understand a *Natural* Motion to be that, *which is made either of Natures own accord, or without any Repugnancy*; and a *Violent* to be that, *which is made either Præternaturally, or with some Repugnancy*. Thus, the Progressive motion of
an Animal, is Natural, because made of Natures own accord; and yet if
the Animal go through a bogg, climb a steep hill, leap, or run, the motion

is

is to be accounted Violent, becaufe though it proceed from an Internal Principle, the Soul of the Animal, yet is it not performed without fome Repugnancy, either internal or external. On the contrary; when a Bullet is fhot through the aer, the motion thereof is violent, becaufe againft the nature of the Bullet, and not performed without fome repugnancy, either internal or external: and yet if the fame Bullet be rowled upon a fmooth plane, the motion thereof is Natural; becaufe though it be caufed by an External Principle, yet is it performed without any Repugnancy either internal or External.

*Art.* 2.
The fame deduced from the *Firft Epicurean* Principle of motion, præmifed: and three confiderable *Conclufions* extracted from thence.

But, that we may take the matter in a higher key, reflecting upon that fo often inculcated Epicurean Principle, That all the motive Virtue of Concretions is originally derived from the mobility inhærent in, and infeparable from Atoms, which compofe them; let us obferve, that forafmuch as that effential mobility of Atoms doth neither ceafe, but is only impeded, when Concretions themfelves begin to obtain a fenfible Quiet; nor is produced anew, but only acquires more liberty, when Concretions begin to be moved: we may thence juftly infer, *that juft fo much motive Force is now, and ever will be in the World, while it is a world, as was in the firft moment of its Creation.* Which really is the fame with that Rule of *Des Cartes* (*princip. philofoph. part.* 2. *art.* 36.) *Deum effe Primariam omnis motus Cauffam, & eandem femper motus quantitatem in univerfo perfeverare.* And Hence may we extract thefe notable *Conclufions.* (1) That, becaufe look how much one Atom, being impacted againft another, doth impel it, juft fo much is it reciprocally impelled by it; and fo the Force of motion doth neither increafe, nor decreafe, but, in refpect of the Compenfation made, remains always the very fame, while it is executed through a free fpace, or without refiftence: therefore, *when Concretions, likewife mutually occurring, do reciprocally impel each other; they are to be conceived, to act upon, or fuffer from each other, fo, as that, if they encounter with equal forces, they retain equal motions on each fide, and if they encounter with unequal forces, fuch a Compenfation of the tardity of one, is made by the fupervelocity of the other, as that accepting both their motions together, or conjunctly, the motion ftill continues the fame.* Which alfo is the fame with that Third Law of Nature, regiftred by *Des Cartes* (*princip. philofoph. pars.* 2. *art.* 40.) *Quod unum Corpus, alteri fortiori occurrendo, nihil amittat de fuo motu: occurrendo, vero minus forti, tantum amittere: quantum in illud transfert.* (2) That forafmuch as Atoms conftantly retain their motive Virtue even in the moft compact and hard Concretions; therefore *can there be no Abfolute Quiet in Nature:* the Atoms unceffant ftriving for liberty, caufing perpetual Commotions in all things, though thofe Commotions be inteftine and infenfible as we have often faid. Which confidered, *Heraclitus* feems to have been more reafonable, in his Denial of all Quiet, but to the dead (*apud plutarch.* 1. *placit.* 23.) than moft have hitherto allowed: He underftanding by the Dead, not only Animals deprived of life, and confequently of motion; but alfo all other things Diffolved, fince then, and only then, the inteftine Commotions of their Component Particles, or Atoms, ceafe. (3) That *Motion is not only much more Natural than Quiet, in the General; but alfo always Natural, in refpect of its Original, forafmuch as it proceeds from Atoms, which are moved by their own Nature, or effentia Gravity: and fometimes Violent, but ever fo only at fecond hand, or from the nature of Concretions, as they are moved with a certain Repugnancy.* And this Rule hath also a parallel in *Des Cartes*, viz. *Non plus Actionis requiri ad motum, quam ad Quietem* (*princip.philosoph.part.2.art.26.*) Nor ought it to seem strange that we admit something to be *Violent* in Nature ; because, though in respect of the Universal Nature, nothing may be accounted Violent : yet, in respect of *Particular Natures*, there may. For, if you conceive it to be *Natural*, that many things in Nature should be *Generated :* you must also conceive it to be equally *Natural*, that as many things in Nature should be *Corrupted* ; and consequently, that they should be moved *violently*, i. e. with Repugnancy to their Particular Natures. Furthermore, notwithstanding the Voluntary motion of an Animal be vulgarly conceived to be Natural ; yet whoever shall consider, that Animal motion is always accompanied with a certain Labour, and attended on by Weariness, which by degrees encreaseth upon Animals, in long, or great and quick motions ; and that strong impaction made against the joynts of one member by the bones and ligaments of another, and of all upon the *Spina Dorsi*, as also of the whole body of the Animal against the ground, on which it treads : we say, whoso duly considers these things, will soon be induced to allow, that such motion is always commixed with some *Violence*. And what hath been here said of Motion, carries the same weight, if applied also to *Quiet* ; forasmuch as Quiet may be understood to be *Violent* in one sense, and *Natural* in another. And, therefore, we shall only add this concerning Quiet ; that it is Natural not only to the whole World, that it should maintain a certain Cohæsion, or Consistence, or Quiet of all its parts ; but also to every single part of the Universe, or every particular body ; because unless the parts sensibly quiesce in the Whole, i. e. be not Dissociated from the Whole, no Concretion or Compage of matter could subsist. We say, *Quiet in the Whole*, not præcisely in *Place* ; because the Whole may be moved, and yet this or that particular part thereof so cohære unto it, or acquiesce in it, as that though it change place together with the whole, yet as to it self, it may be no more moved, nor feel more Repugnancy, than if the whole did acquiesce, and it continued still in the same place therewith.

**Art. 3.**
A short survey of *Aristotles* whole theory concerning the Natural motion of Inanimates : and the Errors thereof.

Now, though the Difficulty is not great, which concerns the motion of Animals ; in respect of that Inequality and Painfulness that accompany, and Lassitude that usually succeeds upon it, all which as we have even now insinuated, import it to be commixt with some Violence : yet that seems to be a very considerable one, which concerns the motion of *Inanimates*, forasmuch as most men, insisting in the Doctrine of *Aristotle*, apprehend it to be *Natural*. It follows therefore, that we henceforth address our Enquiry chiefly to the motion of Inanimates ; as that which may best evince the Impropriety of *Aristotles* Definition of Natural Motion to be that, whose Principle is Internal : wherein that we may be sufficiently circumspect, it behooves us to take a short survey of his whole Theory, touching that subject.

In the first place, He positively affirms, that whatever is moved, or doth move, is moved either *Per se*, or *per Accidens :* subjoyning, that what is moved *per se*, is the subject, or whole ; and what is moved *per Accidens*, is an Accident of the Subject, or Part of the Whole. For Instance ; when a man, in whom are Musick and a Soul, walketh ; the man is moved per se, because he is the subject and the whole : but the Musick, which is in him,

is

is moved per Accidens, because it is an Accident to him; and likewise his soul is moved by Accident, because it is only a Part of him. Again, when He teacheth, that whatever is moved, is moved by *Another*; that ought to be understood of that thing, which is moved *per se*: for, from hence it is, that when in the series of particular movents, He would have us to come at length to one *First Movent*, which is Immoveable, or which is not moved by any other; we are to understand that *Primum Movens* to be *Immoveable per se*, since it may be moved *per Accidens*. Thus, when a stone is moved by a staff, the staff by the hand of a man, the mans hand by his Soul; the soul, indeed, is the First movent and Immoveable: but, understand it to be so, *per se*, because it is at the same moment moved *per Accidens*, i. e. when the hand, arme, and whole body, which contains it, is moved. Moreover, He declares, that whatever is moved *per se*, is moved *juxta Naturam*, according to Nature; such as he affirms that only to be, which is endowed with a soul: yet will He not admit, that what is moved by Another, should always be moved *Præter Naturam*, Præternaturally; but sometimes Un-naturally (as a stone, when it is thrown upward) and sometimes Naturally (as a stone, when it falls Down again.) Now, if you hereupon Demand of Him, What that is, which makes a stone fall Down again; He shall Answer, that what moves it Downward, *per se*, is the Generant it self, or that which first Produced the stone: and that which moves it downward, *per Accidens*, is that which removes the impediment or obstacle to its descent, as the hand of a man, or other thing supporting the stone. And, if you again enquire of him, What is the Difference betwixt the Upward and Downward motion of a stone, how one should be Violent, and the other Natural, since, according to his own Assertion, both are Caused by another: His Return will be, that the Difference lies in this, that the stone is not carried upward, of its own Nature, but Downward; as having the Principle of its Descent, inhærent in it self, but not that of its Ascent. If you urge Him yet farther; since the stone hath in it self the Principle of its Motion, why therefore is it not moved only by it self, but wants Another, or *External Motor*? His Answer will be: that there is a Twofold principle of motion, the one *Active*, the other *Passive*; and in the stone is only the Principle Passive, but in the External Motor is the Active. When yet it may be farther pressed; that since according to his own Doctrine, the Passive principle is the *matter*, and the Active the *Forme*: as to the matter, that cannot be the principle of its motion Downward, no more than of its motion upward; and as for the Forme, if that be neither the Active principle, nor the Passive (as he will by no means admit) certainly there can be none. Which for Him to allow, were plainly to destroy his own great Definition of *Nature*, wherein He acknowledgeth it to be the *Principle of Motion*. But, alas! these are but light and venial Mistakes, in comparison of those gross Incongruities that follow.

When *Aristotle* comes to handle the *Species*, or sorts of Natural Motion, you may remember, that He first Distinguisheth Natural motion in *Direct* and *Circular*; and then subdistinguisheth the *Direct* into (1) that which is from the Circumference toward the Centre, or from the Extrems toward the middle of the world, which He calls *Downward*; and (2) that which is from the Centre toward the Circumference, which He calls *Upward*: assigning the former, or Downward motion, only to *Heavy* things, to the Earth simply, to Water and mixt things, *Secundum quid*; and the Upward only

Art. 4. *Uniformity, or Æquability,* the proper character of a Natural motion: and the want of uniformity, of a Violent.

only to *Light* things, to Fire simply, to Aer *Secundum quid*, and to mixt bodies, according to the greater or less prædominion of Fire in them, over the other Elements. And, as for the *Circular* motion, which is neither toward, nor fromward the Centre or middle of the world, but round a-bout it; He ascribes that only to the Cælestial Orbs, as being things neither Heavy, nor Light. Butt, forasmuch as He doth not make it in the least measure Evident, whether or no all these Things are moved by an Internal Principle, nor whether with some, or without all Repugnancy; and so leaves us still to doubt, whether their Motions be Natural or Violent: are we not constrained, to omit all those Ambages, and Difficulties, that attend upon this His imperfect Doctrine, and (with *Galilæo*) to have recourse to some such *Criterion* or *Character* of *Naturalness* in motions, as seems most consentaneous to truth, because most Evident? Doubtless, as the motion of Atoms, which is most Natural, is most Uniform, or Equal; so also in Concretions, by how much every motion is the more Natural, by so much more doth it appear to be Uniform, or Æqual. And therefore this *Uniformity*, or *Æquability* may be assumed as the truest Character of *Natural* motion: as we may easily conclude, that every Uniform motion is purely Natural, as on the contrary, that every motion, that wants Uniformity, is *Violent*. This may be Confirmed by that common maxime, *that nothing Violent can be Perpetual*; forasmuch as the root of *Perpetuity*, is *Uniformity* (for, nothing in nature can either by growing stronger receive perpetual Increment, or by growing weaker endure perpetual Decrement) and upon consequence, Inæquability, as being opposite to Perpetuity, must be the pathognomonick, or proper and inseparable sign of a Violent thing, and Æquability of a Natural. Hence, as for the *Cælestial motions*, they are argued to be *Natural*, because they are *Uniforme*, and therefore *Perpetual*. And, assuredly, where the wise Creator of the World, would have any motion Perpetual, He ordained it to be Circular: as that, which being equally distant from the Centre in all parts, and wanting both beginning and end, might be continued with one constant tenour, and also uncessantly. And as for *Direct* motions, or such as are competent to Heavy and Light Bodies, whether Elements or mixt, they are on the contrary, to be judged to be *Violent*, in that they are very Unequal, and of little or no Duration. To insist upon that of *Fire*, which perisheth in the same moment wherein it is produced, we need not; nor upon that of *Aer*, which is variously moved, sometimes upward, sometimes downward: because even our sense assures, that their motions are very Unequal. And, as to that Downward motion of *Earth*, and *Water*, and generally of all *mixt* Bodies, commonly accounted *Heavy*; we need only this short observation: that their Motion is not only very short, both as to Time and Space; but also so unequal in it self, and of such vast Acceleration in its progress, as that, if it might be conceived capable of longer Continuation, there is no Body in nature so Compact and Firme, which would not be shivered in pieces, and wholly be Dissolved and Dissipated thereby, in a short time. And who will not readily admit that to be a most evident note of Violence? Since no man can conceive that motion to be Natural, which is comparated not to the Conservation, but inevitable *Destruction* of Nature: but only He, who can admit, that the very Nature, or Formal Constitution of a thing, hath no Repugnancy to Destruction. But, you'l say (we suppose) must then the Principle of all motion, competent to Heavy Bodies, be *External*? Truely, it must; and you know, that we have already declared, that *Aristotle* allows it so to
be,

be. What then, muſt that Exteernal Principle be, as *Ariſtotle* contends, the very *Generant* of the thing move d? Certainly, thats highly Abſurd, ſince the Generant is abſent, and perhaps, long ſince ceaſed to be in *rerum natura:* and nothing either Abſent, or Non exiſtent, can be the Efficient of a Natural Action, ſuch as motion is. If you will have, that to be moved by the Generant, ſignifies no more than *to receive a Virtue or Power of moving it ſelf, from the Generant*, then while you endeavour to ſave *Ariſtotle* from the former *Abſurdity*, you præcipitate him into a groſs *Contradiction* of his own Doctrine: for, ſince the Generant it ſelf ought to be moved by its Generant, and that again to be moved by its Generant, and ſo upward along the whole ſeries of Generants, till you arrive at length at ſome Firſt Generant, from whence that Virtue was firſt derived; you bring *Ariſtotle* to allow a *Firſt* Generant, which impugns his fundamental ſuppoſition of the *Eternity* of the World. Nay, if you admit God to be the Author of the Firſt Generant, it will then follow, that God muſt be the Cauſe of this particular motion, and not the Firſt Generant, no more than the Laſt. Finally, is that the Cauſe, which only removes the *Impediment* to a Heavy bodies Deſcent? Neither is that Reaſonable; for, as *Ariſtotle* himſelf confeſſeth, ſuch a Cauſe is only a Cauſe by *Accident*.

Seeing, therefore, that the Downward motion of a Heavy Body doth not proceed from any Internal Principle, nor from either its Generant, or that Accidental one, which removes the Impediment to its Deſcent, in the ſuppoſed Capacity of an External: let us proceed to enquire, Whether there be not ſome other *External Cauſe*, whereupon we may reaſonably charge that Effect. Which that we may do with the more both of order and plainneſs; it is requiſite, that we firſt remember, how Philoſophers conſtitute divers *ſorts* of Violent, or Externally-cauſed motion. *Empericus* (2. *adverſ. phyſicos*.) makes 4 diſtinct ſpecies thereof, viz. *Pulſion, Traction, Elation, Depreſſion*. And *Ariſtotle* ſometimes ſuperads a fifth, namely *Collſion*; ſometimes diſallowing *Empericus* his Diviſion, affirms that the ſpecies of motion, made by an External principle, are *Traction, Pulſion, Vection*, and *Volutation:* upon good reaſon reducing Elation and Depreſſion to either Traction or Pulſion; foraſmuch as a body may be elevated, or depreſſed by either Traction or Pulſion. But, yet He hath left us rather a Confuſion, than logical Diſcrimination of the ſpecies of Violent motion; for, Colliſion and Pulſion are one and the ſame thing; and Vection may be performed either by Pulſion or Traction, inſomuch as the thing movent doth not forſake the thing pulſed, or drawn, but conſtantly adhæreth unto it: and as for Volutation; it is both Pulſion and Traction at once, as may be eaſily conceived by any man, who ſeriouſly conſiders the manner thereof. Nay, Traction it ſelf may be juſtly reduced to Pulſion; foraſmuch as the movent, which is ſaid to Draw a thing, doth, indeed, nothing but Impel it, by frequently reiterated ſmall ſtrokes, either directly toward it ſelf, or to a lateral region: and yet notwithſtanding, for plainneſs ſake, and the cleerer Demonſtration of our præſent theſis, we judge it convenient, to conſerve the Common Notion, and to determine, that all Motion impreſſed upon one body by another, is performed, in the General either when the movent *Propels* the moveable *from* it ſelf, or *Attracts* it *toward* it ſelf. For, albeit the movent ſometimes propels the thing moved from another body, or attracts it to another; yet can it not poſſibly do that, but it muſt, at the ſame time, either Avert it, in ſome meaſure, from, or Adduce it toward it ſelf. Nevertheleſs, it is not to be denied, but Pulſion is

*Art. 5.*
The Downward motion of Inanimates, derived from an External Principle; contrary to *Ariſtotle*.

always the Chief Species; and for that confideration alone is it, that *Projection* (which is only *Impulfion,* or, as *Ariftotle* emphatically calls it, a *more Violent motion*) is generally accepted as fynonymous to Violent motion; and that Philofophers feldom or never Exemplifie Violent motion, but in *Projectills,* whether they be projected upward, or downward, tranverfly, obliquely, or any way whatever.

*Art. 6.*
That that Ex-
ternal Princi-
ple, is the
Magnetique
*Attraction* of
the Earth.

These things confidered, it follows of pure neceffity, that the Downward motion of Heavy Bodies, being caufed (not by any Internal, but) by an External Force impreffed upon them, muft be effected either by *Impulfion,* or by *Traction.* By *Impulfion* it cannot, becaufe, in the cafe of a ftone thrown Upward, there is nothing External, that can be imagined to impel it Down again, after it hath attained the higheft point of its mountee, unlefs it fhould be the Aer: and if its Defcent did proceed from the impulfe, or depreffive force of the Aer circulated from below upon the upper part of the ftone; then in the projection of the ftone upward, during its Afcent, the motion thereof would, in every degree of its remove from the projicient, be Accelerated in the fame proportion, as its Downward motion is Accelerated, in every degree of its defcent; but Experience teftifies, that its upward motion is more and more Retarded, in every degree of its remove from the projicient, and therefore it cannot be, that the Downward motion thereof fhould be caufed, nay not fo much as advanced by the Aer. Which thing *Gaffendus* (in 1 *Epift. de proport. qua Gravia decidentia accelerantur*) hath copioufly demonftrated; and we our felves, out of him, *in the 9 Article of our 2 Sect. concerning Gravity and Levity, in the 3.Book. precedent.* What, therefore, can remain, but that it muft be by A T-T R A C T I O N ? And, becaufe no other Attractive Force, which might begin and continue the Downward motion of a ftone, can be imagined, unlefs it be that *Magnetique Virtue of the Earth,* whereby it Draws all Terrene Bodies to an Union with it felf, in order to their, and its own better Confervation : we may lawfully Conclude, *that the Caufe of the Downward motion of all Heavy Bodies, is the Magnetique Attraction of the Earth.* Nor need we adferr other Arguments, in this place, to confirm this Pofition; in refpect we have formerly made it the chief fubject of the 2 *Sect. of our Chap. of Gravity and Levity;* whether we, therefore, remit our unfatisfied Reader.

*Art. 7.*
That the *Up-
ward motion*
of *Light
things,* is not
*Accelerated* in
every degree
of their Affent
as *Ariftotle*
præcautioufly
affirmed : but,
the *Downward*
motion of
*Heavy* things
is *Accelerated,*
in every de-
gree of their
Defcent.

From the Caufe of the Downward motion of Heavy bodies, let us advance to the *Acceleration* of them, in every degree of fpace, through which they Fall : there being no confiderable reafon, why we fhould at all enquire into the Acceleration of the *upward* motion of *Light* bodies, in every degree of their Afcent; forafmuch as we know of no man, but *Ariftotle,*that ever durft affirm, that the motion of Fire, and Aer is flower in the beginning, and gradually fwifter and fwifter in the progrefs. And fo fhort was He of proving that his fingular conception, by Experiment, as he ought; that he affumed it upon the credit of only one poor Argument, which is this. " If Fire, and Aer, and other things of the like light and afpiring " nature , faith He (1 *de Cælo.cap.*8.) were Extruded and Impelled up- " ward, by other heavier bodies defcending and crouding toward the mid- " dle of the world, with greater force, as fome have contended; and were " not carried upward by the fpontaneous tendency of their own inhærent " Levity : then would they be moved more fwiftly in the beginning, and " more flowly in the end of their motion; but Fire, and Aer are more " flow in the beginning, and more and more fwift in the progrefs of their " Affent

" Aſſent, therefore are they not moved upward by the Extruſion and Im-
" pulſion, but ſpontaneouſly, or by their own Levity.   And to Confirm
his *Minor* propoſition, that Fire and Aer are Accelerated in every degree of
their Aſſent ; without the ſuffrage of any Experiment, He ſubjoyns only,
" that as a Greater quantity of Earth is moved downward more ſwiftly,
" than a leſs; ſo is a Greater quantity of Fire moved upward more ſwiftly
" than a leſs: which could not be, if either of them were Impelled, or mo-
" ved by an External Force.   But, this is, as the Former, meerly *Petitionary* ; for, why ſhould not a Greater quantity of Earth, or Fire be moved
more ſwiftly than a leſs, both being moved (as we ſuppoſe) by External
force, in caſe the External force be proportionate to the quantity of each ?
Doubtleſs, the force of the ambient Aer, extruding and impelling flame upward, is always ſo much the greater, or more ſenſible, by how much more
Copious the Fire is ; as may be evinced even from the greater Impetus
and waving motion of the flame of a great fire : though it cannot yet be
diſcerned, whether that Undulous or waving motion in a Great flame be
(as He præſumes) more ſwift and rapid, than that more calm and equal one
obſerved in the flame of a Candle.   This (youl ſay) is enough to detect
the incircumſpection of *Ariſtotle*, in aſſuming, upon ſo weak grounds, that
the motion of Light things Aſcending, is accelerated in the progreſs, and
that in the ſame proportion, as that of Heavy things Deſcending is accelerated : but not enough to refute the *Poſition* it ſelf ; and therefore we think
it expedient, to ſuperadd a Demonſtrative Reaſon or two, toward the plenary Refutation thereof.   Seeing it is evident from Experience, that a
Bladder blown up is ſo much the more hardly depreſſed in deep water, by
how much neerer it comes to the bottom ; and a natural Conſequent
thereupon, that the bladder, in reſpect of the Aer included therein, beginning its upward motion at the bottom of the Water, is moved toward the
region of Aer ſo much the more ſlowly, by how much the higher it
riſeth toward the ſurface of the Water, or lower part of the region of Aer incumbent thereupon ; and that the Cauſe thereof is this,
that ſo much the fewer parts of Water are incumbent upon the bladder and
aer contained therein, and conſequently ſo much the leſs muſt that force of
Extruſion be, whereby the parts of Water bearing downward impel them
upward : we may well infer hereupon, that if we imagine that any Flame
ſhould aſcend through the region of Aer ; till it arrived at the region of
Fire, feigned to be immediately above the region of Aer ; that Flame
would always be moved ſo much the ſlower, by how much the higher it
ſhould aſcend, or by how much the neerer it ſhould arive at the region of
Fire.   Becauſe Fire and Aer are conceived to be of the ſame aſpiring nature: and becauſe the ſame Reaſon holds good, in proportion, for the de-
creaſe of Velocity in the aſcenſion of Flame through the Aer, as for that of
the decreaſe of velocity in the aſcenſion of Aer, included in a bladder,
through Water.   And, as for *Ariſtotles* other relative Aſſertion, that *a
Greater quantity of Earth is moved more ſwiftly Downward, than a Leſs*;
manifeſt it is, that He meerly uſurped the conceſſion of this alſo, without,
nay contrary to the ſuffrage of Experiment.   For, an eaſie Experience
doth demonſtrate that a ſtone, or bullet of an hundred pound weight, doth
not fall down more ſwiftly, or ſooner arrive at the ground, than another of
only an ounce weight, both being together præcipitated from the ſame altitude : which may ſeem Paradoxical indeed ; eſpecially to thoſe, who
being educated at the feet of *Ariſtotle*, conceive that Gravity is a Quality
M m m 2                          inhærent

inhærent in bodies accounted Heavy, and that every body must therefore fall down so much the more swiftly and violently, by how much the more of Gravity it possesseth. Having thus totally subverted *Aristotles* erroneous Tenent, that the motion of Light bodies Ascending, is Accelerated in every degree of their Ascension: it follows, that we apply our selves to the consideration of the *Acceleration of the motion of Heavy bodies Descending, in every degree of their Descention.* Wherein the First observeable occurring, is the *Quod sit*, or *that it is so*, which is easily proved from hence, that in all ages it hath been observed, that the motion of Heavy things Descendent, is slower in the beginning, and grows swifter and swifter still toward the end, so as that in fine it becomes highly rapid: experience attesting, that the blow, impulse or impression made upon the Earth, by a thing taln down from on high, is always so much the greater or stronger, by how much the higher the place is from which it fell.

*Art.* 8.
The *Cause* of that Encrease of Velocity in Bodiesdescending; *not the Augmentation of their Specifical Perfection as they approach neerer and neerer to their proper place*: as *Simplicius* makes *Arist.* to have thought.

    The *Second*, is the *Cur sit*, or *Cause* of that velocity Encreasing in bodies Falling; which though enquired into by many of the Ancients, seems yet to have been discovered by none of them. For (1) albeit *Aristotle* Himself was so wary, as not to explicate his thoughts concerning it, yet doth his great Commentator, *Simplicius* tell us (*in Comment.* 87.) that it was His opinion, that a stone, or other thing falling from on high, is Corroborated [ἀπὸ τῆς οἰκείας ὁλότητος] *a Totalitate propria*, and hath its species made more and more perfect, as it comes neerer and neerer to its proper place; and so that a new degree of Gravity acceding to it in every degree of its appropinquation to the Earth, it is accordingly carried more and more swiftly. But, seeing that *Simplicius* hath not expounded, how the whole stone can act upon itself; how it can be Corroborated, or acquire more and more perfection of its species; or how that additament of fresh Gravity should arise unto it: judge you, whether He hath done *Aristotle* any right, in making him the Author of that Opinion, which instead of explaining the matter, leaves it much more obscure than afore. Besides, we have the certificat of Experience, that a descending body is not carried the more swiftly, by reason of any access or additament of Gravity: a stone of an ounce weight, falling as speedily down, as one of an hundred pound.

*Art.* 9.
Nor the Diminution of the quantity of Aer underneath them: as some Others conjectured.

    (2) Others there were (as the same *Simplicius* commemorates) who referred the Cause thereof, to the *Decrease of the quantity of the Aer underneath the stone:* conceiving, that by how much the higher a stone is, by so much the more of Aer is below it, and so much the greater Resistence to the motion of the stone, by how much the greater quantity of the Aer resisting; so that the quantity, and consequently the resistence of the Aer growing less and less, in every degree of the stones descent, the velocity of its motion must be gradually encreased in proportion thereunto. And this after the same manner as weights are carried, sinking in deep water, more slowly neer the top, and more swiftly neer the bottom. But, though we admit, that the subjacent Aer may somewhat resist a stone Descending; yet we deny the resistence to be so great, as to make any sensible difference of velocity in the parts of its motion. And, would you have an Argument to the purpose; be pleased to let fall a stone from the altitude of one fathom; and exactly observe the velocity of its motion. Then let fall the same stone, from the altitude of ten fathoms; and when it hath pervaded nine fathoms

fathoms, obſerve again with what velocity it paſſeth the laſt, or tenth fathom.    This done, conſider, ſince in the latter caſe, the velocity ſhall be
incomparably greater, than in the former ; whether it be not neceſſary, that
that great augmentation of velocity in the ſtone, while it pervadeth the
tenth fathom of ſpace, muſt not ariſe from ſome other, and more potent
Cauſe, than the reſiſtence of the inferior Aer ?  For, in both caſes, the
ſtone carries the ſame proportion of weight ; and in the loweſt fathom
there is the ſame quantity of Aer, and conſequently the ſame meaſure of
reſiſtence.   And, if you weigh the ſtone, firſt in ſome very high place, and
afterward in a low, or very neer the Earth ; ſurely, you cannot expect to
finde it heavier in the low place in reſpect of the leſſer quantity of Aer ſubjacent, than in the high, in reſpect of the greater quantity of Aer there ſu-
ſtaining it.    Laſtly, as for their Argument deſumed from the ſlower ſink-
ing of weights in deep, than in ſhallow Water ; the cauſe thereof is the
ſame with that of the more difficult depreſſion of a Bladder blown up with
Aer, neer the bottom, than neer the top of the Water, which we have lately explained.

*Art. 10.*
Nor the Gra-
dual Diminu-
tion of the
Force impreſt
upon them, in
their projecti-
on upward :
as *Hipparchus*
alleadged.

(3)  A third Conceipt there is (imputed to *Hipparchus*, by the ſame
*Simplicius*) which comparing the Downward motion of a ſtone, cauſed
by its own proper Gravity, with the Upward motion of the ſame ſtone,
cauſed by an External Force impreſſed upon it by the Projicient ; thence
infers, that as long as the force impreſt prævails over the ſtones Gravity, ſo
long is the ſtone carried upward, and that more ſwiftly in the beginning,
becauſe the Force is then ſtrongeſt, but afterward leſs and leſs ſwiftly, becauſe the ſame force impreſt is gradually debilitated, until the ſtones proper Gravity at length getting the upper hand of the force impreſt, the ſtone
begins it motion Downward ; which is ſlower in the beginning, becauſe
the Gravity doth not yet much prævail, but afterwards grows more and
more ſwift, becauſe the Gravity more and more prævails. But this leaves us
more than half way ſhort of the Difficulty ; for, though it be reaſonable to
aſſume, that a certain Compenſation of Velocity is made in both motions, i. e.
that the Decreaſe of Velocity toward the end of the Upward motion, is
made up again by the Encreaſe of Velocity toward the end of the Downward, and that in proportion to the degrees of ſpace : yet foraſmuch as the
motion of a ſtone falling down is conſtantly Accelerated, not only after it
hath been projected Upward, but alſo when it is only dropt down from ſome
high place, to which perhaps it was never elevated, but remained there from
the beginning of the world, as it often happens in deep mines, the earth underneath the ſtones neer the ſurface of it being undermined ; therefore
cannot the ſtones Gravity, gradually prævailing over the Impreſt Force, be,
as *Hipparchus* concludes, the Cauſe of its Encreaſe of Velocity in each degree of its Deſcent.

*Art. 11.*
But, the Mag-
netique Attra-
ction of the
Earth.

Theſe Reaſons thus deluding our Curioſity, let us have recourſe to our
formerly aſſerted Poſition, that *All terrene bodies Deſcend, only becauſe they*
*are Attracted by the magnetique Virtue of the Earth.*  Shall we conceive,
that the magnetique Virtue of the Earth is more potent neer at hand, than
afar off : and thereupon infer, that the downward motion of a ſtone is
therefore more rapid neer the earth, than far from it ; becauſe the magnetick Virtue ſeems to be greater, and ſo the Attraction ſtronger, by how
much neerer the ſtone approacheth to the Earth ?  This certainly is obvious

ous and plausible to our first thought: but insatisfactory to our second.
For, if it were so, then ought the Celerity of the stones motion, in one fathom neer the Earth, to be the same, whether the stone be let fall from the
altitude of only one fathom, or from that of 10, 20, an 100 fathoms, when
we exactly measure the space of time, in which it pervades the one fathom
neer the earth, in the former case, and compare it with that space of time, in
which it pervades the same lowest fathom, in the latter.   It may be farther
observed, that, whether a stone be let fall from a small, or a great altitude,
the motion thereof for the first fathom of its descent, is always of equal
velocity, i. e. it is not more nor less swift for the first fathom of its descent
from the altitude of an 100 fathoms, than from the altitude of only two fathoms: when yet it ought to be more swift for the first fathom of the two,
than for the first of the hundred, if the Attraction of the Earth be more
vehement neer at hand, than far off; in a sensible proportion.   We say,
in a *sensible* proportion; because, forasmuch as the magnetique rays emitted from it, are diffused in round from all parts of the superfice thereof, and
so must be so much the more dense, and consequently more potent, by how
much less they are removed from it: therefore must the Attraction be somewhat more potent at little than at very great distance; but yet there is no
tower or præcipice so high, as to accommodate us with convenience to experiment, whether the power of the Earths magnetique rayes is Greater, to a sensible proportion, in a very low place, than in a very high.

And yet notwithstanding, nothing seems more reasonable than to conceive, that since the magnetique Attraction of the Earth is the true Cause
of a stones Downward motion, therefore it should be also the *true Cause of
the continual Increment of its Velocity, during that motion.*   But how it
should be so; there's the Knot.   Which that we may undo, let us first resume our former supposition (*in the 2. Sect. of our chap. of Gravity and
Levity.*) that a stone were situate in any of the Imaginary spaces; considering that in that case it could not of it self be moved at all: because, holding no Communion with the World (which you may suppose also to be
Annihilated) there could be, in respect thereof, no inferior place or region,
whereto it might be imagined to tend or fall; nor could it have any Repugnancy to motion, because there would be no superior region, to which
it might be conceived to aspire or mount.   Then let us suppose it to be
moved by simple Impulsion, or Attraction, toward any other part of the
Empty, or Imaginary spaces; and without all doubt, it would be moved
thitherward, with a motion altogether Equal or Uniform in all its parts:
because there could be no Reason, why it should be more flow in some parts
of its motion and more swift in others, there being no Centre, to which it
might approach, or from which it might be removed.   Suppose farther,
that, as the stone is in that motion, another Impulse, equal in force to the
former, whereby it was first moved, were impressed upon it; then, assuredly, would the stone be moved forward more swiftly than before, not by reason of any Affection to tend to any Centre, but because the force of the
first impulse persevering, the force of the second impulse is superadded unto it, and the accession of that force must so corroborate the former, as to
augment the Velocity of the stones motion.   And hence comes it, that
to move forward a body already in motion, doth not only prolong, but accelerate the motion thereof.   Imagine moreover, that a third impulse were
incontinently superadded to the second; and then would the motion be yet
more

more fwift than before; the Encreafe of Velocity of neceffity ftill refpon-
ding to the multiplicity of Impulfes made upon the body moved. This
may be familiar to our conceptions, from the Example of a Globe fet upon
a plane; which may be emoved from its place with a very gentle impulfe,
and if many of thofe Impulfes be repeated thickly upon it, as it moves, the
motion thereof will be fo accelerated, as at length to become fuperlatively
rapid. Which alfo feems to be the Reafon, why a clay Bullet is difcharged
by the breath of a man, from a Trunck, with fo great force, as to kill a
Pidgeon at 20, or 30 yards diftance : the Impetus or force impelling the
bullet, growing ftill greater and greater, becaufe in the whole length of the
trunck there is no one point, in which fome of the particles of the mans
breath fucceffively flowing, do not imprefs frefh ftrokes, or impulfes upon
the hinder part of the bullet. The fame alfo may be given, as the moft
probable Caufe, why Long Guns carry or fhot, or bullet farther than fhort,
though yet there be a certain determinate proportion to be obferved betwixt the diametre of the bore, and the length of the barrel or tube, as well
in Truncks, as Guns : experience affuring, that a Gun of five foot, musket
bore, will do as good execution upon Fowl, with fhot, and kill as far, as one
of ten foot, and the fame bore; and confequently that thofe Gunners
are miftaken, who defire to ufe Fowling pieces of above 5, or 6 foot long;
Thefe confiderations premifed, we may conceive, that when a ftone firft
begins to move downward, it then hath newly received the firft impulfe of
the magnetique rays emitted from the Earth : and that if after the impreffion of that firft impulfe, the Attraction of the Earth fhould inftantly
ceafe, and no new force be fuperadded thereunto from any Caufe whatever; in all probability, the ftone would be carried on toward the Earth
with a very flow, but conftantly equal and Uniform pace. But, becaufe
the Attraction of the Earth ceafeth not, but is renewed in the fecond moment by an impulfe of equal force to that firft, which began the ftones motion, and is again renewed in the third moment, in the 4, 5, 6, &c. as it
was in the fecond, therefore is it neceffary, that becaufe the former impulfes, impreffed are not deftroyed by the fubfequent, but fo united as
ftill to corroborate the firft, and all combining together to make one great
force; we fay, therefore is it neceffary, that the motion of the ftone, from
the repeated impulfes, and fo continually multiplied Impetus or Force,
fhould be more fwift in the fecond moment, than in the firft; in the third,
than in the fecond; in the fourth, than the third, and fo in the reft fucceffively; and confequently, that the Celerity fhould be Augmented in one
and the fame tenour, or rate, from the beginning to the end of the motion.

The Third thing confiderable in this Downward motion of Bodies, is the
PROPORTION, or Rate, in which their Celerity is encreafed.
Concerning this, we know of no Enquiry at all made by any one of the
Ancients; only *Hipparchus*, as hath been faid, thought that in the General,
the increment of Velocity in things falling down, was made in the fame reciprocal proportion, as the Velocity of the fame things projected upward.
But, about 90 yeers paft, one *Michael Varro*, an eminent Mathematician (*in
tract. de motu.*) depending meerly upon Reafon; would have the Problem
to be thus folved. What is the Ration, or Proportion of fpace to fpace, the
fame is the Ration of Celerity to Celerity; fo that if a ftone falling down
from the heigth of four fathoms, fhall in the end of the firft fathom acquire
one

*Art.* 12.
That the Pro-
portion, or
Ration of Ce-
lerity to Cele-
rity, encreafing
in the defcent
of Heavy
things; is not
the fame as
the Proporti-
on, or Ration
of Space to
Space, which
they pervade:
contrary to
*Michael Varro*
the Mathema-
tician.

one degree of Velocity, in the end of the second two, in the end of the third three, in the end of the fourth four: it will be moved twice as swiftly in the end of the second fathom, as in the end of the first, thrice as swiftly in the end of the third, and four times as swiftly in the end of the fourth, as of the first. But, this Proportion is deficient, first in this; that though the increment of Celerity, or of its equal degrees, may be compared with the equal moments or parts of space: yet can it not be compared also with the equal moments or parts of Time, without which the mystery can never be explicated. And therefore *Aristotle* did excellently well, in Defining *Swift*, and *Slow*, by *Time*; determining that to be swift, *which percurs a great deal of space in a little time*; and on the contrary, that to be slow, *which is pervading a little of space in a great deal of time*. Again, let us suppose the theorem to be explicable by equal moments of times, and such as are the respites or intervals betwixt the pulses of our Arteries; and that a stone falling down doth pervade the first fathom of space, in the first moment: then, if it pervade the second fathom twice as swiftly as the first (as *Varro* conceives) it must necessarily follow, that the second fathom must be pervaded in the half of a moment; if the third fathom be percurred thrice as swiftly as the first, it must be pervaded in the third part of a moment; and if the fourth fathom be percurred four times as swiftly as the first, it must be pervaded in the fourth part of a moment. And, because, if you conjoyn the half, third, and fourth part of a moment, you shall have a whole moment with one twelfth part of a moment, it will be necessary, that in the second moment, three fathoms (very neer) must be percurred: which indeed is very far from truth. For, because, if we proceed after the same method, so that the fifth fathom be percurred in the fifth part of a moment; the sixth in the sixth part of a moment, and so successively; out of these fragments of time we shall not be able to make up another whole moment, until it be after the stone hath pervaded the eleventh fathom, or thereabout; and so in the third moment seven fathoms shall be pervaded, nor shall we again be able to make up another whole moment, until after the stone hath pervaded the 31 fathom; and so in the fourth moment, it shall pervade 20 fathoms, nor shall we be able to make up another complete moment, until after the stone hath pervaded, neer upon, the 84 fathom, and so in the fifth moment, 53 fathoms shall be percurred, &c. so that proceeding according to a triple proportion, neer upon; you shall consequently, in a very short time, increase it up to Immensity: as is manifest from the short progress through these numbers, 1.2,4,11,31,84,&c. Which is impugned by easie Experience, and not defensible by any Reason whatever.

Art. 13.
But, that the moments or Equal degrees of Celerity, carry the same proportion, as the moments or equal degrees of Time, during the motion: according to the Illustrious Galilæo.

This the brave *Galilæus* well considering, and long labouring his subtle and active thoughts, to explore a fully satisfactory Solution of this dark Riddle; came at length most happily to set up his rest in this. First, He defines *Motion equally Accelerated* to be that, *which receding from quiet, doth acquire equal moments of Celerity, not in equal spaces, but equal Times*. Then proceeding upon Grounds partly *Experimental*, partly *Rational*; He concludes, that the moments, or equal Degrees of Celerity, are as the moments, or equal degrees of Time, or (more plainly) *that the Celerities carry the same proportions as the Times*; so that look how many moments of time pass during the motion, so many degrees of Celerity, are acquired by the thing moved. That the equal spaces, which are percurred continently in single moments of time, do

encrease

encrease in each single moment, according to the progression not of U-
nities, but of Numbers unequal from an Unity: so that if in the first
moment of time, the stone fall down one fathom, in the second moment, it
must fall down three fathom, in the third five, in the fourth seven, in the
fifth nine, in the sixth eleven, and so forward. And, because those
Numbers, which they call *Quadrate* (viz. One is the quadrate of an U-
nity, Fower the quadrate of a Binary, Nine the quadrate of a Ternary,
Sixteen of a Quaternary, and) are made up by the continual addition
of unequal numbers (for, three added to one, make four; five added
to four, make nine; seven, to nine, make sixteen; nine to sixteen, make
twenty five; eleven to twenty five, make thirty six, &c.) thereupon He
infers, that the Aggregates of the spaces percurred from the beginning
to the end of the motion, are as the Quadrates of the times: i. e. assuming any one particular moment of time, so many spaces are found pervaded in the end of that moment, as are indicated in the quadrate number of the same moment.    For Example, when in the end of the first
moment, one fathom of space is pervaded; in the end of the second moment, four fathom shall be pervaded; (viz. three being added to one)
in the end of the third moment, nine fathom (five being added to four)
in the end of the fourth moment, sixteen fathom (seven being added
to nine) and so forward : so that, accordingly, the spaces pervaded
from the beginning to the end of the motion, are among themselves
in a Duplicate Ration of moments (as Geometricians speak) or equal
Divisions of Time, or, all one as the Quadrates of moments are one to
another.

*Galilæus*, we said, herein relyed partly upon Experience, partly
upon Reason.    First, therefore, for his *Experience*; He affirms, that
letting fall a Bullet, from the altitude of 100 Florentine Cubits (i. e. according to exact comparation, 180 feet, Paris measure, and thirty fathom
of ours) He observed it to pervade the whole space, and arrive at the
ground, in the space of five seconds, or ten semiseconds : and according to such a ration, as that in the first semisecond, it fell down one cubit,
in the second semisecond, four cubits; in the third semisecond, nine cubits;
in the fourth sixteen; in the fifth twenty five; in the sixth 36; in
the seventh, forty nine; in the eighth, sixty four; in the ninth, eighty
and one; in the tenth the whole hundred.    And though the good *Mersennus* afterward found a bullet to pervade the same altitude in a much
shorter time; nay, that in the space of five seconds, a bullet fell down
through the space not onely of one hundred and eighty foot, but even of
three hundred, i. e. of fifty fathom: yet doth He fully consent, that the
Acceleration of its motion ariseth exactly according to *Galilæos* progression by the Quadrates of unequal numbers.    So as that if in the first semisecond, it descend one semi-fathom; in the second semisecond, it shall
descend four semifathoms; in the third semisecond, nine semifathoms, &c.
And *Gassendus* likewise, though he wanted the opportunity of experi-
menting the thing, from a Tower of the like altitude; found notwithstanding, from different heights, that the proportion was always the same;
as Himself at large declares (*in Epist.1. de proport. qua gravia decident.
accelerantur.*) Nor need you doubt to find it so your self, if in a Glass
Tube, neer upon two fathom long, divided into an hundred degrees, or
equal parts, marked with figures respectively either cut in, or inscribed

*Art. 14.*
*Galilæo's*
Grounds, Ex-
perience and
Reason.

upon papers (after the manner of those usually starcht on to Weatherglasses, to denote the several degrees) and not perpendicularly erected, but somewhat inclining, you let fall a bullet, and exactly observe the manner of its descent, and rate of Acceleration. For, Heavy bodies are, indeed, moved more slowly in Tubes inclined, than in such as are perpendicularly erected; but yet still with the same proportion of Acceleration.

Secondly, for His *Reason*, it consists in this; that, if the Increment of Velocity be supposed to be Uniforme (and there is no reason, which can persuade to the contrary) certainly, no other proportion can be found out, but that newly exposed: since, with what Celerity, or Tardity soever you shall suppose the first fathom to be pervaded it is necessary that in the same proportion of time following, three fathoms should be pervaded; and in the same proportion of time following, five fathoms should be pervaded; &c. according to the progression of Quadrate Numbers. This, that Great man *Joh. Baptista Ballianus* (whom *Ricciolus* often mentions (*in Almagesto novo*) but never without some honourable attribute) hath demonstrated divers ways *in lib.2. de Gravium motu.*): but the most plain Demonstration of the verity thereof, yet excogitated, we conceive to be this, invented by *Gassendus.*

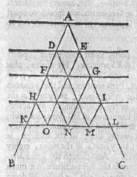

Understand the Lines *L A B* and *A C I* making a rectangular Triangle, by their meeting at the point *A*, to be so divided, on each side, into equal parts, at the points *D E F G H I K L*: (being continued, they may be divided into many more) as that the Lines drawn both betwixt those points, and from them to the points *M N O*, divide the whole space *K A L* into Triangles perfectly alike and equal each to other. This done, Assume the point or Apex *A*, for the beginning of Time, the beginning of space, and the beginning of Velocity: All which are to be here considered in the motion, as beginning together with it. First, then we may account the equal parts of each Line, *A B*, *A C.* for the parts or equal moments of Time, flowing on from the beginning: so that *A E* may represent the first moment, *E G* the second, *G I* the third, *I L* the fourth. Secondly, we may account those equal Triangles, for the equal parts of the space, which are pervaded from the beginning: so that Another perpendicular Line *P Q.* being drawn apart, and representing the fall of a stone, through sixteen fathom, the triangle *A D E*, may refer the first fathom *P R*, which is percurred in the first moment; the three next triangles may refer the three fathoms *R S*, which are percurred in the second moment; the five following triangles, the five fathoms *S T*, which are pervaded in the third moment: and the seven following, the seven fathoms, which are pervaded in the fourth moment. Now from
hence

hence it is manifeft, that the Aggregate fpaces carry the fame proportions, as the Quadrates of Times : when, the Triangle *A D*
*E* (or fpace *P R.*) is one , as the Quadrate of *A E*, that is of
one Time , is one : and the Aggregate *A F G* (or *P S*) is four :
as the Quadrate *A G*, of two, is fower : and the Aggregate *A*
*H I* (or *P T*) is nine : as the Quadrate *A I* of three , is nine ;
and the Aggregate *A K L* (or *P Q.*) is fixteen , as the Quadrate *A L* of four, is fixteen.

Thirdly, we may account the Line *D E* for the firft degree of
Velocity acquired in the end of the firft time ; infomuch, as the
firft time *A E* is not individual, but may be divided into fo many
inftants, or fhorter times, as there are points, or particles in the
line *A E* (or *A D*) fo neither is the degree of Velocity individual, or wholly acquired in one inftant ; but from the beginning
encreafeth through the whole firft time, and may be reprefented
by fo many Lines, as may be drawn parallel to the Line *D E*, betwixt the points of the Lines *A D* and *A E* : fo that , as thofe
Lines do continually encreafe from the point *A* to the Line *D E* ;
fo likewife doth the Velocity continually encreafe from the beginning of the motion, and being reprefented what it is in the intercepted inftants of the firft time, by the intercepted Lines, it may
be reprefented what it is in the laft inftant of the fame firft time,
by the Line *D E* drawn betwixt the two laft points of the Triangle *A D E*. And becaufe the Velocity, thenceforward continuing its Encreafe, may be again fignified, by Greater and Greater
Lines continently drawn betwixt all the fucceeding points of the
remaining Lines, *D B* and *E C* ; hence comes it , that the Line
*F G*, doth reprefent the degree of Velocity acquired , in the end of the
fecond moment : the Line *H I*, the Velocity acquired in the end of the
third moment ; and the Line *K L*, the velocity acquired in the end of
the fourth moment.   And evident it is from hence , how the velocities
refpond in proportions to the Times ; fince, by reafon of the Triangles
of a common angle, and parallel bafes, it is well known, that as *D E* are
to *E A*, fo *F G* to *G A* : *H I* to *I A*, and *K L* to *L A*.  Thus, keeping your
eye upon the Figure, and your mind upon the Analogy ; you fhall fully
comprehend , that in the firft moment of Time, the falling ftone doth acquire one degree of Velocity , and pervades one degree of fpace ; that
in the fecond moment of Time , it acquires another degree of Velocity,
which being conjoynd to the former, makes two , and in the mean while
three fpaces are pervaded ; that in the third moment, it acquires another
degree of Velocity, which conjoyned to the two former makes three, and
in the mean while feven parts of fpace are pervaded ; and fo forward.
You fhall fully comprehend alfo, that the Celerities obtain the fame Ration, as the moments of Time : and that the fpaces pervaded from the
beginning to the end of the motion, have the fame Ration, as the Quadrates of the moments of Time ; which we affumed to Demonftrate, out
of *Gaffendus*. But ftill it concerns you to remember , that we here difcourfe of that Motion, which is Equally, or Uniformly Accelerated ; or
whofe velocity doth continually and uniformly encreafe , nor is there any
moment of the confequent time , in which the motion is not more fwift,
than it was in every antecedent moment, and in which it is not accelerated

N n n 2                    according

P
R
S
T
Q

according to the same Reason. For, the want of this Advertisement in chief, seems to have been the unhappy occasion of that great trouble the Learned Jesuit *Petrus Cazræus* put *Gassendus* to, in his two Epistles, *De Proportione, quâ Gravia decidentia accelerantur.*

**Art. 16.**
**The Physical Reason of that Proportion.**

And this kindly conducts us to the *Physical* Reason of this Proportion, in which the velocity of bodies Descending is observed to encrease. For wholly excluding the supposition of the Aers assistance of the Downward motion of a stone, by recurring above, and so impelling it downward; and admitting the Magnetick Attraction of the Earth to be the sole Cause of its Descent; unto both which the considerations formerly alleadged seem to oblige us: it is familiar for us to conceive, that the Increment of its Celerity, according to the proportion assigned, ariseth from hence. While in the first moment, the earth attracts the stone, one degree of Celerity is acquired, and one degree of space is pervaded. In the second moment, the attraction of the Earth continuing, another degree of celerity is acquired, and three equal spaces are pervaded: one by reason of the degree of celerity in the mean while acquired, and two by reason of the degree of celerity formerly acquired, and still persevering, as that which is doubly æquivalent to the new degree in the mean while acquired; because it is Complete and entire from the very beginning of the 2ᵈ moment, but the other is only acquiring, or in *fieri*, and so not complete till the end of the second moment. Then, according to the same Ration, in the third moment another degree of celerity is acquired, and five spaces (equal) are pervaded; one by reason of the new degree of celerity in the mean while acquired, and fower by reason of the two former persevering, i.e. two in each moment præcedent, or one of a duplicate æquivalency to the new one not yet complete. Then, in the fourth moment another degree of celerity is acquired, and seven spaces are pervaded; one by reason of the fresh degree in the interim acquired, and six by reason of the three former persevering, i. e. two in each præcedent moment. And so of the rest through the whole motion, computing the degrees of encreasing Celerity, by the ration of Quadrate Numbers.

**Art. 17.**
**The Reason of the Equal Velocity of Bodies of very different weights, falling from the same altitude; inferred from the same Theory.**

Now, many are the Physical Theorems, and of considerable importance, which might be genuinely deduced from this excellent and fruitful Physicomathematical speculation; and as many the admired Apparences in nature, that offer themselves to be solved by Reasons more than hinted in the same: but, such is the strictness of our method, and weariness of our Pen, that we can, in the præsent, make no farther advantage of it, than only to infer from thence the most probable Reason of that so famous Phænomenon, *The equal velocity of two stones, or bullets, the one of* 100 *pound, the other of only one ounce weight, descending from the same altitude;* experience constantly attesting, that being dropt down together, or turned off, in the same instant, from the top of a tower; the Lesser shall arrive at the ground, as soon as the Greater. For, this admirable Effect seems to have no other Cause but this; that the Lesser body, as it containeth fewer parts, so doth it require the Impulses or strokes of fewer Magnetical rays, by which the attraction is made: and such is the proportion of the two forces, as that each moveable being considered with what Resistence you please, still is the force in the movent equally sufficient to overcome that resistence, and a few magnetique rays suffice to the

attraction

attraction of a few parts, as well as many to the attraction of many parts. So that the space being equal, which both are to pervade; it follows, that it must be pervaded by both, in equal or the same time. Provided always, that the two bodies assumed, be of the same matter; for, in case they be of divers matters, as the one of Wood, the other of Iron or Lead, that may cause some small Difference in their Velocity. We say, some small Difference; because, if we take two Globes of different materials and weights, but of the same or equal diameters, as (V. G.) one of Lead, the other of Wax: we shall be very far from finding, that the Heavier will be carried down more swiftly than the Lighter, in a proportion to the excess of its Gravity. For, if one be ten times heavier than the other; yet shall not the Heavier therefore, both being turned off, in the same instant, arrive at the ground ten times sooner than the Lighter: but, at the same time as the heavier, arrives at the ground, from the altitude of 10 Fathoms; the lighter shall come within a foot of the earth; so far short doth the lighter come of being nine fathoms behind the Heavier. And the Cause, why the Lighter Globe of Wax, is carried so swiftly, is the same with that, why a bullet of Lead of only an ounce weight, is carried down as swiftly as another bullet of 100 pound. And, what though the Globe of Wax be as great in circumference, as the other of Lead, and somewhat greater; yet seeing still it hath fewer parts to be attracted, it therefore requires fewer magnetical rays to its attraction with equal velocity to the heavier. But, the Cause why it is carried somwhat, though very little, slower than the heavier; is to be derived chiefly from the Aer resisting it underneath, the Aer being more copious in proportion to the virtue Attrahent, in respect of the greatness of its Ambite, or Circumference: and thence is it, that Cork, Pith of Elder, straws, feathers, and the like less compact, and so more light bodies, fall down much more slowly.

From this Experiment, and the Reason of it, we have an opportunity of observing and easily understanding the Distinction of *Gravity* into *Simple* and *Adjectitious*: the *Former* being that, which is competent to a body though unmoved, and whose quantity may be exactly determined by the balance suspending the body in the aer; the *Latter* being proper only to a body moved, and vanisheth as soon as the body attaineth quiet, and whose measure is to be explored both from the quantity of the simple gravity which the body bears during its quiet, and the Altitude from which it falls. Thus, assuming two Bullets, the one of an ounce, the other of 100 pound, Simple Gravity, according to the Scales; the Adjectitious Gravity of the Lesser bullet, acquired by the increment of its velocity during its descent, must be less proportionably to its simple gravity, than the Adjectitious gravity of the Greater bullet, acquired by the increment of its Velocity during its Descent, in the same time, and from the same altitude: because, the space and time of the descent of both being equal, the proportion of the acquired gravity of each must be respondent to the proportion of the simple gravity of each. So that if in the end of the fall of the Lesser bullet of an ounce weight, the Adjectitious Gravity of it shall amount to 10 ounces: the Adjectitious gravity of the Greater of 100 pound weight, shall, in the end of its fall, amount to a thousand pound; nor can the Acquired Gravity of the Lesser ever equal that of the Greater, unless it fall from a far greater Altitude.                                                        Here,

*Art.* 18. Gravity Distinguish't into *Simple*, and *Adjectitious*.

*Art.* 19.
The *Rate* of
that superla-
tive velocity,
with which a
Bullet would
be carried, in
case it should
fall from the
*Moon, Sun,* or
region of the
*Fixed stars,* to
the Earth: and
from each of
those vast
heights, to the
*Centre* of the
Earth.

Here, perhaps, you'l Demand our opinion, concerning that admirable because superlative Velocity, which *Galilæo* and other *Mathematicians* conceive that a bullet would acquire in case it should fall to the Earth from those vast (we might have said Immense) heights of the *Moon, Sun,* and region of the *Fixed* starrs. Of this, therefore, we say in short; (1) That, in this case, Mathematicians are wont to suppose, that there are the same Causes of Gravity and Velocity in those sublime places, as are observed here with us below, or neer the surface of the Earth: and if they be not, certainly our Description and Computation must be altogether vain and fruitless. For, if the Cause of Gravity, and consequently of the Velocity be the Attraction made by the magnetique rays transmitted from the Earth; forasmuch as those magnetique rays must become more Rare, and fewer of them arrive at a body, by how much farther it is removed from the Earth: though, perchance, a bullet might be attracted down from the region of the Moon (and if so, the motion of the bullet would be very slow, for a good while, in respect of the very few magnetique rays, that could arrive to that great height) yet from that far greater height of the region of the Fixt stars, a bullet could not be attracted at all, it being impossible that any magnetique ray should be transmitted so far as half way thither. (2) But, supposing that the magnetique Virtue of the Earth did extend thither; and that a bullet, from whence soever falling, should begin its motion with that speed, and proceed according to the same degrees of Acceleration, which we observe in a stone, or bullet falling from a very high tower: then must it of necessity acquire that incredible Velocity, which our Mathematicians describe. To Particular; conceding the Distances or Intervals betwixt the Earth and each of those Cælestial Orbs, which our modern and best Astronomers generally assign; a bullet would fall from the body, or rather the Limbus of the Moon, to the Earth, in two hours and an half; from the Limbus of the Sun, in eleven hours and a quarter: from the region of the Fixt stars, in 39 hours and a quarter. And so, if we imagine the Earth to be perforated to the Centre, since a bullet would fall from the superfice thereof down to the Centre, in 20 minutes, or the third part of an hour: the same bullet coming from the moon, would pervade the same space from the superfice of the Earth to the Centre of it, in one minute and twenty seconds, or the third part of a minute: coming from the Sun, it would pervade the same semidiametral space of the Earth, in seventeen seconds: and coming from the region of the Fixt stars, it would percur the same semidiametral space of the Earth, in five seconds. So incredibly great would be the Velocity of a bullet falling from such vast Altitudes. And this we think sufficient, concerning the *Downward* motion of Bodies, accounted Heavy.

## SECT. III.

THe Remnant of our præfent Province confifts only in the confideration of the *Upward* motion of Heavy Bodies PROJECTED :
concerning which the principal Enquiries among Philofophers are
(1) *VVhat and whence is that Force, or Virtue motive, whereby bodies projected are carried on,after they are feparated from the Projicient ?* (2)*What
are the Laws of their motion. Direct, and Reflex ?*

Concerning the FIRST, therefore, we obferve, that *Ariftotle* (*in* 8.
*phyfic.cap.ult.*) and moft of his *Sectators* confidently affirm , that a ftone
thrown out of a fling, an arrow fhot from a bow, a bullet difcharged from
a Gun, &c. is moved only by the *Aer* , from the time of its feparation
from the fling, bow, or Gun : and the manner of that motive activity of
the Aer upon the thing projected, They thus explicate.   The Aer (fay
they) which is firft moved by the Projicient,together with the moveable,
doth, at the fame time, both propel the moveable, and impel the Aer immediately beyond it, which being likewife moved, doth in the fame manner propel the moveable, and impel the aer immediately beyond it ; and
that aer being thus moved , doth again impel both the moveable and the
aer next beyond it : and fo confequently the next aer impels both the
moveable and the next aer beyond it ;, until the propulfion and promotion being gradually debilitated , and at length wholly overcome,
partly by the Gravity of the thing moved, partly by the Refiftence of the
occurring Aer,the motion wholly ceafeth, and the thing projected attaineth quiet.

*Art.* 1.
*What,* and
*whence* is that
*Force,* or Vir-
tue *Motive,*
whereby Bo-
dies Projected
are carried on
after their
Difmiffion
from the Pro-
jicient.

And that *Others* contend, that the Body Projected is carryed forward
by a *Force* (as They call it) *Imprefs* ; which they account to be a Quality fo communicated unto the body projected , from the Projicient , as
that not being indelible, it muft gradually decay in the progrefs thereof,
and at length wholly perifh, whereupon the motion alfo muft by degrees
remit its violence, and at length abfolutely vanifh , and the thing projected again recover its native quiet.   But, left we trifle away our præcious
moments , in confuting each of thefe weak Opinions, againft which the
Reafon of every man is ready to object many great abfurdities, efpecially
fuch as the præcedent theory will foon advertife him of : let us præfently
recur to the more folid fpeculations of our mafter *Gaffendus* in his Epi-
ftles (*de motu impreffo a motore tranflato*) and præfenting you the fum-
mary thereof, without further delay fatisfie your Curiofity, and our own
Debt of affifting it.

Firft we are to determine, *that nothing, remaining it felf unmoved, can
move another.*   For, fince our Difcourfe concerns not the Firft Caufe of
all motion, *God,* whofe Power is infinite, who is in all places, who can, only by the force of his Will, create, move, and deftroy all things ; manifeft it is, that nothing *Finite,* efpecially *Corporeal* (and fuch only hath
                                                                              an

an interest in our præsent consideration) can move another thing, unless it self be also moved, at the same time: as *Plato* well observed in his saying, *Neque est Difficile modo, sed etiam plane impossibile, ut quidpiam motum imprimere, sine quapiam sui commotione, valeat*: (*in Timæo.*) And the Reason is this, whatever doth move, doth act; and *e converso*, whatever doth act, doth move; Action and Passion (as Aristotle, 3.physic.3) being the same with motion. Again, the movent and Moveable ought to be together, or to touch each other, because, whether the movent impel, attract, carry, or towle the moveable: necessary it is, that still it should impress some certain Force upon it: and force it can impress none thereupon, unless by touching it. And though it doth touch it, yet if it discharge no force of motion upon it, i. e. remain unmoved it self: there shall be only a meer Contact reciprocal, but no motion, and as the one, so shall the other remain unmoved. Therefore, that the one may move the other: it ought to have that vigour or motion first in it self, which it doth impress upon the other: since if it have none, it can give none. Even sense demonstrates, that by how much more vehement motion the movent it self is in, at the instant it toucheth the moveable, by so much the farther doth it always propel the same: and thence our Reason may necessarily infer, that the movent must it self be in some small motion, in the same instant it gives a small motion to another. Moreover, though *Aristotle* (*in* 8.*Physic cap.*5.) subtly Distinguisheth three Things in motion, viz. the *Movens ut quod*, as (V. G.) a man, the *Movens ut quo*, as a staff: and the *Mobile*, as a stone: and thereupon magisterially teacheth, that the stone is moved, and doth not move; that the staff is moved, and doth move: that the man doth move, and is not moved: yet is it not evident, how far short He comes, of thereby Demonstrating the Immobility of the First Movent, to which He prætended. For whereas He urgeth, that otherwise we must proceed to Infinity; that binds not at all: because the *movens ut quod*, the man is moved by Himself: and sense declares, that the man must move his Arm, or Hand together with the staff, which if you suppose not to be the *movens ut quo*, (the stone being not moved thereby) but the *mobile* it self: is not the movent it self also moved? Suppose also, that the mans Arme, or Hand is the *movens ut quo*, nay if you please, that his whole Body, or the Muscles, or Nerves, or Spirits, are the *movens ut quo*, and deriving the motion from his very Soul, suppose that to be the *movens ut quod*: yet truely can you not conceive, that the Soul, it self remaining Immote, doth move the Arm, or Hand. Nor is the Soul it self then moved onely by Accident (as when a marriner is carried by the motion of his ship) but also *per se*, as when the mariner moves himself, that he may move the Oar, that it may move the ship, in which himself is carried. For, as a ship, in a calm sea, would not be moved it self, nor the mariner be moved with it, by Accident: in case the mariner himself wanted motion, whereby to impel his ship: so neither would the body be moved, nor the Soul be moved therewith by Accident, unless the soul be first agitated within, with a motion whereby the body is moved. Conclude, therefore that nothing can be projected, but the Projicient must not only Touch it, either immediately, or mediately by some Instrument; but also Propel it with the same motion, wherewith it self is, in the same instant, moved,

It is moreover necessary, that the movent be moved, not only in a point, or so far as that point of space, in which it first toucheth the moveable: but also that a while cohæring unto the moveable, it be *moved along with it*: so as we may well conceive them to be made, by that Cohæsion, as it were one and the same body, or one entire moveable, pro tempore; and consequently, that the motion of both the movent and moveable is one intire motion. For, what motion is in the moveable, so long as it remains conjoyned to the movent, is in a manner a certain Tyrocinium, in which the moveable is as it were taught to progress forward in that way, which the movent hath begun, upward, downward, transverse, oblique, circular, and that either slowly, or swiftly, and according as the movent shall guide and direct it, before its manumission or dismission.

Thus, when a man throws a stone with his hand, you may plainly perceive, how the motion thereof begins together with that of his hand: and after it is discharged from his hand, you cannot say, that a new motion is impressed upon the stone, but only that the same motion begun in the hand is continued. And, therefore, it seems also very unnecessary to require the impression of any new and distinct Force upon the stone projected, by the projicient, which should be the Cause of its motion after its Dismission: seeing nothing else is impressed, but the very motion to be continued through a certain space; so that we are not to enquire, what motive Virtue that is, which makes the Persevering motion, but what hath made the motion, that is to persever. In the moveable, certainly, there is none but a Passive Force to motion; nor can the Active Force be required in any thing but the movent: and should we, with the Vulgar, say, that there is an Imprest Force remaining, for some time, in the thing moved, or projected; we could thereby understand no other than the Impetus, or motion it self.

Here might we opportunely insist upon this, that motion is impressed upon a thing moved, only in respect, that the thing moved hath less force of Resistence, than the movent hath of Impulsion: so that the movent, forcing it self into the place of the moveable, compels it to recede, or give way, and go into another place. But it is more material for us to observe, that when a thing projected is impelled, it is first touched by the projicient only in those parts, which are in its superfice or outside and that those outward parts, being pressed by the impulse, do drive inward or press upon the parts next to them; and those again impel the parts next to them, and those again the next to them; till the impulse be by succession propagated quite through the body of the thing projected, to the superficial parts in the opposite side, and then begins the motion of the whole, the parts reciprocally cohæring: as hath been formerly explained, in the example of a long pole, or beam of wood. Which being percussed, but with a very gentle or softly stroke, that one end hath all its parts so commoved successively, as that the stroke may be plainly perceived by a man, that lays his ear close to the other end: which could not be if the impulse were not propagated from parts to parts successively, through the whole substance of the beam. To which it is requisite, that we superad this observable also; that by reason of the force made by Contact, and that short Cohæsion of the moveable to the movent, there is created a certain *Tension*, or stress of all the parts of it, towards the opposite region: and of that by that means, all the parts of the thing pro-

*Art.* 2.
The *Manner* of the Impression of that Force.

O o o                                                   pro-

projected, are difposed or conformed as it were into certain *Fibers*, or direct *Files*; of all which the moft ftrong and powerful is that, which being
trajected through the Centre of Gravity in the thing projected, becomes
as it were the *Axis* to all the circumftant ones.  Our eys afcertain, that
unlefs the Centre of Gravity be in the middle of the thing projected, or
directly obverted to the mark, at which the thing is thrown; the thing
inftantly turns it felf about, and that part, wherein the Centre of Gravity is, always goes foremoft, and as it were carries the reft of the parts, as
that which is the moft Direct and moft Tenfe of all the Fibres.  And
this cannot be effected, but with fome (more or lefs) *Deflection* from the
mark, at which the force, according to the Centre and Axis of Gravity,
was directed; forafmuch as the Centre of Gravity, wherein many Fibres
concur, makes fome Refiftence, and detorting the Fibres, inflecteth them
another way, and fo a new Axis is made *pro tempore*, according to which
the Direction of all the parts in their motion afterward is determined.
Hence is it, that, if you would hit a mark, either with a fling, or ftone-
bow, you muft choofe a ftone, or bullet of an uniform matter and compofition: or, at leaft, turn the heavier part of the body to be thrown, forward; becaufe otherwife, it will Deflect more or lefs, to one fide or other
according to the pofition and inclination of its Centre of Gravity. Moreover, whether foever the thing projected doth tend, all the Fibers conftantly follow the Direction of the Axis, or are made parallels thereunto; fo
that as often as the Centre is changed, fo often doth the Axis, fo often do
all the Fibres change their pofition, and follow the Centre.  Which we
infer chiefly in refpect of the motion of Convolution, or Turning of a
thing projected immediately after its Difmiffion; and of the Curvity of
that Line, which is thereby defcribed, whether afcending, or defcending.
But thefe are onely Tranfient Touches, or Hints; that we might eafily
intimate, why a motion once impreft, is continued rather this way, than
that: and why Feathers, Sponges, and the like Light and Porous bodies,
are incapable of having quick and vehement motions impreft upon them;
becaufe they confift of interrupted Fibres, and fuch as are not Dirigi-
ble with the Centre of Gravity.

*Art.* 3.
That all Moti-
on, in a free
or *Empty*
fpace, muft be
*Uniform*, and
*Perpetual*: and
that the chief
Caufe of the
*Inequality* and
*Brevity* of the
motion of
things project-
ed through
the *Atmofphere*,
is the magne-
tique Attracti-
on of the
Earth.

Here we ask leave, once more to have recourfe to that ufeful fuppofition of a ftone fituate in the immenfity of the Imaginary fpaces.  We
lately faid, as you may remember, that if a ftone placed in the empty Extramundane fpaces, fhould be impelled any way, the motion thereof would be continued the fame way, and that uniformly or equally, and
with tardity or celerity proportionate to the fmartnefs or gentlenefs of the
Impulfe, and perpetually in the fame line; becaufe in thofe empty fpaces
it could meet with no caufe, which by Diverfion might either accelerate, or
retard its motion.    Nor ought it to be Objected, that *nothing Violent
can be Perpetual*; becaufe, in this cafe, there could be no Repugnancy or
Refiftence, but a pure Indifferency in the ftone to all regions, there being
no Centre, in relation whereunto it may be conceived to be Heavy or
Light.   And, therefore, the condition of the ftone would be the very
fame, as to Uniformity and Perpetuity of motion, with that of the Cæleftial Orbs; which being obnoxious to no Retardation, or Acceleration,
but free from all Repugnancy internal, and Refiftence External, conftantly and indefinently maintain that Circular motion, which was, in the firft
moment of their Creation, impreft uupon them, by the Will of the Creator

tor; and that toward one part, rather than any other.   Let us now farther confider; feeing that if upon fome large horizontal plane you fhould
place a fmooth Globe, and then gently impel it; you would obferve it to
be moved therupon equally and indefinently, till it came to the end thereof: why may you not lawfully conjecture, that if the Terreftrial Globe
were of a fuperficie exquifitely polite, or fmooth as the fineft Venice Glafs;
and another fmall Globe as polite were placed in any part of its fuperficie,
and but gently impelled any way, it would be moved with conftant Uniformity quite round the Earth, according to the line of its firft direction;
and having rowled once round the Earth, it would, without intermiffion
again begin, or rather continue another Circuit, and fo maintain a perpetual Circulation upon the furface of the Earth ?   Efpecially, fince there
is no Difficulty to difcourage that conjecture; forafmuch as look how
many parts of the fmall Globe, during the motion thereof, tend toward
the Centre of the Earth, juft fo many are, at the fame time, elevated from
it: fo that a full Compenfation being made in all points of the motion, the
fame cannot but perpetually continue, and in the fame equal tenour, there
being no Declivity, whereby it fhould be Accelerated, no Acclivity, wher-
by it fhould be Retarded, no Cavity, whereby after many accurfes and
recurfes, or reciprocations, it fhould be brought at length to acquiefce.
Moreover, in order to our grand fcope, let us fuppofe, that the fpace,
through which a ftone fhould be Projected, were abfolute Inane, or fuch
as the Imaginary fpaces; and then we muft acknowledge, that it would be
carried in a direct and invariate line, through the fame fpace, and with an
Uniforme and Perpetual motion, until it fhould meet with fome other
fpace, full of magnetique rayes, Aer, or fome other refifting fubftance.   But, here with us, in the Atmofphere, becaufe no fpace is
Inane (fenfibly) but replete as well with Aer, as with millions of magnetique rayes tranfmitted from the Earth; and fo a ftone Projected
muft encounter them in every point of fpace through which it moves:
therefore is it, that it cannot be moved either in a direct Line, or
equally, or long.   For, fince multitudes of magnetique Rayes muft
neceffarily invade and attach it, as foon as it is difcharged from the
Projicient; though at firft fetting forth it break through them, and
fo is fcarce at all Deflected: yet becaufe more and more magnetique
rayes frefhly lay hold of it in every part of fpace, renew the Attraction, and fo more and more infringe and weaken the force of its
motion; hence comes it, that in the progrefs it doth by little and
little Deflect from the Line of Direction, moves flower and flower,
and at length finking down to the Earth, thereon attains its quiet.
Hereupon, when men fhall Demand, *what is that Caufe, which
weakens and at laft quite deftroys the Virtue Impreffed upon a thing Projected*; rightly underftanding, by the Virtue Impreft, the motion begun by the Projicient, and continued by the Projectum: the Anfwer
is manifeft; *viz*. That it is the *Attraction of the Earth*, which firft op-
pofeth, after gradually refracteth, and in fine wholly overcometh the
motion impreft, and fo determineth the Projectum to Quiet.   Hence
alfo may we learn, that *All motion once impreffed, is of it felf Indelible,*
and cannot be Diminifhed, or Determined, but by fome External Caufe,
that is of power to reprefs it.

This

*Art. 4.*
That, in the
Atmosphere,
no body can
be projected
in a *Direct*
line; unless
perpendicu-
larly Upward,
or Downward:
and why.

This confidered, you may pleafe to obferve, that through the Atmofphere, or fpaces circumvironing the Terreftrial Globe, being fo poffeffed by the Aer and fwarms of Magnetique Rayes, *no body can be projected in an abfolute Direct : or perfectly ftreight Line*, unlefs perpendicularly upward or downward.    For, if the projection be made
either obliquely, or parallel to the Horizon ; the projectum fuddainly begins to Deflect from the mark at which it was aimed, and fo defcribes
not a ftreight, but crooked line.    Not that the Deflection or Curvity
is fenfible, at a fmall diftance, especially if the motion be vehement,
fuch as that of an Arrow fhot from a Bowe, or Bullet difcharged from
a Gun : but, that in every point of fpace, and time, the thing Projected is attracted fomewhat Downward ; and there is the fame Reafon
for its Deflection in the firft, as there is for its Deflection in the fecond,
third, fourth, or any following point of fpace, and inftant of time, though
the greater oppofition of the Force impreft makes that Deflection lefs
at the firft.    Nor ought it to incline us to the contrary, that Archers and
Gunners frequently hit the mark, at which they levelled, to fome certain
diftance : becaufe, that Diftance is commonly fuch, as that the Deflection therein is not fenfible, though it be fometimes an hairs-breadth, two,
three, or four, fometimes an inch below the mark.

*Art. 5.*
That the Mo-
tion of a ftone
projected up-
wards ob-
liquely, is
Compofed of
an *Horizontal*
and *Perpendi-
cular* together.

Further you may obferve, that when a ftone is projected, or a bullet
fhot upward, yet not perpendicularly, but obliquely ; the motion thereof is to be confidered, not as fimply perpendicular, or fimply Horizontal,
but as *mixed*, or *compofed of an Horizontal and Perpendicular together :* of a Perpendicular, forafmuch as the Altitude thereof may
be meafured by a Perpendicular line ; of an Horizontal, forafmuch
as it is made according to the Horizon, and the Latitude thereof
may be taken by the plane of the Horizon.    But, becaufe by how
much the more it hath of the perpendicular, fo much the lefs it hath
of the Horizontal ; fo that the Altitude of it may amount to fifty
feet, and the Latitude not exceed one foot : therefore is it manifeft,
that the crooked Line defcribed by this Compafs motion, cannot be
Circular ; and *Galilæo* (*Dialog.* 4.) hath demonftrated that the Line is
*Parabolical*, or fuch as Geometricians defcribe in the ambite of a
Cone, when they fo interfect it obliquely from one fide at the bafe,
that the motion of the interfection is made parallel to the other fide
left whole, for the Area of each refegment is the Geometricians Parabola : and the crooked ambite of the Area, is a Parabolical Line,
and frequently taken for the Parabola it felf.    We remember alfo,
how *Galilæo*, upon confequence, and among other remarkables doth
obferve ; that of all Projections, made by the fame force, the
*Longeft*, and in that refpect the moft *Efficacious*, is that, which is
made to an *half-right* Angle, or by aiming at the *forty fifth* degree of Altitude ; in refpect of the more prolix Parabola which is
defcribed by the Projectum, aimed at that altitude : fince at all
other altitudes the Parabola muft be fhorter ; the fuperior Altitudes
being lefs, and the inferior more open than is requifite.

Now

Now this Compofition of a Perpendicular and Horizontal motion may be moft conveniently Demonftrated unto you , thus.  Being in a fhip, under fayl , if you hold a Ball in your hand ;  the motion of the ball will be onely Horizontal, *viz.* That , whereby the
fhip doth carry you, your hand , and the ball in it.  If the fhip
ftand ftill,  and you throw the ball directly upward ;  the motion of
the ball will be onely Perpendicular :  but if the fhip be moved ,  at
the fame inftant you throw the ball upward ;  then will the motion thereof be Compound, partly Perpendicular, partly Horizontal.  For ,  the ball fhall be carried obliquely , and defcribe a Parabolical line ,  in which it afcends and again falls down again ;  and
in the mean time,  it fhall be promoved Horizontally.  The Perpendicular alone ,  your felf may difcern with your own eye :  becaufe,
the horizontal is common both to the ball and your eye ,  and when
as well the ball ,  as your eye is promoved ,  therefore doth it always
appear imminent over your eye ,  and in the fame perpendicular :  but,
for the Horizontal ,  He onely can deprehend it ,  who ftands ftill on
the fhoare,  or another fhip not carryed on at the fame rate, as that wherein you are.

*Art. 6.*
Demonftrati-
on of that
*Compofition.*

Herein there occur Two things ,  not unworthy our admiration.  The
one is ,  that *though there be two divers Forces or motions impreffed upon
the Ball ,  at the fame time :  the one from the Vibration of your Arm,
the other from the horizontal Tranflation of the fhip :  yet doth neither
deftroy the other ,  but each attains its proper fcope as fully ,  as if
they were impreffed apart.*  For ,  the Ball afcends as high ,  when
the fhip is moved forward ,  as when it ftands ftill :  and whether it
defcribe a Direct ,  or a femiparabolical :  and again ,  it is as much
promoved Horizontally ,  when you divert it upward by projection,
as when you hold it ftill in your hand and fo it be carried onely by
the motion of the fhip :  and confequently whether the motion thereof defcribe a Direct line ,  or a whole Parabola.  Onely this you
are to note :  that a greater Force is required to the projection of a
Ball from the foot to the top of the Maft ,  when the fhip moves forward ,  than when it lies at anchor :  becaufe that femiparabolical line,
which the Ball muft defcribe in the former cafe ,  is fhorter than that
perpendicular one ,  which it muft defcribe in the latter :  and however the vibration or fwing of your arme may feem to you to be e-
qual in both cafes ,  yet is that vibration or force ,  whereby the ball
is carried upward to the top of the Maft ,  when the fhip is in motion,
really greater than that ,  whereby the fame ball is carried to the fame
height ,  when the fhip lies quiet :  becaufe ,  in the former cafe ,  there
is fuperadded to the force of your arme ,  the force which is impreffed
both upon you and your arme (without your apprehenfion) by the
motion of the fhip.  This you fhall plainly perceive ,  if you onely
drop down a ball from the top of the Maft ,  without any fwing or
motion of your arme at all.  For ,  feeing that the ball doth always
fall at the foot of the maft ,  in the fame diftance from it ,  as it was
in the inftant of its dimiffion from the top ;  whether the fhip be moved ,  or quiet :  neceffary it is ,  that fome force be impreft upon
the

*Art. 7.*
That of the
two different
*Forces,* impref-
fed upon a
ball , thrown
upward from
the hand of a
man ftanding
in a fhip, that
is under fayl ;
the one doth
not deftroy
the other : but
each attains
its proper
fcope.

the ball by the motion of the ſhip, or the the ſame motion, where-by both the Maſt it ſelf, and your hand are affected, at the inſtant of its dimiſſion; ſince it muſt deſcribe a ſemiparabolical line, longer than that Direct one, which it would deſcribe, if it fell down the ſhip being quiet. And hence comes it, that if you project a ball from the Poop to the Fore Caſtle of a ſhip, under ſayl, and back again from the Fore-Caſtle to the Poop; you ſhall impreſs a greater force upon it, in throwing it from the Poop to the Fore-Caſtle, than back again from the Fore-Caſtle to the Poop: becauſe, in the former caſe, the force or ſeconding impulſe of the ſhip muſt be ſuperadded to the force of your arme in projection, and ſo make it the ſtronger; and, in the latter caſe, the contrary force of the ſhip doth as much detract from the force of your arme, and ſo make it the weaker. And though the ball be carried over equal ſpaces of the Deck of the ſhip, in both caſes: yet ſhall it not be carried through equal ſpaces in the Aer.

*Art. 8.*

*That the ſpace of time, in which the Ball is Aſcending from the Foot to the Top of the Maſt, is equal to that, in which it is again Deſcending from the Top to the foot.*

Hence may it be Demonſtrated, *that the ſpace of Time which the ball is Aſcending from the foot to the top of the Maſt, is Equal to that in which it is Deſcending again from the top to the foot.* For, were it not ſo, when the ball is projected in a line perpendicular and parallel to the Maſt, the ball would not aſcend and deſcend always at the ſame diſtance from the Maſt, but would either deſert it, or be deſerted by it, the ſhip being in motion. Whence it follows alſo, that in what proportion the velocity of the ball Aſcending doth de-creaſe; in the ſame proportion doth the velocity of the ball again Deſcending encreaſe: ſo that the motion of the ball muſt be of equal velocity, when it is removed from the plane of the ſhip, one fathom aſcending, or deſcending, and likewiſe at the altitude of one foot, aſcending or deſcending. Again, foraſmuch as the force of your arme, projecting the ball, is ſtill equal; but the force ſuperadded thereunto by the motion of the ſhip, may be more or leſs vehement, according as the ſhip is carried with greater or leſs ſpeed: thence it follows, that the Parabolical lines deſcribed by the ball, are reſpe-ctively Greater or Leſs, and the motions of it through the Aer more or leſs ſwift. But, yet all are performed in Equal Time; be-cauſe the times of them all are equal to the ſame time, which is due to the ſimple Aſſent and Deſcent, and with the ſame proportion of parts.

*Art. 9.*

*That, though the Perpendicu-lar motion of a ſtone thrown obliquely up-ward, be Un-equal, both in its aſcent and deſcent: yet is the Horizon-tal of Equal Velocity in all parts of ſpace.*

The *Other*, which deſerves our admiration, is this; that not-withſtanding, of the twofold motion compoſing the Oblique one, that which is Perpendicular, is Unequal, the Velocity there-of being as well diminiſhed in the aſſent, as augmented in the de-ſcent, ſo that; in equal moments of time, leſs ſpaces are pervaded in the aſſent, and greater in the deſcent: yet *is that motion, which is Horizontal, plainly Equal in all its parts, or of equal velocity throughout; ſo that equal ſpaces of the Horizon are pervaded in e-qual times.* The truth of this is conſtant from hence; that if (the ſhip being equally moved on, and the ball being projected in a line parallel to the Maſt) the foot of the Maſt ſhall pervade twenty

paces,

paces, or an hundred foot of horizontal space : the ball shall be horizontally (i. e. toward that region, to which the ship tends) promoved, not more swiftly or slowly in one pace or foot, than in another, but equally in all : for, otherwise, it could not be always imminent over the same part of the ship neer the Mast : nor therefore consist in the same line, or distance from the Mast : which yet it constantly observes. But this easily deceives, that at the end of the balls ascent, or beginning of its descent, the motion is slowest : but then are we to observe, that the Devexity, or Conformity of it to the Horizon is the Greater, as when it comes lower, where the motion is more rapid, the Devexity is less, and its conformity to the Perpendicular greater : so that the whole *Inæquability* doth consist in the Assent and Descent, or *Perpendicular* motion of the ball : while in the mean time there is a perfect *Æquability* in its *Horizontal* advance, or promotion. From hence we collect : that since a thing Projected is moved unequally, insomuch as it tends upward or downward : and not as it progresseth parallel to the Horizon, or Ambite of the Earth : therefore is it, that *the upward and downward motions are both to be accounted Violent : but the Horizontal, or Circular, Natural :* Equality, or Uniformity being the inseparable Character of Natural, and Inequality of Violent motion.

Thus far have we treated of that Returning or Reflex motion of Bodies, whereby, being violently projected upward, they revert or fall down again, by reason of the magnetique Attraction of the Earth : and it now remains onely, that we consider the Reasons of that other species of motion *Reflex* or *Rebounding*, whereby Bodies, being also violently moved or projected any way, are impeded in their course and Diverted from the line of their Direction, by other bodies encountring them. Concerning this Theorem, therefore, be pleased to know, that among all Reflexions, by way of Rebound or Resilition, that is the *Chiefest*, when a body projected, and impinged against another body, is returned from thence *directly*, or in the same line toward the place, from whence it was projected : which always happens, when the Projection is made to right Angles, or in regular line, such as that in which a Heavy body descends upon an horizontal plane. And all other Reflections are in dignity inferior thereunto, as such whereby the thing projected doth not rebound in a direct line toward the same point from whence it was projected, but to some other region by other lines : according as it is projected in lines more or less oblique. Because, with what inclination a body falls upon a plane, with the very same inclination doth it rebound from the plane (especially a Globe, and such as is of an uniform matter, and consequently hath the Centre of magnitude and that of Gravity coincident in the same point) so that by how much the more oblique the projection is, and how much the less is the Angle made of its line with the line of the plane, (called the Angle of *Incidence*) so much the more oblique is the reflexion made, and so much the less the Angle made of its line, with the line of the plane continued (called the Angle of *Reflexion*)

*Art. 10.*
The Reason and Manner of the *Reflexion* or *Rebounding* motion of Bodies, diverted from the line of their direction by others encountring them.

*flexion*) and that so long, as till the line of projection shall become parallel to the plane, and so, no body occurring to or encountring the projectum, no reflexion at all be made.

*Art.* 11.
That the *Emersion of a weight appensed to a string, from the perpendicular, to which it had reduced it self, in Vibration, is a Reflexion Median betwixt No Reflexion at all, and the Least Reflexion assignable; and the Rule of all other Reflexion whatever.*

Know moreover, that betwixt *No* Reflexion at all, and the *Least* Reflexion that is possible, there may be assigned as it were a certain *Medium*; and that is the *Emersion* or Rising up again of a weight appensed to a thread or Lutestring, when performing a vibration or swing from one side to the other, it ascends from the perpendicular Line, to which by descending it had reduced it self. For, in that case, no reflecting body doth occur, a simple Arch is described; and yet there is as a certain Procidence or falling down to the lowest point of the Arch, so also a certain Resilition or rising up again from the lowest point of the Arch, toward the contrary side. Again, having conceived a direct line touching the lowest point of the Arch, so as that the weight suspended by a string, may, in its vibration, glance upon it with its lowest extreme, and onely in a point touch the horizontal line; you shall have on each side an Angle made from the Arch and the line touching it, which is therefore called the Angle of Contingence: and because Geometricians demonstrate, that the Angle of Contingence, which truly differs from a right line, is less than any Rectilinear Angle, however acute; therefore may each of those Angles be said to be Median betwixt the right line, and the Angle either of Incidence, or of Reflexion, how small soever it be; and consequently, the Emersion of the weight in Vibration may as justly be said to be Median betwixt the smallest Reflexion and none at all. However, this Emersion seems to be the Rule of all Reflection whatever; for, as in the *Vibration* of a weight appensed to a string, and describing a simple Arch, the Angle of its *Emersion* is always equal to the Angle of its *Procidence*: so in *Projection* describing an Angular line, the Angle of *Reflection* is always (*quantum ex se est*) equal to the Angle of *Incidence*. We say, *quantum ex se est*; for otherwise, whether it be sensible, or not, because so long as the Projectum is transferred, it is always somewhat depressed toward the earth, for the reason formerly alledged; thence comes it, that the Reflexion can neither be so strong or smart as the Incidence, nor make as great an angle, nor arise to as great an altitude. Which we insinuate, that we might not insist upon this advertisement; that the Æquality of the Angle of the Reflexion to that of the Incidence, may be so much the less, by how much the less the projected body comes to a spherical figure, or doth consist of matter the less uniform.

*Art.* 12.
The Reason of the *Æquality* of the Angles of *Incidence* and *Reflexion.*

For, to attain to that Æquality of the Angles of Incidence and Reflexion, necessary it is, that the body projected be exactly spherical, and of Uniform matter, and so having the Centre of Gravity, and the Centre of magnitude coincident in one and the same point; as we have formerly intimated: it being as well against Reason, as Experience, that bodies wanting those conditions should arise to that æquality which that we may the better understand, let us consider, that as in a Globe, or Ball Falling down, we regard onely that Gravity, which it acquires in its descent, from the magnetique

netique Attraction of the Earth : so in a Globe, or Ball Projected,
we are to regard onely that Impetus or Force, which being imprest
upon it by the Projicient, supplies the place of Gravity, and in respect whereof the Centre of its Gravity may be conceived to be one
with that of its magnitude. Let a Ball, therefore, be projected
Directly or to right Angles, upon a plane ; and, because, in that
case, that Fibre must be the Axis of its Gravity, whose extreme
going foremost is impinged against the plane : thence is it manifest, that the Repreffion must be made, in a direct line, along
that Axis ; the parallel Fibres in equal number on each part invironing that Axis, and so not swaying or diverting the ball more
to one part than to another, by reason of any the least disproportion of quantity on either side. Then, I t the same Ball
be projected Obliquely against the same plane ; and because, in
this case, not that middle Fibre, which constituteth the Axis of
Gravity, but some one or other of the Fibres circumstant about
it, must with one of its extreams strike against the plane : therefore is it necessary, that that same Fibre be represt by that impulse, and by that repreffion compelled to give backward toward
its contrary extream, and thereby in some measure to oppose the
motion begun, which it wholly overcome, and so the ball would
rebound from the plane, the same way it came, if the Fibres on
that side the Axis of Gravity, which is neerest to the plane, were
equal in number to that are on the farther, or contrary side of
it : but, because those Fibres, that are on the farther side, of
on the part of the Centre and Axis, are far more in number, and
so there is a greater quantity of matter, and consequently a greater
force imprest, than on the side neerer to the plane ; therefore doth
the begun motion persever, as prævailing upon the repreffion and
renitency of the Fibre impinged against the plane, and since it cannot be continued in a direct line, because of the impediment arise-
ing from the parts cohærent, it is continued by that way it can,
i. e. by the open and free obliquity of the plane. But, this, of
necessity, must be done with some certain Evolution of the Ball,
and with the contact of the Fibres posited in order both toward the
Axis and beyond it ; and while this is in doing, every Fibre strives
to give back, but, because the farther part doth yet prævail over
the neerer, therefore doth the neerer part still follow the sway, and
conform to the inclination and conduct of the farther, and all the
toucht Fibres change their situation, nor are they any longer capable of returning by the same way they came, because they no longer respect that part from whence they came. We say, with the
Contact of the plane by the Fibres posited toward the Axis and beyond it ; because, since in that Evolution or Turn of the Ball, the
extream of the Axis toucheth the plane, yet neverthelefs no Resilition, or Rebound is therefore caused, in that instant ; and if
there were a refilition, at that time, it would be to a perpendicular, as well the Axis, as all the circumstant Fibres being erected
perpendicularly upon the face of the plane : but the Refilition there
must be beyond it, because the force of the farther part of the Fibres doth yet prævail over that of the neerer.       For, the Force of

P

the

the farther part doth yet continue direct and intire ; but, that of the neerer is reflected, and by the repression somewhat debilitated : and therefore, the Resilition cannot be made, until so much of Repression and Debilitation be made in the further part, as was made at first in the neerer.　　And that must of necessity be done, so soon as ever the plane is touched by some one Fibre, which is distant from the Axis as much beyond, as that Fibre, which first touched the plane, is distant from the Axis on this side : for, then do the two forces become equal, and so one part of the Fibres having no reason any longer to prævail over the other, by counter inclination, the Ball instantly ceaseth to touch the plane, and flies off from it, toward that region, to which the Axis and all the circumstant Fibres are then, i. e. after the Evolution, directed. Now, because the Ball is, after this manner, reflected from the plane, with the same inclination, or obliquity, with which it was impinged against it ; it is an evident consequence, that the Angle of its Reflexion must be commensurable by the Angle of its Incidence : and that each of them must be so much the more *obtuse*, by how much less the line of projection doth recede from a perpendicular ; and contrariwise, so much the more Acute, by how much more the line of projection doth recede from a perpendicular, or how much neerer it approacheth to a parallel with the plane.

*Art.* 13.
Two Inferen-
ces from the
premisses; viz.
(1) That the
oblique Proje-
ction of a
Globe against
a plane, is
composed of a
double Paral-
lel; and (2)
That Nature
suffers no di-
minution of
her right to
the shortest
way by Reflec-
xion.

From these Considerations we may infer *Two observables.* The One, *that the oblique projection of a Globe against a plane, is composed of a double Parallel,* the one with the Perpendicular, the other with the plane : for, the Globe at one and the same time, tends both to the plane, and to that part toward which the plane runs out forward. The Other, *that Nature loseth nothing of her right,* by *the Reflexion of bodies* ; forasmuch as she may nevertheless be allowed still to affect and pursue the shortest, or neerest way : for, because the Angle of Reflexion above the plane, is equal to that Angle, which would have been below the plane, in case the plane had not hinderd the progress of the line of projection beyond it, by reason of the Angles Equal at the Vertex, as Geometricians speak ; therefore, is the Reflex way equal to the Direct, and consequently to the shortest, in which the ball projected could have tended from this to that place.

*Art.* 14.
Wherein the
Aptitude or In-
eptitude of bo-
dies to Re-
flexion doth
consist.

Here, to bring up the rear of this Section, we might advance, a discourse, concerning the *Aptitude* and *Ineptitude* of Bodies to *Reflexion* ; but, the dulness of our Pen with long writing, as well as the Confidence we have of our Readers Collective Abilities, inclining us to all possible brevity, we judge it sufficient onely to advertise, that what we have formerly said, concerning the Aptitude and Ineptitude of Bodies to *Projection,* hath anticipated that Disquisition.　　For, certain it is, in the General, that such Bodies, which are More Compact, Cohærent, and Hard, as they may be, with more vehemence, and to greater distance, *Projected* : so may they, with more vehemence, and to greater distance Rebound, or be *Reflected* ; provided, they be impinged against other bodies
of

of requisite Compactness, Cohærence, and Hardness. And, the Reason, why a Tennis-ball doth make a far greater Rebound, than a Globe of Brass, of the same magnitude, and thrown with equal force; is onely this, that there is not a proportion betwixt the Force imprest by the Projicient, and the Gravity of each of them; or betwixt the Gravity of each, and the Resistence of the Plane. Which holds true also concerning other bodies, of different Contextures.

P 2          CONCLUSION

# CONCLUSION:

*Ingenious Reader,*

I Have kept you long at Sea, I confeſs, and (ſuch was the Unskilfulneſs of my Pen, though ſteered, for the moſt part, according to the lines drawn on thoſe excellent Charts of *Epicurus* and *Gaſſendus*) often ſhipwrackt your Patience. But, be pleaſed to conſider, that our way was very Long and tædious; inſomuch as we had no leſs than the whole of that vaſt and deep Ocean of *Sublunary Corporeal Natures*, to ſayl over: that our paſſage was full of Difficulties, as well in reſpect of thoſe ſundry Rocks of Incertitude, which the great Obſcurity of moſt of thoſe Arguments, whoſe diſcovery we attempted, inevitably caſt us upon; as of thoſe frequent Miſts and Foggs, which the exceeding Variety of mens Opinions, concerning them, ſurrounded and almoſt benighted our judgement withal: and chiefly, that if by the voyage your Underſtanding is brought home not only ſafe, but inriched, though in the leaſt meaſure, with that ineſtimable Wealth, the *Knowledge of Truth*, or what is ſo Like to Truth, as to ſatiſfie your Curioſity as fully; as I have reaſon to congratulate my ſelf, for the happineſs of my Care and Induſtry, in being your Pilot, ſo muſt you to eſteem the adventure of your Time and Attention compenſated with good Advantage. And, now you are on Land agen, give me leave, at parting, to tell you; That all the *Fare* I ſhall ever demand of you, is only a *Candid ſentiment of my Good-will and cordial Devotion to the Commonwealth of Philoſophy*. Which, indeed, doth ſo ſtrongly Animate me on to enterprizes of Publique Utility, though but to thoſe in the Second Form of Scholars; that I can be well contented, not only to neglect opportunities of Temporal advantages to my ſelf, while I am imployed in the ſtudy, how to contribute to the Intellectual promotions of others; but alſo to ſtand in the number of thoſe Active and Free Spirits, who have, through want of Abilities only, miſcarried in their well intended

Endea-

Endeavours for the benefit of Learning; rather than in the lift of thofe Idle, or Envious ones, who having more of Wit, than of Humanity, and wanting nothing but the Inclination to do Good, have buried their Talents, and left the Republique of Arts and Sciences, to fuffer in the want of fuch means of Advancement, as their Capacities might eafily have afforded unto it.

'Tis the Cuftom of the Multitude, you Know, always to eftimate the Counfel of Defigns only by their Succefs; and never allowing for Impediments or finifter Accidents, to account the Goodnefs of an Undertaking to confift wholly in the Felicity of its Event: but, fuch is the juftice of Wifdom, that it configns a Reward to a good Intention; and decrees a Lawrel to be planted on his Grave, who fals in the generous Attempt of any noble Difcovery, as well as one to be placed on his Head, who fhall be fo much beholding to the Favour and Affiftance of his Fortune, as to Accomplifh it. This I put you in mind of, not out of Arrogance, as if I challenged any thing as due to me, befides a lively Refentment of my conftant and fincere Zeale to the Encreafe of Knowledge; but, to poffefs you more fully with the Equity of my Expeĉtation, which aims at no other Reward, but what Detraĉtion it felf dares not difpute my Right unto, and much lefs than what, I prefume, your own Charity would, if I had referred my felf thereunto, have readily affigned me.

But, left I feem to prevent you in your Inclination, or to Extort that from you by force of Argument, which as well your own innate Candor, as judicious Æquanimity, had fufficiently præpared you to offer me of your own accord; I refigne you to your Peace, and the undifturbed enjoyment of thofe Pleafures, which ufually refult from the memory of Difficulties once overcome: Having firft affured you, that your benigne Acceptance of my Services, and Pardon of my Misfortunes (fo I may call all fuch Errors, whofe præcaution was above the power of my humble judgement) in this Voyage; may prove a chief Encouragement to me, to adventure on a *Second*, without which this

F

First must be Imperfect; and that is for a Description of the Nature of that Paradise of the World, that bright shadow of the All-illuminating and yet Invisible Light, that Noble Essence, which we know to be within us, but do not understand because it is within us, and cannot understand without it, the *Humane Soul*; and that, so soon as Quiet and Physick shall have repaired those Decays in the Weather-beaten Vessel of my Body, which long Sitting, frequent Watchings, and constant Solicitude of mind have therein made.

In the mean time, I conjure you, by your own Humanity, to remember and testifie, that in this my Conversation with you, you have found me so far from being Magisterial in any of the Opinions I præsented; that considering my own Humor of Indifferency, and constant Dubiosity (frequently professed, but more expresly, in the First Chapter of this Work, and 1. *Art. of the* 1. *Chap.* 3. *Book.*) it hath somewhat of wonder in it, that I ever proposed them to Others: nor, indeed, can any thing solve that wonder, but my Hopes, thereby secretly to undermine that lofty Confidence of yonger Heads, in the Certitude of Positions and Axioms Physiological; and by my declared Scepticism even in such Notions, as my self have laboured to assert, by the firmest Grounds, and strongest Inducements of Belief, to reduce them to the safer level of

*Quò magis quærimus, magis dubitamus.*

## F I N I S.

Lightning Source UK Ltd.
Milton Keynes UK
UKHW021001211222
414263UK00009B/703